NATO ASI Series

Advanced Science Institutes Series

A series presenting the results of activities sponsored by the NATO Science Committee, which aims at the dissemination of advanced scientific and technological knowledge, with a view to strengthening links between scientific communities.

The series is published by an international board of publishers in conjunction with the NATO Scientific Affairs Division

A	**Life Sciences**	Plenum Publishing Corporation
B	**Physics**	New York and London
C	**Mathematical and Physical Sciences**	Kluwer Academic Publishers Dordrecht, Boston, and London
D	**Behavioral and Social Sciences**	
E	**Applied Sciences**	
F	**Computer and Systems Sciences**	Springer-Verlag
G	**Ecological Sciences**	Berlin, Heidelberg, New York, London,
H	**Cell Biology**	Paris, and Tokyo

Recent Volumes in this Series

Volume 180—The Physics of Submicron Semiconductor Devices
edited by Harold L. Grubin, David K. Ferry, and C. Jacoboni

Volume 181—Fundamental Processes of Atomic Dynamics
edited by J. S. Briggs, H. Kleinpoppen, and H. O. Lutz

Volume 182—Physics, Fabrication, and Applications of Multilayered Structures
edited by P. Dhez and C. Weisbuch

Volume 183—Properties of Impurity States in Superlattice Semiconductors
edited by C. Y. Fong, Inder P. Batra, and S. Ciraci

Volume 184—Narrow-Band Phenomena—Influence of Electrons with Both Band and Localized Character
edited by J. C. Fuggle, G. A. Sawatzky, and J. W. Allen

Volume 185—Nonperturbative Quantum Field Theory
edited by G. 't Hooft, A. Jaffe, G. Mack, P. K. Mitter, and R. Stora

Volume 186—Simple Molecular Systems at Very High Density
edited by A. Polian, P. Loubeyre, and N. Boccara

Volume 187—X-Ray Spectroscopy in Atomic and Solid State Physics
edited by J. Gomes Ferreira and M. Teresa Ramos

Series B: Physics

Nonperturbative Quantum Field Theory

Proceedings of a NATO Advanced Study Institute on
Nonperturbative Quantum Field Theory,
held July 16–30, 1987,
in Cargèse, France

ISBN 978-1-4612-8053-8 ISBN 978-1-4613-0729-7 (eBook)
DOI 10.1007/978-1-4613-0729-7

Softcover reprint of the hardcover 1st edition 1988

Library of Congress Cataloging in Publication Data

NATO Advanced Study Institute on Nonperturbative Quantum Field Theory (1987: Cargèse, Corsica)
Nonperturbative quantum field theory / edited by G. 't Hooft . . . [et al.].
p. cm.—(NATO ASI series, Series B, Physics; vol. 185)
"Proceedings of a NATO Advanced Study Institute on Nonperturbative Quantum Field Theory, held July 16–30, 1987, in Cargèse, France"—T.p. verso.
"Published in cooperation with NATO Scientific Affairs Division."
Includes bibliographical references and index.

1. Quantum field theory—Congresses. 2. Statistical mechanics—Congresses. I. 't Hooft, G. II. North Atlantic Treaty Organization. Scientific Affairs Division. III. Title. IV. Series: NATO ASI series. Series B, Physics; v. 185.
QC174.45.A1N37 1988 88-23927
530.1'43-dc19 CIP

A Division of Plenum Publishing Corporation
233 Spring Street, New York, N.Y. 10013

Nonperturbative Quantum Field Theory

Edited by

G. 't Hooft

University of Utrecht
Utrecht, The Netherlands

A. Jaffe

Harvard University
Cambridge, Massachusetts

G. Mack

University of Hamburg
Hamburg, Federal Republic of Germany

P. K. Mitter

University of Paris VI
Paris, France

and

R. Stora

LAPP
Annecy-le-Vieux, France

Plenum Press
New York and London
Published in cooperation with NATO Scientific Affairs Division

Preface

During the past 15 years, quantum field theory and classical statistical mechanics have merged into a single field, and the need for nonperturbative methods for the description of critical phenomena in statistical mechanics as well as for problems in elementary particle physics are generally acknowledged. Such methods formed the central theme of the 1987 Cargèse Advanced Study Institute on "Nonperturbative Quantum Field Theory."

The use of conformal symmetry has been of central interest in recent years, and was a main subject at the ASI. Conformal invariant quantum field theory describes statistical mechanical systems exactly at a critical point, and can be analysed to a remarkable extent by group theoretical methods. Very strong results have been obtained for 2-dimensional systems. Conformal field theory is also the basis of string theory, which offers some hope of providing a unified theory of all interactions between elementary particles. Accordingly, a number of lectures and seminars were presented on these two topics. After systematic introductory lectures, conformal field theory on Riemann surfaces, orbifolds, sigma models, and application of loop group theory and Grassmannians were discussed, and some ideas on modular geometry were presented.

Other lectures combined traditional techniques of constructive quantum field theory with new methods such as the use of index-theorems and infinite dimensional (Kac Moody) symmetry groups. The problems encountered in a quantum mechanical description of black holes were discussed in detail. A special lecture was devoted to statistics in two dimensions, it made contact with representation theory of braid groups and methods used for integrable systems. Other lectures covered applications of the renormalization group, of Monte Carlo simulation techniques, and combination of high temperature series expansions with perturbation theory. This part of the program may be regarded as a sequel to the 1983 Cargèse summer institute on "Progress in Gauge Field Theory."

Recent developments in conformal quantum field theory have stimulated much research in mathematics, and are leading to a revival of algebraic techniques in quantum field theory. Lectures by mathematicians were presented on noncommutative differential geometry (with application to the quantum Hall effect) and on fixed point formulas. Other mathematical developments were described as parts of the above mentioned lectures by physicists.

In addition to the invited lecturers, participants of the ASI had the opportunity to present own results in the form of short communications . This contributed much to the scientific discussions at the institute and is reflected in these proceedings.

We wish to express our gratitude to NATO whose generous financial contribution made it possible to organize the school, to the CNRS and the Ministry of Foreign Affairs of France for additional support, and to the NSF for travel grants. We thank Maurice Lèvy, the Director of the Institut d'Etudes Scientifiques de Cargèse, as well as the University of Nice, for making available to us the facilities of the Instut. Grateful thanks are due to Marie-France Hanseler for much help with the material aspects of the organization. Last but not least we thank the lecturers and the participants for their enthusiastic involvement which contributed much to the success of the school.

G. 't Hooft
A. Jaffe
G. Mack
P.K. Mitter
R. Stora

CONTENTS

LECTURES*

SEMINARS

*D. Friedan (Chicago) lectured on conformal field theory and modular geometry, but was unable to contribute to these proceedings

OPERATOR METHODS IN STRING THEORY[1]

Luis Alvarez Gaumé

Physics Department, Boston University

I. Introduction

These lectures are based on joint work with C. Gomez, C. Reina, G. Moore and C. Vafa [1,2,3]. The motivation is two-fold. On the one hand, we would like to develop an operator formalism describing high orders of string perturbation theory, as well as conformal field theory on Riemann Surfaces of genus bigger than one. On the other hand, developments in the theory of soliton solutions of the K-P hierarchy (Kadomtsev-Petviashvili equations; for a detailed geometrical account of this theory and references to the literature see for example [4,5]) made it clear that many of the geometrical features of Riemann Surfaces and their moduli spaces can be formulated in terms of the properties of certain two-dimensional quantum field theories [6] , so that the geometrical complexity of a Riemann surface with a field on it can be coded into a state of a standard Fock space of the field theory [1,7,8,9,10,11,12]. This approach gives an important understanding of the action of the Virasoro algebra on the moduli space of surfaces [13,14]. The space of the solutions of the K. P. equation can be described in terms of an infinite dimensional Grassmannian Gr . To any algebraic curve X with a point P selected on it, a local coordinate around P, and a line bundle L over X, we can associate a point in Gr (Krichever construction [4,5]). Moreover. the collection of all those points in Gr is a dense set. It is thus plausible to expect that some subspace of Gr provides an explicit model for the Universal Moduli Space of Friedan and Shenker [15,16] which plays a central role in their non-perturbative approach to string theory. This possibility still remains largely unexplored, mainly because it seems difficult to find integration measures (the non-perturbative analog of the ghost in the bosonic string). After all the progress achieved in string theory over the last few years, we are still scratching the surface of a deep structure whose building blocks seem far from being unravelled. One of the most important clues we seem to be getting from string theory is that the space of classical solutions to a quantum theory of strings is the space of two dimensional conformal theory with a specific value of the central term in the Virasoro algebra.

Conformal theories have been developed along different lines. One approach is to use path integrals to define the theory, and the other is to use operator techniques. The operator formulation is simplest for the sphere and the torus, which has

[1]Supported in part by DOE contract and Alfred P. Sloan foundation fellowship

been discussed in detail in the literature (see for example [16,17]), in fact, most of the substantial work on conformal field theories has been carried out using operator methods for theories formulated in the twice punctured sphere (conformally equivalent to the cylinder). For higher genus surfaces, many results have been obtained using a functional integral formulation, because there is no simple way of constructing a hamiltonian formulation.

It is well known that the energy momentum tensor can be viewed as providing a connection on moduli space. This is quite important in the operator formulation of string theory. One of the main aims in our work was to better understand the Virasoro algebra and moduli space [13]. We can use the Virasoro algebra to move the states in moduli space. Below we describe these issues for the simplest conformal field theories, even though the setup applies in principle to more general situations. One of the spin-offs of the formulation we present is an alternative proof of chiral bosonization completely in operator language. The interested reader is referred to [3] for details. We will not present these results here.

These notes are organized as follows: In section II we review the general ideas of the operator formalism of conformal field theories in arbitrary surfaces, and work out some examples in detail. In section III we turn to the action of the Virasoro algebra and the Polyakov measure for the bosonic string. Section IV contains an outline of the extension of this work to superstrings, together with the conclusions.

Many authors have developed independently an operator formalism for string perturbation theory with a motivation based on string field theory. An incomplete (but representative) list of references can be found in [18,19,20]. There is obviously a good deal of overlap between these approaches and the one presented here; and in many respects they can be considered complementary.

II. Overview and examples

Consider a Riemann surface X with parametrized boundaries, and a quantum field theory on it, whose fields will be collectively denoted by ϕ . If $\mathcal{H}$ is the Hilbert space of the theory in the standard conformal plane, we can construct a state $\mathcal{H}^{\otimes n}$ (one copy per parametrized boundary) associated to the surface and the field theory by performing the Feynman path integral with fixed boundary conditions $\phi_i(t_i), i = 1,2...,n$ at each of the boundaries. This procedure yields a wave functional $\Psi[\phi_1(t_1), \ldots \phi_n(t_n)]$ in $\mathcal{H}^{\otimes n}$. If $\mathcal{P}(g,n)$ denotes the moduli space of Riemann surfaces of genus g with n points on it $P_1, ..., P_n \in X$ and local coordinates about them $t_i(P_i) = 0; 1 \leq i \leq n$; and we fix the field theory, we get a mapping $\Psi : \mathcal{P}(g,n) \to \mathcal{H}^{\otimes n}$. When the theory is not conformally invariant , we have to specify a metric on each surface together with its moduli. For conformal theories however, Ψ only depends on the information parametrized by $\mathcal{P}(g,n)$. Punctured surfaces are of interest in string theory, because a surface with semi-infinite cylinders attached to the boundaries describes string scattering amplitudes, and these semi-infinite cylinders are conformally equivalent to punctured discs. Mathematically, the string scattering amplitude can be understood as measure on the moduli space of punctured surfaces. To illustrate the previous ideas, we consider a surface X divided into two pieces X_1, X_2 by a circle S^1, and a b-c system [17] of spin j on it ($j \in \frac{1}{2}\mathcal{Z}$). We would like to represent the result of functionally integrating over the whole surface as an overlap integral on the Fock space of the circle of two states $\psi_1[f], \psi_2[f]$, where $f(z)$ is the field configuration of $c(z)$ on S^1. To obtain Ψ_1 we functionally integrate the fields over X_1 subject to the boundary conditions $C|_{S^1} = f$, and free boundary conditions on $b(z)$.Standard arguments in functional integration give the following representation for ψ_1:

$$\psi[f] = \int_{c|_{S^1}=f} dbdc \exp(-\int_{X_1}(b\bar{\partial}c + c\bar{\partial}b) + \oint_{S^1} cb) \tag{2.1}$$

One of the basic properties of $\psi[f]$ is its invariance under shifts of f by section of c holomorphic in X_1 :

$$\psi[f + w_n] = \psi[f] \tag{2.2}$$

if for simplicity we choose X_2 to be a disc centered upon the point P chosen on X, and with a local parameter $t, t(P) = 0$; the section $w_n(t)$ may have arbitrary order poles at $t = 0$, but extends holomorphically to X - P. Since the translation of f by w_n is implemented by the operator $\oint_{S^1} w_n(t)b(t) \equiv Q_n$, a more invariant way of expressing (2.2) is $Q_n|\psi >= 0$. A similar argument shows that $|\psi >$ is invariant under shifts of b by sections w_n which extend holomorphically off P. Hence the ray $|\psi >$ is completely defined by

$$Q_n|\psi >= \tilde{Q}_n|\psi >= 0$$

$$Q_n = \oint_{S^1} w_n(t)b(t)$$

$$\tilde{Q}_n = \oint_{S^1} \tilde{w}_n(t)b(t) \tag{2.3}$$

It is a simple consequence of the b-c canonical (anti)-commutation relations that $[Q_n, Q_m] = [\tilde{Q}_n, \tilde{Q}_m] = 0$. The Q_n 's and $\tilde{Q}_n$'s also commute:

$$[Q_n, \tilde{Q}_m] = \oint_{S^1} w_n(t)\tilde{w}_m(t) = 0 \tag{2.4}$$

since $w_n(t)$ (resp $\tilde{w}_m(t)$) is a $1 - j$ (resp. j) differential, the product $w_n(t)\tilde{w}_m(t)$ is a meromorphic differential on X whose only poles are located at P. Then the residue (2.4) vanishes identically. The conditions (2.3) provide the largest set of compatible conditions which can be imposed on a state $|\psi >$; and in most known cases, they uniquely determine the ray represented by $|\psi >$. The exceptions known always involve superconformal ghosts (the $\beta - \gamma$ system in [17]).

a) As a first example, we consider an anticommuting b.c. system of spin j on the sphere with one puncture P. By the Riemann Roch theorem , there are $2j - 1$ global holomorphic $(1 - j)$ differentials: $1, t, t^2, ..., t^{2j-2}$. We can also find $1 - j$ differentials with poles of arbitrary order at P. The w_n 's generate a set $W = \{t^{2j-2}, \ldots, t, 1, t^{-1}, t^{-2}, \ldots\}$. The j-differentials holomorphic off P are generated by $\tilde{W} = \{t^{-2j}, t^{-2j-1}, \ldots\}$. The conserved charges are:

$$Q_n = \oint_0 b(t)t^n = \{b_n, n > -j\} \tag{2.5a}$$

$$\tilde{Q}_n = \oint_0 c(t)t^n = \{c_n, n > j - 1\} \tag{2.5b}$$

with

$$b(t) = \sum_n b_n t^{n-j} \quad , \quad c(t) = \sum_n c_n t^{n+j-1} \tag{2.6}$$

$n \in \mathcal{Z}$ if j is integral , and $n \in \mathcal{Z} + \frac{1}{2}$ if j is half-integral. The state $|\psi>$ is then the $SL_2(C)$ invariant vacuum of the $b-c$ system. We introduce the notation $|\phi_g^1, j>$ for the state obtained on a surface of genus g with n punctures and an anticommuting $b-c$ system of spin j . Thus, the example we have just discussed gives $|\phi_0^1, j>$. When j is an integer, we also have the zero mode algebra $\{b_0, c_0\} = 1$, and the highest weight states of the oscillator algebra are the states $|\pm>$ satisfying

$$c_n|+>= b_n|->= 0, \qquad n \geq 0$$

$$b_0|+>= |->, \qquad c_0|->= |+> \tag{2.7}$$

The relation between $|\pm>$ and $|\phi_0^1>$ is:

$$|+>= c_0 c_1 ... c_{j-1}|\phi_0^1>$$

$$|\phi_0^1>= b_{-j+1}...b_{-1}b_0|+> \tag{2.8}$$

SL_2 invariance of $|\phi_0^1>$ implies

$$L_0|\pm>= -\tfrac{1}{2}j(j-1)|\pm> \tag{2.9}$$

and the natural ghost charge assignment follows from the Atiyah-Patodi-Singer index theorem:

$$Q_{gh}|\phi_0^1>= -(j-\tfrac{1}{2})|\phi_0^1> \tag{2.10}$$

b) For our second example, consider a spin $\frac{1}{2}$ system on a surface X with a single puncture P $\in$X and genus g. If the spin structure is even and non-singular (the kernel of the Dirac operator is zero), the w_n sections are obtained from the Szegö kernel [21] :

$$S(z,w) = \frac{\theta\begin{bmatrix}\alpha\\ \beta\end{bmatrix}(\int_w^z w|\tau)}{\theta\begin{bmatrix}\alpha\\ \beta\end{bmatrix}(0|\tau)} E(z,w) \tag{2.11}$$

where E is the prime form on the surface , and $[\alpha, \beta]$ represents the even spin structure. For z, w in a neighbourhood of P , (2.11) can be written as :

$$S(z,w) = \frac{1}{z-w} + \sum_{n,m=1}^{\infty} B_{nm} z^{n-1} w^{m-1} \tag{2.12a}$$

$$B_{nm} = \frac{1}{(n-1)!(m-1)!} \frac{\partial^{n-1}}{\partial z^{n-1}} \frac{\partial^{m-1}}{\partial w^{m-1}} (S(z,w) - \frac{1}{z-w}) \tag{2.12b}$$

Then the w_n 's are obtained by differentiating $S(z,w)$ with respect to w and setting $w = \mathrm{P}$:

$$w_n(t) = t^{-n} + \sum_{m=1}^{\infty} B_{nm} t^{m-1} \tag{2.13}$$

thus giving the conserved charges :

$$Q_n = \oint_0 w_n(t)c(t) = (c_{n-\frac{1}{2}} + \sum_m B_{nm} c_{-m+\frac{1}{2}}) \tag{2.14}$$

Since $\{c_n, n > 0\}$ are annihilation operators, we can represent c_n by $\partial/\partial b_{-n}$. We obtain a differential equation for $|\phi_g^1, \frac{1}{2}>$:

$$(\frac{\partial}{\partial b_{-n+\frac{1}{2}}} + \sum_m B_{nm} c_{-m+\frac{1}{2}})|\phi_g^1, \tfrac{1}{2} >= 0 \tag{2.15}$$

with solution:

$$|\phi_g^1, \tfrac{1}{2} >= C \exp(\sum_{n,m=1}^{\infty} B_{nm} c_{-n+\frac{1}{2}} b_{-m+\frac{1}{2}})|\phi_0^1, \tfrac{1}{2} > \tag{2.16}$$

If the spin structure is singular or if it is odd, there will be a holomorphic $\frac{1}{2}$ differential on the surface: $h(t)$. Then the associated conserved charge $Q(h) = b(h) \equiv \oint b(t)h(t)$ only involves creation operators. Thus we have to impose the conservation of $Q(h)$ and $\tilde{Q}(h)$; apart from Bogoliubov transformation obtained from the $b - c$ two point function with zero modes removed. If $g(B)$ denotes the Bogoliubov transformation, the state obtained becomes:

$$|\phi_g^1, \tfrac{1}{2} >= g(B) Q(h) \tilde{Q}(h) |\phi_0^1, \tfrac{1}{2} > \tag{2.16}$$

Finally , if we wish to study a Weyl-Majorana fermion, there is a single fermionic field $\psi(t)$ with action $S[\psi] = \int \psi \bar{\partial} \psi$, with commutation relations $\{\psi_n, \psi_m\} = \delta_{m+n,0}$; $n, m \in \mathcal{Z} + \frac{1}{2}$, and the state obtained in this case is :

$$|\phi_g^1, Maj >= C \exp(-\tfrac{1}{2} \sum_{n,m=1}^{\infty} B_{nm} \psi_{-n+\frac{1}{2}} \psi_{-m+\frac{1}{2}})|\phi_0^1, Maj > \tag{2.17}$$

(for an even non-singular spin structure). The $2n$-point functions are easily proved to be expressed in terms of pfaffians:

$$\frac{< 0|\psi(t_1)\psi(t_2)...\psi(t_{2n})|\phi_g^1, Maj >}{< 0|\phi_g^1, Maj >} = Pfaff_{ij} S(t_i, t_j) \tag{2.18}$$

and the partition function involves Pyrm theta functions.

c) Next we explicitly treat the ghost system of the bosonic string. This is a fermionic $b - c$ system with $j = 2$. On any surface X with $g > 1$ there are $3g - 3$ holomorphic quadratic differentials; and we can always find quadratic differentials with poles of arbitrary order at a point P. These statements need extra qualification when P is a Weierstrass point on X, but we will not insist on this issue here. A convenient basis for the meromorphic 2-differentials is (see [3] for details)

$$S_N = (t^N + \sum_{m \geq K} B_{Nm}^{(j)} t^m)(dt)^2, \qquad -\infty < N \leq k-1, \quad k = 3g-3 \tag{2.19}$$

and as a consequence of the Riemann-Roch theorem, the vector fields holomorphic off P must have poles of order larger than k:

$$V_M = (t^M + \sum_{m \geq -k} B_{Mn}^{(1-j)} t^m)(dt)^{-1}, \quad M \leq -(k+1) \tag{2.20}$$

Since the charges:

$$\oint c(t)S_N(t), \quad -1 \leq N \leq k-1 \tag{2.21}$$

only involve creation operators, the state $|\phi_g^1, 2>$ takes the form:

$$|\phi_g^1, 2 >= c(S_0)c(S_1)...c(S_{3g-4})c(S_{-1})\exp(-\sum_{n,m} B_{n+2,m}b_{-n}c_{-m-2})|+>$$

$$c(S_i) = \oint_0 c(t)S_i(t) \tag{2.22}$$

so that the ghost number of $|\phi_g, 2>$ is $3g - 3 + 3/2$, after we assign ghost number 1 (-1) to $c(t)(b(t))$. The extension of (2.22) to surfaces with more punctures is straightforward. We only point out that the ghost number charge of $|\phi_{g,j}^n>$ is $(2j-1)(g-1) + (j - \frac{1}{2})n$.

d) Finally we consider a free scalar on X. This case is important in string theory because the embedding coordinates of the flat bosonic string are free fields from the two dimensional point of view. The state associated to the surface X with a parametrized boundary S^1 and a free single-valued scalar field ϕ can be represented by means of a functional integral

$$\psi[f] = \int_{\phi|_{S^1}=f} d\phi \exp(-\frac{1}{8n}\int_{X_1} d\phi \wedge *d\phi + \frac{1}{8n}\oint_{S^1} \phi * d\phi) \tag{2.23}$$

where $*$ is the duality operation on forms $*dz = -idz$. The state (2.23) is invariant under the shifts of f by functions on S^1 which are boundary values of harmonic functions on X_1 (using the same notation as in (2.1)). As first pointed out by Vafa [9], the explicit construction of $\psi[f]$ is more complicated than in previous cases. The origin of the difficulty is that for generic points P $\in$ X , the Weierstrass gap theorem implies that functions holomorphic off P must have a pole at P of order at least $g+1$. We can however construct multivalued meromorphic functions with poles of order less than $g+1$. Since the shifts involve harmonic functions , we can compensate the multivaluedness by adding some antiholomorphic piece. Explicitly, if $w_i, \bar{w}_i$ are the abelian differentials on X, we introduce the differentials:

$$\eta_n(t) - A_n(Im\tau)^{-1}(w - \bar{w})$$

$$w_i(t) = \sum_{n=1}^{\infty} A_{in}t^{n-1}dt$$

$$\eta_n(t) = \frac{1}{(n-1)!}\frac{\partial}{\partial t}\frac{\partial^n}{\partial y^n}|_n E(t,|y)|_{y=0} \tag{2.24}$$

The harmonic functions of interest are:

$$\begin{aligned}\phi_n(t) &= \int^t (\eta_n(t) - A_n(Im\tau)^{-1}(w - \bar{w})) \\ &= \phi_n^{hol}(t) + \phi_n^{ah}(\bar{t})\end{aligned} \tag{2.25}$$

Expanding $\phi(t)$ in terms of oscillators,

$$\phi(t) = q + iplnt + ip|n\bar{t}\bar{t} + i\sum_{n\neq 0}(\frac{t^n}{n}j_n + c.c.) \tag{2.26}$$

$$[p,q] = +i \quad , \qquad [j_n, j_m] = n\delta_{n+m,0}$$

The conserved charges are:

$$Q_n = \oint (\partial\phi\phi_n^{hol}(t) - \bar{\partial}\phi\phi_n^{ah}(\bar{t}))$$
$$= j_n - \sum_{n=1}^{\infty} (2Q_{nm} + \pi A_n (Im\tau)^{-1} A_m)\frac{j_{-m}}{m} + \pi A_n (Im\tau)^{-1} \bar{A}_m)\frac{\bar{j}_{-m}}{m} \tag{2.27}$$

plus their complex conjugates. Representing j_{-n} as nx_n, and $j_n = \partial/\partial x_n, n \geq 0$ the differential equations $Q_n|\psi >= \bar{Q}_n|\psi >= 0$ yield:

$$|\psi >= c\exp((x,\bar{x})M\binom{x}{\bar{x}}) \tag{2.28}$$

$$M = \begin{pmatrix} Q_{nm} + \frac{1}{2}\pi A_n (Im\tau)^{-1} A_m & -\frac{1}{2}\pi A_n (Im\tau)^{-1} \bar{A}_m \\ -\frac{1}{2}\pi \bar{A}_n (Im\tau)^{-1} A_m & \bar{Q}_{nm} + \frac{1}{2}\pi \bar{A}_n (Im\tau)^{-1} \bar{A}_m \end{pmatrix}$$

The overall constants appearing in front of the states depend on the moduli parameters of the surface. Notice that (2.28) is the first example where the dependence of $|\psi>$ on the moduli parameters is not holomorphic. In the operator language this plays an important role in identifying the correct bosonic string measure.

The previous arguments are easily extended to surfaces with more punctures. Many more details and examples can be found in [3].

III Virasoro algebra and the bosonic string measure

It was shown in [13] that there is a natural action of the Virasoro algebra on the moduli space of punctured surfaces with a local parameter at the puncture. To understand this, let P $\in$ X , and D be the disc centered upon P where the local parameter t vanishing at P is defined. A local meromorphic vector field with poles only at P, $v(t)\partial/\partial t$ can be written in terms of a Laurent expansion:

$$v(t) = \sum a_{n-1} t^n \tag{3.1}$$

These vector fields can be divided into three subsets

i) Vector fields holomorphic in D : $C\|t\|\partial/\partial t$. Among these, those vectors vanishing at P generate local coordinate changes. Those satisfying $v(P) \neq 0$ will also produce infinitesimal motions in the position of the puncture.

ii) Vector fields which extend holomorphically off P. These generate infinitesimal conformal automorphisms of the punctured surface.

iii) Vector fields holomorphic in $D-$P which do not extend holomorphically to either D or X - P. The Riemann Roch theorem shows that this space has dimension $3g-3$; and it is in fact the same as the tangent space to moduli space at the curve X. If we think in terms of the moduli space $\mathcal{P}(g,1)$ of genus g Riemann surfaces with a point and a local parameter, we find that its tangent space at (X,P,t) is isomorphic to i) + ii). Using the Kodaira-Spencer construction, we can generate a new surface S' close to S. Only when v is either in i) or iii) will we get an inequivalent point on $\mathcal{P}(g,1)$. One of the basic properties of the Virasoro algebra in a conformal field theory is that the corresponding variation of $|\psi>$ under the variation of the vector field $v(t)$ is given by the energy momentum tensor:

$$T(v) = \oint T(t)v(t), \qquad T(t) = \sum_{n\in Z} L_n t^{n-2}$$
$$\delta|\psi> = (T(v) + \bar{T}(\bar{v}))|\psi> \tag{3.2}$$

It is easy to show explicitly that (3.2) holds for all the examples considered in the previous section; and it also gives an equation determining the moduli dependence of the constants appearing in front of the states. For instance, if $|\psi>$ is the state associated to a spinor with an even spin structure on the torus, then there is only one vector field in class iii): $t^{-1}d/dt = v$,and $T(v) = L_2$. Then

$$\delta(Cg(B)|0>) = \delta Cg(B)|0> + Cg(B)\delta B|0> = T(v)Cg(B)|0> \tag{3.3}$$

where $g(B)$ is the Bogoliubov transformation (2.16). From (3.3) we obtain a differential equation

$$\frac{\partial}{\partial\tau}\log C \propto \frac{\partial}{\partial\tau}\log\frac{\theta[^{\alpha}_{\beta}](0|\tau)}{\eta(\tau)} \quad ,$$

a well known result for the fermion determinant. More generally, the vector fields in $D-$ P which do not extend to either D or X-P have the form $v_i(t) = t^{-i}d/dt, i = 1, 2, ...3g-3$. Hence $L_{-n}, n \geq 0$ changes the local coordinate, L_1 moves the puncture, and $L_k, k = 2, 3, ..., 3g-3+1$ change the moduli. If $v(t)$ is holomorphic in X-P, then the ray associated to S should not change under the action of $T(v)$. However it is not consistent to require $T(v)|\psi> = 0$, unless the central charge of the Virasoro algebra vanishes in the theory considered, otherwise $T(t)$ does not behave as a quadratic differential under changes of variables. Only when we cover the surface X with patches with $SL_2(C)$ transition functions can we consistently impose $T(v)|\psi> = 0$. To construct the bosonic string measure, let $|W>$ be the state associated to $S \in \mathcal{P}(g,1)$, and the string coordinates $X|^{\mu}$ and ghost fields $b, c, \bar{b}, \bar{c}$. The state $|W>$ has ghost charge $3g-3+3/2$. Let $v_i, i = 1, 3g-3$ be the vector fields representing the tangent vectors to the moduli space $M_g : V_1, ..., V_g$. If $|0>$ is the SL_2 invariant vacuum on the disc, then an obvious candidate for the measure is:

$$\mu(V, \bar{V}) = <0|b(v_1)\bar{b}(\bar{v}_1)...b(v_{3g-3})\bar{b}(\bar{v}_{3g-3})|W> \tag{3.4}$$

To prove that this is indeed a measure on M_g we have to exhibit the independence of $\mu(V, \bar{V})$ with respect to data in $\mathcal{P}(g,1)$ not contained in M_g:

a) We can replace v_i by $v_i + v_0$, where v_0 extends holomorphically to X-P.

b) Independence of the local parameter and the position of the puncture.

The first requirement follows from the definition of $|W>$ in terms of conserved charges, and the second is a consequenceof the properties of the SL_2 invariant vacuum $<0|$. Since $<0|L_{-n} = 0, n = -1, 0, 1, 2, \ldots$, the measure (3.4) is well defined . To finally make sure that (3.4) agrees with the measure obtained using path integral techniques, we will show that $\mu(V, \bar{V})$ is of the form:

$$\mu = (\rho \wedge \bar{\rho})(\det Im\tau)^{-13} \tag{3.5}$$

where ρ is a holomorphic $3g-3$ form on moduli space . To prove this, we only need to show that

$$\partial\bar{\partial}\log\mu(\ldots V_i \ldots) = -13\partial\bar{\partial} tr \log Im\tau \tag{3.6}$$

where $\partial, \bar{\partial}$ are derivatives with respect to the moduli space coordinates. From (3.2) and (3.4) we obtain:

$$\partial\bar{\partial}\log\mu(V_i,\bar{V}_i)=\frac{1}{(<0|\ldots|W>)^2}(<0|\ldots|W><0|\ldots T()\bar{T}()|W> \\ -<0|\ldots T()|W><0|\ldots\bar{T}()|W>) \tag{3.7}$$

where the dots refer to the $b(v_i)$ insertions. Equation (3.7) defines a (1,1)-form. To get a number for the above equation, the energy momentum tensors have to be folded in with vectors. It is clear from (3.7) that $\partial\bar{\partial}\log\tilde{\mu}$ would vanish if the left and right movers did not mix. As shown in section III, the state generated by the string coordinates mixes left and right. Thus, the evaluation of (3.7) reduces to a similar computation for the bosonic contribution to $|W>$. A straightforward computation then proves (3.6).

In (3.4) the ghost charge has been absorbed at a single point on the surface. It is not difficult however to modify the prescription so that the vacuum charge is absorbed at more than one point. Similarly , it is possible to obtain the corresponding measure for scattering amplitudes and derive the physical state conditions as well as the behavior of the measures at the boundary of moduli space; thus making sure that the $3g-3$ form ρ is indeed the Mumford form on M_g (see [4] for details).

IV Conclusions and outlook

The formalism explained in this lecture can also be applied to orbifold and toroidal compactifications of the bosonic string; and in principle it should also help in providing a geometrical interpretation of the discrete series of Virasoro representations [16,23]. There is a natural extension of the operator formalism to super and heterotic strings which is currently being completed [24]. The states are constructed on super-Riemann surfaces with punctures, and follows arguments similar to those in [4], one obtains a well defined supermeasure in supermoduli space. The ambiguities in superstring perturbation theory recently discovered by E. and H. Verlinde [25] (for detailed references and a thorough exposition of these ambiguities see [26]) seem to appear when one tries to construct the super measure at genus g starting with the super-sphere with three punctures (the "vertex") and the sewing operators ("the propagator"). The advantage however, is that one works from the beginning with conformally invariant data and gauge fixing procedure in the space of super-frames on a super-Riemann surface can be circumvented. We believe that this approach will yield some important insights in issues concerning the finiteness an consistency of superstring perturbation theory.

Acknowledgements. I would like to thank the organizers of the school for the opportunity to present these results in such a pleasant atmosphere. I am also grateful to C. Gomez, C. Reina, G. Moore, P. Nelson, G. Sierra and C. Vafa for countless conversations and discussions on many of the issues treated in these lectures.

References

[1] L.Alvarez-Gaumé, C. Gomez and C. Reina, Phys. Lett. **190B**, 55 (1987)

[2] L.Alvarez-Gaumé, C. Gomez and C. Reina, *New methods in string theory*, CERN-TH. 4775/85, Lecture delivered at the " Trieste School on Superstrings", April 1987. World Scientific 1988

[3] L.Alvarez-Gaumé, C. Gomez ,G. Moore and C. Vafa, *Strings in operator formalism* BUHEP-87/49, CERN-TH 4883/87, HUTP-87/A083, IASSNS-HEP-87/68

[4] G. Segal and G. Wilson, Publication Math. de l'I.H.E.S. **61**, 1 (1985)

[5] B.A. Dobrushin, I.M. Krichever and S.P. Novikov, *Topological and algebraic geometric methods in contemporary mathematical physics*, Sov. Sci. Reviews **3**, 1 (1982)

[6] E. Date, M. Jimbo, M. Kashiwara and T. Miwa, J. of Phys. Soc. of Japan **50**, 3806 (1981)

[7] S. Saito, Tokyo Metropolitan Univ. preprints : TMUP-HEL-8613/8615/8701/8707

[8] N. Ishibashi, Y. Matsuo and H. Ooguri, Univ. of Tokyo preprint UT-499 (1986)

[9] C. Vafa, Phys. Lett. **190** , 47 (1987)

[10] E. Witten *Quantum field theory, Grassmannians and algebraic curves* , Commun. Math. Phys. (in press)

[11] G. Segal, talk presented at the conference *Links between geometry and mathematical physics*, held at Schloss Ringberg, March 1987, based on joint work with D. Quillen

[12] C. Vafa, *Conformal theories and punctured Riemann surfaces*, HUTP-87/A 086

[13] A. Beilinson, Y.Manin and Y.A. Schechtman, *Localization and Virasoro and Neveu Schwartz algebras* ,preprint 1986

M. Kontsevich, Funct. Anal. and Appl. **21** 156 (1987) (= Funk. Anal. Pril **21**,78 (1987))

[14] E. Arbarello, C. DeConcini, V. Kac and C. Procesi *Moduli space of curves and representation theory*, Univ of Rome and MIT preprint 1987

[15] D. Friedan and S. Shenker, Phys. Lett. **175B**, 287 (1986); Nucl. Phys. **B281**, 509 (1987)

[16] A. Belavin, A.M. Polyakov and A.B. Zamolodchikov, Nucl. Phys. **241**, 333 (1984)

[17] D. Friedan , E. Martinec and S. Shenker, Nucl. Phys. **271** , 93 (1986)

[18] A. Neveu and P. West, Phys. Lett. **179b**, 235 (1986), *ibid.* **180B**, 34 (1986); **193B** Nucl. Phys. **287** ,601 (1986)

P. West *Multiloop ghost vertices and the determination of the multiloop measure* CERN-Th. 4937/88, 187 (1988); *The group theoretical approach to string amplitudes* CERN-TH preprint, to appear in Commun. Math. Phys., and references therein

M.D. Freeman and P. West, *Ghost vertices for the bosonic string using the group theoretical approach to string amplitudes*, CERN-TH. 49 38/88. A recent review of this approach with many details and references can be found in

P. West *A review of duality string vertices, overlap identities and the group theoretic approach to string theory*, CERN-TH 4819/87

[19] A. Leclair, M.E. Peskin and C.R. Preitschopf *String theory on the conformal plane I, II, III*, SLAC-PUB-4306, 4307, 4464

A. LeClair, Harvard preprint HUTP-87/A033, to appear in Nucl. Phys. B

A. LeClair, *An operator formulation of the superstring*, Princeton preprint PUTP-1080 and references therein

[20] P. di Vecchia, R. Nakayama, J.L.Peterson and S. Sciuto, Nucl. Phys. **B282** , 103 (1987)

[21] J. Fay, *Theta fuctions on Riemann surfaces*, Lecture Notes in Mathematics vol. **352** , Springer Verlag 1973

[22] A. Belavin and V. Knizhnik, Phys. Lett. **168B**, 201 (1986)

[23] D. Friedan, Z. Qiu and S. Shenker, Phys. Rev. Lett. **52** , 1575 (1984)

[24] L.Alvarez-Gaumé, C. Gomez ,P. Nelson , G. Sierra and C. Vafa (in preparation)

[25] E. Verlinde and H. Verlinde , Phys. Lett. **192B**, 45 (1987)

[26] J. Atick, G. Moore and A. Sen, *Some global issues in string perturbation theory*, IASSNS-HEP-87/61

ON THE FIXED POINT FORMULA AND THE RIGIDITY THEOREMS OF WITTEN LECTURES AT CARGÈSE 1987

Raoul Bott
Harvard University
Department of Mathematics
Cambridge, Mass., USA

1. INTRODUCTION

These lectures are meant as an introduction to the interesting ideas which had their origin on the one hand in some string theoretic considerations of Witten and on the other in the more topological considerations of Landweber, Stong, Ochanine, and others [12]. These two diverse points of view have spawned a subject called "Elliptic cohomology" and so, in a sense, these are introductory lectures to that subject also. However my point of view will be rather different. My aim is - *grosso modo* - to show how this whole development fits into the framework of an old fixed point theorem which Atiyah and I proved some 25 years ago, and for which I would in any case like to make some propaganda amongst my physicist friends.

The plan of this somewhat chauvinistic enterprise is then to try and explain the "rigidity" or "vanishing theorems" of Witten entirely in terms of our fixed point formula plus some complex function theory, – at least under the strong additional hypothesis that the S^1 action involved in the theorem only have *isolated singularities.*

Quite recently C. Taubes has – in a *tour de force* of a paper – given a complete proof of the general rigidity theorems by a fixed point theoretic method. Finally, then, – as I came to realize only later in the summer and fall of 1988 under Taubes' tutelage – these remarks are an appropriate introduction also to his rather densely written paper.

Witten's Rigidity Theorem is best thought of as an extension of the "rigidity" of the signature operator d_s on a finite-dimensional oriented manifold M (see section 3 for a definition of d_s) to the loop space $\mathcal{L}(M)$ of M.

The operator d_s has for its index the signature $\tau(M)$ of M and is *rigid* in the *sense that any S^1 action* acts trivially on the virtual S^1-index: *ker* d_s – *coker* d_s. This follows most easily from the *homotopy axiom for de Rham theory,* because both *ker* d_s and *coker* d_s are spaces of harmonic forms on M.

Now Witten argues that d_s "should be definable on $\mathcal{L}(M)$ provided $\mathcal{L}(M)$ is orientable". This, he argues, should be true only if M is spin. Thereafter by beautiful arguments - which, however, are allowed only to physicists - he concludes that the index of this signature operator on $\mathcal{L}(M)$ is *again* rigid, *under* S^1 *motions* of M. Further, to go on, the *ker* and *coker* of d_s on $\mathcal{L}(M)$ are both infinite, but their Fourier coefficients under the natural action of rotating the loops *are finite* and are in fact given by the indexes of certain *twisted* forms $d_s \otimes R_n$ of the signature operator on M; in this manner he finally concludes on the one hand that for *spin-manifolds* the indexes of the $d_s \otimes R_n$ should be rigid, and on the other that apart from a normalization, the power series $\Sigma \quad q^n$ index $(d_s \otimes R_n)$ is modular for the group $\Gamma_0(2)$. For good measure he goes on to argue that the same should be true for a "loop space Dirac operator" and so produces a sequence of representations R'_n for which the index of the twisted Dirac-operator $\partial \otimes R'_n$ is rigid.

In these lectures the loop space will not make its appearance at all. Rather, after reviewing the fixed pont theorem from scratch in section 2, I will explain how it yields an independent proof of the rigidity of d_s and also of the spin operator ∂ in sections 3 and 4. I will then also explain the relationship of these concepts to equivariant characteristic numbers (section 5). Finally I hope to return to the Witten-Taubes theorem proper in section 6, and briefly motivate Witten's formulae as a "push-forward" of the fixed point formulae from $\mathbf{C}^*$ to the torus $\mathbf{C}^*/\{q\}$.

2. THE FIXED POINT FORMULA

To set the stage, consider a map (function)

$$\varphi : X \longrightarrow X, \tag{2.1}$$

of a finite set into itself. This φ then induces a linear map

$$\varphi^* : \mathcal{F}(X) \longrightarrow \mathcal{F}(X), \tag{2.2}$$

in the space of functions on X with values in $\mathbf{R}$, by the rule:

$$\varphi^* f(x) = f\{\varphi(x)\}.$$

Now I claim that:

$$\textit{trace}\ \varphi^* = \text{number of fixed points of } \varphi. \tag{2.3}$$

Indeed the δ-functions, δ_x, give a basis for $\mathcal{F}(X)$ in which φ^* has a diagonal entry of $+1$ precisely at the fixed points of φ. Now the point which Atiyah and I noticed originally is that this formula has a natural generalization in the C^∞ context; however only if φ satisfies the transversality condition that its graph meets the diagonal in $X \times X$ transversally:

The picture that goes with this concept is of course the following one:

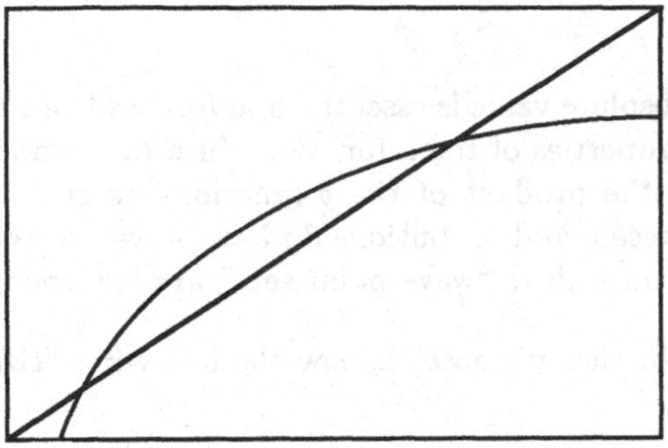

Fig. 1

and transversality is quite equivalent to the demand that at each fixed point p of φ,

(2.4) $$\det(1 - d\varphi_p) \neq 0, \qquad d\varphi_p = \|\frac{\partial \varphi^i}{\partial x_j}\|_p$$

Under these assumptions we argued that the induced linear map φ^* in $\mathcal{F}(X)$ – now the C^∞ functions on X, – "should" have the trace given by:

(2.5) $$\mathit{trace}\ \varphi^* = \sum_{\varphi(p)=p} 1/|\det(1 - d\varphi_p)|.$$

The heuristics for this formula should certainly be convincing to any physicist worth his salt. Consider first a transformation $T : \mathcal{F}(M) \longrightarrow \mathcal{F}(M)$ given by a continuous kernel:

(2.6) $$Tf(x) = \int K(x,y) f(y) |dy|.$$

Clearly the trace of such a T "should be" $\int K(x,x)|dx|$. But the kernel of our φ^* is just the delta-function of the graph of φ:

(2.7) $$K_\varphi(x,y) = \delta(x, \varphi(x)).$$

Hence the trace of φ^* should be the integral $\int_M \delta(x, \varphi(x))$. Now this integral is clearly equal to the sum of the integrals $\int_{U_i} \delta(x, \varphi(x))$ where the U_i are small neighborhoods of the i-th fixed point. But there:

$$\int \delta(x-\varphi(x))|dx| = \int \frac{\delta(u)|du|}{|\det(1-\varphi')|} = \frac{1}{|\det(1-\varphi'_p)|}$$

as the substitution $x - \varphi(x) = u$ shows. QED

By the way, the absolute value is essential and follows from a proper understanding of the functorial properties of the δ-function. In a more modern context one can think of $trace(\varphi^*)$ as the product of the δ-functions carried by the diagonal and the graph of φ, and these two distributions do have a well-defined product precisely because – by assumption – their "wave point sets" are transversal.

Essential for our further purposes is now the following "twisted" version of the fixed point theorem.

Suppose then that E is a vector bundle over M and that our map φ admits a lifting $\widehat{\varphi}$ to E. Precisely this means that

(2.8) $$\widehat{\varphi} : E \longrightarrow E$$

is a map of E into itself which is linear on the fiber E_p of E and which sends $E_{\varphi(p)} \to E_p$ for all $p \in M$. We write $\widehat{\varphi}_p$ for this map and note that it is *only at a fixed point of* φ that $trace(\widehat{\varphi}_p)$ is meaningful. Now, then, given the lifting $\widehat{\varphi}$ of φ we can define an automorphism

(2.9) $$\varphi_E^* : \Gamma(E) \longrightarrow \Gamma(E)$$

on the space of sections of E, by the formula

(2.10) $$(\varphi_E^* s)(p) = \widehat{\varphi}_p \circ s\{\varphi(p)\}.$$

and with this understood our old argument immediately generalizes to the following trace formula for φ_E^*:

(2.11) $$trace\ \varphi_E^* = \sum_p \frac{trace\ \widehat{\varphi}_p}{|\det(1-d\varphi_p)|}.$$

So much for our trace. In the next section I hope to explain why it is a "good" one.

3. THE LEFSCHETZ PRINCIPLE

The index of a linear map $h : V^+ \longrightarrow V^-$ is defined by:

(3.1) $$index\ h = \dim\ Ker\ h - \dim\ Coker\ h.$$

If the dimensions of V^+ and V^- are finite the index h is *independent of* h and is given by:

$$\text{index } h = \dim V^+ - \dim V^-. \tag{3.2}$$

It is this insensitivity of the index to finite dimensional perturbations that makes it a homotopy invariant in certain infinite dimensional contexts.

Similarly the *Lefschetz principle* has its origin in the following extension of (3.1). Namely suppose that T is an endomorphism of the map h. That is, we have a commutative diagram:

$$\begin{array}{ccc} V^+ & \xrightarrow{h} & V^- \\ T\downarrow & & \downarrow T \\ V^+ & \xrightarrow{h} & V^-. \end{array}$$

It follows that T induces a map $H^+(T)$ in *Ker* h and a map $H^-(T)$ in *Coker* $h \approx V^-/hV^+$. The Lefschetz principle asserts that in this situation and in finite dimensions:

$$\text{trace } H^+(T) - \text{trace } H^-(T) = \text{trace } T^+ - \text{trace } T^- \tag{3.3}$$

or, put differently,

$$\text{Supertrace } H(T) = \text{Supertrace } T \tag{3.4}$$

For T the identity this clearly reduces to (3.1) because the trace of the identity map is simply the dimension.

The primary virtue of the trace I defined in the first lecture is that it obeys this *Lefschetz principle in many infinite-dimensional contexts.*

The simplest of these is the following one. Suppose M is compact and that E^+ and E^- are vector bundles over M and that

$$D : \Gamma(E^+) \longrightarrow \Gamma(E^-) \tag{3.5}$$

is an *elliptic* differential operator between them.

Also assume that φ is a map with liftings $\hat{\varphi}^{\pm}$ commuting with D and hence inducing endomorphisms $H^{\pm}(\varphi)$ in $H^+ = \text{Ker } D$ and $H^- = \text{Coker } D$. Under these conditions *our trace satisfies the Lefschetz Principle*: that is,

$$\text{trace } H^+(\varphi) - \text{trace } H^-(\varphi) = \text{trace } \hat{\varphi}_{E^+} - \text{trace } \hat{\varphi}_{E^-}. \tag{3.6}$$

In this way, then, the supertrace of $H(\varphi)$ is computed purely in local terms at the *fixed points*, that is, each point $p = \varphi(p)$, contributes by:

$$\mu_p = \frac{trace\ \widehat{\varphi}_{E^+}(p) - trace\ \widehat{\varphi}_{E^-}(p)}{|\det(1 - d\varphi_p)|}.$$

Example 1. Let M be compact and let $\Omega^* = \oplus \Omega^q$ denote the DeRham complex of q forms on M. Let d be the usual exterior differentiation operator, and let d^* be its adjoint relative to some fixed Riemann structure on M. Next set Ω^{ev} = even-dimensional forms and Ω^{odd} = odd dimensional forms, so that

$$d + d^* : \Omega^{ev} \longrightarrow \Omega^{odd} \tag{3.7}$$

is well defined.

This is an *elliptic* operator, and any *isometry* $\varphi : M \to M$ of M as a Riemann manifold, will naturally lift to Ω^*. The formula (3.6) now gives us information on how the *harmonic forms* $\mathcal{H}^q(M)$ on M move under φ – in terms of the fixed point behaviour of φ.

Indeed,

$$Ker(d + d^*) = \sum_{q \text{ even}} \mathcal{H}^q(M)$$

while

$$coker(d + d^*) = \sum_{q \text{ odd}} \mathcal{H}^q(M).$$

Hence the LHS goes over into:

$$L(\varphi) = \sum (-1)^q trace\ \mathcal{H}^q(\varphi) \tag{3.8}$$

with $\mathcal{H}^q(\varphi)$ the map induced by φ on $\mathcal{H}^q$.

This expression is actually the classical Lefschetz number of φ – and can also be defined quite independently of a Riemann structure by:

$$L(q) = \sum (-1)^q trace\ H^q(\varphi), \tag{3.9}$$

where now $H^q(\varphi)$ denotes the linear map induced on the qth DeRham cohomology:

$$H^q(M) = Ker\ d \text{ in } \Omega^q / \Omega^{q-1}. \tag{3.10}$$

of M.

So much for the LHS.

On the RHS of (3.6) I claim that the fixed point formula *simply yields* ± 1 *at each point.* To see this recall that if A is a finite-dimensional matrix then

$$\det(1 - A) = \Sigma(-1)^q \, trace(\Lambda^q(A) \tag{3.11}$$

where Λ^q denotes its qth exterior power.

Hence at each fixed point p of φ the RHS of (3.6) is given by:

$$\frac{trace\ \widehat{\varphi}_p^+ - trace\ \widehat{\varphi}_p^-}{|\det(1 - d\varphi_p)|} = \frac{\det(1 - d\varphi_p)}{|\det(1 - d\varphi_p)|} = \pm 1. \tag{3.12}$$

In short, (3.6) goes over into the classical expression for the Lefschetz number:

$$L(\varphi) = \sum_{\varphi(p)=p} \pm 1. \tag{3.13}$$

Example 2. The *signature operator* of an *oriented* M is again defined in terms of an operator, $i^{1/2}d + i^{-1/2}d^*$, of our DeRham complex, but now we consider its action in a more subtle splitting of Ω^* than the one induced by the dimension of the forms.

Namely, we use the $*$ operator

$$* : \Omega^q \to \Omega^{n-q} \qquad *^2 = 1 \tag{3.14}$$

furnished by the orientation of M to define $\Omega^\pm$ according to the eigenvalue of $*$. With this understood we set $d_s = i^{1/2}d + i^{-1/2}d^*$ *but acting from* Ω^+ to Ω^-: that is, as the operator:

$$d_s : \Omega^+ \to \Omega^-. \tag{3.15}$$

Now the fixed-point formula reads

$$trace\ \mathcal{H}^+(\varphi) - trace\ \mathcal{H}_s^-(\varphi) = \sum_{\varphi(p)=p} \mu_p \tag{3.16}$$

with

$$\mu_p = \pm \prod_1^{2k} \frac{1 + e^{i\theta_\alpha}}{1 - e^{i\theta_\alpha}},$$

where the θ_α range over the angles by which $d\varphi_p$ rotates the tangent space at p. On the LHS $\mathcal{H}_s^{\pm}(\varphi)$ denotes the map induced by φ on the $*$ invariant and the $*$ anti-invariant harmonic forms.

We will make the sign convention for this formula more precise in the next section. Here let me just record the two other fundamental geometric instances of the fixed point formula:

Example 3. The *Spin Operator*. Here we must assume that M admits as spin-structure so that the Dirac-Weyl operator

(3.17)
$$\partial\!\!\!/ : \Delta^+ \longrightarrow \Delta^-$$

becomes well defined.

The harmonic spinors we denote by $\mathcal{H}^{\pm}(M)$ and the fixed point formula now reads:

$$\mathit{trace}\ \mathcal{H}^+(\varphi) - \mathit{trace}\ \mathcal{H}^-(\varphi) = \sum_{\varphi(p)=p} \mu_p$$

with

(3.18)
$$\mu_p = \pm \prod_1^{2k} \frac{e^{i\theta_\alpha/2}}{1 - e^{i\theta_\alpha}},$$

while the angles θ_α are as before.

Example 4. *The Complex Analytic Case.* Here we assume that M is complex analytic and that φ preserves this structure. Then $\Omega^* = \bigoplus_{a,b}\Omega^{a,b}(M)$ defines a splitting into forms of type (a, b) and $d = \partial + \bar\partial$ splits correspondingly into a ∂ and a $\bar\partial$ operator. These raise a and b respectively by 1, so that fixing a, yields a family of $2d$ separate $\bar\partial$ complexes:

$$\Omega^{a,*} : \Omega^{a,0} \xrightarrow{\bar\partial} \Omega^{a,1} \xrightarrow{\bar\partial} \cdots\ \Omega^{a,2d}$$

and the corresponding fixed point formula for $a = 0$, say, reads

$$\sum (-1)^q\ \mathit{trace}\ \mathcal{H}_{\bar\partial}^{0,q}(\varphi) = \sum \mu_p$$

with

(3.19)
$$\mu_p = \prod \left(\frac{1}{1 - e^{i\theta_\alpha}}\right).$$

For the higher a we multiply μ_p with the ath symmetric function in $e^{i\theta_\alpha}$.

In these formulae the $e^{i\theta_\alpha}$ must however be taken to describe the action of $d\varphi'_p$ on the *holomorphic part* of the tangent space on M.

All these examples are discussed in some detail in [2], and they really just amount to the appropriate linear algebra applied to the geometrically induced liftings of φ at p.

Finally a remark about the twisting of these operators by representations R of the appropriate spinor group.

The fixed point contribution μ'_p of the twisted complex is obtained from the μ_p of the original one by *simply multiplying with the trace of R evaluated at $\widehat{\varphi}_p$*: That is,

$$\mu'_p = \mu_p \cdot trace\ R(\widehat{\varphi}_p). \tag{3.20}$$

Thus for instance on a Spin manifold,

$$d_s = \partial\!\!\!/ \otimes \Delta \qquad \text{where} \qquad \Delta = \Delta^+ + \Delta^-.$$

Correspondingly one has:

$$\frac{\prod e^{i\theta_\alpha/2} \cdot (e^{i\theta_\alpha/2} + e^{-i\theta_\alpha/2})}{\prod 1 - e^{i\theta_\alpha}} = \prod \frac{1 + e^{i\theta_\alpha}}{1 - e^{-i\theta_\alpha}} \tag{3.21}$$

because $\prod_\alpha (e^{i\theta_\alpha/2} + e^{-i\theta_\alpha/2}) = trace\ \Delta(\widehat{\varphi}_p)$.

4. EQUIVARIANT CHARACTERISTIC NUMBERS

From now on let us work in the category of compact oriented manifolds M, on which we assume that a circle S^1 acts *non-trivially*.

In fact let us assume that S^1 acts with *isolated fixed points only*. This is unfortunately *not* a generic assumption. But it is a "morally generic" one in the sense that if one masters this case, then the more general one usually can be mastered by similar methods.

Such an action provides one with a system of integers – one for each fixed point – which describe the action near that point and which we call the "*exponents of the action*" there. Indeed under the action of S^1 the tangent space at p decomposes into a direct sum of invariant 2-planes.

$$T_p = E_1 \oplus \cdots \oplus E_d, \qquad \dim M = 2d \tag{4.1}$$

on each of which $e^{i\theta} \in S^1$ acts by a rotation through $n\theta$. In this way p has associated to it a set of d integers $\{n_\alpha\}$; $\alpha = 1 \cdots d$. We can - by orienting the E_α properly - so arrange it that these "*exponents at p*" are all > 0, and once that is done compare the orientation so induced on T_p with the global one of M. This comparison finally yields $a \pm 1_p$ *canonically attached* to p by the action.

I noted long ago [4] that this system of exponents determined the *ordinary characteristic numbers* of M. Actually, as is explained in [3] they determine *all* the equivariant characteristic numbers also, and the recipe is the following one:

Let $\phi(x_1, \cdots, x_d)$ be a symmetric polynomial in the variables x_1^2, $i = 1 \cdots d$. Then the sum over the fixed points of the action:

$$\phi(M) = \sum_p \pm \frac{\phi(n_1, \cdots, n_d)}{n_1 n_2 \cdots n_d} \tag{4.2}$$

is the characteristic number, $\phi(M)$, associated to ϕ via the standard procedure which represents the kth Pontryagin class p_k by the kth elementary symmetric polynomial in the x_i^2.

REMARKS. Thus $p_1 = \sum x_i^2$, $p_2 = \Sigma_{i<j} x_i^2 x_j^2$, etc. Let me spell this out in greater detail with an example. Let us write L^n for the complex representation of S^1 on C which sends $e^{i\theta}$ to $e^{in\theta}$. Let us choose three numbers $a < b < c$, form the representation $V = L^a \oplus L^b \oplus L^c$ and let $M = P(V)$ be the associated projective space of 1-dimensional subspaces of V. The action on V clearly descends to $P(V)$ and there it will have 3 fixed points given by the three lines L^a, L^b and L^c themselves.

The exponents and "sign" of such a fixed point are then easily computed:

At L^a the exponents are $(b - a, c - a)$ and the sign of L^a counts the sign of $b - a$ and $c - a$ mod 2. One computes similarly at L^b and L^c. Thus if $a, b, c = 0, 1, 3$ – for instance – we can compute $p_1(M)$ and $p_1(M)^2$ by:

$$p_1(M) = \frac{1^2 + 3^2}{1 \cdot 3} - \frac{1^2 + 2^2}{1 \cdot 2} + \frac{3^2 + 2^2}{2 \cdot 3} = 3. \tag{4.3}$$

$$\begin{aligned} p_1(M)^2 &= \frac{(1^2 + 3^2)^2}{1 \cdot 3} - \frac{(1^2 + 2^2)^2}{1 \cdot 2} + (\frac{3^2 + 2^2}{2 \cdot 3})^2 \\ &= 49, \end{aligned} \tag{4.4}$$

and so on.

In short, all the characteristic numbers are determined by the exponents and we could perfectly well take (4.2) as the *definition* of these integers in our discussion. Nevertheless let me make a few remarks here about their topological origin, to set them into the proper perspective.

The point is that in the category of manifolds with S^1 action, there is a natural *cohomology theory* called the *equivariant cohomology* of M, notated $H^i_{S^1}(M)$, and the whole theory of ordinary characteristic classes extends to give invariants in $H^*_{S^1}(M)$ for any bundles $E \to M$ with *S^1-action covering the given one on M.*

Now just as in the ordinary theory, one obtains *characteristic numbers* of a manifold by integrating characteristic *classes* of the tangent bundle to M over the *fundamental class* of M, so in equivariant theory one obtains equivariant characteristic numbers by integrating the equivariant characteristic classes over M. Properly speaking these characteristic numbers are to be thought of as elements in $H^*_{S^1}(pt)$, the equivariant cohomology of a point. In the classical case this cohomology is simply $\mathbf{R}$ in dimension zero, – whence the term characteristic number – however, because $H^*_{S^1}(pt) = \mathbf{R}[[u]]$ is the formal power series generated by an element $u \in H^2$, the equivariant characteristic numbers of the tangent bundles M in general are polynomials in u. Thus for instance if $p_1 \in H^4_{S^1}(M)$ is the first equivariant Pontryagin class of the tangent bundle of M, dim $M = 4k$, say, then

$$\int_M p_1^\ell \in H^{4(\ell-k)}_{S^1}(pt). \tag{4.5}$$

need not be zero for $\ell \geq k$. Rather this integral will be a multiple of $u^{\ell-k} \in H^{4(\ell-k)}(pt)$ and this *multiple we denote by* $p_1^\ell(M)$. In general if $p^\alpha = p_1^{\alpha_1} \cdots p_k^{\alpha_k}$ is a monomial in the equivariant Pontryagin classes of M, we write

$$\int_M p^\alpha = p^\alpha(M) \cdot u^\beta, \qquad \beta = \dim p^\alpha - 4k. \tag{4.6}$$

That these "topological" $p^\alpha(M)$ can be computed by the formula (4.2) for actions of our type is explained in [3], as remarked earlier.

5. VANISHING AND INTEGRALITY PROPERTIES OF THE CHARACTERISTIC NUMBERS

The great virtue of having an S^1-action of the sort we are considering is that the action of $e^{i\theta}$ on M is transversal to the identity for *almost all θ.* Indeed the action clearly fails to be transversal only at the n_α^{th} roots of unity, where n_α is an exponent of the action at some fixed point p. In particular $e^{i\theta}$ acts transversally for θ small enough but not $= 0$. Let us write down the fixed point theorem for the signature operator - say - and study its behaviour as *θ varies on S^1.*

According to (4.1) we obtain for all small $\theta \neq 0$:

$$trace\{H^+(e^{i\theta}) - H^-(e^{i\theta})\} = \sum_{\{p\}} \prod_{\alpha=1}^{d} \left(\frac{1+\lambda^{n_\alpha}}{1-\lambda^{n_\alpha}}\right) \tag{5.1}$$

where I have now written $e^{i\theta} = \lambda$. Note that the LHS of this equation, has the form:

$$LHS = \sum_{n=-N}^{n=N} a_n e^{in\theta}, \quad a_n = a_{-n} \in \mathbf{Z} \tag{5.2}$$

because we are dealing with a finite dimensional *real* representation of S^1. We thus obtain the formula:

$$\sum_{n=-N}^{n=+N} a_n \lambda^n = \sum_{(p)} \pm 1 \prod \left(\frac{1+\lambda^{n_\alpha}}{1-\lambda^{n_\alpha}} \right). \tag{5.3}$$

Here, of course, the $=$ is valid *a priori* only on a dense subset of $e^{i\theta}$'s on S^1. But then - because both sides are analytic in λ except for poles, the *equality must actually hold for all* $\lambda \in \mathbf{C}$. This principle of "*analytic extension*" of the fixed point formula was first used by Lusztig, and is crucial to all that follows.

We conclude that the *poles* on the *two sides must be equal.* On the other hand the LHS has possible poles only at $\lambda = 0$, and ∞ while the RHS has possible poles only on the unit circle!

It follows that the LHS and RHS must both reduce to a constant, or put differently, that the *index* of d_s is *rigid.*

From $a_n = 0$ for $n \neq 0$, it follows that:

$$a_0 = \sum_p \pm 1 \prod_\alpha \left(\frac{1+\lambda^{n_\alpha}}{1-\lambda^{n_\alpha}} \right). \tag{5.4}$$

This a_0 is of course just $\dim H^+ - \dim H^-$, that is, the signature of M which we denoted by $\tau(M)$, so that (5.4) gives us a *large variety of ways of computing* $\tau(M)$.

For instance, setting $\lambda = 0$ we get

$$\tau(M) = \sum_p \pm 1 \tag{5.5}$$

But one can also evaluate $\tau(M)$ in a more adventurous manner by expanding each fixed-point contribution at the pole $\lambda = 1$, and then keeping *only the constant terms.* This procedure must work because the divergent terms have to *cancel after summation* over all the fixed points. This procedure - as I will now show - brings us quite naturally into the realm of Hirzebruch's *L*-polynomials, and their "unstable" extensions.

To see this let us start by studying the behaviour of $(1+\lambda)/(1-\lambda)$ near its pole at $\lambda = 1$. For this purpose set $\lambda = e^{-z}$.

Then the formula:

$$(5.6)\qquad \begin{aligned}\frac{1+e^{-z}}{1-e^{-z}} &= \frac{1}{z}\left\{\frac{z}{1-e^{-z}}\cdot(1+e^{-z})\right\}\\ &= \frac{1}{z}\{L_0 + L_2 z^2 + L_4 z^4 + \cdots\}\end{aligned}$$

defines a power series in z^2 which we call the *L-function.* The coefficients of $L(z)$ are Bernoulli numbers divided by factorials:

$$(5.7)\qquad L(z) = 2 + \sum_{k=1}^{\infty} \frac{(-1)^{k-1}B_{2k}}{(2k)!} z^{2k}.$$

Thus, for example,

$$L(z) = 2 + \frac{1}{6}z^2 - \frac{1}{30\cdot 12}\cdot z^4 + \cdots.$$

From this power series in *one* variable one now generates new ones in several variables by multiplication:

$$(5.8)\qquad L^{(k)}(z_1,\cdots,z_k) = L(z_1)\cdot L(z_2)\cdots L(z_k)$$

and denotes by $L_j^k(z_1,\cdots z_k)$ the homogeneous term of degree j in $L^{(k)}(z_1,\cdots,z_k)$.

For example:

$$(5.9)\qquad \begin{aligned} L^2(z_1,z_2) &= 4 + \underbrace{\frac{1}{3}(z_1^2+z_2^2)}_{(2)} - \underbrace{\frac{1}{30\cdot 6}(z_1^4+z_2^4) + \frac{1}{36}z_1^2 z_2^2}_{(4)} + \cdots\\ &= L_0^{(2)} + L_2^{(2)} + L_4^{(2)} + \cdots.\end{aligned}$$

Finally these L_j^k – which are manifestly symmetric in the z_i^2, can be rewritten in terms of the elementary symmetric functions

$$p_1 = \sum_1^k z_i^2,\quad p_2 = \sum_{i<j} z_i^2 z_j^2,\ \cdots,\ p_k = \sum_{i_\ell\cdots i_k} z_i^2\cdots z_{k_k}^2.$$

Thus for instance:

$$(5.10)\qquad L^{(2)}(z_1,z_2) = 4 + \frac{1}{3}p_1 + \frac{1}{36\cdot 5}\cdot(7p_2 - p_1^2) + \cdots.$$

With all this understood let us evaluate the constant term of a fixed point contribution μ_p at a point with exponents $(n_1, \cdots, n_d)$. Expanding in terms of z we get

$$\mu_p = \frac{\Pi(1+\lambda^{n_\alpha})}{\Pi(1-\lambda^{n_\alpha})} = \frac{1}{n_1 \cdots n_d \cdot z^d} \cdot \{L(n_d z) L(n_2 z) \cdots L(n_d z)\} \tag{5.11}$$

so that the constant term is clearly:

$$\frac{L_d^{(d)}(n_1, n_2, \cdots, n_d)}{n_1 \cdots n_d}. \tag{5.12}$$

It follows that $\tau(M)$ can also be computed by:

$$\sum \pm \frac{L_d^{(d)}(n_1, \cdots, n_d)}{n_1 \cdots n_d} = \tau(M). \tag{5.13}$$

Put differently, $\tau(M)$ is given by the characteristic number $L_d^{(d)}(M)$ – that is, precisely by evaluating the "L-genus" on M.

We can go further by equating the higher order powers in the z-expansion of (5.3). On the LHS one obtains 0, but on the RHS z^j occurs with the coefficient $L_j^{(d)}(M)$. Thus the full force of the rigidity of d_s is expressed by the formula:

$$L_j^{(d)}(M) = \begin{cases} \tau(M), & \text{if } j = d \\ 0, & \text{if } j > d. \end{cases}$$

When M has a spin-structure an entirely similar analysis applied to the Dirac operator is possible and now leads first of all to the formula

$$\sum b_n \lambda^n = \sum (\pm) \frac{\lambda^{\frac{n_1 + \cdots + n_d}{2}}}{\Pi(1-\lambda^{n_\alpha})} \tag{5.14}$$

where now b_n are the "Fourier coefficients" of $\mathcal{H}^+ - \mathcal{H}^-$ as an S^1-module. The poles again have to cancel whence $b_n = 0$ for $n \neq 0$ – *but this time we can evaluate* b_0 *by simply* setting $\lambda = 0$ on the right. This gives $b_0 = 0$! Thus the index of ∂ is rigid with a vengeance. On the other hand the "polar expansion" procedure now produces new identities among the exponents which are derived from the power series $\widehat{A}(z)$ associated to $\frac{\lambda^{1/2}}{1-\lambda^{1/2}}$ by the formula:

$$\frac{e^{-z/2}}{1-e^{-z}} = \frac{1}{z} \cdot \widehat{A}(z). \tag{5.15}$$

Thus up to order 4

$$\widehat{A}(z) = 1 + \frac{z^2}{24} + \frac{7}{2^7 \cdot 3^2 \cdot 5} \cdot z^4 + \cdots \tag{5.16}$$

whence

$$\widehat{A}^{(2)}(z_1, z_2) = 1 + \frac{p_1}{24} + \frac{1}{2^7 \cdot 3^2 \cdot 5}(p_1^2 - 4p_2) + \cdots . \tag{5.17}$$

In any case the vanishing of the LHS of (5.14) now has as a consequence that:

$$\sum \pm \frac{\widehat{A}_j^{(d)}(n_1, \cdots, n_d)}{n_1 \cdots n_d} = 0 \text{ all } j, \tag{5.18}$$

or put differently – that a new array $\widehat{A}_j^{(d)}(M)$ of characteristic number of M vanish – which in particular include the $\widehat{A}$ genus of M.

Thus the identities (5.18) should be thought of as a proof of the celebrated Atiyah-Hirzebruch vanishing theorem [4] for actions of our especially simple type. Their theorem of course asserts that if $\widehat{A}(M) \neq 0$ then M cannot admit nontrivial S^1-action of any sort, and is still the only known obstruction to an S^1 action on simply connected spin manifolds.

Note that when dim $M = 4$ then $\tau(M) = p_1(M)/3$ and $\widehat{A}(M) = p_1(M)/24$. Thus if M is spin, then $\tau(M) = 0$ also. Similarly we see that under these conditions both $p_1^2(M)$ and $p_2(M)$ must vanish because $L_4^{(2)}$ and $\widehat{A}_4^{(2)}$ are linearly independent combinations of those two numbers.

6. REMARKS ON THE NEW VANISHING THEOREMS

I have only time and space for a few remarks of how the *new vanishing* theorems fit into the previous discussion, and what I have to say represents joint work with Taubes – done during the past few months – which reinterprets his earlier work in the present framework.

Let me start by reminding you that every complex torus can be displayed as the quotient of $\mathbf{C}^* = \mathbf{C} - 0$, by the action of an element $q \in \mathbf{C}^*$, $|q| < 1$ under multiplication:

$$T_q = \mathbf{C}^*/(q\lambda \sim \lambda) \qquad \lambda \in \mathbf{C}^* \tag{6.1}$$

Indeed a fundamental domain for this action is the annulus $q \leq |\lambda| \leq 1$,

and multiplication by q then identifies the boundaries to give a torus.

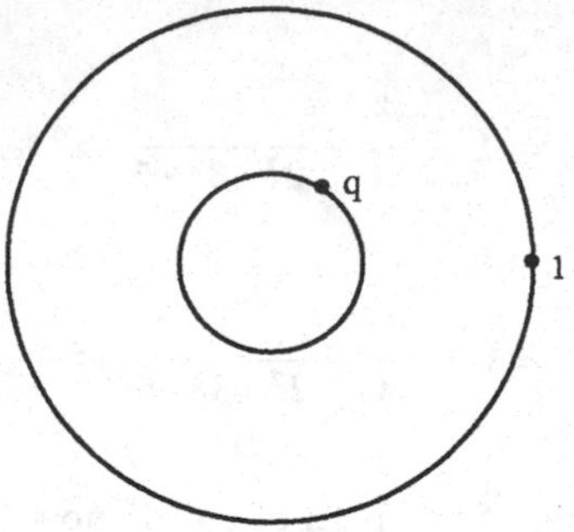

Fig. 2

Let us denote the quotient map

(6.2) $$\mathbf{C}^* \xrightarrow{\pi} T_q$$

by π, and let us try and "push forward" rational functions on $\mathbf{C}^*$ to rational functions on T_q.

For instance, consider

(6.3) $$\varphi(\lambda) = \frac{1+\lambda}{1-\lambda}$$

the basic constituent of the *L*-function.

A first formal attempt at pushing φ forward – i.e., making it invariant under the substitution $\lambda \to q\lambda$ – might be:

(6.4) $$\pi_*\varphi(\lambda) \overset{(?)}{=} \prod_{n \in \mathbf{Z}} \varphi(q^n\lambda).$$

Unfortunately this does not make sense because $\Sigma|q|^{-n}$ and $\Sigma|q|^n$ cannot both be convergent. However note that

(6.5) $$\varphi(\lambda^{-1}) = -\varphi(\lambda).$$

It suggests itself, therefore, that $\pi_*\varphi$ might make sense if we define it by:

(6.6) $$\pi_*\varphi(\lambda) = \varphi(\lambda) \prod_{n=1}^{\infty} \varphi(q^n\lambda) \cdot \varphi(q^n\lambda^{-1}).$$

And indeed - just expanding naively in q and λ now yields a Laurent series for

(6.7) $$\prod_{n=1}^{\infty} \varphi(q^n\lambda)\varphi(q^n\lambda^{-1}) = \sum a_n(q)\lambda^n$$

which converges in the annulus:

$$|q| < |\lambda| < \frac{1}{|q|}. \tag{6.8}$$

Hence $\pi_*\varphi(\lambda)$ is a well defined meromorphic function in this domain, and by virtue of (6.4) is seen to satisfy the identity $\pi_*\varphi(q\lambda) = -\pi_*\varphi(\lambda)$

It follows that $\pi_*\varphi$ ***extends uniquely*** to a meromorphic function in $\mathbf{C}^*$, with the periodic relation:

$$\pi_*\varphi(q\lambda) = -\pi_*\varphi(\lambda), \tag{6.9}$$

and so defines a meromorphic function – not on T_q directly – but certainly on T_{q^2}. As such it has *poles* at 1 and q and *zeroes* at -1 and $-q$.

Now the point ± 1 and $\pm q$ are precisely the points of order 2 on T_{q^2}. Hence, up to a constant, $\pi_*\varphi$ is *uniquely* associated to a torus on *which a nontrivial point of order two has been singled out*, and is one of a family of three such functions on a "bare" torus.

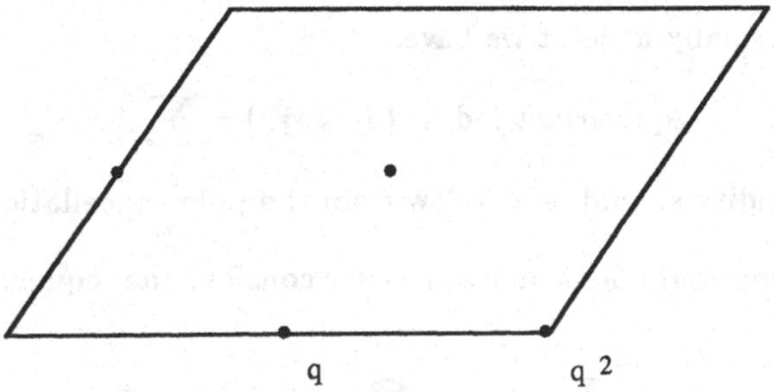

Fig. 3

In any case I want to make Wittens' rigidity theorem plausible, by arguing that his theorem comes about by *pushing* the *rigidity* of d_s *on* M *down to the torus* T_{q^2}.

More precisely, recall that the rigidity of d_s came about from the formula:

$$\textit{index } d_s = \sum_p \pm\mu_p \tag{6.10}$$

where $\mu_p = \varphi(\lambda^{n_1})\cdots\varphi(\lambda^{n_d})$, and can be thought of as a remarkable cancellation of the poles among the rational functions μ_p.

Suppose now that this cancellation persisted under the push-forward. Then we might expect that $\Sigma \pm \pi_* \mu_p$ to again give rise to a constant – i.e., a function of q alone.

But the formula:

$$\pi_*\varphi(\lambda) = \frac{1+\lambda}{1-\lambda} \bigotimes_{n,m} \frac{(1+q^n\lambda)(1+q^m\lambda^{-1})}{(1+q^n\lambda)(1-q^m\lambda^{-1})} \tag{6.11}$$

now suggests that $\pi_* u_p$ should be interpreted as the contribution at p, of a *twist* of d_s by *some representation* W_q. And indeed it is then not too hard to come up with Witten's,

$$W_q = \bigotimes_{n,m\geq 1} \Lambda_{q^n} T \bigotimes S_{q^m} T \tag{6.12}$$

as the unique solution.

Here T is the standard representation of $O(2d)$ and

$$\Lambda_a T = \Sigma a^k \Lambda^k T$$
$$S_b(T) = \Sigma b^k S_k T$$

are the indicated linear combinations of the exterior and symmetric powers of T.

In short, formally at least we have:

$$\text{equivariant index } (d_s \otimes W_q) = \sum \pm \pi_* \mu_p, \tag{6.13}$$

and Witten Rigidity should now follow from the pole cancellation of the $\pi_* u_p$.

To spell things out a little more precisely consider the sequence of representations defined by

$$\sum q^n R_n = \bigotimes_{n,m} \Lambda_{q^n} T \bigotimes \Lambda_{q^m} T. \tag{6.14}$$

The LHS of the formal twist of d_s by W_q in the fixed-point formula is of the form

$$LHS = \sum_n q^n \sum_{k=-N}^{k=N} a_k^n \lambda^k; \quad a_k^n = a_{-k}^n \tag{6.15}$$

with a_k^n the Fourier coefficients of the index of $d_s \otimes R_n$.

The fixed point formula thus takes the form:

$$\sum_{n=0}^{\infty} q^n \sum a_k^n \lambda^k = \sum_p \pm \pi_* \mu_p. \tag{6.16}$$

with the $\pi_*\mu_p$ periodic under the substitution $\lambda \to q^2 \cdot \lambda$!

Hence if we can show that under the summation, all the poles of $\sum_p \pm\pi_*\mu_p$ (just as they did for μ_p) then presumably the LHS would inherit the invariance under the substitution $\lambda \to q^2\lambda$. It would then be a meromorphic function on T_{q^2} with no poles at all – a constant. This is to say that $a_k^n = 0$ for $k \neq 0$, whence the index of $d_s \otimes \mathbf{R}_n$ is rigid. QED

This is then the outline of Taubes' proof in the present context, and we hope to have the details written down shortly. Let me just close here with a few remarks on how the spin-structure enters, and also comment on some of the novel features of this whole argument.

REMARKS. (1) The spin-structure enters at two crucial points of Taubes' pole-cancelling argument. First of all, it is easily seen that under $\lambda \to q\lambda$, $\pi_*\mu_p$ picks up the factor $(-1)^{\Sigma n_\alpha}$. All these factors need to be the same for the "new poles" of $\pi_*\mu_p$ on the circle $|\lambda| = q$ to vanish. But this condition can be *guaranteed only on spin-manifolds.*

Finally to cancel the "even newer" poles on the circles $|\lambda| = q^{m/n_\alpha}$ Taubes uses the fixed point theorem on an ingeniously twisted form of the signature operator d_s^N, associated to the submanifold N of M which is the fixed point set of an element of order n_α in S^1. But again, one can guarantee that such an N is *always orientable* only on a spin-manifold.

What is most amazing to me concerning this argument is that it seems to require the quantum leap from μ_p to $\pi_*\mu_p$ to establish the rigidity of even the *first nontrivial* twist of d_s, namely the twist by $R_1 = T \oplus T$. In any case I know of no more elementary proof of why $d_s \otimes T$ should be rigid on spin-manifolds.

Let me close with just one word of explanation of how all this ties in with "elliptic cohomology". The phenomenon observed by Witten, Landweber, Stong, and Ochanine is that as a function of q, and suitably normalized, the index of $d_s \otimes W_q$ *is always a modular form for the subgroup* $\Gamma_0(2)$ of the modular group associated with tori on which one *nontrivial* point of *order two is singled out.* This can, in our framework, be traced back to the canonical nature of $\pi_*\varphi$ as a function on such a marked torus. Hence, in particular, the coefficients of the Laurent expansion of $\pi_*\varphi$ near $z = 0$, will – after normalization – all be modular forms in $\Gamma_0(2)$.

ACKNOWLEDGEMENT

This work was supported in part by NSF Grant # DMS-86-05482.

REFERENCES

[1] Atiyah, M.F. and Bott, R., A Lefschetz fixed point formula for elliptic complexes I, Ann. of Math. 86, 1967.

[2] Atiyah, M.F. and Bott, R., A Lefschetz fixed point formula for elliptic complexes II, Ann. of Math. 88, 1968.

[3] Atiyah, M.F. and Bott, R., The Moment Map and equivariant cohomology, Topology, Vol. 23 # 1, 1984, 1-28.

[4] Atiyah, M.F. and Hirzebruch, F., Spin manifolds and group actions, "Mémoires dedicated to Georges DeRham", Springer-Verlag, New York-Heidelberg-Berlin, 1970, 18-28.

[5] Atiyah, M.F. and Singer, I.M., The index of elliptic operators I, Ann. of Math. 87, 1967.

[6] Atiyah, M.F. and Segal, G., The index of elliptic operators II, Ann. of Math. 87, 1967.

[7] Atiyah, M.F. and Singer, I.M., The index of elliptic operators III, Ann. of Math. 87, 1967.

[8] Bott, R., Vector fields and characteristic numbers, Mich. Math. Journal 14 (1967), 231-244.

[9] Bott, R. and Tu, L., Differential forms in algebraic topology, Springer-Verlag, New York-Heidelberg-Berlin, 1982.

[10] Landweber, P.S., and Stong, R.E., Circle actions on Spin Manifolds and Characteristic numbers, Rutgers Preprint 1985, to appear in Topology.

[11] Ochanine, S., Genres elliptiques equivariants, Université de Paris Sud Preprint 1986.

[12] Taubes, C., S^1 actions and elliptic genera, Harvard Preprint 1987.

[13] Witten, E., Global Anomalies in String Theory in "Anomalies, Geometry and Topology", ed. by W. Bardeen and A. White, World Scientific, 1985.

[14] Witten, E., The index of the Dirac operator in Loop Space, Princeton Preprint PUPT-1050, to be published in the Proc. of the Conf. on Elliptic Curves and Modular Forms in Alg. Topology (IAS, September, 1986).

NON-COMMUTATIVE GEOMETRY

Alain Connes

Institut des Hautes Etudes Scientifiques (IHES)

F - 91440 Bures-sur-Yvette, France

0 INTRODUCTION

For purely mathematical reasons it is necessary to consider spaces which cannot be represented as point sets and where the coordinates describing the space do not commute. In other words, spaces which are described by algebras of coordinates which are not commutative. If you consider such spaces, then it is necessary to rethink most of the notions of classical geometry and redefine them. Motivated from pure mathematics it turns out that there are very striking parallels to what is done in quantum physics. In the following lectures, I hope to discuss some of these parallels.

1 VECTORBUNDLES, DIFFERENTIALFORMS, CONNEXIONS AND THE DE-RHAM-COMPLEX

Let me start with some simple examples, where purely from mathematics one is forced to consider non-commutative spaces:

Example 1 : (Signature)

Let us consider a simply connected oriented 4k-dimensional manifold M . There is a natural symmetric form on the middle-dimensional cohomology of M, s. /2, p. 572 ff/

$$H^{2k}(M) \times H^{2k}(M) \to H^{4k}(M) \cong \mathbb{R}$$

$$(\omega_1 , \omega_2) \mapsto \omega_1 \wedge \omega_2$$

The signature of M is defined as the signature of this quadratic form (# of positive eigenvalues - # of negative eigenvalues). The signature is computable as the analytical index of the signature operator on M (see for example Bott's lecture at this school). When the manifold M is not simply

Lecture Notes by H.-W. Wiesbrock

connected but still orientable with fundamental group $\Gamma := \pi_1(M)$, you can pass to the universal covering $\tilde{M}$ over M and look at the signature. But by restricting attention to this kind of signature, you lose a lot of information which is crucial in surgery theory for example. There it is fairly obvious that one has to take into account the fundamental grop Γ and the signature has to be more involved than the signature of a quadratic form. The group Γ acts on $H^{2k}(\tilde{M})$ and $H^{2k}(\tilde{M})$ is a finite $\mathbb{C}(\Gamma)$ -module, where $\mathbb{C}(\Gamma)$ is the group ring of Γ .

$$\mathbb{C}(\Gamma) := \{ \sum_{j=1}^{n} a_j \gamma_j \,/\, a_j \in \mathbb{C}, \gamma_j \in \Gamma, n \in \mathbb{N} \}$$

with

$$\left(\sum_{j=1}^{n} a_j \gamma_j \right)\left(\sum_{l=1}^{m} a'_l \gamma'_l \right) = \sum_{\substack{j=1 \\ l=1}}^{n,m} a_j a'_l \gamma_j \gamma'_l$$

you can look at this group ring as a ring of functions defined on Γ :

$$\sum_{j=1}^{n} a_j \gamma_j \mapsto f : \Gamma \to \mathbb{C}, \quad \gamma_j \to a_j$$

with the convolution product:

$$f * g(\gamma) := \sum_{\gamma'\gamma''=\gamma} f(\gamma')g(\gamma'')$$

If Γ is commutative, for example M a torus, $\pi_1(M) = \mathbb{Z}^4$, then by Fourier-transformation the group ring goes over to the algebra of functions on the dual of Γ , $\hat{\Gamma}$. For M a torus, $\hat{\Gamma}$ is the dual torus, with the ordinary product

$$\mathbb{C}(\Gamma) \subset C(\hat{\Gamma})$$

A finite projective module over $C(\hat{\Gamma})$ represents a vectorbundle over $\hat{\Gamma}$ (see below), and $H^{2k}(M)$ thereby defines a vectorbundle over $\hat{\Gamma}$. Now Γ acts invariantly w.r.t. the quadratic form on $H^{2k}(M)$ and you can take the (+) Eigenspace to define a subbundle over $\hat{\Gamma}$ and subtract from it the (-) Eigenspace-bundle (in K-theory of vectorbundles). The invariant you will get in this situation is no longer a real number but a (virtual) vectorbundle, the signature-bundle over the dual $\hat{\Gamma}$, in the case of a torus, over the dual torus. You can interpret this as follows.

Let $\rho \in \hat{\Gamma}$ and take the flat line bundle E_ρ over M , which is defined by the holonomy given by ρ :

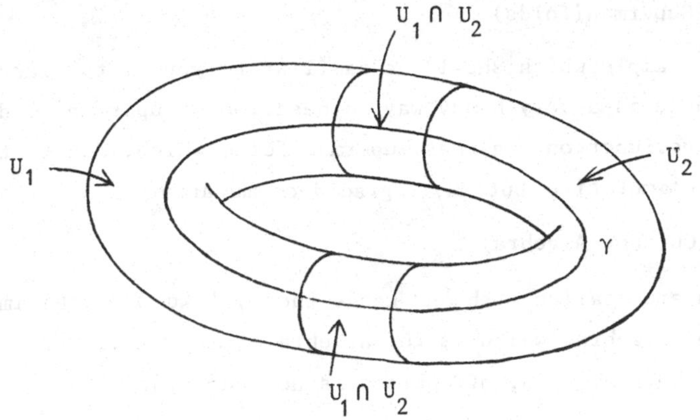

The clutching map in $U_1 \cap U_2$ is given by

$$\varphi|(U_1 \cap U_2)' : \mathbb{C} \to \mathbb{C}$$
$$z \mapsto z$$

and

$$\varphi|(U_1 \cap U_2)'' : \mathbb{C} \to \mathbb{C}$$
$$z \to \rho(\gamma)z$$

For the general construction look at the principal Γ -bundle s /17, p 42 and 160/

$$\begin{array}{ccccc} \tilde{M} & & \tilde{M} \times_\rho \mathbb{C} & & E_\rho \\ \downarrow & \Rightarrow & \downarrow & = & \downarrow \\ M & & M & & M \end{array}$$

and construct by using the homomorphism $\rho : \Gamma \to \mathbb{C}$ the associated $\mathbb{C}$ -fibre-bundle, which is in this case a flat linebundle over M .

Next take the E_ρ -twisted signature-operator on M . The kernel and co-kernel of the twisted operator depend on $\rho \in \hat{\Gamma}$ and thus give rise to the signature-bundle over $\hat{\Gamma}$.

Now assume that Γ is no longer commutative, then you cannot use Fourier transformation to get back to continuous functions on a space and talk about bundles over that space. However, it turns out that it is possible to extend the above construction. In order to do this, you have to understand that even though there is no space which is for example the dual space of a non-commutative group, you can still define vectorbundles, connections, etc. This is the program I want to talk about.

Example 2 : (Supermanifolds)

Another example which should be familiar to you is the case of supermanifolds. There is a very handy way to describe a supermanifold in terms of algebra of superfunctions on that supermanifold, which, due to the fermionic part, is not commutative but $\mathbb{Z}/_2$ graded commutative.

Example 3 : (Current Algebra)

When you are dealing with any gauge theory like QCD, the important thing is the current algebra, which is the algebra of matrix-valued functions on the manifold. This algebra, of course, is non-commutative.

The main idea which I want to persue is that in generalizing the concept of point sets the best way is to encode the information by a suitable algebra.

Example 4 : (Orbifolds)

Another simple example is an orbifold. An orbifold is not just given by a quotient space because you must remember how much folding takes place at each point of the orbifold. The natural way to encode this information is to use the following associated algebra: Let M be a manifold, π a finite group acting on M and consider the algebra

$$C(M) \rtimes \pi \qquad \text{(cross product)}$$

which is defined by $\mathbb{C}$ -valued functions

$$f : M \times \pi \to \mathbb{C}$$

with product

$$f * g(m,\gamma) := \sum_{\gamma' \in \pi} f(m,\gamma')\, g(\gamma' m, \gamma(\gamma')^{-1})$$

As an example, consider the 1-dim. manifold $M = \mathbb{R}$, and $\pi = \mathbb{Z}_2$ which acts on M by $x \to -x$

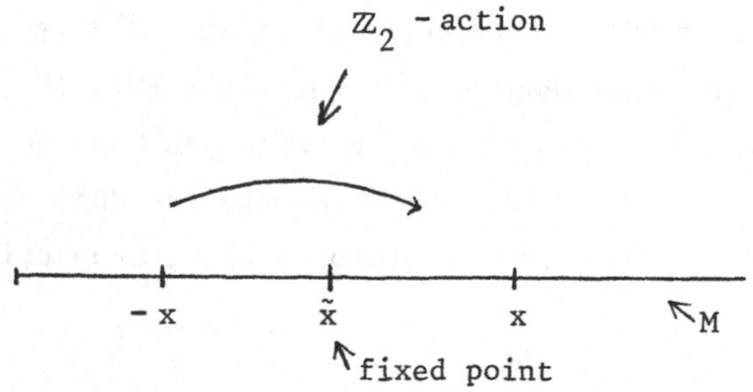

The cross product in this case is isomorphic to the algebra of 2x2 matrix-valued functions on M

$$x \mapsto \begin{bmatrix} f(x,+1) & f(x,-1) \\ f(-x,-1) & f(-x,+1) \end{bmatrix}$$

and it is easily shown, that restricting the domain on $\{x, -x\}$, the spectrum of the algebra you get, i.e. the set of maximal ideals of the algebra, consists of one point, if x is not a fixed point and of two points, if $x = \tilde{x}$ the fixed point. That is, looking at the spectrum associated to an orbifold, x and $-x$ are identified:

spectrum of the algebra

Example 5 : (Foliations)

The bare general case of quotient-spaces is a foliation. This situation was my original motivating example.

Now let us generalize the concept of vectorbundles, connexions et al to the non-commutative situation. To do this, we must first reformulate these concepts in the ordinary case in terms of statements about the associated (commutative) algebra:

Let X be a manifold, E a complex vectorbundle over X and denote by $\mathcal{E} = C(X,E)$ the linear space of all continuous sections of this fibre bundle. $\mathcal{E}$ is a module over $C(X)$, the algebra of $\mathbb{C}$-valued continuous functions on X. The fibre at a point $\tilde{x}$ is obtained by

$$\mathcal{E} \otimes_{C(X)} \mathbb{C}$$

where $C(X)$ acts on $\mathbb{C}$ by

$$fz := f(\tilde{x})z$$

So passing to the module $\mathcal{E}$, you will not lose any information. There is a nice lemma due to Swan, which characterizes the modules over $C(X)$ which come from finite-dimensional vectorbundle over X:

Lemma: (Swan)

There is a one-to-one correspondence between finite dimensional locally trivial vectorbundles over X and finite projective modules over $C(X)$ s. /8/.

Remark:

A module $\mathcal{E}$ is finitely projective iff there exist a module $\mathcal{E}'$, such that $E \otimes E'$ is a finitely generated free module.

There is another way of looking at projective modules: Starting with the free module $[C(X)]^n$, you can describe the finite projective module by the associated projection

$$e \in M_{n,n}(C(X))$$

$M_{n,n}(A)$ - nxn matrices with entries in A, which project down on E. Now

$$M_{n,n}(C(X)) = C(X, M_{n,n}(\mathbb{C}))$$

the continuous $M_{n,n}(\mathbb{C})$-valued functions on X, and e is an idempotent ($e^2 = e$), so that in fact e is a continuous map from X to the Grassmannian.

Geometrically this means that you can reconstruct the vectorbundle on X by pulling back via the map e the universal vectorbundle over the Grassmannian. So for a finite projective module over $C(X)$ we naturally get a vectorbundle over X and vice versa. /3, 18/

The notion of a vectorbundle as a finite projective module can easily be carried over to the non-communtative algebras.

Definition:

A vectorbundle over a non-commutative space with algebra A is given by a finite projective module over A or equivalently by projections in $M_{n,n}(A)$.

The projections are not uniquely defined: Let $e \in M_{n,n}(A)$, $e^2 = e$ and $\tilde{e} \in M_{n+1,n+1}(A)$

$$\tilde{e} := \begin{pmatrix} e & o \\ o & o \end{pmatrix}$$

The e and $\tilde{e}$ describe the same module. It is better to think of the projective module itself.

What we are doing is linear algebra with the algebra A as the ground ring. For example: A trivial bundle over A corresponds to a free module.

With these notions one can define one of the most important cohomology invariants of a space, which is its K-theory and you can extend this invariant to more complicated spaces:

First denote

$$M_\infty(A) := \lim_{\rightarrow} M_{n,n}(A)$$

the algebraic direct limit of $M_{n,n}(A)$ under the imbedding

$$a \to \begin{pmatrix} a & & & 0 \\ & 0 & & \\ & & 0 & \\ 0 & & & 0 \end{pmatrix}$$

This algebra has a natural norm, the completion is called the stable algebra of A and is the C*-tensor product of A and $\mathcal{K}$, $\mathcal{K}$ the algebra of compact operators. To the direct sum of vectorbundles corresponds the direct sum of modules and to the equivalence-relation for vectorbundles which defines K-theory corresponds the algebraic equivalence-relation for projections

$$e \sim f :\Leftrightarrow \exists\, u \quad \text{partial isometry} : e = u^*u \quad \text{and} \quad f = uu^*$$

$$e, f, u \in M_\infty(A) \;, \; e \text{ and } f \text{ projections, s. /13/}$$

The Grothendieck group, which you get by symmetrization of the direct sum, defines $K_o(A)$ for any C^*-algebra A. (It is like $\mathbb{N} \to \mathbb{Z}$).

Remark :

If $A = C(X)$ you have

$$K_o(A) = K^o(X)$$

where $K^o(X)$ is the ordinary K^o-group of vectorbundle-classes over the manifold X. Notice, that upper and lower subscripts refer to the transformation properties of the objects: vectorbundles transform contra-variently w.r.t. X and modules covariantly w.r.t. the algebra A.

In K-theory there is a famous result which makes everything go, the Bott-periodicity Theorem. The first proof was given by using Morse theory in infinite dimensional spaces. However, later it was shown that the appropriate setting for this theorem is the category of Banach-algebras.

Theorem : (Bott-periodicity)

Let A be a Banach-algebra, $Gl_N(A)$ the general linear group of NxN-matrices with entries in A. Then, if N is large enough, you have

$$\pi_n(Gl_N(A)) = \pi_{n+2}(Gl_N(A))$$

and

$$\pi_1(Gl_N(A)) \cong K_o(A)$$

Take for example $A = \mathbb{C}$. Then one gets $(2N \geq n)$

$$\pi_{n-1}(Gl_N(\mathbb{C})) = \begin{cases} 0 & n \text{ odd} \\ \mathbb{Z} & n \text{ even} \end{cases} \quad \text{s. /8,1/}$$

Let me sketch the idea of the proof:

Let $[e] \in K_o(A)$ and the projection $e \in M_{n,n}(A)$ a representant of $[e]$ and look at the loop

$$\gamma : S^1 \to Gl_N(A)$$
$$t \to \exp(2\pi i t e)$$

This gives you a map from $K_o(A) \to \pi_1(Gl_N(A))$ which is in fact an isomorphism.

Now define $\underline{K_1(A) := \pi_o(Gl_N(A))}$

You can also define $K_1(A)$ by mimicking the construction of a suspension to the case of Banach algebras:

$$SA := \{f: \mathbb{R} \to A \;\; \text{If continuous, } \lim_{|x| \to \infty} \|f(x)\| = 0\}$$
$$\tilde{=} C_o(\mathbb{R}) \otimes A$$

and get

$$K_1(A) \tilde{=} K_o(SA) \quad \text{s. /13/}$$

$K_o(A)$ and $K_1(A)$ are now defined for every C*-algebra A and Bott-periodicity works for them. The periodicity shortens the ordinary long exact sequence in cohomology:
If you have an exact sequence of C*-algebras

$$0 \to J \to A \to B \to 0$$

the associated exact sequence in K-theory is

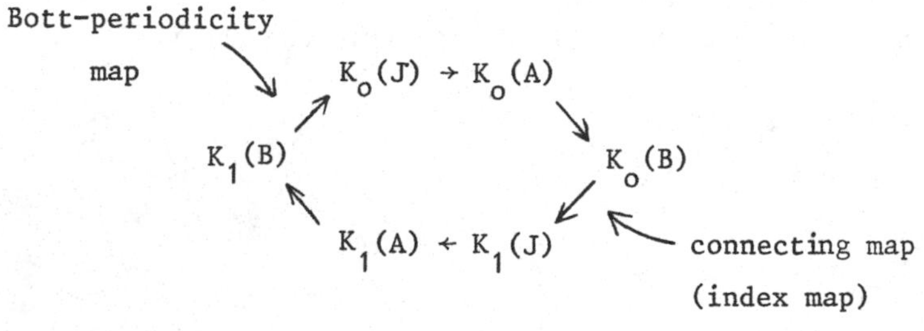

Take for example $U \subset X$ a subspace of the topological space X, $J = C(U)$, $A = C(X)$ and $B = C(X-U)$. In ordinary cohomology you get the long exact sequence

$$\ldots \to H^{q-1}(X,U) \to H^q(U) \to H^q(X) \to H^q(X,U) \to H^{q+1}(U) \to H^{q+1}(X) \to \ldots$$

whereas in K-theory this is shortened to the six-term diagram. This allows a lot of computations in K-theory.

The second main tool is the Chern-character. I will give a description of the Chern-character which fits well into what I want to do later.

Given a bundle E over a manifold X, the Chern-character of the bundle, given for example by the Chern-Weil-homomorphism, is equivalently given by the pairing with an arbitrary closed current (a de-Rham current is an element of the dual space of the differential forms):

$$C \mapsto \langle ch(E), C \rangle$$

But this can be rewritten very simply by use of the projection e which describes the bundle E:

Let $e \in M_{N,N}(C(X)) = C(X, M_{n,n}(\mathbb{C}))$ be a projection associated to E. e is a continuous map from X to the Grassmannian and E is just the pull back of the universal bundle over the Grassmannian. /18,3/ There is a natural connexion for the bundle over the Grassmannian, obtained by projecting a fibre to the nearby fibre by the orthogonal projection. Pulling back this connection via e gives a connection for E and we can compute the Chern-class by applying the Chern-Weil-homomorphism to this connection. /19/ It gives

$$\langle ch(E), C \rangle = \langle e \underbrace{de \ldots de}_{\text{even times}}, C \rangle$$

where $de \in \Omega^1(X) \otimes M_{N,N}(\mathbb{C})$ and $de \ldots de$ is the ordinary $\wedge$-Product for $M_{N,N}(\mathbb{C})$ - valued differential forms. To describe the pairing of the closed current C with a matrix-valued differentialform, it is enough to consider the following situation:

Let $f^0, f^1, \ldots, f^n \in C^\infty(X)$

$$\tau(f^0, f^1, \ldots, f^n) := \langle f^0 df^1 \wedge \ldots \wedge df^k, C \rangle$$

is well defined. Now tensoring the f's with $N^0, \ldots, N^k \in M_{N,N}(\mathbb{C})$ gives a formula reminiscent of Chan-Paton-Factors

$$\tilde{\tau}(f^0 \otimes N^0, \ldots, f^k \otimes N^k) := \tau(f^0, \ldots, f^k)\ \mathrm{Trace}\ (N^0 \ldots N^k)$$

Notice that the projection e associated to the vectorbundle E is not uniquely defined but it turns out (to be proved) that the function

$$e \to \langle e\, de \ldots de, C \rangle$$

although not linear in e, depends only on the K_o-class of e!

What is the right notion of Differentialforms ?

The above notions led to K-theory, which is a cohomology-theory and algebraic topology tells you that there is a corresponding homology-theory due to Alexander-Duality and all that. If you apply this to K-theory you get K-homology. By the work of Atiyah, Brown-Douglas-Fillmore and Kasparov, you can describe K-homology in a beautiful way, provided you introduce some non-commutative spaces, the odd and even spheres Σ^{odd}, Σ^{even} given by non-commutative algebras:

Let $H = H^+ \oplus H^-$ be a $\mathbb{Z}_2$ -graded separable Hilbert space.

Then

$$C(\Sigma^{even}) := \left\{ \begin{pmatrix} a_{11} & 0 \\ 0 & a_{22} \end{pmatrix} \in L(H) \;/\; (a_{11}-a_{22}) \text{ compact} \right\}$$

where $L(H)$ is the algebra of bounded operators on H , and

$$C(\Sigma^{odd}) := \left\{ \begin{pmatrix} a_{11} & a_{12} \\ a_{21} & a_{22} \end{pmatrix} \in L(H) \;/\, a_{12}, a_{22} \text{ compact} \right\}$$

It turns out, that

$$K_o(C(\Sigma^{even})) = \mathbb{Z} \quad ; \quad K_1(C(\Sigma^{even})) = 0$$

$$K_o(C\, \Sigma^{odd})) = 0 \quad ; \quad K_1(C(\Sigma^{odd})) = \mathbb{Z}$$

In ordinary homology-theory you look at the cell-decomposition of the topological space. The basic cycles in the theory are the spheres which are mapped into the space in question. In K-theory this rôle is played by Σ^{even} and Σ^{odd} : s. /16, 13/

<u>Theorem</u>

$$K_1(X) = \mathrm{Hom}\,[\Sigma^{odd}, X]$$

$$K_o(X) = \mathrm{Hom}\,[\Sigma^{even}, X]$$

where K_o, K_1 denote the K-homology, the so-called <u>Steenrod-K-homology</u> of the ordinary space X , and Hom the homotopy classes of maps. A continuous map from $\Sigma^{\substack{even\\odd}}$ to X is a homomorphism of the corresponding algebras

$$\mathrm{Hom}(C(X), C\,(\Sigma^{\substack{even\\odd}}))$$

The pairing of K-theory and K-homology is given as follows:

Let $e \in K_o(C(X)) \cong K^o(X)$, $\pi \in \mathrm{Hom}(C(X), C\,(\Sigma^{even}))$

You can push forward e to $\pi_*(e) \in K_o(C\,(\Sigma^{even}))$

generally of $\mathrm{Hom}(A, C(\Sigma^{\substack{even\\odd}}))$, A any C*-algebra: Given a homomorphism of A to $C(\Sigma^{even})$ you get a representation of $H = H^+ \oplus H^-$ as an A-module.

But due to the properties of $C(\Sigma^{even})$, the two representations of A on H^+ resp. H^- differ only by compact operators! Let me summarize what you get: /6/

a) A $\mathbb{Z}_2$-graded separable Hilbert space $H = H^+ \oplus H^-$ with grading operator

$$\gamma = \begin{pmatrix} 1 & 0 \\ 0 & -1 \end{pmatrix}$$

b) A homomorphism $\varphi : A \to L(H)$ such that

$[\varphi(a),\gamma] = 0 \quad \forall a \in A$

c) An odd operator $F \in L(H)$, i.e. $[\gamma,F]_+ = 0$, such that

$F^2 = 1$

$[F,\varphi(a)]$ is compact $\forall a \in A$

In the above case F is simply given by

$$F = \begin{pmatrix} 0 & 1 \\ 1 & 0 \end{pmatrix}$$

Remark

You can think of the grading operator γ for example as the Γ - operator in Jaffe's lecture or the γ_5 in chiral theories.

The quadruple (H,φ,F,γ) is called a Fredholm-module over A. The odd case is slightly different. There you get

a) A Hilbertspace H and a *-representation

$\varphi : A \to L(H)$

b) $P \in L(H)$ with

$P^2 = \mathrm{Id}$

$[P,\varphi(a)]$ is compact $\forall a \in A$

s. /6, 13/

To see this, notice that restricting the $\pi \in \mathrm{Hom}(A,C(\Sigma^{odd}))$ to H^+ gives a homomorphism of A modulo compact operators (a homomorphism of A to the Calkin algebra of H^+). Define

$$P = \begin{pmatrix} 1 & 0 \\ 0 & -1 \end{pmatrix}$$

and

$$P_+ := \frac{1 + P}{2}$$

Then P_+ is the projection on the (+)-Eigenspace of P .

$$\rho : A \to L(P_+(H))$$
$$a \mapsto P_+\varphi(a)P_+$$

is a homomorphism of A to $L(P_+(H))$ mod. compact operators :

$$\rho(ab)-\rho(a)\rho(b) \quad \text{is compact.}$$

This gives the right thing to define Differentialforms: Let (H,φ,F,γ) be a Fredholm-module of A , (in the odd case (H,φ,P) . A 1-form is obtained as an operator on H

$$da := i[F,\varphi(a)] \in L(H) \quad \text{(compact)}$$

(in the odd case)

$$da := i[P,\varphi(a)] \in L(H)$$

and the general 1-form as a linear combination of

$$\sum_i \varphi(a^j)\, db^i \qquad a^j, b^j \in A$$

To simplify notation, I will drop the φ in the following, looking at H as an A-module. A general k-form is given by

$$\sum_j a_j^0\, da_j^1 \ldots da_j^k \in L(H) \qquad a_j^e \in A$$

The space of k-forms will be denoted by Ω^k . For any algebra A and Fredholm module over A this gives a graded differential algebra. The next notion we need is that of integration of forms:

We are dealing with algebras and the natural generalization of integration for algebras is the concept of trace. Here we have in the even case an additional structure, the $\mathbb{Z}_2$-grading given by γ and we define

$$\int \omega := \text{Trace}_S(\omega) := \text{Trace}\,(\gamma\omega) \quad \text{(supertrace)}$$

in the even case and

$$\int \omega := \text{Trace}\,(\omega)$$

in the odd case where $\omega \in \Omega^k$, whenever the right sides are defined. To be able to use the concept of integration, we refine the definition of a Fredholm-module:

A Fredholm-module (H,φ,F,γ) (in the odd case (H,φ,P)) is called <u>p-summable</u> or <u>of dimension p</u> $p \in R^+$, if and only if

$$[\varphi(a),F] \in L^P(H) \quad (\text{resp. } [\varphi(a),P] \in L^P(H))$$

where $L^P(H)$ is the p.th Schatten class of H .

<u>Remark</u>:

Let $T \in L(H)$ compact, $\{\mu_n(T)\}_{n\in\mathbb{N}}$ the Eingenvalues of $|T|$ counted with multiplicities. Then

$$T \in L^p(H) :\leftrightarrow \sum_n |\mu_n(T)|^p < \infty$$

The right side defines, by taking the p.th root of the sum, a norm $\| \ \|_p$ on $L^p(H)$ and there is a Hölder inequality

$$\|T_1 T_2\| \leq \|T_1\|_{p_1} \|T_2\|_{p_2}$$

if $T_1 \in L^{p_1}(H), T_2 \in L^{p_2}(H)$, s./10/

In particular, for a p-summable Fredholm-module we get

$$\Omega^k \in L^1(H) \qquad k \geq p$$

and the integral of p-forms is well-defined. We get a differential graded algebra Ω^* with differential

$$d : \Omega^k \to \Omega^{k+1}$$

$$\omega \mapsto d\omega := i[F,\omega] = i(F\omega - (-1)^k \omega F) \quad \text{(graded commutator)}$$

and an integral

$$\int : \Omega^* \to \mathbb{C}$$

with

$$\int \omega = \operatorname{trace}(\gamma\omega) \qquad \omega \in \Omega^k \qquad k \geq p$$

$$\int \omega = 0 \qquad\qquad k < p$$

Such a triple $(\Omega^*, d, \int)$ is also called a <u>cycle</u> for A , s/ below.

In the case of commutative algebra $C(M)$, M a manifold, the definition of dimension for Fredholm-modules corresponds to the ordinary definition of the dimension of M .

<u>Example 6:</u>

The fundamental class of a 2n-dimensional compact Spin^c-manifold M in ordinary homology gives naturally a $2n+\varepsilon$-summable Fredholm-module over $C(M)$ and thereby a cycle of $C(M)$ in K-homology:
Let $S = S^+ + S^-$ be the spinor bundle over M and $H = L^2(M,S)$ the Hilbert-space of L^2-spinors. The grading is given by γ_5 and the algebra of continuous functions on M acts on H . Denote $\not{\partial}$ the Dirac-operator and

$$F := \operatorname{sign} \not{\partial}$$

F splits H into the positive and negative energy eigenspaces of $\not{\partial}$.
F is a pseudodifferential operator of order 0 and the commutator of F with any continuous function $a \in C(M)$ is a pseudodifferential operator of order -1 ,

$$[F,a] \approx (1+\Delta)^{-1/2} \quad \text{(in strength)}$$

where Δ is the Laplacian on M. It is an easy exercice to check that (H,F,γ_5) is a p-summable Fredholm-module with $p > \dim M$ because a pseudo-differential operator of order $-p$ is trace class for $p > \dim M$. The dimension of the manifold which is the dimension of the K homology class given by the fundamental class of M, gives the lower limit of p.

Lecture II

In his talk, Harvey considered orbifolds, which are quotients of manifolds by a point-symmetry group. Let me start with a simple example of a space which is described by non-commutative algebra rather than as a point set and which may serve as background for the aformentioned construction of Harvey.

Example 7: (irrational rotation algebra)

Let $M = \mathbb{R}^1$ and the point symmetry given by $\mathbb{Z}+\alpha\mathbb{Z}$, $\alpha \in \mathbb{R}\setminus\mathbb{Q}$ irrationa The "quotient-space" $\mathbb{R}/\mathbb{Z}+\alpha\mathbb{Z}$ doesn't have meaningful topology: First divide $\mathbb{R}^1$ by $\mathbb{Z}$ and you get a torus $T = S^1$. Then you have to divide by $\mathbb{Z}$ acting on T by rotation by the angle α, that is, you want to identify x and y on the same orbit

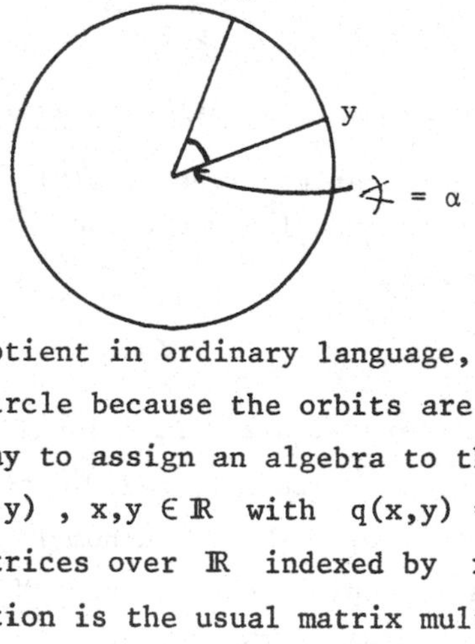

If you think of this as a quotient in ordinary language, the open sets are either empty or the entire circle because the orbits are dense on S^1. However there is a natural way to assign an algebra to this space: Consider the q-numbers $q(x,y)$, $x,y \in \mathbb{R}$ with $q(x,y) = 0$ if $x \neq y \bmod(\mathbb{Z}+\alpha\mathbb{Z})$, that is, as matrices over $\mathbb{R}$ indexed by x,y with $x = y \bmod(\mathbb{Z}+\alpha\mathbb{Z})$. The multiplication is the usual matrix multiplication

$$q_1 \cdot q_2(x,z) := \sum_y q_1(x,y)q_2(y,z)$$

Notice that the summation can be reduced to a countable set, the orbit space of x under $(\mathbb{Z}+\alpha\mathbb{Z})$. The spectrum of the subalgebra defined over one orbit consists of one point. So, for $\alpha \in \mathbb{Q}$, the spectrum of the algebra is exactly $S^1/\alpha\mathbb{Z}$. If $\alpha \in \mathbb{R}\setminus\mathbb{Q}$, the algebra still makes sense and is a natural algebra associated to the above situation. There is a presentation o this algebra as the algebra generated by two matrices U and V:

$$U(x,y) = 0 \qquad x \neq y \qquad (x,y \in S^1)$$
$$U(x,x) = e^{2\pi i x}$$

and

$$V(x,y) = 1 \qquad y = e^{2\pi i \alpha} x$$
$$V(x,y) = 0 \qquad \text{otherwise}$$

and the commutation rule

$$VU = e^{2\pi i \alpha} UV$$

The smooth elements of this algebra are given by

$$\sum_{n,m} a_{nm} U^n V^m$$

with rapid decay of a_{nm}. Denote by A_α the norm closure C* algebra. /12,Th. 1.1/ This algebra is simple. Powers asked for a simple C*-algebra without idempotent. The algebra A_α was a good candidate and Powers tried to show that this algebra has no idempotent and amazingly enough he proved, together with M. Rieffel, in fact, that this algebra contains a projection, which can be described in the simple form

$$U^{-1} f_{-1}(V) + f_o(V) + f_1(V) U$$

with certain functions f_{-1}, f_o, f_1. It turns out that these projections together with the trivial projection 1 generate $K_o(A_\alpha)$.

Pimsner and Voiculescu calculated the $K_o(A_\alpha)$ group s. /4/

$$K_o(A_\alpha) \cong \mathbb{Z}^2 .$$

But there is more information in the K_o-group: as in the case of ordinary manifolds where K-theory contains vector bundle classes of both positive and negative virtual dimension (the latter being virtual bundles), there is also such a notion in the non-commutative case. It turns out that the true bundles are given by the points right to the linc of slope $-{}^1/_\alpha$:

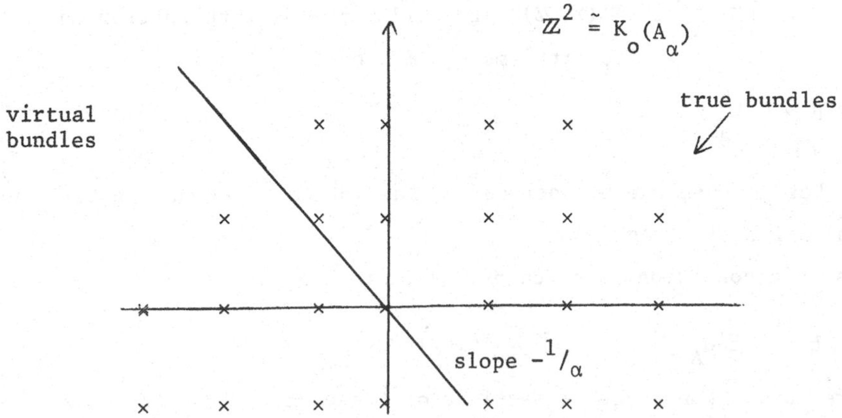

The dimension of "vectorbundles" on this space is defined as follows: Take the (if $\alpha \in \mathbb{R} \setminus \mathbb{Q}$ unique) trace

$$\tau\left(\sum_{n,m} a_{nm} U^n V^m\right) := a_{oo} = \int_{S^1} q(x,x)$$

on the algebra A_α and evaluate the trace of the idempotents associated to the vectorbundles. This defines the dimension of the bundles. You'll find that the dimension is not necessarily an integer, it can also be an irrational number. The dimension of a true bundle must be positive and this gives you the above picture.

For ordinary vectorbundles one has a well defined invariant, the Chern class of the bundle. You can mimick the computation of the Chern-Class of ordinary bundles in the above situation:

You define partial derivations w.r.t. U and V :

$$\partial_1(U) = U \qquad \partial_1(V) = 0$$
$$\partial_2(U) = 0 \qquad \partial_2(V) = V$$

∂_1, ∂_2 commute. Now write down the first Chern-Class-formula for the "vectorbundle"-idempotent e :

$$c_1(e) = \frac{1}{2\pi i} \tau(e(\partial_1 e \partial_2 e - \partial_2 e \partial_1 e))$$

where

$$e(\partial_1 e \partial_2 e - \partial_2 e \partial_1 e)$$

is the curvature. If you compute this expression for the Powers-idempotent you get an integer! /7,p.622/ But this computation gives little understanding of the reason of this integrality. It turns out, that the 1st Chern-Class defined above is the index of an operator and this "explains" why it must be an integer.

Remark:

Let $\begin{pmatrix} a & b \\ c & d \end{pmatrix} \in PSL(2, \mathbb{Z})$, the 2x2 matrices with entries in $\mathbb{Z}$ and det = 1 . You get automorphisms of A_α by

$$U \mapsto U^a V^b$$
$$V \mapsto U^c V^d$$

Let me continue by introducing the notion of connexion for non-commutative algebras:

Usually a connexion is given by a covariant derivative

$$E \overset{\nabla}{\rightarrow} E \otimes_A \Omega^1 \qquad \text{s. /6,p.110/}$$

where E is the space of sections of the vectorbundle, Ω^1 the space of

1-forms on the base-manifold M and $A = C^\infty(M)$. This can be easily transcribed to the case of non-commutative geometry where E is replaced by a finite projective module over A and Ω^1 by the above defined 1-forms. Let us take the simplest example, that is $E \cong A$, the trivial bundle. In this case a connexion is given by

$$d : A \to A \otimes_A \Omega^1 = \Omega^1$$

and the space of connexions is an affine space over $\mathrm{Hom}_A(A, \Omega^1) \cong \Omega^1$ and we identify

$$\begin{aligned} \Omega^1 &\to \{\text{connexion}\} \\ \omega &\mapsto d + \omega \end{aligned}$$

The curvature is given by

$$R = d\omega + \omega^2 .$$

Note, that although the bundle is trivial and the algebra might be commutative, we have to include the ω^2 -term. These are operators on the Hilbert space and in general they don't commute.

I want to give some meaning to this by exploiting the simplest example:

Example 8: (Sato Grassmannian)

Take $A = C^\infty(S^1)$, $H = L^2(S^1, \mathbb{C})$. $\{z^n\}_{n \in \mathbb{Z}}$ is an orthonormal basis of H and define

$$F = \mathrm{sign}\,(i \tfrac{d}{dz})$$

or in terms of eigenfunctions

$$F(z^n) = \mathrm{sign}\,(n)\, z^n .$$

Notice, that we are in the odd case of K-homology. Let us take the trivial bundle $E = A$ and find out what are the flat connexions. In Example 6 we argued, that the 1-forms are Hilbert-Schmidt-operators. So let us take the completion $\overline{\Omega^1}$ of Ω w.r.t. this norm.

Lemma:

The space of connexions on this trivial bundle is isomorphic to the space of Hilbert-Schmidt-operators on H, i.e.

$$\overline{\Omega^1} \cong L^2(H)$$

Proof:

$L^2(H) \cong L^2(S^1 \times S^1)$ by mapping the Hilbert-Schmidt-operator to the associated kernel. Assume $k \in C^\infty(S^1 \times S^1)$, with

$$k(z,\rho) = \sum_{n,m} a_{nm} z^n \rho^m$$

its Fourier expansion. Let us map the Hilbert-Schmidt-operator B with kernel

$$(z,\rho) \to z^n \rho^m$$

to Ω^1 : Let $C \in C^\infty(S^1)$

$$C(z) = \sum_n c_n Z^n$$

the Fourier expansion of C and

$$a_n : S^1 \to \mathbb{C}$$
$$z \to z^n$$

Then

$$(a_n[F,a_m]C)(z) = \sum_\ell (\mathrm{sign}(\ell) - \mathrm{sign}(\ell + m))\, c_\ell z^{\ell+m+n}$$

and

$$BC = (a_{n+1}[F,a_{-m+1}] - a_n[F,a_{-m}])\, C$$

$$B = a_{n+1}\, da_{-m+1} - a_n\, d\, a_{-m}$$

What does it mean for a connexion to be flat? $d+w$ is a flat connexion iff

$$R = dw + w^2 = 0 \quad \Leftrightarrow \quad wF + wF + w^2 = 0$$

(we drop the i). Since $F^2 = 1$ this is equivalent to

$$(F + w)^2 = \mathrm{Id}\ .$$

There is a little Lemma which describes the Grassmannian of Sato in terms of projection operators.

Lemma:

A closed subspace $W \subset H$ is in the Grassmannian, $W \in$ Grass, if and only if

$$I_W + (-I_{W^\perp}) - F = 2\, I_W - I - F$$

is Hilbert-Schmidt. (I_W denotes the orthogonal projection on W , similarly I_W)

Proof:

$$F = I_{H^+} - I_{H^-} \quad \text{and}$$

$$2\, I_W - I - F = 2\,(I_{H^+} + I_{H^-})\,(I_W) - I - F$$
$$= 2\,(I_{H^+} I_W - I_{H^+}) + 2\, I_{H^-} I_W$$

By definition $W \in$ Grass $\Leftrightarrow I_{H^+}\, I_W : W \to H^+$ is Fredholm

and $I_{H^-}\, I_W : W \to H^-$ is Hilbert-Schmidt.

The statement is now obvious /20/. From these Lemmas it easily follows that

Lemma:

There is a one-to-one correspondence between flat connexions of the trivial bundle on S^1 and the Grassmannian of Sato

Proof.

"$\Leftrightarrow$" $d+w$ flat $\Leftrightarrow (F+w)^2 = I$

Let W be the 1-Eigenspace of $F+w$, then

$$w = F+w-F = I_W - I_{W\perp} - F$$

is Hilbert-Schmidt and $W \in$ Grass.

"$\Leftrightarrow$" choose $w = I_W - I_{W\perp} - F$

Carrying these ideas a little bit further, one can describe the Grassmannian as the set of critical points of the action

$$S(w) = \int (wdw + {}^2/_3\, w^3) \qquad w \in \Omega^1$$

where $\int$ denotes the integral (trace), d the derivation in the above Fredholm-module. This is an action of Chern-Simons-type. It gives a Morse function $S : \overline{\Omega^1} \to \mathbb{C}$. One cannot use conventional Morse-theory, the Hessian is a Dirac-type-operator with infinite + and - spectrum. However one can define relative Morse-indices by spectral flow methods. /14, 15/ The indices of S splits the Grassmannian into connected components corresponding to the index of the Fredholm operator $I_{H+} I_W$, s. for example Alvarez-Gaumé's talk.

The gauge-transformations on the trivial bundle are U(1)-valued maps U on S^1, the unitary elements in $C(S^1)$, and they act on $w \in \Omega^1$ by

$$w \mapsto UwU^{-1} + UdU^{-1}.$$

The action S is obviously gauge invariant. Furthermore $\mathrm{Diff}^+(S^1)$, the orientation preserving diffeomorphisms on S^1 acts unitarily on H by

$$(f.\varphi)(z) := \varphi(f^{-1}(z))\,|(f^{-1})'(z)|^{1/2} \qquad f \in \mathrm{Diff}^+ S^1$$

and this action commutes with F modulo Hilbert-Schmidt-operators. /20,p.91/ Thereby it acts on the Fredholm-module and on $\overline{\Omega^1}$ invariantly on S. The central extension of $\mathrm{Diff}^+(S^1)$ leads to the action of the Virasoro-algebra on $\overline{\Omega^1}$ and on the Grassmannian.

To end today's talk, I will give one more example which might explain to you, why I call the above differential forms "first quantized forms".

Example 9 : (Zitterbewegung)

Let $A = C_o^\infty(\mathbb{R}^3)$ the algebra of smooth functions on $\mathbb{R}^3$ with compac support. The natural module, in this case, s. example 6, is given by

$$H = L^2(\mathbb{R}^3, \mathbb{C}^2) \qquad D = \not\partial = i\,\sigma_j \frac{\partial}{\partial x_j}$$

where σ_i are the Pauli-matrices, D the Dirac-operator. The underlying manifold is not compact, D does not have discrete spectrum and we have to take care of the zeromodes in this case!

Let us first consider an operator $\tilde{D}$ which is invertible $a\tilde{D}^{-1} \in L^p(H)$ with bounded commutator $[a,\tilde{D}] \in L(H)$ for $a \in A$. /6,p.68/ It is an easy exercice to show that

$$a \mapsto \begin{pmatrix} a & o \\ o & \tilde{D}^{-1} a \tilde{D} \end{pmatrix}$$

is a homorphism

$$A \to C(\Sigma^{even})$$

The Dirac operator D is not invertible but a simple way to build an invertible operator from D which preserves most of the features like the degree etc. is to use,

$$D_m := D \otimes \begin{pmatrix} 1 & 0 \\ 0 & -1 \end{pmatrix} + m \cdot I \otimes \begin{pmatrix} 0 & 1 \\ 1 & 0 \end{pmatrix}$$

and

$$\tilde{H} = H \otimes \mathbb{C}^2 .$$

Then define

$$F := \operatorname{sign}(D_m) .$$

The Fredholm-module class is independent of $m \neq 0$ by homotopy-invariance.

D_m is the Dirac-Hamiltonian of a free electron moving in $\mathbb{R}^3$ with mas m. If you compute the differentiation w.r.t. F, for example of the coordinate functions x_j, you will obtain the Zitterbewegung.

Lecture III

Cyclic Cohomology

It is quite difficult to handle all Fredholm-modules of a given algebra A. Instead of doing this we weaken the data and pass to the character of a p-summable Fredholm-module and analyse them.

Like the character of group representations the character of a p-summable Fredholm-module, which is an algebra representation, is a $\mathbb{C}$-valued function s. /6/

$\tau : A \times \ldots \times A \to \mathbb{C}$

$(a^o,\ldots,a^n) \mapsto \tau(a^o,\ldots a^n) := \int a^o da^1 \ldots da^n$

Remark:

By using the simple formula

$\frac{1}{2} \operatorname{tr} F(FX + XF) = \operatorname{tr} X$

whenever both sides make sense, it is easily seen that for a p-summable Fredholm-module τ is already defined for $n > p-1$.

τ captures exactly what we need to know about Fredholm-modules to do a lot of computations. The functionals τ have two very important properties.

Proposition:

a) $\tau(a^1, a^2, \ldots, a^n, a^o) = (-1)^n \tau(a^o, a^1, \ldots a^n)$ (cyclicity)

b) $(b\tau)(a^o,\ldots,a^n,a^{n+1}) := \sum_{j=o}^{n} (-1)^j \tau(a^o,\ldots a^{j-1}, a^j a^{j+1}, a^{j+2},\ldots,a^{n+1})$

$+ (-1)^{n+1} \tau(a^{n+1} a^o, a^1, \ldots, a^n)$

$= 0$ (cocycle-condition)

b is the Hochschild Coboundary operator, $b^2 = 0$.

Proof:

a) This property doesn't involve the algebraic structure, but stems from the definition of τ by the trace.

b) This is a reformulation of the fact that the trace is a closed graded trace, i.e. $\int dw = 0$.

To understand what this means, let us consider the simplest case $n = 0$: b) $\Leftrightarrow \tau(a^o a^1) - \tau(a^1 a^o) = 0$ that is, τ a trace. b) is an elementary property of the trace (compare with the character of group representations).

a) arises, as we already saw, if one wants to extend the integral of differential forms to the case of matrix-valued differential forms. Unlike ordinary differential forms, we do not have graded commutativity, but we still maintain cyclicity (using the cyclic property of the trace). Any cyclic cocycle, i.e. a multilinear functional τ satisfying a) and b) of the proposition defines an invariant of vectorbundles resp. of elements in $K_1(A)$ by the generalization of the Chern-Weil-map:

$K_o(A) \overset{\tau *}{\to} \mathbb{C}$

$[e] \mapsto \tau(e,e,\ldots,e)$

e is a projection representing the equivalence class $[e]$ respectively

$$K_1(A) \overset{\tau_*}{\to} \mathbb{C}$$
$$[u] \mapsto \tau(u^{-1},u,u^{-1},..)$$

u is a unitary representant in $[u]$.

Remark:

These maps are well defined, i.e. they don't depend on the special choice of the representants e and u . Moreoever one can prove, that the maps are group homomorphisms.

There is a trivial way to get cyclic cocycles by applying b to a cyclic φ . (But this gives only 0-maps, as one can show using the cyclicity property of τ .) /6,p.104/ If τ comes from an even Fredholm-module (H,φ,F,γ) then

$$\tau_*(e) = \text{ind } F_e^+ \in \mathbb{Z}$$

where F_e^+ is the Fredholm operator

$$F_e^+ = e(H_+ \otimes \mathbb{C}^k) \to e(H_- \otimes \mathbb{C}^k)$$
$$\varphi \mapsto e(F \otimes \mathrm{Id})e(\varphi)$$

when $e \in \mathrm{Proj.}(M_{k,k}(A))$.

Now let me give you the definition of cyclic cohomology: The cochains of the complex are multilinear cyclic functionals on the algebra

$$C_\lambda^n(A) := \{\text{cyclic functionals of } (n+1) \text{ variables}\}$$

(λ refers to cyclicity) and the coboundary is the Hochschild coboundary b . Then the cyclic cohomology group is defined as

$$H_\lambda^n(A) := \frac{Z_\lambda^n(A)}{b(C_\lambda^{n-1}(A))}$$

where $Z_\lambda^n(A)$ is the group of cocycles in $C_\lambda^n(A)$ and $b(C_\lambda^{n-1})$ the group of coboundaries.

Let me describe b in low degrees:

$$b : C_\lambda^o \to C_\lambda^1$$
$$\tau \mapsto ((a^o,a^1) \to (b\tau)(a^o a^1) := \tau(a^o a^1) - \tau(a^1 a^o)$$

and $b\tau = 0 \Leftrightarrow \tau$ a trace, s. above

$$b : C_\lambda^1 \to C_\lambda^2$$
$$\tau \mapsto b\tau$$

with

$$b\tau(a^o,a^1,a^2) = \tau(a^oa^1,a^2) - \tau(a^o,a^1a^2) + \tau(a^2a^0,a^1) \,.$$

If A is a matrix-algebra, it is easy to show that a cyclic 1-cocycle gives a Lie-algebra extension by

$$(A\oplus 1) \times (A\oplus 1) \to (A\oplus 1)$$

$$((a,\alpha),(b,\beta)) \to [(a,\alpha),(b,\beta)] := (ab-ba,\tau(a,b)1)$$

s./11/.

In the case of a commutative algebra $A = C^\infty(M)$, M a compact manifold, and requiring continuous functionals in the cyclic cohomology we get

Theorem:

$$H^k_\lambda(A) = \{\text{closed De-Rham-currents of dimension } k\} \oplus$$

$$H_{k-2}(M,\mathbb{C}) \oplus H_{k-4}(M,\mathbb{C}) \oplus \ldots$$

s. /6, Th. 4.6/

The terms $H_{k-2}(M,\mathbb{C}) \oplus H_{k-4}(M,\mathbb{C}) \oplus .. \oplus H_o(M,\mathbb{C})$ are required by the index theorem. There you have the various components of the $\hat{A}$-genus and they come from lower dimensional homology-classes. So it is absolutely necessary for our cohomology theory to reproduce these lower terms in order for the index to be an integer. Let me give you a picture which will make clear what I'm doing:

Let M be an ordinary manifold

$$\begin{array}{ccccc} & \text{K-theory } (M) & \xrightarrow{\text{Chern}^*} & \text{Cohomology } (M) & \\ \text{pairing} & \updownarrow & & \updownarrow & \text{pairing} \\ & \text{K-homology } (M) & \xrightarrow{\text{Chern}_*} & \text{Homology } (M) & \end{array}$$

The ordinary Chern maps K-theory into the cohomology of the space. What I will construct here is the Chern_* of K-homology (M) to the homology of M such that the pairings on both sides of the diagram coincide. This will reproduce the index-theorems.

Let me explain how to prove this theorem and how to make things work in general.

The properties of cyclic cohomology will be dictated by the situation of Fredholm-modules. There we get these characters τ_n for $n \geq p-1$. In general, one can recover from a cyclic cocycle a cycle $(\Omega,d,\int)$, which is a graded algebra with a graded derivation of degree $1, d^2 = 0$ and $\int : \Omega^n \to \mathbb{C}$ closed graded trace, n - the dimension of the cycle, and a homomorphism $\rho : A \to \Omega^o$. But this cycle need not come from a p-summable Fredholm-module. /6, p.98/

We have seen that in the case of p-summable Fredholm $\tau_n, \tau_{n+2}, \dots$ $n \geq p$ all give cyclic cocycles and the question arises, is the relevant information about the module already contained in τ_n ?

Theorem :

There exists a universal map $S : H_\lambda^n(A) \to H_\lambda^{n+2}(A)$ defined by

$$(S\tau_n)(a^o,\dots,a^{n+2}) := 2\pi i \sum_{j=0}^{n+2} \tau_n(a^o da^1 \dots da^{i-1}) a^i a^{i+1} (da^{j+2} \dots da^{n+2})$$

(purely algebraically), s.t. in the above question

$$\tau_{n+2k} = (S)^k(\tau_n) \in H_\lambda^{n+2k}(A)$$

/6, Th. 1/

This nice result leads to another question: When is a given τ in the image of S ? Or: Given a τ, when is it possible to rewrite τ as $\tau = S^k\tau'$ with a τ' of a lower-dimensional Fredholm-module, which thereby depends on fewer variables?

Theorem 2:

τ is in the image of S, if and only if $\tau \in b(C^{n-1})$ that is, τ is a coboundary of a Hochschild cochain, not necessarily of a cyclic cochain. /6, Cor. 35/

The use of this fact is that the Hochschild cohomology is computable in principle at least, you can use all the developed techniques of homological algebra. In the case of an ordinary manifold we get

Lemma:

For $A = C^\infty(M)$ the Hochschild cohomology $H^k(A,A^*)$ is identical with the space Ω_k of de-Rham currents of dimension k/6, Lemma 45/ .

Remark:

Look at the multilinear functionals

$$\tau : A\times\dots\times A \to \mathbb{C}$$
$$(a^o,\dots,a^k) \mapsto \tau(a^o,\dots,a^k) \ ?$$

These maps give you a map

$$\tilde{\tau}_{a^1,\dots,a^k} : A \to \mathbb{C}$$
$$a^o \to \tau(a^o,a^1,\dots,a^k)$$

where A^* is the dual of A. The maps $\tilde{\tau}$ are the k-cochains of the Hochschild-complex, the coboundary is b, s. the Proposition above. The Hochschild cohomology is defined by

$$H^k(A,A^*) = \frac{\ker b \quad \text{(k-cochains)}}{\operatorname{Im} b \quad \text{((k-1)-cochains}}$$

The remarkable thing about this Lemma is that you get all the de-Rham-currents, the duals of the differential forms, purely algebraically, without involving exterior algebra considerations. We can get rid of the commutativity of the ground ring and recover the currents and the differential for forms by a theorem, not by definition! This will then work for non-commutative algebras where the notion of exterior algebra is not applicable. The Hochschild cohomology is the right generalization of currents in the non commutative case. In the commutative case the above identification is quite simple:

Given a de-Rham current $C \in \Omega_k$ we assign to it the multilinear form

$$\varphi(f^o,f^1,..,f^k) := < C,f^o df^1..df^k >$$

$f^i \in C^\infty(M)$. Notice that I never used properties of locality in the definition of the multilinear functionals defining the cochains of the Hochschild-complex. It is a corollary of the theorem that any element in the Hochschild-cohomology class is equivalent to a local one.

Our aim is the computation of the cyclic cohomology. The cyclic complex is a subcomplex of the Hochschild-complex but not a retraction. The latter would imply that the cohomology groups injected. To show you that this is not so, consider the following simple example.

Example 10:

Let $A = \mathbb{C}$. Then $H^k(\mathbb{C},\mathbb{C}) = 0$, the Hochschild complex is trivial. For the cyclic complex you get:

Let $\varphi \in H^k_\lambda(\mathbb{C})$, then by multilinearity

$$\varphi(a^o,..,a^k) = a^o..a^k \varphi(1,1,..,1)$$

and by cyclicity $H^{2k+1}_\lambda(\mathbb{C}) = 0$ but $H^{2k}_\lambda(\mathbb{C}) = \mathbb{C}$

A generator of $H^2_\lambda(\mathbb{C})$ is given by $\sigma(1,1,1) = 1$ s./6, p.105/.
It turns out that S is just the cup-product with this generator.

Summarizing the situation we have

$$H^n_\lambda(A) \xrightarrow{S} H^{n+2}_\lambda(A) \xrightarrow{I} H^{n+2}(A,A^*)$$

where I is the imbedding. There is a universal operator B , defined purely algebraically by

$$B_\varphi = AB_o\varphi$$

where A is the cyclic antisymmetrization, the projection onto the cyclic part, and

$$(B_o\varphi)(a^o,..,a^{n-1}) = \varphi(1,a^o,..,a^{n-1})-(-1)^n\,\varphi(a^o,..,a^{n-1},1)$$

B_o is the contraction with 1 in the first variable and subtracting a term s.t. it will be zero for cyclic cochains and this operator B continues the above diagram to a long exact sequence

$$\ldots\to H^n_\lambda(A) \overset{S}{\to} H^{n+2}_\lambda(A) \overset{I}{\to} H^{n+2}_\lambda(A,A^*) \overset{B}{\to} H^{n+1}_\lambda(A) \overset{S}{\to} H^{n+3}_\lambda(A) \to \ldots$$

In the case $A = C^\infty(M)$, the I.B operator is exactly the de-Rham boundary for currents. The main lemma in proving the exactness is

Main Lemma: $\dfrac{\operatorname{Im} B \cap \ker b}{b(\operatorname{Im} B)} \cong \dfrac{\ker B \cap \ker b}{b(\ker B)}$ s./6, Lemma 36/

what you get is an exact couple

$$\begin{array}{ccc} & H^*(A,A^*) & \\ {}^{B}\swarrow & & \nwarrow^{I} \\ H^*_\lambda(A) & \xrightarrow{S} & H^*_\lambda(A) \end{array}$$

and you can use all the nice tools of homological algebra: For any exact couple we get a spectral sequence. Let

$$\begin{array}{ccc} A & \overset{i}{\to} & A \\ k\ \nwarrow & & \swarrow\ j \\ & B & \end{array}$$

be an exact sequence of abelian groups (exact couple) and

$d : B \to B \quad d = j \circ k$

Then $d^2 = 0$. Define $B^1 := H(B) := \dfrac{\ker d}{\operatorname{Im} d}$ and $A^1 = i(A)$. Then you get a derived couple by defining

$i^1 : i(A) \to i(A)$ $\qquad$ $j^1 : i(A) \to H(B)$

$(ia) \to i^1(ia) := i \circ i(a)$ $\qquad$ $(ia) \to j^1(ia) := [ja]$

and

$k^1 : H(B) \to i(A)$

$[b] \to k^1([b]) := kb$ s./1, p.155/

It is not difficult to prove that the maps are well-defined and

$$\begin{array}{ccc} A^1 & \xrightarrow{i^1} & A^1 \\ k^1\ \nwarrow & B^1 & \swarrow\ j^1 \end{array}$$

is an exact couple. By this construction you get a sequence of exact couples, the <u>spectral sequence of the exact couple</u>. In the case of cyclic cohomology we get another spectral sequence associated to filtrations of a double complex:

We have an exact sequence of complexes

$$0 \to C_\lambda \to C \to C/_{C_\lambda} \to 0$$

C_λ - the cyclic, C the Hochschild complex. Let $C^{m,n} := C^{n-m}(A,A^*) n,m \in \mathbb{Z}$

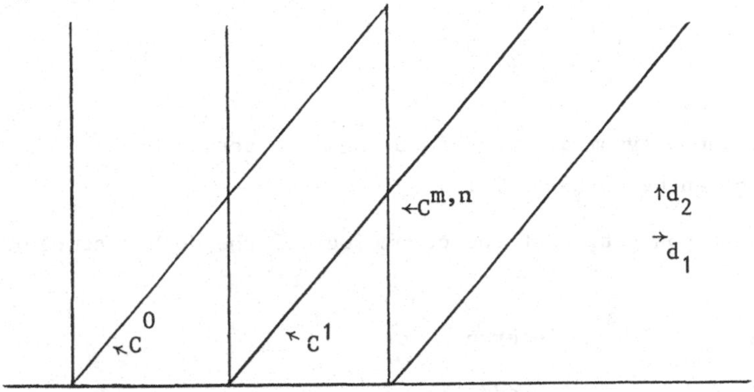

where

$$d_1\varphi = (n-m+1)b\varphi \in C^{n+1,m} \quad , \varphi \in C^{n,m}$$

$$d_2\varphi = \frac{1}{n-m} B\varphi \quad \in C^{n,m+1} \quad , \varphi \in C^{n,m} \quad C = 0 \text{ for } n = m)$$

s. /6, p.123/

by $B^2 = b^2 = Bb + bB = 0$ we get

$$d_1^2 = d_2^2 = d_1 d_2 + d_2 d_1 = 0$$

when you have a double complex of this sort, you get two spectral sequences associated to the two filtrations

$$F_p C = \sum_{n \geq p} C^{n,m}$$

and

$$F^q C = \sum_{m \geq q} C^{n,m}$$

The spectral sequence of the first filtration might not be convergent but for the second we get

<u>Lemma</u>: $H^P(F^q C) = H^n_\lambda(A)$ for $n = P-2q$

s. /6, Th. 40/

where $H^P(F^q C)$ is the cohomology group of the complex $(F^q C, d_1)$

$$H^P(F^q C) = \frac{\ker d_1 \Gamma (\sum_{m \geq q} C^{P,m})}{\operatorname{Im} d_n \Gamma (\sum_{m \geq q} C^{P-1,m})}$$

Furthermore the spectral sequence associated to the second filtration converges and coincides with the spectral sequence of the exact couple.

To prove this, one needs to use the Main Lemma again.

Let me remind you that the characters τ and $S\tau$ for Fredholm-modules contain essentially the same information and in order to get the non-redundant information, we have to "divide" the cyclic cohomology by the action of S . This is performed by defining

$$H^*(A) := \lim_{\to}(H^n_\lambda(A),S) = H^*_\lambda(A) \otimes_{\mathbb{C}_p(S)} \mathbb{C}$$

/6, p. 52/

where $\mathbb{C}_p(S)$ denotes the polynomial ring of S and the action of $p \in \mathbb{C}_p(S)$ on $\mathbb{C}$ is given by $p(S)z := P(1)z$.

As the final result you get, that the cohomology of the double complex is given by

$$H^n(C) = H^{even}(A) \qquad n \text{ even}$$

$$H^n(C) = H^{odd}(A) \qquad n \text{ odd}$$

/6, Th. 40/

Let me go back to the aim : to get some information from the Fredholm-modules, there are three steps in doing that :

1) Compute the cyclic cohomology of the algebra
2) Compute the Chern-character of the Fredholm-module
3) State the Index-Theorem.

As an example of this computation, let me return to

Example 7 : (irrational rotation algebra, II)

In the computation of the currents, i.e. the Hochschild cohomology of A_α , one gets :

If α satisfies the diophantine condition, i.e.

$$|1-e^{2\pi i n\alpha}|^{-1} \underset{n\to\infty}{\approx} \mathcal{O}(n^k)$$

for some k , then

$$H^j(A_\alpha,A^*_\alpha) = \mathbb{C}^2 \qquad j = 1,2$$
$$= 0 \qquad \text{otherwise} \quad \text{s./6, p. 133/}$$

but if α doesn't satisfy the diophantine condition H^1, H^2 are infinite dimensional non-Hausdorff spaces. Compare this result with $\alpha \in \mathbb{Q}$: in general A_α is stably isomorphic to $C_o(\mathbb{R}^2) \rtimes_\alpha \mathbb{R}^2$, i.e.

$$A_\alpha \otimes K \cong C_o(\mathbb{R}^2) \rtimes_\alpha \mathbb{R}^2 \qquad \text{s./4 , p.145 or 12, p.417/}$$

where K is the ideal of compact operator and the action of $\mathbb{R}^2$ on $C_o(\mathbb{R}^2)$ is given by

$$(\xi(\psi))\ (x) := e^{i2\pi\alpha\xi\wedge x}\ \psi(x-\xi)$$

with $\xi \wedge x = \xi_1 x_2 - \xi_2 x_1$.

By looking at the definition of K-theory et al one notices that two stably isomorphic algebras have the same K-groups et al.

If $\alpha \in \mathbb{Q}$, A_α is thereby stably isomorphic to the commutative algebra of smooth functions on a 2-torus and by the above theorem $H^j(A_\alpha, A_\alpha^*)$ is infinite dimensional for $j \leq 2$. But when we pass to cyclic cohomology we get

$$H^{even}(A_\alpha) \cong \mathbb{C}^2$$

$$H^{odd}\ (A_\alpha) \cong \mathbb{C}^2$$

for <u>all</u> α and as a basis

a) even case τ_o, τ_2

denote $f^j = \sum_{n,m} a^j_{n,m}\ U^n V^m \in A_\alpha$

then $\tau_o(f) = a_{o,o}$

and $\tau_2(f^o, f^1, f^2) = \tau_o(f^o(\partial_1 f^1\ \partial_2 f^2 - \partial_2 f^1\ \partial_1 f^2))$

where ∂_j are defined as before, s. Example 7 I .

b) odd case τ_1, τ_1^c

$$\tau_1(f^o f^1) = \tau_o(f^o \partial_1 f^1)$$

$$\tau_1^c(f^o f^1) = \tau_o(f^o \partial_2 f^1) \qquad \text{s./6, p.138/}$$

It is an easy exercise to show that these τ are cyclic cocycles.

Lecture IV

<u>The Quantum Hall Effect</u> (after Bellissard)

There are several different approaches to explaining this effect but I want to stress that of Bellissard which, I feel, explains a lot of things in a conceptually simple fashion. /4, 5/

Let me start by describing the classical Hall effect (1880) :

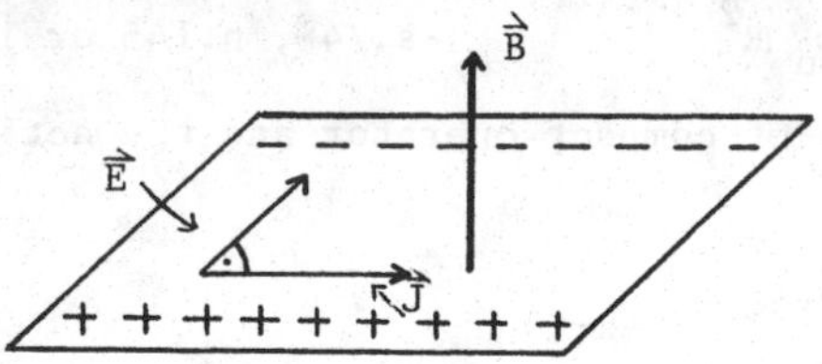

$\vec{B}$ - uniform magnetic field perpendicular to the plane
$\vec{J}$ - electric current along the plane

Hall (1880) observed an electric field $\vec{E}$ perpendicular to $\vec{J}$. The $\pm$ direction of E (sign) depends on the metal used, i.e. whether the electric current is carried by electrons or by holes.

In a stationary state the classical Lorentz-force is

$$Ne\vec{E} + \vec{J} \wedge \vec{B} = 0$$

N-charge carrier density = # of charge per unit of surface, and therefore

$$\vec{J} = \frac{Ne\ \vec{B} \wedge \vec{E}}{|\vec{B}|^2} \quad , \qquad |\vec{J}| = \sigma_H |\vec{E}|$$

with $\sigma_H = \rho_H^{-1}$ the <u>Hall conductivity</u>

$\rho_H = Ne/|\vec{B}|$ the <u>Hall resistivity</u>

The Hall conductivity is classically a linear function of the charge-carrier-density. But if one measures at low temperature and in a very large magnetic field $\vec{B}$ (von Klitzing, Pepper and Dorda) one observes

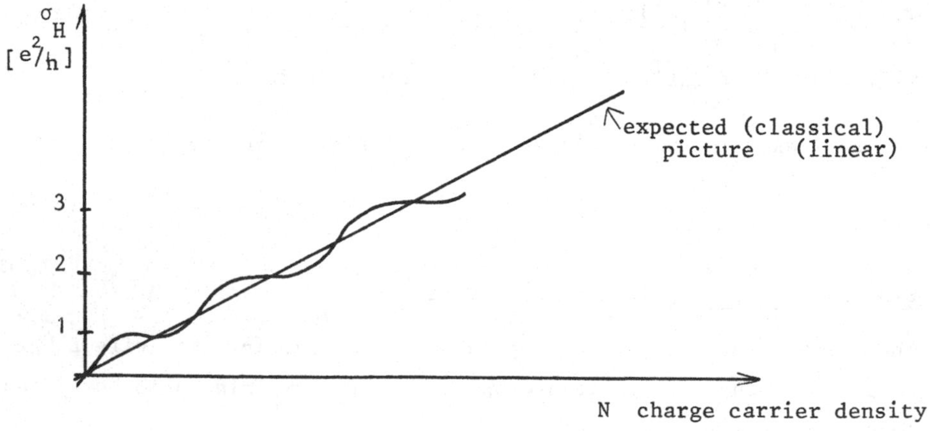

The conductivity becomes stationary for some N , moreover and this is absolutely amazing, the plateaus are exactly at integer multiples of e^2/h with an accuracy better than 10^{-7} . The main surprise was not the Quantum-Hall effect itself but the extremely high accuracy of the result !

I want to give an account for this which is due to Bellissard : /4, 5/

First one has to use Quantum Physics, but we will not use Quantum Field Theory, just the picture of a free Fermi gas, that is a gas of electrons in the 2-dim plane, not interacting with each other. The expectation value of an observable T in this case is given by the thermo-dynamical average w.r.t. the Fermi-Dirac statistic :

$$\langle T\rangle_\beta = \lim_{V\to\infty} \frac{1}{|V|} \text{Trace } \chi_V[(1+e^{\beta(H_\omega-\mu)})^{-1} T]$$

$\beta = (KT_o)^{-1}$ - the inverse temperature
χ_V - characteristic function of $V \subset \mathbb{R}^2$
μ - chemical potential (sometimes called Fermi energy)

$$H_\omega = \frac{1}{2m} ((p_1-eA_1)^2 + (p_2-eA_2)^2) + V_\omega(x)$$
$$= H_o + V_\omega$$

A_1, A_2 is a suitable gauge-potential for the electromagnetic field, take for example

$$A_1 = |\vec{B}|x_2 \qquad A_2 = |\vec{B}|x_1$$

and V_ω is a potential. If we had a perfect lattice, V_ω would be a periodic potential. In general there are some impurities in the metal and the atoms of the lattice have random positions, so that V_ω depends randomly on a parameter ω , which represents the configuration of disorder.

The average is normalized in such a way that

$\langle 1\rangle_\beta$ = charge-carrier-density N

and this fixes the chemical potential μ . To get the Hall conductivity, one has to compute the expectation value of the current. In practice, one induces a current and observes the electric potential E . For doing calculations, one reverses the order, imposes the electric field E and computes the average current in equilibrium. For a stationary state, the result will be the same :

$$\vec{J} = i\, e/\hbar\, [H_\omega, \vec{x}\,]$$

the time evolution, when the electric field E is turned on at time 0 is given by

$$\vec{J}(t) = e^{it/h(H_\omega + \vec{E}\,\vec{x})}\ \vec{J}\ e^{-it/h(H_\omega + \vec{E}\,\vec{x})}$$

the time independent part by

$$\vec{J}_{ave} = \lim_{t\to\infty} \frac{1}{t} \int_o^t < \vec{J}(t) >_\beta \, dt$$

In the case of a perfect lattice and $V_\omega = 0$ one easily computes

$$\vec{J}_o(t) = ie \frac{|\vec{E}|}{|\vec{B}|}\ 1 + e^{it\omega_c}\ \vec{J}_o \quad , \quad \omega_c = e|\vec{B}|/m \qquad \text{cyclotron frequency}$$

and

$$\vec{J}_{o,ave} = ie \frac{|\vec{E}|}{|\vec{B}|} \lim_{V\to\infty} \frac{1}{|V|} \operatorname{trace}(\chi_V (1 + e^{\beta(H_o - \mu)})^{-1})$$

$$= iNe|\vec{E}|/|\vec{B}| \quad .$$

Notice that the Hall conductivity is not quantized as a function of the charge-carrier-density as is observed in experiments !

If one rewrites

$$H_o = \frac{1}{2m}(k_1^2 + k_2^2) \qquad k_j = (p_j - eA_j)$$

with $[k_1, k_2] = i\hbar\ e/c\ 1$

one sees that H_o describes an harmonic oscillator and

$$N = \lim_{V\to\infty} \frac{1}{|V|} \operatorname{trace} \chi_V (1 + e^{\beta(H_o - \mu)})^{-1} = \sum_{n>o} (1 + e^{\beta(\hbar\,\omega_c(n + \frac{1}{2}) - \mu)})^{-1}$$

Passing to the zero temperature limit, $\beta \to \infty$, one gets that N as a function of the chemical potential μ (the Fermi-level or energy) becomes a step function. But μ is adjusted by the normalization $<1>_\beta = N$! Moreover one always has disordered configurations, the effective potential will usually be larger than the spacing between two Landau levels, i.e. $\geq \hbar\,\omega_c$. A priori there is no evidence that the step-behavior will survive in experiment.

The Hamiltonian of the system is given by

$$H_\omega = H_o + V_\omega$$

as above. The physical system is homogeneous in space and the translation by $\alpha \in \mathbb{R}^2$ will result in a physically equivalent disordered configuration $T^\alpha \omega$. Notice that an electric-magnetic field acts on the system, the wave-functions are sections in a $U(1)$-bundle and the translations are given by

$$U(a)\psi(x) = e^{2\pi i|\vec{B}|e/\hbar(x\wedge a)}\psi(x-a)$$

the electromagnetic translations, defined by the gauge-connection (-potential). The homogeneity of the underlying physical situation is reflection by

$$U(a)H_\omega U(a)^* = H_{T^a\omega}$$

and H_ω and $H_{T^a\omega}$ should describe the experiment equivalently. The algebra of observables must contain the subalgebra generated by

$$\{H_{T^a\omega}\}_{a\in\mathbb{R}^2}$$

so that it will be a non-commutative algebra ! The $H_{T^a\omega}$ are unbounded operators and to generate the C^*-algebra of observables, look at the resolvents

$$A_\omega := (z-H_\omega)^{-1} \qquad z \in \mathbb{C} \setminus \mathbb{R} .$$

Rewriting this operator in terms of "matrices" w.r.t. the generalized eigenfunctions of the position operator

$$\langle\delta_x, A_\omega\delta_{x'}\rangle$$

and by using the translation properties we get

$$\langle\delta_x, A_\omega\delta_{x'}\rangle = \langle\delta_o, A_{T^{-x}\omega}\delta_{(x'-x)}\rangle e^{i\pi|\vec{B}|e/\hbar(x\wedge x')}$$

$$= a(T^{-x}\omega, x-x')$$

where a is a function on $\Omega \times \mathbb{R}^2$, Ω the parameter set of the disordered configurations. The algebra of observables can be represented by an algebra of functions on $\Omega \times \mathbb{R}^2$, where $\mathbb{R}^2$ acts on Ω and the operator product becomes /4, p.143/

$$a*b(\omega,x) := \int_{\mathbb{R}^2} dx'\ a(\omega,x')\circ(T^{-x'}\omega, x-x')e^{i\pi|\vec{B}|e/\hbar(x\wedge x')}$$

It turns out that the algebra generated by H_ω is isomorphic to $C_o(\mathbb{R}^2) \rtimes_\alpha \mathbb{R}^2$, $\alpha = |\vec{B}|e/\hbar$, s. Example 7 (II). The two derivations of A in Example 7 (II) are now given by

$$\partial_j a(\omega,x) = x_j a(\omega,x)$$

as an easy calculation shows. Going back to the operator, these derivatives are just the commutator with x_j! Now, the current is given by

$$\vec{J} = 2\pi e/\hbar\ [H,\vec{x}]$$

and the function on $\Omega \times \mathbb{R}^2$ which represents $\vec{J}$ is thereby a derivation of a certain function. To compute the Hall conductivity we have to compute a certain trace involving the current $\vec{J}$.

Passing to the zero-temperature limit $B \to \infty$ leads to

$$(1+e^{B(H_\omega-\mu)})^{-1} \xrightarrow[B\to\infty]{} E_\mu \quad \text{(strongly)}$$

where E_μ is the spectral projection of H_ω on $]-\infty,\mu]$. If μ lies in a gap of the spectrum of H_ω, E_μ is a continuous function of H_ω and an element of the C^*-algebra generated by H_ω . Let us pass to the isomorphic algebra $C_o(\mathbb{R}^2) \rtimes_\alpha \mathbb{R}_2$:

For $\alpha \in \mathbb{R} \setminus \mathbb{Q}$, $\alpha = |\vec{B}|e/\hbar$ there is a unique trace on this algebra, $\vec{J}$ is represented by a derivation of a function and denote by $P_\mu^{'}$ the image of E_μ under this isomorphism.

Bellissard proved (using previous work of Thouless et al.) that by the Kubo formula in the zero-temperature-limit one gets for the Hall conductivity:

$$\sigma_H = e^2/\hbar \; \frac{1}{2\pi i} \; \tau(P_\mu[\partial_1 P_\mu, \partial_2 P_\mu])$$

which is exactly the Chern-character of the module associated to the idempotent P_μ! This might explain the "stability" of the step-behavior for disordered configurations.

What is left to prove now is that σ_H is really quantized, i.e. that $\tau(P_\mu[\partial_1 P_\mu, \partial_2 P_\mu]) \in \mathbb{Z}$.

This is shown by the Index Theory :

Let $H = H^+ + H^- = L^2(\mathbb{R}^2) \oplus L^2(\mathbb{R}^2)$, $\varepsilon = \begin{pmatrix} 1 & 0 \\ 0 & -1 \end{pmatrix}$ the grading and by identifying $\mathbb{C} \cong \mathbb{R}^2$, $z = x_1 + ix_2$

$$F = \begin{pmatrix} 0 & \frac{z}{|z|} \\ \frac{z^*}{|z|} & 0 \end{pmatrix}$$

The module-structure is given by the operator-formalism, i.e.

$$[\rho_\omega(a)\psi](x) := \int_{\mathbb{R}^2} d^2y \; a(T^{-x}\omega, y-x) e^{2\pi i |\vec{B}| e/\hbar \; x \wedge y} \; \psi(y)$$

$(H, \rho_\omega, F, \varepsilon)$ is a 2-summable Fredholm-module for $C_o(\mathbb{R}^2) \rtimes_\alpha \mathbb{R}^2$ and

$$\begin{aligned} \tau(P_\mu[\partial_1 P_\mu, \partial_2 P_\mu]) &= \text{trace}(\varepsilon P_\mu dP_\mu dP_\mu) \\ &= \text{ind}(F^+_{P_\mu}) \in \mathbb{Z} \end{aligned}$$

This argument works for $P_\mu \in C_o(\mathbb{R}^2) \rtimes_\alpha \mathbb{R}^2$ that is when μ is in a gap of the spectrum of H_ω . In general this is not true, the spectrum consists of a continuous part and a pure point part, the corresponding "eigenstates" to free electron states resp. localized states. The first contribute to the conductivity the latter represent bound electrons which don't contribute to the conductivity.

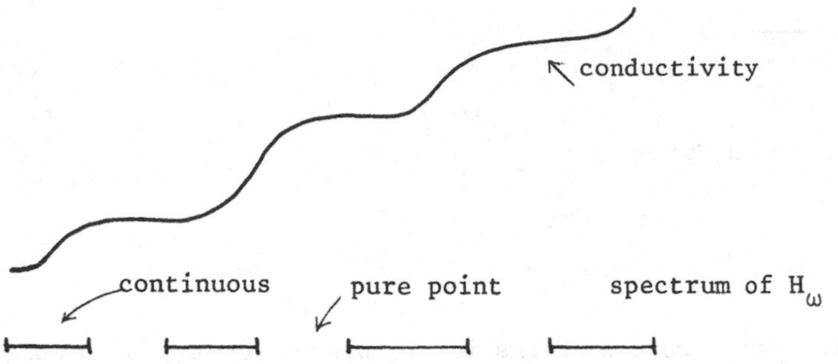

The observed plateus correspond exactly to the region of localized states. Bellissard showed that the proof of the Index Theorem works in a slightly more general situation that is if

$$\|\partial_1 P_\mu\|^2 + \|\partial_2 P_\mu\|^2 < \infty$$

The P_μ need not be in the algebra but need only have finite Dirichlet-integral. When you write down this finiteness condition, you can recognize it as exactly the definition of localized states. What you see is that the Index-Theorem and thereby the quantization of σ_H will hold exactly in the region of localized states and will force the conductivity to be an integer multiple of e^2/h .

Let me stop with this and finish with some remarks and open problems.

1) Can one extend the orbifold string theory to more complicated fibers with spectrum one point.

2) The work of A. Jaffe and coll. defines using $F = Q(1+Q^2)^{-1/2}$ a Fredholm module over the algebra of bosons, compute the ∞ dim. cycle assigned to it :

$$\int \omega_2\omega_1 - (-1)^{\partial_1\partial_2} \omega_2\omega_1 = \int d\omega_1 d\omega_2 \qquad \forall\ \omega_1,\omega_2$$

3) Formulate gauge theories for spaces of non integral Hausdorff dimension, using p-summable Fredholm modules + action :

$$\|dA+A^2\|^2_{H.S.}$$

4) Interpret results in totally int. systems of dim. 1. Using the flat connexions on $S^1 \equiv$ Grassmanian + Action :

+ Action $\int AdA + \frac{2}{3}A^3$.

5) Is string theory for a given target space X yielding an action function on the space of 1-summable Fredholm modules.

6) Higher Schwinger terms $(D+1 > 2)$ and regulator maps :

$$K_{alg}(A) \longrightarrow \mathbb{C}^{x}$$

function on Space

REFERENCES

1) Bott, Tu: "Differential Forms in algebraic topology", Springer
2) Atiyah, Singer: "Index of Elliptic operators III", Ann. of Math. 87
3) Chern: "Complex manifolds without potential theory", Springer
4) Bellissard:"K-theory of C*-algebras in solid state physics", LNP 257
5) Bellissard:Lecture given at the Conference on "Localization in Disordered System", Bad Schandau DDR '86 CPT-86/P. 1949
6) Connes: "Non-Commutative Diff. Geometry", Publ. IHES 62
7) Connes: "A Survey of foliations and operator algebras", Proc. Symposia Pure Math. 38 Part II
8) Karoubi: "K-theory, An Introduction", Springer
9) Pimsner, Voiculescu: "Exact sequences for K-groups and Ext groups of certain cross-product C*-alg.", J. of Op. theory 4 '80
10) Simon: "Trace ideals and their applications", London Math. Soc. LN 35
11) Loday, Quillen: "Cyclic homology and the Lie algebra of matrices", Commun. Math. Helv. 59
12) Rieffel: "C*-algebras associated with irrational rotations", Pac. J. of Math. 93 '81
13) Blackadar: "K-theory for operator algebras" MSRI Publication Springer
14) Witten: "Morse theory and supersymmetry", J. Diff. Geom. 17
15) Proceedings "Bonn Arbeitstagung 1986/1987", MPI Preprints talks of Atiyah '87 and Floer '86/'87
16) Brown, Douglas: "K-homology and index theory, operator algebras and applications", Proc. Symp. Pure Math. 38 Part I '82
17) Dupont: "Curvature and characteristic classes", LNM
18) Wells: "Differential Analysis on complex manifolds", Springer

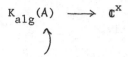

19) Bleeker: "Gauge Theory and Variational principle", Addison Wesley

20) Segal, Pressley: "Loop Groups", Oxford Univ. Press '86

STATISTICS OF FIELDS, THE YANG-BAXTER EQUATION, AND THE THEORY OF KNOTS AND LINKS

Jürg Fröhlich

Theoretical Physics

ETH-Hönggerberg, CH-8093 Zürich

1. Introduction

In these notes we describe an analysis of the statistics problem in quantum field theory. It has been known for some time that in two space-time dimensions there is more to the problem of statistics of local fields than Bose- or Fermi statistics [1]. In three-dimensional space-time, local quantum fields obey Bose- or Fermi statistics, but "extended particles", coupled to the vacuum by fields localized on cones, may exhibit intermediate statistics, so called Θ-statistics [2].

There is a fairly close, and perhaps somewhat surprising, connection between the statistics problem in two-dimensional quantum field theory and the following three topics:

(1) Soliton quantization in two-dimensional quantum field models; [3,4,5,6].

(2) Two-dimensional, conformal quantum field theory; some aspects of the covariant quantization of superstring theory (in particular, the construction of fermion emission vertices); [7,8,9,10].

(3) Polynomial invariants for knots and links [11]. Some of these connections will be sketched briefly, but there is no room here for a detailed treatment. The results discussed in these notes may be viewed as comments on the beautiful and deep lectures by G. Mack, D. Friedan and K.Gawedzki. Some of my results, next to many other things, are also sketched in the notes of B. Schroer.

The first examples of statistics different from standard Bose- or Fermi statistics in the context of simple, two- dimensional quantum field models were encountered in [1]. A related issue, the construction of order- and disorder variables in two-dimensional classical spin systems such as the Ising model, was analyzed by Kadanoff and Ceva [12]. Since the Euclidean description of quantum field theory (see e.g. [13]) was not a well known thing, at the time, Kadanoff and Ceva do, however, not seem to have realized the implications of their work for the statistics problem in two-dimensional quantum field theory. That problem was studied, independently of [12], in [3], where the commutation relations between soliton- and meson fields, sometimes called dual algebra, in simple models exhibiting quantum kinks, like the $\lambda\varphi_2^4$-model, the Ising- or the Potts field theory, were derived and studied. The ideas of Kadanoff and Ceva and the dual algebra played a certain rôle in the work of Jimbo, Miwa and Sato [14] on the scaling limit of the correlation functions of the two-dimensional Ising model. In the context of the Ising model, their work unravelled a connection between the statistics of local fields and monodromy properties of their Euclidean Green functions. That connection was studied in a more general context by the author - and probably other people, as well. Unfortunately, interest in two-dimensional quantum field theory was fading, at that time, and my analysis remained somewhat incomplete and unpublished; but see [4] for a brief sketch, and [5] for results which are relevant in this context. The connection between statistics of fields and monodromy of Euclidean Green functions will be discussed in some detail in these notes.

The work of Kadanoff and Ceva on order-disorder variables was generalized by Wegner [15], who introduced the first lattice gauge theories and the appropriate order- and disorder variables, nowadays called Wilson- and 't Hooft loops. But the field-theoretic interpretation of Wegner's models as lattice approximations to Euclidean-region gauge theories is due to Wilson [16], and the extension of dual algebra to gauge theories, i.e. the 't Hooft commutation relations, is due to 't Hooft [17]. [The 't Hooft commutation relations define the dual algebras generated by Wilson- and 't Hooft loops, in four-dimensional gauge theories, and by Wilson loops and vortex operators in three-dimensional Higgs models. The latter case was actually already considered in 1975/76 in connection with vortex quantization in three- dimensional Higgs models.] The 't Hooft commutation relations have, a priori, nothing immediately obvious to do with the statistics of fields or particles. But it was recognized by Wilczek [18] and systematized by Wu Yong Shi [2] and others that strange statistics can appear in gauge theories in three space-time dimensions: In the presence of a Chern-Simons term in the action (which violates parity) three-dimensional Higgs-theories can have fractionally charged, magnetic vortices among their one-particle states. A magnetic vortex with a fractional electric charge is a particle with intermediate spin and intermediate statistics, i.e. its spin need not be integer or half-integer, and its statistics may be intermediate between Bose- and Fermi statistics (it is called Θ-statistics), depending on the value of its electric charge. For this reason, Wilczek has called them "anyons".

The mathematical connection between 't Hooft type commutation relations and the spin and statistics of anyons has been pinned down in [19], where field theory models of anyons are discussed. The different possibilities for particle statistics in relativistic theories in three space-time dimensions appear to be exhausted by Θ-statistics. This is a consequence of the structure of the three-dimensional quantum mechanical Poincaré group and a spin-statistics connection.

In four- or higher dimensional field theories, statistics of fields and particles is limited to Bose- or Fermi statistics. A deep analysis of statistics within the algebraic approach to local, relativistic quantum physics has been carried out by Doplicher, Haag and Roberts [20]. [Unfortunately, they did not consider the more exotic possibilities appearing in two and three space-time dimensions.]

My work with P.-A. Marchetti on anyons, several stimulating discussions with V. Jones and some aspects of the covariant quantization of superstrings rekindled my interest in the statistics problem in two and three dimensions and its connections with monodromy properties of Euclidean Green functions. In the following, I present a preliminary report on my results. In the next section, some examples of two-dimensional field theory models with soliton states and exotic statistics are briefly reviewed. This serves as a motivation for the model-independent analysis of the statistics problem presented in Sect. 3. Readers not familiar with the elements of constructive quantum field theory may wish to skip Sect. 2 in a first reading and possibly return to it later for motivation. (No attempt at a pedagogical presentation is made in Sect. 2).

In Sect. 4, we describe the relations between statistics of fields, monodromy properties of Euclidean Green functions and representations of the braid groups (or the pure braid groups) on n strings. We then discuss similar issues in the context of two-dimensional conformal quantum field theory. In Sect. 5, we sketch an application to the Wess-Zumino-Witten (WZW) models. In Sect. 6, we sketch connections of the issues discussed in Sects. 3 through 5 with the theory of knots and links. In Sect. 7, we formulate some open problems and conjectures.

Acknowledgements. I wish to thank P.-A. Marchetti and V. Jones for several very stimulating discussions, V. Jones for giving me copies of some of his very interesting unpublished notes on knots and links, T. Kohno for interesting seminars and some preprints, and A. Connes, V. Jones and G. Mack for much needed encouragement.

2. Dual algebras in two-dimensional quantum field models

In this section we consider some Lagrangian quantum field models in two space-time dimensions, with Lagrangian density given by

$$\mathcal{L}(\varphi) = \mathcal{L}_0(\varphi) - \mathcal{L}_I(\varphi), \tag{2.1}$$

$$\mathcal{L}_0(\varphi) = (\partial_t\varphi)^2 - (\partial_x\varphi)^2, \tag{2.2}$$

and $\mathcal{L}_I(\varphi)$ is a local polynomial in the field φ which is bounded from below.

The field φ is a real, scalar field with $N = 1,2,3\cdots$ components, i.e. $\varphi = (\varphi^1, \cdots, \varphi^N)$. The polynomial $\mathcal{L}_I(\varphi)$ is chosen such that $\mathcal{L}(\varphi)$ has a discrete symmetry group, G. It is then possible to choose the coefficients of $\mathcal{L}_I$ in such a way that the physical vacuum of the theory is degenerate, and pure vacuum states are not invariant under G. These claims have been established in constructive quantum field theory, long ago. See [21,22] and refs. given there. It was proven in [3] that, under these circumstances, the theory has non-trivial soliton sectors orthogonal to the vacuum sectors. The soliton sectors are superselection sectors labelled by a multiplicative topological charge. Furthermore, one can construct soliton fields, $s_g(g \in G)$, such that a dense set of states in the soliton sectors can be constructed by applying the fields s_g to a dense set of states in a vacuum sector. The fields constructed in [3] are bounded but in no sense strictly local. Unbounded, local soliton fields have been constructed in [5,6] for models with an abelian, internal symmetry group G. If G is non-abelian the fields are genuinely non-local, in every sense of the word, (although they still correspond to localized *automorphisms of a G-invariant observable algebra). This is an interesting situation which Marchetti and I have studied in the context of some models. [A brief analysis of a situation somewhat more general than the one envisaged here is contained in [6].] So far, we have results for specific models, but a general, model-independent analysis is only beginning to emerge. Suffice it to say that in studying this problem one is very naturally led to loop groups and their representation theory. [The loop group associated with $S0(2)$ made an appearence in [3], and, in retrospect, the study of certain representations of rather general loop groups would have been well motivated by the kinds of problems considered in [3]. An account of our results on models with non-abelian symmetry groups and of some model- independent facts will appear elsewhere.] In this section, we propose to analyze the field algebra generated by φ and by the soliton fields s_g in the case where φ and s_g are local, but in general not relatively local fields. We therefore choose G to be abelian. Without loss, we may assume that $G = Z_n$, for some $n = 2,3,4,\cdots$ and $N = 1$ *or* 2. [For $G = Z_2$, we may choose $N = 1$, for $G = Z_n$, $n \geq 3$, we choose $N = 2$.] If $N = 2$ it is advantageous to work with a complex field $\chi = \varphi^1 + i\varphi^2$. Everyone who knows a little constructive quantum field theory knows that for each $n = 2,3,4,\cdots$, one can choose a polynomial, P_{2k}, of degree $2k$, where $2k$ is the smallest even integer larger than n, such that P_{2k} is Z_n-invariant and bounded from below, and the quantum field theory with Lagrangian

$$\mathcal{L}(\chi) = \mathcal{L}_0(\chi) - :P_{2k}(\chi): \tag{2.3}$$

has a spontaneously broken Z_n-symmetry and an n- fold degenerate vacuum. (In (2.3) the double colons indicate Wick ordering.)

The methods of [3,5,6] then permit us to construct local soliton fields, $s_1, \cdots,$ s_{n-1}, mapping a vacuum sector onto $n-1$ different soliton sectors. The field s_l is obtained from an l-fold product of s_1, with the help of an operator product expansion.

Let x and y be points in two-dimensional space-time, $\mathcal{M}^2$, (with Lorentzian metric given by $\begin{pmatrix}1 & 0\\ 0 & -1\end{pmatrix}$).These points are space-like separated, written $x \times y$, iff $(x-y)^2 < 0$. If $x \times y$ then the inequalities

$$\vec{x} \gtrless \vec{y}, \tag{2.4}$$

where $\vec{x}$ is the space component of x, ($x^0 = t$ is its time component) are Lorentz-invariant. This is because the complement of the closure of the interior of the light cone in $\mathcal{M}^2$ is disconnected. These are the key facts behind the appearance of exotic statistics in two space-time dimensions.

It is shown in [3,4] that $s_k(x)$ and $\chi(y)$ satisfy the following "dual algebra": If $x \times y$ then

$$[s_l(x), s_{l'}(y)] = 0 = [\chi(x), \chi(y)], \tag{2.5}$$

and

$$s_l(x)\chi(y) = \begin{cases} \chi(y)s_l(x), & \text{if } \vec{y} < \vec{x} \\ e^{2\pi i \frac{l}{n}}\chi(y)s_k(x), & \text{if } \vec{y} > \vec{x}. \end{cases} \tag{2.6}$$

Let $\psi_l(x)$ be the field obtained from the product $s_l(x)\chi(y)$, as $y \to x$, (operator product expansion). Then it easily follows from (2.5), (2.6) that, for $x \times y$,

$$\psi_l(x)\psi_l(y) = e^{2\pi i \frac{l}{n}}\psi_l(y)\psi_l(x) \tag{2.7}$$

$$\psi_l(x)\psi_{l'}(y) = e^{2\pi i(\frac{l}{n}'\Theta(\vec{x}-\vec{y})+\frac{l}{n}\Theta(\vec{y}-\vec{x}))}\psi_{l'}(y)\psi_l(x) \tag{2.8}$$

$$\psi_l(x)s_{l'}(y) = e^{2\pi i \frac{l}{n}'\Theta(\vec{x}-\vec{y})}s_{l'}(y)\psi_l(x) \tag{2.9}$$

and

$$\psi_l(x)\chi(y) = e^{2\pi i \frac{l}{n}\Theta(\vec{y}-\vec{x})}\chi(y)\psi_l(x) \tag{2.10}$$

If $n = 2$ and if $\chi(x) = \varphi(x)$ is a real, one-component scalar field then $\psi(x) \equiv \psi_1(x)$ is a real Fermi field, i.e. a Majorana field, according to (2.7). In this case, the dual algebra (2.5) - (2.10) is the one encountered in the $\lambda\varphi_2^4$-model and in the two-dimensional Ising field theory. The Ising field theory is completely characterized by the requirement that $\psi(x)$ be a free Majorana field. [In the $\lambda\varphi_2^4$-model, $\psi(x)$ is not a free field and is very singular in the ultraviolet. See [6].] In these examples, let us denote $\varphi(x)$ by $\Phi_1(x)$ and $\psi(x)$ by $\Phi_2(x)$. Let $x-y$ be space-like and $\sigma = sign(\vec{x}-\vec{y})$. Then by (2.7) and (2.10),

$$\Phi_\alpha(x)\Phi_\beta(y) = R_{\alpha\beta}^{\gamma\delta}(\sigma)\Phi_\gamma(y)\Phi_\delta(x) \tag{2.11}$$

where $R(+)$ and $R(-)$ are 4×4 matrices given by

$$R(+) = \begin{pmatrix} 1 & 0 & 0 & 0 \\ 0 & 0 & 1 & 0 \\ 0 & -1 & 0 & 0 \\ 0 & 0 & 0 & -1 \end{pmatrix}, R(-) = \begin{pmatrix} 1 & 0 & 0 & 0 \\ 0 & 0 & -1 & 0 \\ 0 & 1 & 0 & 0 \\ 0 & 0 & 0 & -1 \end{pmatrix} \tag{2.12}$$

Note that

$$R(+)R(-) = 1 \tag{2.13}$$

[This is no accident, as will be seen in Sect. 3.] The matrices $R(\sigma), \sigma = +or-$, are called statistics matrices. Statistics matrices can be derived from (2.7) - (2.10), for all the examples discussed, so far, i.e. for $G = Z_n$, $n = 2, 3, 4, \cdots$. For $n \geq 3$, we call the fields $\psi_l(x), l = 1, \cdots, n-1$, parafermion fields.

It may be useful to recall that exotic statistics already appears in the theory of the free, massless scalar field, φ, in two space-time dimensions, as discussed e.g. by Streater and Wilde [1]. We define

$$\Phi_\lambda(x) =: \exp i\lambda[\dot{\varphi}(\Theta_x) + \varphi(x)] : \tag{2.14}$$

where $\dot{\varphi} = \partial_t \varphi, \Theta_x(\vec{y}) = \Theta(\vec{y} - \vec{x})$ is the Heavyside step function, and the double colons denote Wick ordering. [A rigorous definition of Φ_λ can be found, for example in [6], but has been known for a long time.] Then the Weyl form of the canonical commutation relations implies that, for $x \times y$,

$$\Phi_\lambda(x)\Phi_{\lambda'}(y) = exp[-i\lambda\lambda'\sigma]\Phi_{\lambda'}(y)\Phi_\lambda(x), \tag{2.15}$$

with $\sigma = \text{sign}\ (\vec{x} - \vec{y})$. This example plays a certain rôle in chiral bosonization and in vertex operator constructions; see e.g. [6] and refs. given there. In [6] the Euclidean Green functions of the fields Φ_λ have been calculated. With a view to what is discussed in Sect. 5, it may be worthwhile to remark that these Green functions give rise to the so-called Burau representation of the braid groups.

For more detailed discussions of soliton quantization and exotic statistics in the framework of simple two-dimensional models we refer the reader to [3-6, 19].

3. Statistics of "local fields" in two-dimensions, and the Yang-Baxter equation

For the benefit of the reader who has skipped Sect. 2 we recall some definitions: Two-dimensional Minkowski space is denoted by $\mathcal{M}^2$, with Lorentzian metric given by $\left(\begin{smallmatrix}1 & 0\\ 0 & -1\end{smallmatrix}\right)$. Two points, x and y, in $\mathcal{M}^2$ are said to be space-like separated, written $x \times y$, if $(x-y)^2 = (x^0 - y^0)^2 - (\vec{x} - \vec{y})^2 < 0$. [Here x^0 is the time component of $x \in \mathcal{M}^2$, and $\vec{x}$ its space component.] Let $\bar{V}_+$ and $\bar{V}_-$ denote the closures of the interiors of the forward and backward light cones and $\bar{V} = \bar{V}_+ \cup \bar{V}_-$. Clearly $\mathcal{M}^2 \backslash \bar{V}$ is disconnected. Therefore if $x \times y$ the inequalities

$$\vec{x} \gtrless \vec{y} \tag{3.1}$$

are Lorentz invariant. We set $\sigma \equiv \sigma_{xy} = sign(\vec{x} - \vec{y})$. Two regions, 0_1 and 0_2, in $\mathcal{M}^2$ are said to be space-like separated, written $0_1 \times 0_2$, if $x \times y$, for every $x \in 0_1$ and every $y \in 0_2$. By $\mathcal{P}_+^\uparrow$ we denote the connected component of the two-dimensional Poincaré group.

We consider a local, relativistic quantum field theory over $\mathcal{M}^2$, in the sense of Wightman [23]. [We must assume that the reader has a vague idea about what a Wightman theory is. We shall not treat any domain problems arising when one deals with unbounded operators, but we shall never implicitly assume more than standard properties of fields in a Wightman theory.] We suppose that the observable fields of the theory are local tensors $J^\alpha_{\mu_1 \cdots \mu_k}(x)$, where $\mu_1, \cdots, \mu_k$ are Lorentz indices, and α labels the different tensor fields of equal rank. By locality of observable fields we mean that

$$[J^\alpha_{\mu_1 \cdots \mu_k}(x), J^\beta_{\nu_1 \cdots \nu_l}(y)] = 0 \tag{3.2}$$

if $x \times y$. Typically, it is reasonable to imagine that among the observable fields of the theory there be the energy-momentum tensor, $T_{\mu\nu}(x)$, of the theory.

In the models discussed in Sect. 2, the observable fields must be invariant under the internal symmetry group, G (see [4]), and it is convenient to choose these fields to be given by e.g. $J(x) =: (\varphi \cdot \varphi) : (x)$ and $T_{\mu\nu}(x)$.

If 0 is an open subset of $\mathcal{M}^2$ we let $\mathcal{A}(0)$ denote the local algebra generated by the operators

$$J^{\alpha}_{\mu_1 \cdots \mu_k}(f) \equiv \int d^2x J^{\alpha}_{\mu_1 \cdots \mu_k}(x) f(x), \tag{3.3}$$

where $f(x)$ is a test function on $\mathcal{M}^2$ vanishing in the complement of 0. [It may be more appropriate to define $\mathcal{A}(0)$ to be the von Neumann algebra generated by the bounded functions of $J^{\alpha}_{\mu_1 \cdots \mu_k}(f)$, supp $f \subset 0$; but we don't have to go into details about this.] Locality can now be expressed by saying that if $0_1 \times 0_2$ then

$$[A, B] = 0, \quad \forall A \in \mathcal{A}(0_1), \forall B \in \mathcal{A}(0_2) \tag{3.4}$$

The theory is Lorentz covariant if $\mathcal{A} = \bigvee_0 \mathcal{A}(0)$ carries a representation, τ, of $\mathcal{P}^{\uparrow}_{+}$ as a group of *automorphisms, with the property that, for $(\Lambda, a) \in \mathcal{P}^{\uparrow}_{+}$,

$$\tau_{(\Lambda,a)}(\mathcal{A}(0)) = \mathcal{A}(0_{(\Lambda,a)}), \tag{3.5}$$

where

$$0_{(\Lambda,a)} = \{x \in \mathcal{M}^2 : \Lambda^{-1}(x - a) \in 0\}.$$

A vacuum sector of the theory is a separable Hilbert space, $\mathcal{H}_0$, which carries an irreducible representation, π_0, of $\mathcal{A}$ with the property that τ is unitarily implementable on $\mathcal{H}_0$, i.e.

$$\tau_{(\Lambda,a)}(A) = U_0(\Lambda, a) A U_0(\Lambda, a)^{-1}, \; ^{*)} \tag{3.6}$$

for all $A \in \mathcal{A}, (\Lambda, a) \in \mathcal{P}^{\uparrow}_{+}$, and U_0 is a unitary representation of $\mathcal{P}^{\uparrow}_{+}$ on $\mathcal{H}_0$. Moreover, the generators $(H_0, \vec{P}_0)$ of the translations $U_0(1, a)$ are supposed to satisfy the relativistic spectrum condition, i.e. spec $(H_0, \vec{P}_0) \subseteq \bar{V}_{+}$. Finally, it is assumed that $(0, \vec{0})$ is a simple eigenvalue of $(H_0, \vec{P}_0)$, i.e. there exists a unique ray, Ω, in $\mathcal{H}_0$ such that $U_0(1, a)\Omega = \Omega, \forall a$, hence

$$U_0(\Lambda, a)\Omega = \Omega, \quad \forall (\Lambda, a) \in \mathcal{P}^{\uparrow}_{+} \tag{3.7}$$

Ω is called vacuum, π_0 a vacuum representation. By the Reeh - Schlieder theorem [23], Ω is separating for $\mathcal{A}(0)$, for all bounded 0, i.e. if $A\Omega = 0, A \in \mathcal{A}(0)$, then $A = 0$. One lesson the reader who has read Sect. 2, or the papers quoted there, should have learned is that not all the physical states of such a theory need to be rays in the vacuum sector $\mathcal{H}_0$. For example, if the theory has soliton states these states do not belong to $\mathcal{H}_0$. In attempting to find all physical states of the theory, it is useful to introduce the concept of covariant representations. A representation, π, of $\mathcal{A}$ on a Hilbert space $\mathcal{H}_\pi$ is called covariant(in these notes - but not everywhere in the

* We identify here $\pi_0(A)$ with A, to keep notations simple.

literature) if τ is unitarily implementable on $\mathcal{H}_\pi$, i.e. there is a unitary representation, U_π, of $\mathcal{P}_+^\uparrow$ on $\mathcal{H}_\pi$ such that

$$\pi(\tau_{(\Lambda,a)}(A)) = U_\pi(\Lambda,a)\pi(A)U_\pi(\Lambda,a)^{-1},$$

for all $A \in \mathcal{A}, (\Lambda,a) \in \mathcal{P}_+^\uparrow$, and, moreover, the generators $(H_\pi, \vec{P}_\pi)$ of $U_\pi(1,a)$ satisfy the relativistic spectrum condition. [Of course, it is not assumed that $\mathcal{H}_\pi$ contain a vacuum.]

One of the basic problems of quantum field theory is this:

Given $(\mathcal{A}, \tau)$, find all covariant representations π, of $\mathcal{A}$.

Assuming that we are able to solve this problem, we then define the physical Hilbert space, $\mathcal{H}$, of the theory by

$$\mathcal{H} = \bigoplus_{\substack{\text{cov. reps.}\\ \pi \text{ of } \mathcal{A}}} \mathcal{H}_\pi, \tag{3.8}$$

and

$$U = \bigoplus_{\substack{\text{cov. reps.}\\ \pi \text{ of } \mathcal{A}}} U_\pi. \tag{3.9}$$

It is generally believed that $(\mathcal{A}, \mathcal{H}, \mathcal{U})$ provides a complete description of the quantum field theory; in particular $\mathcal{H}$ will contain all one-particle states of the theory, etc.

It has turned out that, in practice, one wants to have more than just a complete list of covariant representations of the observable algebra.

One attempts to enlarge the observable algebra $\mathcal{A}$ to a field algebra, $\mathcal{F}$, which acts irreducibly on the physical Hilbert space $\mathcal{H}$. This entails that $\mathcal{F}$ will contain unobservable fields, Φ, which intertwine different covariant representations of $\mathcal{A}$, i.e. make transitions between different sectors, $\mathcal{H}_\pi$ and $\mathcal{H}_{\pi'}, (\pi, \pi'$ covariant representations of $\mathcal{A}$) of the theory. It is a general problem of algebraic quantum field theory to determine whether one can reconstruct a field algebra from a complete list of covariant representations of $\mathcal{A}$. This is quite a vast problem, and the analysis of models shows that, especially in two space-time dimensions, the answer - if there is a completely general answer - is by no means simple. As already mentioned in Sect. 2, the answer will, for example, contain a major part of the theory of loop groups and their representations. [This will be discussed in more detail elsewhere; a special case appears in [3].]

Here we shall assume that the field algebra $\mathcal{F}$ is generated by "local" fields. Examples of such fields are Fermi fields, charged fields, the soliton fields described in Sect. 2, etc. The question is what is meant by "local". One can show that there are at least two notions of locality which are sensible:

(1) Elements of $\mathcal{F}$ are fields, $\Phi(x)$, localized in space-time points $x \in \mathcal{M}^2$.

(2) Elements of $\mathcal{F}$ are fields, $\Phi(\gamma_x)$, localized on space -like curves, γ_x, starting at x and reaching out to space-like $+\infty$, or $-\infty$.

The analysis in these notes shall be restricted to case (1). We now specify what kinds of field algebras, $\mathcal{F}$, we want to admit and what exactly is meant by a local field, in case (1). [It would lead too far to present a general analysis, and my results are not complete, yet.]

We assume that there are finitely many, Poincaré-covariant fields, $\Phi_\alpha(x), \alpha = 1, \cdots, p$, localized in space-time points, $x \in \mathcal{M}^2$, such that $\mathcal{F}$ is generated by $\mathcal{A}$ and by arbitrary polynomials in $\Phi_\alpha(f)$, where f is an arbitrary test function. [In the models studied in Sect. 2, $\mathcal{F}$ is generated by the fields $\chi(x)$ and $s_l(x), l = 1, \cdots, n-1$.]

The rôle of the fields Φ_α is to intertwine the vacuum sector $\mathcal{H}_0$ with superselection sectors $\mathcal{H}_\pi$, where π is a covariant representation of $\mathcal{A}$, i.e. if Ψ is a state in $\mathcal{H}_0$ belonging to the domain of definition of $\Phi_\alpha(f)$ then $\Phi_\alpha(f)\Psi$ belongs to a superselection sector $\mathcal{H}_\pi$ orthogonal to $\mathcal{H}_0$. It is assumed that $\mathcal{F}$ has a common dense domain, $\mathcal{D}$, of definition in $\mathcal{H}$, and that $\mathcal{F}$ acts irreducibly on $\mathcal{H}$.

Let $0_{x_1 \cdots x_n}$ be the space-like complement of the points $x_1, \ldots, x_n$ in $\mathcal{M}^2$; see Fig. 1.

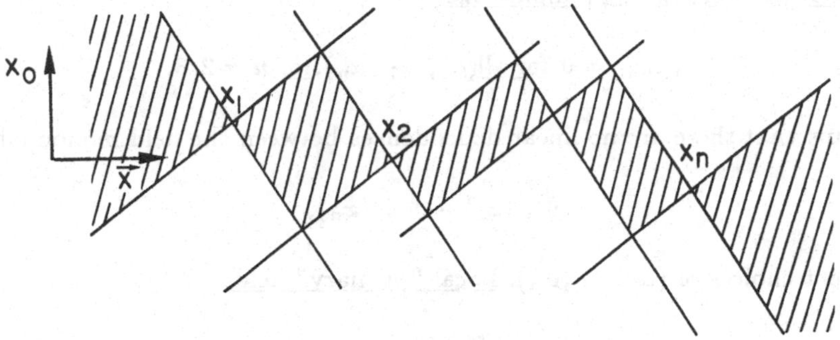

Fig. 1: (The shaded region is $0_{x_1 \cdots x_n}$)

If $x_1, \cdots, x_n$ are pairwise space-like separated $0_{x_1 \cdots x_n}$ consists of of $n-1$ diamonds, a wedge opening to the left and a wedge opening to the right.

By <u>locality</u> of $\Phi_\alpha(x)$ we mean that

(a)

$$[\Phi_\alpha(x), A] = 0, \quad \text{for all} \quad A \in \mathcal{A}(0_x) \tag{3.10}$$

It follows that

$$[\Phi_{\alpha_1}(x_1) \cdots \Phi_{\alpha_n}(x_n), A] = 0, \quad \text{for all} \quad A \in \mathcal{A}(0_{x_1 \cdots x_n}) \tag{3.11}$$

The second property is the following

(b) <u>Reordering Condition:</u> If $x \times y$, <u>then, for every choice of</u> α <u>and</u> β, $\Phi_\alpha(y)\Phi_\beta(x)$ <u>is a linear combination of</u> $\Phi_\gamma(x)\Phi_\delta(y), \gamma, \delta = 1, \cdots, p$.

<u>Motivation:</u> Clearly $\Phi_\alpha(y)\Phi_\beta(x)$ and $\Phi_\gamma(x)\Phi_\delta(y)$ all commute with $\mathcal{A}(0_{xy})$.

Moreover, $\mathcal{F}$ acts irreducibly on $\mathcal{H}$, and locality of the fields Φ_α ought to mean that when forming arbitrary polynomials in $\Phi_{\alpha_1}(x_1) \cdots \Phi_{\alpha_n}(x_n)$ the order of the fields is immaterial if the points $x_1, \cdots, x_n$ are pairwise space-like separated. It is therefore reasonable to assume that $\Phi_\alpha(y)\Phi_\beta(x)$ can be formally expanded in a sum of products of $(D^{n_1}\Phi_{\gamma_1})(x)$ and $(\tilde{D}^{n_2}\Phi_{\gamma_2})(y)$, where D^{n_1} and $\tilde{D}^{n_2}$ are local differential operators of order n_1 and n_2, respectively. It is natural to suppose that the tensorial properties of the fields Φ_α and their ultraviolet dimensions are such that $\Phi_\alpha(y)\Phi_\beta(x)$ can be expanded in a sum of products of $\Phi_\gamma(x)$ and $\Phi_\delta(y)$ and that only monomials of order 2 appear in that expansion. [The reader is supposed to think of Φ_α as some kinds of "primary fields". Assumption (b) looks more natural in the context of conformal quantum field theory.]

Finally we impose the following

(c) Minimality Condition. Let $x_1, \cdots, x_n$ be pairwise space-like separated. Then the equation

$$\sum_{\alpha_1 \cdots \alpha_n} c_{\alpha_1 \cdots \alpha_n} \quad \Phi_{\alpha_1}(x_1) \cdots \Phi_{\alpha_n}(x_n) = 0, \tag{3.12}$$

for complex numbers $c_{\alpha_1 \cdots \alpha_n}$, implies that

$$c_{\alpha_1 \cdots \alpha_n} = 0, \text{for all}(\alpha_1, \cdots, \alpha_n), \text{for } n = 2, 3. \tag{3.13}$$

This means that there are no linear dependences between the field monomials

$$\Phi_{\alpha_1}(x_1) \cdots \Phi_{\alpha_n}(x_n),$$

for different choices of $(\alpha_1, \cdots, \alpha_n)$. Local "primary" fields

$$\{\Phi_\alpha\}_{\alpha=1}^p$$

are fields satisfying $(a), (b)$ and (c). Our basic assumption in this section is that $\mathcal{F}$ is generated by $\mathcal{A}$ and by finitely many (p) local primary fields. I should like to emphasize, once more, that this assumption is special, and that there are two-dimensional field theories with soliton sectors which do not fit into this framework. It may be reasonable to conjecture that it does however accomodate all interesting conformal quantum field theories.

Next, we propose to derive the consequences of assumptions $(a), (b)$ and (c).

From assumption (b) it follows that

$$\Phi_\alpha(y)\Phi_\beta(x) = R_{\alpha\beta}^{\gamma\delta}(x, y)\Phi_\gamma(x)\Phi_\delta(y), \tag{3.14}$$

where repeated indices on the righthand side of (3.14) are to be summed over. $R(x, y) \equiv (R_{\alpha\beta}^{\gamma\delta}(x, y))$ is a $p^2 \times p^2$ square matrix.

Since the fields $\Phi_\alpha(x)$ have been assumed to be Poincaré-covariant, R can only depend on $x - y$ (translation invariance) and must commute with Lorentz transformations, i.e.

$$(S(\Lambda) \otimes S(\Lambda))R(\Lambda^{-1}(x - y)) = R(x - y)(S(\Lambda) \otimes S(\Lambda)), \tag{3.15}$$

where S is a finite-dimensional representation of the Lorentz group. By studying suitable expectation values of $\Phi_\alpha(y)\Phi_\beta(x)$ and of $\Phi_\gamma(x)\Phi_\delta(y)$ one can see that

$R(x-y)$ must be homogeneous in $x-y$ of degree 0. Hence $R(x-y)$ can only depend on $\sigma \equiv sign(\vec{x}-\vec{y})$, in two space-time dimensions, and must be independent of $x-y$, in higher dimensions. The matrix $R(x,y) \equiv R(\sigma)$ is called the statistics matrix of the theory.

Next, we exploit the minimality condition (c).

Clearly, $\sigma = sign(\vec{x}-\vec{y}) = -sign(\vec{y}-\vec{x})$.Hence

$$\begin{aligned}\Phi_\alpha(y)\Phi_\beta(x) &= R^{\gamma\delta}_{\alpha\beta}(\sigma)\Phi_\gamma(x)\Phi_\delta(y)\\ &= R^{\gamma\delta}_{\alpha\beta}(\sigma)R^{\mu\nu}_{\gamma\delta}(-\sigma)\Phi_\mu(y)\Phi_\nu(x).\end{aligned} \tag{3.15}$$

It follows from (c) that

$$R^{\gamma\delta}_{\alpha\beta}(\sigma)R^{\mu\nu}_{\gamma\delta}(-\sigma) = \delta^\mu_\alpha\delta^\nu_\beta, \tag{3.16}$$

i.e

$$\boxed{R(\sigma)R(-\sigma) = 1,} \tag{3.17}$$

in matrix notation. Since in $d \geq 3$ space-time dimensions σ is not Lorentz invariant, and hence R cannot depend on σ, it follows that

$$R^2 = 1, \text{i.e. R is a second root of 1, for } \quad d \geq 3. \tag{3.18}$$

This shows that, within the present framework of local field theory, statistics in $d \geq 3$ space-time dimensions is ordinary Bose - or Fermi-statistics, as expected. [However, for fields localized on space-like cones, more general statistics is possible in three space-time dimensions: $\Theta-$statistics of anyons. See [19].]

Next, we explore the associativity of the field algebra $\mathcal{F}$ and the minimality condition (c). We consider a product of three fields

$$\Phi_\alpha(x)\Phi_\beta(y)\Phi_\gamma(z).$$

We commute them according to the following two schemes:

$$\begin{aligned}&(i) \quad \Phi(x) \quad with \quad \Phi(y), \Phi(x) \quad with \quad \Phi(z), \Phi(y) \quad with \quad \Phi(z).\\ &(ii) \quad \Phi(y) \quad with \quad \Phi(z), \quad \Phi(x) \quad with \quad \Phi(z), \Phi(x) \quad with \quad \Phi(y).\end{aligned}$$

Let

$$\sigma_1 = sign(\vec{y}-\vec{x}), \sigma_2 = sign(\vec{z}-\vec{x})\text{and} \quad \sigma_3 = sign(\vec{z}-\vec{y}).$$

Then scheme (i) yields

$$\begin{aligned}\Phi_\alpha(x)\Phi_\beta(y)\Phi_\gamma(z) &= R^{\mu\nu}_{\alpha\beta}(\sigma_1)\Phi_\mu(y)\Phi_\nu(x)\Phi_\gamma(z)\\ &= R^{\mu\nu}_{\alpha\beta}(\sigma_1)R^{\rho\kappa}_{\nu\gamma}(\sigma_2)\Phi_\mu(y)\Phi_\rho(z)\Phi_\kappa(x)\\ &= R^{\mu\nu}_{\alpha\beta}(\sigma_1)R^{\rho\kappa}_{\nu\gamma}(\sigma_2)R^{\lambda\omega}_{\mu\rho}(\sigma_3)\Phi_\lambda(z)\Phi_\omega(y)\Phi_\kappa(x)\end{aligned} \tag{3.19}$$

Scheme (ii) yields

$$\begin{aligned}\Phi_\alpha(x)\Phi_\beta(y)\Phi_\gamma(z) &= R^{\mu\nu}_{\beta\gamma}(\sigma_3)\Phi_\alpha(x)\Phi_\mu(z)\Phi_\nu(y)\\ &= R^{\mu\nu}_{\beta\gamma}(\sigma_3)R^{\lambda\rho}_{\alpha\mu}(\sigma_2)\Phi_\lambda(z)\Phi_\rho(x)\Phi_\nu(y)\\ &= R^{\mu\nu}_{\beta\gamma}(\sigma_3)R^{\lambda\rho}_{\alpha\mu}(\sigma_2)R^{\omega\kappa}_{\rho\nu}(\sigma_1)\Phi_\lambda(z)\Phi_\omega(y)\Phi_\kappa(x).\end{aligned} \tag{3.20}$$

Comparing (3.19) with (3.20) and using (c) yields

$$\begin{aligned} &R^{\mu\nu}_{\alpha\beta}(\sigma_1)R^{\rho\kappa}_{\nu\gamma}(\sigma_2)R^{\lambda\omega}_{\mu\rho}(\sigma_3) \\ &\qquad = R^{\mu\nu}_{\beta\gamma}(\sigma_3)R^{\lambda\rho}_{\alpha\mu}(\sigma_2)R^{\omega\kappa}_{\rho\nu}(\sigma_1) \end{aligned} \tag{3.21}$$

This is a special case of the Yang-Baxter (*or* $*\,\Delta$) equation. It is advantageous to rewrite (3.21) in matrix notation, in order to see what is going on: Let $V = C^p$. Equation (3.21) is an equation between matrices on $V \otimes V \otimes V$. The first factor corresponds to the field which stands left-most, the second factor to the field in the middle, and the third factor corresponds to the field farthest to the right. We number the factors from 1 to 3, i.e. we write $V_1 \otimes V_2 \otimes V_3$, $V_i = C^p$, for $i = 1, 2, 3$.

We then define

$$R_{12} \equiv R\,|_{V_1 \otimes V_2} \otimes 1\,|_{V_3}$$

$$R_{23} \equiv 1\,|_{V_1} \otimes R\,|_{V_2 \otimes V_3}$$

Then (3.21) is equivalent to the matrix equation

$$\boxed{R_{12}(\sigma_1)R_{23}(\sigma_2)R_{12}(\sigma_3) = R_{23}(\sigma_3)R_{12}(\sigma_2)R_{23}(\sigma_1)} \tag{3.22}$$

Equations (3.17) and (3.22) are the basic equations obeyed by a statistics matrix $R(\sigma)$.

For later purposes, it is convenient to express the contents of equs. (3.17) and (3.22) graphically.

With a product $\Phi_{\alpha_1}(x_1)\cdots\Phi_{\alpha_n}(x_n)$ of local, "primary" fields we associate n points, $(0,1,0), (0,2,0), \cdots, (0,n,0)$, on the 2-axis of R^3. From these points emerge n vertical lines in the positive 3-direction. If $\Phi_{\alpha_i}(x_i)$ is commuted with $\Phi_{\alpha_{i+1}}(x_{i+1})$ the lines above $(0,i,0)$ and $(0,i+1,0)$ are exchanged, more precisely, braided: If $\vec{x}_i - \vec{x}_{i+1} \gtrless 0$ then the ith and $(i+1)^{st}$ line are braided in such a way that the 1-coordinate of a point on the ith line is always $\binom{larger}{smaller}$ than or equal to the 1-coordinate of a point on the $(i+1)^{st}$ line. The n lines end vertically at the points $(0,1,1), (0,2,1), \cdots (0,n,1)$. [In a theory where the fields Φ_α should not be chosen to be symmetric, i.e. $\Phi^*_\alpha \neq \Phi_\alpha$, the lines will bear an orientation: in the $+3$ direction, for Φ_α; in the -3 direction, for Φ^*_α. But in order to keep our notations simple, we shall often consider symmetric, local, "primary" fields, in these notes.] The equations (3.17) and (3.22) can now be represented, graphically, as follows:

(3.17) $\Leftrightarrow$ 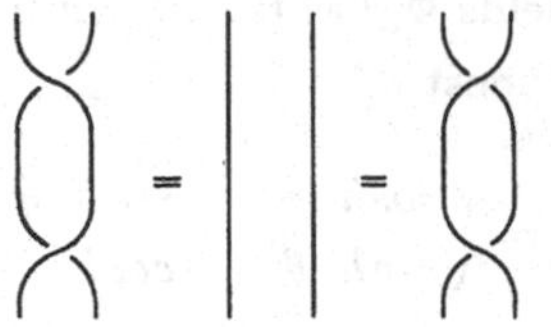 (3.23)

For $\sigma_1 = \sigma_2 = -\sigma_3 = 1$,

(3.22) $\Leftrightarrow$ 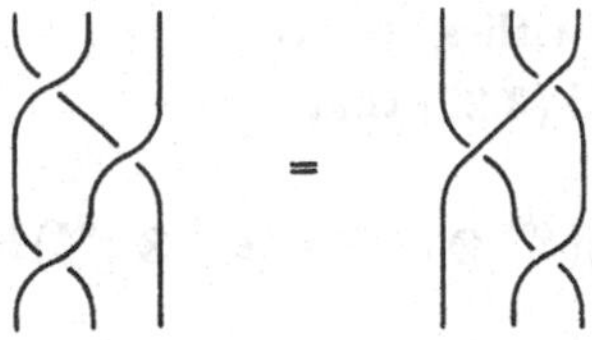 (3.24)

Similar graphical identities hold for the remaining five choices of σ_1, σ_2 and σ_3. [There are, a priori, eight choices of σ_1, σ_2 and σ_3, but only six of them correspond to setting $\sigma_1 = sign(\vec{y} - \vec{x}), \sigma_2 = sign(\vec{z} - \vec{x})$ and $\sigma_3 = sign(\vec{z} - \vec{y})$.]

Identities (3.23) and (3.24) are well known from the theory of knots and links: (3.23) expresses invariance under so-called Reidemeister moves of type II, (3.24) expresses invariance under Reidemeister moves of type III.

Next, we exploit the *invariance of the field algebra. We suppose that there is a matrix μ such that

$$\Phi^*_\alpha(x) = \mu^\beta_\alpha \Phi_\beta(x) \tag{3.25}$$

Since $(\Phi^*_\alpha(x))^* = \Phi_\alpha(x)$, we have

$$\overline{\mu^\beta_\alpha} \mu^\gamma_\beta = \delta^\gamma_\alpha,$$

i.e.

$$\bar{\mu}\mu = 1 \tag{3.26}$$

We define a matrix ${}^TR(\sigma)$ by

$${}^TR^{\gamma\delta}_{\alpha\beta}(\sigma) = R^{\delta\gamma}_{\beta\alpha}(\sigma) \tag{3.27}$$

Since

$$\begin{aligned}(R^{\gamma\delta}_{\alpha\beta}(\sigma)\Phi_\gamma(y)\Phi_\delta(x))^* &= (\Phi_\alpha(x)\Phi_\beta(y))^* \\ &= \Phi^*_\beta(y)\Phi^*_\alpha(x),\end{aligned}$$

it follows, with (3.14), (3.25) and (3.27), that

$$\boxed{\overline{{}^TR(\sigma)}(\mu \otimes \mu) = (\mu \otimes \mu)R(-\sigma)} \tag{3.28}$$

Here $\bar{R}$ is the matrix complex conjugate to R.

We now have to discuss which additional properties the principles of local field theory impose on solutions of equs. (3.17), (3.22) and (3.28) for such solutions to be the statistics matrices of a local field theory in two space-time dimensions.

(i) We assume that the fields $\Phi_\alpha(x)$ transform irreducibly under Lorentz transformations, i.e. for a Lorentz boost,

$$\Lambda = \begin{pmatrix} cosh\ \theta & sinh\ \theta \\ sinh\ \theta & cosh\ \theta \end{pmatrix},$$

$$\Phi^\Lambda_\alpha(x) = e^{\Theta s_\alpha} \Phi_\alpha(\Lambda^{-1}x). \tag{3.29}$$

Let s be the diagonal matrix with $s^\alpha_\alpha \equiv s_\alpha$.

It follows from (3.14) and (3.29) that

$$R(\sigma)(e^{\theta s} \otimes e^{\theta s}) = (e^{\theta s} \otimes e^{\theta s})R(\sigma), \tag{3.30}$$

for all $\theta \in R$.

(ii) We assume that the field theory is a conformal field theory. Let Δ_α be the scaling dimension of Φ_α, and let Δ be the diagonal matrix with $\Delta^\alpha_\alpha \equiv \Delta_\alpha$. Then it follows that

$$R(\sigma)(\lambda^\Delta \otimes \lambda^\Delta) = (\lambda^\Delta \otimes \lambda^\Delta)R(\sigma), \tag{3.31}$$

for arbitrary $\lambda > 0$; (λ parametrizes dilatations).

The proofs of (3.30) and (3.31) are very simple and are left to the reader.

(iii) An important constraint on $R(\sigma)$ is the spin-statistics theorem. For a local field theory in three or more dimensions, this theorem says that fields of integer spin have Bose statistics, while fields of half-integer spin have Fermi statistics. In two space-time dimensions, the connection between spin and statistics is more complicated. Without loss of generality, we may assume that

$$< \Phi_\alpha(x)\Omega, \Phi_\beta(y)\Omega >= \delta_{\alpha\beta} G_\alpha(x,y) \tag{3.22}$$

Positivity of the metric in the physical Hilbert space $\mathcal{H}$ implies that $G_\alpha(x,y)$ is a positive definite distribution. By the spectrum condition, it admits an analytic continuation to the Euclidean region,

$$G_\alpha(x_{-\epsilon}, y_\epsilon) =< \Phi_\alpha(x_\epsilon)\Omega, \Phi_\alpha(y_\epsilon)\Omega >, \tag{3.33}$$

with $x_\epsilon = (i\epsilon, \vec{x}), y_\epsilon = (i\epsilon, \vec{y}), \epsilon > 0$, and $G_\alpha(x_{-\epsilon}, y_\epsilon)$ is positive definite, for arbitrary $\epsilon > 0$. Let $\chi_\epsilon < \pi$ be the largest angle such that $R(\chi_\epsilon)x_\epsilon$ and $R(\chi_\epsilon)y_\epsilon$ remain in the positive-imaginary-time half-space of Euclidean space-time. Here $R(\chi)$ denotes rotation through an angle χ.

Let $U(i\chi)$ denote a Lorentz boost through an imaginary angle $i\chi; U(i\chi)$ is a selfadjoint (positive) operator on $\mathcal{H}$. By the covariance of $\Phi_\alpha(x)$ under Lorentz boosts and the Lorentz invariance of Ω,

$$U(i\chi)\Phi_\alpha(x_\epsilon)\Omega = e^{i\chi s_\alpha}\Phi_\alpha(R(\chi)x_\epsilon)\Omega, \tag{3.34}$$

and the r.h.s. of (3.34) is well defined for all $\chi < \chi_\epsilon$. Clearly

$$< U(i\chi)\Phi_\alpha(x_\epsilon)\Omega, U(i\chi)\Phi_\alpha(y_\epsilon)\Omega > \\ = e^{2i\chi s_\alpha} G_\alpha(x_{-\epsilon}, R(2\chi)y_\epsilon), \tag{3.35}$$

by Euclidean covariance of G_α.

By (3.14), (3.25) and (3.26),

$$e^{2i\chi s_\alpha} G_\alpha(x_{-\epsilon}, R(2\chi)y_\epsilon) = \sum_\rho e^{2i\chi s_\alpha} \mu_\alpha^\gamma \overline{\mu_\kappa^\rho} R_{\gamma\alpha}^{\kappa\rho}(\sigma) < \Omega, \Phi_\rho^*(R(2\chi)y_\epsilon)\Phi_\rho(x_{-\epsilon})\Omega >, \tag{3.36}$$

where χ is chosen so small that

$$\text{Im} \quad (R(2\chi)y_\epsilon)^0 < -\epsilon = \text{Im} \quad x^0_{-\epsilon}.$$

We now first let ϵ tend to 0 and then let χ tend to π. Then (3.35) remains positive definite and the r.h.s. of (3.36) approaches

$$\sum_\rho e^{2i\pi s_\alpha} \mu_\alpha^\gamma \overline{\mu_\kappa^\rho} R_{\gamma\alpha}^{\kappa\rho}(\sigma) G_\rho((0,\vec{y}),(0,\vec{x})) \tag{3.37}$$

Clearly,

$$G_\rho((0,\vec{y}),(0,\vec{x})) \equiv \lim_{\epsilon \searrow 0} G_\rho((-\epsilon,\vec{y}),(\epsilon,\vec{x}))$$

is positive definite. By (3.35), (3.36), the condition on $R(\sigma)$ is that (3.37) be positive-definite. In space-times of dimension three or higher, s_α is integer or half-integer, and $R(\sigma)$ is a square root of 1. In this case the positive- definiteness of (3.37) yields the usual spin statistics theorem. In conformal field theory, the positive definiteness of (3.37) implies that

$$\sum_{\rho:\Delta_\rho=\Delta_\alpha} e^{2i\pi s_\alpha} \mu_\alpha^\gamma \overline{\mu_\kappa^\rho} R_{\gamma\alpha}^{\kappa\rho}(\sigma) \geq 0 \tag{3.38}$$

In many theories, that condition simply implies

$$e^{2i\pi s_\alpha} \mu_\alpha^\gamma \overline{\mu_\kappa^\rho} R_{\gamma\alpha}^{\kappa\rho}(\sigma) > 0,$$

for some $\rho = \rho(\alpha)$. More details about this version of the spin-statistics theorem will be discussed elsewhere.

(iv) The last point that ought to be raised here is whether PCT imposes non-trivial constraints on statistics matrices. As an analysis of the $\lambda\varphi_2^4$-model in the broken symmetry phase, described by the fields $\varphi(x), s(x)$, shows the field algebra of a two-dimensionsal local quantum field theory need not carry a linear representation of PCT. In that case, no simple constraint on $R(\sigma)$ follows from the existence of an anti-unitary PCT operator. If, however, the field algebra does carry a linear representation of PCT then there is a matrix θ such that

$$\Phi_\alpha^{PCT}(x) = \theta_\alpha^\gamma \Phi_\gamma(-x),$$

and simple calculations then show that

$$(\theta \otimes \theta) R(-\sigma) = \overline{R(\sigma)}(\theta \otimes \theta).$$

This and some further matters will have to be discussed in more detail elsewhere. In particular, one would like to have a fairly explicit description of statistics matrices satisfying (3.28) and (i) - (iii) (or (iv)) above. As we know from Sect. 2 and elsewhere, there are certainly non-trivial examples of statistics matrices satisfying those constraints, but a general classification is missing. Since, recently, there has been considerable progress in constructing solutions to the Yang-Baxter equation (starting from the classical Lie algebras), one may be optimistic about the chances for progress.

4. Statistics matrices, representations of braid groups and monodromy of Euclidean Green functions

We start this section by defining the braid groups, B_n, and pure braid groups, P_n, on n strings. We choose n distinct points, $x_1, \cdots, x_n$, in R^2 and consider n maps, $\gamma_1, \cdots, \gamma_n$, from [0,1] to the slab $R^2 \times$ [0,1] in R^3 with the properties

(i) $$\gamma_i(0) = x_i, \quad \gamma_i(1) = x_{\pi(i)},$$

where π is an arbitrary permutation of $\{1, \cdots, n\}$;

(ii) $$\dot{\gamma}_i^3(t) > 0, \text{for all} \quad t \in [0,1];$$

(iii) for $i \neq j$, $\gamma_i(t) \neq \gamma_j(s)$, for all $t \in [0,1]$ and all $s \in [0,1]$.

We now identify all maps, $\gamma_1, \cdots, \gamma_n$, with properties (i) - (iii) related to each other by an ambient isotopy. The resulting object is the braid group on n strings, denoted B_n.

Algebraically, B_n can be defined by choosing the following presentation: B_n is generated by taking arbitrary products of generators $\tau_{i,i+1}, i = 1, \cdots, n-1$, and their inverses, $(\tau_{i,i+1}\tau_{i,i+1}^{-1} = e$, the identity element of B_n), where the generators $\tau_{i,i+1}$ obey the following relations:

$$(a) \quad \tau_{i,i+1} \; \tau_{i+1,i+2} \; \tau_{i,i+1} = \tau_{i+1,i+2} \; \tau_{i,i+1} \; \tau_{i+1,i+2}$$

$$(b) \quad \tau_{i,i+1} \; \tau_{j,j+1} = \tau_{j,j+1} \; \tau_{i,i+1}, \quad \text{for} \mid i - j \mid \geq 2.$$

Geometrically, $\tau_{i,i+1}$ can be represented by the following figure:

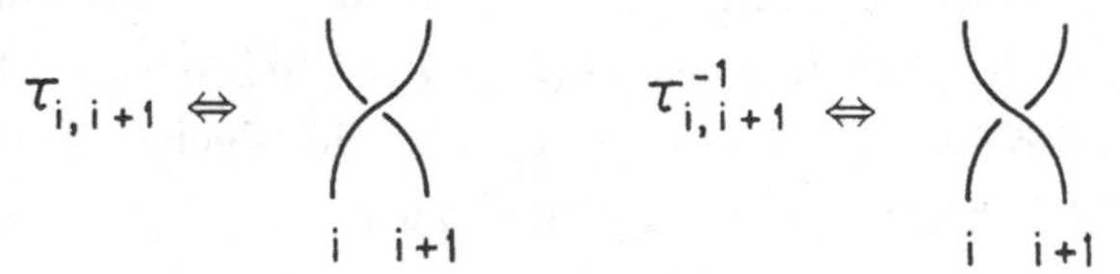

If, in the geometrical definition of B_n, R^2 is replaced by S^2, or by some other compact surface, additional relations between the generators $\tau_{1,2},\cdots,\tau_{n-1,n}$ will emerge in general. In the example of S^2, one has

$$\prod_{i=1}^{n-1} \tau_{i,i-1} \prod_{j=n-1}^{1} \tau_{j,j-1} = e$$

The pure braid group, P_n, on n strings is the subgroup generated by the elements

$$\gamma_{ij} = \tau_{i,i+1} \cdots \tau_{j,j+1}^2 \quad \tau_{j-1,j}^{-1} \cdots \tau_{i,i+1}^{-1} \tag{4.1}$$

Graphically, γ_{ij} corresponds to

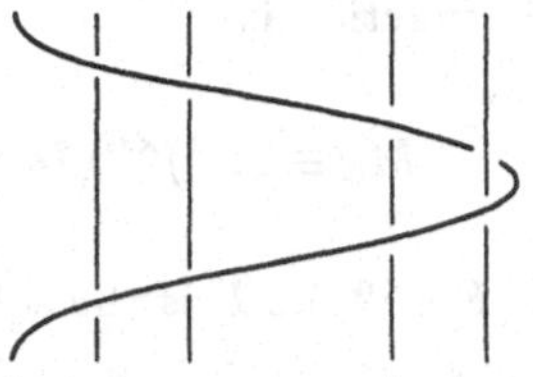

Let $R(\sigma), \sigma = \pm$, be two $p^2 \times p^2$ matrices which are solutions of (3.17) and (3.22). We set $V_i = C^p, i = 1,\cdots,n$, and $V^{(n)} = V_1 \otimes \cdots \otimes V_n$. We define

$$R_{i,i+1}(\sigma) = 1_1 \otimes \cdots \otimes 1_{i-1} \otimes R(\sigma) \otimes 1_{i+2} \otimes \cdots \otimes 1_n, \tag{4.2}$$

with $1_i = 1 \mid_{V_i}$. Representing $\tau_{i,i+1}$ by $R_{i,i+1}(+)$ and $\tau_{i,i+1}^{-1}$ by $R_{i,i+1}(-)$, we see that solutions of (3.17) and (3.22) define representations of B_n, for all $n = 2, 3, \cdots$: Relation (a) follows from (3.22) and (b) follows from (4.2). Thus every statistics matrix of a two-dimensional quantum field theory defines a representation of B_n (hence of P_n), for all $n = 2, 3, \cdots$.

These representations have some significance for the Euclidean description of the quantum field theory. The basic objects in the Euclidean description of quantum field theory are the Euclidean Green- or Schwinger functions,

$$G(\alpha_1, \xi_1, \cdots, \alpha_n, \xi_n), n = 0, 1, 2, \cdots,$$

with, $G(\emptyset) = 1$. Heuristically, they are defined by

$$\begin{aligned} G(\alpha_1, \xi_1, \cdots, \alpha_n, \xi_n) =&< \Omega, \Phi_{\alpha_1}(0, \vec{\xi_1}) e^{-(\xi_2^0 - \xi_1^0)H} \Phi_{\alpha_2}(0, \vec{\xi_2}) \\ &\cdots \Phi_{\alpha_{n-1}}(0, \vec{\xi}_{n-1}) e^{-(\xi_n^0 - \xi_{n-1}^0)H} \Phi_{\alpha_n}(0, \vec{\xi_n}) \Omega >, \end{aligned} \tag{4.3}$$

where H is the Hamilton operator of the theory. [For a precise definition see [13].] By the spectrum condition, i.e. $H \geq 0$, the Euclidean Green functions are well defined if

$$\xi_1^0 < \xi_2^0 < \cdots < \xi_n^0 \tag{4.4}$$

Poincaré covariance of the theory implies Euclidean covariance of the functions

$$G(\alpha_1, \xi_1, \cdots, \alpha_n, \xi_n),$$

so that they are well defined on the smallest Euclidean invariant domain, $\mathcal{E}_n$, containing

$$\{\xi_1, \cdots, \xi_n : \xi_1^0 < \xi_2^0 < \cdots < \xi_n^0\}.$$

Knowledge of the statistics matrix $R(\sigma)$ of the theory permits to construct a huge extension of the domain of definition of the Green functions $G(\alpha_1, \xi_1, \cdots, \alpha_n, \xi_n)$, at the price, though, that the extended Green functions are <u>not</u> single-valued anymore, unless $R(\sigma)^2 = 1$, i.e. for theories with fields that satisfy ordinary Bose-or Fermi statistics. The largest purely Euclidean domain to which $G(\alpha_1, \xi_1, \cdots, \alpha_n, \xi_n)$ may be extended as a multi-valued function is

$$M_n \equiv (E^2)^{\times n} \backslash D \tag{4.5}$$

where D is the "diagonal" of $(E^2)^{\times n}$, i.e.D is the set of points $\{\xi_1, \cdots, \xi_n\}$ for which $\xi_i = \xi_j$, for some $i \neq j$.

The space M_n is not simply connected. In fact, it is easy to see geometrically that

$$\pi_1(M_n) = P_n \tag{4.6}$$

It will be natural to view the Green functions, $G(\alpha_1, \xi_1, \cdots, \alpha_n, \xi_n)$, as single-valued functions on the universal cover $\tilde{M}_n$ of M_n. Because of (4.4) it is actually more natural to extend the definition of $G(\alpha_1, \xi_1, \cdots, \alpha_n, \xi_n)$ to the universal cover, $\tilde{M}_n^{symm.}$, of the space $M_n^{symm.}$, where

$$M_n^{symm.} = M_n / \mathcal{S}_n \tag{4.7}$$

and $\mathcal{S}_n$ is the permutation group of n elements, for each $n = 0, 1, 2, \cdots$. The fundamental group of $M_n^{symm.}$ is

$$\pi_1(M_n^{symm.}) = B_n, \tag{4.8}$$

as one checks easily.

We now indicate how to extend the definition of $G(\alpha_1, \xi_1, \cdots, \alpha_n, \xi_n)$ to $\tilde{M}_n^{symm.}$. First, consider two adjacent arguments, α_i, ξ_i and α_{i+1}, ξ_{i+1}, of G. We choose two oriented curves, γ_i, γ_{i+1}, which map the interval [0,1] into E^2 such that $\gamma_i(0) = \xi_i, \gamma_i(1) = \xi_{i+1}, \gamma_{i+1}(0) = \xi_{i+1}, \gamma_{i+1}(1) = \xi_i$ and $\gamma_i(t) \neq \gamma_{i+1}(s)$, for $t \in (0,1)$ and $s \in (0,1)$. The composition, $\gamma_i * \gamma_{i+1}$, describes an oriented loop with base point ξ_{i+1}. Its orientation is described by a sign, $\sigma = +or-$. Since we are in two dimensions, this sign is invariant under arbitrary, orientation preserving maps of E^2 into itself. The point is now that one can show that, for $(\xi_1, \cdots, \xi_n) \in \mathcal{E}_n$, $\tau \equiv \tau_{i,i+1}$,

$$\begin{aligned} G_\tau(\cdots, \alpha_i, \xi_i, \alpha_{i+1}, \xi_{i+1}, \cdots) = & R_{\alpha_i \alpha_{i+1}}^{\alpha_i' \alpha_{i+1}'}(\sigma) \\ & G_\tau(\cdots, \alpha_i', \gamma_i(t), \alpha_{i+1}', \gamma_{i+1}(t), \cdots) \mid_{t \to 1} \end{aligned} \tag{4.9}$$

The r.h.s. of (4.9) only depends on the element in B_n represented by the two curves $(\gamma_i(t),t),(\gamma_{i+1}(t),t)$. Since the l.h.s. of (4.9) is well-defined, for all $(\xi_1,\cdots,\xi_n)\in\mathcal{E}_n$, one can use (4.9) to extend the definition of G to the domain $\tau_{i,i+1}\mathcal{E}_n$. Although the construction of this extension is not difficult for people who are somewhat familiar with [13] and have read Sect. 3, it takes some space to put this down in words. We must therefore leave it as an exercise to the reader.

Equation (4.9) is the basic fact. Starting from it, one may easily complete the construction of an extension of $G(\alpha_1,\xi_1,\cdots,\alpha_n,\xi_n)$ to $\tilde{M}_n^{symm.}$. Let $b\in B_n$. We present b as a word in the generators $\tau^{\sigma}_{i,i+1},\sigma=\pm1,i=1,\cdots,n-1$:

$$b=\prod_{k=1}^{m}\tau^{\sigma(k)}_{i(k),i(k+1)},\tag{4.10}$$

and define

$$R(b)=\prod_{k=1}^{m}R_{i(k),i(k+1)}(\sigma(k)),\tag{4.11}$$

with $R_{i,i+1}(\sigma)$ given by (4.2). For a given b we choose curves $\gamma_b^{(1)}(t),\cdots,\gamma_b^{(n)}(t)$, mapping [0,1] into E^2, with $\gamma_b^{(i)}(0)=\xi_i$, $\gamma_b^{(i)}(1)=\xi_{\pi(i)}$, for some $\pi=\pi(b)\in\mathcal{S}_n$, such that

$$(\gamma_b^{(1)}(t),t),\cdots,(\gamma_b^{(n)}(t),t),t\in[0,1]$$

is a representative of b. Using (4.9) repeatedly, one may show that

$$\begin{aligned}G_{\underset{\sim}{\alpha}}(\alpha_1,\xi_1,\cdots,\alpha_n,\xi_n)=&R^{\alpha_1'\cdots\alpha_n'}_{\alpha_1\cdots\alpha_n}(b)\\&G_{\underset{\sim}{\alpha}'}(\alpha_1',\gamma_b^{(1)}(t),\cdots,\alpha_n',\gamma_b^{(n)}(t))\,|_{t\to1}\end{aligned}\tag{4.12}$$

This formula permits us to extend the definition of G from $\mathcal{E}_n$ to all of $\tilde{M}_n^{symm.}$, hence to $\tilde{M}_n$. In general, G does not project down to a single-valued function on M_n. For example, the Euclidean Green functions of the fields $\varphi(x)$ and $s(x)$ (meson-and soliton fields) in the $\lambda\varphi_2^4$ - or the Ising$_2$ field theory are double-valued functions on M_n; (see also [6] for an analysis of further examples).

It is not hard to prove that the information contained in equ. (4.12) is completely equivalent to knowing the statistics matrix of the fields $\Phi_\alpha(x),\alpha=1,\cdots,p$.

By equ. (4.12), the Euclidean Green functions $G(\alpha_1,\xi_1,\cdots,\alpha_n,\xi_n)$ carry a representation of B_n, hence of P_n, for $n=0,1,2,\cdots$. This representation uniquely determines the monodromy of the Green functions G: For any element $b\in P_n$,

$$\begin{aligned}G_{\underset{\sim}{\alpha}}(\alpha_1,\gamma_b^{(1)}(0),\cdots,\alpha_n,\gamma_b^{(n)}(0))=&R^{\alpha_1'\cdots\alpha_n'}_{\alpha_1\cdots\alpha_n}(b)\\&G_{\underset{\sim}{\alpha}'}(\alpha_1',\gamma_b^{(1)}(1),\cdots,\alpha_n',\gamma_b^{(n)}(1)).\end{aligned}\tag{4.13}$$

By (4.1), this representation of P_n, in other words the monodromy of G, is trivial iff $R(\sigma)^2=1$, i.e. for ordinary Bose- or Fermi statistics.

By (4.12) and (4.13), we may view the Euclidean Green functions, G, as sections of a complex vector bundle with base space M_n and fiber $V^{(n)}=V_1\otimes\cdots\otimes V_n, V_i=C^p$, for all i. Since the R-matrices in equ. (4.13) only depend on the class $b \in P_n$ represented by the curves $\gamma_b^{(i)}(t), i = 1, \cdots, n$, one can try to interpret these matrices as the holonomy matrices of a flat connection on that vector bundle. Hain and Kohno [24] have shown that if R is any representation of P_n such that $R(\gamma_{ij})$ is sufficiently close to the identity, for all $1 \leq i < j \leq n$, with γ_{ij} given by (4.1), or if R is unipotent, then the matrices $R(b), b \in P_n$, are the holonomy matrices of a flat connection. ω, on that vector bundle which is given by the following formula: A point $\xi = (\xi^0, \xi^1) \in E^2$ is represented by the complex number $z = i\xi^0 + \xi^1$. Then

$$\omega = \sum_{1 \leq i < j \leq n} \Omega_{ij} \ \ \mathrm{d} \ \ \log \ (z_i - z_j), \tag{4.14}$$

where Ω_{ij} are constant matrices acting on $V^{(n)}$. The flatness of ω is equivalent to the equations

$$\boxed{\begin{array}{l} [\Omega_{ij}, \Omega_{ik} + \Omega_{jk}] = [\Omega_{ij} + \Omega_{ik}, \Omega_{jk}] = 0, \\ \text{for } i < j < k, \text{ and} \\ [\Omega_{ij}, \Omega_{kl}] = 0, \text{ for distinct } i, j, k \text{ and } l. \end{array}} \tag{4.15}$$

See [25]. These equations are called infinitesimal pure braid relations. In our case, the matrix Ω_{ij} is given by an endomorphism of $V^{(n)}$ which acts non-trivially only on $V_i \otimes V_j$ and reduces to the identity on all other factors $V_k, k \neq i, j$. It turns out [25] that, in this situation, the equations (4.15) are a special case of the classical Yang-Baxter equations. This is not surprising: Since the statistics matrix $R(\sigma)$ is a solution of the Yang-Baxter equation, the connection Ω whose holonomy is given by $R(\sigma)$ must solve the classical Yang-Baxter equation. [The connection between solutions of the classical $\mathcal{YBE}$ and solutions of the $\mathcal{YBE}$ is analogous to the connection between Lie algebras and Lie groups.] The following result is due to Belavin and Drinfel'd [26]: Let $V_i = V$ carry an irreducible representation π of a simple, complex Lie algebra $\mathfrak{g}$. Let $\{X_\alpha\}_{\alpha=1}^q$ be a basis of $\mathfrak{g}$ which is orthonormal with respect to the Cartan-Killing form on $\mathfrak{g}$. We define

$$\Omega_{ij} = \lambda \sum_{\alpha=1}^{q} 1_1 \otimes \cdots \otimes \pi(X_\alpha) \mid_{V_i} \otimes \cdots \otimes \pi(X_\alpha) \mid_{V_j} \otimes \cdots \otimes 1_n \tag{4.16}$$

Then $\{\Omega_{ij}\}_{1 \leq i < j \leq n}$ solves (4.15). This result will be useful in the next section.

Next, we consider analytic continuations of the Euclidean Green functions, G, to the Bargmann-Hall-Wightman domain [23]. We introduce Euclidean light cone variables:

$$z = i\xi^0 + \xi^1, \bar{z} = -i\xi^0 + \xi^1 \tag{4.17}$$

The axioms of relativistic quantum field theory [23] imply that there are functions,

$$G(\alpha_1, z_1, \bar{z}_1, \cdots, \alpha_n, z_n, \bar{z}_n),$$

which are holomorphic on a certain tube $\mathcal{T}_n$ in C^{2n} (which, using (4.12), is at least as large as the permuted $\mathcal{BHW}$ domain [23]) such that

$$G(\alpha_1, z_1, \bar{z}_1, \cdots, \alpha_n, z_n, \bar{z}_n) = G(\alpha_1, \xi_1, \cdots, \alpha_n, \xi_n). \tag{4.18}$$

The special feature of two-dimensional conformal field theory is that, in two space-time dimensions, the action of the conformal group factorizes in an action transforming only the z-variables and an action transforming only the $\bar{z}$-variables. This enables one to continue the Wightman functions $G(\alpha_1, z_1, \bar{z}_1, \cdots, \alpha_n, z_n, \bar{z}_n)$ to a much larger domain and to treat the z-variables und the $\bar{z}$-variables essentially as independent variables. [It appears that one cannot find a careful axiomatic analysis of two-dimensional conformal field theory, from a Euclidean point of view, in the literature. Unfortunately, this is quite a lengthy enterprize and cannot be presented here. But see [27]; and [28] for some beginnings in a direction related to my own thinking.] If $\bar{z}_1, \cdots, \bar{z}_n$ are kept fixed, we set

$$G(\alpha_1, z_1, \bar{z}_1, \cdots, \alpha_n, z_n, \bar{z}_n) \equiv g(\alpha_1, z_1, \cdots, \alpha_n, z_n) \tag{4.19}$$

One may now attempt to determine the monodromy of the functions g when the variables $z_1 \cdots, z_n$ are varied, but $\bar{z}_1, \cdots, \bar{z}_n$ are kept fixed.

We must look for an equation analogous to (4.12), (4.13), i.e. an equation of the form

$$\begin{aligned} g(\alpha_1, \gamma_b^{(1)}(0), \cdots, \alpha_n, \gamma_b^{(n)}(0)) = \hat{R}_{\alpha_1 \cdots \alpha_n}^{\alpha_1' \cdots \alpha_n'}(b) \\ g(\alpha_1', \gamma_b^{(1)}(1), \cdots, \alpha_n', \gamma_b^{(n)}(1)), \end{aligned} \tag{4.20}$$

where

$$b \in B_n, \text{ and } \gamma_b^{(i)}(\cdot) : [0,1] \to C, i = 1, \cdots, n,$$

are curves with the property that

$$(\gamma_b^{(1)}(t), t) \cdots (\gamma_b^{(n)}(t), t)$$

is a representative of b.

It will not surprize the reader that the matrices $\hat{R}(b)$ are in general not equal to the matrices $R(b)$ defined in (4.11). In fact, it can and does happen that $R(\sigma)^2 = 1$, so that $R(b) = 1$, for all $b \in P_n$, but $\hat{R}(b) \not\equiv 1$; (see [27, 29]).

The interest in computing the matrices $\hat{R}(b)$ comes from the fact that, in conformal field theory, the functions g satisfy differential equations [27,29,$\cdots$]. The matrices $\hat{R}(b)$ then determine the monodromy of those solutions to the differential equations which are candidate Wightman functions.

Let $\hat{R}(\sigma)$ be the matrix with the property that, for b given by (4.10) and $\hat{R}_{i,i+1}(\sigma)$ given by (4.2),

$$\hat{R}(b) = \prod_{k=1}^{m} \hat{R}_{i(k),i(k+1)}(\sigma(k)).$$

One might ask whether $\hat{R}(\sigma)$ is the "statistics matrix" of some chiral operator algebra. If $\hat{R}$ is the Burau representation (see e.g. [25]) of B_n then $\hat{R}(\sigma)$ is the statistics matrix of an algebra of <u>vertex operator</u>, as follows e.g. from the analysis in [6]. More generally, $\hat{R}(\sigma)$ may be the "statistics matrix" of Schroer's <u>exchange algebra</u> [30], but for the moment it is not clear to me how general the notion of exchange algebra is in the context of conformal field theory.

If the functions $G(\alpha_1, \xi_1, \cdots, \alpha_n, \xi_n)$ have an interpretation as single-valued correlation functions of a two-dimensional statistical-mechanical system then the monodromy of $G(\alpha_1, z_1, \bar{z}_1, \cdots, z_n, \bar{z}_n)$ under simultaneous variations of $z_1, \cdots, z_n$ and $\tilde{z}_1, \cdots, \tilde{z}_n$, with $\tilde{z}_i = \bar{z}_i, i = 1, \cdots, n$, must be trivial. This is a powerful constraint [27,29].

5. The example of the Wess-Zumino-Witten models

The Wess-Zumino-Witten (WZW) models [31,32,29] are principal chiral σ-models with a topological Wess-Zumino term in the action. The basic field, $g(\xi)$, of these models is a field with values in a compact, semisimple Lie group G. The Euclidean action is given by

$$\mathcal{S}(g) = \frac{1}{4\lambda^2} \int tr(\partial_\mu g^{-1} \partial^\mu g)(\xi) d^2\xi + k\Gamma(g), \tag{5.1}$$

where $\Gamma(g)$ is the Wess-Zumino (-Witten) term [31,29], and $k \in Z, k \neq 0$. For $\lambda^2 = \frac{4\pi}{k}$, the model turns out to be conformal invariant.

The WZW models form the subject of K. Gawedzki's lectures, so we shall be brief.

For simplicity we consider $G = SU(2)$. Let

$$J(z) = -\frac{1}{2}k \left[(\partial_z g) g^{-1}\right](z) \text{ and } \tilde{J}(\bar{z}) = -\frac{1}{2}k \left[g^{-1} \partial_{\bar{z}} g\right](\bar{z})$$

be the generators of $S\hat{U}(2)_z, S\hat{U}(2)_{\bar{z}}$, respectively. Let $\sigma^1, \sigma^2, \sigma^3$ be the usual Pauli matrices, and t_l^α the spin l representation of $\sigma^\alpha, \alpha = 1, 2, 3$.

In conformal field theory, the energy-momentum tensor, $T(z, \bar{z})$, splits into two tensors, $T(z), \tilde{T}(\bar{z})$, only depending on $z, \bar{z}$, respectively. In the WZW model, $T(z)$ has the Sugawara form

$$T(z) = \frac{1}{2(k+2)} \left\{ \sum_{\alpha=1}^{3} : J_\alpha(z) J_\alpha(z) : \right\}, \tag{5.2}$$

where $J_\alpha = \frac{1}{2} tr(\sigma^\alpha J)$. The double colons on the r.h.s. of (5.2) indicate normal ordering.

Let $\{\Phi_l(z, \bar{z})\}$ be the primary fields, in the sense of [27, 29], of the WZW models. We also use the notations

$$\Phi_T(z) \equiv T(z), \Phi_{J_\alpha}(z) \equiv J_\alpha(z).$$

The conformal Ward identities are (see [27])

$$\begin{aligned} G(T, z, l_1, z_1, \bar{z}_1, \cdots, l_n, z_n, \bar{z}_n) &\text{``}= \langle T(z) \Phi_{l_1}(z_1, \bar{z}_1) \cdots \Phi_{l_n}(z_n, \bar{z}_n) \rangle\text{''} \\ &= \sum_{j=1}^{n} \left\{ \frac{\Delta_{l_j}}{(z - z_j)^2} + \frac{1}{z - z_j} \frac{\partial}{\partial z_j} \right\} G(l_1, z_1, \bar{z}_1, \cdots, l_n, z_n, \bar{z}_n) \end{aligned} \tag{5.3}$$

and the Ward identities for the Kac-Moody symmetry are [29]

$$\begin{aligned} G(J_\alpha, z, l_1, z_1, \bar{z}_1, \cdots, l_n, z_n, \bar{z}_n) &= \text{``}\langle J_\alpha(z)\Phi_{l_1}(z_1,\bar{z}_1)\cdots\Phi_{l_n}(z_n,\bar{z}_n)\rangle\text{''} \\ &= \sum_{j=1}^{n} \frac{t^\alpha_{l_j}}{z-z_j} G(l_1, z_1, \bar{z}_1, \cdots, l_n, z_n, \bar{z}_n). \end{aligned} \tag{5.4}$$

If, on the l.h.s. of (5.3), we insert the Sugawara form (5.2) of the energy-momentum tensor, $T(z)$, and apply (5.4) twice to the resulting expression and finally compute a contour integral in the variable z we obtain the differential equation

$$\begin{aligned} \sum_{i=1}^{n} \frac{\partial}{\partial z_i} g(l_1, z_1, \cdots, l_n, z_n) = \sum_{1\le i<j\le n} \frac{2}{k+2} \frac{(\Omega_{ij})^{l_i' l_j'}_{l_i l_j}}{z_i - z_j} \\ g(l_1, z_1, \cdots, l_i', z_i, \cdots, l_j', z_j, \cdots) \end{aligned} \tag{5.5}$$

where

$$g(l_1, z_1, \cdots, l_n, z_n) = G(l_1, z_1, \bar{z}_1, \cdots, l_n, z_n, \bar{z}_n),$$

with $\bar{z}_1, \cdots, \bar{z}_n$ kept fixed, and

$$\Omega_{ij} = \sum_{\alpha=1}^{3} t^\alpha_{l_i} \otimes t^\alpha_{l_j}. \tag{5.6}$$

Equation (5.5) has been derived in [29]; see also [33]. A similar equation obviously governs the dependence of $G(l_1, z_1, \bar{z}_1, \cdots, l_n, z_n, \bar{z}_n)$ on the variables $\bar{z}_1, \cdots, \bar{z}_n$, for fixed $z_1, \cdots, z_n$. A more compact form for equation (5.5) is

$$\begin{aligned} d\, g(\cdot, z_1, \cdots, \cdot, z_n) = \left\{ \sum_{1\le i<j\le n} \frac{\Omega_{ij}}{k+2} \; d \log \; (z_i - z_j) \right\} \\ \times\; g(\cdot, z_1, \cdots, \cdot, z_n) \end{aligned} \tag{5.7}$$

It has been noticed in Sect. 4 that Euclidean Green functions can be viewed as sections of complex vector bundles with base space M_n. In (4.19), (4.20) it turned out that the functions $g(\cdot, z_1, \cdots, \cdot, z_n)$, too, are sections of a complex vector bundle with base space M_n. Equation (5.7) for the functions g of the WZW models is nothing but the equation for parallel transport on a complex vector bundle with base space M_n and connection ω given by

$$\omega = \sum_{1\le i<j\le n} \Omega_{ij} \;\; d \;\; \log \;\; (z_i - z_j) \tag{5.8}$$

where the matrices Ω_{ij} are given in (5.6). The results of Belavin and Drinfel'd [26] quoted in Sect. 4 show that ω is flat, hence gives rise to a solution of the classical Yang-Baxter equation. The holonomy matrices of ω, corresponding to solutions of the special Yang-Baxter equations (3.22), describe the monodromy of the functions $g(\cdot, z_1, \cdots, \cdot, z_n)$, (as in (4.20)).

Thus from equ. (5.7) we obtain a linear representation of the pure braid group P_n. It is of interest to analyze the properties of that representation. Let us consider the special case where

$$t^{\alpha}_{l_j} = \sigma^{\alpha},$$

for $j = 1, \cdots, n$; i.e., we consider Green functions of primary fields transforming under the fundamental representation of $SU(2)$, in particular $l_1 = \cdots = l_n$. Then equs. (5.7) determine a linear representation of the braid group B_n. This representation factors through the Jones algebra [34] of index $4\cos^2\left(\frac{\pi}{k+2}\right)$, where k is as in (5.2), (the central charge of the Kac-Moody algebra).

[The Jones algebra has generators $\tau_{i,i+1}$ satisfying the relations (a) and (b) stated at the beginning of Sect. 4 and, in addition, the relations

$$(c) \quad \tau^2_{i,i+1} = (q-1)\tau_{i,i+1} + q, \text{ with } q = e^{2\pi i/(k+2)}$$

These representations are unitarizable [34].]

For more details concerning the results discussed here, see [25,33].

One should add that the monodromy of the Euclidean Green functions G under simultaneous variation of the arguments $z_i, \bar{z}_i \cdots$, with $\bar{z}_i = \bar{z}_i$, is trivial, so that the physical fields $\Phi_l(x)$ of the WZW models do not have exotic statistics. [If $G = 0(N)$ then the WZW models have spin fields, and, as in the Ising field theory, non-trivial statistics matrices arise. When $k = 1$ the WZW models can be rewritten in terms of N free, massless Majorana fermions, ψ_k, with

$$J_{kk'}(z) =: \psi_k(z)\psi_{k'}(z) :, \bar{J}_{kk'}(\bar{z}) =: \bar{\psi}_k(\bar{z})\bar{\psi}_{k'}(\bar{z}) :,$$

and

$$g_{kk'}(z,\bar{z}) = \text{const.} : \psi_k(z)\bar{\psi}_{k'}(\bar{z}) : .$$

This is Witten's non-abelian bosonization [31,29]. The statistics problem for the fields $\psi_{k'}\bar{\psi}_k$ and the spin fields is then analogous to the one discussed in Sect. 2 for the Ising_2 field theory.] Finally, we should mention that the primary fields of the WZW models do not transform under all possible representations of G, and that for $g(l_1, z_1, \cdots, l_n, z_n)$ to be non-zero, the labels $l_1, \cdots, l_n$ must satisfy certain constraints. All this is the contents of certain fusion rules which are presumably discussed in K. Gawedzki's lectures.

6. Connections to the theory of knots and links

The purpose of this section is to sketch a connection between the material in Sects. 3-5 and a chapter in topology which was brought to my attention by V. Jones: the theory of knots and links. I am following notes by V. Jones [35] and a recent preprint by Turaev [36].

Suppose $R \in \text{End}\ (V \otimes V)$ is a solution of the Yang-Baxter equs. (3.22), and μ is a diagonal matrix on V such that

$$(\mu \otimes \mu)R = R(\mu \otimes \mu), \tag{6.1}$$

and

$$\left.\begin{aligned} \sum_{\gamma=1}^{k} R^{\alpha\gamma}_{\beta\gamma}\mu_\gamma &= a\, b\, \delta^\alpha_\beta, \\ \sum_{\gamma=1}^{k} (R^{-1})^{\alpha\gamma}_{\beta\gamma}\mu_\gamma &= a^{-1} b\, \delta^\alpha_\beta, \end{aligned}\right\} \tag{6.2}$$

with $k =$ dim V. Then (R, μ, a, b) determine a polynomial invariant of knots and links.

To see this, one represents a link by some braid which is closed by identifying lower and upper end points of the strings. One then associates the matrix

$$R_{i,i+1} = 1_1 \otimes \cdots \otimes 1_{i-1} \otimes R\,|_{V_i \otimes V_{i+1}} \otimes \cdots \otimes 1_n$$

($V_i = V$, for all i) with the generators $\tau_{i,i+1}$ of $B_n, n = 2, 3, 4, \cdots$. This defines a representation $b \in B_n \longmapsto R(b)$, given by (4.11), of B_n, for all $n \geq 2$. Let $w(b)$ be twice the "linking number" (more precisely, the <u>writhe</u>) of the link represented by b. We define

$$T(b) = a^{-w(b)} b^{-n} tr(R(b)\, \underbrace{\mu \otimes \cdots \otimes \mu}_{n \text{ times}}) \tag{6.3}$$

This is an invariant of the link b. An important ingredient in proving this is the invariance of $T(b)$ under Reidemeister moves of type II and III. This invariance is closely related to equs. (3.17) and (3.22).

The quantity $\mathrm{tr}(R(b)\mu \otimes \cdots \otimes \mu)$ can be interpreted as the partition function of a model of classical statistical mechanics defined on the link represented by b (with k states associated with every line (bond) of b).

This approach to finding new invariants for knots and links was pioneered by V.Jones [35]; see also [36] for details concerning the results summarized above, and [37] for a review of knot theory, the Jones polynomial, and statistical mechanics.

Jones' work is also based on constructing solutions to the Yang-Baxter equations, solutions which, in addition, satisfy conditions (6.1) and (6.2).

Although there are statistics matrices satisfying (6.1) and (6.2), the conditions that have to be met by statistics matrices are quite unrelated to the ones met by an R-matrix that gives rise to an invariant for knots and links. In fact, Jones' work is more directly related to the <u>quantum inverse scattering method</u> than to the ideas explained in Sects. 3-5; see [35, 36]. The quantum inverse scattering method is a technique to explicitly diagonalize the transfer matrices of a large class of models of two-dimensional statistical mechanics and to calculate mass spectrum and (factorizable) scattering matrices of a variety of two-dimensional quantum field models.

The point of this short section is to emphasize that there are several areas in quantum field theory, statistical mechanics and in topology, where the Yang-Baxter equation plays an important rôle. One would like to know whether this is an accident or whether there are some fundamental, underlying connections between these subjects.

Pasquier [38] has analyzed models of two-dimensional statistical mechanics (generalized S-0-S models) which appear to tie together the so-called discrete series of two-dimensional conformal quantum field theories (see Friedan's lectures, and refs. given there), the quantum inverse scattering method and the Jones algebra, but it appears that many details remain mysterious. I am tempted to conjecture that in this whole area some basic discoveries remain to be made.

7. Notes and open problems

This final section contains some brief notes and comments on earlier sections and a list of open problems.

Sect. 2, beginning of Sect. 3

The outstanding problem in connection with the results reviewed in Sect. 2 and at the beginning of Sect. 3 is to perform a general analysis of covariant representations of $(\mathcal{A}, \tau)$, where $\mathcal{A}$ is the observable algebra and τ a representation of the Poincaré group as a group of *automorphisms of $\mathcal{A}$, for quantum field theories in two and three space-time dimensions. In particular, the connection between the theory of superselection sectors in two dimensions and the theory of loop groups and their representations should be worked out in more detail.

In three space-time dimensions, it remains to be shown more completely that the field theory of anyons [2,18], as developed in [19], represents all there is that is not covered by conventional Bose- or Fermi statistics. In other words, one would like to show that the general theory of superselection sectors [39] for local theories in three space-time dimensions does not lead to further possibilities besides Bose-, Fermi- and intermediate (-) statistics. This is the problem considered in [19].

Sect. 3

The basic problem posed by the analysis in Sect. 3 is:

(1) What is the most general solution of equations (3.17), (3.22) and (3.28), i.e. what is the most general statistics matrix, that may and does arise in the context of two-dimensional, local, relativistic quantum field theory? In which way must the constructive approach alluded to in Sect. 2, see [3-6], be extended in order to construct examples of field theories with more general statistics matrices?

(2) Let $\{\Phi_\alpha(x)\}_{\alpha=1}^{p}$ be the local "primary" fields of a two-dimensional quantum field theory with statistics matrix $R(\sigma)$. Let 0 be a diamond in M^2, and let $\overline{\mathfrak{F}(0)}$ be the von Neumann algebra generated by the local observable algebra $\mathcal{A}(0)$ and all bounded, selfadjoint functions of $\{\Phi_\alpha(f) :\ \text{supp}\ f \subseteq 0\}, \alpha = 1, \cdots, p$. In which way does the algebraic structure of the net $\{\overline{\mathfrak{F}(0)}\}$ reflect the statistics matrix $R(\sigma)$.

(3) Consider an arbitrary solution, $R(\sigma)$, of equations (3.17), (3.22) and (3.28). Is there an algebra, $\mathfrak{F}$, of operators for which $R(\sigma)$ is, in a suitable sense, the statistics matrix. For example, given a solution $R(\sigma)$ of (3.17), (3.22) and (3.28), are there operators $\Phi_\alpha(\vec{x})$, with $\vec{x} \in Z$, such that

$$\Phi_\alpha(\vec{y})\Phi_\beta(\vec{x}) = R_{\alpha\beta}^{\gamma\delta}(\sigma)\Phi_\gamma(\vec{x})\Phi_\delta(\vec{y}),$$

for arbitrary $\vec{x}, \vec{y}$ in Z, with $\vec{x} \neq \vec{y}; \sigma =$ sign$(\vec{y} - \vec{x})$? If so, what are the algebraic properties of F?

Sect. 4.

One obvious problem in connection with the material in the second part of Sect. 4 is to develop a general Euclidean formulation of conformal quantum field theory in two space-time dimensions.

In particular, it looks plausible that one can prove a general reconstruction theorem that permits one, for example, to reconstruct from the Euclidean Green functions of a conformal theory a unique representation of the conformal group on the physical Hilbert space of radial quantization.

Next, one would like to correctly formulate two-dimensional conformal field theory on Riemann surfaces of arbitrary genus and relate properties of conformal field theory on Riemann surfaces of higher genus to properties of theories in physical space-time. A program in this direction, closely related to ideas worked out independently by the author, has been proposal by G. Segal [28].

All this is clearly intimately related to the covariant quantization of string theory and covariant string perturbation theory. The different spin structures that are encountered on Riemann surfaces of higher genus and the monodromy generated by spin fields in fermionic correlation functions are simple special cases of the general problems discussed in Sect. 4. The general problem is to classify all possible representations of the (pure) braid groups that can arise within the context of general conformal quantum field theory on arbitrary Riemann surfaces.

Sect. 5

In Sect. 5, I have mostly followed refs.7, 25 and 33. After I had started to analyze the representations of braid groups arising in the Wess-Zumino-Witten models, T. Kohno gave two interesting seminars about the work in refs. 24, 25 and 33. I have no results to report that would go beyond those in refs. 25 and 33.

Clearly, these results are not quite the end of the story. One would like to have a more explicit description of the respresentations of B_n and of P_n arising in the WZW models with $G = SU(2)$. One then would like to extend the results sketched in Sect. 5 to general groups G and to the study of Euclidean Green functions of the WZW models on general Riemann surfaces. One would like to make more powerful use of equation (5.7), combined with general theorems of conformal QFT, to obtain explicit information on the properties of the Euclidean Green functions of the WZW models. There are further problems concerning spin fields in the WZW models (e.g. for $G = SO(3)$), and even Witten's non-abelian bosonization [31] ($k = 1$) still poses some interesting questions (e.g. in connection with the functional integral formulation of bosonization in [6]) that are presently investigated.

Sect. 6

An obvious question concerning the material sketched in Sect. 6 is whether knowledge of all solutions of the Yang-Baxter equation (3.22) satisfying (6.1) and (6.2) (for some numbers a and b) would provide a complete set of invariants for knots and links in R^3.

It would be interesting to understand whether there are fundamental mathematical reasons for the formal similarities between "integrable" models of two-dimensional statistical mechanics and quantum field theory, two-dimensional conformal quantum field theories and the theory of knots and links in R^3. In all three areas, the Yang-Baxter equation plays on important rôle, but the rôles played by this equation are different and have different origins. What is the unifying principle? I have only speculations about these questions, but no good answers.

The little I know on knot theory I have learned in conversations with and seminars by V. Jones and H. Morton and from refs. 35, 36 and 37.

I thank these colleagues for stimulating my interests in this area and teaching me a few basic facts.

REFERENCES

1. R.F. Streater and A.S. Wightman, *"PCT, Spin and Statistics and All That", Reading, Mass: Benjamin 1978.* (This book contains many references to the early literature on solitons and statistics in two dimensions.)
   R.F. Streater and I.F. Wilde, *Nuclear Physics* B 24, 561 (1970).
2. Y.S. Wu, *Phys. Rev. Lett.* 52, 2103 (1984); 53, 111 (1984).
3. J. Fröhlich, *Commun. Math. Phys.* 47, 269 (1976); *Acta Phys. Austriaca, Suppl. XV*, 133 (1976).
   J. Bellissard, J. Fröhlich and B. Gidas, *Commun. Math. Phys.* 60, 37 (1978).
   J. Fröhlich, *in :"Invariant Wave Equations"*, Berlin-Heidelberg - New York: Springer-Verlag, Lecture Notes in Physics 73, 1978.
4. J. Fröhlich, *in : "Recent Developments in Gauge Theories"* (Cargese 1979). G.'t Hooft et al. (eds.), New York: Plenum Press, 1980.
5. E.C. Marino and J.A. Swieca, *Nucl. Phys.* B 170 [FS1], 175 (1980); E.C. Marino, B. Schroer and J.A. Swieca, *Nucl. Phys.* B 200 [FS4], 473 (1982); R. Köberle and E. C. Marino, Phys. Lett 126 B, 475 (1983).
6. J. Fröhlich and P.-A. Marchetti, *"Bosonization, Topological Solitons and Fractional Charges in Two-Dimensional Quantum Field Theory"*, submitted to Commun. Math. Phys. and Commun. Math. Phys. *112*, 343-383 (1987)
7. A.A. Belavin, A.M. Polyakov and A.B. Zamolodchikov, *Nucl. Phys.* B 241, 333 (1984); V.G. Knizhnik and A.B. Zamolodchikov, *Nucl. Phys.* B 247, 83 (1984).
8. D. Friedan, E. Martinec and S. Shenker, Phys. Lett. B 160, 55 (1985); *Nucl. Phys.* B 271, 93 (1986).
9. V.G. Knizhnik, *Phys. Lett.* B 160, 403 (1985); B 178, 21 (1986).
10. G. Segal, *"The Definition of Pertubative Conformal Field Theory"*, notes, 1987.
    J. Fröhlich, unpublished notes, Bures-sur-Yvette, 1987.
11. V.F.R. Jones, *Bull. Amer. Math. Soc.* 12, 103 (1985); Notes on a Talk in Atiyah's Seminar (1986); paper in preparation.
12. L. Kadanoff and H. Ceva, *Phys. Rev.* B 11, 3918 (1971).

13. K. Osterwalder and R. Schrader, *Commun. Math. Phys.* 31, 83 (1973); 42, 281 (1975); V. Glaser *Commun. Math. Phys.* 37, 257 (1974); J. Fröhlich, K. Osterwalder and E. Seiler, *Ann. Math.* 118, 461 (1983).

14. M. Sato, T. Miwa and M. Jimbo, *Publ. RIMS*, Kyoto University, 15. 871 (1979).

15. F. Wegner, J. *Math. Phys.* 12, 2259 (1971).

16. K.G. Wilson, *Phys. Rev.* D 10, 2445 (1974).

17. G.'t Hooft, *Nucl. Phys.* B 138, 1 (1978).

18. F. Wilczek and A. Zee, *Phys. Rev. Lett.* 51, 2250 (1983).

19. J. Fröhlich and P.-A. Marchetti, *"Field Theory of Anyons and θ-Statistics"*, to appear.

20. S. Doplicher, R. Haag and J. Roberts, *Commun. Math. Phys.* 23, 199 (1971); 35, 49 (1974). (See also ref. 39.)

21. J. Glimm and A. Jaffe, *"Quantum Physics"* (2nd edition), Berlin-Heidelberg-New York: Springer-Verlag, 1987.

22. J. Imbrie, *Commun. Math. Phys.* 82, 261 (1981), 82, 305 (1981).

23. R.F. Streater and A.S. Wightman, see ref. 1.
R. Jost,*"The General Theory of Quantized Fields"*, Providence, R.I.: Publ. Amer. Math. Soc., 1965.

24. T. Kohno, *Nagoya Math. J.* 92, 21 (1983); *Invent. Math.* 82, 57 (1985).

25. T. Kohno, *"Linear Representations of Braid Groups and Classical Yang-Baxter Equations", to appear in "Contemporary Mathematics: Artin's Braid Group"*, Santa Cruz, 1987.

26. A.A. Belavin and V.G. Drinfel'd, *Funct.Anal.Appl.* 16, 1 (1982).

27. A.A. Belavin, A.M. Polyakov and A.B. Zamolodchikov, *Nucl. Phys.* B 241, 333 (1984).

28. G. Segal, ref. 10.

29. V.G. Knizhnik and A.B. Zamolodchikov, *Nucl. Phys.* B 247, 83 (1984).

30. B. Schroer, *Lecture Notes*, Cargese 1987, to appear in the proceedings; and refs. given there.

31. E. Witten, *Commun. Math. Phys.* 92, 455 (1984).

32. A.M. Polyakov and P.B. Wiegmann, *Phys. Lett.* B 131, 121 (1983).

33. A. Tsuchiya and Y. Kanie, *Lett. Math. Phys.* 13, 303 (1987).

34. V.F.R. Jones, *Invent. Math.* 72, 1983.

35. V.F.R. Jones, unpubl. seminar notes, and paper in preparation.

36. V.G. Turaev, *"The Yang-Baxter Equation and Invariants of Links"*,LOMI Preprint E-3-1987.

37. L.H. Kauffman, *"Statistical Mechanics and the Jones Polynomial"*, Preprint, University of Illinois, 1987.

38. V. Pasquier, *Preprints*, Saclay SPh T/87 - 014, SPh T/ 87 - 62.

39. D. Buchholz and K. Fredenhagen, *Commun. Math. Phys.* 84, 1 (1982).

TOPOLOGICAL ACTIONS IN TWO-DIMENSIONAL QUANTUM FIELD THEORIES*

K. Gawędzki

C.N.R.S., I.H.E.S.

Bures-sur-Yvette, France

ABSTRACT

A systematic approach to 2-dimensional quantum field theories with topological terms in the action is developed using as a mathematical tool the Deligne cohomology. As an application, it is shown how to bosonize the action of free fermions of arbitrary spin on a Riemann surface and how to find the spectrum of the Wess-Zumino-Witten sigma models without recurrence to modular invariance.

LIST OF TOPICS

- Dirac's monopole : an example of a topological action in quantum mechanics.
- Line bundles with connections and Deligne cohomologies.
- Cohomology of topological ambiguities in two-dimensional field theory.
- Bosonized actions for free fermions on a Riemann surface.
- Canonical quantization of the Wess-Zumino-Witten chiral model.
- Spectrum of the Wess-Zumino-Witten model with SU(2) and SO(3) groups.

1. INTRODUCTION

These lectures are devoted to one topological aspect of quantum theory from the plethora of such phenomena which constitute the main topic of this Institute. It appears in theories which lack globally defined actions. The simplest and longest known system of this sort describes a particle in the field of Dirac's magnetic monopole[1]. In Section 2, we shall briefly recall how the lack of global action leads to Dirac's quantization condition for the monopole charge and to other topological effects. This will be done in the spirit of the so-called geometric quantization[2] representing the quantum mechanical states as sections of a line bundle and will

* Extended version of lectures delivered at the Summer School on Nonperturbative Quantum Field Theory, Cargèse 1987.

serve as a natural occasion to introduce cohomological notions known to mathematicians as Deligne cohomology to which Section 3 is devoted.

In 2-dimensional field theories an example of a system without global action is provided by the Wess-Zumino-Witten (WZW) model discussed by Witten in his paper[3] on non-abelian bosonization. Its study was the main motivation for our interest in models without global action. Another system in this category is the free (euclidean) field bosonizing fermions of arbitrary spin on a Riemann surface[4]. In Section 4, we shall discuss path integral quantization of 2-dimensional field theories without global actions on a general 2-dimensional surface using the Deligne cohomology as a mathematical tool. We shall see how the relation of path-integral and canonical quantizations gives rise to natural appearence of line-bundles with connection over the loop spaces of manifolds of field values and to a "stringy" generalization of the notion of parallel transport. In Section 5, the general theory is applied to the bosonized version of free fermionic theories on Riemann surfaces. Section 6 studies with details the WZW model with fields taking values in a simply connected group G . We show how rich symmetries of the classical models give rise to the left-right action of the Kac-Moody group in the space of states of the quantized model represented by sections of the line bundle over the loop group of G . This action decomposes into irreducible representations whose spectrum conjectured by Gepner-Witten in [5] is found explicitly for $G = SU(2)$. Section 7 treats the complications appearing in the case of non-simply connected groups on example of $SO(3)$ model. We find again the result coinciding with the Gepner-Witten spectrum derived originally from the $SU(2)$ one by postulating the modular invariance of the partition function of the model on the torus[5].

ACKNOWLEDGEMENTS

The material presented in these lectures is to large extent a result of joint work with Giovanni Felder and Antti Kupiainen. I would like to thank them for the joy of collaboration. I would also like to acknowledge the discussions with J.-B. Bost and Ch. Soulé from whom we learnt about the Deligne cohomology and with F. Bien who has brought reference[24] to our attention.

2. TOPOLOGICAL ACTIONS IN MECHANICS

We start by discussing the case of quantum mechanics where the topological ambiguities may appear when we attempt to realize Dirac's or Feynman's quantization programmes[6,7] in topologically non-trivial backgrounds. The geometric aspects of this well known phenomenon will be briefly recalled here. Classical mechanics is usually formulated with the help of Lagrangians, but what really enters the classical equations of motion are specific combinations of their derivatives only. To be more concrete, consider a particle moving in magnetic field described by vector potential $\vec{A}(\vec{x})$. The action on the trajectory $\vec{x}(t)$ is given as

$$\frac{1}{2} m \int \dot{\vec{x}}^{\,2} dt + e \int \vec{A} \cdot d\vec{x} \tag{1}$$

whereas the equations of motion involve only magnetic induction $\vec{B}$

$$m\ddot{\vec{x}} = e\vec{B}\wedge\vec{v} \, . \tag{2}$$

In other words, the action contains integral of 1-form $\eta = e\vec{A}\cdot d\vec{x}$ along the trajectory, whereas the equations of motion involve only 2-form $\omega = d\eta = \frac{1}{2}\, e\vec{B}\cdot d\vec{x}\wedge d\vec{x}$.

It is possible to consider classical systems involving a 2-form which is closed but not exact. Such is for example the case for classical particle moving in the field of magnetic monopoles[1,8] and, to make the problem more interesting, infinitely thin Bohm-Aharonov[9] solenoids carrying magnetic fluxes. Away from the singularities at the monopole locations and flux lines, the magnetic induction defines 2-form

$$\omega = \frac{1}{2}\, e\vec{B}\cdot d\vec{x}\wedge d\vec{x} = \frac{1}{2}\sum_{\alpha} e\mu_\alpha \sum_{i,j,k=1}^{3} \varepsilon^{ijk}\, \frac{x^i - x^i_\alpha}{|\vec{x}-\vec{x}_\alpha|^3}\, dx^j dx^k \tag{3}$$

where μ_α are charges of monopoles located at points $\vec{x}_\alpha$. ω is no more exact (vector potentials for magnetic monopoles have singularities along Dirac strings).

Lack of global form η such that $\omega = d\eta$, although no complication on classical level, obstructs the passage to quantum mechanics, where, following Feynman, we should sum the probability amplitudes given as exponentials of (i times) the action over all the trajectories. In some cases however, a global action can still be defined for closed trajectories modulo an ambiguity in $2\pi\mathbb{Z}$ removed by exponentiation. This situation has a nice geometric interpretation : the (topological parts of) the probability amplitudes are the loop holonomies in a line bundle with a hermitian connection of curvature ω . For open trajectories, the parallel transport in the bundle leads to "probability amplitudes" which are no more complex numbers but rather linear maps between the line bundle fibers, but they can still be used to define complex valued probability amplitudes between states represented by sections of the line bundle. That the wave functions should be bundle valued can also be seen from the fact that an important subalgebra of classical physical quantities can be naturally represented by differential operators acting in the space of sections of the line bundle so that the Poisson bracket corresponds to (i times) the commutator, realizing Dirac's quantization programme. This is one of the main constructions of geometric quantization which provided an effective tool in the theory of unitary representations of groups (obtained by quantization of classical group actions carried by orbits of the coadjoint representation of the group[10]). Later, we shall develop some of the concepts of geometric quantization in the infinite-dimensional context of two-dimensional field theories. Here we shall present the elements of the line bundle theory[11] in a cohomological language, specially useful in view of the later discussion.

Suppose that we are given a hermitian line bundle L over a (smooth) manifold M. If $\{O_\alpha\}$ is a sufficiently fine open covering of M then by taking normalized sections s_α of L over O_α, we obtain U(1)-valued transition functions on non-empty intersections $O_{\alpha_0\alpha_1} \equiv O_{\alpha_0} \cap O_{\alpha_1}$: $s_{\alpha_0} = g_{\alpha_0\alpha_1} s_{\alpha_1}$. $(g_{\alpha_0\alpha_1})$ form a 2-cocycle, i.e.

$$g_{\alpha_0\alpha_1} = g_{\alpha_1\alpha_0}^{-1} \; , \quad g_{\alpha_0\alpha_1} g_{\alpha_1\alpha_2} = g_{\alpha_0\alpha_2} \quad \text{on} \quad O_{\alpha_0\alpha_1\alpha_2} \; . \tag{4}$$

A hermitian connection on L can be locally described by real 1-forms η_α on O_α transforming by the gauge transformations

$$\eta_{\alpha_1} = \eta_{\alpha_0} + \frac{1}{i} g_{\alpha_0\alpha_1}^{-1} dg_{\alpha_0\alpha_1} \quad \text{on} \quad O_{\alpha_0\alpha_1} \; . \tag{5}$$

The curvature of the connection corresponding to (η_α) is the real closed 2-form on M equal on each O_α to $d\eta_\alpha$.

If we choose other sections $\tilde{s}_\alpha = \chi_\alpha^{-1} s_\alpha$ of L where χ_α are U(1)-valued functions on O_α then we obtain equivalent local data $(\tilde{g}_{\alpha_0\alpha_1}, \tilde{\eta}_\alpha)$:

$$\begin{aligned} \tilde{g}_{\alpha_0\alpha_1} &= g_{\alpha_0\alpha_1} \chi_{\alpha_1} \chi_{\alpha_0}^{-1} \; , \\ \tilde{\eta}_\alpha &= \eta_\alpha + \frac{1}{i} \chi_\alpha^{-1} d\chi_\alpha \; . \end{aligned} \tag{6}$$

Conversely, the equivalence class $[(g_{\alpha_0\alpha_1}, \eta_\alpha)]$ of local presentations defines the line bundle with connection up to isomorphism (projecting to identity on M). In particular, L can be taken as the set of elements (α,x,z), $x \in O_\alpha$, $z \in \mathbb{C}^1$, with the identification of (α_0,x,z) and $(\alpha_1,x,g_{\alpha_0\alpha_1}(x)z)$ if $x \in O_{\alpha_0\alpha_1}$. If $\{O_\alpha\}$ has the property that all $O_{\alpha_0\ldots\alpha_p}$ are contractible, we may identify the isomorphism classes of the line bundles equipped with hermitian connections and classes $q = [(g_{\alpha_0\alpha_1}, \eta_\alpha)]$.

From the cohomological point of view, the latter are the elements of the cohomology of a homomorphism of sheaves[12] $\mathbb{F}_M^1$

$$U(1)_M \xrightarrow{\frac{1}{i} d \log} \Omega_M^1 \tag{7}$$

where $U(1)_M$ is the sheaf of local smooth U(1) valued functions on M and Ω_M^1 the sheaf of local smooth real 1-forms on M. For a general complex of sheaves of abelian groups

$$\mathcal{F}_0 \xrightarrow{d_0} \mathcal{F}_1 \xrightarrow{d_1} \cdots \xrightarrow{d_{p-1}} \mathcal{F}_p \quad , \tag{8}$$

$d_i \circ d_{i-1} = 0$, and open covering $\{\mathcal{O}_\alpha\}$ of M consider groups $C_i^p \equiv C^p(\mathcal{F}_i)$ of Čech p-cochains with values in $\mathcal{F}_i$. Elements of C_i^p are families $c_i^p \equiv (c^p_{i,\alpha_0\ldots\alpha_p})$, $c^p_{i,\alpha_0\ldots\alpha_p} \in \mathcal{F}_i(\mathcal{O}_{\alpha_0\ldots\alpha_p})$, antisymmetric in indices $\alpha_0,\ldots,\alpha_p$. The Čech coboundary $\delta^p : C_i^p \to C_i^{p+1}$,

$$(\delta^p c_i^p)_{\alpha_0\ldots\alpha_{p+1}} = \sum_{q=0}^{p+1} (-1)^q \, c^p_{i,\alpha_0\ldots\hat{\alpha}_q\ldots\alpha_{p+1}} \quad . \tag{9}$$

From the bi-complex of abelian groups

$$\begin{array}{ccccc} & \downarrow & & \downarrow & \\ \longrightarrow & C_i^p & \xrightarrow{d_i} & C_{i+1}^p & \longrightarrow \\ & \delta^p \downarrow & & \downarrow & \\ \longrightarrow & C_i^{p+1} & \longrightarrow & C_{i+1}^{p+1} & \longrightarrow \\ & \downarrow & & \downarrow & \end{array} \tag{10}$$

we may derive the diagonal complex

$$\longrightarrow E_k \xrightarrow{D_k} E_{k+1} \longrightarrow \tag{11}$$

where

$$E_k = \bigoplus_{\substack{(i,p)\\ i+p=k}} C_i^p \tag{12}$$

and for $j+q = k+1$

$$(D_k((c_i^p)_{i+p=k}))_j^q = (-1)^j \, \delta^{q-1} c_j^{q-1} + (-1)^{j-1} d_{j-1} c_{j-1}^q \quad . \tag{13}$$

The cohomology groups of the complex $\mathbb{F}_M^1$ of sheaves (8),

$$\mathbb{H}^k(\mathbb{F}_M^1) := \frac{\ker D_k}{\operatorname{Im} D_{k-1}} \quad . \tag{14}$$

More exactly, they are given by the inductive limit of the right hand side of (14) over the finer and finer coverings of M , as usually in Čech cohomology. Notice that our local description of isomorphism classes of line bundles with hermitian connection on M identifies them with elements of $\mathbb{H}^1(\mathbb{F}_M^1)$.

Given a local presentation $(g_{\alpha_0\alpha_1}, \eta_\alpha)$ of the bundle and a loop

$\phi : S^1 \to M$ in M, we may express the holonomy $P(\phi) \in U(1)$ corresponding to ϕ (given by the parallel transport around ϕ) in the following way. We break S^1 into a union of intervals b with common points v so that $\phi(b) \subset \mathcal{O}_{\alpha_b}$, see Fig. 1. We also choose for each v an α_v s.t. $v \in \mathcal{O}_{\alpha_v}$. Then

$$P(\phi) = \exp[i \sum_b \int_b \phi^* \eta_{\alpha_b}] \prod_v g_{\alpha_{b_v^-} \alpha_{b_v^+}}(\phi(v))$$

$$= \exp[i \sum_b \int_b \phi^* \eta_{\alpha_b}] \prod_{\substack{v,b \\ v \in \partial b}} g^{-1}_{\alpha_v \alpha_b}(\phi(v)) \tag{15}$$

where on the right hand side the convention is implied which inverts $g^{-1}_{\alpha_v \alpha_b}(\phi(v))$ if v inherits from b negative orientation (i.e. v is the starting point of b). In cohomological language, ϕ allows to pull back any element $q \in \mathbb{H}^1(\mathcal{F}^1_M)$ to an element $\phi^* q \in \mathbb{H}^1(\mathcal{F}^1_{S^1}) \cong U(1)$ and $P(\phi)$ realizes the last isomorphism. Of course, the right hand side of (15) does not depend of the choice of triangulation of S^1, the assignments α_b, α_v or the representative $(g_{\alpha_0 \alpha_1}, \eta_\alpha)$ of q. This is no more true if we apply the same formula to $\phi : I \to M$ where I is a closed interval $[v_i, v_f]$. Then the right hand side of (15) depends on α_{v_i} and α_{v_f} and transforms under the change of this data as an element of $L^*_{\phi(v_i)} \otimes L_{\phi(v_f)}$ giving the parallel transport along ϕ in bundle L which maps fiber $L_{\phi(v_i)}$ into fiber $L_{\phi(v_f)}$. Thus eq. (16) describes the parallel transport in L along both closed and open curves. We should also notice, that although defined for parametrized curves, it does not change under orientation preserving reparametrizations of them.

The knowledge of the curvature of the connection fixes the holonomy of the contractible loops. More generally, if $\phi : \Sigma \to M$ where Σ is a 2-dimensional compact oriented manifold with the boundary composed of loops $(\partial\Sigma)_i$, see Fig.2, then

$$\prod_i P(\phi\restriction_{(\partial\Sigma)_i}) = \exp[i \int_\Sigma \phi^* \omega] \tag{16}$$

where ω is the curvature form.

Given a real closed 2-form ω on M, we may wonder whether there exists a line bundle with a hermitian connection of curvature ω. This is the case if and only if the integrals of ω over 2-dimensional closed surfaces in M are in $2\pi\mathbb{Z}$ (that this is a necessary condition is easily seen from eq. (16)). In the cohomological language this means that the element $[\omega] \in H^2(M, \mathbb{R})$ defined by ω should be in the image of $H^2(M, 2\pi\mathbb{Z})$. We shall call such ω integral. The cohomology group $H^1(M,U(1))$ acts freely on $\mathbb{H}^1(\mathcal{F}^1_M)$ sending $[(g_{\alpha_0\alpha_1}, \eta_\alpha)]$ into $[(\lambda_{\alpha_0\alpha_1} g_{\alpha_0\alpha_1}, \eta_\alpha)]$ for $(\lambda_{\alpha_0\alpha_1})$ being a cocycle defining an element of $H^1(M,U(1))$. The orbits of this action are exactly the sets $Q(M,\omega)$ of isomorphism classes of line bundles with hermitian connection of curvature ω . Thus $H^1(M,U(1))$ enumerates the elements of non-empty $Q(M,\omega)$.

Coming back to our example of the particle moving in the field of magnetic monopoles and flux lines, the integrality of the form ω given by eq. (3) is equivalent to Dirac's quantization condition of magnetic charge

$$e\mu_\alpha \in \frac{1}{2}\mathbb{Z} \ . \tag{17}$$

If this condition is satisfied, the possible choices of line bundle with curvature ω differ by holonomies of loops surrounding flux lines. For small loops these holonomies become $e^{i\theta}$ where $\frac{1}{e}\theta$ are the magnetic fluxes carried by the lines. These holonomy differences are seen in the quantum probability amplitudes (θ vacua [13,14]) although they are ignored by classical trajectories of the particles which is the essence of the Bohm-Aharonov effect.

Another important topological effect appears in the relation between classical and quantum symmetries. If D is a diffeomorphism of M preserving ω , it is a symmetry of classical mechanics. It does not have to give rise to a symmetry of quantum mechanics. One of the obstructions may be that the pull-back of the line bundle L with connection of curvature ω by D can be non-isomorphic to L . If this is not the case, e.g. if $H^1(M,U(1)) = 1$, then D can be lifted to an isomorphism $\hat{D}$ of L (projecting to D) preserving the hermitian structure and the connection. For connected M , the lift is defined up to multiplication by elements of $U(1)$. If G is a group acting on M by diffeomorphisms preserving ω liftable to L , we obtain an action on L of a central extension $\hat{G}$ of G by $U(1)$. This action, in general, gives rise only to projective representations of G on the space of quantum mechanical states. The parallel transports $P(\phi)$ transform covariantly under liftable maps, $P(D\circ\phi) = \hat{D}P(\phi)$ (the right hand side which is an element of $L^*_{D(\phi(v_i))} \otimes L_{D(\phi(v_f))}$ does not depend of the choice of the lift $\hat{D}$ of D) .

For the particle moving in the field of a single monopole of charge μ , the classical mechanics has $SO(3)$ as the symmetry group which lifts to an action on L only for integer μe . For half-integer μe , it lifts to the action of $SU(2)$ on L and gives rise to a projective representation of SO(3) on the space of quantum states : half-integer magnetic charges carry half-integer spins[15].

Similar discussion applies to infinitesimal action of Lie algebras on M and L.

In certain cases, manifold M can be extended to a complex manifold $M^{\mathbb{C}}$ and curvature form ω to a (2,0) holomorphic form $\omega^{\mathbb{C}}$ on $M^{\mathbb{C}}$. It may be convenient then to extend line bundle L with hermitian connection of curvature ω on M to a holomorphic line bundle $L^{\mathbb{C}}$ with holomorphic connection with curvature $\omega^{\mathbb{C}}$ and to consider quantum states extending to holomorphic sections of $L^{\mathbb{C}}$. Isomorphism classes of holomorphic bundles with holomorphic connections may be identified with elements of $\mathbb{H}^1(\mathbb{A}^1_{M^{\mathbb{C}}})$ where $\mathbb{A}^1_{M^{\mathbb{C}}}$ is the sheaf homomorphism

$$O^*_{M^{\mathbb{C}}} \xrightarrow{\frac{1}{i}\,d\log} A^1_{M^{\mathbb{C}}} \tag{18}$$

with $O^*_{M^{\mathbb{C}}}$ denoting the sheaf of local holomorphic nowhere vanishing functions on $M^{\mathbb{C}}$ and $A^p_{M^{\mathbb{C}}}$ the sheaf of local holomorphic (p,0) forms on $M^{\mathbb{C}}$. Groups $\mathbb{H}^1(A^1_{M^{\mathbb{C}}})$ are known as Deligne cohomology (of degree 2), see[16]. More explicitly, elements of $\mathbb{H}^1(\mathbb{A}^1_{M^{\mathbb{C}}})$ are given by classes of $(g_{\alpha_0\alpha_1},\eta_\alpha)$ modulo $(\chi_{\alpha_1}\chi_{\alpha_0}^{-1}, \frac{1}{i}\chi_\alpha^{-1}d\chi_\alpha)$ where $g_{\alpha_0\alpha_1}$ are holomorphic nowhere vanishing functions on $O_{\alpha_0\alpha_1}$ satisfying (4), η_α are holomorphic (1,0) forms on χ_α satisfying (5) and χ_α are nowhere vanishing holomorphic functions on O_α. Parallel transport along path in $M^{\mathbb{C}}$ is defined again by eq. (15) (the holonomy takes values in $\mathbb{C}^*$ now). Let $Q^a(M^{\mathbb{C}},\omega^{\mathbb{C}})$ denote the set of $w \in \mathbb{H}^1(\mathbb{A}^1_{M^{\mathbb{C}}})$ with curvature $\omega^{\mathbb{C}}$. Again, $Q^a(M^{\mathbb{C}},\omega^{\mathbb{C}}) \neq \emptyset$ if and only if $\omega^{\mathbb{C}}$ is integral and for such $\omega^{\mathbb{C}}$, $H^1(M,\mathbb{C}^*)$ acts freely and transitively on $Q^a(M^{\mathbb{C}},\omega^{\mathbb{C}})$.

The gain from extending M to $M^{\mathbb{C}}$ may be the richer symmetry group. For example, in the case of $M = \mathbb{R}^3 \setminus \{0\}$ and ω given by the magnetic induction of a monopole sitting at the origin, we may take $M^{\mathbb{C}} = \{\vec{y} \in \mathbb{C}^3 | \vec{y}^2 \notin]-\infty,0]\}$. This allows to represent $SO(3,\mathbb{C})$ in the space of holomorphic sections of $L^{\mathbb{C}}$. In Appendix 1, we give an explicit local representative of the holomorphic monopole line bundle $L^{\mathbb{C}}$ over $M^{\mathbb{C}}$ which restricted to $\mathbb{R}^3 \setminus \{0\}$ reproduces the monopole bundle with hermitian connection.

3. TOPOLOGICAL AMBIGUITIES IN 2-DIMENSIONAL FIELD THEORY.

In 2-dimensional field theory, closed 3-forms play similar role as closed 2-forms in mechanics. To consider classical equations of motion, we need to know the action locally and only up to an integral of the derivative of a 1 form. Such an information is available if we start from a closed 3-form γ and define (a part of) the local action by taking integrals of local prime 2-forms of γ. On the quantum level, however, one needs the global action to be well defined modulo $2\pi\mathbb{Z}$. To achieve this, we shall employ again the cohomology of sheaf complexes. Consider complex $\mathbb{F}_M^2$

$$\mathcal{U}(1)_M \xrightarrow{\frac{1}{i}\,d\log} \Omega_M^1 \xrightarrow{d} \Omega_M^2 \tag{1}$$

and its second cohomology group $\mathbb{H}^2(\mathbb{F}_M^2)$. The elements of $\mathbb{H}^2(\mathbb{F}_M^2)$ are equivalence classes of systems $(g_{\alpha_0\alpha_1\alpha_2}, \eta_{\alpha_0\alpha_1}, \omega_{\alpha_0})$ such that $g_{\alpha_0\alpha_1\alpha_2}$ are U(1) valued functions on $\mathcal{O}_{\alpha_1\alpha_2\alpha_3}$ (multiplicatively) antisymmetric in indices α_i, $\eta_{\alpha_0\alpha_1} = -\eta_{\alpha_1\alpha_0}$ are real 1-forms on $\mathcal{O}_{\alpha_0\alpha_1}$ and ω_α are real 2-forms on $\mathcal{O}_\alpha$ s.t.

$$g_{\alpha_1\alpha_2\alpha_3} g^{-1}_{\alpha_0\alpha_2\alpha_3} g_{\alpha_0\alpha_1\alpha_3} g^{-1}_{\alpha_0\alpha_1\alpha_2} = 1 \quad \text{on} \quad \mathcal{O}_{\alpha_0\alpha_1\alpha_2\alpha_3}\,, \tag{2}$$

$$\eta_{\alpha_1\alpha_2} - \eta_{\alpha_0\alpha_2} + \eta_{\alpha_0\alpha_1} = \frac{1}{i}\, g^{-1}_{\alpha_0\alpha_1\alpha_2}\, dg_{\alpha_0\alpha_1\alpha_2} \quad \text{on} \quad \mathcal{O}_{\alpha_0\alpha_1\alpha_2}\,, \tag{3}$$

$$\omega_{\alpha_1} - \omega_{\alpha_0} = d\eta_{\alpha_0\alpha_1}\,. \tag{4}$$

The equivalence is defined modulo systems

$$(\chi_{\alpha_1\alpha_2}\chi^{-1}_{\alpha_0\alpha_2}\chi_{\alpha_0\alpha_1}\,,\ \pi_{\alpha_0} - \pi_{\alpha_1} + \frac{1}{i}\,\chi^{-1}_{\alpha_0\alpha_1}\, d\chi_{\alpha_0\alpha_1}\,,\ -d\pi_\alpha) \tag{5}$$

where $\chi_{\alpha_0\alpha_1} = \chi^{-1}_{\alpha_1\alpha_0}$ are U(1) valued function on $\mathcal{O}_{\alpha_0\alpha_1}$ and π_α are real 1-forms on $\mathcal{O}_\alpha$. A class $w = [(g_{\alpha_0\alpha_1\alpha_2}, \eta_{\alpha_0\alpha_1}, \omega_\alpha)]$ defines uniquely a closed 3-form γ on M equal to $d\omega_\alpha$ on each $\mathcal{O}_\alpha$. We shall call γ the curvature of w and denote by $W(M,\gamma)$ the subset of $\mathbb{H}^2(\mathbb{F}_M^2)$ of elements with curvature equal to γ.

If $\phi : \Sigma \to M$ is a map of a connected compact oriented manifold without boundary into M, then, pulling back $w \in \mathbb{H}^2(\mathbb{F}_M^2)$, we obtain an element $\phi^* w \in \mathbb{H}^2(\mathbb{F}_\Sigma^2) \cong U(1)$. To realize the last isomorphism explicitly, let us triangulate Σ by 2-cells c, 1-cells b and vertices v in such a way that $\phi(c) \subset \mathcal{O}_{\alpha_c}$, $\phi(b) \subset \mathcal{O}_{\alpha_b}$, $\phi(v) \in \mathcal{O}_{\alpha_v}$ for some assignment $\alpha_c, \alpha_b, \alpha_v$. If $w = [(g_{\alpha_0\alpha_1\alpha_2}, \eta_{\alpha_0\alpha_1}, \omega_\alpha)]$, then $\phi^* w$ is represented by the number $A(\phi) \in U(1)$,

$$A(\phi) = \exp\Big[i \sum_c \int_c \phi^*\omega_{\alpha_c} - i \sum_{\substack{b,c \\ b\subset\partial c}} \int_b \phi^*\eta_{\alpha_b\alpha_c}\Big] \prod_{\substack{v,b,c \\ v\in\partial b \\ b\subset\partial c}} g_{\alpha_v\alpha_b\alpha_c}(\phi(v))\,, \tag{6}$$

compare expression (2.15) for the loop holonomy in the line bundle case. In (6), $\int_b$ is performed with the orientation of b inherited from c and $g_{\alpha_v\alpha_b\alpha_c}(\phi(v))$ should be inverted if v inherits the negative orientation from c via b. The right hand side of (6) is independent of the choices of the triangulation of Σ, of the assignment $\alpha_c,\alpha_b,\alpha_v$ and of the representative of w (in any covering). It is also invariant under the composition of ϕ with orientation-preserving diffeomorphisms of Σ. We shall call $A(\phi)$ the amplitude of ϕ and interpret it as the exponential of (i times) the global action of ϕ as it is built of integrals of local prime forms of the curvature γ of w and of correction terms.

As in the case of the holonomy of line bundles, amplitudes $A(\phi)$ are partially fixed by curvature γ. If B is a 3-dimensional compact oriented manifold with boundary ∂B composed of connected components $(\partial B)_i$, then

$$\prod_i A(\phi\restriction_{(\partial B)_i}) = \exp[i \int_B \phi^*\gamma] \quad . \tag{7}$$

Witten used (7) to define probability amplitudes in the Wess-Zumino-Witten chiral sigma model where a part of the action is given by a closed but not exact 3-form, see Section 5. More general definition (6) was introduced in [17]. It makes sense also for maps $\phi : \Sigma \to M$ which do not extend to B such that $\Sigma = \partial B$. Similarly, in the line bundle case, the curvature did not fix the holonomies of non-contractible loops.

From relation (7), it follows easily that $w \in W(M,\gamma)$ exists only if integrals of γ over closed 3-dimensional surfaces in M are in $2\pi\mathbb{Z}$ or, in the cohomological language, if $[\gamma] \in H^3(M,\mathbb{R})$ is in the image of $H^3(M,2\pi\mathbb{Z})$ i.e. if γ is integral. It is not difficult to see that this is also a sufficient condition. Moreover, similarly to the case of line bundles with connection, $H^2(M,U(1))$ acts freely on $\mathbb{H}^2(\mathcal{F}_M^2)$ sending $[(g_{\alpha_0\alpha_1\alpha_2},\eta_{\alpha_0\alpha_1},\omega_\alpha)]$ into $[(\lambda_{\alpha_0\alpha_1\alpha_2} g_{\alpha_0\alpha_1\alpha_2},\eta_{\alpha_0\alpha_1},\omega_\alpha)]$ for $(\lambda_{\alpha_0\alpha_1\alpha_2})$ representing an element of $H^2(M,U(1))$, the orbits coinciding with sets $W(M,\gamma)$ for different integral closed 3-forms γ.

In the case of line bundle, $P(\phi)$, as given by eq. (2.15), defined also the parallel transport along open paths. The natural question arises as to whether (6) possesses a meaning for $\phi : \Sigma \to M$ where Σ is a 2-dimensional surface with boundary (being a union of loops). This leads us to the consideration of the loop space LM of all smooth maps from $S^1 \to M$. LM, with the topology of uniform convergence with all derivatives, possesses a natural structure of a (Fréchet) manifold [18] to which we shall refer below. An element $w \in \mathbb{H}^2(\mathcal{F}_M^2)$ defines naturally an isomorphism class Q of line bundles with hermitian connection over LM. We shall describe a local presentation $(G_{A_0A_1}, E_{A_0})$ of Q for an open covering $\{\mathcal{U}_A\}$ of LM constructed as follows. Choose an open covering $\{\mathcal{O}_\alpha\}$ of M and a triangulation of S^1 by intervals b meeting at points v. Choose additionally an assignment α_b, α_v such that the set

$$\{\phi : S^1 \to M \mid \phi(b) \subset \mathcal{O}_{\alpha_b}, \phi(v) \subset \mathcal{O}_{\alpha_v}\} \equiv \mathcal{U}_A \neq \emptyset . \tag{8}$$

$\mathcal{U}_A$ for various triangulations and assignments α_b, α_v cover LM. For $\emptyset = \mathcal{U}_{A_0} \cap \mathcal{U}_{A_1} \equiv \mathcal{U}_{A_0A_1}$, where $\mathcal{U}_{A_i}$ comes from a triangulation of S^1 by intervals b_i and vertices v_i, and assignments $\alpha^i_{b_i}, \alpha^i_{v_i}$, $i = 0,1$, consider the triangulation of S^1 by non-empty intersections $\bar{b} = b_0 \cap b_1$ and vertices $\bar{v}$ of either triangulation. Put $\alpha^0_{\bar{b}} = \alpha^0_{b_0}$, $\alpha^1_{\bar{b}} = \alpha^1_{b_1}$. For a vertex $\bar{v}$ of the new triangulation, set $\alpha^0_{\bar{v}} = \alpha^0_{v_0}$ if $v = v_0$ and $\alpha_{\bar{v}} = \alpha^0_{b_0}$ if $\bar{v}$ is an interior point of interval b_0 otherwise. Similarly define $\alpha^1_{\bar{v}} = \alpha^1_{v_1}$ or $\alpha^1_{b_1}$. The transition functions of the line bundle defined by a system $(g_{\alpha_0\alpha_1\alpha_2}, \eta_{\alpha_0\alpha_1}, \omega_\alpha)$ are given by

$$G_{A_0A_1}(\phi) = \exp[i\sum_{\bar{b}} \int_{\bar{b}} \phi^* \eta_{\alpha^0_{\bar{b}}\alpha^1_{\bar{b}}}] \prod_{\substack{\bar{v},\bar{b} \\ \bar{v}\in\partial\bar{b}}} \frac{g_{\alpha^0_{\bar{v}}\alpha^1_{\bar{v}}\alpha^1_{\bar{b}}}}{g_{\alpha^0_{\bar{v}}\alpha^0_{\bar{b}}\alpha^1_{\bar{b}}}}(\phi(\bar{v})) . \tag{9}$$

The connection 1-forms E_A on $\mathcal{U}_A$ are defined by their action on vectors X_ϕ tangent to LM at loop ϕ. X_ϕ is a vector field on ϕ (i.e. a smooth section of ϕ^*TM).

$$\langle X_\phi, E_A\rangle = \sum_b \int_b \phi^*(X_\phi \rfloor \omega_{\alpha_b}) + \sum_{\substack{v,b \\ v\in\partial b}} X_\phi(v) \rfloor \eta_{\alpha_v\alpha_b}(\phi(v)) . \tag{10}$$

A somewhat tedious but straightforward check shows that $(G_{A_0A_1}, E_A)$ describes locally a line bundle with hermitian connection over LM. If we change $(g_{\alpha_0\alpha_1\alpha_2}, \eta_{\alpha_0\alpha_1}, \omega_\alpha)$ by system (5) then $(G_{\alpha_0\alpha_1}, E_A)$ changes by $(K_{A_1}K^{-1}_{A_0}, \frac{1}{i}K^{-1}_A dK_A)$ where K_A are $U(1)$ valued functions on $\mathcal{U}_A$ defined by

$$K_A(\phi) = \exp[-i\sum_b \int_b \phi^*\pi_{\alpha_b}] \prod_{\substack{v,b \\ v\in\partial b}} \chi_{\alpha_v\alpha_b}(\phi(v)) . \tag{11}$$

Thus eqs. (9) and (10) define a map

$$L : \mathbb{H}^2(\mathcal{F}^2_M) \to \mathbb{H}^1(\mathcal{F}^1_{LM}) . \tag{12}$$

The curvature Ω of Lw is related to the curvature γ of w. More exactly

$$\langle X_\phi, X'_\phi, \Omega\rangle = \int_{S^1} \phi^*(X'_\phi \rfloor X_\phi \rfloor \gamma) \tag{13}$$

for $X_\phi, X'_\phi \in T_\phi LM$. Thus L maps $W(M,\gamma)$ into $Q(LM,\Omega)$.

If line bundle L is obtained by identification of triples (A_0,ϕ,z) and $(A_1,\phi,G_{A_0A_1}(\phi)z)$ for $\phi \in \mathcal{U}_{A_0A_1}$, then the fibers of L over loops ϕ and ϕ' differing by an orientation preserving reparametrization of S^1 are canonically isomorphic : such a reparametrization sends $\mathcal{U}_A$ onto $\mathcal{U}_{A'}$ where A' corresponds to image triangulation of S^1 and $G_{A_0A_1}(\phi) = G_{A'_0A'_1}(\phi')$. Similarly the fibers of L over ϕ and ϕ' differing by an orientation changing reparametrization of S^1 are naturally dual one to another. The fibers of L over constant loops ϕ_0 are canonically identified with $\mathbb{C}^1$ by choosing local representatives (A,ϕ_0,z) of their elements with A defined by a constant assignment of α's .

Now let $\phi : \Sigma \to M$, where Σ is a connected, oriented 2-dimensional compact manifold with boundary composed of loops $(\partial\Sigma)_i$ (parametrized by S^1) . A triangulation of Σ such that $\phi(c) \subset \mathcal{O}_{\alpha_c}$, $\phi(b) \subset \mathcal{O}_{\alpha_b}$, $\phi(v) \in \mathcal{O}_{\alpha_v}$ for some assignments of α's induces triangulations of the boundary circles with (restricted) assigments of α's s.t. $\phi\restriction_{(\partial\Sigma)_i}$ lie in the correponding $\mathcal{U}_{A_i}$'s . For such Σ , the right hand side of (6) is no more triangulation and α-assignment invariant. Instead it picks up a factor $\prod_i G_{A_{i_0}A_{i_1}}(\phi\restriction_{(\partial\Sigma_i)})$ if we change the latter. Thus eq. (6) defines $A(\phi)$ as an element $\otimes_i L_{\phi\restriction(\partial\Sigma)_i}$, generalizing the notion of the parallel transport in L . If we use in eq. (6) an equivalent representative of w , then its right hand side changes by factor $\prod_i K_{A_i}(\phi\restriction_{(\partial\Sigma)_i})$, i.e. transforms by the isomorphism of bundles L obtained from equivalent cocycles.

If $D : M \to M$ preserves the curvature form γ of w , it may, but need not, preserve w except when $H^2(M,U(1)) = 1$. If $D^*w = w$, then the induced map $LD : LM \to LM$, $LD(\phi) = D\circ\phi$ lifts to an isomorphism $\widehat{LD}$ of line bundle L associated to w . $\widehat{LD}$ may be canonically defined up to isomorphisms of L locally represented by U(1) valued functions K_A of eq. (11) with $\chi_{\alpha_1\alpha_2}\chi_{\alpha_0\alpha_2}^{-1}\chi_{\alpha_0\alpha_1} = 1$, $\pi_{\alpha_0} - \pi_{\alpha_1} + \frac{1}{i}\chi_{\alpha_0\alpha_1}^{-1} d\chi_{\alpha_0\alpha_1} = 0$, $d\pi_\alpha = 0$, i.e. isomorphisms multiplying each fiber L_ϕ by the holonomy of ϕ in a flat hermitian line bundle over M defined by $(\chi_{\alpha_1\alpha_2}^{-1}, -\pi_\alpha)$. In particular, if M is simply connected, the lift $\widehat{LD}$ may be canonically chosen in a unique way. In that case actions of groups of diffeomorphisms of M preserving w lifts canonically to bundle L . The amplitudes

$A(\phi)$ of eq. (6) transform covariantly

$$A(D\circ\phi) = \widehat{LD}\, A(\phi) \ . \tag{14}$$

Notice that the right hand side does not depend of the choice of lift $\widehat{LD}$ since the multiplication in L by holonomies of flat bundles over M leaves $A(\phi)$ invariant due to relation (2.16).

Again, similar considerations apply to infinitesimal actions of vector fields on M, i.e. Lie algebras of vector fields on M.

If $M \subset M^{\mathbb{C}}$, where $M^{\mathbb{C}}$ is a complex manifold and γ extends to holomorphic (3,0) form $\gamma^{\mathbb{C}}$ on $M^{\mathbb{C}}$, we may consider classes $[(g_{\alpha_0\alpha_1\alpha_2}, \eta_{\alpha_0\alpha_1}, \omega_\alpha)] \in \mathbb{H}^2(\mathbb{A}^2_{M^{\mathbb{C}}})$ (the Deligne cohomology of degree 3) where $\mathbb{A}^2_{M^{\mathbb{C}}}$ is the sheaf complex

$$\mathcal{O}^*_{M^{\mathbb{C}}} \xrightarrow{\frac{1}{i}\, d \log} A^1_{M^{\mathbb{C}}} \xrightarrow{d} A^2_{M^{\mathbb{C}}} \tag{15}$$

and $d\omega_\alpha = \gamma^{\mathbb{C}}$ (denote by $W^a(M^{\mathbb{C}},\gamma^{\mathbb{C}})$ the set of such classes). $W^a(M^{\mathbb{C}},\gamma^{\mathbb{C}}) \neq \emptyset$ if and only if $\gamma^{\mathbb{C}}$ is integral and $H^2(M^{\mathbb{C}},\mathbb{C}^*)$ enumerates its elements in such case. For $w \in W^a(M^{\mathbb{C}}, \gamma^{\mathbb{C}})$, (9) and (10) define a holomorphic bundle $L^{\mathbb{C}}$ over $LM^{\mathbb{C}}$ with holomorphic connection of curvature $\Omega^{\mathbb{C}}$, a holomorphic (2,0) form on $LM^{\mathbb{C}}$ defined by (13) with γ replaced by $\gamma^{\mathbb{C}}$. The isomorphism class of $L^{\mathbb{C}}$ is an element of $Q^a(LM^{\mathbb{C}},\Omega^{\mathbb{C}})$ which depends only on w. The amplitudes $A(\phi)$ defined by eq. (6) for $\phi : \Sigma \to M^{\mathbb{C}}$ are now in $\mathbb{C}^*$ for Σ without boundary and in $\otimes_i L^{\mathbb{C}}_{\phi\restriction(\partial\Sigma)_i}$ for $\partial\Sigma$ composed of loops $(\partial\Sigma)_i$. Holomorphisms D of $M^{\mathbb{C}}$ which preserve w, induce maps LD of $LM^{\mathbb{C}}$ lifting to isomorphisms of $L^{\mathbb{C}}$. If D is an antiholomorphism of $M^{\mathbb{C}}$ then D^*w is naturally an element of $\mathbb{H}^2(\overline{\mathbb{A}^2_{M^{\mathbb{C}}}})$, where $\overline{A^2_{M^{\mathbb{C}}}}$ is the complex

$$\overline{\mathcal{O}^*_{M^{\mathbb{C}}}} \xrightarrow{\frac{1}{i}\, d \log} \overline{A^1_{M^{\mathbb{C}}}} \xrightarrow{d} A^2_{M^{\mathbb{C}}} \tag{16}$$

So is $-\overline{w}$ represented by $(\overline{g},-\overline{\eta},-\overline{\omega})$. If $D^*w = -\overline{w}$, LD lifts to an antisomorphism of $L^{\mathbb{C}}$.

4. BOSONIZED ACTIONS ON A GENERAL RIEMANN SURFACE.

As the first application of the abstract theory of the previous section, we shall consider the definition of (euclidean) probability amplitudes for (euclidean) bosonic fields defined on a general Riemann surface Σ and corresponding via bosonization[19,4,20] to fermionic fields b,c on Σ with spins $1-\lambda$ and λ respectively, $\lambda \in \frac{1}{2}\mathbb{Z}$. Σ is equipped with a metric g in local complex coordinates given by $g_{z\bar{z}}\, dz d\bar{z}$. The metric connection turns the bundle K^{-1} of holomorphic vectors $a \frac{\partial}{\partial z}$ over Σ into a line bundle with hermitian connection of curvature R satisfying

$$\int_\Sigma R = 2\pi(2 \text{ genus} - 2) \; . \tag{1}$$

Fermionic fields c are sections of a line bundle L over Σ with hermitian connection of curvature $-\lambda R$ and fields b are sections of $K \otimes L^{-1}$ (K is the bundle of holomorphic covectors on Σ). The metric g and (isomorphism class of) L encode the geometric information needed to define the free euclidean field theory of fields b and c (and their complex conjugates) with the action

$$\int_\Sigma b \, d\bar{z} \, \nabla_{\bar{z}} c + \text{c.c.} \tag{2}$$

This is the only information about the fermionic system that we shall use.

Bosonization reexpresses the functional integral of fermionic fields b,c and their complex conjugates by a functional integral of bosonic field φ on Σ with $\varphi \in \mathbb{R}/\frac{1}{2}\mathbb{Z}$. In order to represent correctly the conformal and fermion number anomalies, the euclidean action of φ should have besides the standard free field part $4\pi i \int_\Sigma (\partial\varphi)(\bar{\partial}\varphi)$ a term $(2\lambda-1)i \int_\Sigma R\varphi$. Unfortunately, the latter term is ill-defined on a Riemann surface of genus $g > 0$ as there are smooth field configurations with no smooth real valued φ . What is well defined, however, is the real 3-form

$$\gamma = (1-2\lambda)R \, d\varphi \tag{3}$$

on $M = \Sigma \times \mathbb{R}/\frac{1}{2}\mathbb{Z}$. The additional term in the action tries to integrate a globally non-existent 2-form ω s.t. $\gamma = d\omega$ over the 2-dimensional surface

$$\Sigma \ni \xi \mapsto \phi(\xi) = (\xi,\varphi(\xi)) \in M. \tag{4}$$

As γ is integral,

$$\int_M \gamma = \frac{1}{2} (1-2\lambda)2\pi(2g-2) \in 2\pi \mathbb{Z} \quad , \tag{5}$$

the global action can still be defined up to an ambiguity in $2\pi\mathbb{Z}$, as we know from Section 3. For that, we shall have to choose $w \in W(M,\gamma)$ which we shall do in a special way.

Let us consider first a 2-form $(\frac{1}{2} -\lambda)R$ on Σ . This is a real closed integral form so that there exists a line bundle over Σ with hermitian connection of curvature $(\frac{1}{2} -\lambda)R$ (the isomorphism classes of such bundles form a $2g$-dimensional torus, the Jacobian). We shall construct a natural map ι which assigns to classes of bundles with curvature $(\frac{1}{2} -\lambda)R$ elements of $W(M,\gamma)$:

$$\iota : Q(\Sigma,(\frac{1}{2}-\lambda)R) \to W(M,\gamma) \quad . \tag{6}$$

It is in fact defined by a cup-product operation on the cohomologies of sheaf complexes, but we prefer to give a more down-to-ground definition.

Let $(g_{\alpha_0\alpha_1},\eta_\alpha)$ represent $q \in Q(\Sigma,(\frac{1}{2}-\lambda)R)$ on an open covering $\{\mathcal{O}_\alpha\}$ of Σ . Let $\{\mathcal{O}'_\beta\}$ be an open covering of the circle $\mathbb{R}/\frac{1}{2}\mathbb{Z}$ such that there exist smooth functions $\varphi_\beta : \mathcal{O}'_\beta \to \mathbb{R}$ s.t. $\varphi = \varphi_\beta(\varphi)+\frac{1}{2}\mathbb{Z}$. Notice that $\varphi_{\beta_1}-\varphi_{\beta_0} \in \frac{1}{2}\mathbb{Z}$. Consider the covering $\{\mathcal{O}_{(\alpha,\beta)} = \mathcal{O}_\alpha \times \mathcal{O}'_\beta\}$ of M . Order arbitrarily indices α . Set

$$\omega_{(\alpha,\beta)} = (1-2\lambda)R\,\varphi_\beta \ , \tag{7}$$

$$\eta_{(\alpha_0,\beta_0)(\alpha_1,\beta_1)} = 2(\varphi_{\beta_1}-\varphi_{\beta_0})\eta_{\alpha_0} \quad \text{if} \quad \alpha_0 < \alpha_1 \ , \tag{8}$$

$$g_{(\alpha_0,\beta_0)(\alpha_1,\beta_1)(\alpha_2,\beta_2)} = (g_{\alpha_0\alpha_1})^{2(\varphi_{\beta_2}-\varphi_{\beta_1})} \quad \text{if} \quad \alpha_0 < \alpha_1 < \alpha_2 . \tag{9}$$

It is easy to show that $[(g_{(\alpha_0,\beta_0)(\alpha_1,\beta_1)(\alpha_2,\beta_2)},\eta_{(\alpha_0,\beta_0)(\alpha_1,\beta_1)},\omega_{(\alpha,\beta)})]$ depends only on $q = [(g_{\alpha_0\alpha_1},\eta_\alpha)]$ and defines $\iota q \in W(M,\gamma)$.

Given $w = \iota q$ and a map (4), we may define the amplitude $A(\phi)$ by eq. (3.6). The class q may be chosen in a specific way as the isomorphism class of bundle $K^{-1/2} \otimes L$ where $K^{1/2}$ is a spin bundle, i.e. a line bundle with hermitian connection s.t. $K^{1/2} \otimes K^{1/2}$ is isomorphic to K . L is the line bundle carrying fermionic field c . Indeed, the curvature of $K^{1/2}$ is $-\frac{1}{2}R$ and that of L is $-\lambda R$ so that the curvature of $K^{-1/2}\otimes L$ is equal $(\frac{1}{2}-\lambda)R$, as required. Let us denote the amplitude of ϕ corresponding to this choice of q by $A_{K^{-1/2}\otimes L}(\phi)$. Its definition uses more geometric input that the fermionic theory which we bosonize : namely the isomorphism class of a spin bundle, i.e. a spin structure. We can get rid of this extra input by averaging the amplitude over the spin structures. Let us do this with the weight $\sigma(K^{1/2})$ which is the parity of the spin structure, i.e. -1 to the dimension of the space of sections of $K^{1/2}$ anihilated by covariant antiholomorphic derivative $d\bar{z}\,\nabla_{\bar{z}}$. The complete probability amplitude of the field $\xi \to \phi(\xi) = (\xi, \varphi(\xi))$, with the standard free field term included, is

$$e^{-S(\varphi)} = \exp[-4\pi i \int_\Sigma (\partial\varphi)(\bar{\partial}\varphi)] \sum_{[K^{1/2}]} \sigma(K^{1/2}) A_{K^{-1/2}\otimes L}(\phi) \ . \tag{10}$$

It is clear that it uses as geometric input, besides the metric g , only the isomorphism class of line bundle L with hermitian connection, i.e. the input of the fermionic theory.

Let us cut the surface Σ along the homology basis a_i, b_i, $i = 1,\dots,g$, meeting at one point, reducing the surface to the polygone Σ_c, see Fig. 3. For (smooth) field $\varphi : \Sigma \to \mathbb{R}/\frac{1}{2}\mathbb{Z}$, let us choose a smooth version $\tilde{\varphi} : \Sigma_c \to \mathbb{R}$ of it (i.e. $\varphi(\xi) = \tilde{\varphi}(\xi) + \frac{1}{2}\mathbb{Z}$). Denoting by $P_{K^{-1/2}\otimes L}$ the holonomy in $K^{-1/2} \otimes L$, it is easy to see using the definitions (3.6), (2.15) and (7) to (9) that

$$A_{K^{-1/2}\otimes L}(\phi) = \exp[(1-2\lambda)i \int_{\Sigma_c} R\tilde{\varphi}] \prod_{i=1}^{g} P_{K^{-1/2}\otimes L}(a_i)^{2\int_{b_i} d\varphi} \cdot P_{K^{-1/2}\otimes L}(b_i)^{-2\int_{a_i} d\varphi} . \qquad (11)$$

This shows that (10) reproduces the bosonic field amplitudes defined by Alvarez-Gaumé et al. in[20]. From eq. (10), it is clear that $e^{-S(\varphi)}$ is defined intrinsically, i.e. if D is a diffeomorphism of Σ then

$$e^{-S(\varphi\circ D)} = e^{-S'(\varphi)} \qquad (12)$$

where, if S corresponds to metric g and bundle L over Σ then S' does to D^*g and D^*L. In the work of Alvarez-Gaumé et al.[20] this required an explicite check of modular covariance of expressions defined with the use of a marking of the Riemann surface Σ.

5. WESS-ZUMINO-WITTEN QUANTUM FIELD THEORY.

Wess-Zumino-Witten (WZW) model[3] is another two-dimensional field theory where part of the action is defined by integrals of the local prime forms of a closed but not exact 3-form γ. Fields take values in a compact Lie group G and

$$\gamma = \frac{k}{12\pi} \operatorname{tr}(g^{-1}dg)^3 . \qquad (1)$$

For concreteness, we shall first consider only the case of $G = SU(2)$ and its complexification $G^{\mathbb{C}} = SL(2,\mathbb{C})$ with (3,0) holomorphic form $\gamma^{\mathbb{C}}$ given by the same formula. γ and $\gamma^{\mathbb{C}}$ are integral if and only if $k \in \mathbb{Z}$. As the mapping $g \to g^{-1}$ changes the sign of k, we shall limit ourselves to $k \in \mathbb{N}$. In Appendix 3, we give an explicit local representative $(g_{\alpha_0\alpha_1\alpha_2}, \eta_{\alpha_0\alpha_1}, \omega_\alpha)$ of class $w \in W^a(G^{\mathbb{C}},\gamma^{\mathbb{C}})$ which restricted to G defines an element of $W(G,\gamma)$. Since $H^2(G^{\mathbb{C}},\mathbb{C}^*) = H^2(G,U(1)) = 1$, these classes are unique.

In the euclidean Wess-Zumino-Witten model on a Riemann surface Σ, the field configuration is a map $g : \Sigma \to G$, or more generally $g : \Sigma \to G^{\mathbb{C}}$. Eq. (3.6) allows now to define a part $A(g) \equiv A_\Sigma(g)$ of the probability amplitude of field configuration g coming from $\gamma^{\mathbb{C}}$. It is a non-zero complex number, of modulus one if g takes values in G. The other part of the amplitude comes from the standard action of the chiral σ-model. The complete amplitude is

$$e^{-S_\Sigma(g)} = \exp[\frac{ik}{4\pi} \int_\Sigma tr(g^{-1}\partial g)(g^{-1}\bar{\partial} g)]A_\Sigma(g) \tag{2}$$

and defines global euclidean action $S_\Sigma(g)$ up to $2\pi i \mathbb{Z}$. In (2), we have chosen the special value of the coupling constant of the chiral model, which renders the combined model conformal invariant[3]. Since every mapping g from Σ to $SU(2)$ can be extended to a map $\tilde{g} : B \to SU(2)$ where B is a 3-dimensional compact oriented manifold with boundary $\partial\Sigma$ then due to eq. (3.7), we may rewrite (3) as

$$e^{-S_\Sigma(g)} = \exp[\frac{ik}{4\pi} \int_\Sigma tr(g^{-1}\partial g)(g^{-1}\bar{\partial} g) + \frac{ik}{12\pi} \int_B tr(\tilde{g}^{-1}d\tilde{g})^3] . \tag{3}$$

Let us start with some formal considerations. In quantum field theory, we shall have to calculate functional integrals given formally as

$$\int_{\Sigma G} F(g) e^{-S_\Sigma(g)} D_\Sigma g \tag{4}$$

where $F(g)$ are functionals of g, e.g. $F(g) = \prod_{j=1}^{n} g_{\alpha_j\beta_j}(\xi_j)$, and $D_\Sigma g$ is the Haar measure on the group ΣG of maps from Σ into G.

If $\Sigma = P\mathbb{C}^1$, the functional integral (4) may be given an interpretation in operator formalism. Consider the space of smooth sections ψ of line bundle L over LG associated to (a representative of) the class in $W(G,\gamma)$. $P\mathbb{C}^1 = \mathbb{C}^1 \cup \{\infty\} = D_o \cup D_\infty$, where D_o is the disc $|z| \leq 1$ and D_∞ is the opposite one. Formal examples of sections ψ are provided by functional integrals over fields defined on D_o with boundary values fixed:

$$\psi_F(h) = \int_{D_o G} F(g)\, e^{-S_{D_o}(g)} \delta(g\restriction_{\partial D_o} h^{-1}) D_{D_o} g \tag{5}$$

where $e^{-S_{D_o}(g)}$ is defined by (2) with Σ replaced by D_o and takes values in $L_{g\restriction_{\partial D_o}}$. We may consider formal scalar product of states ψ defined as

$$(\psi_1,\psi_2) = \int_{LG} \overline{\psi_1(h)}\, \psi_2(h)\, D_{S^1} h \tag{6}$$

If we set $\theta g(z) = g(\frac{1}{\bar{z}})$ then it is easy to see that for $g : D_o \to G$ (in any trivilization)

$$\overline{e^{-S_{D_o}(g)}} = e^{-S_{D_\infty}(\theta g)} \quad . \tag{7}$$

Let F_1, F_2 be functionals of $g : \Sigma = P\mathbb{C}^1 \to G$ depending only on fields on D_o. If we denote $(\theta F_1)(g) = \overline{F_1(\theta g)}$ then formally

$$\int_{\Sigma G} (\theta F_1)(g) F_2(g)\, e^{-S_\Sigma(g)} D_\Sigma g$$

$$= \int_{LG} D_{S^1} h \Big(\int_{D_\infty G} (\theta F_1)(g_\infty)\, e^{-S_{D_\infty}(g_\infty)} \delta(g_\infty\restriction_{\partial D_o} h^{-1}) D_{D_\infty} g \Big)$$

$$\cdot \Big(\int_{D_o G} F_2(g_o) e^{-S_{D_o}(g_o)} \delta(g_o\restriction_{\partial D_o} h^{-1}) D_{D_o} g_o \Big)$$

$$= \int_{LG} D_{S^1} h \Big(\overline{\int_{D_o G} F_1(g_o) e^{-S_{D_o}(g_o)} \delta(g_o\restriction_{\partial D_o} h^{-1}) D_{D_o} g_o} \Big)$$

$$\cdot \Big(\int_{D_o G} F_2(g_o) e^{-S_{D_o}(g_o)} \delta(g_o\restriction_{\partial D_o} h^{-1}) D_{D_o} g_o \Big)$$

$$= (\psi_{F_1}, \psi_{F_2}) \quad . \tag{8}$$

This shows that the left hand side should be positive for $F_1 = F_2$ which is the physical positivity condition for the WZW model on Riemann sphere.

Notice that the formal definition (5) may be rewritten as

$$\psi_F(h) = \int_{D_o G} F(gg_h)\, e^{-S_{D_o}(gg_h)} \delta(g|_{\partial D_o}) D_{D_o} g \tag{9}$$

where g_h is a fixed map from D_o into G such that $g_h\restriction_{\partial D_o} = h$. Independence of the integral of the choice of g_h follows from the right invariance of the Haar measure $D_{D_o} g$. If A is a finite set, then the the Haar measure $D_A g$ on AG (i.e. the set of maps from A to G) possesses a richer invariance : for F an analytic function on $AG^{\mathbb{C}}$ and $g' \in AG^{\mathbb{C}}$

$$\int_{AG} F(gg') D_A g = \int_{AG} F(g) D_A g \quad . \tag{10}$$

If we assume this invariance still to hold formally in infinite dimensional case then for F analytic functionals on $D_o G^{\mathbb{C}}$, (9) still makes sense for $h \in LG^{\mathbb{C}}$ and defines formally a holomorphic section of bundle $L^{\mathbb{C}}$ over $LG^{\mathbb{C}}$.

The above formal considerations motivate the choice of space $\Gamma^a(L^{\mathbb{C}})$ of the holomorphic sections of $L^{\mathbb{C}}$ as the space of quantum states of the WZW model. Below we shall show that this space carries a natural representation of a pair of Kac-Moody algebras $\widehat{su}(2)$ with central charge k

(arising by geometric quantization of the classical symmetries of the theory). We shall decompose this representation into irreducible components recovering the spectrum of the latter conjectured by Gepner and Witten in[5].

Classical WZW theory possesses an extraordinarily rich symmetry : its classical (euclidean) equations of motion are $\partial(g^{-1}\bar{\partial}g) = 0$ and if $g_1(g_2)$ is any (anti-)holomorphic map with values in $G^{\mathbb{C}}$ then the transformation $g \to g_1 g g_2^{-1}$ maps local classical solutions into new ones. On the quantum level, this becomes the Kac-Moody symmetry.

It will be important to study this symmetry on the group level rather than on the level of Lie algebras. To this end, we shall first introduce group $\widehat{LG}^{\mathbb{C}}$ which is the central extension of loop group $LG^{\mathbb{C}}$ by $\mathbb{C}^*$. It appears naturally when we lift the classical symmetries to the symmetries of bundle $L^{\mathbb{C}}$ following the rules of geometric quantization. Here, we shall construct directly the final product. As a space, $\widehat{LG}^{\mathbb{C}} = L^{\mathbb{C}^*} \equiv L^{\mathbb{C}} \setminus$ zero section. To describe its multiplication law, let us notice a basic property of the WZW amplitudes e^{-S_Σ}, known as Polyakov formula[21]. If Σ is a Riemann surface with boundary, $g,h : \Sigma \to G^{\mathbb{C}}$, $h\restriction_{\partial\Sigma} \equiv 1$, then

$$e^{-S_\Sigma(gh)} = e^{-S_\Sigma(g)}\, e^{-S_\Sigma(h)}\, e^{\Gamma_\Sigma(g,h)} \tag{11}$$

where

$$\Gamma_\Sigma(g,h) = \frac{ik}{2\pi}\int_\Sigma \mathrm{tr}(g^{-1}\bar{\partial}g)(h\partial h^{-1}) \quad . \tag{12}$$

Notice that $e^{-S_\Sigma(h)}$ is a complex number since the fibers of $L^{\mathbb{C}}$ over constant loops are canonically isomorphic to $\mathbb{C}^1$. Thus (11) is an equality of elements of $\bigotimes_i L^{\mathbb{C}}_{g\,|\,(\partial\Sigma)_i}$. It is easily proven by comparing the t-derivatives of both sides along a homotopy h_t fixed at $\partial\Sigma$ between h and 1. All what is needed is the formula for the derivative of the global action $e^{-S_\Sigma(g_t)}$ over t valid in any trivialization if $\frac{\partial}{\partial t} g_t\restriction_{\partial\Sigma} = 0$:

$$\frac{d}{dt}\, S_\Sigma(g_t) = -\frac{ik}{2\pi}\int_\Sigma \mathrm{tr}\, \partial(g^{-1}\frac{\partial g_t}{\partial t})\, g_t^{-1}\,\bar{\partial}\, g_t \quad . \tag{13}$$

We proove a generalization of (13) admitting any $\frac{\partial g_t}{\partial t}\restriction_{\partial\Sigma}$ in Appendix 4. The multiplication law in $L^{\mathbb{C}}$ is defined by

$$(\lambda_1 e^{-S_{D_o}(g_1)})\cdot(\lambda_2 e^{-S_{D_o}(g_2)}) = \lambda_1\lambda_2\, e^{-S_{D_o}(g_1 g_2)-\Gamma_{D_o}(g_1,g_2)} \tag{14}$$

where $g_i : D_o \to G^{\mathbb{C}}$, $i = 1,2$. That the definition is correct and defines group operation in $\widehat{LG}^{\mathbb{C}}$, follows easily from Polyakov formula (11) and the basic property of Γ_Σ

$$\Gamma_\Sigma(g_1,g_2) = \Gamma_\Sigma(g_1,g_2g_3) - \Gamma_\Sigma(g_1g_2,g_3) + \Gamma_\Sigma(g_2,g_3) \quad . \tag{15}$$

$\mathbb{C}^*$ can be embedded into the fiber of $L^{\mathbb{C}}$ over the constant loop 1 . This way we obtain the exact sequence of groups

$$1 \to \mathbb{C}^* \to \widehat{LG}^{\mathbb{C}} \to LG^{\mathbb{C}} \to 1 \quad . \tag{16}$$

Consider the antiholomorphic involution $j : g \mapsto g^\dagger$ on $G^{\mathbb{C}}$. For $w \in W^a(G^{\mathbb{C}},\gamma^{\mathbb{C}})$,

$$j^* w = -\bar{w} \tag{17}$$

since both sides have the same curvature. The involution Lj of $LG^{\mathbb{C}}$ lifts canonically (G is simply connected) to an antiholomorphic involution. We shall use symbole $\hat{g}^\dagger$ for the image of $\hat{g} \in L^{\mathbb{C}}$ under it. The involution transforms covariantly the probability amplitudes

$$(e^{-S_\Sigma(g)})^\dagger = e^{-S_\Sigma(g^\dagger)} \tag{18}$$

and is antimultiplicative, i.e.

$$(\hat{g}_1 \cdot \hat{g}_2)^\dagger = \hat{g}_2^\dagger \cdot \hat{g}_1^\dagger \quad . \tag{19}$$

The set of points $\hat{g}$ of $LG^{\mathbb{C}}$ s.t. $\hat{g}^\dagger = \hat{g}^{-1}$ is the set of vectors of length one in line bundle L over LG . We shall denote it by $\widehat{LG}$. Due to (19), the multiplication does not lead out of $\widehat{LG}$. This way we get a central extention of LG by $U(1)$:

$$1 \to U(1) \to \widehat{LG} \to LG \to 1 \quad . \tag{20}$$

Central extensions (16) and (20) depend on k . For $k = 1$, we obtain the so-called universal one . The extensions for higher k can be obtained from the universal one by division by $\mathbb{Z}_k$.

Space $\Gamma^a(L^{\mathbb{C}})$ of holomorphic sections of $L^{\mathbb{C}}$ carries two commuting representations ℓ and r of $\widehat{LG}^{\mathbb{C}}$ defined by means of left and right multiplication. If $\hat{g}_1,\hat{g}_2 \in \widehat{LG}^{\mathbb{C}}$ projecting to $g_1,g_2 \in LG^{\mathbb{C}}$ and if $\psi \in \Gamma^a(L^{\mathbb{C}})$ then

$$(\ell(\hat{g}_1)r(\hat{g}_2)\psi)(g) = \hat{g}_1 \cdot \psi(g_1^{-1} g g_2^{\dagger^{-1}}) \cdot \hat{g}_2^\dagger \quad . \tag{21}$$

Denote by $L^\pm G^{\mathbb{C}} \equiv L^\pm$ the subgroups of $LG^{\mathbb{C}}$ composed of boundary values of holomorphic maps of $D_{\substack{o\\ \infty}}$ into $G^{\mathbb{C}}$. $L^\pm$ can be also considered as subgroups of $\widehat{LG}^{\mathbb{C}}$ composed of elements $e^{-S_{D_o}(g)}$, $g \in L^+$, and $e^{S_{D_\infty}(g)}$ i.e. elements dual to $e^{-S_{D_\infty}(g)}$, $g \in L^-$. Notice that for constant loops g_o forming $L^+ \cap L^-$, $e^{-S_{D_o}(g_o)} = e^{S_{D_\infty}(g_o)}$ since $e^{-S_\Sigma(g_o)} = 1$ so that they lift to the same element of $\widehat{LG}^{\mathbb{C}}$ via L^+ and L^- defining an embedding of $G^{\mathbb{C}}$ into $\widehat{LG}^{\mathbb{C}}$. The Lie algebras of L^+ and L^- are spanned by

$\frac{1}{2} i\sigma_j z^n$, $n \geq 0$ and $\frac{1}{2} i\sigma_j z^n$, $n \leq 0$ respectively where σ_j are the Pauli matrices. Define

$$J^j_n\psi = \frac{d}{d\varepsilon}\restriction_{\varepsilon=o} \ell(e^{\frac{1}{2} i\varepsilon\sigma_j z^n})\psi \tag{22}$$

and

$$\bar{J}^j_n\psi = \frac{d}{d\varepsilon}\restriction_{\varepsilon=o} r(e^{\frac{1}{2} i\varepsilon\sigma_j z^n})\psi \ . \tag{23}$$

Straightforward computation based on eq. (14), see Appendix 5, shows that

$$[J^j_n, J^i_m] = \sum_k \varepsilon^{jik} J^k_{n+m} + \frac{k}{2} m\delta^{ji}\delta_{n+m,o} \ . \tag{24}$$

The same relation holds for $\bar{J}$'s . This shows that our representation of $\widehat{LG}^{\mathbb{C}} \times \widehat{LG}^{\mathbb{C}}$ in $\Gamma^a(L^{\mathbb{C}})$ becomes on the level of Lie algebra a pair of commuting representations of the $\widehat{su}(2)$ Kac-Moody algebra with central charge equal k [23].

The representation of $\widehat{LG}^{\mathbb{C}} \times \widehat{LG}^{\mathbb{C}}$, defined somewhat ad hoc, becomes quite natural if we represent states in $\Gamma^a(L^{\mathbb{C}})$ by formal functional integrals (5). Indeed, it is easy to see that formally, for $g_1, g_2 \in L^+$

$$\ell(g_1) r(g_2)\psi_F = \psi_{F'} \tag{25}$$

where

$$F'(g) = F(g_1^{-1} g\, g_2^{\dagger -1}) \ . \tag{26}$$

Besides, in the formal scalar product (8),

$$\ell(g_1)^\dagger = \ell(\theta g_1^\dagger) \ , \ r(g_2)^\dagger = r(\theta g_2^\dagger) \ . \tag{27}$$

The action of $L^+ \times L^+$ together with rules (27) determine the representation of $\widehat{LG}^{\mathbb{C}} \times \widehat{LG}^{\mathbb{C}}$. On the algebraic level, eq. (27) becomes the unitarity rule[23]

$$J^{i\dagger}_n = -J^i_{-n} \ , \quad \bar{J}^{i\dagger}_n = -\bar{J}^i_{-n} \ .$$

We would like to reduce the representation of $\widehat{LG}^{\mathbb{C}} \times \widehat{LG}^{\mathbb{C}}$ in $\Gamma^a(L^{\mathbb{C}})$. To this end, we shall search for lowest weight (LW) vectors in $\Gamma^a(L^{\mathbb{C}})$, the building blocks of the LW subrepresentations[22]. By definition, the LW vectors are states ψ satisfying

$$\ell(g_1) r(g_2)\psi = a_1^{-2j_1} \bar{a}_2^{-2j_2}\psi \tag{29}$$

for $g_1, g_2 \in L^+$, $g_1(0) = \begin{pmatrix} a_1 & 0 \\ * & a_1^{-1} \end{pmatrix}$, $g_2(0) = \begin{pmatrix} a_2 & 0 \\ * & a_2^{-1} \end{pmatrix}$. j_1, j_2 are the left, right (iso-)spins of the LW vector ψ , $0 \leq j_1, j_2 \in \frac{1}{2}\mathbb{Z}$. On constant loops g_o

$$\psi(g_o) = \psi_o(g_o) e^{-S_{D_o}(g_o)} \tag{30}$$

where ψ_o is a holomorphic function on $G^{\mathbb{C}}$. Due to (18), eqs. (21) and (14) imply that for any $g_1, g_2 \in L^+$

$$\psi(g_1 g_2^\dagger) = [\ell(g_1 g_1(0)^{-1}) r(g_2 g_2(0)^{-1})\psi](g_1 g_2^\dagger)$$

$$= (g_1 g_1(0)^{-1}) \cdot \psi(g_1(0) g_2(0)^\dagger) \cdot (g_2(0)^{\dagger -1} g_2^\dagger)$$

$$= \psi_o(g_1(0) g_2(0)^\dagger)\, e^{-S_{D_o}(g_1 g_2^\dagger)} \quad . \tag{31}$$

On the other hand, using constant loops $g_{10} = \begin{pmatrix} a_1 & 0 \\ * & a_1^{-1} \end{pmatrix}$, $g_{20} = \begin{pmatrix} a_2 & 0 \\ * & a_2^{-1} \end{pmatrix}$, we obtain from (29)

$$\psi_o(g_{10}^{-1} g_o g_{20}^{\dagger -1}) = a_1^{-2j_1} \bar{a}_2^{-2j_2} \psi_o(g_o) \quad . \tag{32}$$

Relation (32) means that ψ_o is the LW vector for the left and right regular representation of G in holomorphic functions on $G^{\mathbb{C}}$. Now, the standard result says that the left and right spins of ψ_o have to be equal and ψ_o is unique up to a factor. Indeed, (32) implies that

$$\psi_o(g_{10} g_{20}^{+}) = a_1^{2j_1} \bar{a}_2^{2j_2} \psi_o(1) \quad . \tag{33}$$

Since for $a \neq 0$

$$\begin{pmatrix} a & b \\ c & \frac{1}{a} + \frac{bc}{a} \end{pmatrix} = \begin{pmatrix} a & 0 \\ c & a^{-1} \end{pmatrix} \begin{pmatrix} 1 & 0 \\ \frac{\bar{b}}{\bar{a}} & 1 \end{pmatrix}^\dagger = \begin{pmatrix} 1 & 0 \\ \frac{c}{a} & 1 \end{pmatrix} \begin{pmatrix} \bar{a} & 0 \\ \bar{b} & \bar{a}^{-1} \end{pmatrix} , \tag{34}$$

we infer that $j_1 = j_2 = j$ and for $g_o = \begin{pmatrix} a & b \\ c & d \end{pmatrix}$

$$\psi_o(g_o) = a^{2j} \psi_o(1) \quad . \tag{35}$$

Summarizing, LW vectors ψ in $\Gamma^a(L^{\mathbb{C}})$, see eq. (29), have to have equal spins $j_1 = j_2 = j$ and then, for $g_1, g_2 \in L^+$,

$$\psi(g_1 g_2^\dagger) = \text{const}\ a^{2j}\, e^{-S_{D_o}(g_1 g_2^\dagger)} \quad . \tag{36}$$

where $g_1(0) g_2(0)^\dagger = \begin{pmatrix} a & * \\ * & * \end{pmatrix}$.

By the Birkhoff theorem [22], chapter 8, loops of the form $g_1 g_2^\dagger$, $g_1, g_2 \in L^+$ form an open dense set in $LG^{\mathbb{C}}$. Hence eq. (36), for each j, may have only one solution $\psi \in \Gamma^a(L^{\mathbb{C}})$, up to a multiplicative constant. We shall see that such a solution exists if and only if $j \leq \frac{k}{2}$. The proof of this fact is a variation of the proof of Prop. 11.3.1 of ref.[22].

Let us consider the action of $L^+ \times L^+$ on $LG^{\mathbb{C}}$ given by $g \mapsto g_1 g g_2^\dagger$. We want to describe the orbits of this action. First consider the orbits of $N^+ \times L^+$ in $LG^{\mathbb{C}}$ where $N^+ = \{g_1 \in L^+ \mid g_1(0) = \begin{pmatrix} * & 0 \\ * & * \end{pmatrix}\}$. As proven in [22], Section 8.4, these orbits are of the form

$$\Sigma_n = N^+ e_n L^- = N_n^+ e_n L^- \tag{37}$$

where $e_n = \begin{pmatrix} z^n & 0 \\ 0 & z^{-n} \end{pmatrix} \in LG^{\mathbb{C}}$,

$$N_n^+ = N^+ \cap e_n L_1^+ e_n^{-1} , \tag{38}$$

and $L_1^+ = \{g \in L^+ : g(0) = 1\}$.

Relation (37) follows from the splitting

$$N^+ = N_n^+ (N^+ \cap e_n L^- e_n^{-1}) , \tag{39}$$

elementary to prove. Another elementary splitting is

$$e_n L_1^+ e_n^{-1} = N_n^+ N_n^- \tag{40}$$

where

$$N_n^- = N^- \cap e_n L_1^+ e_n^{-1} \tag{41}$$

$$\text{and } N^- = \{g \in L^- : g(\infty) = \begin{pmatrix} * & * \\ 0 & * \end{pmatrix}\} . \tag{42}$$

Since $U_n \equiv e_n L_1^+ L^-$ is an open (dense) set in $LG^{\mathbb{C}}$ containing e_n and

$$U_n = (e_n L_1^+ e_n^{-1}) e_n L^- = N_n^+ N_n^- e_n L^- , \tag{43}$$

we see that the codimension of Σ_n is the dimension of $N_n^- = \begin{cases} 2n-1 , & n > 0 \\ -2n , & n < 0 \end{cases}$.

Another important information we shall need is that

$$U_n \setminus \Sigma_n \subset \begin{cases} \bigcup\limits_{|m|<n} \Sigma_m , & n > 0 \\ \left(\bigcup\limits_{|m|<n} \Sigma_m \right) \cup \Sigma_{-n} , & n < 0 \end{cases} . \tag{44}$$

This is an easy consequence of the discussion in [22], Chapter 7.3. From the Hartogs theorem it follows that if ψ is a holomorphic section of $L^{\mathbb{C}}$ over $\Sigma_0 \cup \Sigma_1 = U_0 \cup U_1$ then it extends to a holomorphic section over $\Sigma_0 \cup \Sigma_1 \cup \Sigma_{-1} = U_0 \cup U_1 \cup U_{-1}$ since Σ_{-1} is a complex submanifold of the latter set of codimension > 1 . By induction, ψ can be continued to a global section of $L^{\mathbb{C}}$. To see when (36) defines a global section, it is then enough to study when it extends from Σ_0 to Σ_1 .

Consider loops $f_{cn} = \begin{pmatrix} 1 & 0 \\ cz^{-n} & 1 \end{pmatrix}$, $c \in \mathbb{C}^1$, $n = 1,2,\dots$. $f_{cn} \in N_n^-$ and f_{c1} is a general element of N_1^- . If $c \neq 0$ then

$$f_{cn}\, e_n = \begin{pmatrix} z^n & 0 \\ c & z^{-n} \end{pmatrix} = \begin{pmatrix} 1 & \frac{z^n}{c} \\ 0 & 1 \end{pmatrix} \begin{pmatrix} 0 & \bar{c} \\ -\frac{1}{c} & z^n \end{pmatrix}^{\dagger} \equiv g_{1cn} g_{2cn}^{\dagger} \tag{45}$$

where $g_{1cn} \in N^+$ and $g_{2cn} \in L^+$. We shall need to know the asymptotic behavior of $e^{-S_{D_o}(g_{1cn}g_{2cn}^{\dagger})}$ when $c \to 0$. Notice the singularity in

$$g_{1cn} g_{2cn}^{\dagger} = \begin{pmatrix} z^n & \frac{\bar{z}^n z^n - 1}{c} \\ c & \bar{z}^n \end{pmatrix} \tag{46}$$

where $c \to 0$ inside D_o. Using formula (A.4.6) for the derivative of action S_Σ (see Appendix 4), we infer that in any local trivialization

$$\frac{d}{dc} S_{D_o}(g_{1cn}g_{2cn}^{\dagger}) = -\frac{ik}{2\pi} \int_{D_o} \mathrm{tr}\, \partial(g_{2cn}^{\dagger -1} g_{1cn}^{-1} \frac{\partial (g_{1cn}g_{2cn}^{\dagger})}{\partial c}) g_{2cn}^{\dagger -1} g_{1cn}^{-1}$$
$$\bar{\partial}(g_{1cn}g_{2cn}^{\dagger}) + \mathcal{O}(1) \tag{47}$$

where $\mathcal{O}(1)$ contains the boundary terms which are regular when $c \to 0$ as $g_{1c}g_{2c}^{\dagger}$ is regular on ∂D_o. Integrating by parts in (47) and using the fact that $\partial(g_{2cn}^{\dagger -1} g_{1cn}^{-1}\, \bar{\partial}(g_{1cn}g_{2cn}^{\dagger})) = 0$ ($g_{1c}g_{2c}^{\dagger}$ is a classical solution), we obtain

$$\frac{d}{dc} S_{D_o}(g_{1cn}g_{2cn}^{\dagger}) = \frac{ik}{2\pi} \int_{\partial D_o} \mathrm{tr}\, \frac{\partial (g_{2cn}^{\dagger -1} g_{1cn}^{-1})}{\partial c}\, \bar{\partial}(g_{1cn}g_{2cn}^{\dagger}) + \mathcal{O}(1)$$
$$= -\frac{kn}{c} + \mathcal{O}(1)\ . \tag{48}$$

Thus

$$e^{-S_{D_o}(g_{1cn}g_{2cn}^{\dagger})} = c^{kn}\, \mathcal{O}(1)\ . \tag{49}$$

In order to study the holomorphicity of LW state ψ satisfying (36) on U_1, we decompose each element of U_1 as $g_1 f_{c1}\, e_1 g_2^{\dagger}$, where $g_1 \in N_1^+$ and $g_2 \in L^+$. According to (43), this is a unique decomposition. Now for $c \neq 0$

$$\psi(g_1 f_{c1} e_1 g_2^{\dagger}) = \mathrm{const.}\ a^{2j}\, e^{-S_{D_o}(g_1 g_{1c1} g_{2c1}^{\dagger} g_2^{\dagger})}$$
$$= \mathrm{const.}\ a^{2j}\, g_1 \cdot e^{-S_{D_o}(g_{1c1}g_{2c1}^{\dagger})} \cdot g_2^{\dagger} \tag{50}$$

where a is defined by

$$(g_1 g_{1c1} g_{2c1} g_2^{\dagger})(0) = g_1(0) \begin{pmatrix} 0 & -\frac{1}{c} \\ c & 0 \end{pmatrix} g_2(0)^{\dagger} = \begin{pmatrix} a & * \\ * & * \end{pmatrix}.$$

Clearly (see (49)), the right hand side of (50) is analytic at $c = 0$ if and only if $j \leq \frac{k}{2}$. Thus in this and only this case ψ, as given by (36) extends by continuity to a unique holomorphic section of $L^{\mathbb{C}}$. From (36) it is obvious that ψ restricted to Σ_o satisfies (29). We still have to check this for the extension of ψ to the whole loop group. Any element $g \in LG^{\mathbb{C}} \setminus \Sigma_o$ may be written as $g_1 e_n g_2^\dagger$ for $g_1, g_2 \in L^+$ (orbits of $L^+ \times L^-$ on $LG^{\mathbb{C}}$ are of the form $\Sigma_n \cup \Sigma_{-n}$) . Hence g may be approximated by elements $g_1 f_{cn} e_n g_2^\dagger \in \Sigma_o$. Thus (29) follows by continuity. Notice that for $|n| > 2$,

$$\psi(g_1 e_n g_2^\dagger) = \lim_{c \to o} \psi(g_1 f_{cn} e_n g_2^\dagger) \tag{51}$$

$$= \text{const.} \lim_{c \to o} a^{2j} g_1 \cdot e^{-kS_{D_o}(g_{1cn} g_{2cn}^\dagger)} \cdot g_2 = 0 \ , \tag{52}$$

so that LW state ψ vanishes on $\bigcup_{|n| \geq 2} \Sigma_n$. If the spin of the state is less than $\frac{k}{2}$ then ψ vanishes also on Σ_1 and Σ_{-1} .

The spectrum of the LW vectors found here : left-right spins equal taking values $0, \frac{1}{2}, 1 \ldots, \frac{k}{2}$ with multiplicity one, coincides with the spectrum conjectured by Gepner and Witten in[5] . From the general Peter-Weyl type theory for the representations of the Kac-Moody groups[24] it follows that the representation of $\hat{LG}^{\mathbb{C}} \times \hat{LG}^{\mathbb{C}}$ restricted to the subspace of the so-called strongly regular states decomposes into the direct sum of LW representations built on the vectors just described. We do not know if the strongly regular states are dense in $\Gamma^a(L^{\mathbb{C}})$. (This is however plausible as there are no highest weight vectors in $\Gamma^a(L^{\mathbb{C}})$) .

The formal candidates for our LW states are provided by functional integrals of type (5) :

$$\psi_j(h) = \int_{D_o G} N(g(0)^{\otimes_s 2j}) \, e^{-S_{D_o}(g)} \, \delta(g\restriction_{\partial D_o} h^{-1}) D_{D_o} g \tag{53}$$

where $N(g(0)^{\otimes_s 2j})$ stands for a normal ordering of the symmetric tensor power of matrix $g(0)$. Such normal powers are produced from point-splitted expressions by a limiting procedure with an appropriate multiplicative renormalization. How this works in details is described by the operator product expansions or fusion rules [21,5] which we discuss, from the present point of view, in [25].

6. SO(3) WESS-ZUMINO-WITTEN MODEL

In the previous Section, we have studied the canoncially quantized WZW model with a simply connected compact group G taken to be $SU(2)$. Here, we shall examine the new topological aspect appearing if G is not simply connected : the presence of twisted sectors in the space of states

To make the discussion free of group-theory complications, we shall take G to be $SO(3)$ and will denote by $\widetilde{G}$ its simply connected cover $SU(2)$ and by $G^{\mathbb{C}}$ its complexification equal $SL(2,\mathbb{C})/\mathbb{Z}_2$. 3-form γ $(\gamma^{\mathbb{C}})$ on G $(G^{\mathbb{C}})$ given by (5.1) is integral if and only if k is an even integer (the volume of G is half that of $\widetilde{G}$). As before, we may take k positive. A representative of a single $w \in W^a(G^{\mathbb{C}},\gamma^{\mathbb{C}})$ whose restriction to G defines a single element of $W(G,\gamma)$ $(H^2(G^{\mathbb{C}},\mathbb{C}^*) = 1 = H^2(G,U(1)))$ is given in Appendix 3. The loop space $LG^{\mathbb{C}}$ has two components, $L_oG^{\mathbb{C}}$ of contractible loops and $L_tG^{\mathbb{C}}$ of twisted loops, which lift to curves in $\widetilde{G}$ whose ends differ by -1. Line bundle $L^{\mathbb{C}}\restriction_{L_oG^{\mathbb{C}}} \equiv L_o^{\mathbb{C}}$ may be identified with $\widetilde{L}^{\mathbb{C}}/\mathbb{Z}_2$ where $\widetilde{L}^{\mathbb{C}}$ is the line bundle over $L\widetilde{G}^{\mathbb{C}}$ (recall that $\mathbb{Z}_2 \subset \widetilde{G}^{\mathbb{C}} \subset \widehat{L\widetilde{G}^{\mathbb{C}}} \subset L^{\mathbb{C}}$). Thus $\Gamma^a(L_o^{\mathbb{C}})$ may be identified with the subspace $\Gamma^a_{even}(\widetilde{L}^{\mathbb{C}})$. Since the action of $\widehat{L\widetilde{G}^{\mathbb{C}}} \times \widehat{L\widetilde{G}^{\mathbb{C}}}$ on $\Gamma^a(L^{\mathbb{C}})$ commutes with -1, we obtain its representation in $\Gamma^a(L_o^{\mathbb{C}})$ whose LW states correspond to even LW states in $\Gamma^a(\widetilde{L}^{\mathbb{C}})$, i.e. to integer spins $j_1 = j_2 \leq \frac{k}{2}$ appearing with multiplicity one.

We still have to study $L^{\mathbb{C}}\restriction_{L_tG^{\mathbb{C}}} \equiv L_t^{\mathbb{C}}$. We shall define left and right actions of $\widehat{L\widetilde{G}^{\mathbb{C}}}$ on $L_t^{\mathbb{C}}$ (again fixed completely by the rules of geometric quantizations of the classical symmetries of the theory). This actions gives rise to a representation of $\widehat{L\widetilde{G}^{\mathbb{C}}} \times \widehat{L\widetilde{G}^{\mathbb{C}}}$ in $\Gamma^a(L_t^{\mathbb{C}})$. Classifying the LW vectors for this representation, we shall recover the spectrum of the model derived originally by Gepner and Witten from the one for the $SU(2)$ case by use of the modular transformation properties of the characters of the $\widehat{su}(2)$ Kac-Moody algebra[5].

$$\text{Let } e_{1/2} = \begin{pmatrix} z^{1/2} & 0 \\ 0 & z^{-1/2} \end{pmatrix}\mathbb{Z}_2 \in L_tG^{\mathbb{C}}. \text{ Clearly, } L_tG^{\mathbb{C}} = L_oG^{\mathbb{C}}e_{1/2}.$$

We shall represent elements of $L_t^{\mathbb{C}}$ by probality amplitudes of the WZW model. Let $D'_\infty = \{z \in D_\infty \mid |z| \leq 2\}$. $S_2 = \{z \in D_\infty \mid |z| = 2\} \cong S^1$. Let $g : D'_\infty \to G^{\mathbb{C}}$ be such that $g\restriction_{S_2} = e_{1/2}$. Let $\ell \in L^{\mathbb{C}}_{e_{1/2}}$. Then $\ell e^{S_{D'_\infty}(g)}$ ($e^{S_{D'_\infty}(g)}$ is dual to $e^{-S_{D'_\infty}(g)}$) may be naturally considered as an element of $L^{\mathbb{C}}_{g\restriction\partial D_o}$ and every element in $L_t^{\mathbb{C}}$ can be described this way. If $g_1 : D'_\infty \to G^{\mathbb{C}}$, $g_1\restriction_{\partial D'_\infty} = g\restriction_{\partial D'_\infty}$, then $g_1 = gh$ where $h\restriction_{\partial D'_\infty} = 1$. We want to compare $\ell e^{S_{D'_\infty}(g_1)}$ and $\ell e^{S_{D'_\infty}(g)}$. Two cases may arise. If h lifts to $\widetilde{h} : D'_\infty \to \widetilde{G}^{\mathbb{C}}$, $\widetilde{h}\restriction_{\partial D'_\infty} = 1$, then there exists a homotopy h_t s.t. $h_o = 1$, $h_1 = h$ and $h_t\restriction_{\partial D'_\infty} = 1$. Polyakov's formula (5.11) applies and gives

$$\ell e^{S_{D'_\infty}(gh)} = e^{S_{D'_\infty}(h)-\Gamma_{D'_\infty}(g,h)}\ \ell e^{S_{D'_\infty}(g)} \quad . \tag{1}$$

If h lifts to $\tilde{h} : D'_\infty \to \tilde{G}^{\mathbb{C}}$ s.t. $\tilde{h}\restriction_{\partial D_\infty} = 1$, and $\tilde{h}\restriction_{S_2} = -1$ then there exists a homotopy h_t , $h_o = 1$, $h_1 = h$, s.t. $h_t\restriction_{\partial D_\infty} = 1$ and $h_t\restriction_{S_2} = e_{1/2}(e^{2\pi i t})$. In this case, using formula (A.4.6) (Appendix 4) we see that

$$\frac{d}{dt}[S_{D'_\infty}(gh_t)-S_{D'_\infty}(h_t)+\Gamma_{D'_\infty}(g,h_t)]$$
$$= \frac{ik}{4\pi}\int_{S_2} \mathrm{tr}\, h_t^{-1}\frac{\partial h_t}{\partial t}\, g^{-1}dg = -\frac{i\pi k}{2} \quad . \tag{2}$$

(Notice that forms ω_α and $\eta_{\alpha_0\alpha_1}$ given in Appendix 3 do not contribute to (2)). Thus in this case

$$\ell e^{S_{D'_\infty}(gh)} = (-1)^{k/2}\, e^{S_{D'_\infty}(h)-\Gamma_{D'_\infty}(g,h)}\ \ell e^{S_{D'_\infty}(g)} \quad . \tag{3}$$

Similarly,

$$\ell e^{S_{D'_\infty}(hg)} = ((-1)^{k/2})e^{S_{D'_\infty}(h)-\Gamma_{D'_\infty}(h,g)}\ \ell e^{S_{D'_\infty}(g)} \tag{4}$$

where $(-1)^{k/2}$ appears if h lifts to $\tilde{h}$ with $\tilde{h}\restriction_{\partial D_\infty} = 1$ and $\tilde{h}\restriction_{S_2} = -1$.

Before describing the action of $\widetilde{L\tilde{G}^{\mathbb{C}}}$ on $L_t^{\mathbb{C}}$, let us rewrite the one on $\tilde{L}^{\mathbb{C}}$ in a different form. If $g_1 : D_o \to \tilde{G}^{\mathbb{C}}$, $g : D_\infty \to \tilde{G}^{\mathbb{C}}$ and $\tilde{g}_1, \tilde{g}$ are their extensions to $\Sigma = P\mathbb{C}^1$ then

$$(\lambda_1 e^{-S_{D_o}(g_1)})\cdot(\lambda_2 e^{S_{D_\infty}(g)}) = \lambda_1\lambda_2\, e^{S_\Sigma(\tilde{g})}\, e^{-S_{D_o}(g_1)}\cdot e^{-S_{D_o}(\tilde{g})}$$
$$= \lambda_1\lambda_2 e^{S_\Sigma(\tilde{g})}\, e^{-S_{D_o}(g_1\tilde{g})-\Gamma_{D_o}(g_1,\tilde{g})}$$
$$= \lambda_1\lambda_2\, e^{S_\Sigma(\tilde{g})-S_\Sigma(\tilde{g}_1\tilde{g})-\Gamma_{D_o}(g_1,\tilde{g})}\, e^{S_{D_\infty}(\tilde{g}_1 g)}$$
$$= \lambda_1\lambda_2\, e^{-S_\Sigma(\tilde{g}_1)+\Gamma_{D_\infty}(\tilde{g}_1,g)}\, e^{S_{D_\infty}(\tilde{g}_1 g)} \quad . \tag{5}$$

Similarly,

$$(\lambda_2 e^{S_{D_\infty}(g)})\cdot(\lambda_1 e^{-S_{D_o}(g_1)}) = \lambda_1\lambda_2 e^{-S_\Sigma(\tilde{g}_1)+\Gamma_{D_\infty}(g,\tilde{g}_1)}\, e^{S_{D_\infty}(g\tilde{g}_1)} \quad . \tag{6}$$

Eqs. (5) and (6) may serve as a guide-line in the definition of the action of $\widetilde{L\tilde{G}^{\mathbb{C}}}$ on $L_t^{\mathbb{C}}$. Let $g_1, g_2 : D_o \to \tilde{G}^{\mathbb{C}}$, $\tilde{g}_1, \tilde{g}_2$ be their extensions to $D_o \cup D'_\infty$ s.t. $\tilde{g}_1, \tilde{g}_2\restriction_{S_2} = 1$ and $g : D'_\infty \to G^{\mathbb{C}}$, $g\restriction_{S_2} = e_{1/2}$. Let $\Sigma' = D_o \cup D'_\infty$. We shall define

$$(\lambda_1 e^{-S_{D_o}(g_1)})\cdot(\ell e^{S_{D'_\infty}(g)}) = \lambda_1 e^{-S_{\Sigma'}(\widetilde{g}_1)+\Gamma_{D'_\infty}(\widetilde{g}_1,g)}\ell e^{S_{D'_\infty}(\widetilde{g}_1 g)} \tag{7}$$

$$(\ell e^{S_{D'_\infty}(g)})\cdot(\lambda_2 e^{-S_{D_o}(g_2)}) = \lambda_2 e^{-S_{\Sigma'}(\widetilde{g}_2)+\Gamma_{D'}(g,\widetilde{g}_2)}\ell e^{S_{D'_\infty}(g\widetilde{g}_2)} \quad . \tag{8}$$

A straightforward verification shows that this defines commuting left and right actions of $\widehat{L\widetilde{G}^{\mathbb{C}}}$ on $L_t^{\mathbb{C}}$. Notice how in particular -1 acts on $L_t^{\mathbb{C}}$. If $\widetilde{g}_1$ prolongs -1 on D_o to D'_∞, $\widetilde{g}_1\restriction_{S_2} = 1$, then, using (4) and (3), we obtain

$$e^{-S_{D_o}(-1)}\cdot(\ell e^{S_{D'_\infty}(g)}) = e^{-S_{\Sigma'}(\widetilde{g}_1)+\Gamma_{D'_\infty}(\widetilde{g}_1,g)}\ell e^{S_{D'_\infty}(\widetilde{g}_1 g)}$$

$$= (-1)^{k/2} e^{-S_{\Sigma'}(\widetilde{g}_1)+S_{D'_\infty}(\widetilde{g}_1)+\Gamma_{D'_\infty}(\widetilde{g}_1,g)-\Gamma_{D'_\infty}(\widetilde{g}_1,g)}\ell e^{S_{D'_\infty}(g)}$$

$$= (-1)^{k/2}\ell e^{S_{D'_\infty}(g)} = (\ell e^{S_{D'_\infty}(g)})\cdot e^{-S_{D_o}(-1)} \quad . \tag{9}$$

The actions of $\widehat{L\widetilde{G}^{\mathbb{C}}}$ on $L_t^{\mathbb{C}}$ carry over to $\Gamma^a(L_t^{\mathbb{C}})$ inducing a representation $\ell\times r$ of $\widehat{L\widetilde{G}^{\mathbb{C}}}\times\widehat{L\widetilde{G}^{\mathbb{C}}}$ on the sector of twisted states :

$$(\ell(\hat{g}_1)r(\hat{g}_2)\psi)(g) = \hat{g}_1\cdot\psi(g_1^{-1}g g_2^{+-1})\cdot\hat{g}_2^{+} \quad . \tag{10}$$

This representation splits again into irreducible LW representations built over LW states satisfying condition (5.29). We shall find all LW states in $\Gamma^a(L_t^{\mathbb{C}})$. Notice, that due to (9), LW states can involve only integer spins if $k/2$ is even and only half-integer ones if $k/2$ is odd. In fact, the possible spins are more restricted.

As the LW condition of (5.29) fixes the states along orbits $N^+g(N^+)^+$ in $L_tG^{\mathbb{C}}$, we have to study the geometry of the latter. Under the mapping $L_tG^{\mathbb{C}} \ni g \mapsto ge_{1/2} \in L_oG^{\mathbb{C}}$, they become orbits N^+gM^- in $L_oG^{\mathbb{C}}$ where $M^- = \{g \in L^- : g(0) = \begin{pmatrix} * & 0 \\ * & * \end{pmatrix}\}$. Their classification requires an easy refinement of the classification of orbits N^+gL^- in $L\widetilde{G}^{\mathbb{C}}$ given in the previous sections. Namely, orbits N^+gM^- have the form

$$\Sigma_n^1 = N^+\varepsilon_n M^- \quad \text{and} \quad \Sigma_n^2 = N^+e_n M^- \tag{11}$$

where $\varepsilon_n = \begin{pmatrix} 0 & -z^n \\ z^{-n} & 0 \end{pmatrix}\mathbb{Z}_2$, $e_n = \begin{pmatrix} z^n & 0 \\ 0 & z^{-n} \end{pmatrix}\mathbb{Z}_2$ and n is integer (as is easy to see $\Sigma_n^1 \cup \Sigma_n^2 = \Sigma_n/\mathbb{Z}_2$, where Σ_n are the orbits discussed in Section 5). Let $N_o^{+1} = \{g \in N^+ | g(0) = \begin{pmatrix} 1 & 0 \\ * & 1 \end{pmatrix}\}$, $N_n^{+1} = N^+ \cap e_n N_o^{+1} e_n^{-1}$, $N_n^{+2} = N^+ \cap \varepsilon_n N_o^{+1}\varepsilon_n^{-1}$. Then $\Sigma_n^1 = N_n^{+1}\varepsilon_n M^-$ and $\Sigma_n^2 = N_n^{+2}e_n M^-$ and these identities lead to unique decompositions of the elements of the orbits. Moreover, we have splittings :

$$e_n N_o^{+1} e_n^{-1} = N_n^{+1} N_n^{-1} \; , \; \varepsilon_n N_o^{+1} \varepsilon_n^{-1} = N_n^{+2} N_n^{-2} \tag{12}$$

where $N_n^{-1} = N^- \cap e_n N_o^{+1} e_n^{-1}$ and $N_n^{-2} = N^- \cap \varepsilon_n N_o^{+1} \varepsilon_n^{-1}$. Σ_o^1 is an open dense orbit. Since

$$U_n^1 = e_n \Sigma_o^1 = e_n N_o^{+1} \varepsilon_o M^- = N_n^{+1} N_n^{-1} \varepsilon_n M^- \quad , \tag{13}$$

$$U_n^2 = \varepsilon_n \Sigma_o^1 = N_n^{+2} N_n^{-2} e_n M^- \quad , \tag{14}$$

the codimension of Σ_n^1 is equal $\dim N_n^{-1} = \begin{cases} 2n \; , \; n > 0 \\ -2n \; , \; n \leq 0 \end{cases}$ and the codimension of Σ_n^2 equals $\dim N_n^{-2} = \begin{cases} 2n-1 \; , \; n > 0 \\ -2n+1 \; , \; n \leq 0 \end{cases}$. Besides

$$U_n^1 \setminus \Sigma_n^1 \subset \begin{cases} \bigcup_{|m|<n} (\Sigma_m^1 \cup \Sigma_m^2) \cup \Sigma_n^2 \; , \; n > 0 \; , \\ \bigcup_{n<m\leq -n} (\Sigma_m^1 \cup \Sigma_m^2) \; , \; n \leq 0 \; , \end{cases} \tag{15}$$

and

$$U_n^2 \setminus \Sigma_n^2 \subset \begin{cases} \bigcup_{|m|<n} (\Sigma_m^1 \cup \Sigma_m^2) \quad , \quad n > 0 \; , \\ \bigcup_{n<m\leq -n} (\Sigma_m^1 \cup \Sigma_m^2) \cup \Sigma_n^1 \quad , \quad n < 0 \quad , \end{cases} \tag{16}$$

which establishes the hierarchy of the orbits. Proofs of the facts listed above are elementary.

Let us use relation (5.29) to compute a LW state with spins j_1 and j_2 on the open dense orbit $\Sigma_o^1 e_{1/2}^{-1} = N^+ \varepsilon_{1/2} (N^+)^\dagger$ of $N^+ \times N^+$ in $LG_t^{\mathbb{C}}$. Let $\tilde{\varepsilon}_{1/2} : D'_\infty \to G^{\mathbb{C}}$ be such that $\tilde{\varepsilon}_{1/2}\restriction_{\partial D_o} = \varepsilon_{1/2}$, $\tilde{\varepsilon}_{1/2}\restriction_{S_2} = e_{1/2}$. For a LW state ψ , represent

$$\psi(\varepsilon_{1/2}) = \ell e^{S_{D'_\infty}(\tilde{\varepsilon}_{1/2})} \tag{17}$$

for $\ell \in L_{e_{1/2}}$. Then for $g_1, g_2 \in N^+$, $g_1(0) = \begin{pmatrix} a_1 & 0 \\ * & a_1^{-1} \end{pmatrix}$, $g_2(0) = \begin{pmatrix} a_2 & 0 \\ * & a_2^{-1} \end{pmatrix}$

$$\begin{aligned} \psi(g_1 \varepsilon_{1/2} g_2^\dagger) &= a_1^{2j_1} \bar{a}^{2j_2} (\ell(g_1) r(g_2) \psi)(g_1 \varepsilon_{1/2} g_2^\dagger) \\ &= a_1^{2j_1} \bar{a}_2^{2j_2} g_1 \cdot (\ell e^{S_{D'_\infty}(\tilde{\varepsilon}_{1/2})}) \cdot g_2^\dagger \\ &= a_1^{2j_1} \bar{a}_2^{2j_2} e^{-S_{\Sigma'}(\tilde{g}_1) + \Gamma_{D'_\infty}(\tilde{g}_1, \tilde{\varepsilon}_{1/2})} (\ell e^{S_{D'_\infty}(\tilde{g}_1 \tilde{\varepsilon}_{1/2})}) \cdot g_2^\dagger \\ &= a_1^{2j_1} \bar{a}^{2j_2} e^{-S_{\Sigma'}(\tilde{g}_1) - S_{\Sigma'}(\tilde{g}_2^\dagger) + \Gamma_{D'_\infty}(\tilde{g}_1, \tilde{\varepsilon}_{1/2}) + \Gamma_{D'_\infty}(\tilde{g}_1 \tilde{\varepsilon}_{1/2}, \tilde{g}_2^\dagger)} \\ &\qquad \ell e^{S_{D'_\infty}(\tilde{g}_1 \tilde{\varepsilon}_{1/2} g_2^\dagger)} \end{aligned} \tag{18}$$

where $\widetilde{g}_i$ extends g_i to Σ' so that $\widetilde{g}_i\restriction_{S_2} = 1$, $i = 1,2$. For (18) to determine ψ on the dense open orbit $\Sigma^1_o e^{-1}_{1/2}$ it is necessary and sufficient that the right hand side does not change when $g_1 \mapsto g_1\begin{pmatrix} a & 0 \\ 0 & a^{-1}\end{pmatrix}$, $g_2 \mapsto \pm g_2\begin{pmatrix} \bar{a} & 0 \\ 0 & \bar{a}^{-1}\end{pmatrix}$. As the action of -1 on the sections was already studied, we can consider only the + sign case. Using equation (5.13), we obtain after an easy algebra

$$\frac{d}{da}\,[S_{D'_\infty}(\widetilde{g}_1\widetilde{\varepsilon}_{1/2}\widetilde{g}_2^{\dagger})-S_{\Sigma'}(\widetilde{g}_1)-S_{\Sigma'}(\widetilde{g}_2^{\dagger})+\Gamma_{D'_\infty}(\widetilde{g}_1,\widetilde{\varepsilon}_{1/2})+\Gamma_{D'_\infty}(\widetilde{g}_1\widetilde{\varepsilon}_{1/2},\widetilde{g}_2^{\dagger})]$$

$$= -\frac{ik}{2\pi}\int_{\partial D_o} \operatorname{tr} g_1^{-1}\frac{dg_1}{da}\,\varepsilon_{1/2}d\varepsilon_{1/2}^{-1} = -\frac{k}{a}\,. \tag{19}$$

On the other hand,

$$\frac{d}{da}\log a_1^{2j_1}\bar{a}_2^{2j_2} = 2(j_1+j_2)\,\frac{1}{a} \tag{20}$$

so that the right hand side of (18) is a-independent if and only if

$$j_1+j_2 = \frac{k}{2}\,. \tag{21}$$

For a pair of spins satisfying this condition, integer for even $\frac{k}{2}$ and half-integer for odd $\frac{k}{2}$, eq. (18) defines a holomorphic section of $L^{\mathbb{C}}$ over $\Sigma^1_o e^{-1}_{1/2}$ up to a constant factor. Remains to be seen if we can extend this section to a holomorphic section over entire $L_t G^{\mathbb{C}}$. Obstructions can arise only when extending the section to orbits of codimension 1 , i.e. to $\Sigma^2_o e^{-1}_{1/2}$ and $\Sigma^2_1 e^{-1}_{1/2}$.

By eq. (14), any element of $U^2_o e^{-1}_{1/2}$ may be uniquely written as $g'_1\begin{pmatrix} 1 & b \\ 0 & 1\end{pmatrix} e^{-1}_{1/2}\, g'^{\dagger}_2$ where $g'_1 \in N^{+2}_o$, $g'_2 \in N^+$. It belongs to $\Sigma^2_o e^{-1}_{1/2} \subset U^2_o e^{-1}_{1/2}$ if and only if $b = 0$, for $b \neq 0$ being an element of $\Sigma^1_o e^{-1}_{1/2}$. Indeed,

$$\begin{pmatrix} 1 & b \\ 0 & 1\end{pmatrix} e^{-1}_{1/2} = \begin{pmatrix} 1 & 0 \\ b^{-1} & 1\end{pmatrix}\varepsilon_{1/2}\begin{pmatrix} -\bar{b}^{-1} & -z \\ 0 & -\bar{b}\end{pmatrix}^{\dagger} \equiv g_{1b}\varepsilon_{1/2}g^{\dagger}_{2b}\,. \tag{22}$$

Now, $\psi(g'_1 g_{1b}\varepsilon_{1/2}g^{\dagger}_{2b}g'^{\dagger}_2) \equiv \psi(g_b)$ is given by the right hand side of (18) with $g'_1 g_{1b} \mapsto g_1$, $g'_2 g_{2b} \mapsto g_2$. Writing it in a somewhat more convenient way, we obtain

$$\psi(g_b) = a_1^{2j_1}\bar{a}_2^{2j_2}\, e^{-S_{\Sigma'}(\widetilde{g}_{1b})-S_{\Sigma'}(\widetilde{g}^{\dagger}_{2b})+\Gamma_{D'_\infty}(\widetilde{g}_{1b},\widetilde{\varepsilon}_{1/2})+\Gamma_{D'_\infty}(\widetilde{g}_{1b}\widetilde{\varepsilon}_{1/2},\widetilde{g}^{\dagger}_{2b})}$$

$$g'_1\cdot(\ell e^{S_{D'_\infty}(\widetilde{g}_{1b}\widetilde{\varepsilon}_{1/2}\widetilde{g}^{\dagger}_{2b})})\cdot g'^{\dagger}_2 \tag{23}$$

where $g'_1(0) = \begin{pmatrix} a_1 & 0 \\ * & a_1^{-1}\end{pmatrix}$ and $g'_2(0)\begin{pmatrix} -\bar{b}^{-1} & 0 \\ 0 & -\bar{b}\end{pmatrix} = \begin{pmatrix} a_2 & 0 \\ * & a_2^{-1}\end{pmatrix}$. Hence

$a_1^{2j_1} a_2^{-2j_2} = \mathcal{O}(b^{-2j_2})$ when $b \to 0$. To study the asymptotics of the exponential expression on the right hand side of (23), we apply formula (A.4.6) to infer that

$$\frac{d}{db}\,[S_{D'_\infty}(\tilde{g}_{1b}\tilde{\varepsilon}_{1b}\tilde{g}_{2b}^{+}) - S_{\Sigma'}(\tilde{g}_{1b}) - S_{\Sigma'}(\tilde{g}_{2b}^{+}) + \Gamma_{D'_\infty}(\tilde{g}_{1b},\tilde{\varepsilon}_{1b}) + \Gamma_{D'_\infty}(\tilde{g}_{1b}\tilde{\varepsilon}_{1/2},\tilde{g}_{2b}^{+})]$$

$$= -\frac{ik}{2\pi}\int_{\partial D_o} \mathrm{tr}\,[\,\frac{\tilde{g}_{2b}^{+-1}}{\partial b}\,\bar{\partial}\tilde{g}_{2b}^{+}\tilde{g}_{1b}^{-1}\,\frac{\partial \tilde{g}_{1b}}{\partial b}\,\tilde{\varepsilon}_{1/2}\tilde{g}_{2b}^{+}\,d(g_{2b}^{+-1}\,\varepsilon_{1/2}^{-1})]$$

$$+\ \mathcal{O}(1) = \frac{k}{b} + \mathcal{O}(1) \tag{24}$$

so that

$$\psi(g_b) = \mathcal{O}(b^{k-2j_2}) \tag{25}$$

and $\psi(g_b)$ is analytic at $b = 0$ since $j_2 \le \frac{k}{2}$.

Similarly, every element of $U_1^2 e_{1/2}^{-1}$ may be uniquely written as

$$g_1'\begin{pmatrix} 1 & 0 \\ cz^{-1} & 1 \end{pmatrix} e_{1/2}g_2'^{+}$$ where $g_1' \in N_1^{+2}$ and $g_2' \in N^{+}$. It belongs to $\Sigma_1^2 e_{1/2}^{-1}$ if and only if $c = 0$, for $c \neq 0$ being an element of $\Sigma_o^1 e_{1/2}^{-1}$. Since

$$\begin{pmatrix} 1 & 0 \\ cz^{-1} & 1 \end{pmatrix} e_{1/2} = \begin{pmatrix} 1 & \frac{z}{c} \\ 0 & 1 \end{pmatrix} \varepsilon_{1/2} \begin{pmatrix} \bar{c} & 0 \\ 1 & \frac{1}{\bar{c}} \end{pmatrix}^{+} \equiv g_{1c}\varepsilon_{1/2}g_{2c}^{+}, \tag{26}$$

$\psi(g_1'g_{1c}\varepsilon_{1/2}g_{2c}^{+}g_2'^{+}) \equiv \psi(g_c)$ is given by the right hand side of (23) with $c \to b$ and a_1, a_2 defined by relations $g_1'(0) = \begin{pmatrix} a_1 & 0 \\ * & a_1^{-1} \end{pmatrix}$, $g_2'(0)\begin{pmatrix} \bar{c} & 0 \\ 1 & \frac{1}{\bar{c}} \end{pmatrix} = \begin{pmatrix} a_2 & 0 \\ * & a_2^{-1} \end{pmatrix}$. Thus $a_1^{2j_1} a_2^{-2j_2} = \mathcal{O}(c^{2j_2})$. On the other hand, the left equality of (24) still holds (with $c \mapsto b$) and produces a regular contribution. Hence

$$\psi(g_c) = \mathcal{O}(c^{2j_2}) \tag{27}$$

and ψ is analytic at $c = 0$. By induction based on the Hartogs theorem, ψ extends now to a global holomorphic section, for which (5.29) follows by continuity.

Summarizing, in the twisted sector the LW states correspond with multiplicity 1, to spins j_1, j_2, $j_1 + j_2 = \frac{k}{2}$, j_1, j_2 integer for $\frac{k}{2}$ even and half-integers if $\frac{k}{2}$ is odd. This coincides with the Gepner-Witten result[5].

For both SU(2) and SO(3) case, the spectrum of the LW states found here gives immediately modular invariant partition functions of the theory on the torus. These are the two infinite series from all possible modular invariants built from bilinears in $\widehat{su}(2)$ Kac-Moody algebra characters, see[26,27].

APPENDIX 1

We shall describe explicitly the magnetic-monopole line bundle. The monopole curvature 2-form

$$\omega = \frac{1}{2} e\mu \sum_{i,j,k=1}^{3} \varepsilon^{ijk} \frac{x^i}{|\vec{x}|^3} dx^j dx^k , \tag{1}$$

see (1.3), extends to a (2,0)-form $\omega^{\mathbb{C}}$ on $M^{\mathbb{C}} = \{\vec{y} \in \mathbb{C}^3 | \vec{y}^{\,2} \notin]-\infty,0]\}$ given by (1) with $\vec{y} \mapsto \vec{x}$ and $|\vec{y}| \equiv \sqrt{\vec{y}^{\,2}}$ (with the square root cut along the negative axis). Cover $M^{\mathbb{C}}$ by $\{\mathcal{O}_{-1}, \mathcal{O}_o, \mathcal{O}_1\}$,

$$\begin{aligned} \mathcal{O}_{-1} &= \{\vec{y} \mid \mathrm{Re}\, \frac{y_1}{|\vec{y}|} > 0\} , \\ \mathcal{O}_0 &= \{\vec{y} \mid \frac{y_2^2+y_3^2}{\vec{y}^{\,2}} \notin]-\infty,0]\} , \\ \mathcal{O}_1 &= \{\vec{y} \mid \mathrm{Re}\, \frac{y_1}{|\vec{y}|} < 0\} . \end{aligned} \tag{2}$$

For $e\mu \in \frac{1}{2}\mathbb{Z}$, put

$$\begin{aligned} \eta_{-1} &= e\mu\left(\frac{y_1}{|\vec{y}|} -1\right) \frac{y_3 dy_2 - y_2 dy_3}{y_2^2+y_3^2} = \eta_o , \\ \eta_1 &= e\mu \left(\frac{y_1}{|\vec{y}|} +1\right) \frac{y_3 dy_2 - y_2 dy_3}{y_2^2+y_3^2} , \end{aligned} \tag{3}$$

$$g_{-10} = 1 , \; g_{01} = \left(\frac{i y_2 + y_3}{\sqrt{\frac{y_2^2+y_3^2}{\vec{y}^{\,2}}}\; |\vec{y}|} \right)^{2e\mu} .$$

$(g_{\alpha_0\alpha_1}, \eta_\alpha)$ defines the unique element of $Q^a(M^{\mathbb{C}},\omega^{\mathbb{C}})$ and, by restriction to $\mathbb{R}^3 \setminus \{0\}$, the unique element of $Q(M,\omega)$, i.e. the monopole line bundle.

APPENDIX 2

The group of orientation preserving reparametrizations of S^1, Diff_+S^1, acts on the loop space LM by $\phi \mapsto \phi\circ\tau^{-1}$, $\tau \in \mathrm{Diff}_+S^1$. The canonical isomorphism of the fibers of the line bundle L over loops ϕ and $\phi\circ\tau^{-1}$ may be used to lift this action to the action of Diff_+S^1 on L preserving the hermitian structure and the connection given by (3.10). Here we shall show that the connection projects from L to the line bundle L/Diff_+S^1 over LM/Diff_+S^1. One has to check that if (τ_t) is a curve in Diff_+S^1, $\tau_o = \mathrm{id}$, then its lift $\ell(t) = \tau(t)\ell_o$ to L, $\ell_o \in L$, is horizontal.

Let in local presentation

$$\ell_o = (\mathcal{U}_A,\phi,z) \tag{1}$$

where $\mathcal{U}_A$ is given by (3.8), $\phi \in \mathcal{U}_A \subset LM$ and $z \in \mathbb{C}$. Then

$$\ell(t) = (\mathcal{U}_{A_t},\phi\circ\tau_t^{-1},z) \tag{2}$$

where if A corresponds to the triangulation of S^1 by intervals b and vertices v and to assignment α_b,α_v then A_t is related to the triangulation by $\tau_t(b)$ and $\tau_t(v)$ with the unchanged assignment of α's. Changing the trivialization by the transition function G_{A_tA} given by (3.9), we obtain

$$\ell(t) = (\mathcal{U}_A,\phi\circ\tau_t^{-1},G_{A_tA}(\phi\circ\tau_t^{-1})z)\,. \tag{3}$$

We have to show that the covariant derivative

$$\frac{D}{dt}\,\ell(t) = [G_{A_tA}(\phi\circ\tau_t^{-1})^{-1}\frac{d}{dt}G_{A_tA}(\phi\circ\tau_t^{-1}) - i\langle X_t,E_A\rangle]\,\ell(t) = 0 \tag{4}$$

where

$$X_t = \frac{\partial}{\partial t}(\phi\circ\tau_t^{-1})\,, \tag{5}$$

see (3.10). It is enough to show (4) for $t = 0$. For simplicity, let us assume that $\tau_t(v) > v$ for each vertex v of the triangulation of S^1 (in the natural order on the circle). The general case proceeds the same way. Denote by b_v^+ and b_v^- two intervals b bordering on v, b_v^+ later than b_v^-, see Fig. 4. From eq. (3.9), we compute

$$G_{A_tA}(\phi\circ\tau_t^{-1}) = \exp[i\sum_v \int_v^{\tau_t(v)} (\phi\circ\tau_t^{-1})^*\eta_{\alpha_{b_v^-}-\alpha_{b_v^+}}] \cdot \prod_v \frac{g_{\alpha_v\alpha_{b_v^-}\alpha_{b_v^+}}(\phi\circ\tau_t^{-1}(v))}{g_{\alpha_v\alpha_{b_v^-}\alpha_{b_v^+}}(\phi(v))}\,. \tag{6}$$

On the other hand, by definition (3.10),

$$\langle X_o, E_A\rangle = -\sum_b \int_b \frac{\partial\tau_o}{\partial t} \lrcorner\, \phi^*\omega_{\alpha_b} - \sum_{\substack{v,b\\ v\in\partial b}} \langle \frac{\partial\tau_o}{\partial t}(v), \phi^*\eta_{\alpha_v\alpha_b}\rangle$$

$$= \sum_v \langle \frac{\partial\tau_o}{\partial t}(v), \phi^*(\eta_{\alpha_v\alpha_{b_v^+}} - \eta_{\alpha_v\alpha_{b_v^-}})\rangle \quad . \qquad (7)$$

Thus

$$G_{A_tA}(\phi)^{-1}\frac{d}{dt}\, G_{A_oA}(\phi\circ\tau_o^{-1}) - i\langle X_o, E_A\rangle$$

$$= i\sum_v \langle \frac{\partial\tau_o}{\partial t}(v), \phi^*(\eta_{\alpha_{b_v^-}\alpha_{b_v^+}} - \frac{1}{i}\, g^{-1}_{\alpha_v\alpha_{b_v^-}\alpha_{b_v^+}}\, dg_{\alpha_v\alpha_{b_v^-}\alpha_{b_v^+}}$$

$$- \eta_{\alpha_v\alpha_{b_v^+}} + \eta_{\alpha_v\alpha_{b_v^-}})\rangle = 0 \qquad (8)$$

by (3.3).

APPENDIX 3

In order to describe a representative of $w \in W^a(SL(2,\mathbb{C}), \frac{k}{12\pi}\,tr(g^{-1}dg)^3 \equiv \gamma^{\mathbb{C}})$ let us parametrice

$$SL(2,\mathbb{C}) = \{z_o + i\sum_{i=1}^{3} z_i\sigma_i \mid z_o, z_i \in \mathbb{C} \ , \ z_o^2 + z^2 = 1\} \ . \qquad (1)$$

We shall use the (stereographic) coordinates $\vec\xi, \vec\zeta$ on the subsets $z_o \notin\,]-\infty,-1]$ and $z_o \notin [1,\infty[$ respectively, $\vec\xi, \vec\zeta \in \mathbb{C}^3$, $\vec\xi^{\,2}, \vec\zeta^{\,2} \notin\,]-\infty,-1]$,

$$\frac{1-\vec\xi^{\,2}}{1+\vec\xi^{\,2}} = z_o = \frac{\vec\zeta^{\,2}-1}{\vec\zeta^{\,2}+1} \ , \qquad (2)$$

$$\frac{2\xi_i}{1+\vec\xi^{\,2}} = z_i = \frac{2\zeta_i}{\vec\zeta^{\,2}+1} \ . \qquad (3)$$

An easy computation gives

$$\gamma^{\mathbb{C}} = \frac{8k}{\pi}\,\frac{1}{(1+\vec\xi^{\,2})^3}\, d\xi^1 d\xi^2 d\xi^3 = -\frac{8k}{\pi}\,\frac{1}{(\vec\zeta^{\,2}+1)^3}\, d\zeta^1 d\zeta^2 d\zeta^3 \ . \qquad (4)$$

Cover $SL(2,\mathbb{C})$ by

$$\mathcal{O}_{-2} = \{|\vec\xi^{\,2}| < 1\} \ , \quad \mathcal{O}_{-1} = \{\mathrm{Re}\,\frac{\xi_1}{|\vec\xi|} > 0\} \ ,$$

$$O_0 = \{\frac{\xi_2^2+\xi_3^2}{\vec{\xi}^2} \notin]-\infty,0]\} \quad , \tag{5}$$

$$O_1 = \{\mathrm{Re}\,\frac{\xi_1}{|\vec{\xi}|} < 0\} \quad , \quad O_{-2} = \{|\vec{\zeta}^{\,2}| < 1\}$$

where $|\vec{\xi}| = \sqrt{\vec{\xi}^{\,2}}$. Let

$$\omega = \int_0^1 \frac{4kt^2}{\pi(t^2\,\vec{\xi}^2+1)}\,dt \sum_{i,j,k} \varepsilon^{ijk}\xi^i d\xi^j d\xi^k \tag{6}$$

and ω' be given by the same formula with $\vec{\zeta} \to \vec{\xi}$. Notice that they are well defined for $\vec{\xi}^2, \vec{\zeta}^2 \notin]-\infty,-1]$ and that $d\omega = \gamma = -d\omega'$ there. Define for integer k

$$\omega_{-2} = \omega \,, \quad \omega_{-1} = \omega \,, \quad \omega_0 = \omega \,, \quad \omega_1 = \omega \quad , \quad \omega_2 = -\omega' \,,$$

$$\eta_{-2,-1} = 0 \,, \eta_{-2,0} = 0 \,, \; \eta_{-2,1} = 0 \,, \; \eta_{-1,0} = 0 \,, \; \eta_{0,1} = 0 \,,$$

$$\eta_{-1,2} = \frac{k}{2}\,(\frac{\xi_1}{|\vec{\xi}|} - 1)\,\frac{\xi_3 d\xi_2 - \xi_2 d\xi_3}{\xi_2^2+\xi_3^2} = \eta_{0,2} \,,$$

$$\eta_{1,2} = \frac{k}{2}\,(\frac{\xi_1}{|\vec{\xi}|} + 1)\,\frac{\xi_3 d\xi_2 - \xi_2 d\xi_3}{\xi_2^2+\xi_3^2} \quad , \tag{7}$$

$$g_{-2,-1,0} = 1 \,, \quad g_{-2,0,1} = 1 \,, \quad g_{-1,0,2} = 1 \,,$$

$$g_{0,1,2} = \left(\frac{i\xi_2+\xi_3}{\sqrt{\frac{\xi_2^2+\xi_3^2}{\vec{\xi}^2}}\;|\vec{\xi}|}\right)^k \,.$$

It is easy to check that $(g_{\alpha_0\alpha_1\alpha_2}, \eta_{\alpha_0\alpha_1}, \omega_\alpha)$ determines the unique element of $W^a(SL(2,\mathbb{C}),\gamma^{\mathbb{C}})$ and, by restriction to $SU(2)$ (i.e. to $\vec{\xi}$ and $\vec{\zeta}$ real), the unique element of $W(SU(2),\gamma)$. Notice the relation between (7) and the local data of the monopole bundle of Appendix 1.

For even k , define another representative of $w \in W^a(SL(2,\mathbb{C}),\gamma^{\mathbb{C}})$ by putting

$$\tilde{\omega}_{-2} = \omega \,, \quad \tilde{\omega}_2 = -\omega' \quad ,$$

$$\tilde{\omega}_{-1} = \omega + \frac{k}{8|\vec{\xi}|^3} \sum_{i,j,k} \varepsilon^{ijk}\xi^i d\xi^j d\xi^k = \tilde{\omega}_1 = \tilde{\omega}_0 \quad ,$$

$$\tilde{\eta}_{-2,-1} = \frac{k}{4}\,(\frac{\xi_1}{|\vec{\xi}|} - 1)\,\frac{\xi_3 d\xi_2 - \xi_2 d\xi_3}{\xi_2^2+\xi_3^2} = \tilde{\eta}_{-2,0} = \tilde{\eta}_{-1,2} \quad ,$$

$$\tilde{\eta}_{-2,1} = \frac{k}{4}\left(\frac{\xi_1}{|\vec{\xi}|}+1\right)\frac{\xi_3 d\xi_2 - \xi_2 d\xi_3}{\xi_2^2+\xi_3^2} = \tilde{\eta}_{0,2} \quad ,$$

$$\qquad\qquad\qquad\qquad \| $$

$$\qquad\qquad\qquad\qquad \tilde{\eta}_{1,2}$$

$$\tilde{\eta}_{-1,0} = 0 = \tilde{\eta}_{0,1} \quad ,$$

$$\tilde{g}_{-2,-1,0} = 1 = \tilde{g}_{0,1,2} \quad ,$$

$$\tilde{g}_{-2,0,1} = \left(\frac{i\xi_2 - \xi_3}{\sqrt{\frac{\xi_2^2+\xi_3^2}{\vec{\xi}^2}}\ |\vec{\xi}|} \right)^{k/2} \quad ,$$

$$\tilde{g}_{-1,0,2} = \left(\frac{i\xi_2 + \xi_3}{\sqrt{\frac{\xi_2^2+\xi_3^2}{\vec{\xi}^2}}\ |\vec{\xi}|} \right)^{k/2} \quad .$$

The multiplication by -1 in $SL(2,\mathbb{C})$ corresponds to $\vec{\xi} \to -\frac{\vec{\xi}}{\vec{\xi}^2}$, $\vec{\zeta} \to -\frac{\vec{\zeta}}{\vec{\zeta}^2}$. A straightforward check shows that $(\tilde{g},\tilde{\eta},\tilde{\omega})$ is invariant under this transformation so that it defines the unique element in $W^a(SL(2,\mathbb{C})/\mathbb{Z}_2,\gamma^{\mathbb{C}})$ and in $W(SO(3),\gamma)$.

APPENDIX 4

We shall compute here $\frac{d}{dt} S_\Sigma(g_t)$ for a t-dependent map $g_t : \Sigma \to G^{\mathbb{C}}$ and Σ any oriented 2-dimensional compact surface with boundary. First, we shall differentiate the $A_\Sigma(g_t)$ part of the amplitude, see (5.2), as given by formula (3.6)

$$-\frac{d}{dt}\log A_\Sigma(g_t) = \frac{d}{dt}\Big[-i\sum_c \int_c g_t^*\omega_{\alpha_c} + i \sum_{\substack{b,c \\ b\subset\partial c}} \int_b g_t^*\eta_{\alpha_b\alpha_c}\Big]$$

$$- \sum_{\substack{v,b,c \\ v\in\partial b \\ b\subset\partial c}} g_{\alpha_v\alpha_b\alpha_c}(g_t(v))^{-1}\frac{d}{dt} g_{\alpha_v\alpha_b\alpha_c}(g_t(v)) \quad . \tag{1}$$

Let $\tilde{g}_t : \Sigma \to \Sigma \times G^{\mathbb{C}}$ be obtained from g_t by putting $\tilde{g}_t(\xi) = (\xi, g_t(\xi))$. Denote by $\tilde{X}_t$ any vector field tangent to $\Sigma \times G^{\mathbb{C}}$ s.t. $X_t(\tilde{g}_t(\xi)) = \frac{d}{dt}\tilde{g}_t(\xi)$. For a form χ on $G^{\mathbb{C}}$ denote by $\tilde{\chi}$ its pull-back to $\Sigma \times G^{\mathbb{C}}$ by the projection on the second factor. We have

$$\frac{d}{dt}\int g_t^*\chi = \frac{d}{dt}\int \tilde{g}_t^*\tilde{\chi} = \int g_t^* L_{\tilde{X}_t}\tilde{\chi} \quad . \tag{2}$$

where $L_{\widetilde{X}_t}\widetilde{\chi} = \widetilde{X}_t \lrcorner d\widetilde{\chi} + d(\widetilde{X}_t \lrcorner \widetilde{\chi})$ is the Lie derivative of $\widetilde{\chi}$ with respect to $\widetilde{X}_t$. Hence

$$-\frac{d}{dt}\log A_\Sigma(g_t) = -i\sum_c \int_c \widetilde{g}_t^* L_{\widetilde{X}_t}\widetilde{\omega}_{\alpha_c} + i\sum_{b,c}\int_b \widetilde{g}_t^* L_{\widetilde{X}_t}\widetilde{\eta}_{\alpha_b\alpha_c}$$

$$- \sum_{v,b,c} \langle \widetilde{X}_t(\widetilde{g}_t(v)), \widetilde{g}^{-1}_{\alpha_v\alpha_b\alpha_c}\, d\widetilde{g}_{\alpha_v\alpha_b\alpha_c}\rangle$$

$$= -i\sum_c \int_c \widetilde{g}_t^*(\widetilde{X}_t \lrcorner \widetilde{\gamma}^{\mathbb{C}}) - i\sum_c \int_c \widetilde{g}_t^* d(\widetilde{X}_t \lrcorner \widetilde{\omega}_{\alpha_c})$$

$$+ i\sum_{b,c}\int_b \widetilde{g}_t^*(\widetilde{X}_t \lrcorner d\widetilde{\eta}_{\alpha_b\alpha_c}) + i\sum_{b,c}\int_b \widetilde{g}_t^* d(\widetilde{X}_t \lrcorner \widetilde{\eta}_{\alpha_b\alpha_c})$$

$$- \sum_{v,b,c} \langle \widetilde{X}_t(\widetilde{g}_t(v)), \widetilde{g}^{-1}_{\alpha_v\alpha_b\alpha_c}\, dg_{\alpha_v\alpha_b\alpha_c}\rangle$$

$$= -i\int_\Sigma \widetilde{g}_t^*(X_t \lrcorner \widetilde{\gamma}^{\mathbb{C}}) - i\sum_{b,c}\int_b \widetilde{g}_t^*(\widetilde{X}_t \lrcorner (\widetilde{\omega}_{\alpha_c} - d\widetilde{\eta}_{\alpha_b\alpha_c}))$$

$$+ i\sum_{v,b,c} \langle \widetilde{X}_t(\widetilde{g}_t(v)), \widetilde{\eta}_{\alpha_b\alpha_c} - \frac{1}{i} g^{-1}_{\alpha_v\alpha_b\alpha_c}\, dg_{\alpha_v\alpha_b\alpha_c}\rangle$$

$$= -i\int_\Sigma \widetilde{g}_t^*(\widetilde{X}_t \lrcorner \widetilde{\gamma}^{\mathbb{C}}) - i\sum_{b,c}\int_b \widetilde{g}_t^*(\widetilde{X}_t \lrcorner (\widetilde{\omega}_{\alpha_c} - \widetilde{\omega}_{\alpha_b} - d\widetilde{\eta}_{\alpha_b\alpha_c}))$$

$$-i\sum_{b\subset\partial\Sigma}\int_b \widetilde{g}_t^*(\widetilde{X}_t \lrcorner \widetilde{\omega}_{\alpha_b})$$

$$+ i\sum_{v,b,c} \langle \widetilde{X}_t(\widetilde{g}_t(v)), \widetilde{\eta}_{\alpha_b\alpha_c} - \widetilde{\eta}_{\alpha_v\alpha_c} + \widetilde{\eta}_{\alpha_v\alpha_b} - \frac{1}{i} g^{-1}_{\alpha_v\alpha_b\alpha_c}\, dg_{\alpha_v\alpha_b\alpha_c}\rangle$$

$$-i\sum_{\substack{v,b\\ v\in\partial b\\ b\subset\partial\Sigma}} \langle \widetilde{X}_t(\widetilde{g}_t(v)), \widetilde{\eta}_{\alpha_v\alpha_b}\rangle$$

$$= -i\int_\Sigma \widetilde{g}_t^*(\widetilde{X}_t \lrcorner \widetilde{\gamma}^{\mathbb{C}}) - i\sum_{b\subset\partial\Sigma}\int_b \widetilde{g}_t^*(\widetilde{X}_t \lrcorner \widetilde{\omega}_{\alpha_b})$$

$$-i\sum_{\substack{v,b\\ v\in\partial b,\, b\subset\partial\Sigma}} \langle \widetilde{X}_t(\widetilde{g}_t(v)), \widetilde{\eta}_{\alpha_v\alpha_b}\rangle \tag{3}$$

where we have used the defining properties of $(g_{\alpha_0\alpha_1\alpha_2}, \eta_{\alpha_0\alpha_1}, \omega_\alpha)$. Now

$$-i\int_\Sigma \widetilde{g}_t^*(\widetilde{X}_t \lrcorner \widetilde{\gamma}^{\mathbb{C}}) = -\frac{ik}{4\pi}\int \mathrm{tr}(g_t^{-1}\frac{\partial g_t}{\partial t})(g_t^{-1}dg_t)^2$$

$$= \frac{ik}{4\pi}\int \mathrm{tr}(g_t^{-1}\frac{\partial g_t}{\partial t})d(g_t^{-1}dg_t)$$

$$= \frac{ik}{4\pi}\int_{\partial\Sigma}\mathrm{tr}(g_t^{-1}\frac{\partial g_t}{\partial t})(g_t^{-1}dg_t) - \frac{ik}{4\pi}\int_\Sigma \mathrm{tr}\,\partial(g_t^{-1}\frac{\partial g_t}{\partial t})(g_t^{-1}\bar{\partial} g_t)$$

$$- \frac{ik}{4\pi}\int \mathrm{tr}\,\bar{\partial}(g_t^{-1}\frac{\partial g_t}{\partial t})(g_t^{-1}\partial g_t) \quad . \tag{4}$$

The contribution of the sigma model action $S_\Sigma(g_t)$ (see (5.2)) gives

$$
\begin{aligned}
& -\frac{ik}{4\pi}\frac{d}{dt}\int_\Sigma tr(g_t^{-1}\partial g_t)(g_t^{-1}\bar\partial g_t) \\
& = \frac{ik}{4\pi}\int_\Sigma tr(g_t^{-1}\frac{\partial g_t}{\partial t})(g_t^{-1}\partial g_t)(g_t^{-1}\bar\partial g_t) \\
& - \frac{ik}{4\pi}\int_\Sigma tr(g_t^{-1}\partial(\frac{\partial g_t}{\partial t}))(g_t^{-1}\bar\partial g_t) \\
& + \frac{ik}{4\pi}\int_\Sigma tr(g_t^{-1}\partial g_t)(g_t^{-1}\frac{\partial g_t}{\partial t})(g_t^{-1}\bar\partial g_t) \\
& - \frac{ik}{4\pi}\int_\Sigma tr(g_t^{-1}\partial g_t)(g_t^{-1}\bar\partial(\frac{\partial g_t}{\partial t})) \\
& = -\frac{ik}{4\pi}\int_\Sigma tr\,\partial(g_t^{-1}\frac{\partial g_t}{\partial t})(g_t^{-1}\bar\partial g_t) \\
& + \frac{ik}{4\pi}\int_\Sigma tr\,\bar\partial(g_t^{-1}\frac{\partial g_t}{\partial t})(g_t^{-1}\partial g_t)\ .
\end{aligned}
\tag{5}
$$

Eqs. (3), (4) and (5) together yield

$$
\begin{aligned}
\frac{d}{dt}S_\Sigma(g_t) &= -\frac{ik}{2\pi}\int_\Sigma tr\,\partial(g_t^{-1}\frac{\partial g_t}{\partial t})(g_t^{-1}\bar\partial g_t) \\
&+ \frac{ik}{4\pi}\int_{\partial\Sigma} tr(g_t^{-1}\frac{\partial g_t}{\partial t})(g_t^{-1}dg_t) - i\sum_{b\subset\partial\Sigma}\int_b \tilde g_t^*(\tilde X_t \lrcorner\, \tilde\omega_{\alpha_b}) \\
&- i\sum_{\substack{v,b\\ v\in\partial b,\, b\subset\partial\Sigma}} \langle \tilde X_t(\tilde g_t(v)), \tilde\eta_{\alpha_v\alpha_b}\rangle\ .
\end{aligned}
\tag{6}
$$

Of course (6) implies (5.13) but it provides also the boundary terms occurring when $\frac{\partial g_t}{\partial t}$ does not vanish on $\partial\Sigma$.

Notice that due to (3.10), eq. (6) may be rewritten as

$$
\begin{aligned}
\frac{D}{dt}\exp[-S\,(g_t)] = \{&\frac{ik}{2\pi}\int_\Sigma tr\,\partial(g_t^{-1}\frac{\partial g_t}{\partial t})(g_t^{-1}\bar\partial g_t) \\
&- \frac{ik}{4\pi}\int_{\partial\Sigma} tr(g_t^{-1}\frac{\partial g_t}{\partial t})(g_t^{-1}dg_t)\}\exp[-S_\Sigma(g_t)]
\end{aligned}
\tag{7}
$$

where $\frac{D}{dt}$ is the covariant derivative.

APPENDIX 5

In order to prove (5.24), it is enough to check the commutation relations in the Lie algebra of $\widehat{LG}^{\mathbb{C}}$ as ℓ and r are two commuting representations of $\widehat{LG}^{\mathbb{C}}$. The only non-trivial relation is the commutator

$$[\frac{d}{d\varepsilon}\Big|_{\varepsilon=0} e^{S_{D_\infty}(e^{\frac{1}{2}i\varepsilon\sigma_j z^n})}, \frac{d}{d\varepsilon}\Big|_{\varepsilon=0} e^{-S_{D_o}(e^{\frac{1}{2}i\varepsilon\sigma_\ell z^m})}] \tag{1}$$

with $n < 0$ and $m \geq 0$. This is equal to

$$\frac{d}{d\varepsilon^2}\Big|_{\varepsilon=0} e^{S_{D_\infty}(g_1^{-1})} \cdot e^{-S_{D_o}(g_2^{-1})} \cdot e^{S_{D_\infty}(g_1)} \cdot e^{-S_{D_o}(g_2)} \tag{2}$$

where

$$g_1 = e^{\frac{1}{2}i\varepsilon\sigma_j z^n}, \quad g_2 = e^{\frac{1}{2}i\varepsilon\sigma_\ell z^m}. \tag{3}$$

Setting $\tilde{g}_1 = e^{\frac{1}{2}i\varepsilon\sigma_j f}$, where f extends z^n on D_∞ to D_o, we may rewrite expression (2) :

$$\begin{aligned}(2) &= \frac{d}{d\varepsilon^2}\Big|_{\varepsilon=0} e^{S_\Sigma(\tilde{g}_1^{-1})+S_\Sigma(\tilde{g}_1)} e^{-S_{D_o}(\tilde{g}_1^{-1})} \cdot e^{-S_{D_o}(g_2^{-1})} \cdot e^{-S_{D_o}(\tilde{g}_1)} \cdot e^{-S_{D_o}(g_2)} \\
&= \frac{d}{d\varepsilon^2}\Big|_{\varepsilon=0} \exp[\Gamma_\Sigma(\tilde{g}_1^{-1},\tilde{g}_1)-\Gamma_{D_o}(\tilde{g}_1^{-1},g_2^{-1})-\Gamma_{D_o}(\tilde{g}_1,g_2) \\
&\qquad e^{S_{D_o}(\tilde{g}_1^{-1}g_2^{-1})} \cdot e^{S_{D_o}(\tilde{g}_1 g_2)} \\
&= \frac{d}{d\varepsilon^2}\Big|_{\varepsilon=0} \exp[\Gamma_{D_o}(\tilde{g}_1^{-1},\tilde{g}_1)-\Gamma_{D_o}(\tilde{g}_1^{-1},g_2^{-1})-\Gamma_{D_o}(\tilde{g}_1,g_2) \\
&\qquad - \Gamma_{D_o}(\tilde{g}_1^{-1}g_2^{-1},\tilde{g}_1 g_2)]\, e^{-S_{D_o}(\tilde{g}_1^{-1}g_2^{-1}\tilde{g}_1 g_2)}\end{aligned} \tag{4}$$

where we have used (5.11) and (5.14). The expression in brackets equals

$$\begin{aligned}&- \frac{ik\varepsilon^2}{8\pi}\int_{D_o} tr(\sigma_j\sigma_\ell)(\bar{\partial}f)(\partial z^m) + O(\varepsilon^2) \\
&= - \frac{ik\varepsilon^2}{4\pi} m\delta^{j\ell} \int_{\partial D_o} z^{n+m}\frac{dz}{z} + O(\varepsilon^2) \\
&= \varepsilon^2 \frac{k}{2} m\, \delta^{j\ell}\, \delta_{n+m,o} + O(\varepsilon^2) .\end{aligned} \tag{5}$$

On the other hand,

$$\tilde{g}_1^{-1} g_2^{-1} \tilde{g}_1 g_2 = e^{\frac{1}{2} i\varepsilon^2 \sum_k \varepsilon^{j\ell k} \sigma_k f z^m} (1+O(\varepsilon^2)) \quad .$$

Now it follows from (5.11) and (5.13) $(S(e^{O(\varepsilon^2)}) = O(\varepsilon^4))$ that if $n+m \geq 0$ then

$$\frac{d}{d\varepsilon^2}\Big|_{\varepsilon=o} e^{-S_{D_o}(\tilde{g}_1^{-1} g_2^{-1} \tilde{g}_1 g_2)} = \frac{d}{d\varepsilon^2}\Big|_{\varepsilon=o} e^{-S_{D_o}(e^{\frac{1}{2} i\varepsilon^2 \sum_k \varepsilon^{j\ell k} \sigma_k z^{n+m}})} \tag{6}$$

and if $n+m < 0$ then

$$\frac{d}{d\varepsilon^2}\Big|_{\varepsilon=o} e^{-S_{D_o}(\tilde{g}_1^{-1} g_2^{-1} \tilde{g}_1 g_2)} = \frac{d}{d\varepsilon^2}\Big|_{\varepsilon=o} e^{S_{D_\infty}(\frac{1}{2} i\varepsilon^2 \sum_k \varepsilon^{j\ell k} \sigma_k z^{n+m})} \quad . \tag{7}$$

Gathering (5), (6) and (7), we obtain for $n+m \gtrless 0$

$$(1) = \frac{d}{d\varepsilon^2}\Big|_{\varepsilon=o} e^{\left\{\begin{matrix} -S_{D_o} \\ S_{D_\infty} \end{matrix}\right. (\frac{1}{2} i\varepsilon^2 \sum_k \varepsilon^{j\ell k} \sigma_k z^{n+m})} + \frac{k}{2} m\delta^{j\ell} \delta_{n+m,o} \tag{8}$$

what was to be shown.

REFERENCES

1. P.A.M. Dirac, Proc. R. Soc. London A133, 60 (1931).
2. D.J. Simms, N.M. Woodhouse, Lecture Notes in Physics, Vol.53, Springer (1976).
3. E. Witten, Commun. Math. Phys. 92, 455 (1984).
4. L. Alvarez-Gaumé, J.-B. Bost, G. Moore, P. Nelson, C. Vafa, Phys. Lett. B178, 105 (1986).
5. D. Gepner, E. Witten, Nucl. Phys. B278, 493 (1986).
6. P.A.M. Dirac, The principles of quantum mechanics, 4th edition, Oxford University Press (1958).
7. R.P. Feynman, A.R. Hibbs, Quantum mechanics and path integrals, Mc. Graw-Hill.
8. T.T. Wu, C.N. Yang, Phys. Rev. D14, 437 (1976).
9. Y. Aharonov, D. Bohm, Phys. Rev. 115, 485 (1959).
10. A. Kirillov, Eléments de la théorie des représentations, Editions Mir (1974).
11. B. Kostant, in Lecture Notes in Math., Vol.170, pp.87-208, Springer (1970).
12. R. Godement, Topologie algébrique et théorie des faisceaux, Hermann (1964).
13. C.G. Callan, R.F. Dashen, D.J. Gross, Phys. Let. 63B, 334 (1976).
14. R. Jackiw, C. Rebbi, Phys. Rev. Lett. 37, 172 (1976).
15. T.R. Ramadas, Commun. Math. Phys. 93, 355 (1984).

16. H. Esnault, E. Viehweg, Deligne-Beilinson cohomology, publication of Max-Planck-Institut für Mathematik, Bonn.

17. O. Alvarez, Commun. Math. Phys. 100, 279 (1985).

18. R. Hamilton, Bull. Am. Math. Soc. 7, 65 (1982).

19. D. Friedan, E. Martinec, S. Shenker, Nucl. Phys. B271, 93 (1986).

20. L. Alvarez-Gaumé, J.-B. Bost, G. Moore, P. Nelson, C. Vafa, Bosonization on higher genus Riemann surfaces, Harvard-CERN preprint.

21. V.G. Knizhnik, A.B. Zamolodchikov, Nucl. Phys. B247, 83 (1984).

22. A. Pressley, G. Segal, Loop groups, Clarendon Press (1986).

23. V.G. Kac, Infinite dimensional Lie algebras, Cambridge University Press (1985).

24. V.G. Kac, D.H. Peterson, in Arithmetics and geometry, Vol.2, pp.141-166, Birkhäuser (1983).

25. G. Felder, K. Gawedzki, A. Kupiainen, The spectrum of Wess-Zumino-Witten models, IHES preprint.

26. D. Gepner, Nucl. Physics. B287, 111 (1987).

27. A. Cappelli, C. Itzykson, J.-B. Zuber, Nucl. Phys. B280 [FS 18], 445 (1987) and The A-D-E classification of minimal and $A_1^{(1)}$ conformal invariant theories, to appear in Commun. Math. Phys.

ISSUES IN SUPERSTRING THEORY

Michael B. Green

Department of Physics
Queen Mary College
University of London
Mile End Road, London E1 4NS, U.K.

1. INTRODUCTION

String theory has major links with many areas of modern mathematics, the theory of critical phenomena in two-dimensional statistical mechanical systems and a quantum theory of all the fundamental forces, including gravity. It is the latter prospect that sparked the current interest in the particle physics community but the links with mathematics and two-dimensional statistical mechanics have also been very fruitful.

String theory describes the relativistic dynamics of one-dimensionally extended relativistic particles moving in a curved D-dimensional space-time. A crucial feature of the string spectrum (see fig. 1.1) is that it includes a massless spin-2 mode (along with other massless modes) which is identified with the graviton, the fluctuation of the metric of space-time. This signifies that the theory necessarily describes the gravitational force along with other forces and therefore incorporates general relativity and hence must determine the geometry of the space-time in which the string is moving. The present understanding of string theory is based on a perturbation expansion around a fixed, classical spacetime background. This perturbation expansion, which defines what we mean by string theory, is a semiclassical approximation analogous to the expansion of Einstein's theory in small fluctuations of the metric around any given classical space-time solution. In superstring theory the individual terms in this perturbation expansion have certain extraordinary features which suggest that the usual problems of conventional perturbative quantum gravity are avoided. In particular, there are no infinities or chiral anomalies when the classical background is chosen consistently.

For reasons of brevity I shall refer to the perturbation expansions around any of the many consistent choices for the classical background as different string "theories". They are more properly described as different semiclassical expansions of the same theory.

One of the most basic features of string theory is that it contains a new dimensional constant, the string tension T, which is associated with a fundamental distance scale

$$l_S = \sqrt{\frac{ch}{\pi T}}. \tag{1.1}$$

This sets the scale of the frequencies of the infinite number of excited modes of vibration

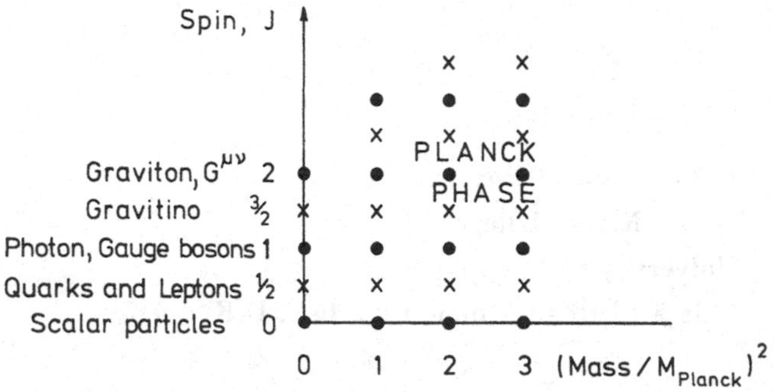

Figure 1.1. A typical superstring spectrum showing the angular momentum and mass of the states. The dots indicate bosonic states and the crosses indicate fermionic ones (each dot and cross generally represents a number of degenerate states). There are an infinite number of massive excitations corresponding to the infinite number of modes of higher and higher frequency. The separation between successive levels is determined by the string tension $\Delta m^2 = T\pi h/c^3 \sim (10^{19} Gev/c^2)^2$. It is important that the massless states always include states with spins corresponding to the graviton (spin 2), gravitino (spin 3/2), photon and gauge bosons (spin 1), leptons and quarks (spin 1/2), scalar bosons (spin 0). These massless fields are typical of those required in conventional grand unified field theories with supersymmetry, which approximates string theory at "low" energies ($E \ll \sqrt{\pi hcT}$). Supersymmetry also results in an equal number of fermion and boson states at each mass level.

of the string. Since a mode of given frequency corresponds (via the de Broglie relation) to a quantum state of given energy and hence of given mass, the string describes an infinite number of quantum states with masses (and spins) that increase indefinitely (fig. 1.1). It is natural to identify l_S with the Planck distance,

$$l_{Planck} = \sqrt{\frac{Gh}{c^3}} \sim 10^{-35} meter \tag{1.2}$$

(where G is Newton's constant) which is the distance scale that enters into any theory of quantum gravity. This means that $T \sim c^4/\pi G$ so the separation between the masses of the excited string states is huge $\Delta m^2 \sim T\pi h/c^3 \sim (10^{19} Gev/c^2)^2$. This infinitely rich spectrum of states in string theory is associated with an infinite extension of conventional Yang–Mills and gravitational gauge invariances. In the "low energy" approximation (at energies much less than the Planck energy, $E \ll \sqrt{\pi chT} \sim 10^{19} Gev$) the massive states decouple, leaving an effective theory of massless pointlike particles corresponding to the massless states in fig. 1.1. Since the massless states include the spin-2 graviton and spin-1 Yang–Mills particles (as well as the spin-3/2 gravitino characteristic of supersymmetry), superstring theory reduces in this low energy limit to a conventional supersymmetric Yang–Mills theory interacting with supergravity. Different string "theories" (*i.e.*, theories with different choices of background) differ in the details of the spectrum of massless states (such as the Yang–Mills gauge group and the representations of the leptons) and so correspond to different possible unifications of the non-gravitational forces. Certain of these low energy limits come remarkably close to describing the observed world of elementary particles and their forces.

Despite these attractive features of the individual terms in string perturbation theory there should be more to a quantum theory of gravity than a semiclassical expansion, so our understanding of the theory is far from complete. At present we are in the rather extraordinary situation of knowing these very interesting perturbation expansions without knowing the fundamental structure that is being expanded (*i.e.*, the analogue of the curvature scalar in the Einstein–Hilbert theory). One would expect the perturbation expansion to be inadequate for understanding many issues in superstring theory (as it is in many ordinary field theories, such as QCD). In particular, in a quantum theory of gravity space-time coordinates should not be introduced into the theory as fundamental input, as in string perturbation theory, but rather should emerge from the dynamics unified with the particles and forces.

One might expect a semi-classical approximation to the space-time background to break down at the Planck scale and therefore that the string perturbation may be inconsistent even though it appears consistent at any finite order. In other words, just at the scale at which the extended nature of strings becomes important - leading, for example, to the good ultraviolet properties of individual string Feynman diagrams - the simple picture of a single string moving in a fixed background may not be sensible. This suggests that strings are incorrect variables with which to describe string theory. Indications of a phase transition in the theory around the Planck scale support the view that new coordinates are needed to describe the theory in the "Planck phase" (fig. 1.1).* In the perturbative description of string theory the background metric of space-time, $G_{\mu\nu}$, takes a fixed nonzero value (approximately the Minkowski metric) characteristic of the spontaneous breakdown of general coordinate invariance. A more fundamental treatment would describe the theory in the unbroken phase in which the spacetime metric would not be singled out from the rest of the string states (Witten's topological models of gravity are possible examples of this kind of theory). Needless to say an intensive search is being undertaken for such a fundamental setting to string theory. Many paths are open since the only criterion for deciding on a correct approach (apart from aesthetic ones) is that it should reproduce the (presumably divergent) string perturbation expansion in a (possibly inconsistent!) approximation.

I will begin with a survey of the developments in string perturbation theory and then comment on a few of the ideas that have been suggested about how to formu-

---

* The original developments in string theory began in 1968 (with the work of Veneziano) with strings representing the hadrons - the strongly interacting particles, namely, the baryons, such as the proton and neutron, and the mesons. For hadrons the string tension has a value of around $10^{-19} M_{Planck}$ and there is a phase transition at a correspondingly low temperature to a phase of deconfined quarks and gluons, the constituents of QCD. Nowadays, the stringlike spectrum observed in hadronic experiments is supposed to be explained by QCD.

late a nonperturbative theory that reproduces string perturbation theory. I have not attempted to give explicit references to the huge amount of recent work in string theory.†

2. STRING PERTURBATION THEORY

The quantum mechanical description of string theory is based closely on the Feynman path integral formulation of quantum mechanics. Recall that for relativistic point particles the amplitude for a particle to move from a point labelled x_1^μ (where $\mu = 0, 1, \ldots, D-1$ is a world-sheet vector index) to a point x_2^μ is given by an average over all world lines $x^\mu(\tau)$ connecting these points (where τ is the parameter labelling the path),

$$A(x_1 \to x_2) = \sum_{world\ lines} e^{\frac{i}{\hbar} S[x(\tau)]}. \tag{2.1}$$

The action S is the length of the world line, a parametrization-independent quantity.

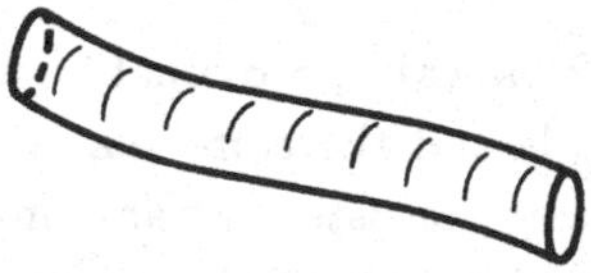

Figure 2.1. A world surface joining initial and final closed strings. The path integral includes sums over all possible surfaces joining initial and final surfaces.

The analogous quantity in string theory describes the amplitude for a string to move from a given initial curve $X_1^\mu(\sigma)$ to a curve $X_2^\mu(\sigma)$ and is given as a sum over all surfaces, or world sheets, connecting the curves.‡ These two-dimensional surfaces may be parametrized by σ and τ but no physical quantity should depend on the parametrization used. The simplest parametrization-independent action is the *area* of the world-sheet embedded in space-time, which is the Nambu action,

$$S' = T \int d\sigma d\tau \sqrt{\det \partial_\alpha X^\mu \partial_\beta X_\mu}, \tag{2.2}$$

where the indices $\alpha, \beta = 0, 1$ label the world sheet coordinates and $G_{\mu\nu}(X)$ is the metric of space-time (the embedding space).

---

† For a compilation of recent work in string theory see, for example, *Proceedings of the 1988 ICTP Spring School on Superstrings*, M.B. Green, M. Grisaru, R. Iengo, E. Sezgyn and A. Strominger eds. (World Scientific, to be published).

‡ I will discuss only closed-string theories. These are theories with closed, orientable world sheets. The "type 1" theories which also contain open strings (*i.e.*, strings with end points) are more complicated since they require non-orientable world sheets as well as world sheets with boundaries. They appear to be unrealistic as unified theories.

It proves to be very convenient to describe these world sheets not only by their embedding, $X^\mu(\sigma,\tau)$, in space-time, but also by an intrinsic metric, $g^{\alpha\beta}(\sigma,\tau)$, of the sheet. This leads to the so-called 'Polyakov' method of calculating string theory amplitudes in which the sum over histories involves an integral over all intrinsic metrics as well as all embeddings of the world sheet, so that (2.1) is replaced by

$$A(X_1^\mu(\sigma) \to X_2^\mu(\sigma)) = \int_1^2 Dg^{\alpha\beta} DX^\mu D\psi e^{-S/h}. \tag{2.3}$$

The variables $\psi(\sigma,\tau)$ in this expression represent the additional fermionic variables that are present in the various kinds of superstring theories. It has also been assumed that it is sensible to continue τ to imaginary value, $\tau \to i\tau$ (a 'Wick rotation') so that $e^{iS/h} \to e^{-S/h}$. The obvious reparametrization invariant action is now given by

$$S_0 = -\frac{T}{2}\int d\sigma d\tau \sqrt{g} g^{\alpha\beta}(\sigma,\tau)\partial_\alpha X^\mu(\sigma,\tau)\partial_\beta X^\nu(\sigma,\tau)G_{\mu\nu}(X). \tag{2.4}$$

(where $g \equiv \det g_{\alpha\beta}$). The expression (2.4) can be viewed as the action for a *two-dimensional* field theory. The coordinates $X^\mu(\sigma,\tau)$ are simply D scalar two-dimensional fields (the Lorentz index μ playing the rôle of an internal symmetry). The action S_0 is the action of two-dimensional gravity coupled to D scalar fields and hence it is manifestly reparametrization invariant. The classical equation of motion for the intrinsic metric relates it to the induced metric, $g_{\alpha\beta} = f(\sigma,\tau)\partial_\alpha X^\mu \partial_\beta X^\nu G_{\mu\nu}(X)$ (where f is an arbitrary function). Substituting this in S_0 gives back S' (2.2) and hence the two expressions are equivalent for the classical theory. However, the procedure which includes the intrinsic metric makes the analysis of the quantum theory significantly clearer.

In addition to S_0 a number of other terms can obviously be considered in the action. One of these involves an antisymmetric background field $B_{\mu\nu}(X)$,

$$S_1 = -\frac{T}{2}\int d\sigma d\tau \epsilon^{\alpha\beta}\partial_\alpha X^\mu \partial_\beta X^\nu B_{\mu\nu}(X). \tag{2.5}$$

Another possible term involves a scalar field, $\Phi(X)$, which corresponds to a 'dilaton' field in the embedding space,

$$S_2 = \frac{1}{4\pi}\int d\sigma d\tau \sqrt{g} R^{(2)}(g)\Phi(X), \tag{2.6}$$

where R^2 is the two-dimensional curvature (made out of $g^{\alpha\beta}$ in the usual way) which is a total derivative. Therefore, for constant $\Phi(X) = \phi$ the integrand in S_2 is a total derivative and hence the integral is determined by the topology of the two-dimensional surface alone. For a closed surface with l handles S_2 is then given simply in terms of the Euler characteristic of the surface,

$$S_2 = \int d^2\sigma R^{(2)}\phi/4\pi = 2(1-l)\phi. \tag{2.7}$$

The factor of e^{-S_2} in the sum over histories therefore gives a factor of $e^{-2\phi}$ for each handle which means that the string coupling constant (the coupling between three closed strings) is determined by the dilaton expectation value to be given by $g = e^{-\phi}$.

I will ignore other possible terms in the action which involve fermionic fields in the embedding space. I will also concentrate on S_0 for illustrative purposes in much of this talk.

3. CONFORMAL FIELD THEORY

The action S_0 is manifestly invariant under reparametrixations $\tau \to \tau + \xi^0(\sigma,\tau)$ and $\sigma \to \sigma + \xi^1(\sigma,\tau)$, where ξ^α is an arbitrary world-sheet vector. It is also invariant under Weyl transformations, $g^{\alpha\beta} \to e^{\phi(\sigma,\tau)}g^{\alpha\beta}$ where ϕ is an arbitrary function. The reparametrization invariance can be used to gauge away two components of the metric tensor and so choose a gauge in which the intrinsic metric tensor (which is a 2×2 symmetric tensor) has the form

$$g^{\alpha\beta} = e^{\phi(\sigma,\tau)}\eta^{\alpha\beta}, \tag{3.1}$$

where $\eta_{\alpha\beta} = \mathrm{diag}(-1,1)$ is the Minkowski world-sheet metric which becomes the unit matrix after a Wick rotation to imaginary τ. Evidently the action S_0 simplifies considerably in this *conformal* gauge so that, defining the complex coordinates,

$$z = e^{\tau+i\sigma}, \qquad \bar{z} = e^{\tau-i\sigma}, \tag{3.2}$$

we have

$$S_0 = -\frac{T}{2}\int d^2z\partial_\alpha X^\mu \partial^\alpha X^\nu G_{\mu\nu}. \tag{3.3}$$

In this gauge, and in a flat space-time background ($G_{\mu\nu}(X) = \eta_{\mu\nu}$), S_0 is simply the action for D free two-dimensional boson fields. The dilation mode $\phi(\sigma,\tau)$ cancels out of the classical action due to the Weyl symmetry and the action S_0 is invariant under arbitrary conformal transformations, $z \to \tilde{z}(z)$ and $\bar{z} \to \tilde{\bar{z}}(\bar{z})$. As usual, the choice of such a gauge requires the imposition of constraints which are simply the $g^{\alpha\beta}$ equations of motion evaluated in this gauge,

$$T_{\alpha\beta} \equiv \frac{-2}{\sqrt{g}T}\frac{\partial S_0}{\partial g^{\alpha\beta}} = 0, \tag{3.4}$$

where $T_{\alpha\beta}$ is the two-dimensional energy-momentum tensor. It is convenient to express $T_{\alpha\beta}$ in terms of its complex components $T_{zz}, T_{\bar{z}\bar{z}}, T_{z\bar{z}}$ (where $T_{z...} = T_{0...} + iT_{1...}$, etc.).

Weyl invariance of the action is expressed by the tracelessness of the energy-momentum tensor, $T^{\alpha}_{\ \alpha} = 0$, *i.e.*,

$$T_{z\bar{z}} = T_{\bar{z}z} = 0. \tag{3.5}$$

In this case conservation of $T_{\alpha\beta}$, $\partial_\alpha T^{\alpha\beta} = 0$ implies

$$T_{zz} = T(z), \qquad T_{\bar{z}\bar{z}} = \bar{T}(\bar{z}). \tag{3.6}$$

$T(z)$ and $\bar{T}(\bar{z})$ are the generators of the residual conformal transformations that preserve the form of the metric in the conformal gauge. The modes of T,

$$L_n = \frac{1}{2\pi i}\oint \frac{dz}{z} z^n T(z), \tag{3.7}$$

are the *Virasoro* operators. They satisfy the infinite dimensional Virasoro algebra obtained by taking the moments of the conformal algebra satisfied by $T(z)$,

$$[L_m, L_n] = (m-n)L_{m+n} + \frac{c}{12}\delta_{m+n,0}(m^3 - m), \tag{3.8}$$

where c is an important constant. The modes of $\bar{T}$ define another set of generators, $\bar{L}_n$, which satisfy a similar algebra and commute with L_n. The *central extension* with coefficient c is a quantum mechanical term which corresponds to an anomaly in the classical algebra of the generators of conformal transformations.

For closed strings in the simplest string theory, defined by the action S_0 (with $G_{\mu\nu} = \eta_{\mu\nu}$) with no other terms, the X^μ equations of motion have the general solution

$$X^\mu(z,\bar{z}) = (X^\mu_L(\bar{z}) + X^\mu_R(z))/2, \tag{3.9}$$

where the left and right handed polarizations, X_L and X_R, are referred to as *left-movers* and *right-movers*. They have periodic expansions in terms of normal modes,

$$X^\mu_R(z) = x^\mu + ip^\mu \ln z + i\sum_{n\neq 0}\frac{1}{n}\alpha^\mu_n z^n, \tag{3.10}$$

(where $\alpha^\mu_{-n} \equiv \alpha^{\mu\dagger}_n$) with a similar expression for $X_L(\bar{z})$ in terms of left polarized modes, $\tilde{\alpha}^\mu_n$. In the quantum theory the commutation relations derived from the action S_0 result in harmonic oscillator commutation relations for the modes,

$$[\alpha^\mu_m, \alpha^\nu_n] = m\delta_{m+n,0}, \tag{3.11}$$

with similar relations for $\tilde{\alpha}^\mu_n$. In this simple theory the L_n's and $\bar{L}_n$'s are simple quadratic expressions,

$$L_n = \frac{1}{2}\sum_{-\infty}^{\infty}\alpha^\mu_{-m+n}\alpha_{m\mu} \qquad \bar{L}_n = \frac{1}{2}\sum_{-\infty}^{\infty}\tilde{\alpha}^\mu_{-m+n}\tilde{\alpha}_{m\mu} \tag{3.12}$$

(where $\alpha^\mu_0 = p^\mu$). In this case each component of X^μ contributes 1 to the value of c in (3.8) so that the total value is $c = D$, the dimension of space-time.

The description of the theory in the conformal gauge is streamlined by introducing ghost fields to take care of the Fadeev–Popov determinant in the functional integral. The gauge-fixing ghost, $c^\alpha(\sigma,\tau)$ is a world-sheet vector while the conjugate antighost, $b_{\alpha\beta}(\sigma,\tau)$ is a tensor. The ghost terms in the action, S_{ghost}, give a contribution to the energy-momentum tensor which satisfies an algebra like (3.8) but with an anomaly term with coefficient $c_{ghost} = -26$. We see that the total anomaly will vanish when $c = 26$ which happens in $D = 26$ space-time dimensions for the bosonic theory in flat space-time. The condition for the absence of the anomaly also constrains the spectrum of states so that the gauge particles are massless (and the ground state is a tachyon - a state of negative (mass)^2 - which indicates the bosonic theory has problems).

The simplest generalization of the bosonic theory includes free world-sheet fermionic fields in addition to the bosonic $X^\mu(\sigma,\tau)$. A single Majorana fermion contributes $c = \frac{1}{2}$ to the Virasoro anomaly. In the Neveu–Schwarz–Ramond *spinning string* model the fermions also have a *space-time vector* index, *i.e.*, $\psi^\mu_a(\sigma,\tau)$ (where $a = 1,2$ is the world-sheet spinor index) describes D anticommuting Majorana *world-sheet spinors*. The fields ψ^μ contribute Dirac terms to the action S which make it supersymmetric on the world sheet ($S_\psi \sim \int d\sigma d\tau \bar{\psi}^\mu \rho \cdot \partial \psi^\nu \eta_{\mu\nu}$ in the conformal gauge) The Virasoro algebra is now extended to the *superconformal* (or super-Virasoro) algebra which involves anticommuting generators, F_n, in addition to the Virasoro generators. In this case the total contribution to c from X^μ and ψ^μ is $c = 3D/2$. The Fadeev–Popov ghosts for the world-sheet supersymmetry $(\beta^a_\alpha, \gamma^a)$ together with the b, c ghosts give rise to an anomaly, $c_{ghost} = -15$ so that the total anomaly vanishes when $c = 15$. This happens in $D = 10$ space-time dimensions in the supersymmetric theory in flat space-time. Spinor fields may be double-valued so that they can satisfy either periodic or antiperiodic boundary conditions. The spectrum of states now includes space-time fermions as well as space-time bosons but, in its original form, still has a problematic tachyon ground state. However, in calculating amplitudes it is important to sum over all possible boundary conditions (*i.e.*, sum over all *spin structures*). This sum (the GSO projection) eliminates the tachyon leaving the massless ground states shown in fig. 1.1. The projected theory not only has world-sheet supersymmetry but also turns out to be supersymmetric in the space-time in which the string is moving so that there are an equal number of space-time boson states and space-time fermion states at every mass level in fig. 1.1. Theories with space-time supersymmetry are superstring theories and it is such theories which are remarkably consistent.

More generally, instead of considering actions with a specific field content, string theories may be defined algebraically by studying unitary representations of the Virasoro algebra defined by (3.8) without reference to any particular action. The unitarity condition is the requirement that $L_n = L^\dagger_{-n}$ which amounts to reality of the energy-momentum tensor. Different values of c correspond to different theories. For any c we can define *highest-weight* representations, labelled by h_i, by

$$L_n|h_i\rangle = 0, \quad h_i > 0, \qquad L_0|h_i\rangle = h_i|h_i\rangle. \tag{3.13}$$

The classical constraint equations (3.4) are transcribed in the quantum theory into the statement that the physical string states are highest-weight states.

This discussion is also of relevance to the analysis of the behaviour of two-dimensional statistical mechanical systems at a critical point corresponding to a continuous phase transition (*i.e.*, a second-order or higher-order critical point). The special feature of two dimensions is that the global scale invariance characteristic of critical systems is augmented to the infinite symmetry generated by the L_n and $\bar{L}_n$. Therefore the classification of representations h_i of the Virasoro algebra for all allowed values of c amounts to a classification of possible critical behaviour in two-dimensional statistical systems. The absence of the reparametrization constraints, (3.3), in the description of critical systems means that the restriction to highest-weight states is not present in this case.

Transversality of physical states

The various anomalies in the algebraic structure of the quantum theory referred to earlier are typical of quantum mechanical effects arising from the path integral measure which are not present in the classical theory. They often spell disaster for the consistent interpretation of the theory. The situations in which anomalies cancel therefore play a special rôle. We saw that in the bosonic theory anomalies in reparametrization and Weyl symmetries are absent if $c = 26$ while in theories which have world-sheet supersymmetry they are absent if $c = 15$. String theories with such restrictions on c are called 'critical' string theories and are the ones of interest as possible unifying theories. The special feature of the critical values of c is that the physical states in the theory are 'transverse'. The unphysical timelike modes created by the negative-normed oscillators, $\alpha_n^{0\,*}$, cancel with the longitudinal modes. This is analogous to the situation in quantum electrodynamics or massless Yang–Mills theories, in which there are only $D-2$ physical polarizations of the photon corresponding to the transverse components of the vector potential. The huge gauge invariance of string theory ensures that all the states of the theory are created by transverse oscillators (oscillators with $D-2$ components) for the special values of c.

The absence of anomalies in the critical theories is characterized in the conformal gauge by the existence of BRST (Becchi, Rouet, Stora and Tyupin) symmetry for the action that includes Fadeev–Popov ghosts. This global symmetry encodes all the local symmetries of the theory. The BRST charge, Q, which generates this symmetry, satisfies the nilpotency condition $Q^2 = 0$ only for the critical theories (which is one way of discovering the critical values of c). Furthermore, when ghost modes are included the constraints on the physical states, (3.13), are replaced by the single equation

$$Q|\phi\rangle = 0. \tag{3.14}$$

The solutions of this are cohomology classes of the operator Q. These are just the transverse physical states which are defined modulo gauge transformations $|\phi\rangle \to |\phi\rangle + Q|\lambda\rangle$ (for arbitrary $|\lambda\rangle$).

In models with larger values of c than the critical values the timelike modes do not decouple and there are physical states of negative norm. This is a disaster since

⋆ Returning to the Minkowski signature metric $\eta^{\mu\nu} = \text{diag}(-1, 1, \ldots)$

positivity of the norms of the states is crucial to the quantum interpretation of these theories. Subcritical theories are those with smaller c. The Weyl anomalies induce longitudinal components in the states (the timelike modes again decouple). Although they are not yet properly studied, it is likely that the presence of longitudinal modes implies that these theories do not contain massless gauge particles (fig. 1.1). Polyakov has pursued subcritical theories as models of hadronic interactions and in studying the three dimensional Ising model. I shall only be discussing string theories with critical values of c.

Curved space ($G_{\mu\nu} \neq \eta_{\mu\nu}$).

So far we have seen that certain (unreasonably large) dimensions of space-time are singled out for strings propagating in a flat space-time background. More general situations involve curved backgrounds. This, for example, is one way of obtaining theories in lower dimensions starting from one of the above flat-space theories. The extra spatial dimensions may be curled up so that the effective dimensionality of the theory may then be more realistic. When $G_{\mu\nu} \neq \eta_{\mu\nu}$ the action (3.3) is that of an interacting two-dimensional theory - a *nonlinear sigma model*. Although the action is conformally invariant as a classical two-dimensional theory ($T^{\alpha}_{\ \alpha} = 0$) there will generally be quantum anomalies in the tracelessness of the energy-momentum tensor caused by the interactions. Although the theory cannot be solved exactly with a general $G_{\mu\nu}$ it can be studied in perturbation theory in powers of the inverse string tension, $\alpha' = 1/\pi T$. The condition that the energy-momentum tensor be traceless is more or less equivalent to the statement that the renormalization group β functions vanish. In a nonlinear sigma model, such as that defined by the action $S_0 + S_1 + S_2$ in the gauge $g_{\alpha\beta} = e^{\phi}\eta_{\alpha\beta}$, there is a beta *functional* that describes the renormalization of each of the background fields, $G_{\mu\nu}$, $B_{\mu\nu}$, Φ, ... The conditions for conformal invariance are therefore

$$\beta^G[G_{\mu\nu}, B_{\mu\nu}, \Phi, \ldots] = 0, \quad \beta^B[G_{\mu\nu}, B_{\mu\nu}, \Phi, \ldots] = 0, \quad \beta^{\Phi}[G_{\mu\nu}, B_{\mu\nu}, \Phi, \ldots] = 0, \quad (3.15)$$

with further equations corresponding to other background fields (such as the fermionic ones in superstring theories). The perturbative evaluation of the β functions therefore leads to a set of consistency conditions on the background fields. For example, ignoring all fields other than $G_{\mu\nu}$ and Φ the $\beta^G = 0$ equation gives

$$R_{\mu\nu} - \nabla_\mu \nabla_\nu \Phi + \frac{\alpha'}{2} R_{\mu\rho\omega\gamma} R^{\rho\omega\gamma}{}_{\nu} + \ldots = 0. \quad (3.16)$$

The ... indicate an infinite number of terms of higher order in α'. In this equation $R_{\mu\rho\omega\nu}$ is the Riemann curvature of the embedding space formed out of the metric $G_{\mu\nu}$ in the usual way and $R_{\mu\nu}$ is the Ricci tensor. Equation (3.16) is a generalization of Einstein's equation for the space-time in which the string is moving! At energies much smaller than $1/\sqrt{\alpha'}$ only the lowest-order term is relevant and Einstein's equation is recovered. The higher order terms alter Einstein's equation at very high energies (short distances). This is one of the clearest ways of seeing that string theory is a theory that contains general relativity. Remarkably, the quantum mechanical description of a string moving through a curved space *requires* that space to satisfy a generalization of Einstein's equation.

The connection between string theory and general relativity goes beyond this. It is possible to identify the massless spin-2 particle excitation of the string with fluctuations of the background metric, $G_{\mu\nu}$. This connection shows that string theory is not simply a theory of particles moving through an inert space-time, but also geometrical properties of space-time are determined by the conditions for conformal invariance.

Although (3.16) is given as a perturbation expansion in powers of α' there are classes of curved spaces which are known to be solutions of this equation (together with the other $\beta = 0$ equations) to all orders in α'. Consider, for example, a superstring theory which is initially defined in ten space-time dimensions and assume that four dimensions are flat and six spatial dimensions are curved. In this case the solutions to the $\beta = 0$ equations are six-dimensional spaces known as *Calabi–Yau* spaces.* The resulting four-dimensional theory has space-time supersymmetry as well as other features which suggest that it may describe a kind of Grand Unified Theory.

Figure 3.1. (a) A world sheet for the scattering of two incoming strings to give two outgoing strings. (b) The conformal symmetry of string theory means that the parameter-space of (a) can be mapped to a standard sphere, with the external particles mapped onto the dots.

String tree ampltudes

The Polyakov–Feynman path history method can be used to evaluate string scattering amplitudes. For example, the four-string on-shell tree amplitude is associated with a sum over world-sheets of the form shown in fig. 3.1 (a). This picture shows that the strings incoming from infinity interact *locally* by touching at a point and joining to form an intermediate string which then breaks into the strings which propagate out to infinity. The parameter space of this world sheet has the topology of a sphere. The integration over metrics is trivially eliminated by choosing a particular conformal gauge, for example, that of a standard (Riemann) sphere. The incoming and outgoing strings are mapped into infinitesimal punctures labelled by dots in fig. 3.1 (b). Each puncture is represented by the insertion of a *Vertex operator*, $V(k_r, z_r, \bar{z}_r)$ (where k_r^μ is the momentum carried by the external particle), which is a conformal field. Conformal invariance dictates that the vertex operator has unit conformal weight which requires

* These are Kähler spaces with vanishing first Chern class.

the external states to be on their mass shells, *i.e.*, they carry momentum k_r satisfying $k_r^2 = -m^2$. The amplitude is then given by a correlation function of vertex operators,

$$A(k_1, \ldots, k_M) \sim g^{M-2} \int DXD\psi e^{-S} \prod_{r=1}^{M} \int d^2 z_r V(k_r, z_r, \bar{z}_r). \qquad (3.17)$$

In the simplest models the integrations over the embeddings (X^μ, ψ^μ, ...) can be performed since they are gaussian in this gauge and the resulting expression for the amplitude is then given as an integral over the z_r. In superstring theory the world sheet becomes a *super*-world sheet with ordinary coordinates z, $\bar{z}$ and anticommuting coordinate θ. Superstring amplitudes are then given by an integral over the positions of the M external states over the super world sheet,

$$A(k_1, \ldots, k_M) = g^{M-2} \int \prod d^2 z_r d\theta_r F(\{k_r, z_r, \theta_r\}). \qquad (3.18)$$

When the on-shell external scattering particles are massless this expression reduces at low energy to the sum of the Feynman diagrams for the scattering of these states in a standard field theory of (super)-Yang–Mills coupled to (super)gravity. It is one of the attractive features of string theory that all the Feynman diagrams of any given order are packaged together into one expression.

One-loop amplitudes - modular invariance

The study of one-loop scattering amplitudes raises some important new issues. These are the lowest order radiative corrections to the tree amplitudes.

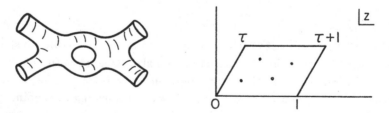

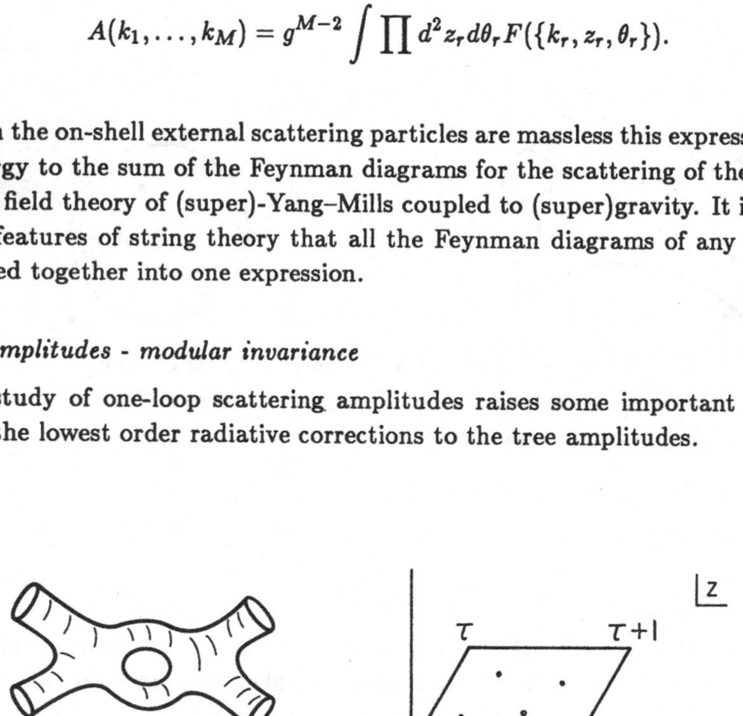

Figure 3.2. (a) A toroidal world sheet representing the one-loop correction to the four-string scattering amplitude. (b) Representation of the torus as a parallelogram. The complex *modulus* τ labels tori which are not equivalent under reparametrizations and conformal transformations. The external particles are mapped onto the dots.

In this case the world-sheet is a torus (fig. 3.2 (a)). In contrast to the case of the sphere, not all tori are equivalent up to reparametrizations and Weyl transformations. In other words the integration over all intrinsic metrics $g_{\alpha\beta}(z, \bar{z})$ cannot simply be performed by choosing a particular conformal gauge slice. There is a one (complex) parameter class of gauge inequivalent metrics on a torus which are labelled by the modulus, τ (fig. 3.2), which is the complex structure of the torus. The expression for the amplitude is therefore given by an integral over τ as well as over the positions of the external particles over the parallelogram in fig. 3.2. There are large reparametrizations of the toroidal world-sheet which are not continuously deformable to the infinitessimal ones considered earlier. For example, if the torus in fig. 3.2 (a) is cut along a circle and the two edges reglued after a relative rotation of 2π, a twist is introduced into the parametrization which cannot be undone continously. All transformations of this type are expressed in terms of τ as combinations of the two transformations

$$\tau \to \tau + 1, \qquad \tau \to -\frac{1}{\tau}. \tag{3.19}$$

Successive transformations of these types leads to a general transformation of the form

$$\tau \to \tau' = \frac{a\tau + b}{c\tau + d}, \tag{3.20}$$

where a, b, c, d are integers satisfying $ad - bc = 1$. This defines the group $SL(2, Z)$, or the *modular* group. Reparametrization invariance of string theory requires invariance under modular transformations as well as infinitesimal reparametrizations.

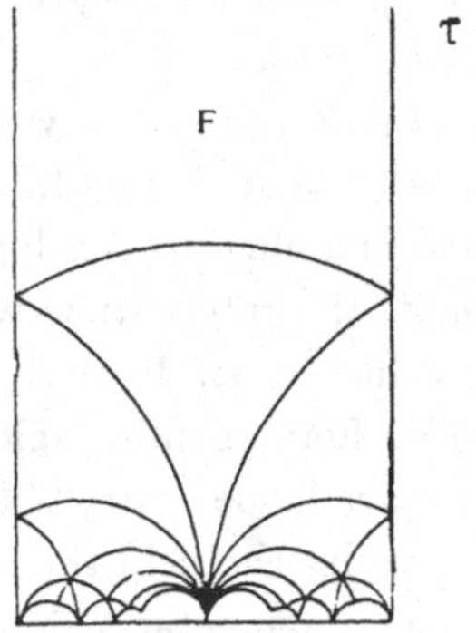

Figure 3.3. The τ integration region. Modular symmetry relates the integrand in the infinite number of regions in the upper half τ plane. The region marked F is the only region (up to $\tau \to \tau+1$ transformations) which is disconnected from the Im$\tau = 0$ axis.

This means that not all values of τ should be included in the integration over metrics since those related by (3.20) are equivalent. Figure 3.3 shows how the τ plane divides up into fundamental regions related by modular transformations. The τ integration is taken over only one of these regions, such as the one marked F. These comments generalize to multi-loop amplitudes which are associated with more moduli analogous to τ. The analogue of the upper-half τ plane is generally known as Teichmüller space while the modular group generalizes to the mapping class group. The space of moduli is therefore a fundamental domain of the mapping class group.

The simplest one-loop amplitude is the lowest-order contribution to the cosmological constant, Λ, which is a measure of the curvature of the universe in the absence of matter. This is determined by the vacuum energy density which is given by loop diagrams with no external particles. Experimentally the cosmological constant is known to be zero or extremely close to zero, $\Lambda < 10^{-120} M^4_{planck}$, when expressed in the natural units. In conventional quantum field theory the characteristic value of Λ generated by quantum fluctuations of the vacuum is $\Lambda \sim 1 \times M^4_{planck}$, one of the worst mistakes in the history of science!

In the simplest string theory, the 26-dimensional bosonic string theory, Λ is given, at one loop, by

$$\Lambda = \int_F \frac{d^2\tau}{(\mathrm{Im}\tau)^2} (\mathrm{Im}\tau)^{-12} e^{4\pi \mathrm{Im}\tau} \left[\prod_{n=1}^{\infty} (1 - q^{2n})(1 - \bar{q}^{2n}) \right]^{-24}, \tag{3.21}$$

where $q = e^{i\pi\tau}$. The measure $d^2\tau/(\mathrm{Im}\tau)^2$ is modular invariant as is the rest of the integrand. The power -24 arises from the fact that the bosonic theory is only conformally invariant in $D = 26$ space-time dimensions. In a more general theory the corresponding expression is of the form

$$\Lambda = \int \frac{d^2\tau}{(\mathrm{Im}\tau)^2} \sum \mathcal{N}^{i\bar{j}} \chi^i(\tau) \chi^{\bar{j}}(\bar{\tau}), \tag{3.22}$$

where χ^i is the character of the highest-weight representation of the Virasoro algebra associated with conformal weight h^i.

Several features of (3.21) and (3.22) are worthy of note:

It might have seemed plausible that Λ in (3.21) should equal the sum of the cosmological constants of the infinite number of pointlike fields propagated by the string. That result would have followed if the integration was over the whole semi-infinite rectangle $-1/2 < \mathrm{Re}\tau < 1/2$, $0 < \mathrm{Im}\tau < \infty$. In that case the answer would have had an obvious infinite factor since each fundamental region contributes the same nonzero amount to Λ. In fact, the integral has to be restricted to a single fundamental region, giving an answer that effectively divides the field theory result by an infinite factor.

The integral in (3.21) over the fundamental region F has no ultraviolet divergences since these can be shown to arise from the region $\mathrm{Im}\tau \to 0$, which is excluded. This comment also generalizes to loop amplitudes with external particles. This is in marked contrast to point particle theories of gravity where the ultraviolet divergences make the perturbation theory diagrams infinite in a nonrenormalizable manner. The expression (3.21) does, nevertheless, diverge due to the $\mathrm{Im}\tau \to \infty$ endpoint. This divergence is related to the presence of a tachyon in the bosonic theory which is absent in superstring theories (which have $\Lambda = 0$) as well as in certain non-supersymmetric theories (such as one associated with the symmetry $O(16) \times O(16)$) which have finite values of Λ.

In superstring theories the integrand of Λ vanishes point by point in τ space. Unfortunately, we do not expect the experimental vanishing of the cosmological constant to be explained by this mechanism since this would require supersymmetry to be unbroken in the real world.

Moore has pointed out the possibility, even in non-supersymmetric string theories, of a symmetry (known as Atkin–Lehner symmetry) of the integrand which is nonlocal in τ which causes the integrated cosmological constant to vanish. At present only one such example is known - at one loop in a two-dimensional theory - and it is not clear that any others exist. Furthermore, as mentioned above we do not expect that the observed value $\Lambda = 0$ should be determined by perturbation theory.

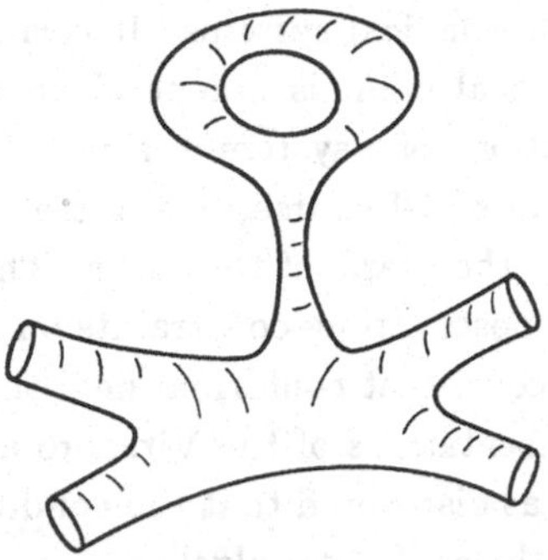

Figure 3.4. Divergences may arise in loop amplitudes due to the sum over worldsheets in which a toroidal tadpole disappears into the vacuum with a very thin neck. The propagation of a massless scalar (dilaton) particle in the neck causes the divergence. The coefficient of the divergence is proportional to the dilaton vacuum expectation value (which is also the cosmological constant).

Even though there are no ultraviolet divergences, string loop amplitudes may still diverge due to the emission of a closed-string massless scalar state (the 'dilaton') into the vacuum. In the sum over histories there are world sheets in which a closed string disappears into the vacuum via a toroidal tadpole. Since there is no momentum transferred through the neck ($k = 0$) the dilaton propagator ($1/k^2$) is infinite. The tachyon state also gives rise to divergences. Divergences of this type (which are proportional to the dilaton vacuum expectation value) suggest that the theory has been defined around an incorrect vacuum state. These divergences are also manifested as anomalies in the BRST symmetry. Fischler and Susskind and others have shown how such anomalies in one-loop diagrams can be cancelled by modifying the tree diagrams. This mechanism redefines the vacuum state to include a condensate of dilatons which gives rise to a compensating tree diagram anomaly.

In superstring theories space-time supersymmetry ensures that the dilaton expectation value (which is equal to Λ) vanishes so that the one-loop amplitudes are finite. The simplest examples are the ten-dimensional type 2 superstring theories which have been known to have finite one-loop amplitudes since 1982.

None of the published modular-invariant expressions for one-loop closed-string amplitudes is quite correct since they are real expressions while the amplitude has overlapping branch cuts! In practice, the standard analysis of the unitarity properties of these amplitudes breaks them into pieces which do not have overlapping branch cuts. This, however, apparently does violence to modular invariance. A proper analysis requires an understanding of the '$i\epsilon$' prescription that defines causal amplitudes. This has not yet been carried out.

An important feature of (3.22) is that it is given by a sum of terms which are products of holomorphic and antiholomorphic functions of the modulus (*i.e.*, functions of τ and functions of $\bar{\tau}$). This 'holomorphic factorization' is a very general feature that originates from the fact that the left-polarized and right-polarized modes contribute independent factors to the amplitudes (apart from subtleties with the zero modes). This generalizes to multiloop amplitudes and has deep geometrical significance.

The question of modular invariance is also of great significance in critical behaviour of two-dimensional statistical mechanical systems. It corresponds to the fact that the behaviour of a system at a critical point is independent of the boundary conditions. For example, the partition function* of a system in a rectangular box is invariant under interchange of the axes of the box. Of course, in the statistical mechanics context no integration is performed over τ, the shape of the torus. The condition that Λ is an integral of a modular function imposes strong constraints on which representations occur in the sum over i in (3.22). A consistent conformal field theory is therefore one with a particular combination of representations of the Virasoro algebra. In an important recent development E. Verlinde has discovered that the conditions for modular invariance are related to local features of the conformal algebra.

Multiloop amplitudes

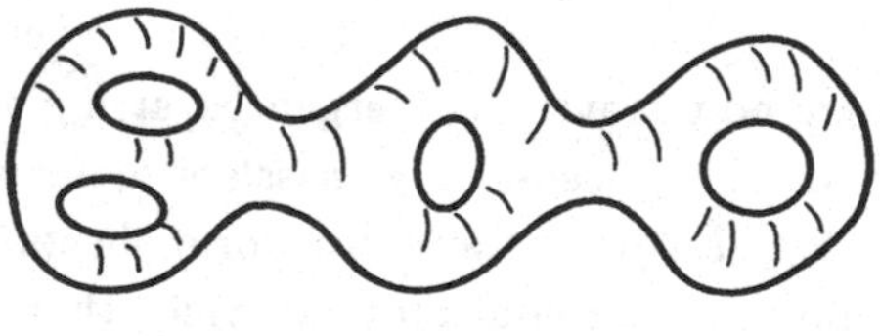

Figure 3.5. A g-loop amplitude is described by a world sheet of genus l.

The multiloop amplitudes for closed string theories are defined by integrating over world sheets of higher genus. The genus l contribution has a coefficient of g^{2l-2} which arises from the e^{-S_2} term in the functional integral (where S_2 was defined in (2.7)). Amplitudes with M external on-shell amplitudes also have a power of g^M. The fact that there is precisely one string diagram at each order in perturbation theory is an incredible simplification compared to standard pointlike quantum field theories, in which there is a dramatic escalation of diagrams as the order increases.

The multiloop diagrams are Riemann surfaces of nontrivial topology. The integration over all possible intrinsic metrics, $g_{\alpha\beta}(\sigma,\tau)$, now reduces to a finite-dimensional integral over $3l-3$ complex moduli (analogous to the one-loop modulus, τ) when $l>1$. This number is given by the Riemann–Roch theorem (roughly speaking each added

* The partition function in statistical mechanics is the analogue of the cosmological constant discussed earlier.

handle is made by gluing two circular holes - each hole is specified by a complex position and a radius). The l-loop cosmological constant for the bosonic theory in a flat space-time background has a form that generalizes (3.21),

$$\Lambda^l = g^{2l-2} \int dy_1 \wedge \ldots \wedge dy_{3g-3} \wedge d\bar{y}_1 \wedge \ldots \wedge d\bar{y}_{3g-3}$$
$$(\det \mathrm{Im}\mathrm{T})^{-13} \mathrm{M}(\mathrm{y}_1, \ldots \mathrm{y}_{3g-3}) \bar{\mathrm{M}}(\bar{\mathrm{y}}_1, \ldots \bar{\mathrm{y}}_{3g-3}), \tag{3.23}$$

where y_i, $\bar{y}_i$ are the moduli and $\mathbf{T}$ is the 'period matrix' which is a matrix function of the moduli (that replaces τ in the one-loop case). The $\wedge$'s signify that the integrand is a $6g-6$ form on moduli space.

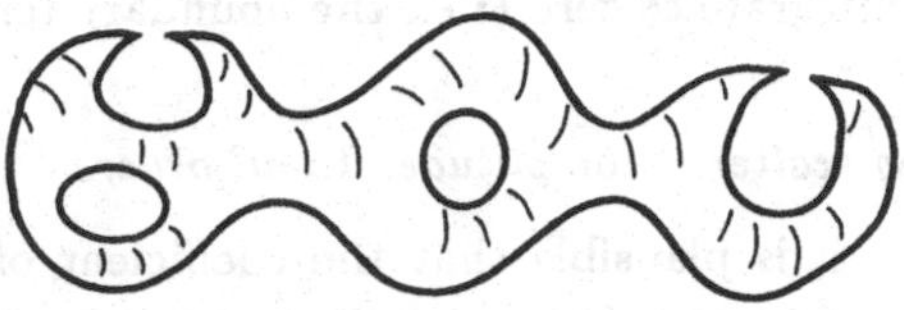

Figure 3.6. The amplitudes with external states can be obtained from the expression for the integrand of the cosmological constant as a function over moduli space. Various boundaries in moduli space correspond to handles pinching and splitting. The series of terms obtained by expanding in powers of the radius of the neck correspond to amplitudes with external states of increasing mass.

The expression (3.23) (which was obtained by Belavin and Knizhnik whose recent premature death is a very great loss to physics) again exhibits the holomorphic factorization (factorizing into the functions M and $\bar{M}$) that was encountered in the one-loop case. The integrand is also invariant under modular transformations (transformations of the mapping-class group) and the y_i's are integrated over one fundamental region of this group. As for the one-loop amplitudes, the divergences of this expression can be associated with the emission of massless states and tachyons into the vacuum rather than ultraviolet divergences (fig. 3.4). Amplitudes with external on-shell particles attached can be obtained from knowledge of the integrand of (3.23) as a density on moduli space. For example, there is a boundary in moduli space at which a given handle becomes very thin, forms a node and then breaks apart. This is illustrated in fig. 3.6. The amplitudes are recovered by expanding the integrand of (3.23) in powers of the radius of a neck that is being pinched.

Much effort has gone into constructing the multiloop amplitudes by various other techniques. These include: The direct evaluation of the functional integral over world-sheet metrics and embeddings with vertex operators representing the external states; sewing together the external legs of tree diagrams to make loops; group theory techniques for determining the form of the amplitudes; ....

There has also been very much effort invested in generalizing this analysis to the case of closed superstring amplitudes, using the formalism with world-sheet supersymmetry. In this case there are $(2l-2)$ complex *super*moduli which are anticommuting (Grassmann) coordinates - the supersymmetric analogues of the moduli. The cosmological constant is now expressed as an integral over a density in supermoduli space which again factorizes into holomorphic and antiholomorphic factors. The amplitudes can again be recovered by the pinching of handles of the super-Riemann surface in a manner analogous to fig. 3.6. The limit of zero radius for a handle is now defined by a boundary in the space of moduli and supermoduli.

There has been some confusion in the literature resulting from the fact that the parametrization of the supermoduli is not unique. After integrating over the supermoduli the expression for the integrand of the cosmological constant should vanish for all values of the ordinary moduli, y_i. However, different parametrizations of the supermoduli give rise to expressions which differ by total derivatives of y_i. In a consistent theory such total derivatives should integrate to zero (*i.e.*, the boundary terms should vanish).

Finiteness of superstring scattering amplitudes to all orders

As in the one-loop case, it is plausible that the coefficient of the divergence of a multiloop amplitude is proportional to the cosmological constant. In this case it is necessary, and possibly sufficient, to have $\Lambda = 0$ for the amplitude to be finite. Atick, Moore and Sen have given a somewhat indirect proof of the vanishing of the cosmological constant to all orders in the case of heterotic theories.

The formulation of string theory based on the light-cone parametrization (in which $\tau = X^0 + X^{D-1}$) leads to a very useful parametrization for the supermoduli. Mandelstam has used the light-cone gauge to construct the amplitudes for the ten-dimensional type 2 theories and claims that they are finite to all orders in perturbation theory. His parametrization avoids certain singular terms that are problematic in other parametrizations. Mandelstam's expression for the amplitudes is similar to the expression derived from another light-cone analysis based on the space-time supersymmetric formalism by Restuccio and Taylor (but the presence of contact interactions in that formalism makes the analysis rather complicated). The analyses of Restuccio and Taylor and of Kallosh and Morozov (see later) also point to the finiteness of the heterotic theories so it seems that the perturbative analysis of superstring theories is in good shape.

4. MODEL BUILDING

We have seen that consistent string theories can be built from representations of the Virasoro algebra with $c = 26$ or $c = 15$, in the case of the super-Virasoro algebra. In closed string theories the left polarized modes are to some extent independent of the right polarized modes. Denoting the left-moving and right-moving Fock spaces by $|L, p\rangle$ and $|R, p\rangle$, respectively, a general closed-string state has the form

$$|L\rangle \otimes |R\rangle. \tag{4.1}$$

The only link between the left-movers and right-movers is that they have the same space-time momentum, p^μ. Otherwise $|L\rangle$ and $|R\rangle$ are independent states that generally belong to different representations of the Virasoro algebra. This means that there are two classes of supersymmetric string theories.

The first class is that of **Type 2** theories in which both left and right-movers have supersymmetry and hence $c = 15$. There are two possible models of this type one of which is non-chiral (type 2a) while the more interesting one is chiral (type 2b).

The second class of theories has $c = 15$ for the right polarized modes which therefore correspond to a superstring theory and $c = 26$ for the left polarized modes which therefore correspond to the bosonic theory. This is the class known as **heterotic** theories. Since the left-moving and right-moving modes of the heterotic string are different, theories of this type are chiral with respect to the world sheet.

The value of c may be composed from smaller values. In particular, if we wish to consider string theories in D flat space-time dimensions (with $D = 4$ being of particular interest) the total c is given by

$$26 = c_i + D, \qquad \text{or} \qquad 15 = c_i + \frac{3}{2}D, \tag{4.2}$$

where the residual c_i represents the part of c which must be provided by 'internal' symmetry. In $D = 4$ space-time dimensions c_i must obviously take the values

$$c_i = 22, \qquad \text{or} \qquad c_i = 9. \tag{4.3}$$

These values may be obtained from tensor products of representations of Virasoro algebras with smaller values of c such that the value of the sum of the c's add up to 22 (for bosonic strings) or 9 (for superstrings). The particular combinations of representations that are permitted are highly constrained by requiring modular invariance of the loop amplitudes.

The earliest four-dimensional string models were those constructed (by Candelas, Horowitz, Strominger and Witten) by curling up six dimensions in the ten-dimensional models. There are powerful restrictions on these six compactified dimensions which require the resulting four-dimensional theory to be supersymmetric in space-time. The known spaces of this type are the Calabi–Yau spaces referred to earlier, as well as spaces which are flat everywhere except at points. The latter spaces are known as orbifolds. There are many attractive features of these models as far as their possible relevance to observed physics.

A more general study of all possible string theories requires the classification of all possible modular invariant unitary conformal field theories with arbitrary c. This also amounts to a classification of all possible unitary critical two-dimensional statistical mechanical models. This is a subject of intense research and there has been very much recent progress. since the original work by Belavin, Polyakov and Zamolodchikov (which was based on Polyakov's conformal bootstrap of the early 1970's).

Only models with $c > 0$ and $h_i \geq 0$ are unitary. For $c < 1$ there is a complete classification (discovered by Friedan, Qui and Shenker) which consists of an infinite series of models with different discrete values of c. These are labelled by an integer, m (≥ 1),

$$c = 1 - \frac{6}{(m+2)(m+3)}. \tag{4.4}$$

These discrete values of c start at $c = 1/2$ (the $m = 0$, $c = 0$ case is trivial) and condense at $c = 1$ as $m \to \infty$. For each of these values of c the dimensions weights of the highest-weight representations are given by

$$h_{pq} = \frac{[(m+3)p - (m+2)q]^2 - 1}{4(m+2)(m+3)}, \tag{4.5}$$

where $p = 1, 2, \ldots, m+1$ and $q = 1, 2, \ldots, p$. The constraints of modular invariance were first considered by Cardy, since when Cappeli, Itzykson and Zuber, as well as Nahm, Gepner and others have given the complete classification of all modular invariant theories with $c < 1$. The simplest theory ($c = 1/2$) is the Ising model, for which there are three highest-weight representations (with $h = 0, 1/2$ and $1/16$). The next theory ($c = 7/10$) is the tricritical Ising model. More generally the representations for a given c package together in different ways to give modular invariant partition functions of the form (3.22). Therefore there are generally several different critical statistical models with a given value of c.

For $c \geq 1$ there are unitary representations of the Virasoro algebra for all values of c, for each of which there are *infinite* numbers of values of h_i. However there is no systematic classification of the modular invariant theories. There are certain discrete infinite series of models which are generalizations of (4.4) incorporating extra symmetries. For example, the series of models with

$$c = \frac{3}{2}\left(1 - \frac{8}{(m+2)(m+4)}\right), \tag{4.6}$$

($m \geq 1$) have two-dimensional supersymmetry. This is an infinite series of models with $c < 3/2$. Obviously, any model with $c < 1$ must have ocurred on the list (4.4) so that the model at $c = 7/10$ is now seen to be a supersymmetric model.

Many other discrete series of models are known incorporating extra symmetries described by infinite dimensional Kac–Moody algebra. A unified group-theoretic construction of these series, known as the "coset" construction, has been given by Goddard, Kent and Olive. This leads to a beautiful description of modular invariants in terms of the characters of a finite number of highest-weight representations of the Kac–Moody algebra. A particularly interesting series of models from the point of view of superstring theory is the series with extended, $N = 2$, two-dimensional supersymmetry with c values

$$c = \frac{3m}{m+2}, \tag{4.7}$$

for which $1 \leq c < 3/2$. These models are specially interesting because the two world-sheet spinors in these theories are interpreted in string theory as a single (four-dimensional) space-time spinor associated with space-time supersymmetry. We saw earlier that a consistent superstring theory requires $c_i = 9$ which is larger than any terms in this series. (For example, the Calabi–Yau spaces that describe the six compactified dimensions in the early work on superstring theory can be considered to be two-dimensional theories with $c = 9$ and $N = 2$ world-sheet supersymmetry.) It is therefore of interest to consider the tensor products of models in the $N = 2$ discrete series. For example the tensor product of 5 models with $c = 9/5$ gives a superstring theory which has been analysed to some extent by Gepner who also considers many other possibilities. These discrete statistical mechanical models are the same as superstring theories in which six dimensions have been compactified on particular spaces which preserve supersymmetry in the remaining flat four dimensions, namely the Calabi–Yau spaces and orbifolds referred to earlier.

This demonstrates explicitly how the analysis of two-dimensional statistical mechanics can lead to an understanding of the compactified space in which the string is moving. It is particularly interesting that certain Calabi–Yau spaces can be described in terms of two-dimensional statistical models.

Comments

There are very many string theories in four dimensions. This classification is based on considering properties of the free theory along with the modular constraints deduced from the one-loop cosmological constant. In addition, finiteness of one-loop amplitudes (or vanishing of the one-loop cosmological constant) singles out the models with space-time supersymmetry (*i.e.*, superstrings) as the only possibly consistent ones. There are good arguments which suggest that these conditions are sufficient to give consistent amplitudes to all orders in string perturbation theory although certain problems can arise. (For example, certain conformal field theories are consistent with modular invariance at one loop but supersymmetry is broken - technically, Fayet–Illiopoulos D terms are generated - in a manner that makes the two-loop cosmological constant non-zero due to one-loop effects.)

Many methods are used to construct such string theories. Among these are: Calabi–Yau spaces; orbifolds; world-sheet fermions; bosonized fermions; current algebra; asymmetric orbifolds; .... Almost all known models are connected to each other by continuous parameters. Many of these models can also be related to the ten-dimensional models by scaling the radii of compactified dimensions to infinity.

For each model there is a determined spectrum of massless particles and couplings between them. There is therefore some hope of selecting a physically interesting string theory on the basis of experimental input, such as the number of generations of quarks and leptons, the absence of very rapid proton decay, ... For the moment models of the heterotic type ($c = 15$ for the right-movers and $c = 26$ for the left-movers) are the only ones which are able to accomodate the standard model with the observed particle representation content. Type 2 models ($c = 15$ for both left-movers and right-movers) come tantalizingly close but, in our present way of thinking about phenomenology (which might be wildly wrong), they are unable to accomodate the complete standard model.

Some of the consistent heterotic models have a spectrum of massless states and interactions that corresponds quite closely to that of the standard model $SU(3) \times SU(2) \times$

$U(1)$. Unfortunately, since space-time supersymmetry is unbroken in perturbation theory, more detailed phenomenology is not possible until nonperturbative effects are understood. Similarly, the implications of string theory for the early history of the universe and other cosmological phenomena must await a deeper understanding of the theory.

The apparently vast array of string "theories" should be viewed as different semi-classical approximations to a more fundamental theory. This means that they are different solutions of the same set of equations, such as the equations for (super)conformal invariance, (3.15).

The construction of particular embedding space-time geometries by assembling different two-dimensional statistical mechanical models illustrates again the interplay between properties of the world sheet and the space-time in which it is embedded. It seems that the properties that we need for an effective space-time are also incorporated in the world-sheet. These include: coordinate invariance, chirality and supersymmetry. This interplay between the geometry of the world sheet and space-time ought to be a clue for understanding a more fundamental formulation of string theory.

5. SPACE-TIME SUPERSYMMETRY

The formalism based on world-sheet supersymmetry is not manifestly covariant with respect to space-time supersymmetry. This obscures the essential fact that superstring theory is supersymmetric in space-time. It ought, therefore, to be fruitful to formulate superstring theory in a covariant manner which would make space-time supersymetry manifest. In conventional supersymmetric field theories, for example, many features of perturbation theory (such as non-renormalization theorems) are much clearer when described in terms of superspace coordinates. In string theory these space-time superspace coordinates are denoted

$$X^\mu(\sigma,\tau), \qquad \Theta^a(\sigma,\tau), \tag{5.1}$$

where the index a denotes a chiral space-time spinor which has $2^{(D-2)/2}$ components and is a Grassmann variable (it is an anticommuting fermionic variable). In ten dimensions this spinor can be chosen to be real (*i.e.*, it is a Majorana spinor) and it has sixteen components. The coordinates (5.1) are scalars with respect to transformations of the world sheet. Space-time supersymmetry is defined by the transformations

$$\delta X^\mu = i\bar{\epsilon}\gamma^\mu\Theta(\sigma,\tau), \qquad \delta\Theta = \epsilon, \tag{5.2}$$

where γ^μ are the ten-dimensional Dirac matrices and ϵ^a is a constant spinor parameter. In type 2 theories there are two D-dimensional spinor coordinates, Θ_1^a and Θ_2^a.

There is a classical action based on these coordinates which has a very different form from the actions discussed earlier. In flat space it is given by the sum of three terms (for the heterotic theory)

$$S = S_1 + S_2 + S_3, \tag{5.3}$$

where S_1 is the obvious supersymmetric generalization of (2.4) in which $\partial_\alpha X^\mu$ is made supersymmetric,

$$S_1 = -\frac{T}{2}\int d\sigma d\tau \sqrt{g} g^{\alpha\beta}\Pi^\mu_\alpha \Pi_{\alpha\mu}. \tag{5.4}$$

In this expression $\Pi^\mu_\alpha = \partial_\alpha X^\mu - i\bar{\Theta}\gamma^\mu\partial_\alpha\Theta$ is manifestly supersymmetric. Notice that S_1 defines an interacting theory since it contains terms which are cubic and quartic in the coordinates. The second term in the action is

$$S_2 = -iT\int d\sigma d\tau \epsilon^{\alpha\beta}\partial_\alpha X^\mu \bar{\Theta}\gamma_\mu \partial_\beta \Theta, \tag{5.5}$$

which can also be shown to be supersymmetric. However, this is not totally obvious and only works in certain space-time dimensions, namely, $D = 3, 4, 6$, or 10. Such a restriction on the space-time dimension of a classical string theory is a new feature. Only the ten-dimensional theory gives a quantum theory with transverse states, of relevance for a theory with low-energy gauge symmetries. However, the other theories may well be the supersymmetric analogues of the subcritical bosonic theories studied initially by Polyakov. The third term in the action, S_3, describes the remaining internal coordinates of the heterotic theories and is irrelevant for the discussion of space-time supersymmetry.

Although the theory defined by this action appears to be interacting, the combination $S_1 + S_2$ has an intriguing local fermionic symmetry (analogous to one discovered by Siegel in the case of the supersymmetric point particle) which is not possessed by S_1 or S_2 alone. This symmetry enables the quantum theory to be analyzed in the light-cone parametrization ($\tau = (X^0 + X^9)$, $(\gamma^0 + \gamma^9)\Theta = 0$), in which the theory is a free theory. Although it is not ultimately satisfactory to fix a non-covariant parametrization such as the light-cone parametrization, it is very useful for many explicit calculations. For example, this is the easiest way to analyse the free string spectrum. Furthermore, the one-loop amplitudes are obtained in a very simple form and their finiteness is almost trivial to see. Whereas in the formalism with world-sheet supersymmetry the amplitudes are given as a sum over terms with different boundary conditions on the world-sheet fermions, in the space-time supersymmetric formalism there is only one term in the amplitude since there are no world-sheet fermions.

Recently there have been some interesting results using a variant of the light-cone parametrization. This is a parametrization (originally used by Carlip) in which $g^{\alpha\beta} = e^\phi \eta^{\alpha\beta}$ and $(\gamma^0 + \gamma^9)\Theta = 0$. The theory is still conformally invariant in this parametrization so that many of the attractive features of conformal field theory are retained. Kallosh and Morozov have used this gauge to give a proof of the nonrenormalization of multiloop amplitudes with less than four external external on-shell massless particles to all orders in perturbation theory. Their proof uses a very simple power-counting argument, originally used for the one-loop amplitudes. Grisaru, Nishino and Zanon have also discussed the theory in a curved background in this gauge.

Unfortunately, covariant quantization of this formalism has proved to be very difficult. This is due in part to the fact that the action incorporates a peculiar mixture of phase-space constraints. However, recently a number of authors have made some progress. Siegel has recast the action in terms of a first-order formalism in which the troublesome second-class constraints are absent. The quantum theory is still complicated since the Fadeev–Popov ghost field action has further gauge invariances. In fact there must be an infinite number of Fadeev–Popov ghosts in the covariant quantum theory. This suggests an intriguing new structure which is currently the subject of study. One hopeful direction is the formalism of Siegel and Zweibach based on the supergroups $Osp(D+n, n|2n)$ (where the optimal value of n has yet to be determined). In another approach, Nissinov, Pacheva and Solomon introduce new "harmonic" coordinates.

Progress in this area raises the prospect of a deeper understanding of the interplay between space-time and world-sheet supersymmetry. There are, for example, some intriguing connections between this formulation of superstring theory and twistor space (as discussed originally by Witten and in a different context by Fairlie and Manogue, Shaw and Bengtsson). There are also connections with octonions and Jordan algebras (as discussed by Corrigan and Hollowood, Gürsey and others).

The covariant superspace action has also been generalized to describe super-"membranes" moving through space-time. This subject is too far from the theme of this article for me to review here.

6. PROBLEMS WITH PERTURBATION THEORY

We have seen that string theory is defined by a series of string diagrams analogous to Feynman diagrams. Certain mathematical features of these diagrams are very striking, in particularly the freedom from chiral anomalies and the finiteness of the diagrams of superstring theories. However, it is obviously not satisfactory to have the theory defined in terms of perturbation theory.

There are several arguments that sensible phenomenology can never emerge from string perturbation theory. For example, supersymmetry is unbroken to all orders - this is crucial for the diagrams to be finite. However, if supersymmetry is of any relevance in the physical world it is very obviously a broken symmetry. As long as supersymmetry is unbroken the string ground states (which are supposed to describe the observed elementary particles) have precisely zero mass. In order to understand the observed non-zero masses, which are tiny when expressed in Planck units, it is important to understand the non-perturbative mechanism that breaks supersymmetry. Since any familiar symmetry-breaking mechanism generates an unacceptably large cosmological constant a key problem is to understand why the observed cosmological constant is so small.

For perturbation theory to have any chance of being a good approximation the coupling constant must be a small parameter. However, in string theory the string coupling constant is not a free parameter. It is determined by the constant part of the expectation value of the dilaton field, $g = e^{-\phi}$, which follows from the topological term in the string action, S_2 (2.6). The value of ϕ should be determined dynamically and there is no reason for it to be very large.

There are several indications that the separation into diagrams of different genus is somewhat artificial. The distinction between a surface with l handles, with one infinitesimally thin one, and a surface with $l-1$ handles is very slight. More explicitly, the conditions for conformal invariance are modified by degeneration of handles. For example, the 'Fischler–Susskind mechanism' demonstrates how the conformal anomalies of loop diagrams of different genus are linked.

On purely aesthetic grounds it is unsatisfactory to be forced to choose one particular conformal field theory out of the huge number of possibilities as the semiclassical ground state of the perturbation expansion. If such an approximation is of relevance to real physics it is essential to understand the theoretical principle that selects one particular ground state.

The choice of a particular fixed background space-time appears to be at variance with the logic of the uncertainty principle at the Planck distance or smaller scales. At larger length scales strings behave like point particles, so there appears to be no scale at which strings should be the relevant degrees of freedom. This suggests that the perturbative expansion of string theory should diverge. Bosonic string perturbation theory has indeed been shown to be divergent by Gross and Periwal and is at best an asymptotic series. The same is probably true for superstring theories. This means that nonperturbative effects are certainly of importance.

The space-time metric, $G_{\mu\nu}$, is one particular mode out of the infinite number of string modes. In the perturbation expansion it is given a nonzero expectation value ($G_{\mu\nu} = \eta_{\mu\nu}$ for the flat space-time dimensions). Picking a non-zero value for a field in this manner is a signal that the theory is defined in a broken-symmetry phase - the phase below the Planck energy. A deeper understanding of string theory should involve an understanding of the degrees of freedom in the symmetric phase above the Planck energy.

7. BEYOND PERTURBATION THEORY

As yet there is no single compelling approach to a formulation of string theory that goes beyond the perturbation expansion. The only clues in discovering such an underlying formulation are that it should reproduce the perturbation expansion in some approximation (maybe a poor approximation) and it should presumably be based on an elegant principle. I will briefly mention a number of approaches.

String field theory

One rather obvious strategy, by analogy with second-quantization in relativistic point-particle physics, is to introduce fields defined on the space of string configurations, $\Phi[X^\mu(\sigma)]$. In string quantum field theory such fields create and destroy complete strings. [In the case of superstring theories the fields also depend on the fermionic world-sheet coordinates.] Such fields can be expanded as a sum over a complete basis with coefficients which are fields depending only on the zero modes, x^μ. These coefficients are the fields associated with each of the infinite number of states described by the string.

Several different formulations of open string field theory have been proposed. The most geometrical of these is the open-string theory of Witten* based on an action of the form

$$\int (A * QA + \frac{2}{3} A * A * A) \tag{7.1}$$

where the string field A can be thought of as a one-form in string space just as the vector potential in electrodynamics is a one-form in space-time. The BRST charge, Q, is an external derivative in this space (*i.e.*, $Q \sim d$) and the outer product $*$ of two string fields and integration operation are suitably defined. The action (7.1) is the string analogue of the integral of the Chern–Simons form that arises in ordinary three-dimensional gauge theories. The spectrum of the free theory is determined by the field equation

$$QA = 0, \tag{7.2}$$

which shows that the physical states are the cohomology classes of Q, as we saw earlier, (3.14).

It has been shown (by D'Hoker, Giddings, Martinec and Witten) that the perturbation expansion of the theory defined by (7.1) reproduces all the correct string world-sheet diagrams - Riemann surfaces with boundaries defined by the world lines of string endpoints. The string Feynman rules build the diagrams by sewing together flat sections of world sheet. In fact, the string Feynman rules define a simple triangulation for the space of moduli of Riemann surfaces with boundaries.

However, one of the unappealing features of this approach is that it does violence to the reparametrization-invariance of the world sheet since it selects out a particular parametrization. The process of assembling the world sheet from flat segments obscures the geometrical features of the complete string diagrams. Furthermore, the theory depends on a fixed background space-time metric since the definition of Q contains the metric. In fact Q is defined by conformal field theory and so (7.1) is appropriate for describing an expansion around a particular classical background.

In the case of closed-string field theory there are other severe problems. The calculation of the cosmological constant, Λ (the vacuum amplitude), described in an earlier section poses a particularly obvious problem for closed-string field theory. At lowest order Λ is determined by the toroidal world-sheet (parametrized by the complex modulus τ) which is a Riemann surface with no boundaries or punctures. None of the obvious candidates for closed-string field theory give Feynman rules which are able to correctly reproduce this diagram.

Suggestions have been made (by HIKKO and by Horowitz, Lykken, Rohm and Stominger) for how to formulate background-independent string field theory in which the

---

* A different approach, pioneered by Siegel has been pursued by Siegel and Zweibach, Hata, Itoh, Kugo, Kunitomo and Ogawa (HIKKO), Neveu and West and others.

kinetic term in (7.1) (the term involving Q) arises as an approximation from expanding the string field around a particular classical solution. At present such suggestions have a rather formal status and do not reveal the hoped-for compelling geometrical principle.

The superstring analogue of the bosonic open-string action, (7.1), was also proposed by Witten. This is based on the formalism with world-sheet supersymmetry and uses separate string fields for fermions and bosons which is analogous to the component formulation of ordinary supersymmetric field theory. Recently, it has been suggested (by Wendt) that further "contact' terms are needed in Witten's action to ensure the gauge invariance of this theory.

World-sheet dynamics

A different kind of approach towards finding a fundamental formulation of string theory is based on generalization of the geometrical properties of conformal field theories on (super)Riemann surfaces. Recall that any l-loop string amplitude is determined as an integral over the $3l-3$ ordinary moduli and $2l-2$ supermoduli of the closed surface of genus l, together with the punctures corresponding to the positions of the external on-shell states. The full perturbation expansion of the amplitude is given as a sum of integrals over the moduli spaces of the surfaces for all l. The integrations over moduli are the only remnants of the integration over the world-sheet metrics and the embeddings. This means that all of the properties of the embedding space("space-time") as well as the symmetry properties of any string theory are encoded in the properties of (super)moduli space. The analytic properties of the l-loop partition function as a density on (super)moduli space are of vital significance in determining properties of the theory, and in some sense they determine the nature of the embedding space-time. This is one of the most attractive features of string theory.

The moduli space for a diagram of a given l can be viewed as a boundary of the larger space of moduli of surfaces of larger l. This motivated Friedan and Shenker to suggest a setting for string theory based on an infinitely large space - 'Universal Moduli Space' - which in some sense contains the union of the moduli space of all Riemann surfaces of arbitrary genus, including disconnected surfaces. This huge space should define a completion of conformal field theories defined over the finite genus Riemann surfaces (rather as the real numbers are a completion of the integers). This proposal is designed to avoid the artificial separation of the theory into surfaces of different genus (so that a surface with a degenerating neck and the surface obtained after the neck has been severed are both included). The conventional string perturbation series emerges as a sum of boundary contributions to the integration over universal moduli space. The hope is that, at least in its supersymmetric version, the geometry of universal moduli space very constrained. Thus, whereas there may be many consistent modular conformal field theories at finite genus, there may be fewer consistent theories defined over universal moduli space.

One possible mathematical setting for these ideas is the Grassmannian. This can be thought of as the space of all possible Fock states of a field defined on a circular boundary of an arbitrary Riemann surface. The Grassmannian describes arbitrary surfaces of any genus with nodes and cusps as well as infinitely many other states which cannot be associated with any Riemann surface. Although the Grassmannian is a rather concrete expression of universal moduli space it is not yet clear if it is the appropriate setting for nonperturbative string theory. It is, however, an obviously interesting technical approach towards constructing the diagrams of arbitrary finite genus. It also has interesting connections to the KP equations which are nonlinear differential equations with soliton solutions.

A rather different approach to the search for a fundamental setting for string theory is based on the fact that the set of conformal field theories, of relevance to perturbative string theory, is a subspace of all possible two-dimensional field theories. Such general theories are parametrized by an infinite number of coupling parameters, generalizations of $G_{\mu\nu}, B_{\mu\nu}$ and Φ in the theories described earlier. These coupling constants, g^i, define an infinite-dimensional manifold $\mathcal{M}$. The subspace of conformal field theories is determined by the renormalization group equations

$$\beta^i(g)\frac{dg^i}{dt} = 0, \tag{7.3}$$

where the beta function β^i is a vector field in $\mathcal{M}$ and $t = \ln a$ where a is a spatial cutoff. There are many solutions of these equations

$$g = g_r^*, \tag{7.4}$$

which are the fixed points in $\mathcal{M}$ defining the conformal field theories (including all the ones defined in the partial classification described earlier). Each fixed point is therefore characterized by a particular value of the central extension c of the Virasoro algebra, together with a list of highest-weight representations with weights $\{h_i\}$.

The $\{g^i\}$ are analogous to the component fields of a string field and the infinite number of conditions on g^i, specified by (7.3), are similar to the string field theory equations of motion. This suggests that the space $\mathcal{M}$ is an appropriate fundamental setting for string theory, reducing to the usual conformal field theory formulation only in the classical approximation. Renormalization group trajectories interpolate between the fixed points, therefore interpolating between different classical string vacuum states. This suggests a quantum picture which includes tunneling between classical ground states, much as in ordinary quantum mechanics.

A prerequisite to studying this idea is an understanding of the connectivity of conformal field theories in the larger space. An important general result is that of Zamolodchikov who has shown that a certain funtion $C(g)$ which interpolates between the values of c (so that $C(g_r^*) = c_r$) satisfies

$$C(g_2) < C(g_1), \tag{7.5}$$

where g_1, g_2 are two points at t_1 and t_2 ($t_2 > t_1$) on any renormalization group trajectory. This means that for a renormalization group trajectory linking two conformal field theories, c_0 (the value of c at $t = 0$) is always greater than c_∞ (the value of c at $t = \infty$). These ideas might lead to an understanding of the nonperturbative stability of string theories as well as an understanding of which, if any, conformal field theory is a good approximation to a complete theory.

In this and other approaches the Riemann surface picture of world sheets should emerge as an approximation to the more fundamental underlying theory.

Ideas about the Planck phase

Earlier I indicated that strings are probably not the appropriate coordinates for a fundamental description of string theory. This point of view is supported by the evidence for the existence of a phase transition at the "Hagedorn" temperature, T_0, which is around the Planck temperature ($T_{Planck} = M_{Planck}c^2/k$). Such a phase transition is a very general thermodynamic feature of a system (such as string theory) with a density of single-particle states which increases exponentially with mass. In fact, Hagedorn had described transitions of this type in the context of his hadronic bootstrap which was a precursor to the original string theory. Originally the Hagedorn temperature was thought to be an "ultimate" temperature, reached only at infinite energy density. Some early work by Frautschi and by Carlitz suggested that the transition can occur at finite energy density and recently Kagan and Sathapalian have given a very simple explanation of this transition and pinpointed a relevant order parameter.

Atick and Witten have emphasised the strong analogy between this transition and the deconfining phase transition in four-dimensional QCD. In its low temperature phase QCD may be thought of as a theory of confined quarks and gluons analogous to a string theory, with an exponentially increasing density of states. Above the deconfining temperature the system is in its symmetric phase (the phase in which chiral symmetry is unbroken) and behaves like a gas of quarks and gluons with a free energy density that increases like $T^D = T^4$ asymptotically - the behaviour characteristic of $D = 4$-dimensional field theory. Just as QCD is understood from a fundamental point of view in terms of the coordinates of its high-temperature symmetric phase, one would expect that knowledge of the high temperature phase of string theory would provide the insight lacking in the string phase. Gross and others have emphasized the fact that there are indications from the asymptotic behaviour of string perturbation theory diagrams that there should be a phase with infinitely larger symmetry. In this case the masses of all the string states may in some sense be generated by spontaneous breakdown of this huge symmetry, in the same way as gauge boson masses are generated in the electro-weak theory.

Atick and Witten have used the analogy with QCD to argue that certain features of the high temperature phase may be deduced from the asymptotic behaviour of the series of string perturbation diagrams that describe the low temperature phase. In particular, they argue that the Hagedorn transition is strongly first order and that the free energy density in the high temperature phase increase only like T^2. This slow increase (which looks like the behaviour of a *2-dimensional* field theory!) is related to the absence of the ultaviolet catastrophe in string theory. This analysis is problematic in detail since the very notion of equilibrium and the definition of temperature can only be approximate in a theory of gravity due to the instability of an infinite gravitating system. Nevertheless it has pinpointed some dramatic ways in which string theory differs from any conventional field theory. Starting from an entirely different viewpoint Klebanov and Susskind have also concluded that there is evidence that string theory arises from a theory with the number of degrees of freedom appropriate to a 2-dimensional field theory (one space and one time).

In a separate development Witten has interpreted the mathematical results of

Donaldson concerning the topology of four-dimensional manifolds in terms of relativistic quantum field theory (generalizing non-relativistic results of Atiyah). He starts from a four-dimensional field theory with general coordinate invariance but no dependence on a space-time metric. Since the metric has vanishing expectation value the theory is one in which general covariance is unbroken. In such a phase there is no notion of distance and therefore no possibility of propagation of signals. All correlation functions are pure numbers - Donaldson's topological invariants that characterize the manifold.

These mathematically motivated ideas may eventually turn out to be connected to the ideas about the high temperature phase of string theory which should also be a phase in which general covariance is unbroken.

8. SUMMARY

String theory is in a phase of great activity with significant developments on various fronts.

String perturbation theory - conformal field theory - has lead to a partial classification of possible consistent theories in four space-time dimensions. Among the large number of apparently consistent models there are some which are remarkably close to explaining the observed phenomenology of elementary particle physics - the standard model with the correct anomaly-free assignment of representations for the quarks and leptons. This success indicates that string theory may well have the elements needed for a unified understanding of all the forces.

It is expected that space-time supersymmetry is unbroken and all observed particles are predicted to be massless to any order in superstring perturbation theory. However, the observed world is one in which supersymmetry is broken (if it is present at all) and particle masses are not zero (although they are miniscule on the scale of the Planck mass). It is therefore essential to understand non-perturbative effects in order to determine the observed mass spectrum. It is quite likely that nonperturbative effects will only be properly understood in terms of a reformulation of string theory - possibly along the lines of one of the proposals outlined earlier. Certainly, such a reformulation is essential for understanding the logic of string theory and how it gives rise to a dynamical theory of space-time unified with the elementary particles.

LECTURES ON HETEROTIC STRING AND ORBIFOLD COMPACTIFICATIONS

Jeffrey A. Harvey

Joseph Henry Laboratories
Princeton University
Princeton, NJ 08544

ABSTRACT

These lectures are concerned with the structure of the heterotic string and with special soluble compactifications of string theory. The first lecture discusses the connection between anomaly cancellation in the low-energy limit of consistent strings and the chiral current algebra structure of the heterotic string. The second lecture gives a brief introduction to toroidal and orbifold compactifications in string theory. The third lecture is based on work done in collaboration with G.Moore and C.Vafa and concerns "quasicrystalline" orbifold compactifications. These compactifications exhibit symmetries related to those of quasicrystals at irrational values of the background fields. They are most naturally formulated using certain ideas from number theory.

LECTURE I. AXION STRINGS AND HETEROTIC STRINGS

There are three special aspects of heterotic string. All appear in other string theories, but their role in heterotic strings is particularly striking or important. The first is the existence of different conformal structures for the left- and right-moving degrees of freedom. The second is the existence of two-dimensional chiral currents which can propagate along the string and which are responsible for the gauge degrees of freedom. The third is the crucial role that modular invariance plays in determining the allowed representations of the affine Lie algebras associated with these currents. In this lecture I will discuss the connection between the first two of these aspects of the heterotic string and its low-energy structure.

It is remarkable that combining the left-moving bosonic string and the right-moving superstring degrees of freedom produces a consistent and rich new type of string. However, little light has yet been shed on why this procedure seems to be work so well, or whether it is favored by nature as preliminary phenomenological investigations seem to indicate. I will not have anything to say about this, but I will try to explain how one can guess the structure of the heterotic string simply from the cancellation of anomalies at low energies and the assumption that there is a purely closed string theory which has this low-energy limit. This connection was mentioned briefly in.[1]

The starting point is a physical interpretation[3] of the anomaly descent equations.[2] The chiral anomaly in a global $U(1)$ current in $2n+2$ space-time dimensions in the presence of a background gauge field A is given by the $2n+2$ form $\omega_{2n+2} = \frac{1}{(2\pi)^n} TrF^{n+1}$ where $F = dA + [A, A]$ is the field strength associated with the gauge field A. The descent equations determine the $2n$ dimensional gauge anomaly ω^1_{2n} in terms of ω_{2n+2} via

$$\begin{aligned} \omega_{2n+2} &= d\omega_{2n+1} \\ \delta\omega_{2n+1} &= d\omega^1_{2n} \end{aligned} \tag{1}$$

Here ω_{2n+1} is the Chern-Simons form and δ indicates variation with respect to a gauge transformation $\delta_\Lambda A = d\Lambda + [A, \Lambda]$. This result has a deep mathematical basis, and is also of great practical use in reducing the algebra needed to determine anomalies in higher dimensional theories. of algebra.

To construct a model which provides a physical interpretation of these descent equations consider the following model (in four dimensions for simplicity). We will have a global, chiral $U(1)$ symmetry and a local, non-chiral $U(1)$ gauge symmetry. The Lagrangian is given by

$$L = -\frac{1}{4}F_{\mu\nu}F^{\mu\nu} + \bar{\psi} i \not{D}\psi + \partial_\mu\Phi^*\partial^\mu\Phi + \bar{\psi}(\Phi_1 + i\Phi_2\gamma_5)\psi - V(\Phi) \tag{2}$$

where $\Phi = \Phi_1 + i\Phi_2$ is a complex scalar field, ψ is a Dirac fermion minimally coupled to the $U(1)$ gauge field, and we will assume that the minimum of the potential $V(\Phi)$ occurs for $\langle\Phi\rangle = v$ with v nonzero. The vacuum expectation value of Φ breaks the chiral $U(1)$ symmetry but preserves the $U(1)$ gauge symmetry. Since $\pi_1(U(1)) = Z$ there exist topologically stable vortex solutions, albeit with divergent energy per unit length (because unlike vortices in a superconductor there is no long range gauge field coupled to the scalar field). For such a vortex the general form of the classical scalar field will be

$$\Phi = f(\rho)e^{i\theta(\phi)} \tag{3}$$

with $f(\rho) \to v$ as $\rho \to \infty$ and with θ having a nonzero winding number around the vortex. The phase of Φ will be a massless Goldstone boson which may be thought of as an axion field. We will consider the physics of fermions coupled to such a vortex or "axion string" and then later will imagine replacing the axion string by a macroscopic superstring.

It is often useful to make a chiral transformation on the fermion fields to remove the phase in the coupling of Φ to the fermions. Because of the chiral anomaly, this results in a couping of the form

$$\frac{e^2}{32\pi^2}\int \theta \epsilon_{\mu\nu\lambda\rho} F^{\mu\nu} F^{\lambda\rho} d^4x \tag{4}$$

in the effective Lagrangian. However, in the field of a vortex, θ is not single valued, so it is more correct to integrate by parts and write this coupling as

$$-\frac{1}{32\pi^2}\int \partial_\mu \theta K^\mu d^4x \tag{5}$$

where

$$K_\mu = 2e^2 \epsilon_{\mu\nu\lambda\rho} A^\nu F^{\lambda\rho} \tag{6}$$

Now we see that the spatial variation of θ gives rise to a current in a background electromagnetic field. Varying with respect to A^μ we see that the current is given by

$$J_\mu = \frac{e}{8\pi^2}\epsilon_{\mu\nu\lambda\rho}\partial^\nu \theta F^{\lambda\rho} \tag{7}$$

This demonstrates the well known fact that there are new contributions to the current in the presence of a spatially varying axion field. This current takes on an interesting form when the variation of the axion field carries non-trivial topological information. In particular, consider an axion string which runs along the z-axis and has unit winding number. Then we can take $\theta = \phi$ in cylindrical coordinates. If there is an electric field of strength E in the z-direction, then we see that there is a radial component of the current given by $J_\rho = -eE/4\pi^2\rho$. So it appears as if the axion string acts as a sink for electric charge in the presence of a background electric field. Of course charge should be conserved, even in the presence of a vortex , so the vortex must have some way to carry the charge that is flowing onto it. The fact that the charge flowing onto the vortex is proportional to E provides an important clue for how this can happen.

Recall that in 1+1 dimensions a $U(1)$ gauge field coupled to chiral charged fermions has an anomaly given by $\partial^a J_a = \frac{1}{2\pi}\epsilon^{ab}\partial_a A_b$ We can think of this as saying that charge is not conserved in the presence of background fields by an amount that is linear in the background field. This sounds very much like the proper two-dimensional description of the charge which flows onto our axion string. From the $1+1$ dimensional point of view, the charge inflow appears to violate conservation of charge since the charge appears from outside of the $1+1$ dimensional system. In fact one might predict that this system can only make sense by having chiral charged fermion zero modes which are trapped on the axion string and which allow for a consistent description of what is going on both from the four-dimensional point of view outside the string and also from the two-dimensional point of view on the string. In fact it is quite easy to show that there are precisely the right type of fermion zero modes for this interpretation to make sense. It is straightforward to solve the Dirac equation explicitly and to exhibit the required zero-mode solutions.[3] Their existence is also guaranteed by an index theorem analyzed in Ref. 4.

If you track down all the factors of 2 and π that appear in connecting the charge inflow due to the topology of the axion field and the charge appearing due to the two-dimensional anomaly, you find that it all hangs together beautifully. I will not go into the details here, they can be found in Refs. 3, 5,6, and 7. This example can be easily generalized to include gravitational anomalies and more general types of topological defects in higher dimensions.

The moral of this example is that it is sometimes possible to deduce the microscopic structure of zero modes which live on a topological defect by looking at low-energy or long distance couplings in the topologically non-trivial background provided by the defect.

Let us now try to apply this idea to superstrings. We will try to determine the microscopic structure by looking at the known low-energy couplings which must arise from any consistent string theory. Green and Schwarz[8] showed that gauge and gravitational anomalies can be cancelled in ten-dimensional supergravity coupled to gauge fields provided that the gauge group is $SO(32)$ or $E_8 \otimes E_8$. The key to their anomaly cancellation mechanism was the presence of an antisymmetric tensor field $B_{\mu\nu}$. In order to cancel anomalies they had to take B to vary under gauge and local Lorentz transformations in such a way that the generalized field strength $H = dB + (\omega_{3Y} - \omega_{3L})$ was gauge invariant. Here ω_{3Y} and ω_{3L} are the Yang-Mills and Lorentz Chern-Simons three forms. The kinetic energy term for B is given by $\int {}^*HH$ where ${}^*H$ is the dual of H. Expanding this out results in a coupling of the form

$$\int {}^*dB(\omega_{3Y} - \omega_{3L}) \tag{8}$$

This coupling will play a role which is analogous to the role played by the coupling of the axion field to F^2 in the previous example. Although it is possible to present the analysis working purely in ten-dimensional Minkowski space, this analogy is even stronger if one imagines compactifying six of the dimensions and then focuses on the part of $B_{\mu\nu}$ where all indices are four-dimensional. Then $B_{\mu\nu}$ acts precisely like the axion field in the previous example. To see this explicitly we use $\partial^\mu H_{\mu\nu\lambda} = 0$ to write $H_{\mu\nu\lambda} = \epsilon_{\mu\nu\lambda\rho}\partial^\rho\sigma$ for some scalar field σ. The coupling (8) is then equivalent to the axion-gauge field coupling (4) plus its gravitational counterpart with σ playing the role of the axion field.

Now suppose we want to invent a string theory which gives rise to such a coupling in the low-energy limit. Of course we know this happens for Type I open strings with gauge group $SO(32)$, but we want to obtain $E_8 \otimes E_8$ as well. Let us try to do it with a theory which only contains closed strings. Then it makes perfectly good sense to consider a macroscopic closed string since the string cannot break apart into open strings. Since the string eventually closes it is not topologically stable, but if it closes far away, or outside our horizon, we can think of it locally as being analogous to the vortex we considered earlier, that is we can consider the classical configuration of fields outside the string and see what effects we can deduce from the low-energy couplings we assume to be present.

In the case of axion strings the crucial ingredients were the coupling of the axion to the Chern-Simons form and the fact that the axion string acted as a source for the axion string. We have seen that σ couples to the $E_8 \otimes E_8$ and Lorentz Chern-Simons forms with opposite sign as a consequence of the low-energy anomaly cancellation. To show that a macroscopic heterotic string acts as a source of σ requires a simple vertex operator calculation which is given in Ref.1. The result is that for a string running along the z-axis

$$[\partial_x, \partial_y]\sigma = 2\pi\delta(x)\delta(y) \tag{9}$$

which is equivalent to saying that σ is proportional to the angular coordinate about the string, just as in the case of axion strings.

We can now vary the effective low-energy action to obtain the gauge and Lorentz currents in the presence of background gauge and gravitational fields. One finds as before that the string acts as a sink of gauge and Lorentz "charge", yet the full theory must be gauge and Lorentz invariant. So one concludes that there must be currents on the string which can carry both gauge and Lorentz quantum numbers, and the relative minus sign in the coupling of σ to the Chern-Simons terms dictates that the currents must flow in opposite directions along the string in order to correctly conserve charge. This of course is precisely the structure found in the heterotic string. The structure was not originally deduced in this way, but I think it is amusing that it could have been.

Once one has these currents there is still much to be done to formulate a consistent string theory. This has been discussed extensively elsewhere.[9] I think it is worth emphasizing however that the chiral current algebras which appear in the heterotic string have a very physical interpretation which rests on the consistency between the string point of view and the low-energy effective field theory point of view. Demanding consistency between these two points of view has been a very powerful tool in string theory, particularly in deriving information about string compactifications based on low-energy supersymmetry.[10]

LECTURE II. TOROIDAL AND ORBIFOLD COMPACTIFICATIONS

In this lecture I would like to describe certain very simple compactifications of string theory. Since string theory can be formulated simply and consistently in ten space-time dimensions, any attempt to use string theory to describe the real world must figure out what to do with the extra six dimensions. The approach which has been pursued most vigorously over the last few years is to look for six-dimensional compact spaces K which solve the classical string equations of motion. This is equivalent to demanding that the non-linear sigma model which describes string propagation on K be conformally invariant. If K is a Ricci-flat Kahler manifold (Calabi-Yau space) then it is known that one can construct such a conformally invariant sigma model by perturbing K in a well-defined way.[12] It is truly remarkable that the spectrum of low-energy fluctuations about such a background looks a great deal like the real world in as much as it contains a non-trivial gauge group with fermions in chiral representations. It was soon realized that K did not have to be a manifold, it could be an "orbifold" of a manifold, where one identifies points on the manifold under the action of some discrete group of symmetries of the manifold, and, in certain cases, this led to soluble string theories which had many of the nice features of Calabi-Yau compactifications.[13]

More generally, the role of the "extra" dimensions in string theory is to provide a (super) conformal field theory with the correct conformal anomaly to cancel the conformal anomaly coming form the ghost system and the free conformal field theory constructed out of the four-dimensional Minkowski space coordinates of the string. Thus, subject to certain restrictions, one can construct "compactifications" of string theory out of (super) conformal field theories which often seem to have nothing at all to do with geometry. However as recent results have shown,[11] this is often an illusion, and there is often a geometrical interpretation of the theory. Precisely what generalization of geometry is given by general (super) conformal field theories is an interesting open question.

In any event, orbifolds are an important class of soluble conformal field theories. They show that string theory can be formulated on spaces that are more general than manifolds, and in the case of asymmetric orbifolds, they show that the geometrical structure can be quite subtle. They are also a useful tool in trying to probe the more general structures of conformal field theory. In this lecture I will discuss the simplest types of orbifolds which appear as compactifications of the bosonic string. I will not discuss interactions on orbifolds. Ref.14 contains a more detailed review of orbifolds.

Toroidal Compactification

The simplest way to compactify D dimensions of the bosonic string is to take the internal space to be a D-dimensional torus

$$T^D = \mathrm{R}^D/\Lambda \tag{10}$$

where Λ is a rank D lattice. Thus we take the string to propagate not in 26-dimensional Minkowski space, but on the tensor product of (26-D)-dimensional Minkowski space with a D-dimensional torus. Since the torus is flat, this theory is still conformally invariant, and it is simple to solve for the spectrum of a string propagating on this space. I will consider only the closed bosonic string. Then the string coordinates on T^D satisfy a free two-dimensional wave equation $(\partial_\sigma^2 - \partial_t^2)X = 0$ where σ, t are coordinates on the string world sheet. The string coordinates are periodic in σ up to translation by a lattice vector which defines the lattice Λ. We can therefore split the string coordinates into left- and right-moving parts

$$X = X_L^i + X_R^i \qquad i = 1..D \tag{11}$$

where

$$X_L^i = x_L^i + p_L^i(t+\sigma) + \frac{i}{2}\sum_{n\neq 0}\frac{\tilde{\alpha}_n^i}{n}e^{-2in(t+\sigma)} \tag{12}$$

$$X_R^i = x_R^i + p_R^i(t-\sigma) + \frac{i}{2}\sum_{n\neq 0}\frac{\alpha_n^i}{n}e^{-2in(t+\sigma)} \tag{13}$$

and

$$(p_L, p_R) = (p/2 + w, p/2 - w) \tag{14}$$

where $w \in \Lambda$ are the winding numbers and $p \in \Lambda^*$ are the momenta. Here Λ^* denotes the dual lattice of Λ. We can think of the elements (p_L, p_R) as elements of a Lorentzian lattice $\Gamma_{D,D}$ with signature $(+^D, -^D)$ by defining the inner product of two elements to be $\langle p_L, p_R | p'_L, p'_R\rangle = p_L \cdot p'_L - p_R \cdot p'_R$. It is then easy to check using (14) that $\langle p_L, p_R | p_L, p_R\rangle$ is an even integer and that any vector which has an integer inner product with every (p_L, p_R) is also in the lattice. In other words, the vectors (p_L, p_R) define points in an even, self-dual, Lorentzian lattice $\Gamma_{D,D}$ with signature $(+^D, -^D)$.

String propagation on T^D gives rise to a conformal field theory with left- and right-moving central charges equal to D. The Virasoro generators are constructed as usual out of the moments of the stress tensor associated to the D free scalar fields X^i. In particular one has

$$L_0 = \frac{p_R^2}{2} + N_R - \frac{D}{24} \tag{15}$$

and

$$\tilde{L}_0 = \frac{p_L^2}{2} + N_L - \frac{D}{24} \tag{16}$$

where

$$N_R = \sum_{n=1}^{\infty} \alpha_{-n}^i \alpha_n^i \qquad N_L = \sum_{n=1}^{\infty} \tilde{\alpha}_{-n}^i \tilde{\alpha}_n^i \tag{17}$$

and the Hamiltonian of this system is simply $H = L_0 + \tilde{L}_0$.

A very natural object to consider is the partition function

$$Z(q, \bar{q}) = Tr q^{L_0} \bar{q}^{\tilde{L}_0} \tag{18}$$

From a mathematical point of view, $q$and $\bar{q}$ are just a formal complex variables and $Z(q, \bar{q})$ is the character of this particular representation of the left- and right-moving Virasoro algebras. From a physical point of view we write $q = e^{2\pi i \tau}$ and then $Z(q, \bar{q})$ gives the amplitude for a string to propagate for a time $Im\tau$ while being translated in σ by an amount $Re\tau$. $Z(q, \bar{q})$ can be calculated as a path integral over the torus which depends only on the conformal equivalence class of the torus and is manifestly coordinate independent. From this viewpoint τ is the modular parameter of the torus and $Z(q, \bar{q})$ should be invariant under modular transformations which correspond to global diffeomorphisms of the torus. These transformations have the form

$$\tau \longrightarrow \frac{a\tau + b}{c\tau + d} \qquad a, b, c, d \in Z \quad ad - bc = 1. \tag{19}$$

To check modular invariance explicitly we first need to work out $Z(q, \bar{q})$. This is straightforward in the operator formalism. The string Fock space vacuum is defined by

$$\alpha_n^i|0\rangle = \tilde{\alpha}_n^i|0\rangle = 0 \qquad n > 0. \tag{20}$$

States in the Fock space are created by acting on this vacuum with creation operators α_{-n}^i and $\tilde{\alpha}_{-n}^i$ for $n > 0$. To count the number of states with say N_L eigenvalue equal to n_L we must count the partitions of n_L into integers. This is done using the fact that

$$\frac{1}{\prod_{n=1}^{\infty}(1-q^n)} = \sum_{N=0}^{\infty} P(N)q^N \tag{21}$$

where $P(N)$ is the number of partitions of N. Including the left- and right-moving creation operators and the $-D/24$ offset in L_0 and $\tilde{L}_0$ we see that the oscillator contribution to Z is given by

$$Z_{osc} = \frac{1}{\eta(q)^D \bar{\eta}(q)^D} \tag{22}$$

where $\eta(q) = q^{1/24}\prod(1-q^n)$ is the Dedekind eta function. The string Fock states are also labeled by the eigenvalues of p_L and p_R which also must be summed over in calculating Z. Thus the full answer is

$$Z(q,\bar{q}) = \frac{\sum_{p_L,p_R} q^{p_R^2/2}\bar{q}^{p_L^2/2}}{\eta(q)^D\bar{\eta}(q)^D} \equiv \frac{\Theta_\Gamma(\tau,\bar{\tau})}{\eta(q)^D\bar{\eta}(q)^D} \tag{23}$$

To check that this is invariant under modular transformations it suffices to show that it is invariant under the generators of the modular group which are given by

$$T: \quad \tau \to \tau + 1 \tag{24}$$

$$S: \quad \tau \to -1/\tau \tag{25}$$

The transformation law for the Dedekind eta function can be found in any book on modular forms and is given by

$$\eta(-1/\tau) = (-i\tau)^{1/2}\eta(\tau) \qquad \eta(\tau+1) = e^{2\pi i/24}\eta(\tau) \tag{26}$$

The numerator of (23) is a generalization of the usual theta function associated to a positive definite form. These generalized theta functions were first studied by Siegel.[15] The theta function in (23) is associated to the indefinite quadratic form which is determined by the inner products of basis vectors for the Lorentzian lattice $\Gamma_{D,D}$. The fact the $\Gamma_{D,D}$ is even ensures that Θ_Γ is invariant under T and one can show using Poisson summation that under S it obeys the transformation law

$$\Theta_\Gamma(-1/\tau,-1/\bar{\tau}) = (\tau)^{D/2}(\bar{\tau})^{D/2}\Theta_\Gamma(\tau,\bar{\tau}) \tag{27}$$

Combining (26)and (27)we see that the partition function is invariant under modular transformations as claimed.

It was realized by Narain[16] that one could generalize these toroidal string compactifications by taking (p_L, p_R) to lie in a general even, self-dual Lorentzian lattice. It turns out that any such lattice $\Gamma_{D,D}$ can be transformed by a $SO(D,D)$ rotation into D copies of the unique lattice $\Gamma_{1,1}$. On the other hand the mass spectrum for the bosonic string depends on the combination $p_L^2 + p_R^2$ and is only invariant under $SO(D)_L \otimes SO(D)_R$ rotations. Thus the moduli space of these Narain compactifications is given by

$$M = \frac{SO(D,D)}{SO(D) \otimes SO(D)} \tag{28}$$

modulo certain discrete identifications. These generalized toroidal compactifications thus depend on D^2 parameters. It was shown in Ref. 17 that these parameters could be interpreted as giving constant expectation values to the background metric G_{ij} and torsion B_{ij} on the torus. These generalized toroidal compactifications will be utilized in the third lecture in the construction of some exotic string compactifications.

For toroidal compactifications of the heterotic string this construction is generalized by the addition of constant background gauge fields in the Cartan subalgebra of $E_8 \otimes E_8$. The appropriate even, self-dual, Lorentzian lattice is $\Gamma_{D,16+D}$ and the moduli space is modified appropriately.

One of the most interesting features of these Narain compactifications is that they can lead to enhanced gauge symmetries at certain point in the moduli space M. For example,choose a simply laced Lie algebra G of rank D. Take p_L, p_R to lie in the weight lattice of G with $p_L - p_R$ in the root lattice of G. This leads to the existence of left- and right-moving current algebras on the string world sheet which allow one to organize the string states into representations of $G_L \otimes G_R$ and which also leads to an enhanced gauge symmetry of $G_L \otimes G_R$ in the low-energy theory.

Orbifolds

We can think of toroidal compactifications as saying that we identify points in R^D under the action of a finite group of translations given by the lattice Λ. To construct an orbifold we generalize this procedure to include the identification of points under a group which also contains rotations. The general procedure is to consider the quotient space $O^D = \mathsf{R}^D/S$ where S is a space group. Elements of S are of the form $g = (\theta, v) \in S$ where θ is an element of $O(D)$ and v is a vector in R^D. S acts on the coordinates on R^D as $(\theta, v)x = \theta x + v$. In special cases O^D will still be a manifold. This occurs when no element of S leaves a subspace of R^D fixed. For example, the Mobius strip and the Klein bottle can be obtained as quotients of the two-torus by freely acting symmetries. The classification of such O^D is discussed in detail in the book by Wolf.[18] Here we will be interested in the case where S does not act freely and O^D is not a manifold, but an orbifold.

The original motivation for considering such spaces was the role that they play in the construction of a certain class of Calabi-Yau spaces. In that context on blows up the singularities to obtain a smooth manifold. In string theory this resolution of the singularities is unnecessary and one can obtain finite sensible results for physical quantities even though the underlying space has singular points. Furthermore, we will see in the last lecture that one can construct what are known as left-right asymmetric orbifolds where the underlying geometry is further obscured. In general the orbifold procedure can be viewed as a procedure for manufacturing a new conformal field theory starting from a conformal field theory with symmetry and can often be applied without any underlying geometrical interpretation. For now we will restrict ourselves to the simpler symmetric orbifolds.

To classify possible orbifolds one must first classify the possible space groups. In general this problem is enormously complicated. In two and three dimensions the problem has been solved in connection with the classification of crystallographic symmetries. In four dimensions a complete classification is also available.[19] In higher dimensions only partial results are available. However most of the ideas are already present in lower-dimensional examples in the bosonic string so I will restrict myself to one and two-dimensional examples for the moment. Also, for simplicity I will only consider the case when S can be written as the semi-direct product of the translation group, or lattice, Λ, and a point group P consisting of rotations unaccompanied by translations. Thus I will ignore the possibility of "glide" symmetries. Then the orbifold $O^D = \mathbb{R}^D/S$ can also be viewed as $O^D = T^D/P$.

In one dimension the only possible point group transformations are the identity and a reflection $x \to -x$. In two dimensions there are only seventeen possible space groups, and the point group elements must have order $1, 2, 3, 4$ or 6. To see how this classification comes about let $\{e_1, e_2\}$ be a basis for Λ. Suppose the rotation

$$\Theta(\phi) = \begin{pmatrix} cos\phi & sin\phi \\ -sin\phi & cos\phi \end{pmatrix}$$

is an automorphism of Λ. We can then transform Θ to the lattice basis $\{e_1, e_2\}$, $\Theta \to A\Theta A^{-1}$ with $A \in GL(2, \mathbb{R})$. Then since Θ takes lattice points to lattice points, and the lattice is the integer span of the basis $\{e_1, e_2\}$, $A\Theta A^{-1}$ must be an element of $GL(2, \mathbb{Z})$. We thus obtain the crystallographic condition that $Tr(\Theta) = 2cos\phi \in \mathbb{Z}$ which implies that Θ has order 1,2,3,4 or 6.

To construct the Hilbert space for string propagation on an orbifold one constructs a Hilbert space H_g for each $g \in P$ which is the Hilbert space for strings twisted by g, i.e. the boundary conditions on the string fields are that the string closes only up to a transformation by g. In each such subspace of the total Hilbert space one then projects onto those states which are invariant under the action of P. If one were only considering the propagation of point particles then there would be no need for the Hilbert spaces H_g. One would simply take the original Hilbert space and project onto group invariant states. In string theory the necessity for the spaces H_g is geometrically obvious, since we must consider strings which are closed on the orbifold, and this includes strings which close only up to group transformations on the original space. This construction is also necessary to ensure modular invariance of the resulting theory. To see why, consider calculating the partition function for the orbifold theory following the above prescription. We will have

$$Z = \sum_g Tr_{H_g} \mathsf{P} q^{L_0} \bar{q}^{\bar{L}_0} \tag{29}$$

where P is the projection operator onto group invariant states $\mathsf{P} = \frac{1}{|P|} \sum_{h \in P} h$ (I have assumed here that the point group P is abelian, for the non-abelian case see Ref.13). We can write (29)in the form $Z = \sum_{g,h} Z_{g,h}$ where

$$Z_{g,h} = \frac{1}{|P|} Tr_{H_g} h q^{L_0} \bar{q}^{\bar{L}_0} \tag{30}$$

But $Z_{g,h}$ can also be thought of as the result of performing the path integral over a torus with modular parameter τ with boundary conditions on the string coordinates given by $X(\sigma+\pi, t) = gX(\sigma, t)$, $X(\sigma, t+\pi) = hX(\sigma, t)$. Since modular transformations mix the σ and t directions on the torus it is necessary to sum over all possible boundary conditions in order to obtain a modular invariant answer.

It is helpful to see how this actually works in a specific example. One of the simplest cases to work out is the orbifold obtained by dividing an arbitrary torus T^D by the Z_2 automorphism generated by $\theta : x \to -x$. We start with the Hilbert space H_1 which describes strings obeying the usual boundary conditions on the torus T^D. The projection onto group invariant states within H_1 gives

$$Z_{1,1} + Z_{1,\theta} = \frac{1}{2} Tr_{H_1} (1+\theta) q^{L_0} \bar{q}^{\bar{L}_0} \tag{31}$$

Clearly $Z_{1,1}$ is modular invariant since it only differs from the torus partition function by a factor of two. To evaluate $Z_{1,\theta}$ we note that θ acting on a state obtained by acting on the state $|p_L, p_R\rangle$ with an even (odd) number of creation operators gives $+(-)| - p_L, -p_R\rangle$. We thus find

$$Z_{1,\theta} = \frac{1}{2} \frac{q^{-D/24} \bar{q}^{-D/24}}{\prod(1+q^n)^D \prod(1+\bar{q}^n)^D}. \tag{32}$$

Using the fact that

$$\prod(1-q^n)\prod(1+q^n)=\prod(1-q^{2n})=q^{-1/12}\eta(2\tau) \tag{33}$$

we can rewrite $Z_{1,\theta}$ as

$$Z_{1,\theta}=\frac{1}{2}\left(\frac{\eta(\tau)\bar{\eta}(\tau)}{\eta(2\tau)\bar{\eta}(2\tau)}\right)^D . \tag{34}$$

Now it is clear that $Z_{1,\theta}$ is not modular invariant under $\tau\to-1/\tau$. Using (26)we find that this transformation takes

$$Z_{1,\theta}\to\frac{1}{2}2^D\left|\frac{\eta(\tau)}{\eta(\tau/2)}\right|^{2D} . \tag{35}$$

If we now use $\prod(1-q^{n-1/2})\prod(1-q^n)=\prod(1-q^{n/2})$ we can write (35) as

$$\frac{1}{2}2^D q^{D/48}\bar{q}^{D/48}\left|\frac{1}{\prod(1-q^{n-1/2})}\right|^{2D} \tag{36}$$

I claim that this is equal to $Z_{\theta,1}$, i.e. $Z_{1,\theta}(-1/\tau)=Z_{\theta,1}(\tau)$. To check this we must construct the "twisted " Hilbert space H_θ.

H_θ is constructed using string coordinates obeying the twisted boundary conditions $X(\sigma+\pi,t)=-X(\sigma,t) \quad mod\ \Lambda$. Solving the two-dimensional wave equation with these boundary conditions gives

$$X=x_{cm}+\frac{i}{2}\sum_{n\in\mathbb{Z}+1/2}\left(\frac{\alpha_n}{n}e^{-2in(t-\sigma)}+\frac{\tilde{\alpha}_n}{n}e^{-2in(t+\sigma)}\right) \tag{37}$$

where $x_{cm}=-x_{cm}\ mod\ \Lambda$ are the fixed points of θ. The number of such fixed points is given by $det(1-\theta)=2^D$. Classically, the twisted string sits at one of the fixed points of θ. We can build quantum states about any one of these 2^D classical solutions, this explains the factor of 2^D in $Z_{\theta,1}$. We again construct states in these Fock spaces by acting with the creation operators α_{-n} with n positive. Using (21) as before, this explains the factor involving $\prod(1-q^{n-1/2})$. The last factor I need to explain is the factor of $q^{D/48}\bar{q}^{D/48}$. There are many ways of deriving this factor. The fastest is to view it as arising form the zero point energy of the infinite number of harmonic oscillators appearing in the mode expansion of X.[20] When X obeyed periodic boundary conditions this zero point energy was

$$D\sum_{n\in Z}\frac{n}{2}=\frac{D}{2}\left(\sum n^{-s}\right)_{s=-1}=\frac{D}{2}\zeta(-1)=-D/24 \tag{38}$$

thus explaining the factor of $D/24$ in (15). For X anti-periodic we get

$$D\sum_{n\in Z+1/2}\frac{n}{2}=\frac{D}{4}\sum_{nodd}n \tag{39}$$

and if we carry these formal manipulations to their logical extreme we have

$$\sum_{n\ odd}n=\sum n-\sum_{n\ even}n=\sum n-2\sum n=-\sum n=1/12 \tag{40}$$

which gives a total zero point energy of $D/48$. A more satisfactory derivation of this number involves constructing the Virasoro algebra in the twisted Hilbert space and noting that the usual Virasoro algebra is obtained only when L_0 is shifted by this amount.

It is now straightforward to calculate $Z_{\theta,\theta}$ and one finds

$$Z_{\theta,\theta}=\frac{1}{2}2^D q^{D/48}\bar{q}^{D/48}\left|\frac{1}{\prod(1+q^{n-1/2})}\right|^{2D} \tag{41}$$

The modular behavior of the terms in Z are just what one would expect from the path integral point of view. Under $\tau\rightarrow\tau+1$, $Z_{1,1}$ and $Z_{\theta,1}$ are invariant while $Z_{1,\theta}$ and $Z_{\theta,\theta}$ are interchanged. Under $\tau\rightarrow-1/\tau$, $Z_{1,1}$ and $Z_{\theta,\theta}$ are invariant while $Z_{1,\theta}$ and $Z_{\theta,1}$ are interchanged. The general modular structure of orbifold compactifications is that each term in the partition function is only invariant under some congruence subgroup Γ' of the full modular group Γ. The various terms in the partition function are permuted under the action of Γ/Γ'.

So far I have only considered simple abelian orbifolds of the bosonic string. A richer structure emerges when on generalizes this construction to chiral superstrings like the heterotic string. Modular invariance is no longer automatic, and demanding space-time supersymmetry restricts the possible point groups which are allowed. For a discussion of these issues see Ref.13. Also, I have not said anything about how one calculates correlation functions for orbifold compactifications of string theory. This is a very pretty application of the techniques of conformal field theory. Details can be found in Ref. 21.

LECTURE III. QUASICRYSTALLINE ORBIFOLDS

In the last lecture I sketched the general framework behind toroidal compactifications and their orbifold generalizations. In constructing orbifolds the point group was always taken to act symmetrically on the left- and right-moving string degrees of freedom. However more general constructions are allowed. We can twist the theory by any symmetry, whether or not it acts the same on left and right movers, as long as modular invariance is preserved.[22] We thus need to classify more general symmetries of generalized toroidal compactifications. In doing this we will discover some symmetries which are quite closely related to symmetries which arise in the theory of quasicrystals and which can be most conveniently described using some elementary number theory. A more detailed presentation of the results discussed in this lecture can be found in Ref. 23.

If we consider a toroidal compactification given by a particular even, self-dual Lorentzian lattice $\Gamma_{D,D}$ then an automorphism g of $\Gamma_{D,D}$ must preserve $p_L^2 - p_R^2$. To be a symmetry of the Hamiltonian

$$H = N + \tilde{N} + p_L^2 + p_R^2 - 2 \tag{42}$$

it must preserve $p_L^2 + p_R^2$. Therefore g will actually lie in $O(d)_L \otimes O(d)_R$. The action of g (taking g to act without shifts) is given by

$$g \mid p_L, p_R\rangle = \mid \theta_L p_L, \theta_R p_R\rangle \tag{43}$$

where $\theta_{L,R} \in O(d)_{L,R}$. We would like to know when automorphisms of $\Gamma_{D,D}$ exist with $\theta_L \neq \theta_R$. We can then use these automorphisms to construct asymmetric orbifolds. We will see that θ_L and θ_R then have to satisfy stringent conditions.

Let us consider an almost trivial example. It is not completely trivial however since it will illustrate certain general features of previously considered asymmetric orbifolds. We will compactify one dimension of the bosonic string. Then there is no torsion since B must be antisymmetric. The compactification is determined by a single number, R, the radius of the circle. We then have $p = m/R$ and $w = nR$ where m, n are arbitrary integers. A general lattice point is thus of the form

$$(p_L, p_R) = (m/2R + nR, m/2R - nR) = me_1 + ne_2 \tag{44}$$

where $e_1 = (1/2, 1/2)/R$ and $e_2 = (1, -1)R$ are a basis for $\Gamma_{1,1}$. Now any symmetry g we twist by must lie in $O(1) \otimes O(1)$ which is just $Z_2 \otimes Z_2$. An asymmetric twist would be of the form $(\theta_L, \theta_R) = (1, -1)$. This will be an automorphism if we can solve

$$(p'_L, p'_R) = (m'/2R + n'R, m'/2R - n'R) = (p_L, -p_R) = (m/2R + nR, -m/2R + nR) \tag{45}$$

where m, n, m', n' are integers. This implies $m' = 2nR^2$ and $n' = m/2R^2$. Taking $m = n = 1$ we find that we must have $R = 1/\sqrt{2}$.

There are several interesting things about this example. First there is an enhanced discrete symmetry for a special point in the moduli space (here just a special value of R) and when we twist by this symmetry we would expect that the massless mode in the low-energy field theory that corresponds to changing R disappears. This is true since this mode corresponds to the vertex operator

$$\partial_z X_L \partial_{\bar{z}} X_R \tag{46}$$

and this vertex operator is odd under the twist so the corresponding state is projected out of the spectrum.

Actually what is happening at this value of R is well known. The $U(1)$ current algebra generated by $\partial_z X$ and $\partial_{\bar{z}} X$ is enlarged to a $SU(2)_L \otimes SU(2)_R$ current algebra with the extra currents corresponding to $e^{ip_L X}$ and $e^{ip_R X}$ with $p_L^2 = 2$ and $p_R^2 = 2$. This is reflected in the presence of additional massless scalars transforming as $(3, 3)$ under $SU(2)_L \otimes SU(2)_R$. The twist with $\theta_R = -1$ breaks the symmetry down to $SU(2)_L \otimes U(1)_R$. This example is quite typical. At special points one finds both an enhanced space-time gauge symmetry G and an enhanced discrete world sheet symmetry corresponding to the Weyl group W of G. Twisting by elements of W breaks G down to some subgroup and fixes some of the moduli. There are usually additional massless scalars which are not removed by the twist however and one has to determine whether or not these scalars become new moduli.

This particular example is actually not consistent because the asymmetric twist does not preserve modular invariance. If we think about the twisted Hilbert space then by the analysis in Lecture II the zero point of L_0 for the right-movers will be shifted up by $1/16$ as a result of the twisted boundary conditions, while the zero point of $\bar{L}_0$ will still be at $-1/24$. As a result there will not be any states in this sector that satisfy the physical state condition that $L_0 = \bar{L}_0$ which follows from invariance of the theory under constant shifts in σ. One can show that this implies a breakdown of modular invariance by considering transformations such as $\tau \rightarrow \tau + N$ for an order N twist,[13,24] and, that for abelian orbifolds, the existence of physical states in all twisted sectors is sufficient to guarantee modular invariance.

In the heterotic string one already has more gauge symmetry than is needed or convenient for phenomenological applications and most compactifications have large numbers of massless scalars corresponding to a large vacuum degeneracy. It would thus be interesting to construct asymmetric orbifolds where the enhanced discrete world-sheet symmetry does not correspond to enhanced space-time gauge symmetry, and the massless moduli are eliminated. Although I will not discuss the heterotic string in this lecture, the constructions I will describe are partially successful in achieving this goal.

Another motivation for the construction I will describe comes from the analysis of conformal theories with $c = 1$. As described by E.Verlinde at this conference, all known $c = 1$ conformal field theories can be obtained as string propagation on a circle, a Z_2 left-right symmetric orbifold of the circle, or a left-right symmetric orbifold based on a discrete subgroup of $SU(2)$ which acts as a quantum symmetry of the theory at radius $R = 1/\sqrt{2}$. Their are various ideas as to how one might classify conformal field theories with $c \geq 1$. One possible approach involves first classifying "rational " conformal field theories and then trying to approximate 'irrational" theories by the rational ones. There are various possible definitions of what rational should mean, I will take it to mean that the partition function is of the form

$$Z(\tau, \bar{\tau}) = \sum_{i}^{N} f_i(\tau) \bar{g}_i(\tau) \tag{47}$$

where f_i and g_i are modular functions for some subgroup of the modular group (one should also allow multiplier systems which cancel between f_i and g_i). For $c = 1$ this idea could work since the only irrational theories appear along the orbifold or circle lines at irrational values of R^2 and are continuously related to rational theories. This approach would run into difficulties in its most naive application if there were conformal field theories which were irrational and isolated from other conformal field theories. These are precisely the sorts of theories that we will construct.

We can construct lattices $\Gamma_{D,D}$ with discrete symmetries but without enhanced gauge symmetry by ensuring that operators of the form e^{ipX} with $(p_L^2, p_R^2) = (2, 0)$ and their right-moving counterparts are absent. This will be true if p_L^2 and p_R^2 only take on irrational values. To construct such lattices we will borrow an idea that arises in the theory of quasicrystals. There one uses the fact that a $p + q$ dimensional lattice Λ^{p+q} can have a crystallographic symmetry which is not allowed for any p or q dimensional lattice. Then one can form subsets of the projection of Λ^{p+q} onto a p dimensional subspace which define the vertices of an aperiodic tiling. As an example, Z_5 symmetry is allowed in four dimensions, but as we saw earlier, not in two. We can thus construct a four-dimensional lattice with Z_5 symmetry (such as the root lattice of $SU(5)$) and an appropriate projection of this onto a two-dimensional subspace will give rise to the Z_5 symmetric aperiodic tiling of the plane constructed by Penrose. We will construct analogs of these in string theory and then show that many massless states are removed because p_L^2 and p_R^2 will turn out to be irrational if they are non-zero.

We want to construct even, self-dual Lorentzian lattice $\Gamma_{D,D}$ with symmetries $\theta = (\theta_L, \theta_R)$ with θ crystallographic in $2d$ dimensions but where θ_L and θ_R are not separately crystallographic in d dimensions. Now if the sets of vectors $\{p_L\}$ or $\{p_R\}$ form a lattice then this is clearly impossible. A necessary and sufficient condition for the left-moving and right-moving momenta to form a lattice is the rationality of the background metric, torsion, and (for the heterotic string) gauge fields.[25,26] Thus we will need toroidal compactifications where the background fields take on irrational values.

Lattices of the type we wish to construct must satisfy several constraints. A very powerful constraint comes from modular invariance. If the twist θ leaves no sublattice of $\Gamma_{D,D}$ invariant and is left-right symmetric ($\theta_L = \theta_R$), then in the twisted sector of the resulting orbifold there is a vacuum degeneracy given by the number of fixed points of θ which by the Lefschetz fixed point theorem is

$$det(1-\theta_L) = \sqrt{det(1-\theta_L)det(1-\theta_R)} = \sqrt{det(1-\theta)} \tag{48}$$

When $\theta_L \neq \theta_R$ the geometrical interpretation is less obvious but modular invariance still requires a degeneracy of

$$\sqrt{det(1-\theta)} \tag{49}$$

and to have a sensible Hilbert space interpretation this must be an integer.[22] If θ has an invariant lattice $I \subset \Gamma_{D,D}$ then this is generalized to

$$\sqrt{\frac{det(1-\theta)}{|I^*/I|}} \tag{50}$$

where I^* is the dual lattice of I and $|I^*/I|$ is the index of I^* in I. Here I will only consider the situation with $I = 0$. When one generalizes the construction I will describe to the heterotic string it is crucial to allow for the possibility of invariant lattices.

If θ acts crystallographically then it must have finite order, $\theta^m = 1$. We must have integer vacuum degeneracy in the sectors twisted by all powers of θ, so when $I = 0$, the condition we must satisfy is

$$det(1-\theta^r) = n_r^2 \quad n_r \in \mathbf{Z}, \quad r = 1 \cdots m-1 \tag{51}$$

Now let us consider what kind of θ can satisfy this condition with θ_L and θ_R acting acrystallographically. Since θ has order m, if we diagonalize θ it will be of the form $diag(\xi^{i_1}, \ldots \xi^{i_{2d}})$ where $\xi = e^{2\pi i/m}$. If θ is crystallographic then we can make a similarity transformation to a lattice basis

$$\theta \mapsto A^{-1}\theta A \quad A \in GL(2d, \mathrm{R}) \tag{52}$$

and in this basis θ must have integer entries. So we require that

$$Tr\theta = \sum_{j=1}^{2d} \xi^{i_j} \in \mathrm{Z} \tag{53}$$

We can ensure that θ_L and θ_R act acrystallographically if no proper subset of the ξ^{i_j} have an integer sum (this is clearly more restrictive than is required and will be relaxed later on).

One can proceed by trial and error and try to find solutions to all these criterion. However we will save a great deal of work by putting this problem in the proper framework and using some simple number theoretical results. This leads us to discuss cyclotomic fields.

<u>Cyclotomic fields and Construction of Lattices</u>

Given an mth root of unity $\xi = e^{2\pi i/m}$ we construct the mth cyclotomic field $\mathrm{Q}[\xi]$ as an extension of the rationals Q. Elements $\alpha \in \mathrm{Q}[\xi]$ are linear combinations of powers of ξ with rational coefficients

$$\alpha = \sum_{i=1}^{m} a_i \xi^{i-1} \quad , a_i \in \mathrm{Q} \tag{54}$$

It is trivial to check that this defines a field. It is also useful to view $\mathrm{Q}[\xi]$ as a finite dimensional vector space over Q. The dimension of this vector space is called the degree of $\mathrm{Q}[\xi]/\mathrm{Q}$ and is equal to $\phi(m)$ where $\phi(m)$ (the Euler totient function) is the number of integers less than m which are relatively prime to m. A fundamental object in the study of these fields is the mth cyclotomic polynomial which is given by

$$\Phi_m(X) = \prod_{(i,m)=1} (X - \xi^i) \tag{55}$$

Here ξ^i for $(i, m) = 1$ are the primitive mth roots of m (those roots for which $(\xi^i)^a = 1$ implies a is a multiple of m).

One can show that $\Phi_m(x)$ is irreducible in $\mathrm{Z}(x)$ and that the coefficients of $\Phi_m(x)$ lie in Z. This later fact can be proved by induction using the useful relation

$$X^m - 1 = \prod_{d|m} \Phi_d(X) \tag{56}$$

which allows one to calculate the Φ_m recursively. In particular this implies that $\sum_{(i,m)=1} \xi^i \in \mathrm{Z}$ and since $\mathrm{Q}[\xi]/\mathrm{Q}$ has degree $\phi(m)$ no proper subset of the primitive mth roots has an integer sum.

This suggests that we consider a twist by θ in (d,d) dimensions where the order m of θ satisfies $2d=\phi(m)$ and the eigenvalues of θ are the $\phi(m)$ primitive mth roots of unity. We then satisfy the demand that $Tr\theta \in Z$ and θ_L and θ_R be acrystallographic. We still have to choose θ so that (51) is satisfied. Now $det(1-\theta)=\Phi_m(1)$. One can show using induction and the prime factorization of m that

$$\Phi_m(1)=\begin{cases} p, & m=p^r\ ;\ \text{p prime} \\ 1, & m\neq p^r \end{cases} \tag{57}$$

so using (51) we can immediately rule out $m=1,2,3,4,5,7,8,9,11,13,etc.$. If m is of the form $2p^r$ for p prime and $r\in Z$ then $\phi(m)=\phi(2)\phi(p^r)=\phi(p^r)=\phi(m/2)$. Therefore θ^2 will have eigenvalues consisting of the primitive $(p^r)^{th}$ roots of unity and so $det(1-\theta^2)=p$. This eliminates $m=6,10,14,18,22,\cdots$. So if we stick with the rule that only primitive roots are allowed we are left with the possibilities $(m,d)=(12,2),(15,4),(20,4),(21,6),(24,4),(28,6),(30,4)$ etc.. One can show that it is not necessary to consider any further powers of θ in (51) so these values for (m,d) satisfy all the constraints we have demanded. For example consider $d=2$ and $m=12$. Then $\theta=diag(\xi,\xi^{11},\xi^5,\xi^7)$, $\theta^2=diag(\xi^2,\xi^{10},\xi^{10},\xi^2)$ and $\theta^3=diag(\xi^3,\xi^9,\xi^3,\xi^9)$ so θ^r is left-right symmetric if $(12,r)\neq 1$ and has eigenvalues which are a permutation of those of θ if $(12,r)=1$.

We still have to show that we can actually construct lattices with these symmetries and that constructing orbifolds based on these symmetries preserves modular invariance. This is not as difficult as it might seem. Let's work out a $\Gamma_{2,2}$ with Z_{12} symmetry as an illustration. The general method will be to start with a vector v and to construct a basis for the lattice consisting of v rotated by powers of the rotational symmetry we are trying to obtain. In our example we start with $v=(v_L,v_R)$ and take as a basis $e_1=v, e_2=\theta v, e_3=\theta^2 v$, and $e_4=\theta^3 v$. Here θ is the Z_{12} rotation $\theta=(\theta_L,\theta_R)$ with θ_L a rotation by $2\pi/12$ and θ_R a rotation by $10\pi/12$. Now this is a basis for a four-dimensional lattice but we have to show that it has Z_{12} symmetry. Clearly θ takes $e_1\mapsto e_2, e_2\mapsto e_3, e_3\mapsto e_4$, but we have to check its action on e_4. It is easy to check that $\theta^4=\theta^2-1$ (this is a reflection of the fact that primitive 12th roots of unity are roots of $\Phi_{12}(x)=x^4-x^2+1$. Thus the action of θ on e_4 is $e_4\mapsto e_3-e_1$. Therefore θ takes lattice points to lattice points and is an automorphism of this lattice.

A choice of v determines a two-parameter family of lattices with Z_{12} symmetry (we can obviously choose $v_L=(v_{1L},0), v_R=(v_{1R},0)$ and then v_{1L} and v_{1R} are the two parameters). We have to show that we can choose these parameters so that the resulting lattice is even and self-dual. Let $a = v^2 = v_L^2 - v_R^2$ and $b = v \cdot \theta \cdot v$. Then $e_i \cdot e_j = Q_{ij}$ with

$$Q = \begin{pmatrix} a & b & a/2 & 0 \\ b & a & b & a/2 \\ a/2 & b & a & b \\ 0 & a/2 & b & a \end{pmatrix} \tag{58}$$

Matrices of this form can always be block diagonalized using the change of basis

$$f_i = (e_i + e_{2d+1-i})/\sqrt{2}\; i = 1, \ldots d$$

$$f_i = (e_i - e_{2d+1-i})/\sqrt{2}\; i = d+1, \ldots 2d \tag{59}$$

Using this one easily finds $det(Q) = (3a^2 - 4b^2)^2/16$ so choosing $a = 2$ and $b = 2$ yields an even self-dual lattice with Z_{12} symmetry. It is also clear from the construction that the Z_{12} symmetry acts without mixing left- and right-movers as required to construct an asymmetric orbifold.

This rotation method can be easily generalized to arbitrary (m, d) satisfying our criterion. A choice of v will determine a d-parameter family of lattices with Z_m symmetry. A basis for these lattices consists of $v, \theta v, \cdots \theta^{2d-1} v$. To show that this lattice has Z_m symmetry we use the fact that $\theta^{\phi(m)}$ is related to lower powers of θ with integer coefficients through the cyclotomic polynomial.

The difficult part that remains is to show that one can choose v so that the resulting lattice is even, self-dual. We do not yet have a proof that this is always possible. One can construct such lattices explicitly as long as d is not too large. The calculations are simplified by noting a few simple facts. First, if $Q_{ij} = e_i \cdot e_j$ is the quadratic form we want then it is clear from its construction that Q_{ij} has the form $Q_{ij} = a_{|i-j|+1}$ where $a_i \quad i = 1 \cdots 2d$ is the first row of Q. Second, we can block diagonalize Q into two $d \times d$ blocks by making the change of basis (59).

These facts plus access to a symbolic manipulation program allow one to construct further examples fairly easily. For instance

$$a = (2, 1, 0, 0, 0, 0, -1, -1, 0, 0, 0, 0) \tag{60}$$

defines an even, self-dual $\Gamma_{6,6}$ with Z_{21} symmetry. It would be nice to have a general proof that an appropriate choice of v always exists. It seems that it should since the condition (51) is the only obvious obstruction to the existence of even, self-dual lattices with these symmetries.

Moduli and Massless Scalars

Now that I have demonstrated that lattices with "quasicrystalline" symmetries exist, it is necessary to show that one can construct orbifolds based on these symmetries that are modular invariant. I will show how this works in an example and then discuss some of the structure of these theories. For concreteness I will mainly focus on the Z_{12} orbifold in $d = 2$ as a compactification of the bosonic string.

First consider the question of modular invariance for the Z_{12} orbifold. It suffices to check matching of the left- and right-moving vacuum energies in the sector twisted by θ. It is easy to generalize the previous calculation of the vacuum energy for a Z_2 twist to a twist of a complex boson by $e^{2\pi i\eta}$. The result is that the vacuum energy is shifted by an amount $\eta(1-\eta)/2$. This gives

$$E_L - E_R = \frac{1 \times 11 - 5 \times 7}{2 \times 12 \times 12} = \frac{-1}{12} \tag{61}$$

as the difference between left and right-moving energies so we see that level matching is satisfied. It is a bit surprising that level matching works here since in general this constraint is quite restrictive for asymmetric orbifolds. It should also be emphasized that it is independent of the condition (51) which was a condition required for the existence of even, self-dual lattices with certain symmetries. For twists of the form considered here it is possible to show that level matching is always satisfied. The proof is a bit technical and utilizes the structure of the group of units in $\mathbb{Z}/m\mathbb{Z}$. The details can be found in Ref. 23.

A generic feature of asymmetric orbifolds is that many of the usual massless moduli are removed, but as discussed earlier, there are usually other massless scalars left after twisting. For the Z_{12} orbifold we will construct it turns out that there are no massless scalars at all so this must correspond to an isolated solution of the bosonic string.

One key to demonstrating the absence of massless scalars is the fact that the only non-zero values of p_L^2 and p_R^2 are irrational. To show this recall that in our construction of the lattice we began with a vector $v \in \Gamma_{2,2}$ of the form $(v_{L1}, 0, v_{R1}, 0)$. v_{L1}^2 and v_{R1}^2 are then determined from Q in terms of $cos(2\pi/12) = (\xi + \xi^{-1})/2$. It is then easy to show from this construction that p_L^2 is of the form

$$v_{L1}^2 \sum_{k,l=1}^{4} n_k n_l cos(k-l)2\pi/12 \tag{62}$$

for $n_k \in \mathbb{Z}$. Now $deg(\mathbb{Q}[\xi + \xi^{-1}]/\mathbb{Q}) = 2$. Therefore the quadratic form in (62) can be written as

$$Q_L^0 + (\xi + \xi^{-1})Q_L^1 \tag{63}$$

In particular

$$Q_L^1 = \frac{1}{6}\begin{pmatrix} 4 & 3 & 2 & 0 \\ 3 & 4 & 3 & 2 \\ 2 & 3 & 4 & 3 \\ 0 & 2 & 3 & 4 \end{pmatrix} \tag{64}$$

and can be seen to be positive definite by block diagonalizing as described earlier. Hence p_L^2 can never be rational if it is non-zero. Since $p_L^2 - p_R^2 \in 2Z$ the same is true of p_R^2.

Now let us look at the spectrum of this orbifold. Since there are no states with $p_L^2 = 2$ or $p_R^2 = 2$, in the untwisted sector the only possible scalars are of the form

$$\alpha_{-1}^i|0\rangle_R \otimes \alpha_{-1}^j|0\rangle_L \quad i,j = 1,2 \tag{65}$$

but these are clearly projected out by the twist.

In the twisted sectors we have to check that there are no massless scalars constructed by acting on the twisted vacuum with fractionally moded oscillators. In the sector twisted by θ the left-moving vacuum energy is

$$E_L = -1 + \frac{11}{288} \tag{66}$$

and we have oscillators with modings that are multiples of 1/12. Hence there are no states with $E_L = 0$. In fact in all the twisted sectors there is a similar mismatch between the vacuum energy and the moding of the oscillators so that no massless states exist. There are of course massive states as required by modular invariance.

So we see that the irrationality of p_L^2 and p_R^2 coupled with the left-right asymmetric twist is very effective in reducing the number of massless scalars. In other quasicrystalline orbifolds there are usually massless scalars appearing in twisted sectors. These scalars have no good reason to be moduli, that is have flat potentials, but a calculation which we have not done is required to test this claim.

So far I have only discussed quasicrystalline orbifolds in the context of bosonic string compactifications. It is certainly possible to generalize these ideas to compactifications of the heterotic string. However, the chiral structure of the heterotic string coupled with these left-right asymmetric twists, makes it more difficult to construct modular invariant theories. Several possibilities are discussed in Ref.23. Again, the number of massless scalars is reduced compared with what one expects in rational orbifolds, and the resulting gauge group can be small enough to be phenomenologically interesting. On the other hand, these models generically break space-time supersymmetry, so their phenomenological prospects are uncertain at the present time.

Comments and Conclusions

Quasicrystalline orbifolds have some interesting implications for the structure of conformal field theories. It has been suggested that there might exist rational conformal field theories which are dense in the space of all conformal field theories and which could be used to approximate a general conformal field theory.[26] For toroidal compactifications and orbifolds rational can be viewed in the sense of (47), or equivalently it means that the background fields take on rational values. It is clear that quasicrystalline orbifolds cannot be reached from rational theories by perturbing by a marginal operator since the Z_{12} theory has no marginal operators. Still, one could hope to approximate these theories by other rational theories without marginal operators, even though the theories were not continuously connected. It is hard to rule out this possibility without a classification of conformal field theories. Still, if there is a rational approximation it would seem likely for it to be a rational orbifold. This possibility can be ruled out as follow. We will take approximation here to mean a uniform approximation of the operator product algebra of the conformal fields. The basic point is that there exists a twist field for the quasicrystalline orbifold which cannot be the twist field of any rational orbifold because for a rational orbifold the twist fields must have dimensions consistent with the crystallogrpahic condition in d dimensions whereas more general twist fields are allowed for quasicrystalline orbifolds. One might try instead to approximate the twist field by some other primary field such as the exponential of a linear combination of free bosons (by definition the primary fields of an orbifold are combinations of the twist fields, $\partial_z X$ and e^{ipX}). Then one can work out the operator product expansion for the product of two twist fields and see that this does not agree with the operator product of two exponentials. Unfortunately to make this argument carefully one must worry about primary fields that combine twists and shifts and the possibility of twisting by quantum symmetries which arise only for certain values of the background fields. For the Z_{12} quasicrystal orbifold the argument can be made carefully.[23] In general the argument would be quite complicated, but the principle is simple: there exist symmetries at irrational values of the background fields which do not appear at any rational values, and orbifolds based on these symmetries cannot be rationally approximated because the symmetry cannot be approximated.

These orbifolds also raise some interesting questions in the fermionic formulation of orbifolds.[27] It is known that one can often rewrite these theories either in terms of bosonic or fermionic degrees of freedom. However if one wishes to work in terms of free fermions than the fermions obey twisted boundary conditions of finite order only if the corresponding background data in the bosonic theory are rational.[28] For irrational values it is not really clear whether the fermionic formulation makes sense or is useful. If one then further twists the fields as we have done here it becomes even less clear how one would treat these theories in a fermionic formulation.

One further comment I would like to make regards the extent to which these fermionic formulations or their bosonic counterparts correspond to truly lower-dimensional string theories or whether they should be viewed as compactifications of higher dimensional theories. It has been claimed that some of these theories are really lower dimensional string theories which are not continuously connected to theories in ten dimensions. This may be true in some cases (as in the Z_{12} quasicrystalline orbifold in $d = 2$) but it is not sufficient to simply rewrite the string variables in a fermionic formulation. One must investigate the moduli space or the space of perturbations generated by the marginal operators in the theory and show that it is not connected to higher dimensional theories. In some of the theories considered here there are no marginal operators and the theory is an isolated point in the space of conformal field theories. In various fermionic formulations there are often a variety of marginal operators, and the structure of the resulting moduli space has not been carefully explored.

In any event, the orbifolds described here may seem a bit "cute" and contrived, but they illustrate some of the subtleties that we will have to come to grips with in the search for a systematic understanding of classical string vacua.

ACKNOWLEDGEMENTS

I would like to thank C. Callan, L.Dixon, G.Moore, C.Vafa, and E.Witten for collaboration on much of the material discussed here. I am also grateful to the organizers of the Cargese school for their hospitality and for the invitation to speak. This work was supported in part by NSF grant PHY80-19754 and a NSF PYI award PHY86-58033.

REFERENCES

1. E.Witten, *Phys. Lett.* **153B** (1985), 243
2. M.F.Atiyah and I.M Singer, Proc. Nat. Acad. Sciences 81, April 1984; B.Zumino, Lectures at Les Houches Summer School (1983); R.Stora, Cargese Lectures (1983)
3. C.Callan and J.Harvey, *Nucl.Phys.* **B250** (1985),
4. E.Weinberg, *Phys. Rev.* **D24** (1981), 2669.
5. P.Ginsparg, Applications of topological and differential geometric methods to anomalies in quantum field theory, in *New Perspectives in Quantum Field Theory (XVI GIFT seminar)*, eds. J.Abad et al., World Scientific (1986).

6. S.Naculich, Axionic Strings: Covariant Anomalies and Bosonization of Chiral Zero Modes Princeton preprint PUPT-1053

7. D.B.Kaplan and A. Manohar, Anomalous Vortices and Electromagnetism, HUTP-87/A031, CTP #1469.

8. M.Green and J.Schwarz, *Phys. Lett.* **149B** (1984), 117

9. D.Gross, J.Harvey, E.Martinec and R.Rohm, *Phys.Rev.Lett.* **54** (1985), 502, *Nucl.Phys.* **B256** (1985), 253, *Nucl.Phys.* **B267** (1986), 75.

10. See M.Dine and N.Seiberg, preprint IASSNS/HEP-87/50 and references therein

11. D.Gepner, Princeton Preprints PUPT-1056 and PUPT-1066 (1987)

12. P.Candelas, G.Horowitz, A.Strominger, and E.Witten, *Nucl.Phys.* **B258** (1985), 46; D.Zanon, M.Grisaru and A.Van de Ven, *Phys. Lett.* **173B** (1986), 423; D.Gross and E.Witten, *Nucl. Phys.* **B277** (1986), 1; D.Nemeschansky and A.Sen, *Phys. Lett.* **178B** (1986), 365.

13. L.Dixon, J.Harvey, C. Vafa, and E.Witten, *Nucl.Phys.* **B261** (1985), 620 , *Nucl.Phys.* **B274** (1986), 285.

14. L.Dixon, Lectures at the 1987 ICTP Summer Workshop, PUPT-1074 (1987).

15. C.L.Siegel, Indefinite quadratische Formen und Funktionen Theorie I & II, *Math. Ann.* **124** (1951), (52).

16. K.Narain, *Phys.Lett.* **169B,** (1986), 41

17. K.Narain, M.H.Sarmadi, and E.Witten, *Nucl. Phys.* **B279** (1986), 93

18. J.A. Wolf, *Spaces of Constant Curvature*, New York, McGraw Hill (1967).

19. H.Brown et al., *Crystallographic Groups of Four-Dimensional Space*, New York, Wiley (1978).

20. L.Brink and H.K.Nielsen, *Phys. Lett.* **45B** (1973), 332

21. L.Dixon,D.Freidan, E.Martinec, and S.Shenker, *Nucl.Phys.* **B282** (1987), 13 M.Bershadskii and A.Radul, *Int. J. Mod. Phys.* **A2** (1987),165 S.Hamidi and C.Vafa, *Nucl.Phys.* **B279** (1987), 465.

22. K.Narain, M.H.Sarmadi, and C.Vafa, *Nucl.Phys.* **B288** (1987), 551.

23. J.Harvey, G.Moore, and C.Vafa, Princeton preprint PUPT-1068 (1987)

24. L.Dixon, J.Harvey, C.Vafa, and E.Witten, *Nuc. Phys.* **B274** (1986), 285 C.Vafa, *Nucl. Phys.* **B273** (1986), 592.

25. G.Moore, HUTP-A013, Nucl.Phys.B, in press.

26. D.Freidan and S.Shenker, unpublished

27. See the reviews by J.Schwarz, Caltech preprint CALT-68-1432 (1987), and A.N. Schellekens, CERN preprint CERN-TH.4807/87.

28. J.Bagger, D.Nemeschansky, N.Seiberg, and S.Yankielowicz, *Nucl. Phys.* **B289** (1987), 53

QUANTUM GRAVITY AND BLACK HOLES

G. 't Hooft

Institute for Theoretical Physics
Princetonplein 5, P.O.Box 80.006
3508 TA Utrecht, The Netherlands

ABSTRACT

At energies beyond the Planck mass gravitational interactions become fundamentally non-perturbative. The fact that the gravitational field of an ultra-energetic massless particle can be given in a simple closed form and vanishes nearly everywhere can be used to shed some light on the problem of "quantum gravity". It is argued that black holes must play an essential role in any successful theory for quantum gravity. A detailed introduction is given to the phenomenon of Hawking radiation but its derivation has shortcomings. We suggest a theory in which the set of states that build up the Hilbert space of black holes is entirely generated by the geometry of its horizon.

1. INTRODUCTION

The question how to reconcile the notion of General Relativity with that of Quantum Mechanics has bewildered physicists for more than half a century. And in spite of the recent claim that "Superstring theories" bear the promise of a complete unification of all interactions, including gravity, much of this problem is still shrouded in mystery. In these lectures we intend to show that there are not only mysteries but also *paradoxes*: questions to which known laws of physics seem to give conflicting answers. The apparent phenomenon of *Hawking radiation* [1] by a black hole provides us with such a paradox: is there a Schrodinger equation for phenomena at the Planck length scale or not? How should one catagorize states in Hilbert space?

In our first lecture we show that some questions *can* be answered by application of known laws of physics. We find that the scattering amplitude for particles can be computed exactly in the limit where the Mandelstam variable t stays much smaller than the Planck mass squared but s is in the order of the Planck mass squared or larger. The outcome shows some resemblance to the well-known Veneziano amplitude, but is in many important ways different. Indeed, at large s and small t, string theories cannot even be approximately right.

When s becomes very much larger than the Planck mass squared we expect black hole formation. It is here that the real difficulties arise. We quickly review the black hole solution to Einstein's equations in the next lecture, after which we show a detailed derivation of the phenomenon of Hawking radiation. As stated earlier, the result seems to defy common sense, because it suggests that purely quantum mechanical wave functions spontaneously turn into probabilistic mixtures of different states. Differently from other researchers in this field [1,2] we observe that

(i) such an outcome would suggest a probabilistic mixture of different Hamiltonians, whereas only one Hamiltonian can be acceptable if we want to maintain energy- and momentum conservation,

and (ii) the calculation ignored one extremely important effect: the gravitational interactions between in- and outgoing matter at the horizon of the black hole. This is a large s , small t interaction, which becomes dominant when in- and outgoing particles are separated by time scales larger than $\mathcal{O}(M \log M)$ in Planck units. This is to be compared to the life time of a black hole, which is of order M^3 .

Some speculative ideas on how to construct a black hole Hilbert space are given in the end.

2. THE GRAVITATIONAL FIELD OF A FAST, LIGHT PARTICLE

When a particle at rest is much lighter than the Planck mass, then at distance scales larger than the Planck length it generates a space-time metric very well approximated (in this chapter we take units such that G = 1), by

$$g_{\mu\nu} = \begin{bmatrix} -1+\frac{2m}{r} & & & 0 \\ & 1+\frac{2m}{r} & & \\ & & 1+\frac{2m}{r} & \\ 0 & & & 1+\frac{2m}{r} \end{bmatrix} , \qquad (2.1)$$

(compared to the usual Schwarzschild metric, eqs (4.1), (4.5), the r coordinate here is shifted by an amount $2m$). In order to find the field of a fast moving particle we rewrite this in a covariant way introducing the covariant velocity u_μ :

$$u_\mu = \frac{1}{\sqrt{1-\vec{v}^2}} \begin{bmatrix} 1 \\ \vec{v} \end{bmatrix} , \tag{2.2}$$

such that $u^2 = -1$. We can write (2.1), which holds for $\vec{v} = 0$, as

$$g_{\mu\nu} = \eta_{\mu\nu} \left[1+ \frac{2m}{r} \right] + \frac{4m}{r} u_\mu u_\nu ; \tag{2.3}$$

here r is now defined by

$$r = \sqrt{x^2 + (x.u)^2} . \tag{2.4}$$

Next, let us take the limit

$$u_\mu \to \infty \quad , \quad m \to 0 \quad , \quad m u_\mu \to p_\mu \quad \text{(fixed)} \quad , \quad p^2 \to 0 . \tag{2.5}$$

In this limit,

$$r \to |x.u| . \tag{2.6}$$

The infinitesimal line element ds becomes

$$ds^2 = g_{\mu\nu} dx_\mu dx_\nu | \to ds_0^2 + \frac{4m}{r} (u_\mu dx^\mu)^2 \to ds_0^2 + \frac{4(p_\mu dx)^2}{p.x|} , \tag{2.7}$$

where ds is the flat metric. Now, using the definition (2.4) for r, we introduce two new sets of coordinates, $z^{\pm}_\mu$:

$$z^{\pm}_\mu = x_\mu \pm 2p_\mu \log r , \tag{2.8}$$

so that

$$dz^2_{\pm} = dx^2 \pm 4(p_\mu dx^\mu)^2/(p.x) . \tag{2.9}$$

Notice that we are reproducing (2.7) apart from the absolute value signs. Apparently we have

$$\begin{aligned} &\text{at } (x.p) > 0 : \quad ds^2 \to dz_+^{\ 2} , \\ &\text{at } (x.p) < 0 : \quad ds^2 \to dz_-^{\ 2} . \end{aligned} \tag{2.10}$$

thus, both before and behind the plane $(x.p) = 0$ we have flat space, if the coordinates z_μ are used there. However, at the plane $(x.p) = (z_+.p) = (z_-.p) = 0$ itself these two flat spaces are glued together in such a way that

$$z^+_\mu = z^-_\mu + 2p_\mu \log \tilde{z}^2 , \tag{2.11}$$

where $\tilde{z}_\mu$ is the transverse part of the coordinate z_μ (which is the same for z^+_μ as for z^-_μ). Because of the non-trivial $\tilde{z}$-dependence we have a δ-distributed Riemann curvature there.

As far as we are aware, this space-time was first described by Aichelburg and Sexl [3]. Our physical interpretation is that the plane $(z.p) = 0$ is a *shock wave*, carried along by the particle. Notice that multiplying the argument of the logarithm in (2.11) by a constant just corresponds to a relative shift between the coordinates z^+_μ and z^-_μ and hence does not affect the physical features of this space-time. Furthermore one can show that the effects of the first derivatives of the shift with respect to $\tilde{z}$ are also locally unobservable (they can be removed by a relative Lorentz transformation of the coordinates $z^\pm$). The locally observable amplitude of the shock wave therefore decreases as $1/\tilde{z}^2$.

The fact that energetic massless particles are surrounded by a space-time that is flat nearly but not quite everywhere is extremely important for understanding the problems in quantum gravity and some possible alleys towards their resolution.

3. GRAVITON DOMINANCE IN ULTRA-HIGH ENERGY SCATTERING [4]

Consider now two (electrically neutral) particles with rest masses $m^{(1)}, m^{(2)} \ll M_{\text{Planck}}$. Let us first use a coordinate system in which the ingoing particle (1) is at rest or moves slowly. Let the second particle arrive from the right along the trajectory

$$\begin{bmatrix} x^{(2)} \\ y^{(2)} \end{bmatrix} \overset{=}{\text{def}} \tilde{x}^{(2)} = 0 \quad ; \quad z^{(2)} = -t^{(2)} , \tag{3.1}$$

with energy

$$\tfrac{1}{2}p_-^{(2)} = p_0^{(2)} = -p_3^{(2)} = \mathcal{O}(1/Gm^{(1)}) , \tag{3.2}$$

where G is Newton's constant in units where $\hbar = c = 1$. Since we take $m^{(i)} \ll M_{\text{Planck}}$ the velocity of particle (2) can be regarded to be that of light. The energy (3.2) is so tremendous that we can no longer ignore the gravitational field of particle (2). This field is due to the curvature found in the previous section. Two flat regions of space-time, $R^{(4)}_{(+)}$ $(t>-z)$ and $R^4_{(-)}$ $(t<-z)$, are glued together at the null plane $z=-t$, such that on this plane (see eq. 2.11),

$$
\begin{aligned}
\tilde{x}_{(+)} &= \tilde{x}_{(-)} \quad , \\
z_{(+)} &= z_{(-)} + 2Gp_o^{(2)}\log(\tilde{x}^2/C) \ , \\
t_{(+)} &= t_{(-)} - 2Gp_o^{(2)}\log(\tilde{x}^2/C) \ ,
\end{aligned}
\tag{3.3}
$$

where C is an irrelevant constant.

For simplicity we now take particle (1) to be spinless. In $R^4_{(-)}$ its wave function is

$$
\begin{aligned}
\psi^{(1)}_{(-)} &= e^{i\tilde{p}^{(1)}\tilde{x} + ip_3^{(1)}z - ip^{(1)}t} = \\
&= e^{i\tilde{p}^{(1)}\tilde{x} - ip_+^{(1)}u - ip_-^{(1)}v} \ ,
\end{aligned}
\tag{3.4}
$$

where $u = (t-z)/2$ and $v = (t+z)/2$ are lightcone coordinates.

Immediately after the shock wave went by we have the shifted wave function in $R^4_{(+)}$:

$$
\psi^{(1)}_{(+)} = e^{i\tilde{p}^{(1)}\tilde{x} - ip_+^{(1)}(u+2Gp_o^{(2)}\log(\tilde{x}^2/C))} \ , \quad \text{at} \quad v = 0 \ . \tag{3.5}
$$

This we can expand in plane waves,

$$
\psi^{(1)}_{(+)} = \int A(k_+,\tilde{k})\, dk_+ d^2\tilde{k}\; e^{i\tilde{k}\tilde{x} - ik_+u - ik_-v} \ , \tag{3.6}
$$

with

$$
k_- = (\tilde{k}^2 + m^{(1)2})/k_+ \ . \tag{3.7}
$$

Clearly,

$$
\begin{aligned}
&A(k_+,\tilde{k}) = \\
&\delta(k_+ - p_+^{(1)}) \frac{1}{(2\pi)^2} \int d^2\tilde{x}\; e^{i(\tilde{p}^{(1)} - \tilde{k}^{(1)})\tilde{x} - 2iGp_+^{(1)}p_o^{(2)} \log(\tilde{x}^2/C)} \ .
\end{aligned}
\tag{3.8}
$$

The integral here is elementary:

$$\int d^2\tilde{x}\, e^{i\tilde{k}\tilde{x} - iB \log x} = \frac{\pi\Gamma(1-iB)}{\Gamma(iB)} \left[\frac{4}{\tilde{k}^2} \right]^{1-iB} . \tag{3.9}$$

In our case (see eq. (3.2)), $B = 2Gp_+^{(1)} p_o^{(2)} = -2G(p^{(1)}.p^{(2)}) = Gs$, where s is the usual Mandelstam variable. Notice furthermore that

$$dk_+ d\tilde{k} = \frac{k_+}{k_o}\, dk_3 d\tilde{k} . \tag{3.10}$$

Thus, concentrating only on particle (1,) we get

$$_{out}\langle\vec{k}^{(1)}|\vec{p}^{(1)}\rangle_{in} =$$

$$\frac{k_+}{4\pi k_o}\, \delta(k_+ - p_+)\, \frac{\Gamma(1-iGs)}{\Gamma(iGs)}\, C^{iGs} \left[\frac{4}{(\tilde{p}-\tilde{k}^2)}\right]^{1-iGs} . \tag{3.11}$$

No particle production or Bremsstrahlung is seen in any coordinate frame (as long as particle (1) is electrically neutral and $m^{(1)} \ll M_{Planck}$), so the scattering is elastic. There is an exchange of momentum

$$q = k^{(1)} - p^{(1)} . \tag{3.12}$$

The Dirac delta in (3.11) is just energy conservation. Defining the Mandelstam variable $t = -q^2$, we find the elastic scattering amplitude to be (apart from the canonical factor $(k_+/k_o)\delta(\Sigma\, k - \Sigma\, p)$,

$$U(s,t) = \frac{\Gamma(1-iGs)}{4\pi\Gamma(iGs)}\, C^{iGs} \left[\frac{4}{-t} \right]^{1-iGs} , \tag{3.13}$$

from which the cross section follows,

$$\sigma(\tilde{p}^{(1)} \to \tilde{k}^{(1)})d^2\tilde{k} = \frac{4}{t^2} \left| \frac{\Gamma(1-iGs)}{\Gamma(iGs)} \right|^2 d^2k = 4G^2 \frac{s^2}{t^2}\, d^2\tilde{k} . \tag{3.14}$$

This resembles Rutherford scattering, except for the factor $s^2 \approx p_+^{(1)2}\, p_-^{(2)2}$. Such an extra energy dependence is of course to be expected from a theory of gravity. Apparently the cross section is just as if a single graviton were exchanged, but the amplitude, eq. (3.13), is more complicated. From the derivation it must be clear that the phase factor C^{iGs} is meaningless.

How exact is the amplitude (3.13)? Exchange of other massless particles would alter it. For instance, it is easily seen that electric charges $e^{(1)}$, $e^{(2)}$, just cause a shift:

$$Gs \to Gs - e^{(1)}e^{(2)}/4\pi , \tag{3.15}$$

in eqs (3.13) and (3.14) (this is found by considering the electric shock wave from a charged particle which is quite analogous to the gravitational shock wave). In many respects, eq. (3.13) can be seen to be an "eikonal approximation" [5]. However, we claim that other quantum field theoretic effects, for instance those due to exchange of scalar or massive particles, will be swamped by eq. (3.13) at sufficiently large Gs or small Gt. This is not only because of the obvious divergence at $t \to 0$. Take a particle (1) with a definite impact parameter b with respect to particle (2). If b is large we only have effects from the graviton (and other massless particles). But if b is small, we have the divergence of the logarithm in eqs. (2.11) and (3.3): particle (1) is being shifted along in the direction of $p_\mu^{(2)}$. Whatever it does, these effects will only be seen at much later times by any observeres in $R^4_{(+)}$.

Because of the above we suspect that the poles in our amplitude, which occur at

$$s = M^2_{\text{Planck}} \ (-Ni + k\alpha) \ , \tag{3.16}$$

where N is a positive integer, k is any integer, and α is the finestructure constant, may indicate the presence of new physical states. Their properties are subjects of further investigation.

Finally one may ask what happens if Gs becomes much larger than one. In that case it seems more appropriate to consider both particles (1) and (2) and their gravitational fields in the c.m. frame. The computation of the general relativistic effects when both shock waves collide is complicated however [6]. In general a spacelike singularity in space-time is expected. If the impact parameter is less than the Schwarzschild radius corresponding to the c.m. energy then one obviously expects a black hole to form, and classical gravitational waves will be emitted (in particle terms these are just coherent many-graviton modes). The corresponding computations are technically difficult, but should follow completely from the well known laws of general relativity, although we hasten to add to this that questions of a proper formulation of the initial conditions and the actual existence of solutions are far from settled mathematically [7]. In any case, non-trivial quantum field theoretical phenomena are well hidden behind the horizon.

4. THE BLACK HOLE

So at $s \gg M^2_{\text{Planck}}$ we expect black hole formation to affect our amplitude (3.13), perhaps creating new poles (which, by estimates of the black hole life time one might expect to occur at $s/M^2_{\text{Planck}} \to A - iB$, with A large and B of order one.) At first sight this may also seem to be a doable calculation, for which only classical physics is needed. Black holes just look like solitons, and all we have to do is compute their first non-trivial quantum corrections. Unfortunately, this turns out to be impossible with our present knowledge. We would expect black holes to form a spectrum of states, extending the spectrum of all known, much lighter particles into the regime of ultra-high energies. After all, there should be no fundamental difference whatsoever between the "ordinary" particles and black holes, both carrying a gravitational field described by Einstein's equations, and both associated with De Broglie waves [8].

To see what is going on we first briefly resume the elementary mathematical aspects of black holes [10].

A spherically symmetric solution of Einstein's equation after matter has moved to the center ($T_{\mu\nu} = 0$) can be written as

$$ds^2 = - F(r)\, dt^2 + G(r)\, dr^2 + H(r)\, (d\theta^2+\sin^2\theta\, d\varphi^2) \; , \qquad (4.1)$$

but we still have the freedom to redefine r: $r \to r'$, such that after the redefinition,

$$H(r) = r^2 \; . \qquad (4.2)$$

The equations $R_{\mu\nu} = 0$ give three equations for F and G, but of these one is redundant because of the automatic Bianchi identity

$$2\partial_\mu R_{\mu\nu} = \partial_\nu R \; . \qquad (4.3)$$

One finds successively

$$\partial_r(FG) = 0 \quad \to \quad F(r).G(r) = \text{Const.} \; , \qquad (4.4)$$

which constant can be put equal to one by rescaling t, and

$$\partial_r(rF) = 1 \quad \to \quad F(r) = 1/G(r) = 1 - 2M/r \; . \qquad (4.5)$$

Here, $2M$ is an arbitrary integration constantant. But one may observe that the function $F(r)$ corresponds directly to the gravitational red-shift, so that it can easily be identified as the gravitational potential, which is asymptotically

$$\sqrt{F(r)} \rightarrow 1 - M/r , \tag{4.6}$$

and one may conclude that

$$M = Gm , \tag{4.7}$$

where m is the black hole mass.

At the points

$$r = 2M , \tag{4.8}$$

this metric is singular, but this singularity is an artifact of the coordinates chosen. Consider the new time coordinate

$$\tilde{t}_+ = t + 2M \log(r-2M) , \tag{4.9}$$

then in the coordinates $(\tilde{t}_+, r, \theta, \varphi)$ the singularity disappears. These are the so-called "ingoing" Eddington-Finkelstein coordinates. The lines $\tilde{t}_+ + r =$ constant (at constant angles θ and φ) are the geodesics of infalling light rays. The region $0 < r < 2M$ can be reached from the outside. $\tilde{t}_+$ is real there, but t is complex. In that region (which will be called region III later), the local future light cone points entirely towards the singularity at $r = 0$.

One may also consider the "outgoing" Eddington-Finkelstein coordinates, replacing t by

$$\tilde{t}_- = t - 2M \log(r-2M) ; \tag{4.10}$$

in these coordinates the region $0 < r < 2M$ can be reached going backwards in time. The local future lightcone points outwards. We will call this region IV. Here also t is complex. However, if this metric is regarded as a solution of Einstein's equations of a black hole formed by collapse of matter at $t = t_1$, then region IV is unphysical: to reach it one would have to cross the region $-\infty < t < t_1$, where the black hole was not yet formed, and matter was present so that the vacuum Einstein equations were not valid there.

In spite of the fact that region IV is not present in such a "physical" black hole, it is still worthwhile to consider coordinates that show both regions III and IV. These are the so-called Kruskal coordinates (x,y,θ,φ), where x and y are defined by

$$\left[\frac{r}{2M} - 1 \right] e^{r/2M} = -xy \ , \tag{4.11}$$

$$e^{t/2M} = -x/y \ . \tag{4.12}$$

By differentiating one finds

$$4 \frac{dx\ dy}{xy} = \frac{1}{4M^2} \left[\frac{dr^2}{(1-2M/r)^2} - dt^2 \right] , \tag{4.13}$$

$$ds^2 = -2A(r)\ dx\ dy + r^2\ d\Omega^2 \ , \tag{4.14}$$

where

$$A(r) = \frac{16M^3}{r}\ e^{-r/2M} \ , \qquad d\Omega^2 = d\theta^2 + \sin^2\theta\ d\varphi^2 \ . \tag{4.15}$$

Notice that in these coordinates the singularity at $r + 2M$ totally disappears. The lines x = const., Ω = const., and the lines y = const., Ω = const., are light rays.

At every r,t we have two solutions for x and y (differing by a sign), so the regular region $r > 2M$ occurs twice in (x,y) space (to be called regions I and II). We indicate the regions I to IV in fig. 1.

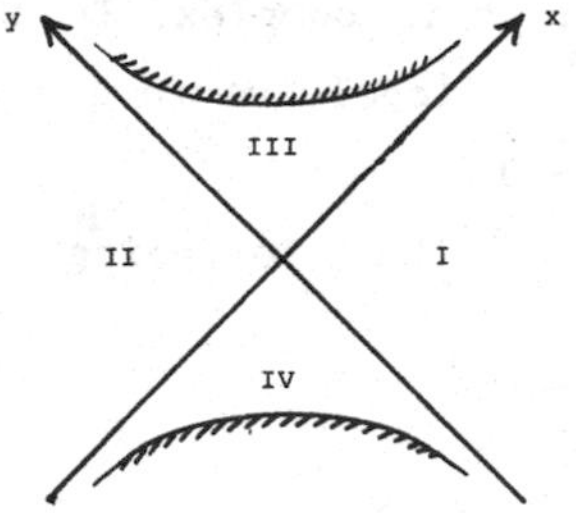

Fig. 1. The Kruskal coordinates.

The central region, $|x| \ll 1$, $|y| \ll 1$, is very important for understanding the quantum mechanics of the black hole. It is there where objects sent in in the far past, and particles that will emerge in the distant future meet each other. To describe that region, the *curvature* of space-time seems to be only of secondary importance. If we replace it by flat space we have the so-called *Rindler space*.

Consider a flat Minkowski space-time described in coordinates t, z and $\tilde{x} = (x,y)$. Let us then consider the new coordinates τ, ζ, and $\tilde{x}$, given by

$$z = \varsigma \cosh \tau \ ,$$
$$t = \varsigma \sinh \tau \ , \qquad (4.16)$$
$$\tilde{x} = \tilde{x} \ .$$

Of these coordinates, τ can be considered to be a *time* coordinate, because any shift of the form $\tau \to \tau + \tau_1$ is nothing but a Lorentz transformation, and hence leaves the laws of physics, as phrased in these "Rindler coordinates", invariant: the laws of physics do not change with time. A stationary observer in these coordinates has ς = const., which is a curved trajectory in Minkowski space; hence such an observer feels a gravitational field which is constant in time. This field becomes infinitely strong at $\varsigma = 0$. Clearly,. Rindler space is a model for a gravitational field. ς and τ play the role of the Schwarzschild coordinates r and t. The Kruskal coordinates x and y correspond to the Minkowski lightcone coordinates $t \pm z$. By comparing (4.12) and (4.16) at the origin of x-y space we see that τ corresponds to $t/4M$.

5. FIELD THEORY IN RINDLER SPACE

Consider now a quantum field theory, possibly with interacting particles, in a flat Minkowski background metric. Usually its Hamiltonian, H_M, can be written as an integral over a Hamiltonian density:

$$H_M = \int \mathcal{H}(\vec{x}) \, d^3\vec{x} \ . \qquad (5.1)$$

Now a time boost in Rindler space is generated by the operator

$$H_R = \int (\mathcal{H}(\vec{x}) \, z - \mathcal{P}_z(\vec{x}) \, t) \, d^3\vec{x} \ , \qquad (5.2)$$

to be recognized as the generator of Lorentz transformations in Minkowski space. usually we consider it at $t = 0$,

$$H_R = \int \mathcal{H}(\vec{x}) z \, d^3\vec{x} \ . \qquad (5.3)$$

In contrast with the usual Hamiltonian (5.1) this is obviously not bounded from below. Let us write

$$H_R = H_1 - H_2 \ , \qquad (5.4)$$

with

$$H_1 = \int_{z>0} \mathcal{H}(\vec{x}) z \, d^3\vec{x} \,,$$

$$H_2 = \int_{z<0} \mathcal{H}(\vec{x}) \, |z| \, d^3\vec{x} \,. \tag{5.5}$$

We quickly find that quite generally

$$[H_1, H_2] = 0 \,. \tag{5.6}$$

Consider namely two operators F_1 and F_2, defined by

$$F_i = \int f_i(\vec{x}) \, \mathcal{H}(x) \, d^3\vec{x} \,, \tag{5.7}$$

then

$$[F_1, F_2] \propto \int f_1(\vec{x}) \, \vec{\partial} f_2(\vec{x}) \, \ldots + \int f_2(\vec{x}) \, \vec{\partial} f_1(\vec{x}) \quad \ldots \tag{5.8}$$

With $f_1 = z\theta(z)$ and $f_2 = -z\theta(-z)$ this vanishes.
Physically this result is understandable: H_1 governs the evolution within region I and H_2 the evolution in region II. No information can be transmitted between the two regions.

At first sight this situation is quite pleasing: all physics in region I is described by H_1 and H_1 alone. Unfortunately, there is a divergence. This divergence is seen more clearly if we substitute in (4.16),

$$\zeta = e^{\sigma} \,. \tag{5.9}$$

A Lagrangian of the form

$$\mathcal{L} \, d^3\vec{x} \, dt = (-1/2(\vec{\partial}\varphi)^2 + 1/2(\partial_t\varphi)^2 - \tfrac{1}{2}m^2\varphi^2) \, d^3\vec{x} \, dt \,, \tag{5.10}$$

then takes the form

$$\left[-1/2(\partial_\sigma\varphi)^2 + 1/2(\partial_\tau\varphi)^2 + e^{2\sigma}(-1/2(\tilde{\partial}\varphi)^2 - \tfrac{1}{2}m^2\varphi^2)\right] d^2\tilde{x} \, d\sigma \, d\tau \,. \tag{5.11}$$

At $\sigma \to -\infty$ the first two terms survive and describe plane waves. The problem is that there is an asymptotic region at $\sigma \to -\infty$ into which wave packets may disappear at $\tau \to +\infty$ or from which wave packets may come at $\tau \to -\infty$. If we would try to formulate scattering against the Rindler horizon we would need a boundary condition at $\sigma = -\infty$. We do not know what it is. *Hawking radiation* seems to be the "answer" at present: there is thermal radiation emerging from the region $\sigma = -\infty$. Let us now consider the derivation of this phenomenon.

Everything can be understood as a feature of the central region in Kruskal space, and indeed all we need is the Rindler coordinate transformation (4.16). We will only consider non-interacting scalar particles; other cases are not really different.

A scalar field Φ in Minkowski space $(\vec{r},t)$ can be written as

$$\Phi(\vec{r},t) = \int \frac{d^3\vec{k}}{\sqrt{2k_o(\vec{k})V}} \left[a(\vec{k})\, e^{i\vec{k}\vec{r} - ik_o t} + a^\dagger(\vec{k})\, e^{-i\vec{k}\vec{r} + ik_o t}\right] , \qquad (5.12)$$

$$\dot{\Phi}(\vec{r},t) = \int \frac{d^3\vec{k}}{\sqrt{2k_o(\vec{k})v}} \left[-ik_o\, a(\vec{k})\, e^{ikx} + ik_o\, a^\dagger(\vec{k})\, e^{-ikx}\right] . \qquad (5.13)$$

Here, $V = (2\pi)^3$, and we have

$$[a(\vec{k}), a^\dagger(\vec{k}')] = \delta^3(\vec{k}-\vec{k}') , \qquad (5.14)$$

and

$$[\dot{\Phi}(\vec{r}), \Phi(\vec{r}')] = -i\, \delta^3(\vec{r}-\vec{r}') , \qquad (5.15)$$

etc.

First we make the transition to lightcone coordinates,

$$u = (t-z)/2 , \quad v = (t+z)/2 ,$$

$$k_+ = k_o + k_3 , \quad k_- = k_o - k_3 . \qquad (5.16)$$

Then in Rindler time these evolve as

$$v \to v\, e^{\tau} ,$$

$$u \to u\, e^{-\tau} . \qquad (5.17)$$

And we define new annihilation operators a_1:

$$a(\vec{k})\,\sqrt{k_o} = a_1(\tilde{k},k_+)\,\sqrt{k_+}\ , \tag{5.18}$$

which, because

$$\frac{\partial k_+}{\partial k_3}\Big|_{\tilde{k}} = \frac{k_+}{k_o}\ , \tag{5.19}$$

are now normalized by

$$\left[a_1(\tilde{k},k_+),a_1^{\dagger}(\tilde{k}'.k')\right] = \delta^2(\tilde{k}-\tilde{k}')\ \delta(k_+-k')\ . \tag{5.20}$$

So we can write

$$\Phi(\vec{r},t) = A(\vec{r},t) + A^{\dagger}(\vec{r},t)\ ; \tag{5.21}$$

$$A(\tilde{r},u,v) = \int_{k_+>0} \frac{d\tilde{k}\ dk_+}{\sqrt{2Vk_+}}\ a_1(\tilde{k},k_+)\ e^{i\tilde{k}\tilde{r}\ -\ ik_+u\ -\ ik_-v}\ . \tag{5.22}$$

Now in Rindler time, $\tilde{k}$ is constant and

$$\begin{aligned} k_+ &\rightarrow k_+\ e^{\tau}\ ,\\ k_- &\rightarrow k_-\ e^{-\tau} \end{aligned} \tag{5.23}$$

If we Fourier transform the field Φ of eq. (5.21) with respect to τ then we may expect to get annihilation and creation operators corresponding to definite amounts of energy for the Rindler observer. Therefore we now choose to Fourier transform a_1 with respect to log k_+:

$$a_1(\tilde{k},k_+)\sqrt{k_+} = (2\pi)^{-1/2}\int_{-\infty}^{\infty} d\omega\ a_2(\tilde{k}.\omega)\ e^{-i\omega\ \ln(k_+/\mu)}\ , \tag{5.24}$$

where

$$\mu^2 = \tilde{k}^2 + m^2 = k_+k_-\ , \tag{5.25}$$

and the new annihilation operators a_2 are normalized as

$$[a_2(\tilde{k},\omega),a_2^\dagger(\tilde{k}',\omega')] = \delta^2(\tilde{k}-\tilde{k}')\ \delta(\omega-\omega') \ . \qquad (5.26)$$

The inverse of eq. (5.24) is

$$a_2(\tilde{k},\omega) = \int_0^\infty dk_+\ (2\pi k_+)^{-1/2}\ a_1(\tilde{k},k_+)\ e^{i\omega\ \ln(k_+/\mu)} \ . \qquad (5.27)$$

The Rindler Hamiltonian is

$$H_R = \int z\ \mathcal{H}_M(\vec{r},0)\ d^3\vec{r} \ , \qquad (5.28)$$

with

$$\mathcal{H}_M(\vec{r},t) = \tfrac{1}{2}\dot{\Phi}^2 + 1/2(\vec{\partial}\Phi)^2 + \tfrac{1}{2}m^2\Phi^2 \ . \qquad (5.29)$$

A straightforward calculation now yields:

$$H_R = \int d^2\tilde{k} \int_{-\infty}^{\infty} d\omega\ \omega\ a_2^\dagger(\tilde{k},\omega)\ a_2(\tilde{k},\omega) \ , \qquad (5.30)$$

so indeed in all respects, a_2 behaves as an annihilation operator corresponding to a Rindler energy ω.

Nevertheless, a_2 is *not* the annihilation operator we want to work with. We would like to split the operator H_R into two parts:

$$H_R = H_1 - H_2 \ , \qquad (5.31)$$

with

$$H_1 = \int z\theta(z)\ \mathcal{H}_M(\vec{r},0)\ d^3\vec{r} \ ; \quad H_2 = -\int z\theta(-z)\ \mathcal{H}_M(\vec{r},0)\ d^3\vec{r} \ . \qquad (5.32)$$

And now it is important to note that H_1 and H_2 do *not* split the integral (5.30) in the same manner.

To see what happens, let us write the field A in eq. (5.21) in terms of a_2:

$$A(\vec{r},t) = \int_{-\infty}^{\infty} d\omega \int \frac{d^2\tilde{k}}{\sqrt{4\pi V}}\ K(-\omega,\mu u,\mu v)\ e^{i\tilde{k}\tilde{r}}\ a_2(\tilde{k},\omega) \ . \qquad (5.33)$$

Here, u and v are the coordinates (5.16) and K is an integration kernel, which turns out to be

$$K(\omega,\alpha,\beta) = \int_0^\infty \frac{dx}{x}\, x^{i\omega}\, e^{-ix\alpha - i\beta/x} \tag{5.34}$$

We need some properties of K. For $\alpha < 0$ and $\beta > 0$ the integrand in (5.34) converges rapidly if $\mathrm{Im}(x) \geq 0$. Therefore we may rotate the integration contour by

$$x \to x\, e^{i\varphi} , \qquad 0 \leq \varphi \leq \pi . \tag{5.35}$$

Taking $\varphi = \pi$ gives us the identity

$$K(\omega,\alpha,\beta) = \int_0^\infty \frac{dx}{x}\, x^{i\omega}\, e^{-\pi\omega}\, e^{ix\alpha + i\beta/x} = e^{-\pi\omega}\, K^*(-\omega,\alpha,\beta) \;, \quad \alpha < 0 \;, \quad \beta > 0 \;. \tag{5.36}$$

When $\alpha > 0$ and $\beta < 0$ we have, using a similar contour shift,

$$K(\omega,\alpha,\beta) = e^{\pi\omega}\, K^*(-\omega,\alpha,\beta) \;, \qquad \alpha > 0 \;, \; \beta < 0 \;. \tag{5.37}$$

We now split the integral (5.33) into two integrals for positive ω. If $\vec{r}$ is in region I we have $u < 0$ and $v > 0$. Therefore,

$$A(\vec{r},t) = \int_0^\infty d\omega \int \frac{d^2\tilde{k}}{\sqrt{4\pi V}}\, e^{i\tilde{k}\tilde{r}} \Big(K(-\omega,\mu u,\mu v)\, a_2(\tilde{k},\omega) + $$

$$+ e^{-\pi\omega}\, K^*(-\omega,\mu u,\mu v)\, a_2(\tilde{k},-\omega)\Big) \;, \tag{5.38}$$

and

$$A^\dagger(\vec{r},t) = \int_0^\infty d\omega \int \frac{d^2\tilde{k}}{\sqrt{4\pi V}}\, e^{i\tilde{k}\tilde{r}} \Big(K^*(-\omega,\mu u,\mu v)\, a_2^{\dagger}(-\tilde{k},\omega) + $$

$$+ e^{-\pi\omega}\, K(-\omega,\mu u,\mu v)\, a_2^{\dagger}(-\tilde{k},-\omega)\Big) \;. \tag{5.39}$$

Combining these now into the field Φ (see 5.21)), we see

$$\Phi(\vec{r},t) = \int_0^\infty d\omega\, \frac{d^2\tilde{k}}{\sqrt{4\pi V}}\, e^{i\tilde{k}\tilde{r}} \Big[K(-\omega,\mu u,\mu v)\big(a_2(\tilde{k},\omega) + e^{-\pi\omega}\, a_2^{\dagger}(-\tilde{k},-\omega)\big) + $$

$$K^*(-\omega,\mu u,\mu v)\big(a_2^{\dagger}(-\tilde{k},\omega) + e^{-\pi\omega}\, a_2(\tilde{k},-\omega)\big)\Big] \tag{5.40}$$

This prompts us to define an operator $a_I(\tilde{k},\omega)$ as follows,

$$a_{\rm I}(\tilde{k},\omega)\sqrt{1-e^{-2\pi\omega}} = a_2(\tilde{k},\omega) + e^{-\pi\omega}\, a_2^{\dagger}(-\tilde{k},-\omega) \ ; \tag{5.41}$$

Clearly, if $\tilde{r},t$ are in region I then $\Phi(\tilde{r},t)$ only depends on $a_{\rm I}$ and its Hermitean conjugate. Similarly, in region II we have $a_{\rm II}$

$$a_{\rm II}(\tilde{k},\omega)\sqrt{1-e^{-2\pi\omega}} = a_2(\tilde{k},-\omega) + e^{-\pi\omega}\, a_2^{\dagger}(-\tilde{k},\omega) \ . \tag{5.42}$$

The normalization factors are needed to get the commutation rules

$$\left[a_{\rm I}(\tilde{k},\omega)\ ,\ a_{\rm I}^{\dagger}(\tilde{k}',\omega')\right] = \delta^2(\tilde{k}-\tilde{k}')\ \delta(\omega-\omega') \ ; \tag{5.43}$$

similarly for $[a_{\rm II}\ ,\ a_{\rm II}^{\dagger}]$, and furthermore we have:

$$[a_{\rm I}\ ,\ a_{\rm I}] = [a_{\rm II}\ ,\ a_{\rm II}] = [a_{\rm I}\ ,\ a_{\rm II}] = [a_{\rm I}\ ,\ a_{\rm II}^{\dagger}] = 0 \ . \tag{5.44}$$

And now indeed,

$$\begin{aligned} H_1 &= \int_0^\infty d\omega \int d^2\tilde{k}\ \omega\ a_{\rm I}^{\dagger}a_{\rm I} + C \quad ; \\ H_2 &= \int_0^\infty d\omega \int d^2\tilde{k}\ \omega\ a_{\rm II}^{\dagger}a_{\rm II} + C \ , \end{aligned} \tag{5.45}$$

where C is common, irrelevant constant coming from the re-ordering process. It cancels in $H_{\rm R}$, eq. (5.31).

From the commutation rules (5.43)-(5.44) we see that all observables in region II commute with all $a_{\rm I}$, $a_{\rm I}^{\dagger}$, and *vice versa*. Therefore, not the operators a_2, $a_2^{\dagger}$, but $a_{\rm I}$ and $a^{\dagger}_{\rm I}$ are the proper annihilation and creation operators for Rindler observers in region I, and $a_{\rm II}$, $a^{\dagger}_{\rm II}$ in region II. Transformations such as (5.41) and (5.42) involving a and $a^{\dagger}$ are called "Bogolyubov transformations".

Let now $|\Omega\rangle$ be the vacuum state as defined by an observer in Minkowski space, i.e.

$$a|\Omega\rangle = a_1|\Omega\rangle = a_2|\Omega\rangle = 0 \ , \text{ for all } \tilde{k},\omega \ . \tag{5.46}$$

It is now opportune to introduce as a basis for Hilbert space those states which at each set of values of $\pm\tilde{k}$ and ω have definite values for $n_{\rm I} \underset{\rm def}{=} a^{\dagger}{}_{\rm I}(\tilde{k},\omega)\ a_{\rm I}(\tilde{k},\omega)$ and for $n_{\rm II} \underset{\rm def}{=} a^{\dagger}{}_{\rm II}\ a_{\rm II}$. At each $(\pm\tilde{k},\omega)$ we label these states as $|n_{\mathrm{I}}, n_{\mathrm{II}}>$. Clearly,

$$\prod_{\tilde{k},\omega} |0,0> \neq |\Omega> . \tag{5.47}$$

To express $|\Omega>$ in our Rindler basis we use, from (5.41) and (5.42),

$$a_{\mathrm{I}}(\tilde{k},\omega)\ |\Omega> - e^{-\pi\omega}\ a_{\mathrm{II}}{}^{\dagger}(-\tilde{k},\omega)\ |\Omega> = 0\ ; \tag{5.48}$$

$$a_{\mathrm{II}}(\tilde{k},\omega)\ |\Omega> - e^{-\pi\omega}\ a_{\mathrm{I}}{}^{\dagger}(-\tilde{k},\omega)\ |\Omega. = 0\ , \tag{5.49}$$

so that, when acting on $|\Omega>$, we have

$$a^{\dagger}{}_{\mathrm{I}}a_{\mathrm{I}} = e^{-\pi\omega}\ a^{\dagger}{}_{\mathrm{I}}a^{\dagger}{}_{\mathrm{II}} = e^{-\pi\omega}\ a^{\dagger}{}_{\mathrm{II}}a^{\dagger}{}_{\mathrm{I}} = a^{\dagger}{}_{\mathrm{II}}a_{\mathrm{II}}\ . \tag{5.50}$$

Consequently, $|\Omega>$ only consists of states with

$$n_{\mathrm{I}} = n_{\mathrm{II}}\ ; \tag{5.51}$$

$$|\Omega> = \sum_n f_n\ |n,n>\ . \tag{5.52}$$

We find f_n from (5.48):

$$\sum_n f_n\ \sqrt{n}\ |n-1,n> = e^{-\pi\omega} \sum_n f_n\ \sqrt{n+1}\ |n,n+1>\ ; \tag{5.53}$$

$$f_{n+1} = e^{-\pi\omega}\ f_n\ . \tag{5.54}$$

Conclusion:

$$|\Omega> = \prod_{\tilde{k},\omega} \sqrt{1-e^{-\pi\omega}} \sum_{n=0}^{\infty} e^{-\pi n\omega}\ |n,n>_{\pm\tilde{k},\omega}\ , \tag{5.55}$$

where the square root is a normalization factor. Notice that (5.51) implies

$$H_{\mathrm{R}}|\Omega = 0\ , \tag{5.56}$$

or: $|\Omega>$ is Lorentz invariant. More surprising perhaps is that there are very many other Lorentz invariant states. These, as all elements $|n,m>$ of our basis, must have divergent expectation values for their energy and momentum.

The probability that a Rindler observer in region I, while looking at the states $|\Omega\rangle$, observes n_I particles with energy ω and transverse momentum $\tilde{k}$ in region I is

$$P_{n_I} = \sum_{n_{II}} \langle\Omega|n_I,n_{II}\rangle\langle n_I,n_{II}|\Omega\rangle = |f_{n_I}|^2 = (1-e^{-2\pi\omega})^{-1}\, e^{-2\pi n_I\omega} \tag{5.57}$$

This we can write as

$$P_n = e^{-\beta(E-F)} \ , \tag{5.58}$$

where $E = n\omega$ is the energy, and $\beta = 1/T$ can be interpreted as a *temperature*. F is then the free energy. One concludes that the Rindler observer detects particles radiating in all directions at a temperature which is in his units of energy

$$T_R = 1/2\pi \ . \tag{5.59}$$

Now consider the central region in the Kruskal space of a black hole. For the distant observer this is just Rindler space. Now as a matter of fact all observers must see matter here, namely the ingoing objects that produced the black hole, at a very early time. If at later times nothing more is thrown into the hole then the Rindler observer will see the *incoming* particles approach the state $|0,0\rangle$, which, as we remember, also corresponds to a highly energetic state in the local Minkowski space. But all these particles are *ingoing* particles. In the local Minkowski frame no particles are seen coming out. To describe those we need the state $|\Omega\rangle$. An observer at late times in the Kruskal region I will think he sees this state $|\Omega\rangle$. This is why we expect him to see radiation corresponding with a temperature T. Notice that when Rindler space is replaced by Kruskal space we must insert a factor $4M$ in the unit of time (see the end of sect. 4). Therefore, in natural units, the temperature of the expected radiation from a black hole is

$$T_H = 1/8\pi M \ . \tag{5.60}$$

This is the Hawking temperature [1].

6. THE GRAVITATIONAL BACK REACTION

There is something very peculiar about this result of the previous section. It namely suggests that any observer who looks at a black hole long after the last objects have been thrown in, will see a thermodynamic mixture of quantum mechanical states, to be described by a density matrix,

$$\rho = N \prod_{\tilde{k},\omega} \sum |n_{\mathrm{I}}\rangle \, e^{-\beta_{\mathrm{I}}\omega} \langle n_{\mathrm{I}}| \ , \tag{6.1}$$

where N is a normalization factor. After all, we must assume that the labels n_{II} corresponding to particles in region II are irrelevant to him. Only with this matrix ρ we can reproduce the probabilities (5.57). But what if we started with one pure quantum mechanical state describing imploding matter? Does then also the density matrix (6.1) evolve?

Suppose now that a black hole is simply the most compact, and in some sense the most general, object with a given total energy. Suppose that in all other respects black holes may be assumed to obey the ordinary rules of quantum mechanics. This then would imply that when the state of all particles that made up the black hole in some implosion process were completely specified, then also the state of all outgoing particles should be well determined, probably as a complicated linear superposition of many different "decay modes". In particular, it should not be a mixture of different states in a density matrix ρ unless

$$\mathrm{Tr}\ \rho^2 = \mathrm{Tr}\ \rho = 1 \ , \tag{6.2}$$

which means that it can be seen as a single pure state, if the initial state was pure as well.

Apparently, this is not what one finds when applying standard quantum field theory in the vicinity of the black hole horizon. One finds a thermal spectrum (6.1), to be normalized by

$$\mathrm{Tr}\ \rho = 1 \ , \tag{6.3}$$

so that eq. (6.2) cannot be obeyed. What is wrong in standard quantum field theory near a horizon?

First of all we note that the difference between pure states and mixed states will be more and more difficult to detect as the black hole becomes larger. The final state could be pure but so complicated that for all practical purposes it can be handled as a thermodynamical mixture. By the time a black hole is so large that observers can be sent in nobody will ever detect the difference.

For small black holes however the question of quantum mechanical purity is extremely important. It seems then that (6.1) can only be an "approximation". Is there a way to replace it by a single realtistic wave function?

What was ignored in the standard derivation was the gravitational interaction between ingoing and outgoing matter. But this interaction is crucial. Suppose a particle goes in with momentum p_1 at time $t = t_1$. After a long time, $t = t_2 >> t_1$, we look again at the black hole and observe a Hawking particle with momentum p_2 coming out. At some time t_o, roughly halfway between t_1 and t_2 the two particles must have met, that is, they were at the same distance from the horizon, one entering, the other leaving. Particles and shock waves collide at $t = t_o$. The center-of-mass energy with which this collision takes places can be easily estimated:

$$E^2_{c.m.} = -s = \mathcal{O}\left[p_1 p_2 \, e^{(t_2-t_1)/4M}\right] , \tag{6.4}$$

increasing far beyond control when

$$t_2 - t_1 >> 4M . \tag{6.5}$$

Now if in (6.4) both p_1 and p_2 are large then the effect this has on the metric is so complicated that exact solutions probably do not exist. If only p_1 (or only p_2) is large then we have a δ-distributed Riemann curvature at the past horizon and, remarkably, the metric can be found analytically [10]. It corresponds to two Schwarzschild solutions, shifted with respect to each other at the past horizon by a shift δy in the Kruskal coordinate y. By imposing Einstein's equations

$$R_{\mu\nu} = 0 , \tag{6.6}$$

one finds for δy the equations

$$\frac{\partial}{\partial y} \delta y = 0 , \quad (\Delta_\Omega - 1) \, \delta y(\theta,\varphi) = C \, p_1 \, \delta^2(\Omega - \Omega_{in}) , \tag{6.7}$$

where Ω_{in} is the set of angles θ,φ at which the incoming particle entered, and C is a numerical constant.

$$\Delta_\Omega = \frac{\partial^2}{\partial\theta^2} + \cot\theta \frac{\partial}{\partial\theta} + \frac{1}{\sin^2\theta} \frac{\partial^2}{\partial\varphi^2} \tag{6.8}$$

is the angular Laplacian. The Green function corresponding to eq. (6.7) can be found:

$$\delta y(\Omega_1) = C' \int_0^\infty \frac{\cos \frac{1}{2}\sqrt{3}\, s\, ds}{\sqrt{\cosh s - \cos(\Omega_1 - \Omega_2)}}\, p_1(\Omega_2) \ . \tag{6.9}$$

where C' is another numerical constant. For Ω_1 close to Ω_2 this diverges as $\log(\Omega_1 - \Omega_2)$, just as in the flat space case.

Indeed, in reality collisions between the particles p_1 and p_2 with these energies do not take place, or rather, they are not seen. The ingoing particle does not see the outgoing particle but experiences a surrounding vacuum. However, as soon as we introduce a detector at $t = t_2$ that can distinguish different modes of outgoing particles, we have trouble at $t = t_0$, because the detector has split up the wave function in pieces that may contain these objects, hitting the ingoing particles with tremendous center-of-mass energies.

A black hole should be in a pure state if all particles that produced the hole in some distant past were in a pure state. We now notice that when observations are made at much later times, we are tempted to use elements of Hilbert space that are extremely singular in the past, and decomposition of the ingoing states into outgoing modes will be made extremely difficult because of this. Since the standard derivation of the Hawking effect ignores these gravitational self-interactions, we believe that the resulting density matrix ρ of eq. (6.1) cannot be trusted completely.

Consider a black hole that was formed by a collapse at $t = t_1$, and an observer at $t = t_2$ has decomposed the wave function as just described, by looking at particles emerging from the hole at various angles. If we follow these particles back in the past, we see that for a while they stick to the "past horizon", but then at $t \approx t_1$, they are released. By that time they have energies roughly described by eq. (6.4), and, coming from different directions they collide. The Schwarzschild radius corresponding to (6.4) is large, and hence a space-time singularity at $t = t_1$ is now unavoidable [11]. Thus there will be a time-reflected black hole in the past. This is sometimes called a "white hole".

Considering some Cauchy surface at $t = t_0$, we see that quantum mechanical superpositions must be allowed between states that contain different kinds of black hole singularities both in the future and the past. We also see that the distinction between "primordial" black holes and black holes that have been formed a relatively short time ago disappears, and our view upon black holes is entirely symmetric under time-reversal.

Since the arguments presented in this section are essentially independent of the assumptions mentioned earlier we believe that they provide further support to the idea that a black hole can decay entirely into "ordinary" particles. The concept of a naked remnant singularity [12] does not fit very well in this picture.

But the most difficult problem that the present discussion confronts us with is that not all configurations of particles surrounding a black hole late in its evolution should be accepted as independent, mutually orthogonal elements of Hilbert space. Most of these, in certain linear combinations, would produce the space-time singularity of a white hole in the (not very distant) past, whereas the real black hole should not have such a singularity.

How then should Hilbert space elements be labeled? The remainder of this section is an attempt.

Let us compare two states in Hilbert space. The second is the same as the first, except for one extra particle going in at $t = t_1$. That portion of the horizon $\mathcal{J}$ that corresponds to times $t < t_1$ in these two states is not in exactly the same position. The displacement can be accurately calculated. Let us write eq. (6.9) as

$$\delta u(\Omega) = f(\Omega,\Omega')\ p_{in}(\Omega') \ , \tag{6.10}$$

where p_{in} is the momentum of a particle coming in at solid angles Ω', in some suitable unitys; u is a Kruskal coordinate (the other Kruskal coordinate will be called v), but for simplicity we will soon replace u and v by Rindler coordinates.

Let now p_{out} be the momentum of an observed outgoing particle. Then naturally the displacement (6.10) will give its wave function an extra factor

$$e^{-ip_{out}(\Omega)\ \delta u(\Omega)} = e^{-ip_{out}(\Omega)\, f(\Omega,\Omega')\ p_{in}(\Omega')} \ . \tag{6.11}$$

If we suppose that the shift δu is all information we'll ever get back from the ingoing particle, then (6.11) must give the amplitude for the process. What have we found out? The functions $p_{in}(\theta,\varphi)$ describe the ingoing momenta, and $p_{out}(\theta,\varphi)$ the outgoing momenta. Suppose we can treat these as operators depending on θ and φ. The angles θ and φ are continuous; we'll quickly replace them by a dense but discrete lattice in Ω space. This will be necessary in order to make our expressions well-defined. Thus, any Dirac delta of the form $\delta(\Omega-\Omega')$ will have to be thought of as a Kronecker delta on some dense lattice.

In the usual Hilbert space of particles we expect

$$[p_{in}(\Omega) \ , \ p_{in}(\Omega')] = 0 \ , \qquad (6.12)$$

and the same for the outgoing particles. The conjugated operators are $v_{in}(\Omega)$, $u_{out}(\Omega)$, with

$$[p_{in}(\Omega), \ v_{in}(\Omega')] = -i\delta(\Omega,\Omega') \ , \qquad (6.13)$$

Now we see that eq. (6.11) suggests

$$u_{out}(\Omega) = - \int f(\Omega,\Omega') \ p_{in}(\Omega') \ d^2\Omega' \ , \qquad (6.14)$$

$$v_{in}(\Omega) = \int f(\Omega,\Omega') \ p_{out}(\Omega') \ d^2\Omega' \ , \qquad (6.15)$$

obtaining

$$\langle p_{out}(\Omega')|p_{in}(\Omega)\rangle = N \ e^{-i \int p_{out}(\Omega) \ f(\Omega,\Omega') \ p_{in}(\Omega') \ d\Omega \ d\Omega'} \ , \qquad (6.16)$$

where N is a normalization factor.

Notice that (6.16) is an entirely acceptable unitary "scattering matrix". Unfortunately, the Hilbert space generated by (6.12) and (6.13), in which this matrix acts, is quite unnatural. It resembles a bit the Fock space of in- and out-particles in a mixed coordinate-momentum representation where the transverse coordinates Ω and the longitudinal momenta p_r are specified for each poarticle. What is unusual about it is that at *every* pair of values for the angles θ and φ we must have *exactly one* particle!

As physicists we might not be too much worried about this situation. After all, we already suspected that the black hole will be surrounded by particles, possibly swimming in a Dirac sea.

Suppose we had a dense lattice in Ω-space. Why not reshuffle those particles a little bit so that there is exactly one for each point in this Ω lattice? The answer is that it is not so easy to link this special Hilbert space to the space of real particles in the real world. In particular the situation far away from the black hole will be difficult to handle. Also, any such procedure will depend delicately upon the Ω cut-off procedure used.

There is another reason why the need for an Ω cut-off should not surprise us. Tiny values $\Delta\Omega$ can only be detected by particles with large transverse momentum. But these particles will not only shift the horizon as given by (6.10) but also cause shifts in the transverse direction. A difficulty here is that these shifts will produce more complicated forms of curvature in space-time. As long as we cannot handle this situation precisely we will stick to more crude *Ansätze* for a lattice cut-off.

A simple one-dimensional model for a quantum mechanically "coherent" black hole is constructed in ref. [13]. A single Dirac particle bounces back and forth against the horizon. It is found to display a discrete set of energy levels of the form

$$\omega_N \approx 2\pi N \;/\; \ln(Ng/m^2) \;, \tag{6.17}$$

for large N, where g is a "gravitational" coupling constant.

7. CONCLUSION

Our present picture of the combined laws of quantum mechanics and general relativity is still beset by inconsistencies. In all other areas of physics we had frequently the good fortune that experimental observations eventually gave us the crucial pieces of information to sort out such problems, and invariably it would turn out that the correct answers given by nature are completely rational and logical. There is little doubt that a completely satisfactory description of quantum gravity including black holes should be possible. The question now is how to get there with at best only extremely indirect experimental evidence (such as the set of gauge groups, fermionic representations of these, and coupling constants). In this lecturer's opinion more *Gedanken*-experiments should be done, after which simply all conceivable answers should be compared. Our "scattering experiment" in section 3 is an example. We found that the result can be computed unambiguously. As for black holes, we think that some spectrum of states should exist. Perhaps it begins with the poles (3.16). Unfortunately these are not the poles given by string theory as was once hoped [11]. We remarked that the geometry of the horizon of a black hole could in principle generate its Hilbert space. This is the geometry of two-dimensional Riemann spaces, so there *could* be an important role for two-dimensional worlds in a future theory of quantum gravity.

REFERENCES

[1] S.W. Hawking, Comm. Math. Phys. 43 (1975) 199 ; J.B. Hartle, S.W.Hawking, Phys. Rev. D13 (1976) 2188 ; W.G. Unruh, Phys. Rev. D14

(1976) 870 ; S.W. Hawking and G. Gibbons, Phys. Rev. D15 2738 (1977); S.W. Hawking, Comm. Math. Phys. 87 (1982) 395.

[2] R.M. Wald, Comm. Math. Phys. 45 (1975) 9; Phys. Rev. D20 (1979) 1271.

[3] P.C. Aichelburg and R.U. Sexl, Gen. Rel. Grav. 2 (1971), 303.

[4] This result was reported in: G. 't Hooft, Phys. Lett. B (1987), to appear.

[5] H. Abarbanel and C. Itzykson, Phys. Rev. Letters, 23 (1969) 53; M. Levy and J. Sucher, Phys. Rev. 186 (1969) 1656.

[6] P.D. D'Eath, Phys. Rev. D18 (1978) 990.

[7] See the lectures by T. Damour and B. Carter in "Gravitation in Astrophysics", Cargese 1986, Plenum, NATO ASI Series B, vol. 156,B. Carter and J.B. Hartle, eds.

[8] G. 't Hooft, Nucl. Phys. B256 (1985) 727.

[9] There are meny excellent textbooks on general relativity. For instance: R. Adler, M. Bazin, M. Schiffer, "Introduction to General Relativity", McGraw-Hill 1965; P.A.M. Dirac, "General Theory of Relativity", Wiley Interscience 1975; S. Weinberg, "Gravitation and Cosmology: Principles and Applications of the General Theory of Relativity", J. Wiley & Sons; S.W. Hawking, G.F.R. Ellis, "The largescale structure of Space-time", Cambridge Univ. Press 1973;S. Chandrasekhar, "The Mathematical Theory of Black Holes", Clarendon Press, Oxford Univ. Press; R.M. Wald, "General Relativity", Univ. ofChicago Press, 1984.

[10] T. Dray and G. 't Hooft, Nucl. Phys. B253 (1985) 173; Comm. Math.Phys. 99 (1985) 613; Class. Quantum Grav. 3 (1986) 825.

[11] G. 't Hooft, Physica Scripta T15 (1987) 143.

[12] See for instance: M.A. Markov, "on the stability of Elementary Black Holes", Trieste Preprint IC/78/41, 1978; Progr. Theor. Phys., Suppl. Extra Number, 85 (1965); ZHETF, 51, 878 (1966); id., "Micro-Macro Symmetric Universe", Acad. of Sciences of the USSR, Inst. of Nucl. Research, Moscow, Prepr. 1985.

[13] G. 't Hooft, talk presented at the Moscow Seminar on Quantum Gravity, USSR Acad. of Sciences, May 1987.

SUPERSYMMETRIC QUANTUM FIELDS AND INFINITE DIMENSIONAL ANALYSIS*

Arthur Jaffe and Andrzej Lesniewski
Harvard University, Cambridge, MA 02138 USA

I. INTRODUCTION

Supersymmetric quantum fields are interesting from the point of view both of physics and of mathematics. Their interest for physics (see, e.g. [35]) stems primarily from the fact that they may provide a natural framework for a unified theory of elementary interactions. In particular, recent developments in superstring theory [15] suggest that such a theory has the potential to unify gravitational forces with the strong, weak and electromagnetic forces.

From the point of mathematics, supersymmetric quantum field theory provides the framework for a theory of analysis in infinitely many dimensions. One aspect of this theory, and the one we have extensively studied, is the formulation of the theory of Dirac operators on a class of infinite dimensional spin manifolds.

A particular feature of certain two-dimensional models of supersymmetric quantum field theory concerns their local regularity properties. They may be ultraviolet finite. In that case the divergences usually associated with quantum field theory cancel identically. Finite models are specified by their Lagrangians. In other words, the parameters of the Lagrangian does not require modification, and there are no hidden constants required to define the equations of motion.

However, even finite models require regularization as an intermediate step toward their construction. One hopes that after removal of the regularization, finite models have limits which are independent of the regularization, at least within some large class of regularizations.

* Supported in part by the National Science Foundation under Grant DMS/PHY 86–45122.

We have studied a family of finite supersymmetric theories. In these notes we summarize some of our results [18–26] on the mathematics of supersymmetric quantum fields. These examples have infinitely many degrees of freedom, coupled through a nonlinear system of equations of motion. It is hard to believe that they could be exactly soluble. Other examples of infinite dimensional analysis arise in constructive field theory and in statistical mechanics. However, all these other examples either involve infinite renormalizations, or else they are either finite dimensional or exactly soluble.

The basic methods of our constructions rest heavily on establishing *a priori* estimates. These estimates are infinite dimensional generalizations of elliptic partial differential inequalities such as Gårding's inequality. It is essential, however, that the constants in these *a priori* estimates are dimension independent. For this reason Sobolev spaces are not an appropriate framework for such infinite dimensional analysis. Rather, trace class properties of the heat kernel $\exp(-\beta H)$ plays a fundamental role in regularization. We call our procedure "heat kernel regularization," see [18].

Elliptic estimates are a cornerstone of finite dimensional analysis. Our approach is to approximate our infinite dimensional manifold (loop-space) by a sequence of finite dimensional manifolds. We establish estimates for each of these finite dimensional approximations, with constants that do not depend on the dimension. The finite dimensional approximation is the regularization alluded to above; it is also known as an "ultraviolet cutoff." The approximating theory satisfies classical elliptic inequalities; it is easy to prove estimates for the approximations – but with dimension-dependent constants. The brunt of our work goes into developing methods to improve these estimates so that the constants are uniform in the dimension. We use these uniform, *a priori* estimates to establish the existence of a probability measure on loop space. Similarly, we use the estimates to define a set of operators on loop space, including covariant Dirac operators and Laplacians. Our uniform estimates carry over to the infinite dimensional case. We thereby obtain, for example, a Gårding inequality in infinite dimensions. A byproduct of these estimates is the construction of finite, supersymmetric quantum field models.

Having constructed these models, we find that there is a local Hamiltonian H related to a local supercharge Q by $H = Q^2$. In fact Q can be interpreted as a Dirac operator on loop space. Let us contrast this with the finite dimensional case. On a compact manifold M, a Dirac operator has the form $\sum_{i=1}^{R} \gamma_i \nabla_i$ where γ_i are elements of a Clifford algebra and where ∇_i are covariant derivatives on M. In our infinite dimensional generalization, the fermion field $\psi(x)$ on Fock space can be interpreted as an infinite dimensional Clifford algebra. The relations $\gamma_i\gamma_j + \gamma_j\gamma_i = 2\delta_{ij}$

are replaced by the canonical anticommutation relations $\psi(x)^*\psi(y) + \psi(y)\psi(x)^* = \delta(x-y)$. The covariant derivative ∇_i is replaced by a covariant, Frechet derivative, $\nabla_{\varphi(x)} = -i\delta/\delta\varphi(x) + f(\varphi(x))$, where f can be interpreted as a connection on our infinite dimensional manifold. Then the supercharge Q has the general form

$$Q = \int_{T^1} \psi(x)\nabla_{\varphi(x)}\, dx \quad .$$

Here T^1 denotes the circle underlying a cylindrical space-time. This form of Q justifies the interpretation of Q as Dirac operator. The function f is linear in a potential function V, and we study polynomial potentials. A complete definition of Q is given in Sections V.5 and V.8 and below.

The models we study in these notes are known in the formal physics literature as the $N = 2$ Wess-Zumino models, see for example, [5] and [10]. The idea that the supercharge of a supersymmetric field theory model is a Dirac operator on loop space is due to Witten [36–38]. The actual construction of all these objects was first given in [18–26].

While the local singularities in these models are no worse than those encountered in constructive field theory models that involve renormalization, see [12], we need to sharpen the known inequalities in order to gain the desired control over the limit of infinite dimensions. With these estimates we establish the independence of the limiting theories within a wide class of regularizations, as desired above. Furthermore, we prove the continuity of the topological data under the infinite dimensional limit. Establishing this continuity is the basic technical content of our work [18–26] described here. We describe here the index theory for the operators Q; in Section III we give background on this topic. In future work, we hope to develop a more extensive topological characterization of the infinite dimensional manifolds (loop spaces) which we construct here.

The main technical tool which we use to study the analysis of Dirac theory on loop space is the heat kernel

$$e^{-\beta H} = e^{-\beta Q^2} \quad . \tag{I.1}$$

We develop functional integral representations for (I.1), by introducing the path space over loop space. This yields a path space representations for

$$\mathrm{Trace}(Ae^{-\beta H}) \qquad \text{and} \qquad \mathrm{Supertrace}(Ae^{-\beta H}) \ , \tag{I.2}$$

as explained in Chapter VI. Here A denotes a function of the quantum field. These trace formulas require the existence of non-Gaussian measures on the function space over the space-time torus

$$\{\varphi: T^2 \to \mathbf{C}\} \quad .$$

Here, constructive quantum field theory provides the framework and tools to study these measures.

II. SUPERSYMMETRIC QUANTUM FIELDS AND LOOP SPACE

In this chapter we present an elementary introduction to supersymmetry in the context of the models we study. The discussion in this chapter — in distinction to the remainder of these notes — is formal. Here we do not assure that there exist operators with the properties we describe. Rather, we emphasize the algebraic relations that such a structure would entail. In later chapters we make precise our mathematical understanding of the state-of-the-art in using constructive field theory to realize these structures.

II.1 Supersymmetry

The self-adjoint supercharge Q is the infinitesimal generator of a supersymmetry transformation. Let Γ be a self-adjoint grading operator, i.e., an operator which satisfies

$$\Gamma^2 = I \,, \qquad \Gamma = \Gamma^* \,, \qquad \Gamma Q + Q\Gamma = 0 \ . \tag{II.1.1}$$

For an operator A, define

$$A^\Gamma = \Gamma A \Gamma \quad , \tag{II.1.2}$$

so it is always possible to write

$$A = A_+ + A_- \quad , \tag{II.1.3}$$

where

$$(A_\pm)^\Gamma = \pm A_\pm \quad . \tag{II.1.4}$$

Explicitly,

$$A_\pm = \frac{1}{2}\left(A \pm A^\Gamma\right) \quad . \tag{II.1.5}$$

In terms of these objects, the infinitesimal supersymmetry transformation is

$$dA = [iQ, A_+] + \{iQ, A_-\} \quad , \tag{II.1.6}$$

where $\{A, B\} = AB + BA$ denotes the anticommutator. The definition (III.1.6) can also be restated as

$$dA = \Gamma[i\Gamma Q, A] \quad . \tag{II.1.7}$$

It follows that

$$d^2 A = -[H, A] \quad , \tag{II.1.8}$$

where

$$H = Q^2 \tag{II.1.9}$$

is the Hamiltonian.

A more concrete realization of this algebra arises from a quantum field theory of a boson φ and a fermion ψ. Let us take, for example, a field theory where φ is complex and ψ is a two-component Dirac spinor, both acting on a two-dimensional, cylindrical space-time. (This example will be a central feature of later sections.) Then the realization of the algebra depends crucially on the functional form of Q. The simplest example arises from the free field. In that case, Q is linear in ψ and linear in φ. In later sections we introduce a potential function V. We retain linearity in ψ but introduce a nonlinear dependence on φ. In that case V' (= the derivative of V) serves as the connection f described above.

Let us write down the simplest case, the massless free field. Then we require that

$$d\varphi = \frac{1}{\sqrt{2}} (\psi_1 + \psi_2) \quad , \tag{II.1.10}$$

$$\begin{aligned} d\psi_1 &= \frac{i}{\sqrt{2}} (\pi^* - \partial_x \varphi) \, , \\ d\psi_2 &= \frac{i}{\sqrt{2}} (\pi - \partial_x \varphi^*) \, , \end{aligned} \tag{II.1.11}$$

where ψ_i denotes the ith spinor component, and where π is the canonical momentum. This can be realized by taking for Q the operator

$$Q = \frac{1}{\sqrt{2}} \int \bar{\psi}_2(\pi^* - \partial_x\varphi) + \bar{\psi}_1(\pi - \partial_x\varphi^*) \, dx + \text{ h.c.} \quad , \tag{II.1.12}$$

where h.c. means hermitian conjugation. Then $H = Q^2$ is the free Hamiltonian $H_0 = H_{0,b} + H_{0,f}$; it is the sum of bosonic and fermionic, zero-mass, free Hamiltonians. The zero particle state Ω_0 (i.e., the ground state of H_0) satisfies

$$Q\Omega_0 = 0 \quad . \tag{II.1.13}$$

It is characteristic of supersymmetric theories that $H = Q^2 \geq 0$. We say that supersymmetry is unbroken if there exists $\Omega \in \mathcal{H}$ such that

$$Q\Omega = 0 \quad . \tag{II.1.14}$$

Clearly, Ω is then a ground state of the theory. If no such Ω exists, then we say that supersymmetry is broken.

II.2 The Quantum Field Hilbert Space

The standard Hilbert space for a free quantum field is Fock space. For a system of boson and fermion fields, one takes the Hilbert space $\mathcal{H}$ to be the tensor product

$$\mathcal{H} = \mathcal{H}_b \otimes \mathcal{H}_f$$

of a bosonic Fock space with a fermionic Fock space.

In general, however, the quantum field Hilbert space is not Fock space. Furthermore, the Hilbert space for a nonlinear field is not in general a tensor product. In fact, the existence problem for quantum fields entails construction of this Hilbert space as part of the existence proof. This construction generally involves limiting procedures which yield the desired Hilbert space as a limit of tensor products of Fock spaces. The limiting procedure involves a "renormalization" and a change to a representation of the canonical commutation/anti-commutation relations which is unitarily inequivalent to the Fock representation.

It turns out, however, that in a finite space-volume (i.e., a field theory over a torus) and in the special case of one-space (torus = circle) and one time dimension, the situation simplifies: with polynomial interactions Fock space suffices. This is the class of examples we study in these notes. In particular, the restriction to polynomial interactions includes the assumption that the scalar field takes values in $\mathbb{C}$ or $\mathbb{R}$.

II.3. Loop Space

There are two standard representations of a Fock-Hilbert space. The first representation is known as the Fock representation. The idea is to represent $\mathcal{H}_b$ and $\mathcal{H}_f$ as sums of n-particle spaces,

$$\mathcal{H}_b = \bigoplus_{n=0}^{\infty} \mathcal{H}_{b,n} , \qquad \mathcal{H}_f = \bigoplus_{n=0}^{\infty} \mathcal{H}_{f,n} . \tag{II.3.1}$$

Each n-particle space $\mathcal{H}_{b,n}$ or $\mathcal{H}_{f,n}$ is an element of the n-fold linear tensor algebra over the one-particle space $\mathcal{H}_{b,1}$ or $\mathcal{H}_{f,1}$. The one particle space is $L_2(\mathbb{R}) \oplus L_2(\mathbb{R})$, one summand for particles, the other for antiparticles. In the bosonic case we take the symmetric tensor algebra to construct $\mathcal{H}_{b,n}$. In the fermionic case, we take the antisymmetric tensor algebra to construct $\mathcal{H}_{f,n}$.

The second representation, known also as the Schrödinger representation, is unitarily equivalent to the Fock representation. In the Schrödinger representation states are represented as functions on the space of time-zero fields. The time-zero bosonic field is a map φ from the circle to $\mathbb{C}$ (or $\mathbb{R}$). Hence the space of the bosonic fields is the loop space $\Lambda\mathbb{C}$ of $\mathbb{C}$ (or $\mathbb{R}$), namely

$$\Lambda\mathbb{C} = \{\varphi : S^1 \to \mathbb{C}\} \ . \tag{II.3.2}$$

In order to define a well-behaved measure on the loop space, however, we need to enlarge the category of maps φ. We require a *generalized* loop space, which we also denote by (II.3.2); this is the space (II.3.2), with the maps φ in the class of generalized functions (distributions).

The free field Fock-Hilbert space is defined as the L_2 functions on generalized loop space; thus for the free field of mass $m > 0$,

$$\mathcal{H}_b \cong L_2(\Lambda\mathbb{C},\, d\nu_0) \ . \tag{II.3.3}$$

The measure $d\nu_0 = d\nu_0(m)$ is a Gaussian, probability measure (on the space $\mathcal{D}'(S^1)$ of periodic distributions with mean zero and covariance $G = (2\sqrt{-d^2/dx^2 + m^2})^{-1}$. The construction of $d\nu_0$ is well understood; see, for example, [12].

Under the unitary equivalence (II.3.3), the zero particle (vacuum) state in $\mathcal{H}_b$, namely to solution to $H_0\Omega_0 = 0$, is equivalent to the function 1 in $L_2(\Lambda\mathbb{C})$. In work we describe here, we find a family of probability measures $d\nu_V$ on the loop space $\Lambda\mathbb{C}$ which are parameterized by a polynomial potential function V. Each of these measures gives rise to an equivalence

$$\mathcal{H}_b \cong L_2(\Lambda C,\, d\nu_V) \quad . \tag{II.3.4}$$

For such a measure, the function $1 \in L_2(\Lambda C)$ corresponds to a state defined by the space of ground states of a supersymmetric quantum field theory on $\mathcal{H}_b$ with Hamiltonian H. A major application of constructive quantum field theory to infinite dimensional analysis of the manifold ΛC (or $\Lambda \mathbb{R}$) is the construction of the measures $d\nu_V$ which establish the various topologies on loop space. The case $V = 0$ reduces to the measure $d\nu_0$.

One of the most useful tools to investigate loop space ΛC is the functional integral representation of the heat kernel. Basically, this boils down to the definition of a measure $d\mu_V$ on the path space over loop space. The construction of $d\mu_V$ is carried out in the framework of constructive field theory. We describe this in detail in Sections V and VI. The measure $d\nu_V$ is the conditional expectation of $d\mu_V$ on the time zero hyperplane.

III. ELEMENTS OF INDEX THEORY

In this section we derive an elementary stability theorem for the index. We deal with a setting which is an abstraction of the Dirac operator setting which we study here and in our related papers. While stability theorems can be found in the mathematics literature, we present here two simple, direct proofs applicable in our special case. General references for this section are [9] and [27].

III.1. Fredholm Operators

Let $\mathcal{H}_1$, $\mathcal{H}_2$ be separable, complex Hilbert spaces. We consider densely defined linear operators $A: \mathcal{H}_1 \to \mathcal{H}_2$. The domain of A is denoted $\mathrm{Dom}(A)$. The graph of A is the set of pairs $\{\psi,\, A\psi\}$ in $\mathcal{H}_1 \oplus \mathcal{H}_2$ for $\psi \in \mathrm{Dom}(A)$. An operator A is closed, if its graph is a closed subset of $H_1 \oplus H_2$.

A densely defined operator A: $\mathcal{H}_1 \to \mathcal{H}_2$ has a uniquely defined adjoint $A^*: \mathcal{H}_2 \to \mathcal{H}_1$ with domain $\mathrm{Dom}(A^*)$, and given for $\psi \in \mathrm{Dom}(A)$ by

$$\langle \chi,\, A\psi \rangle_{\mathcal{H}_2} = \langle A^*\chi,\, \psi \rangle_{\mathcal{H}_1} \quad , \tag{III.1.1}$$

whenever $\langle \chi,\, A\psi \rangle_{\mathcal{H}_2}$ is continuous in χ.

The kernel of A is denoted $\mathrm{Ker}(A)$ and consists of the null space of A. If A is a closed operator, then $\mathrm{Ker}(A)$ is a closed subspace of $\mathcal{H}_1$.

The cokernel of A, denoted $\mathrm{Coker}(A)$, is defined by

$$\mathrm{Coker}(A) = \mathcal{H}_2/\mathrm{Range}(A) \quad .$$

Proposition III.1. *If* $\mathrm{Range}(A)$ *is closed, then* $\mathrm{Coker}(A)$ *and* $\mathrm{Ker}(A^*)$ *are canonically isomorphic.*

Proof. Suppose $\chi \in \mathrm{Ker}(A^*)$. Then for all $\psi \in \mathrm{Dom}(A)$

$$0 = \langle A^*\chi, \psi\rangle_{\mathcal{H}_1} = \langle \chi, A\psi\rangle_{\mathcal{H}_2} ,$$

and $\chi \perp \mathrm{Range}(A)$. If $\mathrm{Range}\ A$ is closed, then

$$\mathcal{H}_2 = \mathrm{Range}(A) \oplus (\mathrm{Range}(A))^\perp \quad . \tag{III.1.2}$$

It follows that $\chi \in \mathrm{Coker}(A)$. Conversely, if $\chi \in \mathrm{Coker}(A)$, and $\mathcal{H}_2$ is the sum (III.1.2), then $\chi \in (\mathrm{Range}\ (A))^\perp = \mathrm{Ker}(A^*)$.

Definition III.2. *An operator* $A: \mathcal{H}_1 \to \mathcal{H}_2$ *is Fredholm if*

(i) $\dim \mathrm{Ker}(A) < \infty$, $\dim \mathrm{Ker}(A^*) < \infty$,

and

(ii) $\mathrm{Range}(A)$ *is closed.*

Definition III.3. *The index* $i(A)$ *of a Fredholm operator* A *is*

$$i(A) = n_+ - n_-$$

where $n_+ = \dim \mathrm{Ker}(A)$, $n_- = \dim \mathrm{Ker}(A^*)$.

III.2. Dirac Operators

Let the Hilbert space $\mathcal{H}$ be a direct sum

$$\mathcal{H} = \mathcal{H}_+ \oplus \mathcal{H}_- \quad . \tag{III.2.1}$$

In our example, $\mathcal{H}_\pm$ are eigenspaces of a grading operator Γ on $\mathcal{H}$, *cf.* (II.1.1). We say that Q is odd with respect to the grading operator, if Q is off-diagonal with respect to the decomposition (III.2.1). In other words, an odd operator Q on $\mathcal{H}$ has the form

$$Q = \begin{pmatrix} 0 & Q_- \\ Q_+ & 0 \end{pmatrix} \tag{III.2.2}$$

where $Q_+ : \mathcal{H}_+ \to \mathcal{H}_-$ and $Q_- : \mathcal{H}_- \to \mathcal{H}_+$. We require that Q be densely defined.

We are interested in a subset of the odd operators which have the essential features of our Dirac operators. Thus we define

$$\mathrm{DIR}(\mathcal{H}) = \mathrm{DIR}(\mathcal{H}_+ \oplus \mathcal{H}_-) \tag{III.2.3}$$

as follows:

Definition III.2.1. *The operators* DIR($\mathcal{H}$) *are operators* Q *on* $\mathcal{H} = \mathcal{H}_+ \oplus \mathcal{H}_-$ *which are* (i) *odd with respect to the grading* Γ, (ii) *self-adjoint, and* (iii) *have a compact resolvent.*

Properties (i) and (ii) mean that Q has the form (III.2.2) where Q_+ is closed and where $Q_- = Q_+^*$. Property (iii) concerns the resolvent $(Q - i)^{-1}$ of Q. It can be derived from $I = (Q - i)(Q - i)^{-1}$, yielding the representation

$$(Q - i)^{-1} = \begin{pmatrix} i\Delta_+ & Q_-\Delta_- \\ Q_+\Delta_+ & i\Delta_- \end{pmatrix} , \tag{III.2.4}$$

where

$$\Delta_+ = (Q_+^* Q_+ + I)^{-1} \quad \text{and} \quad \Delta_- = (Q_-^* Q_- + I)^{-1} . \tag{III.2.5}$$

The compactness of $(Q - i)^{-1}$ is equivalent to compactness of each of the four component operators (III.2.4).

Theorem III.2.2. *If* $Q \in \mathrm{DIR}(\mathcal{H})$, *then* Q_+ *and* Q_- *are Fredholm.*

Proof. Since Δ_+ and Δ_- are compact operators, their eigenspaces corresponding to eigenvalue 1 are finite dimensional. But these are just $\mathrm{Ker}(Q_+)$ and $\mathrm{Ker}(Q_-)$, respectively. Thus to show that Q_+ or Q_- is Fredholm, we need only to show that $\mathrm{Range}(Q_+)$ or $\mathrm{Range}(Q_-)$ is closed. Let $\chi_n \in \mathrm{Range}(Q_+)$ be a convergent sequence, with limit point χ. We claim that $\chi \in \text{Range}(Q_+)$. If $\chi_n = Q_+\psi_n$, we can write

$$y_n = \begin{pmatrix} 0 \\ \chi_n \end{pmatrix} \qquad x_n = \begin{pmatrix} \psi_n \\ 0 \end{pmatrix} , \tag{III.2.6}$$

and $y_n = Qx_n$. It is no loss of generality to assume that $\psi_n \perp \text{Ker}(Q_+)$.

Secondly, we remark that on the domain $\text{Dom}(Q_-Q_+) \oplus \text{Dom}(Q_+Q_-)$, the operator $(Q-i)^{-1}(Q-i) = I$ can be expressed in terms of

$$(Q-i)^{-1} = \begin{pmatrix} i\Delta_+ & \Delta_+Q_- \\ \Delta_-Q_+ & i\Delta_- \end{pmatrix} . \tag{III.2.7}$$

Comparing with (III.2.4), we find that

$$Q_+\Delta_+ = \Delta_-Q_+ \quad \text{on} \quad \text{Dom}(Q_+) , \tag{III.2.8}$$

and

$$Q_-\Delta_- = \Delta_+Q_- \quad \text{on} \quad \text{Dom}(Q_-) . \tag{III.2.9}$$

Since the left sides of (III.2.8–9) are bounded (and in fact compact) operators, the right sides are also and extend by continuity to $\mathcal{H}_+$ and $\mathcal{H}_-$, respectively. Thus the equation $x_n = (Q-i)^{-1}(Q-i)x_n$ yields

$$\psi_n = Q_-\Delta_-\chi_n + \Delta_+\psi_n , \tag{III.2.10}$$

where $Q_-\Delta_-$ and Δ_+ are compact.

It is sufficient to prove that the $\{\psi_n\}$ are bounded. In fact, $\{\chi_n\}$ is convergent, so it follows from (III.2.10) that $\{\psi_n\}$ has a convergent subsequence with limit ψ. But Q is assumed self-adjoint, so Q_+ is a closed operator from $\mathcal{H}_+$ to $\mathcal{H}_-$. Hence $\psi \in \text{Dom}(Q_+)$ and $Q_+\psi = \chi$, showing that $\text{Range}(Q_+)$ is closed.

We now show that $\{\psi_n\}$ is bounded. Let us assume, on the contrary, that $\{\psi_n\}$

has an unbounded sequence, which we again denote $\{\psi_n\}$. Since $\{\chi_n\}$ is convergent, $\chi_n/\|\psi_n\|$ converges to zero. Define the bounded sequence $\eta_n = \psi_n/\|\psi_n\|$. Then (III.2.10) can be written

$$\eta_n = Q_-\Delta_-\chi_n/\|\psi_n\| + \Delta_+\eta_n \quad . \tag{III.2.11}$$

Since Δ_+ is compact, the sequence η_n has a convergent subsequence to a unit vector η satisfying

$$\eta = \Delta_+\eta = (Q_-Q_+ + I)^{-1}\eta \quad . \tag{III.2.12}$$

By the spectral theorem, this is possible only if $\eta \in \mathrm{Ker}(Q_+)$. But since we assumed $\psi_n \perp \mathrm{Ker}(Q_+)$, this means $\eta = 0$, a contradiction. Thus the $\{\psi_n\}$ are bounded, and the range of Q_+ is closed. Hence Q_+ is Fredholm.

The argument to show that $\mathrm{Range}(Q_-)$ is closed is the same, so we conclude that both Q_+ and Q_- are Fredholm.

III.3.1. Homotopy Invariance of the Index

We are actually interested in studying a continuous family of operators in $\mathrm{DIR}(\mathcal{H})$, where $\mathcal{H} = \mathcal{H}_+ \oplus \mathcal{H}_-$. We introduce an (incomplete) metric topology on $\mathrm{DIR}(\mathcal{H})$ defined by norm resolvent convergence. For $Q_1, Q_2 \in \mathrm{DIR}(\mathcal{H})$, define the distance $d_{\mathcal{H}}$ between Q_1 and Q_2 to be

$$d_{\mathcal{H}}(Q_1, Q_2) = \|(Q_1 + i)^{-1} - (Q_2 + i)^{-1}\|_{\mathcal{H}} \quad . \tag{III.3.1}$$

(When confusion does not arise, we can drop the subscript $\mathcal{H}$.) If Q_1, Q_2 have norms bounded by 1, the topology defined by (III.3.1) is equivalent to the topology defined by the distance $\|Q_1 - Q_2\|$. However, for unbounded operators the topology (III.3.1) can still be used. The space $\mathrm{DIR}(\mathcal{H})$ breaks up into various connected components with respect to the metric (III.3.1). The main property that we wish to establish concerns the mapping from $Q \in \mathrm{DIR}(\mathcal{H})$ to the index:

$$Q = \begin{pmatrix} 0 & Q_- \\ Q_+ & 0 \end{pmatrix} \mapsto i(Q_+) = -i(Q_-) \quad . \tag{III.3.2}$$

Theorem III.3.1. *The index map (III.3.2) is constant on connected components of* $\mathrm{DIR}(\mathcal{H})$.

We now prove the theorem with a method patterned after Gilkey's proof for bounded Fredholm operators [9]. In the following section we show that with a stronger topology on $\mathrm{DIR}(\mathcal{H})$, a much easier proof can be given. The basic strategy is to proceed in three steps.

Step 1. Choose a reference operator $D \in \mathrm{DIR}(\mathcal{H})$. Use D to define a space $\hat{\mathcal{H}} \supset \mathcal{H}$, with a grading. For every $Q \in \mathrm{DIR}(\mathcal{H})$ there is a natural operator $\hat{Q} \in \mathrm{DIR}(\hat{\mathcal{H}})$.

Step 2. The next step is to prove local stability of bounded invertibility in $\mathrm{DIR}(\hat{\mathcal{H}})$. We show that $\hat{D}$ has a bounded inverse. Furthermore, if $d_{\mathcal{H}}(Q, D)$ is sufficiently small, then $\hat{Q}$ also has a bounded inverse.

Step 3. As a consequence of the local stability of bounded invertibility on $\mathrm{DIR}(\hat{\mathcal{H}})$ we infer the local constancy of the index on $\mathrm{DIR}(\mathcal{H})$.

Step 1. The Construction of $\hat{\mathcal{H}}$

Consider a fixed $D \in \mathrm{DIR}(\mathcal{H})$. Define

$$\hat{\mathcal{H}}_\pm = \mathrm{Ker}(D_\mp) \oplus \mathcal{H}_\pm \qquad \text{and} \qquad \hat{\mathcal{H}} = \hat{\mathcal{H}}_+ \oplus \hat{\mathcal{H}}_- \ .$$

Also let $P_\pm$ be the orthogonal projections of $\mathcal{H}_\pm$ onto $\mathrm{Ker}(D_\pm)$.

Now choose $Q \in \mathrm{DIR}(\mathcal{H})$. With respect to the decomposition

$$\hat{\mathcal{H}} = \mathrm{Ker}(D_-) \oplus \mathcal{H}_+ \oplus \mathrm{Ker}(D_+) \oplus \mathcal{H}_-$$

we define an operator $\hat{Q}$ by

$$\hat{Q} = \begin{pmatrix} 0 & 0 & 0 & P_- \\ 0 & 0 & P_+^* & Q_- \\ 0 & P_+ & 0 & 0 \\ P_-^* & Q_+ & 0 & 0 \end{pmatrix} = \begin{pmatrix} 0 & \hat{Q}_- \\ \hat{Q}_+ & 0 \end{pmatrix} , \tag{III.3.3}$$

where

$$\hat{Q}_+ = \begin{pmatrix} 0 & P_+ \\ P_-^* & Q_+ \end{pmatrix} , \qquad \hat{Q}_- = \begin{pmatrix} 0 & P_- \\ P_+^* & Q_- \end{pmatrix} . \tag{III.3.4}$$

Clearly $\hat{Q} \in \mathrm{DIR}(\hat{\mathcal{H}})$. Furthermore, $\hat{Q}$ is invertible if and only if both $\hat{Q}_\pm$ are invertible, in which case

$$(\hat{Q})^{-1} = \begin{pmatrix} 0 & \hat{Q}_+^{-1} \\ \hat{Q}_-^{-1} & 0 \end{pmatrix} \quad .$$

Lemma III.3.2. *The operators* $\hat{D}_\pm$ *(and therefore* $\hat{D}$*) have bounded inverses.*

Proof. We claim that for a general $Q \in \mathrm{DIR}(\mathcal{H})$,

$$\mathcal{H}_\pm = \mathrm{Ker}(Q_\pm) \oplus (\mathrm{Ker}(Q_\pm))^\perp = \mathrm{Ker}(Q_\pm) \oplus \mathrm{Range}(Q_\mp) \quad . \tag{III.3.5}$$

In fact $\mathrm{Ker}(Q_+) = (\mathrm{Range}(Q_-))^\perp$, so $\mathcal{H}_+ = \mathrm{Ker}(Q_+) \oplus (\mathrm{Range}(Q_-))^{\perp\perp}$. Since Q is Fredholm, Range (Q_-) is closed and $\mathrm{Range}(Q_-) = (\mathrm{Range}(Q_-))^{\perp\perp}$. The argument for $\mathcal{H}_-$ is identical.

Now we set $Q = D$, the operator used to define $\hat{\mathcal{H}}$. Since

$$\hat{D}_+ : \mathrm{Ker}(D_-) \oplus \mathcal{H}_+ \to \mathrm{Ker}(D_+) \oplus \mathcal{H}_- \quad ,$$

we can write the matrix representation of $\hat{D}_+$ with respect to

$$\hat{D}_+ : \mathrm{Ker}(D_-) \oplus \mathrm{Ker}(D_+) \oplus \mathrm{Range}(D_-) \to \mathrm{Ker}(D_+) \oplus \mathrm{Ker}(D_-) \oplus \mathrm{Range}(D_+)$$

as

$$\hat{D}_+ = \begin{pmatrix} 0 & P_+ & 0 \\ P_-^* & 0 & 0 \\ 0 & 0 & D_+ \end{pmatrix} \quad .$$

Since $P_+ \restriction \mathrm{Ker}(D_+)$, $P_-^* \restriction \mathrm{Ker}(D_-)$ and $D_+ \restriction (\mathrm{Ker}(D_+))^\perp$ each have kernel zero, it follows that $\mathrm{Ker}(\hat{D}_+) = 0$. Similarly $\mathrm{Ker}(\hat{D}_-) = 0$. Hence $\hat{D}_\pm$ are invertible. Furthermore, $\hat{D}_\pm$ are surjective and closed. From the closed graph theorem we infer that $\hat{D}_\pm^{-1}$ are bounded.

Lemma III.3.3. *The map* $\hat{}\ : \mathrm{DIR}(\mathcal{H}) \to \mathrm{DIR}(\hat{\mathcal{H}})$ *is continuous and*

$$d_{\hat{\mathcal{H}}}(\hat{Q}, \hat{Q}') \le 6 d_{\mathcal{H}}(Q, Q') \quad . \tag{III.3.5}$$

Proof. Write $\hat{Q} = \hat{Q}_0 + \hat{Q}_1$, where $\hat{Q}_j \in \mathrm{DIR}(\hat{\mathcal{H}})$,

$$(\hat{Q}_0)_+ = \begin{pmatrix} 0 & 0 \\ 0 & Q_+ \end{pmatrix} , \qquad \text{and} \quad \|\hat{Q}_1\| = 1 .$$

Then, with the resolvent $R(\hat{Q}) = (\hat{Q} + i)^{-1}$, clearly

$$\|R(\hat{Q}_0) - R(\hat{Q}'_0)\| = d_{\mathcal{H}}(Q, Q') . \tag{III.3.6}$$

We require the identity

$$R(\hat{Q}) = (I + R(\hat{Q}_0)\hat{Q}_1)^{-1} R(\hat{Q}_0) ,$$

and the fact that $I + R(\hat{Q}_0)\hat{Q}_1$ has an inverse with norm bounded by 2. To see this, use the resolvent equation,

$$R(\hat{Q}) = R(\hat{Q}_0) - R(\hat{Q})\hat{Q}_1 R(\hat{Q}_0) .$$

Thus on the domain of $\hat{Q}_0$,

$$R(\hat{Q})R(\hat{Q}_0)^{-1} = I - R(\hat{Q})\hat{Q}_1 ,$$

and

$$(I + R(\hat{Q}_0)\hat{Q}_1)^{-1} = R(\hat{Q})R(\hat{Q}_0)^{-1} = I - R(\hat{Q})\hat{Q}_1 ,$$

which is bounded in norm by $1 + \|\hat{Q}_1\| \le 2$.

Let $K(\hat{Q}_0, \hat{Q}_1) = (I + R(\hat{Q}_0)\hat{Q}_1)^{-1}$. Then we can write

$$\begin{aligned} d_{\hat{\mathcal{H}}}(\hat{Q}, \hat{Q}') &= \|R(\hat{Q}) - R(\hat{Q}')\| \\ &= \|K(\hat{Q}_0, \hat{Q}_1)(R(\hat{Q}_0) - R(\hat{Q}'_0)) + (K(\hat{Q}_0, \hat{Q}_1) - K(\hat{Q}'_0, \hat{Q}_1))R(\hat{Q}'_0)\| \\ &\le 2\|R(\hat{Q}_0) - R(\hat{Q}'_0)\| + \|K(\hat{Q}_0, \hat{Q}_1) - K(\hat{Q}'_0, Q_1)\| . \end{aligned} \tag{III.3.7}$$

Using (III.3.6), the first term is bounded by $2d_{\mathcal{H}}(Q, Q')$, as desired. For the second term we use the resolvent identity to write

$$K(\hat{Q}_0,\hat{Q}_1) - K(\hat{Q}'_0,\hat{Q}_1) = -K(\hat{Q}_0,\hat{Q}_1)(R(\hat{Q}_0) - R(\hat{Q}'_0))\hat{Q}_1 K(\hat{Q}'_0,Q_1)\,. \quad \text{(III.3.8)}$$

Since K is bounded in norm by 2 and $\hat{Q}_1$ is bounded in norm by 1, the norm of (III.3.8) is bounded by $4d_{\mathcal{H}}(Q,Q')$. Hence the bound (III.3.5) holds.

Step 2. Stability of Bounded Invertibility on DIR($\hat{\mathcal{H}}$)

Lemma III.3.4. *If* $d_{\mathcal{H}}(Q,D)$ *satisfies*

$$d_{\mathcal{H}}(Q,D) < (6 + 6\|\hat{D}^{-1}\|)^{-1} \quad , \qquad \text{(III.3.9)}$$

then $\hat{Q}$ *has a bounded inverse.*

Proof. We would like to write

$$\hat{Q}^{-1} = (I - iR(\hat{Q}))^{-1}R(\hat{Q}) \quad . \qquad \text{(III.3.10)}$$

In fact, by Lemma III.3.2,

$$\|R(\hat{Q}) - R(\hat{D})\| \le 6d_{\mathcal{H}}(Q,D) \quad ,$$

so $(I - iR(\hat{Q}))^{-1}$ has a convergent Neumann series for $d_{\mathcal{H}}(Q,D)$ sufficiently small. In particular,

$$(I - iR(\hat{Q}))^{-1} = (I - iR(\hat{D}) - i(R(\hat{Q}) - R(\hat{D})))^{-1}$$

$$= \sum_{n=0}^{\infty} [i(I - iR(\hat{D})^{-1}(R(\hat{Q}) - R(\hat{D}))]^n (I - iR(\hat{D}))^{-1} \quad ,$$

where $(I - i(R(\hat{D}))^{-1} = I + i\hat{D}^{-1}$ is bounded. This series converges for $\|R(\hat{Q}) - R(\hat{D})\|(1 + \|\hat{D}^{-1}\|) < 1$, which is true for

$$6d_{\mathcal{H}}(Q,D)(1 + \|\hat{D}^{-1}\|) < 1 \quad .$$

Hence (III.3.10) expresses $\hat{Q}^{-1}$ as a product of two bounded operators and the proof is complete.

Step 3. Local Constancy of the Index on DIR($\mathcal{H}$)

Lemma III.3.5. *If* $d_{\mathcal{H}}(Q,D)$ *satisfies (III.3.9), then* $i(Q_+) = i(D_+)$.

Proof. According to Lemma III.3.4, the operator $\hat{Q}$ has a bounded inverse. Thus $\mathrm{Ker}(\hat{Q}_+) = \mathrm{Ker}(\hat{Q}_-) = 0$. Let us define j_+ as the injection

$$j_+ : \mathcal{H}_+ \to \hat{\mathcal{H}}_+$$

and π_- as the orthogonal projection

$$\pi_- : \hat{\mathcal{H}}_- \to \mathcal{H}_- \quad .$$

Thus $\mathrm{Ker}(j_+) = 0$ and $\mathrm{Ker}(j_+^*) = \mathrm{Ker}(D_-)$. Also $\mathrm{Ker}(\pi_-) = \mathrm{Ker}(D_+)$ and $\mathrm{Ker}(\pi_-^*) = 0$. But

$$Q_+ = \pi_- \hat{Q}_+ j_+ \quad \text{and} \quad Q_+^* = j_+^* \hat{Q}_+^* \pi_-^* \quad .$$

Therefore, by inspection,

$$i(Q_+) = \dim(\mathrm{Ker}(D_+)) - \dim(\mathrm{Ker}(D_-)) = i(D_+) \quad ,$$

to complete the proof.

Theorem III.3.1 is an immediate consequence of Lemma III.3.5.

III.4. The Heat Kernel Representation of the Index

There is a standard heat kernel representation of the index. The heat kernel is a map from $\mathrm{DIR}(\mathcal{H})$ to a semigroup on $\mathcal{H}$,

$$Q \mapsto \exp(-\beta Q^2) = \exp(-\beta H) \tag{III.4.1}$$

where $\beta \in \mathbf{R}_+$. There is a very convenient representation of the index in terms of the heat kernel, at least when the heat kernel obeys some basic regularity properties. In our examples we use a Feynman-Kac representation of the heat kernel which allows us to establish strong regularity properties for the heat kernel.

Throughout this section we assume that $\mathcal{H} = \mathcal{H}_+ \oplus \mathcal{H}_-$, that $Q \in \mathrm{DIR}(\mathcal{H})$, and that $\Gamma = P_{\mathcal{H}_+} - P_{\mathcal{H}_-}$ is the grading operator, where $P_{\mathcal{H}_\pm}$ are the orthogonal projections of $\mathcal{H}$ onto $\mathcal{H}_+$ and $\mathcal{H}_-$, respectively. We also let $H = Q^2$. Let $I_p(\mathcal{H})$

denote the pth Schatten class of "l_p operators" with the usual norm

$$\|T\|_p = (\mathrm{Tr}_{\mathcal{H}}(T^*T)^{p/2})^{1/p} .$$

Theorem III.4.1. *If* $\exp(-H) \in I_p$ *for some* $p < \infty$, *then*

$$i(Q_+) = \mathrm{Tr}(\Gamma e^{-\beta H}) \quad , \tag{III.4.2}$$

whenever $\beta \geq p$.

Proof. The condition $\exp(-H) \in I_p$ ensures that $\exp(-\beta H)$ is trace class for $\beta \geq p$. In particular, $\exp(-\beta H)$ is compact so eigenvalues have finite multiplicity. The eigenspace of the eigenvalue 0 of H contributes $i(Q_+)$ to the trace (III.4.2). If $E > 0$ is an eigenvalue of H with eigenspace $\mathcal{H}_E$, then we can decompose $\mathcal{H}_E = \mathcal{H}_{E,+} \oplus \mathcal{H}_{E,-}$ where $\mathcal{H}_{E,\pm} \subset \mathcal{H}_\pm$ are isomorphic subspaces. In fact, if P_E is the projection onto $\mathcal{H}_E$, then

$$P_{E,\pm} = \frac{E^{1/2} \mp Q}{2E^{1/2}} P_E \tag{III.4.3}$$

are the projections onto $P_{E,\pm}$. Since $\{\Gamma, Q\} = 0$, and $\Gamma = \Gamma^{-1}$, it follows that

$$\Gamma P_{E,+}\Gamma = P_{E,-} \quad ,$$

and $\mathcal{H}_{E,\pm}$ are isomorphic. Thus for $E > 0$,

$$\mathrm{Tr}(P_E \Gamma \exp(-\beta H) P_E) = 0 \quad ,$$

and (III.4.2) is valid.

The representation (III.4.2) suggests another topology on $\mathrm{DIR}(\mathcal{H})$, namely the I_p topology. While this topology is not defined on the entire space $\mathrm{DIR}(\mathcal{H})$, it is nevertheless extremely useful.

Definition III.4.2. *Let* $DIR_p(\mathcal{H})$ *denote the subset of* $\mathrm{DIR}(\mathcal{H})$ *such that* $\exp(-Q^2) \in I_p$. *For* Q, $Q' \in \mathrm{DIR}_p(\mathcal{H})$, *define the metric*

$$d_p(Q, Q') = \|e^{-Q^2} - e^{-Q'^2}\|_p \quad . \tag{III.4.4}$$

Theorem III.4.3. *The index map*

$$i \ : \ \mathrm{DIR}_2(\mathcal{H}) \to \mathbf{Z} \ ,$$

is constant on components of $\mathrm{DIR}_2(\mathcal{H})$ *connected with respect to the metric (III.4.4).*

Remark. The formulation and proof of continuity is especially simple for $p = 2$ (the Hilbert-Schmidt case). This case is applicable to our quantum field models. However, similar results can be established which involve the metric d_p. The case $p = \infty$ is related to the result on continuity of the index in the resolvent norm, established in Section III.3.

Proof. For any self-adjoint operator A,

$$|\mathrm{Tr}(A)| \le \|A\|_1 \ . \qquad \text{(III.4.5).}$$

Also, for $A, B \in I_2$, $AB \in I_1$ and the Schwarz inequality $\|AB\|_1 \le \|A\|_2 \|B\|_2$ holds. Take $\beta = 2$, and write

$$i(Q_+) - i(Q'_+) = \mathrm{Tr}(\Gamma(e^{-2Q^2} - e^{-2Q'^2})) \ . \qquad \text{(III.4.6)}$$

Hence by (III.4.5),

$$|i(Q_+) - i(Q'_+)| \le \left\| \Gamma(e^{-2Q^2} - e^{-2Q'^2}) \right\|_1 \le \left\| e^{-2Q^2} - e^{-2Q'^2} \right\|_1 \ .$$

Note that

$$e^{-2Q^2} - e^{-2Q'^2} = e^{-Q^2}(e^{-Q^2} - e^{-Q'^2}) + (e^{-Q^2} - e^{-Q'^2})e^{-Q'^2} \ .$$

Hence by the Schwarz inequality

$$|i(Q_+) - i(Q'_+)| \le \left(\left\| e^{-Q^2} \right\|_2 + \left\| e^{-Q'^2} \right\|_2 \right) d_2(Q, Q'). \qquad \text{(III.4.7)}$$

In particular, if $Q(s)$, $s \in [0,1]$ is a continuous path from Q to Q' in $\mathrm{DIR}_2(\mathcal{H})$, the index, which is always integer, remains constant on the path.

IV. THE $N = 2$ SUPERSYMMETRIC QUANTUM MECHANICS

In this section we give the details of a very simple family of quantum mechanical examples. This supersymmetric quantum mechanics will occur later as the endpoint of the homotopy of the finite quantum field theory models described in Section V.

We consider the complex Clifford Algebra C_4 whose generators we denote by γ_j, $j = 0, 1, 2, 3$. The generators obey the usual algebra

$$\gamma_i\gamma_j + \gamma_j\gamma_i = 2\delta_{ij} \ . \tag{IV.1}$$

Define

$$\Gamma = \gamma_0\gamma_1\gamma_2\gamma_3 \ . \tag{IV.2}$$

As a consequence of (IV.1),

$$\Gamma^2 = 1 \ . \tag{IV.3}$$

There is a representation of C_4 in End($\mathbb{C}^4$) in which the generators γ_j are given by the self-adjoint matrices

$$\gamma_0 = \begin{pmatrix} 0 & I \\ I & 0 \end{pmatrix} , \quad \gamma_j = \begin{pmatrix} 0 & i\sigma_j \\ -i\sigma_j & 0 \end{pmatrix} , \quad j = 1, 2, 3 , \tag{IV.4}$$

where

$$\sigma_1 = \begin{pmatrix} 0 & 1 \\ 1 & 0 \end{pmatrix} , \quad \sigma_2 = \begin{pmatrix} 0 & -i \\ i & 0 \end{pmatrix} , \quad \sigma_3 = \begin{pmatrix} 1 & 0 \\ 0 & -1 \end{pmatrix} \tag{IV.5}$$

are the usual generators of the Lie algebra $su(2)$. In this representation,

$$\Gamma = \begin{pmatrix} I & 0 \\ 0 & -I \end{pmatrix} \ . \tag{IV.6}$$

We also define

$$\psi_1 = \frac{1}{2}(\gamma_0 - i\gamma_3) , \quad \psi_2 = \frac{1}{2}(\gamma_1 - i\gamma_2) \ . \tag{IV.7}$$

and denote by ψ_j^* the conjugate of ψ_j.

We consider the Hilbert space

$$\mathcal{H} = L_2(\mathbb{R}^2) \otimes \mathbb{C}^4 \cong \bigoplus_{j=1}^{4} \mathcal{H}_j \ , \tag{IV.8}$$

where $\mathcal{H}_j = L_2(\mathbb{R}^2)$. The Clifford algebra C_4 acts naturally on $\mathcal{H}$. With this action, Γ is a grading operator. Let

$$\mathcal{H} = \mathcal{H}_+ \oplus \mathcal{H}_- \tag{IV.9}$$

be the corresponding polarization of $\mathcal{H}$.

Let $\bar{\partial} = (1/2)(\partial/\partial x + i\partial/\partial y)$ be the Cauchy-Riemann operator and let ∂ be its conjugate. For an arbitrary polynomial $V(z)$ in $z = x + iy$ we define the operator

$$\bar{\partial}_V = i\psi_2\bar{\partial} + i\psi_1\partial V \quad , \tag{IV.10}$$

with domain $\mathrm{Dom}(\bar{\partial}_V) = \mathcal{D}_0 \equiv \mathcal{S}(\mathbb{R}^2) \otimes \mathbb{C}^4$. By $\bar{\partial}_V^*$ we denote the formal conjugate of $\bar{\partial}_V$. We set

$$D_V = \bar{\partial}_V + \bar{\partial}_V^* \ , \tag{IV.11}$$

and

$$\Delta_V = D_V^2 \ . \tag{IV.12}$$

Here $\mathrm{Dom}(D_V) = \mathrm{Dom}(\Delta_V) = \mathcal{D}_0$. We call these operators the Dirac and Laplace operators, respectively. Whenever there is no danger of confusion, we suppress the subscript V in D_V and Δ_V. Note

$$\Gamma D_V = -D_V \Gamma \ ,$$

so D_V is odd with respect to Γ.

Theorem IV.1. (i) *The operators D and Δ are essentially self-adjoint. (We denote their closures again by D and Δ, respectively.)*

(ii) *The resolvents of D and Δ are compact.*

Proof. Let D_+ denote the component of the decomposition of D defined by (III.2.2). A computation yields

$$D_+ = \begin{pmatrix} (\partial V)^* & i\partial \\ i\bar{\partial} & -\partial V \end{pmatrix} \ , \tag{IV.13}$$

and thus as $D_- = (D_+)^*$,

$$D_+D_- = (-\partial\bar{\partial} + |\partial V|^2)I\ , \tag{IV.14}$$

$$D_-D_+ = (-\partial\bar{\partial} + |\partial V|^2)I + K\ , \tag{IV.15}$$

where

$$K = \begin{pmatrix} 0 & -i\partial^2 V \\ i(\partial^2 V)^* & 0 \end{pmatrix}\ . \tag{IV.16}$$

Consequently,

$$\Delta = (-\partial\bar{\partial} + |\partial V|^2)I + \begin{pmatrix} K & 0 \\ 0 & 0 \end{pmatrix}\ . \tag{IV.17}$$

Standard results in the theory of Schrödinger operators imply that $-\partial\bar{\partial} + |\partial V|^2$ is essentially self-adjoint and has compact resolvent. Furthermore, for any $\epsilon > 0$ there is $C_\epsilon < \infty$ such that

$$|\partial^2 V| \le \epsilon|\partial V|^2 + C_\epsilon\ . \tag{IV.18}$$

Using the standard perturbation theory of Kato [27], we infer from (IV.16–IV.18) that Δ is essentially self-adjoint and has compact resolvent. The statements about D follow, since D is symmetric and $D^2 = \Delta$.

With the notation of Section III, we have

Corollary IV.2. $D \in \mathrm{DIR}(\mathcal{H})$.

Theorem IV.3. (Vanishing Theorem [21]). *The kernel of D is a subspace of $\mathcal{H}_+$, i.e.,*

$$\mathrm{Ker}(D_-) = 0\ . \tag{IV.19}$$

Remark. The standard way to prove a vanishing theorem is to write an operator as a sum of squares. We do the same here.

Proof. Since $D_- = D_+^*$, it follows that $\mathrm{Ker}(D_-) = \mathrm{Ker}(D_+D_-)$. We use (IV.14) and note that any solution u of the elliptic equation

$$-\partial\bar{\partial}u + |\partial V|^2 u = 0 \ , \qquad \text{(IV.20)}$$

with the C^∞ potential $|\partial V|^2$, is smooth. Hence we may multiply by $\bar{u}$ and integrate by parts to conclude that $\|\bar{\partial}u\|^2 + \|\partial V u\|^2 = 0$. Thus $\bar{\partial}u = (\partial V)u = 0$. Hence $u = 0$, as claimed.

Theorem IV.4. (Index Theorem).

$$i(D_+) = \dim \operatorname{Ker}(D_+) = \deg V - 1. \qquad \text{(IV.21)}$$

Proof. First consider the case

$$V(z) = V_0(z) = \lambda z^n \ , \qquad \lambda \neq 0 \ .$$

We then establish for the general case that D_V and D_{V_0} lie in the same connected component of $\mathrm{DIR}(\mathcal{H})$. The theorem then is a consequence of Theorem III.3.1.

Lemma IV.5. $\dim \operatorname{Ker}((D_{V_0})_+) = n - 1$.

Essentially we solve the zero-eigenvalue problem for $\Delta_{V_0} = (D_{V_0})^2$. We omit the elementary but lengthy proof of this lemma. For the details, see [21].

We now return to the proof of the theorem. Write the lower degree perturbation as

$$v(z) = V(z) - V_0(z) = \sum_{j=0}^{n-1} \lambda_j z^j \ . \qquad \text{(IV.22)}$$

Lemma IV.6. *The map*

$$\mathbf{R}^n \ni (\lambda_0, \ldots, \lambda_{n-1}) \to D_V \in \mathrm{DIR}(\mathcal{H}) \qquad \text{(IV.23)}$$

is continuous in the metric $d_{\mathcal{H}}$ *defined in (III.3.1). In particular* $i((D_V)_+) = i((D_{V_0})_+)$.

Proof. Let V, V' be two polynomials of degree n with the same leading term. Let $c \neq 0$ be a real constant, and define the resolvent $R_{ic}(D_V) = (D_V + ic)^{-1}$. Using the resolvent equation

$$R_{ic}(D_V) - R_{ic}(D_{V'}) = -R_{ic}(D_V)[i\psi_1\partial(\delta v) - i\psi_1^*\bar{\partial}(\delta v)^*]R_{ic}(D_{V'}) \,, \qquad \text{(IV.24)}$$

where $\delta v = v - v'$. Using $\|R_{ic}(D)(\Delta + c^2)^{1/2}\| = \|(\Delta + c^2)^{1/2}R_{ic}(D)\| = 1$ we obtain

$$\|R_{ic}(D_V) - R_{ic}(D_{V'})\| \leq \|R_{c^2}(\Delta_V)^{1/2}\psi_1\partial(\delta v)R_{c^2}(\Delta_{V'})^{1/2}\|$$

$$+ \|R_{c^2}(\Delta_V)^{1/2}\psi_2\bar{\partial}(\delta v)^*R_{c^2}(\Delta_{V'})^{1/2}\| \,, \qquad \text{(IV.25)}$$

where $R_{c^2}(\Delta) = (\Delta + c^2)^{-1}$. Let us study the first term on the right-hand side of (IV.25), the analysis of the second term being similar. Set

$$T = (-\partial\bar{\partial} + n^2|\lambda|^2|z|^{2(n-1)})I \quad . \qquad \text{(IV.26)}$$

Then for any $\epsilon > 0$ there is $C(\epsilon, \lambda_0, \ldots, \lambda_{n-1}) \geq 0$ such that

$$\Delta_V \geq (1-\epsilon)T - C(\epsilon, \lambda_0, \ldots, \lambda_{n-1}) \qquad \text{(IV.27)}$$

and similarly for $\Delta_{V'}$. Take $\lambda_j', j = 0, \ldots, n-1$ close to λ_j. Then $C(\epsilon, \lambda_0', \ldots, \lambda_{n-1}')$ can be taken close to $C(\epsilon, \lambda_0, \ldots, \lambda_{n-1})$, say $|C(\epsilon, \lambda_0, \ldots, \lambda_{n-1}) - C(\epsilon, \lambda_0', \ldots, \lambda_{n-1}')| \leq 1$. Set $c^2 = C(\epsilon, \lambda_0, \ldots, \lambda_{n-1}) + 2$. Then

$$\|R_{c^2}(\Delta_V)^{1/2}(T + I)^{1/2}\| < \infty \quad , \qquad \text{(IV.28)}$$

$$\|(T + I)^{1/2}R_{c^2}(\Delta_{V'})^{1/2}\| < \infty \quad , \qquad \text{(IV.29)}$$

and we can bound the first term in (IV.25) by

$$\text{Const}\|R_1(T)^{1/2}\psi_1\partial(\delta v)R_1(T)^{1/2}\| =$$

$$\text{Const}\|R_1(T)^{1/2}\partial(\delta v)R_1(T)^{1/2}\| = o(1) \quad ,$$

as $\lambda_j' \to \lambda_j$, $j = 0, \ldots, n-1$. This completes the proof of the lemma and hence of the theorem.

V. THE LOOP SPACE $S^1 \to \mathbb{C}$ AND QUANTUM FIELD THEORY

The Hilbert space $\mathcal{H}$ of our theory is a tensor product of the bosonic Hilbert space $\mathcal{H}_b$ and the fermionic Hilbert space $\mathcal{H}_f$, namely $\mathcal{H} = \mathcal{H}_b \otimes \mathcal{H}_f$. In both cases we assume that the one-particle space is built over the circle (one torus) T^1 of length ℓ.

V.1. The Bosonic Fock Space

The one particle space of the complex scalar field is

$$W = L_2(T^1) \oplus L_2(T^1) \ .$$

The Fock space $\mathcal{H}_b$ is a symmetric tensor algebra over W with the natural inner product yielding on the n-fold tensor product $\|f \otimes \ldots \otimes f\| = \|f\|^n$, $f \in W$. In the Fourier space (momentum representation) we define annihilation operators $a_\pm(p)$ on $\mathcal{H}_b$ so that $a_\pm \Omega_0^b = 0$, $\Omega_0^b = (1, 0, \ldots, 0, \ldots)$, and

$$\begin{aligned} &[a_\pm(p), a_\pm(q)] = [a_\pm(p), a_\mp(q)] = [a_\pm(p), a^*_\mp(q)] = 0, \\ &[a_\pm(p), a^*_\pm(q)] = \delta_{pq} \ , \end{aligned} \tag{V.1.1}$$

where $p \in \hat{T}^1 \equiv \frac{2\pi}{\ell} \mathbb{Z}$ and δ_{pq} is the Kronecker delta. The time zero field is defined by

$$\varphi(x) = (2\ell)^{-1/2} \sum_{p \in \hat{T}^1} \mu(p)^{-1/2} \left(a^*_+(p) + a_-(-p)\right) e^{-ipx} \ , \tag{V.1.2}$$

where $\mu(p) = (p^2 + m^2)^{1/2}$, and $m > 0$. The canonical momentum is

$$\pi(x) = i(2\ell)^{-1/2} \sum_{p \in \hat{T}^1} \mu(p)^{1/2} \left(a^*_-(p) - a_+(-p)\right) e^{-ipx} \ . \tag{V.1.3}$$

The scalar field satisfies the commutation relations

$$\begin{aligned} &[\varphi(x), \varphi(y)] = [\pi(x), \pi(y)] = [\pi^*(x), \varphi(y)] = 0 \ , \\ &[\pi(x), \varphi(y)] = -i\delta(x - y) \ , \end{aligned} \tag{V.1.4}$$

where $\delta(x - y)$ is the Dirac measure.

V.2. The Schrödinger Representation (Loop Space)

Another, unitarily equivalent, representation of $\mathcal{H}_b$ is given by the Schrödinger (or loop space) representation. Let $\mathcal{D}(T^1)$ denote the space of smooth maps (loops) from T^1 to $\mathbb{C}$, with the topology defined by uniform convergence of each derivative. Let $\mathcal{D}'(T^1)$ denote the topological dual, i.e., the space of complex distributions on T^1. Let $d\nu_{G_\ell}$ denote the Gaussian measure on $\mathcal{D}'(T^1)$ with mean zero and covariance $G_\ell = \frac{1}{2\mu}$, where

$$\mu = (-d^2/dx^2 + m^2)^{1/2} \quad . \tag{V.2.1}$$

It is then well known that $\mathcal{H}_b \cong L^2(\mathcal{D}'(T^1), d\nu_{G_\ell})$, see e.g. [12]. Under this isomorphism $\varphi(x)$ becomes a multiplication operator, and $\pi(x)$ becomes $-i\delta/\delta\varphi(x) + i\mu\varphi$, where $\delta/\delta\varphi(x)$ is the Frechet derivative.

V.3 The Fermionic Fock Space

The fermionic Fock space $\mathcal{H}_f$ is the anti-symmetric tensor algebra over $L^2(T^1) \oplus L^2(T^1)$. The annihilation operators are $b_\pm(p)$, $p \in \hat{T}^1$ and they satisfy

$$\begin{aligned} &\{b_\pm(p), b_\pm(q)\} = \{b_\pm(p), b_\mp(q)\} = \{b_\pm(p), b_\mp^*(q)\} = 0 \ , \\ &\{b_\pm(p), b_\pm^*(q)\} = \delta_{pq} \ , \end{aligned} \tag{V.3.1}$$

where $\{\cdot\ ,\ \cdot\}$ is the anti-commutator. The time zero Fermi fields are defined by

$$\begin{aligned} \psi_1(x) &= (2\ell)^{-1/2} \sum_{p\in\hat{T}^1} \mu(p)^{-1/2} \left(\nu(-p) b_-^*(p) + \nu(p) b_+(-p)\right) e^{-ipx} \ , \\ \psi_2(x) &= (2\ell)^{-1/2} \sum_{p\in\hat{T}^1} \mu(p)^{-1/2} \left(\nu(p) b_-^*(p) - \nu(-p) b_+(-p)\right) e^{-ipx} \ , \end{aligned} \tag{V.3.2}$$

where $\nu(p) = (\mu(p) + p)^{1/2}$. Let $\bar{\psi}_1(x) \equiv \psi_2^*(x)$, $\bar{\psi}_2(x) \equiv \psi_1^*(x)$, corresponding to $\bar{\psi} = \psi^* \left(\begin{smallmatrix} 0 & 1 \\ 1 & 0 \end{smallmatrix}\right)$. Then

$$\begin{aligned} &\{\psi_\mu(x), \psi_\nu(y)\} = 0, && \mu, \nu = 1, 2, \\ &\{\bar{\psi}_\mu(x), \psi_\mu(y)\} = 0, && \mu = 1, 2, \\ &\{\bar{\psi}_1(x), \psi_2(y)\} = \{\bar{\psi}_2(x), \psi_1(y)\} = \delta(x - y) \ . \end{aligned} \tag{V.3.3}$$

Relations (V.3.3) mean that for $f \in L^2(T^1)$,

$$\psi_\mu(f) \equiv \int_{T^1} \psi_\mu(x) f(x)\, dx \qquad \text{and} \qquad \bar{\psi}_\mu(f) \equiv \int_{T^1} \bar{\psi}_\mu(x) f(x)\, dx$$

generate an infinite dimensional, complex Clifford algebra. It also follows from (V.3.3) that $\psi_\mu(f)$ is a bounded operator and $\|\psi_\mu(f)\| = \|f\|_{L^2}$.

V.4 The Operators N_τ

For $0 \leq \tau \leq 1$ we define the operators

$$N_{\tau,b} = \sum_{j=\pm} \sum_{p\in\hat{T}^1} \mu(p)^\tau a_j^*(p) a_j(p) \ ,$$

$$N_{\tau,f} = \sum_{j=\pm} \sum_{p\in\hat{T}^1} \mu(p)^\tau b_j^*(p) b_j(p) \ , \tag{V.4.1}$$

on dense subspaces of $\mathcal{H}_b$ and $\mathcal{H}_f$, respectively. Let

$$N_\tau = N_{\tau,b} \otimes I_{\mathcal{H}_f} + I_{\mathcal{H}_b} \otimes N_{\tau,f}$$

be defined on $\mathcal{H}$. Clearly the number operator is $N = N_0$ and the free field Hamiltonian is $H_0 = N_1$. For $0 < \tau < 1$ these N_τ operators interpolate between N and H_0. It clearly causes no confusion to suppress the tensor products with I.

Let us also introduce the involution

$$\Gamma = \exp(i\pi N_f) \tag{V.4.2}$$

on $\mathcal{H}$, where $N_f = N_{0,f}$ is the fermionic number operator. Γ induces a grading on H, as described in Section II.

Proposition V.4.1. *For* $\tau, \beta > 0$ *the operators* $\exp(-\beta N_{\tau,b})$ *and* $\exp(-\beta N_{\tau,f})$ *are trace class and*

$$\mathrm{Tr}_{\mathcal{H}_b}(\exp(-\beta N_{\tau,b})) = \prod_{p\in\hat{T}^1} (1 - \exp(-\beta\mu(p)^\tau))^{-2} \ , \tag{V.4.3}$$

$$\mathrm{Tr}_{\mathcal{H}_f}(\exp(-\beta N_{\tau,f})) = \prod_{p\in\hat{T}^1} (1 + \exp(-\beta\mu(p)^\tau))^{2} \ , \tag{V.4.4}$$

$$\mathrm{Tr}_{\mathcal{H}_f}\left(\Gamma \exp(-\beta N_{\tau,f})\right) = \prod_{p\in\hat{T}^1} \left(1 - \exp(-\beta\mu(p)^\tau)\right)^2 \ . \qquad \text{(V.4.5)}$$

Proof: See [23].

As a corollary to (V.4.3) and (V.4.5) we obtain the following identity

$$\mathrm{Tr}_{\mathcal{H}}\left(\Gamma \exp(-\beta N_\tau)\right) = 1 \quad , \qquad \text{(V.4.6)}$$

valid for $\tau, \beta > 0$.

V.5 The Regularized Dirac Operator

Let V be a polynomial of degree $n \geq 2$. The Dirac operator Q is defined as a bilinear form on $\mathcal{H}$,

$$Q = \frac{1}{\sqrt{2}} \int_{T_1} \psi_1(\pi - \partial_x\varphi^* - i\partial V(\varphi)) + \psi_2(\pi^* - \partial_x\varphi - i\partial V(\varphi)^*))\, dx + \text{h.c.} \ , \qquad \text{(V.5.1)}$$

The domain $\mathcal{D}_0$ of Q we choose consists of Fock states with finite number of particles and $\mathcal{D}(T^1)$-valued wave functions. Notice that Q has the structure of a Dirac operator on an infinite dimensional manifold (loop space $\mathcal{D}(T^1)$) with circle action. The terms $\partial_x\varphi$ and $\partial_x\varphi^*$ are generators of the circle action $\varphi(x) \to \varphi(x+y)$, $y \in T^1$. Also $(\partial V(\varphi), \partial V(\varphi)^*)$ is the connection of a flat bundle over $\mathcal{D}(T^1)$. Defining (V.5.1) as an operator on $\mathcal{H}$ requires careful definition of its domain. We first smooth the form Q, and then we exhibit cancellation of local singularities. Finally we justify removing the smoothing.

We use the following smooth approximation to the periodic Dirac measure based on a cutoff function χ satisfying

(i) $0 \leq \chi \in \mathcal{S}(\mathbb{R})$,

(ii) $\int_{-\infty}^{\infty} \chi(x)\, dx = 1$,

(iii) $\chi(-x) = \chi(x)$,

(iv) $\hat{\chi}(p) \geq 0$,

(v) $\mathrm{supp}\ \hat{\chi}(p) \subseteq [-1, 1]$.

We set

$$\chi_\kappa(x) = \kappa \sum_{n\in\mathbb{Z}} \chi(\kappa(x - n\ell)) \ , \tag{V.5.2}$$

where $\kappa > 0$. We define regularized fields by convoluting with χ_κ on T^1,

$$\varphi_\kappa(x) = \chi_\kappa * \varphi(x), \quad \psi_{\mu,\kappa}(x) = \chi_\kappa * \psi_\mu(x) \ . \tag{V.5.3}$$

It is convenient to write $V(\varphi)$ in the form

$$V(\varphi) = \frac{1}{2}\, m\varphi^2 + P(\varphi) \ . \tag{V.5.4}$$

The regularized supercharge $Q(\kappa)$ is defined as a bilinear form on $\mathcal{H}$,

$$Q(\kappa) = Q_0 + Q_{i,\kappa} \ , \tag{V.5.5}$$

where

$$Q_0 = \frac{1}{\sqrt{2}} \int_{T^1} (\psi_1(\pi - \partial_1\varphi^* - im\varphi) + \psi_2(\pi^* - \partial_1\varphi - im\varphi^*))\, dx + \text{h.c.} \ , \tag{V.5.6}$$

and

$$Q_{i,\kappa} = -\frac{i}{\sqrt{2}} \int_{T^1} (\psi_1 \partial P(\varphi_\kappa) + \psi_2 \partial P(\varphi_\kappa)^*)\, dx + \text{h.c.} \ . \tag{V.5.7}$$

Proposition V.5.1. *The form $Q(\kappa)$ uniquely defines an essentially self-adjoint operator with domain D_0, such that its square equals*

$$H(\kappa) \equiv Q(\kappa)^2 = H_0 + \int_{T^1} (m\varphi^* \partial P(\varphi_\kappa) - (\bar{\psi}_1\psi_1)_\kappa \partial^2 P(\varphi_\kappa) + h.c.)\, dx$$

$$+ \int_{T^1} dx\, |\partial P(\varphi_\kappa)|^2 \ . \tag{V.5.8}$$

Here $(\bar{\psi}_\mu\psi_\mu)_\kappa \equiv \frac{1}{2}(\bar{\psi}_{\mu,\kappa}\psi_\mu + \bar{\psi}_\mu\psi_{\mu,\kappa})$. Thus $Q(\kappa)$ and $H(\kappa)$ extend uniquely to self-adjoint operators which we also denote $Q(\kappa)$ and $H(\kappa)$.

Proof: See [23], [24].

V.6 The Zero Momentum Limit

Set

$$\varphi_0 = \ell^{-1/2}\hat{\varphi}(0), \qquad \psi_{\mu,0} = \ell^{-1/2}\hat{\psi}_\mu(0) \quad , \tag{V.6.1}$$

where $\hat{\varphi}(p) = \ell^{-1/2} \int_{T^1} \varphi(x)e^{ipx}\, dx$. Define

$$Q(0) = Q_0 + Q_{i,0} \quad , \tag{V.6.2}$$

where

$$Q_{i,0} = -\frac{i}{\sqrt{2}}\,\ell(\psi_{1,0}\partial P(\varphi_0) + \psi_{2,0}\partial P(\varphi_0)^*) + \text{h.c.} \quad . \tag{V.6.3}$$

We also set $H(0) = Q(0)^2$. Here $H(0)$ is the Hamiltonian of a theory where the only interacting mode is the zero mode.

V.7 Fundamental A Priori Elliptic Estimates

We state now the crucial part of our construction, the fundamental *a priori* estimates. These estimates generalize certain classical elliptic estimates for differential operators on $L_2(\mathbb{R}^M)$ to operators on L_2 of an infinite dimensional (loop) space. The estimates are proved in [19]. For example, a fundamental *a priori* estimate in partial differential equations is Gårding's inequality which bounds an elliptic operator from below by a power of the Laplace operator. Our first estimate generalizes Gårding's inequality to an infinite dimensional setting:

THEOREM V.7.1. *Choose* $\tau \in [0,1)$. *Then there exist constants* $\zeta > 0$ *and* $C < \infty$ *which are independent of* κ *and for which*

$$\zeta N_\tau \leq H(\kappa) + C \quad . \tag{V.7.1}$$

The fact that the bound (V.7.1) is uniform in κ is characteristic of the *a priori* bounds established here. We use such estimates to establish the existence of the $\kappa \to \infty$ limit. Such a philosophy is standard in the constructive field theory [12]. The $H(\kappa)$ with $\kappa < \infty$ are operators with a finite number of degrees of freedom (plus an infinite number of uncoupled degrees of freedom). It is important that the constants in our estimates are independent of the number of degrees of freedom $= O(\kappa)$. We next state the continuity and convergence of the finite dimensional approximations to the semigroups

$$\beta \to \exp(-\beta H(\kappa)) , \qquad \beta \geq 0 . \tag{V.7.2}$$

THEOREM V.7.2. *For $\beta > 0$ fixed, the map*

$$\kappa \to \exp(-\beta H(\kappa)) \tag{V.7.3}$$

is norm-continuous for $0 \leq \kappa$. Furthermore, the family

$$\{\exp(-\beta H(\kappa))\}$$

is norm-convergent as $\kappa \to \infty$.

We denote the limiting semigroup by $T(\beta)$, $\beta \geq 0$, namely $\exp(-\beta H(\kappa)) \to T(\beta)$. In order to express $T(\beta)$ in terms of an infinitesimal generator H, we require strong continuity of $T(\beta)$. The consequence of strong continuity at $\beta = 0$ is the representation $T(\beta) = e^{-\beta H}$, with H a self-adjoint operator on $\mathcal{H}$. The delicate question whether H exists is more subtle in the infinite dimensional setting than in finite dimensions. For example, no vector in the smooth domain $\mathcal{D}_0$ of C^∞ wave functions with a finite number of particles is in the domain of H.

Theorem V.7.3. *The semigroup $T(\beta)$ is strongly continuous at $\beta = 0$,*

$$\operatorname*{st\,lim}_{\beta \to 0} T(\beta) = I . \tag{V.7.4}$$

Corollary to Theorems V.7.2, 3. *The limiting Hamiltonian H satisfies the Gårding estimate (V.7.1)*

$$\zeta N_\tau \leq H + C . \tag{V.7.5}$$

In our examples, the Dirac operator $Q(\kappa)$ is related to $H(\kappa)$ by $H(\kappa) = Q(\kappa)^2$. We wish to construct a limiting Q as well as a limiting H, and we desire $H = Q^2$. The supercharge is a Dirac operator on loop space, while H is a Laplace operator. We require continuity of $Q(\kappa)$ in κ, as well as convergence of $Q(\kappa)$ in the following manner as $\kappa \to \infty$. Let $\delta Q = (Q(\kappa) - Q(\kappa'))^-$, where $^-$ denotes the operator closure.

Theorem V.7.4. *Let $\beta > 0$. Then $\text{Range}(e^{-\beta H(\kappa)}) \subset \text{Dom}(\delta Q)$ and*

$$\|e^{-\beta H(\kappa')}\delta Q e^{-\beta H(\kappa)}\| = o(1) \tag{V.7.6}$$

as $|\kappa - \kappa'| \to 0$, and as κ, $\kappa' \to \infty$.

V.8. The Dirac Operator on Loop Space

We will show now how the *a priori* estimates of Section V.8 yield the fundamental continuity in the cutoff.

Theorem V.8.1. *The resolvent*

$$S_\kappa = (Q(\kappa) + i)^{-1} \tag{V.8.1}$$

is norm-continuous in κ and norm-convergent as $\kappa \to \infty$. The limiting operator $S = \lim_{\kappa\to\infty} S_\kappa$ is the resolvent of a self-adjoint operator Q such that $Q^2 = H$.

Proposition V.4.1 ensures that N_τ has a compact resolvent. The estimate (V.7.1) shows that $H(\kappa)$ is relatively compact with respect to N_τ, and hence it too has a compact resolvent. Since $Q(\kappa)^2 = H(\kappa)$, and $Q(\kappa)$ is self-adjoint, also $Q(\kappa)$ has a compact resolvent. Norm limits of compact operators are compact, so we have by Definition III.2.1,

Corollary V.8.2. *For $0 \le \kappa \le \infty$, $Q(\infty) \equiv Q$,*

$$Q(\kappa) \in \text{DIR}(\mathcal{H}) \quad . \tag{V.8.2}$$

Moreover,

$$i(Q_+) = i(Q(0)_+) \quad . \tag{V.8.3}$$

Proof of Theorem V.8.1. We note first that as a consequence of Theorem V.7.3, the mapping

$$\kappa \to R_\kappa = (H(\kappa) + I)^{-1} \tag{V.8.4}$$

is norm-continuous for $0 \leq \kappa \leq \infty$, where $H(\infty) = H$ is the generator of the limiting group $T(\beta)$.

Choose $\epsilon > 0$. We claim that for κ, κ' sufficiently large,

$$\|S_\kappa - S_{\kappa'}\| \leq 7\epsilon \quad . \tag{V.8.5}$$

Let $E_\kappa = E_\kappa(\lambda)$ denote the spectral projection onto the subspace $H(\kappa) \leq \lambda$. We choose $\lambda = \epsilon^{-2}$. Since $H(\kappa) = Q(\kappa)^2$, E_κ commutes with S_κ. We study $\delta S = S_\kappa - S_{\kappa'}$. Then

$$\delta S = E_{\kappa'}\delta S E_\kappa + (I - E_{\kappa'})\delta S E_\kappa + \delta S(I - E_\kappa) \quad . \tag{V.8.6}$$

We claim that for κ, κ' large,

$$\|\delta S(I - E_\kappa)\| \leq 3\epsilon \ , \qquad \|(I - E_{\kappa'})\delta S E_\kappa\| \leq 3\epsilon \ . \tag{V.8.7}$$

In fact using

$$\|R_\kappa^{-1/2} S_\kappa\| = 1 \quad , \tag{V.8.8}$$

we obtain

$$\|S_\kappa(I-E_\kappa)\| = \|R_\kappa^{-1/2} S_\kappa R_\kappa^{1/2}(I-E_\kappa)\| \leq \|R_\kappa^{1/2}(I-E_\kappa)\| \leq (\lambda+1)^{-1/2} \leq \epsilon \ . \tag{V.8.9}$$

Furthermore, for κ, κ' sufficiently large, we infer from the continuity of (V.8.4) that

$$\|R_\kappa^{1/2} - R_{\kappa'}^{1/2}\| \leq \epsilon \ . \tag{V.8.10}$$

Here we use the fact that norm convergence of resolvents implies norm convergence of the square root. Thus

$$\begin{aligned}\|R_{\kappa'}^{1/2}(I - E_\kappa)\| &\leq \|R_\kappa^{1/2}(I - E_\kappa)\| + \|(R_{\kappa'}^{1/2} - R_\kappa^{1/2})(I - E_\kappa)\| \\ &\leq \epsilon + \epsilon = 2\epsilon \quad ,\end{aligned}$$

and by (V.8.8),

$$\begin{aligned}\|S_{\kappa'}(I - E_\kappa)\| &\leq \|R_{\kappa'}^{-1/2} S_{\kappa'} R_{\kappa'}^{1/2}(I - E_\kappa)\| \\ &\leq \|R_{\kappa'}^{1/2}(I - E_\kappa)\| \leq 2\epsilon \ .\end{aligned} \tag{V.8.11}$$

It follows from (V.8.9), (V.8.11) that

$$\|\delta S(I - E_\kappa)\| \le 3\epsilon \ ,$$

which is the estimate on the last term in (V.8.6). The estimate on $(I - E_{\kappa'})\delta S E_\kappa$ is similar. Hence

$$\|\delta S\| \le 6\epsilon + \|E_{\kappa'}\delta S E_\kappa\| \ . \tag{V.8.12}$$

We now use the facts that $\|S_\kappa\| \le 1$, and that the resolvent identity

$$\delta S = S_\kappa - S_{\kappa'} = S_{\kappa'}(Q(\kappa') - Q(\kappa))S_\kappa \tag{V.8.13}$$

holds as a bilinear form. Thus we also have the form identity

$$E_{\kappa'}\delta S E_\kappa = S_{\kappa'}E_{\kappa'}\delta Q E_\kappa S_\kappa \ , \tag{V.8.14}$$

where $\delta Q = (Q(\kappa') - Q(\kappa))^-$.

By Theorem V.7.4, with $\beta > 0$, and with κ, κ' sufficiently large,

$$\begin{aligned} \|E_{\kappa'}\delta Q E_\kappa\| &= \|E_{\kappa'}e^{\beta H(\kappa')}e^{-\beta H(\kappa')}\delta Q e^{-\beta H(\kappa)}e^{\beta H(\kappa)}E_\kappa\| \\ &\le e^{2\lambda\beta}\|e^{-\beta H(\kappa')}\delta Q e^{-\beta H(\kappa)}\| \le e^{2\lambda\beta}o(1) \le \epsilon \ . \end{aligned} \tag{V.8.15}$$

From (V.8.12–15), we infer $\|\delta S\| \le 7\epsilon$ as claimed.

This completes the proof of convergence of S_κ as $\kappa \to \infty$. The same type of argument shows that $\{S_\kappa\}$ is continuous in κ for $\kappa < \infty$. Since $\|\delta S\| = \|\delta S^*\|$, the continuity and convergence of S_κ^* follows. This completes the proof of the norm continuity S_κ.

We now proceed to show that $S = \lim_{\kappa\to\infty} S_\kappa$ is the resolvent of a self-adjoint operator Q. The main technical issue is to show that S is invertible, namely that $\mathrm{Ker}(S) = 0$. A similar issue arose in the proof of self-adjointness of H, and it was solved by showing that the semigroup $T(\beta) = \lim_{\kappa\to\infty} \exp(-\beta H(\kappa))$ was strongly continuous at $\beta = 0$, c.f. Theorem V.7.3. In this case we have no heat kernel representation for Q, but we use the existence of a dense domain for H. In fact

$$S^*S = \lim_{\kappa\to\infty} S_\kappa^* S_\kappa = \lim_{\kappa\to\infty} (H(\kappa) + I)^{-1} = (H + I)^{-1} \ . \tag{V.8.16}$$

Since $0 \leq H = H^*$, the null space of $(H+I)^{-1}$ is zero. Thus the null space of S is trivial and S is invertible. It then follows by Theorem 4 of [11] that $S = (Q+i)^{-1}$ is the resolvent of a self-adjoint operator Q. Furthermore,

$$S^*S = (Q^2+I)^{-1} = (H+I)^{-1} \quad ,$$

so $H = Q^2$ and the proof of Theorem V.8.1 is complete.

V.9 The Index Theorem

In the preceding section we showed that the *a priori* estimates of Section V.7 are sufficient to establish a homotopy between $Q(\infty)$ and $Q(0)$ with $i(Q(\kappa)_+)$ constant. Here we evaluate $i(Q(0)_+)$.

Theorem V.9.1. (Index theorem) *Let $V(\varphi)$ be a polynomial of degree $n \geq 2$, and let Q be the corresponding Dirac operator. Then $i(Q_+) = n-1$.*

Proof: By Corollary V.8.2, we need only consider the case $\kappa = 0$. Decompose the Fock space as an orthogonal sum $\mathcal{H} = \mathcal{H}_0 \oplus \mathcal{H}_0^{\perp}$, where $\mathcal{H}_0$ is the subspace spanned by the zero-momentum modes. According to this decomposition, the free supercharge Q_0 can be written

$$Q_0 = \begin{pmatrix} Q_0^0 & 0 \\ 0 & Q_0^{\perp} \end{pmatrix} \quad , \tag{V.9.1}$$

where the zero modes contributes

$$\begin{aligned} Q_0^0 = \frac{1}{\sqrt{2}} \Big(&(\hat{\psi}_1(0) + \hat{\bar{\psi}}_1(0))\hat{\pi}(0) + (\hat{\psi}_2(0) + \hat{\bar{\psi}}_2(0))\hat{\pi}^*(0) \\ &+ im\hat{\varphi}(0)(\hat{\bar{\psi}}_1(0) - \hat{\psi}_1(0)) + im\hat{\varphi}^*(0)(\hat{\bar{\psi}}_2(0) - \hat{\psi}_2(0)) \Big) \quad . \end{aligned} \tag{V.9.2}$$

Introduce new variables $z = \hat{\varphi}(0)$ and

$$\begin{aligned} \Psi_1 &= -\frac{1}{2}\Big(b_-^*(0) + b_+(0) - b_-(0) + b_+^*(0)\Big), \quad \bar{\Psi}_2 = \Psi_1^* \ , \\ \Psi_2 &= -\frac{1}{2}\Big(b_-^*(0) - b_+(0) + b_-(0) + b_+^*(0)\Big), \quad \bar{\Psi}_1 = \Psi_2^* \ . \end{aligned} \tag{V.9.3}$$

Then we verify

$$\begin{aligned} &\{\Psi_\mu, \Psi_\nu\} = 0, && \mu,\nu = 1,2, \\ &\{\Psi_\mu, \bar{\Psi}_\mu\} = 0, && \mu = 1,2, \\ &\{\Psi_1, \bar{\Psi}_2\} = \{\Psi_2, \bar{\Psi}_1\} = 1 \quad . \end{aligned} \tag{V.9.4}$$

In terms of these variables

$$Q_0^0 = i\bar{\Psi}_1(\partial/\partial z) + i\Psi_2(\partial/\partial \bar{z}) + i\Psi_1 mz - i\bar{\Psi}_2 m\bar{z} \quad . \tag{V.9.5}$$

The operator $Q_{i,0}$ can also be expressed as an operator on $\mathcal{H}_0$, namely

$$Q_{i,0} = i\ell^{1/2}\Psi_1\partial P(\ell^{-1/2}z) - i\ell^{1/2}\bar{\Psi}_2\partial P(\ell^{-1/2}z)^* \quad .$$

Therefore

$$\begin{aligned} Q_0^0 + Q_{i,0} = & \, i\bar{\Psi}_1(\partial/\partial z) + i\Psi_2(\partial/\partial \bar{z}) \\ & + i\ell^{1/2}\Psi_1\partial V(\ell^{-1/2}z) - i\ell^{1/2}\bar{\Psi}_2\partial V(\ell^{-1/2}z)^* \quad . \end{aligned} \tag{V.9.6}$$

Comparing (V.9.6) with Section IV, we find that $Q_0^0+Q_{i,0}$ is exactly the supercharge of our model of supersymmetric quantum mechanics. In addition, we verify by explicit calculation that $\Gamma\restriction_{\mathcal{H}_0}$ is identical to the operator Γ of Section IV. Thus the index calculation for $(Q_0^0 + Q_{i,0})_+$ reduces to the calculation of Section IV where we established $n_+ = n-1$, $n_- = 0$. On the full Fock space

$$Q(0) = \begin{pmatrix} Q_0^0 + Q_{i,0} & 0 \\ 0 & Q_0^\perp \end{pmatrix} \quad . \tag{V.9.7}$$

Clearly $Q_0^\perp$ has a unique ground state which is an element of $\mathcal{H}_+$; it has no ground state in $\mathcal{H}_-$. This completes the proof of the theorem.

VI. FUNCTIONAL INTEGRAL REPRESENTATION OF THE HEAT KERNEL

The main technical device to prove the *a priori* estimates of Section V.7 is the use of functional (or path space) integrals. Their merit is that they allow to replace operator (non-commutative) estimates by measure theoretical (commutative) estimates. The bridge between the operators and path space measures is provided by Feynman-Kac formulas, which give path space representatives of various states on the field algebra. Our analysis requires several basic states on the field algebra:

(i) *The trace state:*

$$\langle \cdot \rangle_\beta = (\mathrm{Tr}_{\mathcal{H}} e^{-\beta H})^{-1} \mathrm{Tr}_{\mathcal{H}}(\cdot \, e^{-\beta H}) \quad , \tag{VI.0.1}$$

where $\beta > 0$. If $H = H_0$, the free Hamiltonian, then we call the corresponding state the free trace state and denote it by $\langle \cdot \rangle_{\beta,0}$.

(ii) *The graded (or super) trace state:*

$$\langle \cdot \rangle_\beta^\Gamma = (Tr_{\mathcal{H}} \Gamma e^{-\beta H})^{-1} \mathrm{Tr}_{\mathcal{H}}(\cdot \Gamma e^{\beta H}) \quad , \tag{VI.0.2}$$

where Γ is the grading operator. The corresponding free state is denoted by $\langle \cdot \rangle_{\beta,0}^\Gamma$.

(iii) *The free vacuum state:*

$$\langle \cdot \rangle_0 = \langle \Omega_0 , \cdot \Omega_0 \rangle \tag{VI.0.3}$$

where Ω_0 is the Fock vacuum vector. This state is the $\beta \to \infty$ limit of (VI.0.1) in the case $H = H_0$.

We have constructed non-Gaussian measures on path spaces such that the states (VI.0.1) – (VI.0.2) are represented as integrals with respect to these measures. We then use the state (VI.0.1) to prove Theorems V.7.1, 2 and 4. The state (VI.0.3) plays an important role in the proof of Theorem V.7.3. Finally, we use the state (VI.0.2) to establish a functional integral representation for the index $i(Q_+)$.

VI.1 States on the Field Algebra and Boundary Conditions on Path Space

We consider the two-dimensional, Euclidean space-time $B \times T^1$, where either $B = T^1_\beta$ (a one-torus of circumference β) or $B = \mathsf{R}$. We define $C_\ell = (-\Delta + m^2)^{-1}$ if $B = \mathsf{R}$, and similarly we define $C_{\ell,\beta}$ if $B = T^1_\beta$. Let $\not{D} = i\gamma_0^E \partial_0 + i\gamma_1^E \partial_1$ be the Euclidean Dirac operator on $B \times T^1$. Here γ_j^E are the Euclidean Dirac matrices which we choose to be $\gamma_0^E = i\sigma_1$, $\gamma_1^E = -i\sigma_2$ with σ_j given by (IV.5). We define S_ℓ and $S_{\ell,\beta}$ as $(\not{D} + m)^{-1}$ for $B = \mathsf{R}$ and $B = T^1_\beta$, respectively. By $\tilde{\not{D}}$ we denote the Dirac operator on the torus twisted in the time direction by π. Functions in the domain of $\tilde{\not{D}}$ satisfy

$$f(x_0 + \beta, x_1) = -f(x_0, x_1) \quad . \tag{VI.1.1}$$

We define $\tilde{S}_{\ell,\beta} = (\tilde{\not{D}} + m)^{-1}$. Note that $\tilde{S}_{\ell,\beta}(x)$ is antiperiodic in the time direction and periodic in the space direction.

The bosonic path space is $\mathcal{S}'(B \times T^1)$. Let $d\mu_{C_\ell}$ be a Gaussian measure on $\mathcal{S}'(\mathsf{R} \times T^1)$ with covariance C_ℓ, and let $d\mu_{C_{\ell,\beta}}$ be a Gaussian measure on $\mathcal{S}'(T^1_\beta \times T^1)$ with covariance $C_{\ell,\beta}$.

We can also define a "fermionic path space" [4]. Let $\mathcal{H}_\alpha(B \times T^1)$ be the Sobolev space of order α with the norm $\|f\|_{\mathcal{H}_\alpha} = \|(-\Delta + m^2)^{\alpha/2} f\|_{L_2}$. Let $\mathcal{G}$ be an infinite dimensional Grassmann algebra with generators $\Psi_\mu(f)$, $\bar{\Psi}_\mu(f)$, $\mu = 1, 2$, $f \in \mathcal{H}_{-1/2}(B \times T^1)$. In other words $\mathcal{G}$ is the algebra of polynomials in $\Psi_\mu(f)$, $\Psi_\nu(g)$, $f, g \in \mathcal{H}_{-1/2}(B \times T^1)$. Let S be any of the fermionic Green's operators defined above. A Gaussian Berezin integral $\int \cdot \, d\mu_S$ on $\mathcal{G}$ is a linear functional $\mathcal{G} \to \mathbf{C}$ defined on the monomials as follows: it is zero if the number of Ψ's in the monomial is not equal to the number of $\bar{\Psi}$'s, otherwise

$$\int \Psi_{\mu_1}(f_1)\bar{\Psi}_{\nu_1}(g_1)\ldots\Psi_{\mu_n}(f_n)\bar{\Psi}_{\nu_n}(g_n)\, d\mu_S = \det\{S_{\mu_j\nu_k}(f_j, g_k)\} \,. \tag{VI.1.2}$$

We extend (VI.1.2) to all of $\mathcal{G}$ by linearity. Note that $\int \cdot \, d\mu_S$ extends also to certain infinite series in Ψ and $\bar{\Psi}$.

Theorem VI.1.1.

(i) *The free ungraded trace state is given by the Gaussian measure* $d\mu_{C_{\ell,\beta}} \otimes d\mu_{\tilde{S}_{\ell,\beta}}$ *on path space.*

(ii) *The free graded trace state is given by the Gaussian measure* $d\mu_{C_{\ell,\beta}} \otimes d\mu_{S_{\ell,\beta}}$ *on path space.*

(iii) *The free vacuum state is given by the Gaussian measure* $d\mu_{C_\ell} \otimes d\mu_{S_\ell}$ *on path space.*

Proof. Since the states (VI.0.1) – (VI.0.3) with $H = H_0$ are Gaussian and have mean zero, it is sufficient to check that the covariance operators agree. Let us prove e.g., that

$$\begin{aligned} &\langle e^{-sH_{0,b}}\varphi(f)e^{-(t-s)H_{0,b}}\varphi^*(g)e^{tH_{0,b}}\rangle_{\beta,0} \\ &\quad = C_{\ell,\beta}(f \otimes \delta_s, g \otimes \delta_t), \qquad \text{for} \qquad f, g \in \mathcal{H}_{-1}(T^1) \,. \end{aligned} \tag{VI.1.3}$$

The proofs of the other statements are similar. We represent $\mathcal{H}_b$ as

$$\mathcal{H}_b \cong \bigotimes_{p,j} \mathcal{H}_b(p, j) \,, \tag{VI.1.4}$$

where $\mathcal{H}_b(p,j)$ is the subspace of $\mathcal{H}_b$ spanned by polynomials in the creation operator $a_j^*(p)$, $p \in \hat{T}^1$, $j = \pm$, applied to Ω_0^b. Straightforward calculations show that

$$\begin{aligned}&\mathrm{Tr}_{\mathcal{H}_b(p,j)}(a_\pm^*(p)e^{-\sigma H_{0,b}(p,j)}a_\pm(p)e^{-(\beta-\sigma)H_{0,b}(p,j)})\\&\quad=\Big(1-\exp(-\beta\omega(p))\Big)^{-2}\exp(-(\beta-\sigma)\omega(p)) \quad ,\end{aligned}$$

and

$$\begin{aligned}&\mathrm{Tr}_{\mathcal{H}_b(p,j)}(a_\pm(p)e^{-\sigma H_{0,b}(p,j)}a_\pm^*(p)e^{-(\beta-\sigma)H_{0,b}(p,j)})\\&\quad=\Big(1-\exp(-\beta\omega(p))\Big)^{-2}\exp(-\sigma\omega(p)) \quad ,\end{aligned}$$

where $\sigma = t - s$ and $H_{0,b}(p,j) = \omega(p)a_j^*(p)a_j(p)$. The left-hand side of (VI.1.3) can be thus written as

$$\sum_{p\in\hat{T}^1}\hat{f}(p)\hat{g}(p)(2\omega(p))^{-1}\Big(1-\exp(-\beta\omega(p))\Big)^{-1}\Big(\exp(-\sigma\omega(p))+\exp(-(\beta-\sigma)\omega(p))\Big)$$

$$=\sum_{n\in\mathbb{Z}}\sum_{p\in\hat{T}^1}\hat{f}(p)\hat{g}(p)(2\omega(p))^{-1}\exp(-|\sigma+n\beta|\omega(p))$$

$$=\sum_{n\in\mathbb{Z}}\sum_{p\in\hat{T}^1}\hat{f}(p)\hat{g}(p)\,\frac{1}{2\pi}\int_{-\infty}^{\infty}\frac{1}{E^2+p^2+m^2}\,e^{i(\sigma+n\beta)E}\,dE$$

$$=C_{\ell,\beta}(f\otimes\delta_s, g\otimes\delta_t)\,,$$

and the claim follows. The fermionic calculation can be found in [23].

VI.2 The Non-Gaussian Measures $d\mu_V$

The states (VI.0.1) and (VI.0.2) corresponding to non-free Hamiltonians H can also be represented by a measure on path space. Each interaction polynomial V determines two unique measures $d\tilde{\mu}_V$, $d\mu_V$. Now, however, $d\tilde{\mu}_V$, $d\mu_V$ are neither product measures nor are they Gaussian. As in the free case, the measure $d\tilde{\mu}_V$ is periodic in space; it is antiperiodic in time in the fermionic coordinates and periodic in time in the bosonic coordinates. On the other hand, the measure $d\mu_V$ is periodic both in the fermionic and the bosonic degrees of freedom and both in space and in time.

The measures $d\tilde{\mu}_V$ and $d\mu_V$ are constructed as limits of approximating measures of the form

$$d\tilde{\mu}_V^{(\kappa)},\ d\mu_V^{(\kappa)} = \begin{pmatrix}\text{Infinite dimensional}\\ \text{Gaussian measure}\end{pmatrix} \otimes \begin{pmatrix}\text{Non} - \text{Gaussian measure}\\ \text{in } O(\kappa) \text{ variables}\end{pmatrix} .$$

The limits $\kappa \to \infty$ yield $d\tilde{\mu}_V$ and $d\mu_V$. These limits are studied using estimates on convergence of these path space integrals. They yield Hilbert-Schmidt (or Schatten class I_2) convergence of the corresponding heat kernels $\exp(-\beta H(\kappa))$ and hence convergence of their trace states.

Let $\Phi \in \mathcal{S}'(\mathsf{R} \times T^1)$ and let $\Phi_\kappa(x)$ be a regularized approximation to $\Phi(x)$:

$$\Phi_\kappa(x) = \chi_\kappa \star \Phi(x) \equiv \int_{T^1} \chi_\kappa(x_1 - x_1')\Phi(x_0, x_1')\, dx_1' \quad , \qquad \text{(VI.2.1)}$$

where χ_κ is given by (V.5.2). We set

$$A_\ell^{(\kappa)}(\Phi) = \int_{[0,\beta]\times T^1} (m\Phi\partial P(\Phi_\kappa)^* + m\Phi^*\partial P(\Phi_\kappa) + |\partial P(\Phi_\kappa)|^2)\, dx\ . \qquad \text{(VI.2.2)}$$

Let $K_\ell^{(\kappa)}(\Phi)$ be the operator on $\mathcal{K}_{1/2} = \mathcal{H}_{1/2} \oplus \mathcal{H}_{1/2}$ whose integral kernel is given by

$$K_\ell^{(\kappa)}(\Phi)(x,y)$$

$$= \frac{1}{2}\int_{T^1} S_\ell(x-z)\chi_\kappa(z_1 - y_1)\left(\partial^2 P(\Phi_\kappa(z)) + \partial^2 P(\Phi_\kappa(y))\right)\Lambda_+\, dz_1$$

$$+ \frac{1}{2}\int_{T^1} S_\ell(x-z)\chi_\kappa(z_1 - y_1)\left(\partial^2 P(\Phi_\kappa(z))^* + \partial^2 P(\Phi_\kappa(y))^*\right)\Lambda_-\, dz_1 \ , \qquad \text{(V.2.3)}$$

where

$$\Lambda_+ = \begin{pmatrix}1 & 0\\ 0 & 0\end{pmatrix} , \qquad \Lambda_- = \begin{pmatrix}0 & 0\\ 0 & 1\end{pmatrix}$$

are the chiral projections, and where $z = (y_0, z_1)$.

As a consequence of Theorem VI.1.1(i) and standard approximation arguments we obtain the following identity:

$$\langle \Omega_0, e^{-\beta H(\kappa)} \Omega_0 \rangle = \int \det(I - K_\ell^{(\kappa)}(\Phi)) \exp\{-A_\ell^{(\kappa)}(\Phi)\}\, d\mu_{C_\iota}(\Phi) \ , \qquad \text{(VI.2.4)}$$

where det is the Fredholm determinant, see [24]. This identity as well as its generalizations are used in [19] to construct the limit $\kappa \to \infty$ of the heat kernel $\exp\{-\beta \mathcal{H}(\kappa)\}$ and to establish the properties of H and Q required in Theorems V.7.1–4. We refer to [19] for details and content ourselves here with the following remarks.

One of the difficulties in establishing the $\kappa \to \infty$ limit of (VI.2.4) is that both $\det(I - K_\ell^{(\kappa)}(\Phi))$ and $A_\ell^{(\kappa)}(\Phi)$ become singular, as $\kappa \to \infty$. However, if we write

$$\det(I - K_\ell^{(\kappa)}(\Phi)) = \det_3(I - K_\ell^{(\kappa)}(\Phi)) \exp\left\{-\mathrm{Tr}\, K_\ell^{(\kappa)}(\Phi) - \frac{1}{2} \mathrm{Tr}\, K_\ell^{(\kappa)}(\Phi)^2\right\}, \qquad \text{(VI.2.5)}$$

then we notice that all the singularities of

$$A_\ell^{(\kappa)}(\Phi) + \mathrm{Tr}\, K_\ell^{(\kappa)}(\Phi) + \frac{1}{2}\, \mathrm{Tr}\, K_\ell^{(\kappa)}(\Phi)^2 \equiv \mathcal{A}_\ell^{(\kappa)}(\Phi) \qquad \text{(VI.2.6)}$$

cancel against each other! This property is known as *ultraviolet finiteness* of a supersymmetric field theory model. Moreover, $\det_3(I - K_\ell^{(\kappa)}(\Phi))$ has an integrable, $\kappa \to \infty$ limit and thus we obtain

$$\langle \Omega_0, e^{-\beta H}\, \Omega_0 \rangle = \int \det_3\, (I - K_\ell(\Phi)) \exp\{-\mathcal{A}_\ell(\Phi)\}\, d\mu_{C_\iota}(\Phi) \ . \qquad \text{(VI.2.7)}$$

VI.3. Path Integral Representation of the Index

Let $A_{\ell,\beta}^{(\kappa)}(\Phi)$ and $K_{\ell,\beta}^{(\kappa)}(\Phi)$ be defined by (VI.2.2) and (VI.2.3), respectively, with C_ℓ replaced by $C_{\ell,\beta}$ and with S_ℓ replaced by $S_{\ell,\beta}$. As a consequence of (III.4.2) and Theorem VI.1 (iii) we obtain

Theorem VI.3.1. [23] *The index* $i(Q_+)$ *has the following path integral representation*

$$i(Q_+) = \int \det\, (I - K_{\ell,\beta}^{(\kappa)}(\Phi)) \exp\{-A_{\ell,\beta}^{(\kappa)}(\Phi)\}\, d\mu_{C_{\ell,\beta}}\, (\Phi) \ , \qquad \text{(VI.3.1)}$$

where $0 \leq \kappa < \infty$.

Formula (VI.3.1) can be used to give an alternative proof of the index theorem V.9.1, see [23]. Taking the $\kappa \to \infty$ limit as explained in the previous section, we obtain the following

Theorem VI.3.2. *With the self-explanatory notation,*

$$i(Q_+) = \int \det_3 (I - K_{\ell,\beta}(\Phi)) \exp\{-\mathcal{A}_{\ell,\beta}(\Phi)\} \, d\mu_{C_{\ell,\beta}}(\Phi) \quad . \tag{VI.3.2}$$

VII. CONJECTURES AND OPEN QUESTIONS

VII.1. The Real Case

In these notes we have described a supersymmetric model of a complex scalar field interacting with a Dirac field. This gives rise to a theory of Dirac operators on the loop space $\Lambda\mathbb{C} = \{\text{maps} : S^1 \to \mathbb{C}\}$ and to a function space measure $d\mu$ on $L_2(\mathcal{S}'(T^2))$ over the torus T^2.

The corresponding real loop space is

$$\Lambda\mathbb{R} = \{\text{maps} : S^1 \to \mathbb{R}\} \ .$$

In this case we can also find a supersymmetric field theory which is a polynomial perturbation of the free field. The real case differs in several other ways from the complex case. The major difference is the fact that the real problem is not finite. It involves a renormalization, although this is a renormalization of a straightforward kind, namely, normal (Wick) ordering of the supercharge Q. However, this means that the resulting Dirac operator Q on loop space and its Hamiltonian (Laplacian) $H = Q^2$ depend on one parameter in addition to the polynomial V. This additional parameter is the Wick ordering mass.

A second qualitative feature of the real model is that the fermion is Majorana rather than Dirac (which can be interpreted as two copies of a Majorana fermion). The function space integral over the fermions gives rise to a Pfaffian in place of a determinant, and in fact to a relative Pfaffian which is the formal ratio of two Pfaffians of Dirac operators on a torus.

Conjecture 1. The real polynomial supersymmetric field models exist.

The construction of these models is work in progress [26]. In order to deal with the analytic properties of the Pfaffians which arise, we have developed an analytic theory of regularized, relative Pfaffians of two antisymmetric operators A, B, see [25]. In the finite dimensional case, the relative Pfaffian satisfies

$$\mathrm{Pf}(A,B) = \frac{\mathrm{Pf}(A^{-1} - B)}{\mathrm{Pf}(A^{-1})} .$$

The relative Pfaffian has the property

$$\mathrm{Pf}(A,B)^2 = \det(I - AB) .$$

We also define the regularized, relative Pfaffian $\mathrm{Pf}_n(A,B)$. It satisfies $\mathrm{Pf}_1(A,B) = \mathrm{Pf}(A,B)$ while for $n = 2, 3, \ldots$,

$$\mathrm{Pf}_n(A,B)^2 = \det_n(I - AB) .$$

The relative Pfaffian $\mathrm{Pf}(A,B)$ has an expansion in terms of minors which has a limit so long as A and B are both Hilbert-Schmidt. Furthermore, $\mathrm{Pf}_n(A,B)$ extends to antisymmetric operators A, B such that A^n and B^n are both Hilbert-Schmidt.

We believe that the relative Pfaffian is a useful analytic concept in the study of loop groups [30], in the study of conformal field theories, in the study of the theory of Pfaffian bundles over an infinite dimensional Grassmannian [40], etc., as well as in the study of supersymmetric quantum field theory.

In the real case, the index can also be computed by a homotopy of Q onto quantum theory of one bosonic and one fermionic degree of freedom. In this case one finds

$$i(Q_+) = \pm[(1 + \deg V) \bmod 2] .$$

In fact, the quantum mechanics model involved has the Hamiltonian

$$H = \frac{1}{2}(P^2 + (V')^2 + V''\sigma_3) ,$$

arising from a supercharge

$$Q = \frac{1}{\sqrt{2}} (p\sigma_1 + V'\sigma_2) ,$$

where $\sigma_1, \sigma_2, \sigma_3$ are defined by (IV.5). This Hamiltonian describes a supersymmetric quantum mechanics studied extensively by Witten [36] and other authors.

VII.2 The Vanishing Theorem

Exactly how many vacuum states do exist? Let $\mathcal{H}_0$ denote the eigenspace $E = 0$ of H. Clearly the representation

$$H = \begin{pmatrix} Q_+^* Q_+ & 0 \\ 0 & Q_-^* Q_- \end{pmatrix} \qquad \text{(VII.2.1)}$$

shows that the number of zero energy vacuum states in

$$\dim \mathcal{H}_0 = n_+ + n_- \quad , \qquad \text{(VII.2.2)}$$

while the index

$$i(Q_+) = n_+ - n_- \quad . \qquad \text{(VII.2.3)}$$

In particular, the index bounds the number of vacuum states,

$$\dim \mathcal{H}_0 \geq | \, i(Q_+)| \quad . \qquad \text{(VII.2.4)}$$

Conjecture 2. (Vanishing Theorem) For the models on the loop spaces ΛC and ΛR,

$$\dim \mathcal{H}_0 = |i(Q_+)| \quad . \qquad \text{(VII.2.5)}$$

The above analysis can be made for each cutoff κ, i.e., for each point in the homotopy between $Q = Q(\kappa = \infty)$ and $Q(\kappa = 0)$. This family of intermediate Q's are each Dirac operators on a finite dimensional manifold. In the finite dimensional case, $Q(\kappa = 0)$ satisfies the conjecture (VII.2.5) [21, 36], as described in Section IV in the complex case. We expect that this vanishing theorem, which in the complex case says $n_- = 0$, is true for every $\kappa \leq \infty$. In other words, we are conjecturing for the examples ΛC and ΛR, that $n_\pm(\kappa)$ are individually invariant under variation of κ, as well as under variation of their difference.

Another equivalent statement of the vanishing theorem is:

Conjecture 2′. For the loop spaces $\Lambda\mathbf{C}$ and $\Lambda\mathbf{R}$, and the Dirac operators with polynomial potentials described above,

$$n_+(\kappa)\, n_-(\kappa) = 0 \tag{VII.2.6}$$

for all $\kappa \leq \infty$.

VII.3. Breaking Supersymmetry

In a finite volume, i.e., for field theories on the space of closed loops, supersymmetry breaking is defined by

$$\dim \mathcal{H}_0 = 0 \quad . \tag{VII.3.1}$$

In other words, the vacuum state(s) of the theory have energy $E > 0$. In fact, since for $E > 0$, the operators (III.4.3) project onto isomorphic subspaces of eigenvalues of Q, namely $\mathcal{H}_{E,\pm}$. Thus (VII.3.1) and a supersymmetric field theory ensures the existence of at least two vacuum states. We thus summarize the situation:

If the index $i(Q_+)$ vanishes, and the vanishing theorem (Conjecture 2) holds, then supersymmetry is broken, the vacuum energy $E > 0$, and the ground state of the theory is degenerate.

Conjecture 3. Supersymmetry is broken for all the real, polynomial models with potentials V of odd degree. In particular, supersymmetry is broken for φ^4, φ^8, ..., φ^{4j}, ... models, but unbroken for φ^{4j+2} models.

Note that in these theories, Conjecture 3 is known to hold for the $\kappa = 0$ endpoint of the homotopy for $Q(\kappa)$. Furthermore, the breaking of supersymmetry is a global property of the potential V. It is determined by whether

$$\lim_{t\to\infty} \operatorname{sign}(V(t)\, V(-t)) = \pm 1 \quad ,$$

rather than by the local properties of V.

VII.4. The Infinite Volume Limit

The infinite volume limit corresponds to taking the limit $T^1 \to \mathbf{R}$ as the radius ℓ of the circle increases to infinity. In this limit, the loop space $\Lambda\mathbf{C}$ or $\Lambda\mathbf{R}$ is replaced by a function space of maps from $\mathbf{R}$ to $\mathbf{C}$, or to $\mathbf{R}$, respectively.

The infinite volume limit is standard in the study of quantum field models, since one wishes to ensure action of the Lorentz (or Poincaré) group on the fields. This necessitates the passage to an infinite space (and space-time) volume.

The most detailed analysis of the infinite volume limit comes from a technique known as the cluster expansion [12]. This method has been successful under two conditions:

(C1). The ground state of the finite volume model is unique.

(C2). The gap $m = m(\ell)$ in energy between the ground state and the rest of the spectrum of H is bounded away from zero as $\ell \to \infty$.

In addition, a third condition of a technical nature has been imposed in the examples studied. This condition plays the role of assuring (C1) and (C2), as described below.

(C3). The model can be parameterized as a small perturbation of a free field or of a direct sum of free fields.

It is possible to use the cluster expansion even in cases where normally one would expect (C1) and (C2) not to be valid, for example, in field theory models with multiple phases (multiple ground states) [13, 14]. In this case, one introduces boundary conditions on a large volume which have the effect of assuring both (C1) and (C2). However, different boundary conditions give rise to different limits $\ell \to \infty$ (the different ground states).

In the supersymmetric situation a new set of issues must be dealt with, in addition to ordinary phase transitions and symmetry breaking. In particular the breaking of supersymmetry may be independent of the other phenomena. Let us discuss these issues within the framework of the $\Lambda\mathbb{C}$ and $\Lambda\mathbb{R}$ models.

Presumably the easiest problem is to develop a cluster expansion for the real, $N = 1$ models with potentials V of even degree. These theories should have unbroken supersymmetry, a uniform mass gap, and a unique vacuum (since we expect that the vanishing theorem holds). Let

$$V_{\substack{even\\odd}}(\varphi) = \frac{1}{2} m\varphi^2 + \lambda P_{\substack{even\\odd}}(\varphi) \quad , \qquad |\lambda| << 1 \quad , \tag{VII.4.1}$$

where $P_{\substack{even\\odd}}$ is a real polynomial of even degree or odd degree respectively.

Conjecture 4A. The cluster expansion can be developed for the models with potentials V_{even}. Supersymmetry is unbroken in the $\ell \to \infty$ limit, the ground state is unique, and the mass gap is positive.

We remark that the corresponding case for P_{odd} is presumably more difficult. In the physics literature, there is a suggestion that these models have a zero mass fermion (a "Goldstino"). In a finite volume we know that H has purely discrete spectrum. (The Hamiltonian has a compact resolvent.) However, the gap in the spectrum is presumably not uniform in ℓ.

The models of the form (VII.4.1) with potential V_{odd} are, because of the small values of λ, again presumably in a classical region. For these models, we conjecture that an infinite volume exists with broken supersymmetry. Here one expects two ground states in a finite volume (Conjecture 4A) which remain degenerate in the infinite volume limit.

Conjecture 4B. The real models with potential (VII.4.1) and V_{odd} have infinite volume limits with a Goldstone fermion. They do not have a mass gap and they have a degenerate vacuum.

Models with large coupling constants presumably behave qualitatively differently. In purely bosonic theories (without supersymmetry) one has regions of parameter space where multiple phases exist. A multiple phase quantum field is characterized by the vacuum being a mixture, rather than a superposition. (A mixture is characterized by a density matrix, rather than a vector as ground state.) A density matrix can also be represented as a vector expectation in a larger Hilbert space, essentially one copy for each vacuum. In this representation, all operators are diagonal. This is a "superselection rule" in the language of physics, and describes a "phase transition."

Consider the case of $P(\varphi)$ bosonic fields with a generic interaction of the form

$$\prod_{i=1}^{n} \frac{1}{2} (\varphi - c_i)^2 + \sum_{i=1}^{n-1} a_i \varphi^i \quad ,$$

where c_i are well separated and the lower degree perturbation $\sum |a_i|$ is sufficiently small. Here Imbrie [17] completely characterized the phase diagram in the space of couplings $\{a_i\}$ as topologically the corner of an $n-1$ cube: There is one point, near $a_i = 0; i = 1, 2, \ldots, n-1$ for which n phases coexist. There are exactly $n-1$

lines with $(n-1)$ phases coexisting, exactly $n-2$ surfaces with $n-2$ phases, etc. This was proved using a cluster expansion.

What happens in the supersymmetric case? We presume that if the zeros of the potential V' are well separated (recall $|V'|^2$ enters the energy) and if the derivatives of V' are large at the zeros, then a cluster expansion can be developed. The question is: how to pick boundary conditions that favor a particular phase?

Conjecture 4C. For well-separated classical minima of the energy with large classical masses, the supersymmetric models have a phase transition. In each phase there is a complete model of the Wightman axioms. A phase diagram like the one for pure bosonic theories can be established.

We note that the existence of a phase transition is completely independent of supersymmetry breaking. Supersymmetry may be broken or it may be unbroken. For example, if we start from a complex, finite volume theory with $\deg V - 1$ ground states and unbroken supersymmetry, we could obtain a theory in infinite volume with one ground state and also with $H = Q^2$, i.e., unbroken supersymmetry. Starting from a real, odd V with broken supersymmetry we can obtain also an infinite volume limit with one vacuum. However, the vacuum must be translation invariant and hence

$$H = \lim_{\ell \to \infty} (H_\ell - E_\ell)$$

where $E_\ell = \inf \text{spectrum } (H_\ell)$, and one expects $E_\ell = O(\ell)$ as $\ell \to \infty$. Thus we do not expect that Q_ℓ (which satisfies $Q_\ell^2 = H_\ell$) has a limit as $\ell \to \infty$.

In infinite volume, one says that supersymmetry is unbroken if a local operator Q exists for which $H = Q^2$, and such that $Q^\Gamma = -Q$.

Finally, let us comment on the critical theory for the $V(\varphi)$ models. In the

$$V(\varphi) = \sum_{j=0}^{n} a_j \varphi^j$$

models, the parameter space is $\mathbf{a} = (a_0, \ldots, a_n)$ where $\mathbf{a} \in \mathbb{C}^{n+1}$ or $(m, \mathbf{a}) \in \mathbb{R}^{n+2}$. The phase transition surfaces are the subsets of parameter space for which the infinite volume theories have degenerate ground states. Of interest is the surface on which exactly $(n-1)$ phases coexist (corresponding to the $(n-1)$ minima of $|\partial V|^2$). This is called the $(n-1)$-surface or point. (The word "point" corresponds to the fact that in the sub-region of the theory without fermions described by Imbrie, there is exactly one point on the $(n-1)$-surface.) The multicritical point is the boundary of the $(n-1)$ surface.

Of special interest are the scaling limits of the multicritical theories. The scaling limits are achieved by passing to the multicritical surface, while at the same time scaling the coupling constants a_j (which have dimension of mass) in such a way to keep a particular mass gap fixed.

It has been conjectured by Zamolodchikov [41] that in the real case, this scaling limit corresponds to the exactly soluble models of Belavin, Polyakov and Zamolodchikov [3], which arise from the $c < 1$ representations of the Virasoro algebra. In the case of the spin $1/2$ Ising model, solved in [29, 31], the S-matrix has a specially simple form. For the supersymmetric models, this is apparently not the case. (We are grateful to S. Shenker for bringing Reference 41 to our attention, and for discussion of his unpublished work on scaling limits.) In any case, the investigation of the supersymmetric models in both the complex and real case, both in the scaling limits (which include a conformally invariant, zero mass point) and in a neighborhood of these limits form a very interesting set of problems.

VII.5. Existence of Multicomponent Models

The loop spaces $\Lambda\mathbb{C}$ and $\Lambda\mathbb{R}$ have natural generalizations to $\Lambda\mathbb{C}^M$ and $\Lambda\mathbb{R}^M$. In this case we wish to construct quantum field models and Dirac operators on loop space which correspond to perturbations of M copies of the corresponding supersymmetric free fields. These perturbations are defined by potentials

$$V\colon \mathbb{C}^M \to \mathbb{C} \qquad \text{and} \qquad V\colon \mathbb{R}^M \to \mathbb{R} \tag{VII.5.1}$$

for the complex and real models, respectively.

The question arises for which class of (e.g., polynomial) potentials V the constructive methods can be implemented. In particular, it is clear that problems can arise if $|V|$ does not increase in all directions at infinity.

A polynomial V of the form (VII.5.1) is *well-behaved* if

$$|\partial_i V| \ , \ i = 1, 2, \ldots, M$$

is unbounded along every generic line to ∞.

Conjecture 5. Every model arising from a well-behaved potential V exists in finite volume.

The index theory of the multicomponent models is clearly more subtle than for the one-component models. In particular, a general vanishing theorem cannot be true in the real case. We expect that a vanishing theorem holds in the complex case if V is a holomorphic polynomial.

VII.6. Sigma Models

Clearly it is interesting to study the loop space

$$\Lambda M = \{\text{maps}: \ S^1 \to M\} \quad ,$$

where M is a Riemannian manifold. The corresponding supersymmetric field theories are supersymmetric σ-models corresponding to the manifold M. In fact such models are very interesting from a topological point of view as has been indicated by Witten [37]. From an analytic point of view, these models appear renormalizable. However, it is well known that a class of supersymmetric σ-models is finite in perturbation theory [1].

One often wants a regularization of σ-models preserving their good properties. A supersymmetric regularization is not known outside perturbation theory. We propose that a useful regularization can be obtained by imbedding the manifold M in a Euclidean space $\mathbf{R}^m$. Then we take our polynomial model on $\mathbf{R}^m$ with a potential which localizes the field near the manifold M.

Let us consider a very simple example, a σ-model on the $(m-1)$-sphere $\varphi_1^2 + \varphi_2^2 + \ldots + \varphi_m^2 = 1$. Let us choose $r^2 = \sum_{i=1}^m \varphi_i^2$. Clearly one cannot choose a polynomial function $V(\varphi)$ which localizes φ near $r = 1$. Such a polynomial would necessarily satisfy $V(\varphi) = g(r)$ and

$$|\nabla V|^2 = \left(\frac{dg}{dr}\right)^2 = (r^2-1)^2 \cdot (\text{Poly}\,(r))^2$$

so that $g(r)$ is a polynomial in r of odd degree, and is not polynomial in φ. On the other hand

$$g(r) = (r^2-1)^2 \quad , \qquad V(\varphi) = (\varphi^2-1)^2 \tag{VII.6.1}$$

localizes φ in the neighborhood of $M \cup \{0\}$, as

$$|\nabla V|^2 = 4\varphi^2(\varphi^2-1)^2 \quad .$$

This is a "φ^6 model." Scaling V by λ and taking the limit $\lambda \to \infty$ we have

Conjecture 6. The $\lambda \to \infty$ of the model with potential λV, where V is given by (VII.6.1), gives the σ-model for the m-sphere. The probability that φ lies at zero decouples in this limit. The model (VII.6.1) provides (up to Wick ordering) a finite supersymmetric regularization of the σ-model.

Clearly the index theory and vanishing theorems have to be reinvestigated for this $\lambda \to \infty$ limit. The topology of M will determine whether a vanishing theorem holds. ($n_\pm$ are given in the quantum mechanics case as sums of Betti numbers of M and the index is the Euler-Poincaré characteristic of M [10], [37]). Thus we expect the index theory of the finite λ models to reflect these complications of the limit. It is clearly interesting to investigate these approximations.

Other interesting directions have been recently suggested in [39].

VII.7. Cohomology of Infinite Dimensional Manifolds

The models of Dirac operators on loop space give rise to an interesting cohomology theory. They fit into the framework of Connes' noncommutative differential geometry, with the p-summability condition replaced by a weaker condition: for any $\beta > 0$,

$$\mathrm{Tr}\,(e^{-\beta H}) \; < \; \infty \; . \tag{VII.7.1}$$

The framework of cyclic cohomology of Connes [7], see also Loday and Quillen [28], can be given concrete realization through these examples. In principle, the sequence of cohomology groups has an infinite support and thus is inherently infinite dimensional. Thus we should find an infinite sequence of invariants associated with these models, the simplest invariant being the index. We are presently working in this direction [20].

REFERENCES

1. Alvarez-Gaumé, L. and Ginsparg, P., Finiteness of Ricci Flat Supersymmetric Nonlinear σ-Models, *Commun. Math. Phys.* **102**, 311–326 (1985).

2. Atiyah, M. and Singer, I., The Index of Elliptic Operators. III, *Ann. Math.* **87**, 546–604 (1968).

3. Belavin, A. A., Polyakov, A. M., and Zamolodchikov, A. B., Infinite Conformal Symmetry in Two-Dimensional Quantum Field Theory, *Nucl. Phys* **B241**, 333–380 (1984).

4. Berezin, F. A., *The Method of Second Quantization*, New York, Academic Press (1966).

5. Cecotti, S. and Girardello, L., Functional Measure, Topology and Dynamical Supersymmetry Breaking, *Phys. Lett.* **110B**, 39–43 (1982).

6. Cecotti, S. and Girardello, L., Stochastic and Parastochastic Aspects of Supersymmetric Functional Measures: A New Nonperturbative Approach to Supersymmetry, *Ann. Phys.* **143**, 81–89 (1983).

7. Connes, A., Noncommutative Differential Geometry, Publ. *Math. I.M.E.S.* **62**, 41–44 (1986).

8. Getzler, E., The Degree of the Nicolai Map in Supersymmetric Quantum Mechanics, *J. Funct. Anal.* **73** (1987).

9. Gilkey, P., *Invariance Theory, the Heat Equation and the Atiyah-Singer Index Theorem*, Wilmington, Del.: Publish or Perish (1984).

10. Girardello, L., Imbimbo, C., and Mukhi, S., On Constant Configurations and the Evaluation of the Witten Index, *Phys. Lett.* **132B**, 69–74 (1983).

11. Glimm, J. and Jaffe, A., Singular Perturbations of Self-Adjoint Operators, *Comm. Pure Appl. Math.* **22**, 401–414 (1963).

12. Glimm, J. and Jaffe, A., *Quantum Physics*, New York, Berlin, Heidelberg: Springer Verlag (1987).

13. Glimm, J., Jaffe, A., and Spencer, T., Phase Transitions for φ_2^4 Quantum Fields, *Commun. Math. Phys.* **45**, 203–216 (1975).

14. Glimm, J., Jaffe, A., and Spencer, T., A Convergent Expansion About Mean Field Theory, Parts I and II, *Ann. Phys.* (N.Y.) **101**, 610–630 and 631–669 (1976).

15. Green, M., Schwarz, J., and Witten, E., *Superstring Theory, Vol. I and II*, Cambridge University Press, Cambridge (1987).

16. Gross, L., On the Formula of Mathews and Salam, *J. Funct. Anal.* **25**, 162–209 (1977).

17. Imbrie, J., Phase Diagrams and Cluster Expansions for Low Temperature $P(\phi)_2$ Models, Parts I and II. *Comm. Math. Phys.* **82**, 261–343 (1981).

18. Jaffe, A. Heat Kernel Regularization and Infinite Dimensional Analysis, in *Constructive Quantum Field Theory*, Canadian Mathematical Proceedings, J. Feldman and L. Rosen, eds., Amer. Math. Soc., Providence, 1988.

19. Jaffe, A. and Lesniewski, A., *A priori* Estimates for $N = 2$ Wess-Zumino Models on a Cylinder, *Comm. Math. Phys.*, to appear.

20. Jaffe, A. and Lesniewski, A., Quantum K-Theory, in preparation.

21. Jaffe, A., Lesniewski, A., and Lewenstein, M., Ground State Structure in Supersymmetric Quantum Mechanics, *Ann. Phys.* **178**, 313–329 (1987).

22. Jaffe, A., Lesniewski, A., Osterwalder, K., On Convergence of Inverse Functions of Operators, *J. Funct. Anal.*, to appear.

23. Jaffe, A., Lesniewski, A., and Weitsman, J., Index of a Family of Dirac Operators on Loop Space, *Comm. Math. Phys.* **112**, 75–88 (1987).

24. Jaffe, A., Lesniewski, A., and Weitsman, J., The Two-Dimensional, $N = 2$ Wess-Zumino Model on a Cylinder, *Comm. Math. Phys.*, to appear.

25. Jaffe, A., Lesniewski, A., and Weitsman, J., Pfaffians on Hilbert Space, *J. Funct. Anal.*, to appear.

26. Jaffe, A., Lesniewski, A., and Weitsman, J., The Loop Space $S^1 \to \mathbf{R}$ and Supersymmetric Quantum Fields, preprint **HUTMP B-212**.

27. Kato, T., *Perturbation Theory for Linear Operators*, Berlin, Heidelberg, New York: Springer (1984).

28. Loday, J.-L. and Quillen, D., Cyclic Homology and the Lie Algebra Homology of Matrices, *Comment. Math. Helv.* **59**, 565–591 (1984).

29. McCoy, B., Tracy, C., and Wu, T. T., Two-Dimensional Ising Model as an Explicitly Solvable Relativistic Field Theory: Explicit Formulas for n-Point Functions, *Phys. Rev. Lett.* **38**, 793–796 (1977).

30. Pressley, A. and Segal, G., *Loop Groups*, Oxford, Claredon Press (1986).

31. Sato, M., Miwa, T., and Jimbo, M., Holomorphic Quantum Fields I. Publ. *Res. Inst. Math. Sci.*, Kyoto University **14**, 223–267 (1978).

32. Seiler, E., Schwinger Functions for the Yukawa Model in Two Dimensions with Space-Time Cutoff, *Comm. Math. Phys.* **42**, 163–182 (1975).

33. Seiler, E. and Simon, B., Nelson's Symmetry and All That in the Yukawa_2 and φ_3^4 Field Theories, *Ann. Phys.* **97**, 470–518 (1976).

34. Shenker, S. (private communication).

35. Wess, J. and Bagger, J., *Supersymmetry and Supergravity*, Princeton, Princeton University Press (1983).

36. Witten, E., Dynamical Breaking of Supersymmetry, *Nucl. Phys.* **B188**, 513–554 (1981).

37. Witten, E., Constraints on Supersymmetry Breaking, *Nucl. Phys.* **B202**, 253–316 (1982).

38. Witten, E., Supersymmetry and Morse Theory, *J. Diff. Geom.* **17**, 661–692 (1982).

39. Witten, E., Elliptic Genera and Quantum Field Theory, *Comm. Math. Phys.* **109**, 525–536 (1987).

40. Witten, E., Quantum Field Theory, Grassmannian and Algebraic Curves, *Comm. Math. Phys.*, to appear.

41. Zamolodchikov, A. B., Conformal Symmetry and Multicritical Points in Two-Dimensional Quantum Field Theory, *Soviet J. Nucl. Phys.* **44**, 529–533 (1986).

SOLUTION OF THE LATTICE ϕ^4 THEORY IN 4 DIMENSIONS [1)]

M. Lüscher

DESY

Hamburg, Federal Republic of Germany

ABSTRACT

Some recent analytical and numerical studies of the one-component ϕ^4 theory on a 4-dimensional hypercubic lattice are reviewed. Taken together, the results obtained provide a complete solution of the model in the sense that most low energy amplitudes can be calculated with reasonable accuracy in those parts of the phase diagram, where the ultra-violet cutoff Λ satisfies $\Lambda \geqslant 2m$ ($\Lambda = 1/a$, a: lattice spacing, m: physical particle mass). Further topics discussed include the issue of "triviality" and a possible upper bound on the Higgs meson mass.

1. INTRODUCTION

Although the one-component ϕ^4 theory has so far not found any direct application in elementary particle physics, it has been used for many years as a guinea-pig to test and develop new ideas in quantum field theory. Among today's motivations to study the lattice regularized ϕ^4 theory are the following.

1) Lectures given at the Nato Advanced Study Institute on "Non-Perturbative Quantum Field Theory", Cargèse (1987)

(a) There is overwhelming evidence /1-17/ that this model is "trivial" in 4 dimensions, i.e. that its continuum limit is a free field theory. As I will explain later (sect. 2), "trivial" field theories can nevertheless serve as accurate mathematical models for interacting elementary particles. However, "triviality" implies an upper bound on the interaction strength and one of the questions one would like to answer is, where exactly this bound lies and whether a non-perturbative (strong interaction) sector is excluded, in particular.

(b) In the limit of vanishing gauge coupling, the SU(2) Higgs model (which is an important part of the standard electro-weak theory) reduces to three copies of Maxwell fields and the 4-component ϕ^4 theory. By studying the latter, one thus hopes to get some insight into how the Higgs model behaves, especially when the scalar self-coupling is large and perturbation theory is not reliable. In particular, it is possible, at least for small gauge coupling, that the "triviality" of the scalar sector implies the "triviality" of the full Higgs model /18,19/, and this would then give rise to an upper bound on the Higgs meson mass /20-31/.

(c) Because of its simplicity, the ϕ^4 theory is an ideal laboratory to test improved numerical simulation algorithms /8,32,50-52/, to learn how the systematical errors in these calculations can be controlled and to develop new methods to extract the more elusive quantities of physical interest (such as scattering amplitudes) from the numerical data /33/. To a large extent, the present excitement in this field is due to the fact that <u>accurate</u> numerical simulations are feasible with the available computer power and, as we shall see, that detailed analytical "predictions" exist, which can be immediately compared with the "experimental" results.

In these lectures, I would first like to expand a little on points (a) and (b) above and I will then proceed to explain in outline how the one-component ϕ^4 theory in the symmetric phase can be solved analytically /34/. Of course, by a solution I do not mean that an exact and explicit formula for (say) the scattering matrix can be given, but that most low energy quantities can be calculated with respectable accuracy by combining renormalized perturbation theory with data obtained from the "high temperature" expansion. It is important that these expansions are only used in regions of the parameter space where they really apply, i.e. no analytic extrapolations are performed and an effort is made to estimate the systematic errors which arise when truncating the expansions at a finite order.

The analytic solution of the ϕ^4 theory can be extended to the broken symmetry phase of the model /35/, but before explaining how this goes (sect. 6), I shall review the numerical work of Montvay and Weisz /33/ on the 4-dimensional Ising model, which is a limiting case of the ϕ^4 theory. Their results agree very well with the analytic solution. In addition, they have made a detailed finite size analysis, which enabled them, for the first time in a numerical simulation, to determine a scattering matrix element (the S-wave scattering length). The conclusion from this beautiful "experiment" is that within errors the analytic solution of the ϕ^4 theory in the symmetric phase is correct and that a complete quantitative understanding of the model has hence been achieved. Simulations in the broken symmetry phase are already on the way and hopefully result in a similar confirmation of the analytic solution.

2. THE MEANING OF "TRIVIALITY"

The action of the lattice ϕ^4 theory may be written in the form

$$S = a^4 \sum_x \left\{ \frac{1}{2}(\partial_\mu \phi_0)^2 + \frac{1}{2} m_0^2 \phi_0^2 + \frac{g_0}{4!} \phi_0^4 \right\}, \tag{2.1}$$

where "a" denotes the lattice spacing, $\phi_0(x)$ $(x/a \in \mathbb{Z}^4)$ is a real scalar field and $\partial_\mu \phi_0$ the nearest neighbor lattice derivative of ϕ_0. For stability we require $g_0 \geqslant 0$ and we also assume that the bare mass parameter m_0^2 is in the range where the reflection symmetry $\phi_0 \to -\phi_0$ is not spontaneously broken (the discussion below is however equally valid in the broken symmetry phase).

Let us now define a wave function renormalization constant Z_R, a renormalized mass m_R and a renormalized coupling g_R through

$$(2.2)\quad \Gamma^{(2)}(p,-p) = -Z_R^{-1}\{m_R^2 + p^2 + O(p^4)\} \quad (p \to 0),$$

$$(2.3)\quad \Gamma^{(4)}(0,0,0,0) = -Z_R^{-2}\, g_R,$$

where $\Gamma^{(n)}(p_1,\dots,p_n)$ denotes the n-point vertex function of ϕ_o. The renormalized parameters m_R, g_R are well-defined functions of a, m_o^2 and g_o, which (by dimensional analysis) are of the form

$$(2.4)\quad m_R = \frac{1}{a}\, r(a^2m_o^2, g_o),$$

$$(2.5)\quad g_R = s(a^2m_o^2, g_o).$$

Using Lebowitz' inequality, one may show that $g_R \geq 0$ and, by definition, we also have $m_R > 0$ throughout the symmetric phase region.

If it exists at all, the continuum limit of the lattice theory is obtained by fixing m_R, g_R and sending the cutoff mass $\Lambda = 1/a$ to infinity. This assumes, in particular, that for given m_R, g_R and arbitrarily large Λ, bare parameters $m_o^2(\Lambda)$, $g_o(\Lambda)$ exist such that eqs. (2.4), (2.5) hold. In a "trivial" theory, this precondition is only fulfilled if $g_R = 0$. In other words, for all $g_R > 0$, eqs. (2.4), (2.5) imply an upper bound on the cutoff Λ of the form

$$(2.6)\quad \ln(\Lambda/m_R) \leq f(g_R),$$

where $f(g_R)$ is continuous and

$$(2.7)\quad \lim_{g_R \to 0} f(g_R) = \infty.$$

Thus, if one insists on taking the cutoff to infinity, one also has to scale g_R to zero so that in the end one is left with a free field theory.

The lattice ϕ^4 theory is most likely trivial /1-17/, but a completely rigorous proof of triviality is still missing. The solution of the one-component model, which I shall discuss later, also implies triviality and moreover yields an estimate for the function $f(g_R)$, which enters the triviality bound (2.6).

An obvious question is, whether a trivial theory is necessarily useless for the description of interacting elementary particles. The answer is definitely no here, because the bound (2.6) is often not very restrictive from a practical point of view. For example, in case of the ϕ^4 theory, we shall see that

$$(2.8) \qquad f(g_R) \underset{g_R \to 0}{\sim} 16\pi^2/3g_R,$$

and for $g_R = 1$ (which is sufficiently small for (2.8) to apply), the triviality bound hence becomes

$$(2.9) \qquad \Lambda/m_R \leq 7\cdot 10^{22}.$$

Thus, even for reasonably large couplings, the cutoff Λ can be pushed to very high values which may be orders of magnitude beyond the experimentally accessible energy region. In such an instance, the presence of the cutoff has no practical relevance, i.e. at low energies E, the theory behaves effectively like a continuum theory. Of course, cutoff effects are not totally absent, but since they are of order E^2/Λ^2 /36/, they are usually completely negligible.

Still, a trivial theory can only be a valid description of elementary particles and their interactions up to some finite energy scale and thus cannot by itself be a fundamental theory. It is however conceivable that trivial theories arise by integrating out the high energy degrees of freedom of an underlying ultra-violet stable theory. In that case, the triviality bound (2.6) provides an upper bound on the energy scale where "new physics" has to set in.

3. THE ϕ^4 THEORY AS A LIMIT OF THE SU(2) HIGGS MODEL

The Higgs sector of the standard model of electro-weak interactions is described by the (euclidean) action

$$S = S_G + S_H, \tag{3.1}$$

$$S_G = \int d^4x \; \frac{1}{4} W^a_{\mu\nu} W^a_{\mu\nu}, \tag{3.2}$$

$$S_H = \int d^4x \; \{ D_\mu \phi^\dagger \cdot D_\mu \phi + \frac{\lambda}{2} (\phi^\dagger \cdot \phi - \frac{v^2}{2})^2 \}, \tag{3.3}$$

where ϕ is an SU(2) doublet and

$$W^a_{\mu\nu} = \partial_\mu W^a_\nu - \partial_\nu W^a_\mu + g \epsilon^{abc} W^b_\mu W^c_\nu, \tag{3.4}$$

$$D_\mu \phi = (\partial_\mu + g W^a_\mu \frac{\sigma^a}{2i}) \phi \tag{3.5}$$

(σ^a are the Pauli matrices and the indices a,b,c,... run from 1 to 3). For convenience, I here use a continuum notation, but everything what follows, with obvious modifications, also applies to the standard lattice version of the model (e.g. ref. /28/).

In the Higgs phase, i.e. for large positive v^2, the model describes a triplet of heavy vector bosons ("W bosons") and a neutral scalar particle (the "Higgs boson"). At tree level of perturbation theory, the masses of these particles are

$$m_W = \frac{1}{2} g v, \tag{3.6}$$

$$m_H = \sqrt{\lambda}\, v. \tag{3.7}$$

The physical values of g and v are approximately given by

(3.8) $g \simeq 0.65,$

(3.9) $v \simeq 250$ GeV.

The Higgs self-coupling λ, on the other hand, proved to be a very elusive parameter so that today its value is essentially unknown (experimental bounds on the Higgs meson mass are given in ref. /37/, for example).

It is conceivable that λ is in fact quite large. In this case, the Higgs particle would be heavy, perhaps $m_H \simeq 1$ TeV, and the perturbation expansion in powers of λ would become unreliable. Thus, non-perturbative methods are required to determine the properties of the Higgs model in this situation and one obvious possibility then is to apply the numerical simulation technique to the latticised model (see /38,39/ for reviews and /28-30/ for recent papers in this field). These simulations are done with the complete model including all fields and interactions as listed at the beginning of this section. They are therefore rather complicated and it is not easy to obtain solid results in a short time.

At this point, it is useful to note that the gauge coupling g is actually rather small (the relevant expansion parameter is $g^2/4\pi \simeq 1/30$). Thus, as has been proposed by Dashen and Neuberger some time ago /25/, the solution of the Higgs model at large λ may be attempted by first expanding in powers of g at fixed λ, v and then evaluating the coefficients in this expansion by numerical simulation or any other non-perturbative method.

To lowest order in g, the gauge field W^a_μ and the Higgs field ϕ decouple. Furthermore, the gauge action (3.2) reduces to the action for a triplet of non-interacting Maxwell fields and the Higgs action (3.3) becomes the action of an O(4) symmetric φ^4 theory:

(3.10) $$S_H = \int d^4x \, \{ \tfrac{1}{2} \partial_\mu \varphi \cdot \partial_\mu \varphi + \tfrac{\lambda}{8} (\varphi \cdot \varphi - v^2)^2 \},$$

(3.11) $$\phi = \frac{1}{\sqrt{2}} \begin{pmatrix} \varphi_2 + i\,\varphi_1 \\ \varphi_4 - i\,\varphi_3 \end{pmatrix}, \qquad \varphi_\alpha \text{ real.}$$

Since the limit $g \to 0$ is taken in the Higgs phase, the parameters in (3.10) are such that the O(4) symmetry is spontaneously broken. The associated Goldstone bosons are the former W bosons with a longitudinal spin polarization (the transversely polarized W bosons become the "photons", which are described by the gauge action).

The Higgs particle corresponds to a radial excitation of the scalar field and remains massive for $g = 0$. However, since it can decay into any even number of Goldstone bosons, it is actually a resonance with a decay width given by

(3.12) $$\Gamma_H/m_H = 3\lambda/32\pi + O(\lambda^2).$$

Thus, for large λ the Higgs particle is presumably a broad resonance.

Besides the Higgs mass m_H, there is another physical scale F in the φ^4 theory, which is associated with the dynamics of the Goldstone bosons. Suppose the vacuum expectation value of φ_α is in the 4-direction and let

(3.13) $$A^a_\mu = \varphi_4\,\partial_\mu\varphi_a - \varphi_a\,\partial_\mu\varphi_4, \qquad a = 1,2,3,$$

be the conserved currents, which generate the spontaneously broken symmetries. Then, F is defined by the matrix element

(3.14) $$\langle 0|\,A^a_\mu(0)\,|p,b\rangle = i p_\mu\,\delta^{ab}\,F,$$

where $|p,b\rangle$ denotes the state of a single Goldstone boson with momentum p and symmetry label b. The normalizations are such that

(3.15) $$\langle q,a|p,b\rangle = 2|\vec{q}|(2\pi)^3\,\delta^{ab}\,\delta(\vec{q}-\vec{p})$$

and $p_\mu = (i|\vec{p}|,\,\vec{p})$ (the time derivative in eq. (3.13) is with respect to euclidean time). Eq. (3.14) defines F non-perturbatively and there is also no normalization ambiguity, because the normalization of the currents A^a_μ is fixed by the associated Ward identities. Incidentally, by a simple application of these identities, it is possible to show that /40/

(3.16) $$F = \langle 0|\varphi_4|0\rangle,$$

provided φ_α is renormalized in such a way that the Goldstone pole in the two-point function of φ_a has unit residue. In particular, $F = v + O(\lambda)$.

So far I have discussed what happens at g = 0. If the gauge coupling is now switched on again, the most important effect is that the gauge bosons and the Goldstone bosons become massive and combine to form the W vector bosons as indicated above. To first order in g, the vector boson mass is proportional to g and one may actually show that /25/

(3.17) $$m_W^2 = \frac{1}{4} g^2F^2 + O(g^4 \ln g^2).$$

The proof of this nice formula is based solely on the O(4) Ward identities at g = 0 and it is therefore an exact result valid for all values of v^2 and λ. It also holds literally on the lattice (the lattice artefacts only show up at order g^4). Essentially, eq. (3.17) should be considered a form of the Goldstone theorem.

Closed expressions to first order in g^2 could perhaps also be derived for other physical quantities such as the WW scattering amplitude, but I would now like to proceed to discuss another issue, which is how triviality gives rise to an upper bound on the Higgs meson mass.

In view of eq. (3.7), a possible definition of a renormalized Higgs self-coupling λ_R at g = 0 is

(3.18) $$\lambda_R = m_H^2/F^2.$$

The triviality bound (2.6) for the (lattice regularized) φ^4 theory with action (3.10) then reads

(3.19) $\ln(\Lambda/m_H) \leq f(m_H^2/F^2)$.

At least for small λ_R and presumably in the whole range of λ_R, the function $f(\lambda_R)$ is monotonically increasing when λ_R is made smaller so that (3.19) may be rewritten in the form

(3.20) $m_H^2/F^2 \leq f^{-1}(\ln(\Lambda/m_H))$.

Finally, using eq. (3.17) to eliminate the scale F, one obtains

(3.21) $m_H^2/m_W^2 \leq \frac{4}{g^2} f^{-1}(\ln(\Lambda/m_H))$.

Since g and m_W are measured, eq. (3.21) provides an upper bound on the Higgs mass if we require that Λ is greater than (say) $2m_H$ (for lower values of Λ, the low energy properties of the Higgs model would be strongly influenced by non-universal cutoff effects). Of course, it may also be sensible to require that Λ is beyond the Planck scale or some other huge mass, in which case the bound (3.21) would be more stringent.

To extract actual numbers from eq. (3.21), one needs the function $f(\lambda_R)$, which is defined in the (pure) φ^4 theory with action (3.10). Unfortunately, only the asymptotic form of $f(\lambda_R)$ for $\lambda_R \rightarrow 0$ is known presently, but there is little doubt that $f(\lambda_R)$ will soon be determined in the full range by the analytic method, which I shall explain later for the one-component model, and by numerical simulations (see /31,41/ for first attempts in this direction). Finally, I would like to remark that in the derivation of the bound (3.21), we have neglected the correction term in eq. (3.17) and, of course, we have also discarded the influence of the fermions and the other fields in the standard model, which are not included in the Higgs action (3.1)-(3.3).

4. SOLUTION OF THE ONE-COMPONENT MODEL IN THE SYMMETRIC PHASE

I now sketch how the lattice ϕ^4 theory defined in sect. 2 can be solved analytically in the symmetric phase region. A more detailed discussion is given in ref. /34/.

For what follows, it is convenient to rewrite the action (2.1) in the form

$$S = \sum_x \{ -\varkappa \sum_{\mu=0}^{3} (\phi(x)\phi(x+\hat{\mu}) + \phi(x)\phi(x-\hat{\mu})) + \phi(x)^2 + \lambda (\phi(x)^2 - 1)^2 \}, \tag{4.1}$$

where $\varkappa \geq 0$, $0 \leq \lambda \leq \infty$ and the lattice spacing "a" has been set equal to one for convenience, i.e. I shall use lattice units from now on. The relation between the old and the new notation is

$$\phi_0 = \sqrt{2\varkappa}\, \phi, \tag{4.2}$$

$$m_0^2 = (1-2\lambda)/\varkappa - 8, \tag{4.3}$$

$$g_0 = 6\lambda / \varkappa^2. \tag{4.4}$$

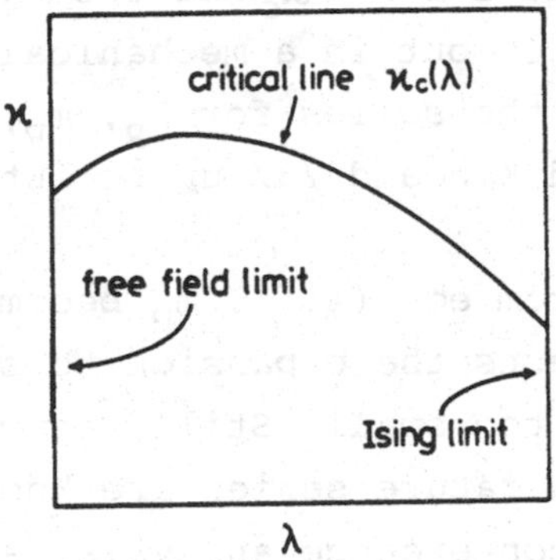

Fig. 1. Qualitative plot of the phase diagram of the lattice model with action (4.1). For $\lambda \to \infty$, the theory reduces to the Ising model.

The phase diagram of the model (4.1) is displayed in Fig. 1. There are two phases separated by a second order critical line $\varkappa = \varkappa_c(\lambda)$. Here we are interested in the region $\varkappa < \varkappa_c(\lambda)$, which corresponds to an unbroken reflection symmetry $\phi \to -\phi$.

From the action (4.1) one derives in the usual way the correlation functions $\langle\phi(x_1)\ldots\phi(x_n)\rangle$ and the n-point vertex functions $\Gamma^{(n)}(p_1,\ldots,p_n)$ in momentum space. Suppose now we define Z_R, m_R, g_R as before through eqs. (2.2), (2.3). The immediate goal in what follows then is, to calculate these quantities as a function of the bare parameters $\varkappa$ and λ. As we shall see later, the solution of this problem also leads to a reasonably accurate determination of the low energy properties of the model, at least in the region $\Lambda \geqslant 2m_R$ ($\Lambda = 1$ in lattice units).

For $\varkappa = 0$, the field variables at different points of the lattice decouple and the model becomes soluble. Relying on this fact, it is easy to derive an expansion of Z_R, m_R, g_R in powers of $\varkappa$, e.g. for m_R we have

$$m_R = \frac{1}{\sqrt{\varkappa}} \sum_{\nu=0}^{\infty} m_R^{(\nu)}(\lambda)\, \varkappa^{\nu}. \tag{4.5}$$

This expansion has been known for a long time in statistical mechanics, where it is called the "high temperature expansion". It is convergent for $\varkappa < \varkappa_c$ and the expansion coefficients can be worked out in a mechanical way to a high order. In particular, the series for Z_R, m_R, g_R have been tabulated by Baker and Kincaid /3/ up to 10th order.

As one can see from eq. (4.5), m_R becomes large for $\varkappa \to 0$ so that one expects the expansion to be practically useful when m_R is not too small. Still, since the first 10 terms in the high temperature series are known, it is possible to perform a careful convergence analysis, and one then finds that the truncation error stays reasonably small up to $\varkappa = 0.95\,\varkappa_c$ which corresponds to $m_R \simeq 0.5$ (estimates for $\varkappa_c$ are given in ref. /34/). Some results at $\varkappa = 0.95\,\varkappa_c$ obtained in this way are listed in Table 1. The data show that Z_R is surprisingly close to $1/2\varkappa$, which is the lowest order term in the weak coupling perturbation expansion of this quantity. The renormalized coupling g_R is monotonically rising with λ and reaches a maximal value of about 41 at $\lambda = \infty$. This is actually not such a big value, at least, it is only about 2/3 of the tree level unitarity bound and renormalized perturbation theory should in general still be applicable at these values of the coupling (the "natural" expansion parameter in perturbation theory is $\alpha_R = g_R/16\pi^2$).

With the help of the high temperature expansion we have thus been able to solve the theory in the region $\kappa \lesssim 0.95\,\kappa_c$, which corresponds approximately to $\Lambda/m_R \lesssim 2$ (see Fig. 2). To get closer to the critical line, i.e. closer to the continuum limit, we shall use the renormalization group equations. One of these equations is usually written as

$$(4.6)\qquad -\Lambda\left(\frac{\partial g_R}{\partial \Lambda}\right)_\lambda = \beta,$$

where β is the Callan-Symanzik β-function. In lattice units, $\Lambda = 1$ by definition and the proper form of eq. (4.6) then is

$$(4.7)\qquad m_R\left(\frac{\partial g_R}{\partial m_R}\right)_\lambda = \beta.$$

Table 1. Values of Z_R, m_R, g_R as a function of λ at $\kappa = 0.95\,\kappa_c$ as calculated from the high temperature expansion

λ	κ	$2\kappa Z_R$	m_R	g_R
0.00	0.1188	1.0	0.649	0.0
0.01	0.1206	1.0000(2)	0.639(1)	3.57(5)
0.10	0.1298	0.9990(5)	0.599(6)	16(1)
1.00	0.1267	0.990(2)	0.54(1)	34(4)
∞	0.0710	0.973(4)	0.49(1)	41(6)

This equation describes the evolution of g_R as one moves towards the critical line at fixed λ. Similar equations exist for Z_R and κ. Thus, if we knew the β function (and the other Callan-Symanzik coefficients), we could easily calculate Z_R, m_R, g_R at (say) point B of Fig. 2 by integrating the renormalization group equations using the known values of these quantities at point A as initial data.

The crucial observation now is, that as we have noted above, the coupling g_R is already in the perturbative domain along the line $\varkappa = 0.95\,\varkappa_c$ where the integration of the renormalization group equations is started. Thus, we may employ renormalized perturbation theory to calculate $\beta(g_R)$, at least during the initial steps of the integration. In fact, since β is positive, eq. (4.7) drives g_R to smaller values as m_R decreases and perturbation theory hence becomes an ever better approximation the closer one is to the critical line. Thus, in this way it is possible to compute Z_R, m_R, g_R as a function of $\varkappa, \lambda$ everywhere in the white area below the critical line in Fig. 2.

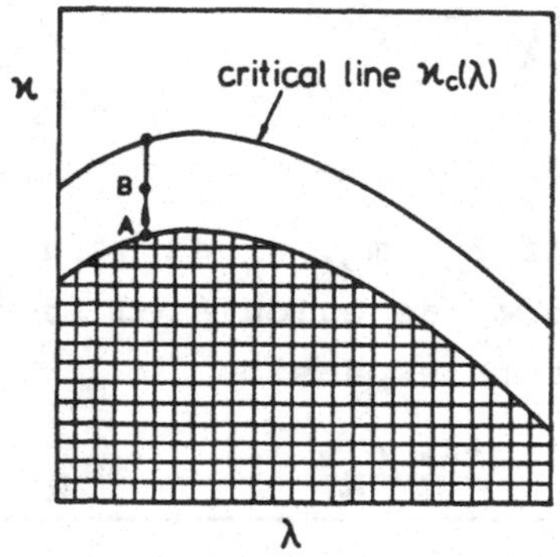

Fig. 2. Same as Fig. 1, but showing the region where the high temperature expansion applies (cross-hatched area)

For illustration, some results obtained by integrating the renormalization group equations are listed in Table 2. The errors quoted derive from the errors in the initial data at $\varkappa = 0.95\,\varkappa_c$. As can be seen from Table 1, the errors are maximal for $\lambda = \infty$, in particular, the estimated accuracy in g_R is never worse than 15 %. The data in Table 2 smoothly join the high temperature curves at the matching point $\varkappa = 0.95\,\varkappa_c$

Table 2. Values of $\varkappa$, Z_R, g_R as a function of m_R at $\lambda = \infty$ (Ising model)

m_R	$\varkappa$	$2\varkappa Z_R$	g_R
0.40	0.0722(2)	0.973(7)	35(5)
0.20	0.0741(4)	0.975(9)	24(2)
0.10	0.0746(4)	0.976(9)	18(1)
0.05	0.0747(4)	0.974(9)	15.0(8)
0.01	0.0748(4)	0.972(9)	10.6(4)

and they also agree well with the Monte Carlo data of ref. /33/ (cf. sect. 5).

In the limit $\Lambda/m_R \to \infty$, λ fixed, the coupling g_R eventually goes to zero according to the implicit asymptotic formula

$$m_R/\Lambda = C_1 (\beta_1 g_R)^{17/27} e^{-1/\beta_1 g_R} \{1 + O(g_R)\}, \tag{4.8}$$

where C_1 is a constant (depending on λ) and $\beta_1 = 3/16\pi^2$ is the one-loop coefficient of the β function. Eq. (4.8) is just the asymptotic form of the general solution of the renormalization group equation (4.7), which one obtains when the initial value of g_R is sufficiently close to the origin (which we have argued to be the case). The triviality of the ϕ^4 theory is essentially a consequence of the scaling law (4.8). This can be seen more clearly from Fig. 3, where I have plotted the curves of constant g_R in the $\varkappa,\lambda$-plane.[2)] Along these curves, only the cutoff Λ (in units of m_R) changes while the low energy physics is fixed. Now it turns out that the maximal value of Λ/m_R is attained in the Ising limit ($\lambda = \infty$) where the curves end, and the triviality bound (2.6) is thus given by

2) I shall later explain how to obtain the curves in the broken symmetry phase $\varkappa > \varkappa_c$.

(4.9) $$\ln(\Lambda/m_R) \leq \frac{1}{\beta_1 g_R} - \frac{17}{27}\ln(\beta_1 g_R) + C + O(g_R),$$

where

(4.10) $$C = -\ln C_1(\infty) = -1.5(2).$$

The correction terms in eq. (4.9) are negligible for $g_R \leq 10$ and for the larger values of g_R, the triviality bound can be read off from Table 2.

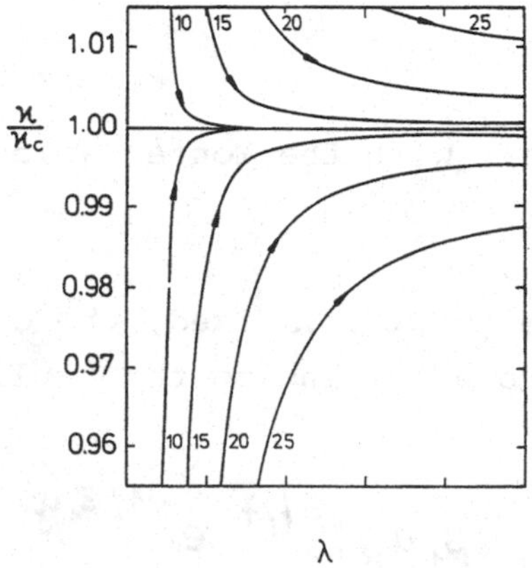

Fig. 3. Curves of constant coupling $g_R = 10,15,20,25$ in the plane of bare parameters. The arrows are in the direction of increasing cutoff. All curves end at the Ising line (for $g_R = 10$, the distance to the critical line is so small that it cannot be resolved in this drawing)

An important result of the discussion so far is that the coupling g_R is always less than about 2/3 of the tree level unitarity bound when $\Lambda \geqslant 2m_R$. In other words, whenever the cutoff is sufficiently high for the theory to have essentially cutoff independent low energy properties, the coupling is necessarily small and renormalized perturbation theory should be applicable. In particular, the scattering matrix for low energy processes can be computed in this way to a respectable accuracy and a complete solution of the ϕ^4 theory in the symmetric phase has thus been achieved.

5. NUMERICAL SIMULATION OF THE ISING MODEL

For the analytical solution of the ϕ^4 theory, the Ising limit is the most difficult case, because g_R assumes its largest values there. On the other hand, numerical simulations of the Ising model are relatively easy and a significant comparison between "theory" and "experiment" is hence feasible.

The numerical work, which I am now going to describe, has been done by Montvay and Weisz /33/. They chose two values of $\varkappa$ in the symmetric phase, approximately corresponding to $m_R=0.5$ and $m_R=0.2$. The lattices considered were of the form $L^3 \times T$ with (ordinary) periodic boundary conditions in all directions and

$$T = 12, \quad L = 4,6,\ldots,12 \quad \text{for } m_R \simeq 0.5,$$
(5.1)
$$T = 24, \quad L = 8,10,\ldots,20 \quad \text{for } m_R \simeq 0.2$$

(T is time, L is space). On each lattice, several million field configurations were generated using a standard Metropolis algorithm. The reason for having L vary over a range of values is that in this way a detailed finite size analysis is possible, as I will explain later. In particular, the results quoted below are, within errors, infinite volume values.

It is not easy to determine m_R in a Monte Carlo simulation, because the finite extent of the lattice implies that the momentum p in eq. (2.2) is quantized with a lowest non-zero value, which may not be sufficiently small to suppress the $O(p^4)$ terms (for the lattices (5.1), for example, one has $p^2 > m_R^2$ if $p \neq 0$). A more readily accessible quantity

Table 3. Results from a numerical simulation of the Ising model /33/ and comparison with the analytic solution /34/. In the Monte Carlo calculation, $\varkappa$ is given, while for the analytic calculation m_R is taken as the independent parameter

	Monte Carlo	analytic solution
$\varkappa$	0.07102	0.0710(2)
m_R	0.4923(5)	0.4923
$2\varkappa Z_R$	0.970(3)	0.973(7)
g_R	44(4)	42(7)
$\varkappa$	0.07400	0.0740(3)
m_R	0.2148(5)	0.2148
$2\varkappa Z_R$	0.962(7)	0.975(9)
g_R	25(2) [3)]	24(2)

is the physical mass m, which, for all L, is defined through the exponential decay of the two-point function of $\phi(x)$ in the time direction. As I will discuss shortly, the L-dependence of m is weak and well understood. Furthermore, for $L = \infty$ we have

$$m_R = 2 \sinh m/2 \; (1 + O(g_R^2)), \tag{5.2}$$

where the $O(g_R^2)$ correction has been calculated and was found to be negligible ($< 10^{-4}$ for $g_R < 44$).

Some results obtained by Montvay and Weisz, for the two values of $\varkappa$ considered, are listed in Table 3, where I have used eq. (5.2) to eliminate m in favour of m_R. Within the quoted errors, the agreement with the analytic solution is

3) This number includes an analytically calculated finite size correction of $\Delta g_R = 1.7$ at L = 18. The error quoted is statistical only.

perfect. Thus, the qualitative assumptions on which the analytic solution is based (for example, that renormalized perturbation theory may be applied when $g_R \lesssim 41$) appear to be justified and little doubt remains that the solution is in fact correct.

I would now like to digress a little and discuss the lattice size dependence of the particle mass m(L) and the lowest two-particle energy W(L), which I shall define later. First note that m(L) is an eigenvalue of the transfer matrix and is hence independent of T by definition (in a Monte Carlo simulation, T must however be large so that the exponential decay of the two-point function of ϕ can be followed over a significant distance). As a function of L, the finite size mass shift

$$\delta_1 = [m(L) - m(\infty)]/m(\infty) \tag{5.3}$$

decays exponentially according to

$$\delta_1 = -\frac{3}{2m^2}\int_{-\pi}^{\pi} \frac{d^3q}{(2\pi)^3\, 2\omega(q)}\; e^{-\omega(q)L}\, F(q) + O(e^{-\bar{m}L}), \tag{5.4}$$

where $\omega(q)$ denotes the energy of a single particle with momentum q, F(q) an elastic forward scattering amplitude and $\bar{m} > m$. All quantities m, $\omega(q)$ and F(q) on the right hand side of eq. (5.4) are defined and evaluated at $L = \infty$. I have first presented this formula at Cargèse 1983 /42/ and since then provided a detailed proof /43/ (the lattice corrections have been discussed by Münster /48/).

It is of course possible to compute $\omega(q)$ and F(q) in renormalized perturbation theory. Taking the first order expressions and inserting the values of m_R, g_R at $\varkappa = 0.07102$ as given by Table 3, one obtains curve "a" in Fig. 4. The agreement with the Monte Carlo data at L = 6,8,10,12 (the points with the small error bars in Fig. 4) is very good although perhaps a bit fortuitous given that only the first order perturbative formulae were used and that the error term in eq. (5.4) was also neglected. Anyway, eq. (5.4) certainly gives the right order of magnitude for δ_1 and there is no doubt that the finite size effects are negligible compared to the statistical errors beyond say z = 6.

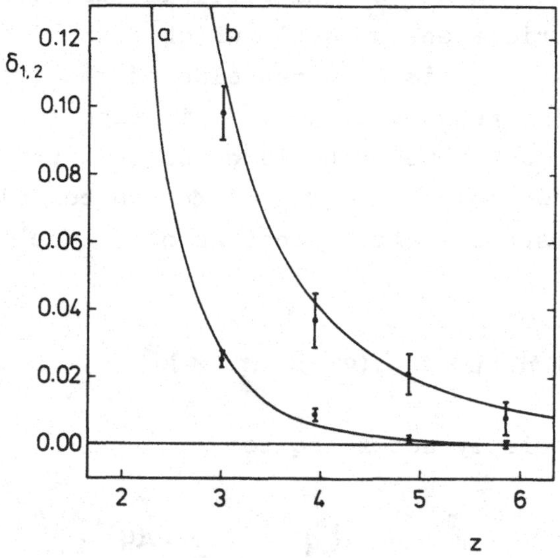

Fig. 4. Finite size energy shifts δ_1 and δ_2 as a function of $z = m(L)L$. Curves a and b correspond to eqs. (5.4) and (5.5) evaluated at $\varkappa = 0.07102$ ($m(\infty) \simeq 0.49$ and $a_0 \simeq -0.68$ at this point)

Another quantity considered by Montvay and Weisz is the two-particle energy W(L), which is the lowest energy above the vacuum in the sector of even states under $\phi \rightarrow -\phi$. The corresponding energy eigenstate describes two particles, which, being confined to the finite lattice, are in a stationary scattering state. Thus, W(L) is essentially equal to 2m with a finite size correction δ_2 given by

$$\delta_2 = [W(L) - 2m(L)]/2m(\infty)$$

$$= -\frac{2\pi a_0}{m^2 L^3}\{1 + c_1 \frac{a_0}{L} + c_2 \frac{a_0^2}{L^2}\} + O(L^{-6}), \qquad (5.5)$$

$$(5.6)\quad c_1 = -2.837297,$$

$$(5.7)\quad c_2 = 6.375183,$$

where a_o denotes the S-wave scattering length.[4] A proof of eq. (5.5) in the framework of quantum field theory has been given recently /43/, but for the case of non-relativistic hard spheres in a periodic box, the formula has actually been derived much earlier by Huang and Yang /44/ (see also refs. /45,46/).

If we use renormalized perturbation theory to two loops to compute a_o at $\varkappa = 0.07102$, eq. (5.5) yields curve b in Fig. 4. Again the Monte Carlo data of Montvay and Weisz agree very well with the theoretical prediction. In fact, the scattering length a_o could have been extracted from the data by fitting them with eq. (5.5). It has thus been demonstrated that a calculation of S-matrix elements through numerical simulation is possible in certain cases, the main difficulties being that very accurate data are required and that several lattices of variable size must be considered. Of course one hopes to apply the method to other models such as the Higgs model or even QCD.

A plot similar to Fig. 4 could also be produced at $\varkappa = 0.07400$, which corresponds to $m(\infty) = 0.21$ approximately. The picture would look less impressive in this case, because the errors are larger and because the maximal value of z would only be around 4 for the lattices considered. Within these limitations, the agreement between theory and experiment is however equally good.

6. SOLUTION OF THE ONE-COMPONENT MODEL IN THE BROKEN SYMMETRY PHASE

For $\varkappa > \varkappa_c(\lambda)$, the reflection symmetry $\phi \rightarrow -\phi$ of the action (4.1) is spontaneously broken and the field ϕ acquires

4) It is possible to develop a full-fledged scattering theory for euclidean lattice field theories /49/ and a_o is hence a completely well-defined quantity for all $\varkappa, \lambda$.

a non-zero vacuum expectation value

(6.1) $v = \langle \phi \rangle > 0.$

If we define Z_R and m_R as in the symmetric phase (eq. (2.2)), the renormalized vacuum expectation value is given by

(6.2) $v_R = v \, Z_R^{-1/2}$

and a renormalized coupling may be introduced through

(6.3) $g_R = 3m_R^2/v_R^2$

(to first order in g_R, this definition is equivalent to eq. (2.3)).

As in the symmetric phase, the first goal now is to compute Z_R, m_R and g_R as a function of $\varkappa$ and λ. However, since there is no known practical expansion for $\varkappa \to \infty$, which could play the rôle the high temperature expansion did in our analysis of the symmetric phase, a different strategy is needed.

The basic idea is as follows /35/. As we have discussed in sect. 4, the renormalized coupling g_R in the symmetric phase scales to zero as one approaches the critical line in such a way that the limit

(6.4) $$C_1(\lambda) = \lim_{\varkappa \to \varkappa_c} m_R \, (\beta_1 g_R)^{-17/27} \, e^{1/\beta_1 g_R}$$

exists (cf. eq. (4.8)). Similarly, a constant $C_1'(\lambda)$ may be defined by approaching $\varkappa_c(\lambda)$ from the broken symmetry phase. Both constants are defined at the critical line and it is therefore not surprising that they can be given an interpretation in terms of the critical (massless) theory. It then turns out that $C_1'(\lambda)$ is actually proportional to $C_1(\lambda)$ with a proportionality constant, which is exactly given by

(6.5) $C_1'(\lambda) = e^{1/6} \, C_1(\lambda)$

for our choice of renormalization conditions.

$C_1(\lambda)$ can be calculated for all λ to a reasonable estimated accuracy from the solution of the model in the symmetric phase. Thus, $C_1'(\lambda)$ is also known and may be used as initial datum for the integration of the renormalization group equations along the lines λ = constant in the broken symmetry phase starting at $\varkappa = \varkappa_c$ (see Fig. 5). Since the β-function (and the other Callan-Symanzik coefficients) are only known in perturbation theory, the integration must be stopped when g_R

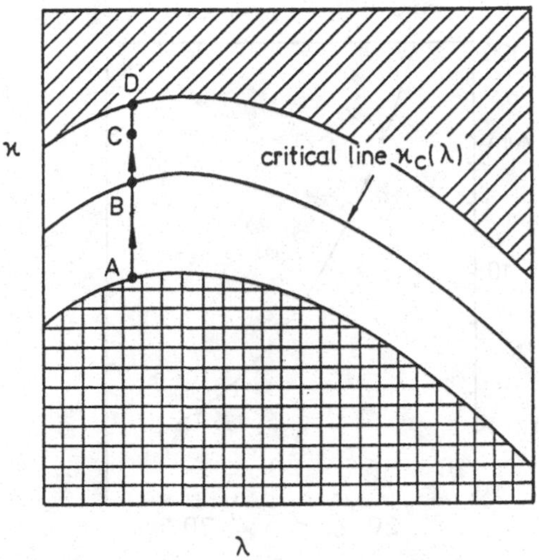

Fig. 5. Qualitative plot of the phase diagram of the lattice ϕ^4 theory. The integration of the renormalization group equations is started at e.g. point A at the boundary of the high temperature region (cross-hatched area) and follows the line λ = constant towards point B, where C_1 and C_1' are determined. The integration can then be carried on to (say) point C in the broken symmetry phase.

becomes large (point D in Fig. 5). Thus, in this way the theory can only be solved in a narrow band above the critical line, but as it turns out, this band includes the whole region $\Lambda \gtrsim 2m_R$. The shaded area in Fig. 5, where the theory remains unsolved, is therefore not a very interesting region since there, similarly to the high temperature region in the symmetric phase, the physics at scales of m_R is strongly influenced by non-universal cutoff effects.

The most conspicuous feature of the solution of the model in the broken symmetry phase obtained along these lines is that, concerning the scaling behaviour, there is practically no difference to what happens in the symmetric phase. In particular, for $\Lambda \geq 2m_R$, the renormalized coupling g_R does not exceed a maximal value of about 2/3 of the tree level unitarity bound and renormalized perturbation theory should

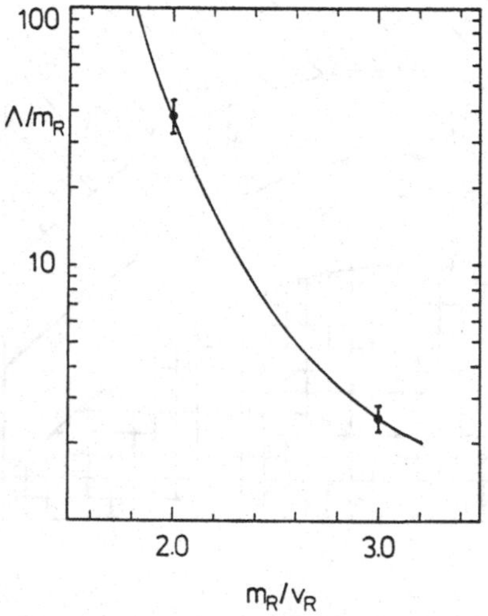

Fig. 6. Maximal value of the ultra-violet cutoff Λ in units of m_R for given m_R/v_R. The size of the estimated errors in the calculation is indicated at two representative points.

hence give an essentially correct description of the particle interactions at low energies in this region. Furthermore, as shown by Fig. 3, the flow of the curves of constant coupling g_R in the plane of bare parameters also looks similar on both sides of the critical line, a marked difference being that the interval of $\varkappa$ corresponding to $\Lambda \geq 2m_R$ is about a factor of 3 smaller in the broken symmetry phase.

In Fig. 6, I have plotted the triviality bound (2.6), where, instead of the coupling g_R, I have taken the ratio

m_R/v_R as the independent variable (cf. eq. (6.3)). One expects that a similar result will be obtained for the O(4) symmetric ϕ^4 theory and, as discussed in sect. 3, this will then lead to an upper bound on the Higgs meson mass. Fig. 6 applies to the one-component model and is therefore not amenable to such an interpretation. However, it is interesting to note that if we insert $m_R = m_H$, v_R = 250 GeV and assume $\Lambda \geqslant 2m_H$ for the purpose of illustration, the bound $m_H \leqslant 800$ GeV is obtained, which is actually not far from what other people have found earlier /20-31/.

7. CONCLUDING REMARKS

The analytic solution of the one-component ϕ^4 theory in the broken symmetry phase has not yet been checked by a large scale numerical simulation, but such calculations are on the way, one computing the effective action /50/ and another one /51/ employing the highly efficient up-dating algorithm of Swendson and Wang /52/. Work is also in progress on the physically more interesting O(4) symmetric model, which I expect to be soluble in the same way as the one-component model. In particular, the Goldstone modes in the broken symmetry phase should not give rise to any great difficulties for the analytic approach. Still, a substantial amount of labour remains to be done, especially so since the Baker-Kincaid tables /3/ are only for the one-component model and the high temperature series for Z_R, m_R and g_R must hence be newly derived /53/.

ACKNOWLEDGEMENTS

The bulk of these lectures is based on joint work with Peter Weisz, whom I would like to thank for the enjoyable collaboration. I am also indebted to Istvan Montvay for discussions on various aspects of the Monte Carlo simulation technique and I finally thank the organizers of the Cargèse Summer School for their kind invitation.

REFERENCES

1. E. Brézin, J.C. Le Guillou and J. Zinn-Justin: Field theoretical approach to critical phenomena, in: Phase

transitions and critical phenomena, vol. 6, eds. C. Domb, M.S. Green (Academic Press, London, 1976)
2. K.G. Wilson and J. Kogut: Phys. Reports 12 (1974) 75
3. G.A. Baker and J.M. Kincaid: J. Stat. Phys. 24 (1981) 469
4. D.S. Gaunt, M.F. Sykes and S. Mc Kenzie: J. Phys. A12 (1979) 871
5. B. Freedman, P. Smolensky and D. Weingarten: Phys. Lett. 113B (1982) 481
6. I.A. Fox and I.G. Halliday, Phys. Lett. 159B (1985) 148
7. C.B. Lang: Block spin approach to the fixed-point structure of lattice ϕ^4 theory, in: Advances in lattice gauge theory (Tallahassee conference 1985), eds. D.W. Duke, J.F. Owens (World Scientific, Singapore, 1985)
8. I.T. Drummond, S. Duane and R.R. Horgan: Nucl. Phys. B280 [FS18] (1987) 25
9. O.G. Mouritsen and S.J. Knak Jensen: Phys. Rev. B19 (1979) 3663
10. M. Aizenmann: Phys. Rev. Lett. 47 (1981) 1
11. M. Aizenmann: Comm. Math. Phys. 86 (1982) 1
12. M. Aizenmann and R. Graham: Nucl. Phys. B225 [FS9] (1983) 261
13. J. Fröhlich: Nucl. Phys. B200 [FS4] (1982) 281
14. C. Aragão de Carvalho, S. Carraciolo and J. Fröhlich: Nucl. Phys. B215 [FS7] (1983) 209
15. K. Gawędski and A. Kupiainen: Phys. Rev. Lett. 54 (1985) 92
16. K. Gawędski and A. Kupiainen: Comm. Math. Phys. 99 (1985) 197
17. J. Feldman, J. Magnen, V. Rivasseau and R. Sénéor: Comm. Math. Phys. 109 (1987) 437
18. A. Hasenfratz and P. Hasenfratz: Florida preprint, FSU-SCRI-86-30
19. I. Montvay: The lattice regularized standard Higgs model, in: Lattice Gauge Theory '86 (Brookhaven conference), eds. H. Satz et al. (Plenum, New York 1987)
20. L. Maiani, G. Parisi and R. Petronzio: Nucl. Phys. B136 (1978) 115
21. D.J.E. Callaway: Nucl. Phys. B233 (1984) 189
22. M.A.B. Bég, C. Panagiotakopoulos and S. Sirlin: Phys. Rev. Lett. 52 (1984) 883
23. A. Bovier and D. Wyler: Phys. Lett. 154B (1985) 43
24. A. Sirlin and R. Zucchini: Nucl. Phys. B266 (1986) 389
25. R. Dashen and H. Neuberger: Phys. Rev. Lett. 50 (1983) 1897
26. P. Hasenfratz and J. Nager: Bern preprint, BUTP-86-20
27. B. Grądkowski and M. Lindner: Phys. Lett. 178 (1986) 81
28. W. Langguth, I. Montvay and P. Weisz: Nucl. Phys. B277 (1986) 11
29. W. Langguth and I. Montvay: DESY-87-20, to appear in Z. Phys. C
30. A. Hasenfratz and T. Neuhaus: Florida preprint FSU-SCRI-87-29
31. A. Hasenfratz, K. Jansen, C.B. Lang, T. Neuhaus and H. Yoneyama: Florida preprint, FSU-SCRI-87-52
32. G. Mack: Multigrid methods , this volume
33. I. Montvay and P. Weisz: DESY-87-056, to appear in Nucl. Phys. B

34. M. Lüscher and P. Weisz: Nucl. Phys. B290 [FS20] (1987) 25
35. M. Lüscher and P. Weisz: DESY-87-075, to appear in Nucl. Phys. B
36. K. Symanzik: Cutoff dependence in lattice ϕ^4 theory, in: Recent developments in gauge theories (Cargèse 1979), eds. G. 't Hooft et al. (Plenum, New York 1980)
37. S.J. Freedman, J. Napolitano, J. Camp and M. Kroupa: Phys. Rev. Lett. 52 (1984) 240
38. J. Jersák: Lattice Higgs models, in: Lattice gauge theory - a challenge in large scale computing (Wuppertal 1985), eds. B. Bunk et al. (Plenum, New York 1986)
39. H.G. Evertz: PhD Thesis, Aachen (1987)
40. K. Symanzik: Comm. Math. Phys. 16 (1970) 48
41. M.M. Tsypin: Lebedev Physical Institute preprint No. 280 (Moscow 1985)
42. M. Lüscher: On a relation between finite size effects and elastic scattering processes, in: Progress in gauge field theory (Cargèse 1983), eds. G. 't Hooft et al. (Plenum, New York 1984)
43. M. Lüscher: Comm. Math. Phys. 104 (1986) 177
44. K. Huang and C.N. Yang: Phys. Rev. 105 (1957) 767
45. H.W. Hamber, E. Marinari, G. Parisi and C. Rebbi: Nucl. Phys. B225 [FS9] (1983) 475
46. G. Parisi: Phys. Reports 103 (1984) 203
47. M. Lüscher: Comm. Math. Phys. 105 (1986) 153
48. G. Münster: Nucl. Phys. B249 (1985) 659
49. J.C.A. Barata and K. Fredenhagen: DESY-87-149, to appear in the Proceedings of the International Symposium on Field Theory on the Lattice (Seillac 1987), Eds. A. Billoire et al., Nuclear Physics B, Proceedings Supplement
50. J. Kuti and Y. Shen: Talk presented at the International Symposium on Field Theory on the Lattice (Seillac 1987)
51. K. Jansen, J. Jersák, I. Montvay, G. Münster, T. Trappenberg and U. Wolff: to be published
52. R.H. Swendsen and J.-S. Wang: Phys. Rev. Lett. 58 (1987) 86
53. M. Lüscher and P. Weisz: to be published

MULTIGRID METHODS IN QUANTUM FIELD THEORY

Gerhard Mack

II. Institut für Theoretische Physik der Universität Hamburg

Abstract

The multigrid formulation of Euclidean quantum field theory (= classical statistical mechanics) on the continuum or on a lattice of small lattice spacing is helpful both for analytical and numerical investigations. Basically it amounts to simultaneous performance of a sequence of renormalization group transformations. It permits Monte Carlo simulations for critical and nearly critical systems without critical slowing down. Combined with analytic tools - the theory of polymer systems on the multigrid - it offers a chance to perform computer simulations for continuum systems without UV-cutoff, rather than their lattice approximations.

I. Generalities: Renormalization group and multigrid

II. Multigrid Monte Carlo simulations for $\lambda\phi^4$-theory. Possible generalization to gauge theories

III. Towards computer simulation of continuum theories without UV-cutoff via simulation of polymer systems on the multigrid

IV. Analytic tools: Polymer systems on the multigrid. Renormalization and UV-convergence

I made an effort to make section II as selfcontained as possible.

The multigrid reformulation of Euclidean quantum field theory as used here was introduced by A. Pordt and the author some years ago [1]. The numerical multigrid investigations of ϕ^4-theory were done in collaboration with S. Meyer [2]. The polymer approach was developed jointly with A. Pordt, K. Pinn, and H.-J. Timme [1,3,4,5]. Financial support from Deutsche Forschungsgemeinschaft is gratefully acknowledged.

I. Generalities: Renormalization Group and Multigrid

To solve an Euclidean quantum field theory or a classical statistical mechanical system one should compute its partition function $Z(\psi)$ as a function of an external source or field ψ. $\ln Z(\psi)$ will then serve as generating function for the correlation functions. This requires evaluation of the many- or ∞-dimension integral that defines $Z(\psi)$.

Example: We may consider $\lambda\phi^4$-theory (= Landau Ginzburg ferromagnet) or, more generally, a theory of one real field $\phi(z)$ on some lattice ($\equiv$ *base*) of lattice spacing a, with action[1]

$$\mathcal{H}(\phi) = \mathcal{H}_{free}(\phi) + V(\phi) \tag{0.1a}$$

$$\mathcal{H}_{free}(\phi) = \frac{1}{2}\int_{z\in base} [\nabla_\mu \phi(z)]^2 = \frac{1}{2}(\phi, -\Delta\phi) \tag{0.1b}$$

$$V(\phi) = \int_{z\in base} \mathcal{V}(\phi(z)) \; .$$

$\mathcal{H}_{free}$ is the action of a massless free field theory, with propagator

$$v = (-\Delta)^{-1} \quad , \quad v(x,y) = \text{integral kernel of } v \; .$$

$\mathcal{V}$ might equal $\frac{1}{2}m_u^2\phi(z)^2 + \lambda_u\phi(z)^4$ or similar. It is customarily called the interaction. But it is better to think of the propagator $v(x,y)$ as mediating interaction between points x, y of the lattice.

The partition function in the presence of an external field ψ is

$$Z(\psi) = \int d\mu_v(\phi) e^{-V(\phi+\psi)} \; . \tag{0.2}$$

$d\mu_v(\phi)$ is the Gaussian measure for a free field theory with propagator v, viz

$$d\mu_v(\phi) = \mathcal{N}\mathcal{D}\phi \; e^{-\mathcal{H}_{free}(\phi)} \quad , \quad \mathcal{D}\phi = \prod_{z\in base} d\phi(z) \; ;$$
$$\mathcal{N} = \text{normalization factor.} \tag{0.3}$$

$\ln Z(\psi)$ is generating function of the free-propagator-amputated Greens functions [1, Appendix].

1. Effective Hamiltonians and ~ Observables

A crucial notion in Wilson's renormalization group [7] theory is that of effective Hamiltonian (or action) and observables. Given a classical statistical mechanical system on the continuum or on a lattice $(a\mathbf{Z})^d$ with arbitrarily small lattice spacing a - for instance an Euclidean field theory or a ferromagnet - with fields (spins) $\phi(x)$, Hamiltonian $\mathcal{H}(\phi)$, observables $\mathcal{O}_i = \mathcal{O}_i(\phi)$, it

gets mapped into a lattice theory

[1] In classical statistical mechanics, $\mathcal{H}$ is called the Hamiltonian. We use lattice notations as follows. On a lattice Λ of lattice spacing a, we write e_μ for the (lattice) vector of length a in μ-direction; $e_\mu = -e_{-\mu}$. In d dimensions

$$\int_{z\in\Lambda} = a^d \sum_{x\in\Lambda} \quad , \quad \nabla_\mu f(z) = \frac{1}{a}[f(z+e_\mu) - f(z)] \quad , \quad -\Delta = \nabla_\mu\nabla_{-\mu}$$

and the δ-function is $\delta(z) = a^{-d}\delta_{z,0}$. The fundamental lattice from which we start will be called *base*.

on a lattice of (larger) lattice spacing a' with lattice fields $\Phi(x')$, Hamiltonian $\mathcal{H}'(\Phi) = \mathcal{H}_{eff}(\Phi)$, and observables $\mathcal{O}'_i(\Phi)$, such that all expectation values of corresponding observables are exactly equal in the two theories

$$\langle \mathcal{O}_i \rangle_{\mathcal{H}} = \langle \mathcal{O}'_i \rangle_{\mathcal{H}'} \ . \tag{1.1}$$

This is achieved by a *block spin transformation.* It can be defined by specifying the new field $\Phi(x')$, called block spin, as a function of the original field ϕ. For instance[1]

$$\Phi(x') = \underset{x \underline{\in} x'}{\text{av}} \phi(x) \equiv (C\phi)(x') \ , \tag{1.2}$$

or the same formula with a constant factor, to absorb wave function renormalization later on.

Given that a'/a is an integer > 1, points x' in Λ' can be identified with blocks of $(a'/a)^d$ points in Λ, and the definition (1.2) involves the average of ϕ over such a block. Typically one is ultimately interested in a' of the order of the correlation length.

The effective Hamiltonian $\mathcal{H}'$ is defined by

$$e^{-\mathcal{H}'(\Phi)} = \int \mathcal{D}\phi \ e^{-\mathcal{H}(\phi)} \prod_{x'} \delta\Big(\underset{x \underline{\in} x'}{\text{av}} \phi(x) - \Phi(x') \Big) \tag{1.3a}$$

Similarly, effective observables are defined by

$$\mathcal{O}'_i(\Phi) \ e^{-\mathcal{H}'(\Phi)} = \int \mathcal{D}\phi \ \mathcal{O}_i(\phi) \ e^{-\mathcal{H}(\phi)} \prod_{x'} \delta\Big(\underset{x \underline{\in} x'}{\text{av}} \phi(x) - \Phi(x') \Big) \tag{1.3b}$$

Equality of expectation values (1.1) is obvious from the definitions.

It is worth emphasizing that the block spin variables need not be fields of the same kind as those in the original theory [6]. For instance, in the case of an Ising model with variables $\sigma(z) = \pm 1$ one may work with real block spins.

Now I will rewrite the effective Boltzmannian. This will lead to a *twogrid* reformulation of the theory.

We proceed by splitting the field ϕ into a term which is determined by the block spin Φ, and a remainder ζ, called the fluctuation field, which has vanishing block averages

$$\phi(x) = \int_{x' \in \Lambda'} \mathcal{A}(x, x')\Phi(x') + \zeta(x) \tag{1.4}$$

The kernel $\mathcal{A}$ should be invariant under translations by multiples of the block lattice spacing a', and its block averages must equal

$$\int_{x \underline{\in} y'} \mathcal{A}(x, x') = \delta(x' - y') \qquad (x', y' \in \Lambda'). \tag{1.5}$$

Given the kernel $\mathcal{A}$, the split is unique.

It is an important problem how to choose the kernel $\mathcal{A}$ properly. I will follow

[1] The notation $x \underline{\in} x'$ shall mean that x is in the cube (in Λ) specified by the point x' (in Λ').

Kupiainen and Gawędzki [8,9]. Consider the free Hamiltonian first. Inserting the split (1.4) we obtain

$$\mathcal{H}_{free} = \frac{1}{2}(\Phi, -\Delta_{eff}\Phi) + \frac{1}{2}(\zeta, -\Delta\zeta) + (\zeta, -\Delta\mathcal{A}\Phi) \tag{1.6}$$

The scalar product in the first term involves summation $\int_{x'}$ over the block lattice. The effective Laplacian on the block lattice is defined by $\Delta_{eff} = \mathcal{A}^*\Delta\mathcal{A}$; it is an integral operator with kernel

$$\Delta_{eff}(x', y') = \int_{x\in\Lambda} \mathcal{A}(x, x')\Delta_x\mathcal{A}(x, y') \tag{1.7}$$

Since ζ has zero block average, the last term in (1.6) will vanish if we choose $\mathcal{A}$ so that its Laplacian

$$-\Delta_x\mathcal{A}(x, y') \text{ is constant as a function of } x \text{ on blocks } x' \tag{1.8}$$

If we adopt this choice there will be no coupling between block spin Φ and fluctuation field ζ in $\mathcal{H}_{free}$. Let us insert the resulting expression for $\mathcal{H} = \mathcal{H}_{free} + V$ in the defining formula (1.3a) for the effective Hamiltonian, and shift the integration variable ϕ by $-\mathcal{A}\Phi$. This gives

$$e^{-\mathcal{H}'(\Phi)} = \int \mathcal{D}\zeta\ e^{-\mathcal{H}(\mathcal{A}\Phi+\zeta)} \prod_{x'\in\Lambda'} \delta\Big(\operatorname*{av}_{x\in x'}\zeta(x)\Big). \tag{1.9}$$

The δ-function may be regarded as limit of a Gaussian $\exp -\frac{\kappa}{2}(\zeta, C^*C\zeta)$ (C^* = adjoint of averaging operator C, cp. eq. (1.2)). Therefore eq. (1.9) can also be written as

$$e^{-\mathcal{H}'(\Phi)} = e^{\frac{1}{2}(\Phi,\Delta_{eff}\Phi)} \int d\mu_\Gamma(\zeta)\ e^{-V_\Phi(\zeta)} \quad \text{with} \quad V_\Phi(\zeta) = V(\mathcal{A}\Phi + \zeta) \tag{1.10}$$

apart from a constant factor. This formula involves the Gaussian measure $d\mu_\Gamma(\zeta)$ which is the functional measure of a free field theory with propagator Γ,

$$\Gamma = \lim_{\kappa\to\infty}(-\Delta + \kappa C^*C)^{-1}. \tag{1.11}$$

It follows from eq. (1.3a) that the partition function Z of the original theory equals

$$Z = \int \mathcal{D}\Phi\ e^{-\mathcal{H}'(\Phi)} = \int \mathcal{D}\Phi\ e^{\frac{1}{2}(\Phi,\Delta_{eff}\Phi)} \int d\mu_\Gamma(\zeta)\ e^{-V_\Phi(\zeta)} \tag{1.12}$$

We may imagine that V depends parametrically on an external field or source, so that $\ln Z$ is generating function for correlation functions.

Eq. (1.12) represents the rewriting of the partition function of the original theory as partition function of a statistical mechanical system on a *twogrid*. There are two lattices Λ, Λ' involved. Their (disjoint) union is the twogrid. Later we shall generalize to a multigrid, see figure 1 below.

The lattice Λ has lattice spacing a, and the lattice Λ' has larger lattice spacing $a' = La$ (L integer, ≥ 2). With each point of the twogrid, an integration variable is associated: fluctuation field variables $\zeta(x)$ for $x \in \Lambda$, and block spin variables $\Phi(x')$ for $x' \in \Lambda'$.

What is the advantage of this rewriting, and of its generalization to a multigrid? To explain this, we need to recall Wilson's renormalization group philosophy.

Comparing expression (1.10) for the effective Boltzmann factor $e^{-\mathcal{H}'(\Phi)}$ with eq. (0.2) for a partition function, we see that $e^{-\mathcal{H}'(\Phi)}$ is obtained in the form of partition function of an *auxiliary Euclidean field theory with a free propagator* Γ *with infrared cutoff* a'^{-1}. In contrast with the original nearly critical theory, this auxiliary system will have a correlation length of the order of one block lattice spacing a' only, i.e. it is not a nearly critical system if a'/a is not too large. This is the crucial point in Wilson's philosophy. We will discuss below what can be established to support it. Summing up, the ζ-integration will amount to treatment of a noncritical system.

Noncritical systems are much more manageable than critical ones. They can be treated analytically by convergent expansions, at least if the coupling is not too strong. Such expansions are furnished by the theory of polymer systems. This will be discussed in section IV. Noncritical systems can also be treated numerically by Monte Carlo simulations much more effectively than nearly critical ones, because of the absence of critical slowing down. The number of sweeps needed to obtain independent configurations on a lattice of lattice spacing a goes with the correlation length ξ like $(\xi/a)^2$ approximately, see G. Parisi's lectures at this school.

To treat the full theory, one needs to perform also the integration over the block spin Φ which lives on the block lattice Λ'. The effective theory on the block lattice has the same long distance behaviour as the original theory, and therefore the same correlation length. So it is still a critical theory. But the number of lattice points (integration variables) has decreased by $(a/a')^d$ in d dimensions, and the quantity $(\xi/a')^2$ which governs critical slowing down in Monte Carlo simulations, is less than $(\xi/a)^2$ by a factor $(a/a')^2$. So there is a net gain. If $(\xi/a')^2$ is still too large, a further split of the block spin Φ into a fluctuation part and a block block spin can be done, repeatedly if necessary. This leads to a multigrid instead of the twogrid and will be discussed in the next subsection.

What has been said so far about critical slowing down is not the full story yet, and we will come back to this question in section II when we deal with Multigrid Monte Carlo simulations.

The rest of this subsection I-1 is devoted to a discussion of the precise meaning of the assertion that the auxiliary statistical mechanical system, whose partition function is the effective Boltzmannian, has correlation length of order one block lattice spacing a'.

First, this is true to all orders in perturbation theory in the interaction V. It follows that $\mathcal{H}'$ has good locality properties, to all orders in perturbation theory. These findings follow from the decay properties of the kernel $\mathcal{A}$, the fluctuation field propagator Γ, and the kernel of the effective Laplacian Δ_{eff}. They are as follows [8]. Because of translation invariance properties, it suffices to consider $\mathcal{A}(x,0), \Gamma(x,0)$ for $x \in \Lambda$, and $\Delta_{eff}(x',0)$ for $x' \in \Lambda'$. All these quantities decay exponentially with x or x', with decay length of order a'. This can be proven by finding their Fourier representations, and shifting the path of the p-integration into the complex plane [8,9]. Another method, which is also applicable in situations without translation invariance (propagators in external gauge fields, for instance) was introduced by Balaban [10]. It is based on generalized random walk expansions. In section II,

I will describe a method to evaluate the kernels $\mathcal{A}$ and Δ_{eff} numerically, and present the results. It turns out that the decay length is rather less than a'. Γ could also be computed when $\mathcal{A}$ is known, using the following formula which gives the split of the free propagator v of the original theory

$$\begin{aligned} v &= \mathcal{A}^* u \mathcal{A} + \Gamma \quad ; \\ u &= (-\Delta_{eff})^{-1} = C v C^* \end{aligned} \tag{1.13}$$

This formula follows by considering representation (1.12) of the partition function Z for the special case $V(\phi) = i(q, \phi)$ and evaluating both sides by use of the known formula [14]

$$\int d\mu_v(\phi)\, e^{i(q,\phi)} = e^{-\frac{1}{2}(q,vq)} \tag{1.14}$$

for the Fourier transform of a Gaussian measure. Γ is not needed for numerical purposes, though.

We return to the discussion of the supposedly noncritical behaviour of the auxiliary statistical mechanical system with partition function (1.10). I will now explain what is known nonperturbatively. Nonperturbative analytical studies were limited to the case of sufficiently weak coupling [9,11,12]. The following picture emerged.

To all orders of perturbation theory, the effective Hamiltonian $\mathcal{H}'(\Phi)$ has good locality properties. Outside of perturbation theory this is no longer true for arbitrary block spin configuration Φ.

Suppose that we have defined effective Hamiltonians $\mathcal{H}'_{\Lambda'}(\Phi)$ for arbitrary shape and size of Λ' by a formula like (1.10). $\mathcal{H}'_{\Lambda'}$ will depend on $\Phi(x')$ for $x' \in \Lambda'$. Then we may consider the part $\mathcal{H}'(X|\Phi)$ of the effective Hamiltonian that is "spread throughout the subset X of Λ' ". It is uniquely defined by validity of the following formula for arbitrary Λ'

$$\mathcal{H}'_{\Lambda'}(\Phi) = \sum_{X : X \subseteq \Lambda'} \mathcal{H}'(X|\Phi)$$

We assume that additive constants are so chosen that $\min \mathcal{H}'_{\Lambda'}(\Phi) = 0$. $\mathcal{H}'(X|\Phi)$ will only depend on the restriction of Φ to X. Good locality properties of $\mathcal{H}'$ mean roughly that $\mathcal{H}'(X|\Phi)$ becomes very small when X is large. The size of X is measured by the length (in units of a') of the shortest tree whose vertices are all points of X.

Consider a large, or infinitely extended lattice Λ' again. Define an energy density $\mathcal{E}(x'|\Phi)$ by ($|X|$ = no. points in X)

$$\mathcal{E}(x'|\Phi) = \sum_{X : x' \in X} |X|^{-1} \mathcal{H}'(X|\Phi)$$

so that $\mathcal{H}'(\Phi) = \int_{x'} \mathcal{E}(x'|\Phi)$. Given a configuration Φ on Λ', we may divide Λ' into *small field region* and *large field region* as follows. A site x' will be in the large field region together with a neighbourhood of diameter of order $a' \ln e\gamma^{-1}$ if

$$e = \mathcal{E}(x'|\Phi) > \gamma$$

where γ is a suitably chosen number appreciably bigger than 1 so that $e^{-\gamma}$ is small. The small field region is what is left over. [The name "small field region" is merely historical. It depends on the model wether $\Phi(x')$ is really small in the small field region. In some models, x' with $\Phi(x') = 0$ would not be in the small field region.]

The configuration Φ is a random variable with probability distribution proportional $e^{-\mathcal{H}'(\Phi)}$. Appealing to ergodicity, we may look at a generic ("typical") configuration Φ on an infinite lattice Λ' instead. The restriction of Φ to finite parts of Λ' which are translates of each other will then produce configurations with appropriate probability distribution.

For a generic configuration Φ, most of Λ' will be small field region. The complement decomposes into connected components, which we call large field islands. In weakly coupled models, these large field islands are typically small (tree length of one or few lattice spacings) while large islands are very rare, i.e. dilute.

Given Φ, let us look at the auxiliary statistical system whose random variables are the fluctuation field values $\zeta(x)$. One finds that the correlations are short range, of order one block lattice spacing a', in the small field region determined by Φ, and $\mathcal{H}'$ has good locality properties there. That is, $\mathcal{H}'(X|\Phi)$ will decay exponentially with the tree length of X (in units of a'), in an appropriate norm.

In the large field region, the situation may be different. There could sometimes be correlations throughout individual large field islands; in large islands correlations could be over large distances, and long range correlations in ζ can induce nonlocalities in the effective Hamiltonian. But large islands will be very rare.

In the presence of a cutoff which bounds the fields so that all the lattice is small field region, perturbation theory would converge. The contributions from the large field region are therefore the truely nonperturbative ones. In the weakly coupled models that were studied they are very small. [It is an advantage of the RG-methods that nonperturbative contributions to the RG-flow are separated from nonperturbative long distance effects like spontaneous symmetry breaking.]

These results of analytical investigations clarified in a satisfactory way in what sense Wilson's philosophy is valid in weakly coupled models. Results of numerical investigations by K. Pinn indicate [13] that the large field region can be more important, and locality problems associated with it may aggravate in non weakly coupled models, such as a nearly critical 2-dimensional Ising model with real block spins. In any case, one must conclude that the truncation of effective Hamiltonians, which one needs to make in standard Monte Carlo renormalization group approaches, is a subtle and potentially dangerous aspect of these approaches. In analytical studies, the problem is overcome by use of polymer representations of the effective Boltzmannian in the large field region, rather than expansions of the effective Hamiltonian[1]. It may ultimately prove necessary to do the same in Monte Carlo renormalization group work (cp. end of section III).

In the multigrid approach there is no need to truncate effective Hamiltonians.

2. Iteration of Renormalization Group Transformations: Multigrid Transform

Ultimately one is interested in the effective Hamiltonian $\mathcal{H}' = \mathcal{H}_{eff}$ on a lattice with lattice spacing a' of the order of the physical correlation length, maybe even larger. In our numerical work on ϕ^4-theory, we compute the effective Hamiltonian as a function of the magnetization, i.e. for a single block which contains the whole lattice.

If the original UV-cutoff a^{-1} is large, this means that also a'/a will be large. It is

[1] In this way one avoids having to take the logarithm of a function which may have zeros very near to the real axis - cp. sections III, IV.

then advantageous to perform the desired block spin transformation $a \to a'$ through a sequence of block spin transformations as described in subsection I-1. For instance

$$\begin{aligned} a \equiv a_N \to a_{N-1} \to \ldots \to a_1 \to a_0 \equiv a' = L^N a \\ a_{j-1}/a_j = L = \text{small integer} > 1 \end{aligned} \tag{2.1}$$

For the choice (1.2) of block spin as block average, this decomposition into steps is particularly simple because the average of an average is an average

$$\operatorname*{av}_{x \in x''} \phi(x) = \operatorname*{av}_{x' \in x''} \operatorname*{av}_{x \in x'} \phi(x). \tag{2.2}$$

Therefore we only need to iterate the procedure of subsection I-1. In this way a sequence of Hamiltonians is introduced

$$\mathcal{H} \equiv \mathcal{H}^N \to \mathcal{H}^{N-1} \to \ldots \to \mathcal{H}^1 \to \mathcal{H}^0 \equiv \mathcal{H}_{eff} \tag{2.3}$$

They live on a sequence of lattices Λ_j of lattice spacing a_j ($j = N, \ldots, 0$). By doing simple dimensional analysis, one may also regard them as living on lattices of lattice spacing 1 instead. The resulting sequence of Hamiltonians is called a (discrete) *Renormalization Group flow* (RG-flow). The Hamiltonian $\mathcal{H}^j$ will depend on a (block spin) field Φ^j which lives on sites of the lattice Λ_j

$$\Phi^j(x) = \operatorname*{av}_{z \in x} \phi(z) \quad \text{for} \quad x \in \Lambda_j \qquad (z \in \mathit{base}). \tag{2.4}$$

A great advantage of the (numerical) multigrid approach, aside from absence of critical slowing down, is that it yields at the same time probability distributions of the original field $\phi = \Phi^N$ and of the block spins Φ^j at different length scales a_j, $j = 0, \ldots, N-1$. By looking at conditional probabilities, one may extract from this also information about the RG-flow. A special trick to get at conditional probabilities will be mentioned in section II.

Repeating the procedure of subsection I-1 will lead to a successive decomposition of the field ϕ on the original lattice $\Lambda_N = \mathit{base}$, as follows.

Let us consider the first block spin transformation $a_N \to a_{N-1}$, as described in the last subsection. Rename ϕ into Φ^N (original field), Φ into Φ^{N-1} (block spin), ζ into φ^N (fluctuation field). The split (1.4) of the field ϕ reads

$$\Phi^N(x) = \int_{x' \in \Lambda_{N-1}} \mathcal{A}(x, x')\, \Phi^{N-1}(x') + \varphi^N(x) \tag{2.5}$$

with a kernel $\mathcal{A}(x, x') \equiv \mathcal{A}^{[N,N-1]}(x, x')$ which is defined for $x' \in \Lambda_{N-1}$ and satisfies the following requirement for $j = N-1$

$$\int_{z \in y'} \mathcal{A}(z, x') = \delta(x' - y') \qquad \text{for} \qquad x', y' \in \Lambda_j\ ,\ z \in \mathit{base} \tag{2.6}$$

We may choose the kernel as discussed before [8], so that there will be no coupling between the fluctuation field φ^N and block spin Φ^{N-1} in the free Hamiltonian $\mathcal{H}_{free}$. This is achieved by requiring that we have for $j = N-1$, $z \in \mathit{base}$

$$-\Delta_z \mathcal{A}(z, x) = \text{constant as a function of } z \text{ on blocks } y \in \Lambda_j \text{ if } x \in \Lambda_j. \tag{2.7}$$

Now we repeat the procedure. We split the field Φ^{N-1} on Λ_{N-1} into a fluctuation field φ^{N-1} which lives on Λ_{N-1}, and a part which is determined by the block spin field Φ^{N-2} on Λ_{N-2}. Upon inserting the result into eq. (2.5) we obtain the following formula for the special case $j = N - 2$

$$\phi(z) = \int_{x\in\Lambda_j} \mathcal{A}(z,x)\ \Phi^j(x) + \zeta^j(z) \tag{2.8a}$$

$$\zeta^j(z) = \sum_{k=j+1}^{N} \int_{x\in\Lambda_k} \mathcal{A}(z,x)\ \varphi^k(x) \qquad (z \in base) \tag{2.8b}$$

The kernel $\mathcal{A}(z,x)$ is newly defined for argument $x \in \Lambda_{N-2}$ as a convolution of $\mathcal{A}^{[N,N-1]}(z,y)$ with the kernel $\mathcal{A}^{[N-1,N-2]}(y,x)$ which appeared in the decomposition of Φ^{N-1} $(y \in \Lambda_{N-1})$. Thus a recursion relation of the following form holds for $j = N - 2$

$$\mathcal{A}^j(z,x) = \int_{y\in\Lambda_{j+1}} \mathcal{A}^{j+1}(z,y)\ \mathcal{A}^{[j+1,j]}(y,x)\ . \tag{2.8c}$$

The kernels obey condition (2.6). This ensures that Φ^j is block average (2.4) of ϕ.

For arguments x on base space one introduces a trivial kernel

$$\mathcal{A}(z,x) = \delta(z-x) \qquad \text{if} \qquad x \in \Lambda_N = base \tag{2.9}$$

And so it goes. Repeating the procedure yields a split of the field $\phi(z)$ in the form (2.8) for all $j = N, N-1, ..., 1, 0$. The field φ^k lives on lattice Λ_k and obeys the constraint that its average over blocks $x' \in \Lambda_{k-1}$ is zero

$$\operatorname*{av}_{x\in x'} \varphi^k(x) = 0 \qquad \text{for all} \quad x' \in \Lambda_{k-1}\ ,\ k = N, ..., 1. \tag{2.10}$$

The multigrid Λ is the disjoint union of all the lattices[1] Λ_j, $j = 0, ..., N$ (later on we will admit $N = \infty$).

$$\Lambda = \Lambda_0 + \Lambda_1 + \Lambda_2 + ... + \Lambda_N \qquad (N \le \infty).$$

It may be regarded as a lattice in one more dimension, see figure 1.

The extra direction will be called "vertical" or scale direction. The lattice Λ has a somewhat unusual geometry. Let us call $x \in \Lambda_j$ a nearest neighbour of $y \in \Lambda_k$ $(k \le j)$ if x, y are either nearest neighbours in the same lattice $\Lambda_j = \Lambda_k$, or x is contained in the block $y \in \Lambda_{j-1} = \Lambda_k$. Next nearest neighbours are nearest neighbours of nearest neighbours etc. With this definition, the number of k-th nearest neighbours increases exponentially with k, while in a standard d-dimensional lattice it would increase like k^{d-1}. We will have occasion to consider also a multigrid with infinitely many layers $\Lambda_0, \Lambda_1, ...$ $(N = \infty)$, in order to deal with continuum theories, and with either periodic boundary condition or infinite extension in the horizontal directions.

This infinite multigrid admits a semigroup S of symmetries (maps) compounded from

1) translations by $\vec{n} \cdot a_0$, $\vec{n}$ = integer vector

2) rotation by $\frac{\pi}{2}$ around axes of the cubic lattice Λ_0

3) dilatations by L^{-k}, $k = 0, 1, 2, ...$ $\qquad (a_{j-1}/a_j = L)$.

[1] I use the notation $+$, $\sum$ for union of disjoint sets throughout.

One might want to call Λ a "semicrystal" since it has long range order and a semigroup of symmetries.

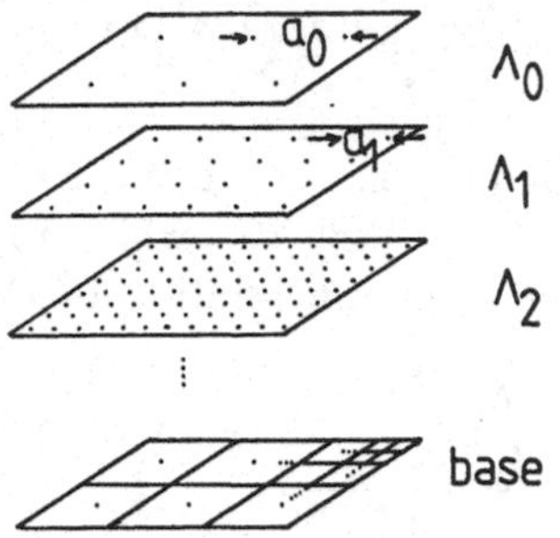

Fig. 1a and b. *The multigrid.* A point $x \in \Lambda_j$ of the multigrid may be regarded as a cube of side length a_j in base space Λ_N. The cubes are positioned in such a way that smaller cubes fit into larger ones (i.e. do not intersect their boundaries).
a Base space with lattices of side length $a_j = L^{-j}a_0$ superimposed, $j = 0, 1, 2, \ldots$
b Every cube in one of the lattices Λ_j that is superimposed on base space is represented by a dot in this pictorial representation of the multigrid $\Lambda = \Lambda_0 + \Lambda_1 + \Lambda_2 + \ldots$ The dots in layer Λ_j represent cubes of side length a_j. (Drawing for 2 dimensions)

Let us return to the split (2.8) of the field. It is convenient to introduce a field φ on the multigrid Λ by setting

$$\begin{aligned} \varphi^0 &\equiv \Phi^0 \\ \varphi(x) &\equiv \varphi^j(x) \qquad \text{if} \qquad x \in \Lambda_j,\ j = 0, \ldots, N \end{aligned} \tag{2.11}$$

and to introduce special notations for summation over the multigrid, cp. footnote[1]. The decomposition (2.8) of the field reads now, for $j = 0$

$$\phi(z) = \int_{x\in\Lambda} \mathcal{A}(z,x)\ \varphi(x) \equiv (\mathcal{A}\varphi)(z) \tag{2.13}$$

[1] Multigrid notation:

$$\int_{x\in\Lambda} (\ldots) = \sum_{j=0}^{N} \int_{x\in\Lambda_j} (\ldots) = \sum_{j=0}^{N} a_j^d \sum_{x\in\Lambda_j} (\ldots) \tag{2.12}$$

in d dimensions.

The field φ obeys the constraints (2.10) for all layers except $j = 0$. These constraints determine $\varphi(x)$ at block centers $x = x_c \in \Lambda_j$, x_c = center of block $x' \in \Lambda_{j-1}$, in terms of the other variables (take L odd ...). There are as many independent variables $\varphi(x)$ as there were independent variables $\phi(z)$. In other words, φ is uniquely determined by ϕ. It is a linear function of ϕ. Therefore its probability distribution is determined by the probability distribution of ϕ.

Let us first consider massless free field theory on base space, with free propagator $v = (-\Delta)^{-1}$. Its expectation values will be written $\langle \cdot \rangle_0$. By definition

$$\langle \phi(z_1)\phi(z_2)\rangle_0 = v(z_1, z_2)$$

The kernels $\mathcal{A}$ can be chosen in such a way that there will be no coupling between fields attached to different layers Λ_k of the multigrid in the free Hamiltonian

$$\mathcal{H}_{free}(\phi) = -\frac{1}{2}\int_{z \in \text{base}} \phi(z)\Delta\phi(z)$$

when the split (2.8) of the field is inserted. Using property (2.6) of kernels and remembering that φ^j have vanishing block average (2.10) for $j \neq 0$, it is readily verified that this decoupling property obtains when the Laplacian $\Delta_z \mathcal{A}(z, x)$ has the constancy property (2.7) for all j. As a result (scalar product $(\cdot\,,\cdot)$ involves $\int_{x \in \Lambda_j}$)

$$\mathcal{H}_{free}(\phi) = \frac{1}{2}(\Phi^j, -\Delta^{(j)}_{eff}\Phi^j) + \sum_{k=j+1}^{N} \frac{1}{2}(\varphi^k, -\Delta^{(k)}_{eff}\varphi^k) \tag{2.14a}$$

$$= \frac{1}{2}\sum_{k=0}^{N}(\varphi^k, -\Delta^{(k)}_{eff}\varphi^k) \tag{2.14b}$$

The second formula (b) is the special case $j = 0$ of the first (a), and

$$\Delta^{(j)}_{eff}(x, y) = \int_{z \in \text{base}} \mathcal{A}(z, x)\ \Delta_z \mathcal{A}(z, y) \qquad \text{for} \qquad x, y \in \Lambda_j \tag{2.15}$$

Since $\varphi(x)$ are linear functions of ϕ, they will have a Gaussian probability distribution given by that of ϕ, in a free field theory. Because of the absence of coupling between fields attached to different layers $\Lambda_j \neq \Lambda_k$, there is no correlation between them. Therefore

$$\langle \varphi(x)\varphi(y)\rangle_0 = \mathrm{v}^j(x, y)\ \delta_{jk} \qquad \text{for} \qquad x \in \Lambda_j,\ y \in \Lambda_k \tag{2.16}$$

and this equation defines v^j as a positive semidefinite operator on the Hilbert space of square summable functions on Λ_j. The decomposition (2.13) of the field $\phi(z)$ induces a decomposition of the free propagator $v(z_1, z_2) = \langle \phi(z_1)\phi(z_2)\rangle_0$. According to eq. (2.16) it takes the form

$$v(z_1, z_2) = \sum_{j=0}^{N} \iint_{x_1, x_2 \in \Lambda_j} \mathcal{A}(z_1, x_1)\mathrm{v}^j(x_1, x_2)\mathcal{A}(z_2, x_2). \tag{2.17}$$

v^j may be recovered from v recursively by using eq. (2.6).

The decay properties of the kernels $\mathcal{A}(z, x)$ and propagators v^j which were discussed for the twogrid in subsection I-1 generalize to the multigrid as follows [8,9].

For $x \in \Lambda_j$, the kernel $\mathcal{A}(z,x)$ decays exponentially with distance $|z-x|$ with decay length a_j. The propagator $\mathrm{v}^0(x_1,x_2)$ on the coarsest lattice Λ_0 is a massless propagator which is given by the average of the massless propagator $v(z_1,z_2)$ over blocks $z_1 \in x_1$, $z_2 \in x_2$; $x_1,x_2 \in \Lambda_0$. The propagators $\mathrm{v}^j(x_1,x_2)$ on the other layers Λ_j $(j \neq 0)$ have an infrared cutoff. They decay exponentially with distance $|x_1-x_2|$ with decay length $a_{j-1} = La_j$. In units of the appropriate lattice spacing, the decay length of $\mathcal{A}(z,x)$ is 1, and the decay length of the propagators v^j is L (for $j \neq 0$).

In numerical work one is forced to restrict attention to a finite volume, this will provide an infrared cutoff for the last layer Λ_0 [In section II we will use a multigrid whose last layer Λ_0 contains only one point]. With such an infrared cutoff for the last layer, the multigrid transformation which has been described transforms a massless free field theory on base space Λ_N into a noncritical statistical mechanical system on the multigrid Λ, with no coupling or correlations between layers, and correlation length of about L (e.g. $= 2$ or 3) lattice spacings on each layer Λ_j, except the last layer Λ_0.

Let us now turn to field theory with interaction

$$\mathcal{H} = \mathcal{H}_{free} + V \quad ; \quad V(\phi) = \int_{z\in base} \mathcal{V}(\phi(z)) \tag{2.18}$$

By definition, the effective Hamiltonian $\mathcal{H}^j$ at length scale a_j is a function of the block spin $\Phi^j(x)$ on lattice Λ_j and is given by

$$e^{-\mathcal{H}^j(\Phi^j)} = \int \mathcal{D}\phi \left[\prod_{x\in\Lambda_j} \delta(\Phi^j(x) - \operatorname*{av}_{z\in x} \phi(z))\right] e^{-\mathcal{H}(\phi)} \tag{2.19}$$

Inserting the split (2.8) of the field ϕ on base space results in

$$\mathcal{H}^j(\Phi) = \mathcal{H}^j_{free}(\Phi) + V^j(\Phi) \tag{2.20a}$$

with

$$\mathcal{H}^j_{free}(\Phi) = \frac{1}{2}(\Phi, -\Delta^{(j)}_{eff}\Phi) \equiv -\frac{1}{2}\iint_{x,y\in\Lambda_j} \Phi(x)\Delta^{(j)}_{eff}(x,y)\Phi(y) \tag{2.20b}$$

and

$$e^{-V^j(\Phi^j)} = \mathcal{N}_j \int [\prod_{x\in\Lambda_{\geq j+1}}{}' d\varphi(x)]\, e^{-\sum_{k=j+1}^N \mathcal{H}^k_{free}(\varphi^k)} \exp\left[-V_{\Phi^j}(\sum_{k=j+1}^{N} \mathcal{A}\varphi^k)\right] \tag{2.20c}$$

Here we used the abbreviation

$$V_{\Phi^j}(\zeta) \equiv \int_{z\in base} \mathcal{V}\left(\int_{x\in\Lambda_j} \mathcal{A}(z,x)\Phi^j(x) + \zeta(z)\right). \tag{2.20d}$$

$\mathcal{N}_j$ is a suitable constant normalization factor. We use the notation

$$\Lambda_{\geq j} = \Lambda_j + \Lambda_{j+1} + \ldots + \Lambda_N \tag{2.21}$$

The prime on $\prod' d\varphi(x)$ stands to indicate that the fields φ satisfy a constraint (2.10), therefore one should only integrate over independent variables, e.g. those attached to sites $x \in \Lambda_k$ which are not centers of blocks $x' \in \Lambda_{k-1}$. φ^k is the restriction of φ to layer Λ_k, and

$$\mathcal{A}\varphi^k(z) = \int_{x\in\Lambda_k} \mathcal{A}(z,x)\varphi(x). \tag{2.22}$$

Using the standard notation $d\mu_\Gamma(\zeta)$ for a Gaussian measure with covariance Γ, eq. (2.20c) for the effective interaction Hamiltonian V^j can also be written as

$$e^{-V^j(\Phi^j)} = \int [\prod_{k=j+1}^{N} d\mu_{\mathrm{v}^k}(\varphi^k)] \exp\left[-V_{\Phi^j}\left(\sum_{k=j+1}^{N} \mathcal{A}\varphi^k\right)\right] \tag{2.20e}$$

I used the freedom to adjust normalization factors $\mathcal{N}_j$, this only changes $\mathcal{H}$ by an additive constant. Formulae (2.20) are valid for all j, in particular $j = 0$. To obtain partition functions and expectation values for the full theory, it remains to integrate over $\Phi^0 = \varphi^0$. In shorthand notation (2.13)

$$\begin{aligned} Z &= \int [\prod_{x\in\Lambda}{}' \, d\varphi(x)] \; e^{-\mathcal{H}(\mathcal{A}\varphi)} \\ \langle \mathcal{O} \rangle &= Z^{-1} \int [\prod_{x\in\Lambda}{}' \, d\varphi(x)] \; \mathcal{O}(\mathcal{A}\varphi) \, e^{-\mathcal{H}(\mathcal{A}\varphi)} \end{aligned} \tag{2.23}$$

Eq. (2.20e) generalizes formula (1.10) for the effective Hamiltonian.

The stated purpose of such a rewriting was to exhibit the effective Boltzmann factor $e^{-\mathcal{H}^j(\Phi)}$ as partition function of an auxiliary noncritical system in the situation where a_j may be much larger than the original lattice spacing $a \equiv a_N$. Let us discuss this point.

Whatever j, an integration over φ^0 never appears in (2.20e). The Gaussian measures $d\mu_{\mathrm{v}^k}(\varphi^k)$ $(k \neq 0)$ describe free field theory on lattices Λ_k with propagators (free correlation functions) v^j which decay exponentially, with correlation length $\approx L$ lattice spacings a_j. Therefore, in the limit of vanishing interaction, the statistical system whose partition function is (2.23) is noncritical. There are no correlations between layers in this limit.

It remains true to all orders in perturbation theory, and can be established more generally, at least for weakly coupled models, that the correlations between field variables $\varphi(x)$ and $\varphi(y)$ which are attached to sites x, y of the same layer Λ_j decay exponentially with decay length of order of L lattice spacings a_j. Also the correlations between $\varphi(x)$, $x \in \Lambda_j$, and $\varphi(y)$ with $y \in \Lambda_k$ with $k > j$ fall off exponentially with the "horizontal" distance $|x - y|$ with decay length of order a_j. But the number of points within this distance of x increases exponentially with $k - j$ (like $L^{d(k-j)}$ in d dimensions). Therefore there could still be strong correlations with very many variables, unless their strength decays quickly enough with $|k-j|$, i.e. in the "vertical" or "scale"- direction.

Rigorous work on iterated RG-transformations and phase cell cluster expansions for weakly coupled models [1,4,5,8,9,11,15,16,17] supports the view that sufficiently fast decay of correlations in the vertical direction, i.e. between (relatively) "low frequency fields" φ^j and "high frequency fields" φ^k, $k \gg j$, is a prerequisite for renormalizability of the interaction (in a particular regularization scheme which is specified by a choice of block spin and kernels $\mathcal{A}$). This insight supplements Wilson's observation [7] that the chief effect of interactions of a low frequency field Φ^j with high frequency fields φ^k $(k > j)$ *upon integration* of φ^k is to renormalize the effective (running) coupling constants which determine the relevant local part in the effective Hamiltonian $\mathcal{H}^j$. We will have occasion to come back to this point in section IV. A further prerequisite for renormalizability is that the bare coupling constant does not become infinite for finite values of the UV-cutoff a_N^{-1}, when some renormalized coupling constants are held fixed. This is violated in ϕ^4-theory in 4 dimensions and puts a limit on how far one can raise the UV-cutoff a_N^{-1}, see Lüscher's lectures at this school.

In conclusion, the noncriticality of the auxiliary statistical mechanical system on the multigrid $\Lambda_{\geq j}$ whose partition function is the effective Boltzmannian, has a kinematical aspect - the decay of propagators v^j which comes from the presence of an infrared cutoff a_{j-1} - but is also tied in with the workings of renormalization.

We spoke about correlations between layers. They come about because the interaction couples fields on different layers. For instance, a ϕ^4-interaction becomes

$$\begin{aligned}\phi(z)^4 &= \left[\sum_j \int_{x\in\Lambda_j} \mathcal{A}(z,x)\varphi(x)\right]^4 \\ &= \sum_{j_1}\cdots\sum_{j_4}\int_{x_1\in\Lambda_{j_1}}\cdots\int_{x_4\in\Lambda_{j_4}} \mathcal{V}_4(x_1,\dots,x_4)\varphi(x_1)\dots\varphi(x_4) \qquad (2.24)\end{aligned}$$

with

$$\mathcal{V}_4(x_1,\dots,x_4) = \int_{z\in base}\prod_{a=1}^4 \mathcal{A}(z,x_a)\ .$$

$\mathcal{V}_4$ decays exponentially in the horizontal direction (with decay length a_j, $j = \min(j_1,j_2)$ in $|x_1 - x_2|$, for instance), but it couples arbitrary layers $\Lambda_{j_1}\dots\Lambda_{j_4}$.

In the multigrid language, removal of the UV-cutoff means that one lets the number N of layers become infinite, so that the UV-cutoff $a_N^{-1} = L^N a_0^{-1} \to \infty$. Existence of the continuum limit without UV-cutoff becomes a kind of ∞ volume limit in this way, whose existence is assured by sufficient decay of correlations in the vertical directions. The catch is that this has to be brought about in the presence of interactions that are nonlocal in the vertical direction.

II. Multigrid Monte Carlo Simulations

Multigrid relaxation methods were introduced more than 10 years ago to solve partial differential equations efficiently [18]. This method to solve partial differential equations is also useful to compute propagators in external gauge fields, which are needed in simulations of gauge theories with fermions [31]. Here I describe joint work with S. Meyer [2] on the application of a multigrid method to simulate statistical systems. I got interested in this problem after Parisi's lectures in the Cargèse school 1983 [18] which mentioned the possibility. An earlier proposal by Hildegard Meyer-Ortmanns, J. Goodman and A.D. Sokal [19] is simpler than ours, but it ignores the need to minimize couplings between low- and high frequency fields in order to make the whole system really noncritical (see below).

1. Multigrid Simulation Procedure for $\lambda\phi^4$-Theory in 4 Dimensions

The simulation procedure must produce configurations ϕ of a real field (with one or more components) on a hypercubic lattice $\Lambda_N \equiv base$ of lattice spacing $a = a_N$, with probability distribution proportional to $e^{-\mathcal{H}(\phi)} \prod_{z\in\Lambda_N} d\phi(z)$,

$$\mathcal{H}(\phi) = \int_{z\in base} \left[-\frac{1}{2}\phi(z)\Delta\phi(z) + \frac{1}{2}m_u^2\phi(z)^2 + \lambda_u\phi(z)^4 \right] \tag{1.1}$$

We introduce a multigrid

$$\Lambda = \Lambda_0 + \Lambda_1 + ... + \Lambda_N \tag{1.2}$$

which consists of a sequence of $N+1$ lattices Λ_j of decreasing lattice spacing a_j, see figure 1, with $a_j/a_{j-1} = L$, L a small integer (2 or 3). In the numerical work performed so far, we used 3 or 4 lattices $\Lambda_0, ..., \Lambda_N$, with Λ_0 consisting of a single point so that the side length of all lattices equals a_0. One takes the fullest advantage of the multigrid by choosing L as small as possible, i.e. $L = 2$, because the correlation lengths on lattices Λ_j ($j \neq 0$) are of order L lattice spacings. We chose $L = 3$ to make the computer program more transparent. It is not strictly necessary to have L independent of j, we took $a_0/a_1 = 4$ and $a_2/a_1 = 3$ for work on a 12^4 lattice.

The points $x \in \Lambda_j$ specify cubes of sidelength a_j in $\Lambda_N = base$. These cubes are also denoted by x. The requirement $z\underline{\in}x$ shall mean that the point z in base space Λ_N is in cube x.

The multigrid procedure is defined by specifying for each $j = 0, ..., N-1$ block spins Φ^j which live on Λ_j, and a split of the field ϕ on Λ_N into a "fluctuation field" ζ^j, and a "background field" Ψ^j which is determined by the block spin Φ^j. We chose block averages as block spin

$$\Phi^j(x) = \operatorname*{av}_{z\underline{\in}x} \phi(z) \tag{1.3}$$

and a linear decomposition of ϕ

$$\phi(z) = \Psi^j(z) + \zeta^j(z) \tag{1.4a}$$

$$\Psi^j(z) = a_j^4 \sum_{x\in\Lambda_j} \mathcal{A}(z,x)\Phi^j(x) \quad , \quad z \in base \tag{1.4b}$$

The fluctuation field has block average 0, and the kernel $\mathcal{A}(z,x)$ is defined for $z \in base$ and all $x \in \Lambda = \Lambda_0 + ... + \Lambda_{N-1}$, it has property (I-2.6), viz

$$\operatorname*{av}_{z\underline{\in}y} \mathcal{A}(z,x) = a_j^{-4}\delta_{xy} \quad \text{for} \quad x,y \in \Lambda_j \tag{1.4c}$$

One sets $\Phi^N = \phi$, $\zeta^N = 0$, $\mathcal{A}(z,x) = a_N^{-4}\delta_{zx}$ for $x \in \Lambda_N$. Given the kernel $\mathcal{A}$, equations (1.3) and (1.4) define ζ^j as function of ϕ. The multigrid transform φ of ϕ is a field on the multigrid Λ and is defined so that the following formula holds

$$\zeta^j(z) = \sum_{k=j+1}^{N} a_k^{-4} \sum_{x\in\Lambda_k} \mathcal{A}(z,x)\varphi(x) \tag{1.5a}$$

This is accomplished by the definition

$$\varphi(x) = \operatorname*{av}_{z\underline{\in}x} \zeta^{j-1}(z) \quad \text{for} \quad x \in \Lambda_j,\ j \neq 0 \tag{1.5b}$$

We complete the definition by setting

$$\varphi(x) = \Phi^0(x) \qquad \text{for} \qquad x \in \Lambda_0 \ .$$

As a result of eq. (1.5b) and (1.5a) with $j = 0$ we find that the multigrid transform φ of ϕ determines

$$\phi = \mathcal{A}\Phi^0 + \zeta^0$$

viz

$$\phi(z) = \sum_{j=0}^{N} a_j^{-4} \sum_{x \in \Lambda_j} \mathcal{A}(z,x)\varphi(x) \ . \tag{1.5c}$$

Considering ϕ as determined by φ turns $\mathcal{H}(\phi)$ into Hamiltonian of a statistical mechanical system on the multigrid Λ, with basic random variables $\varphi(x)$, $x \in \Lambda$. The idea is that this system should be *not nearly critical, i.e. a given variable* $\varphi(x)$ *should have nonnegligible correlations with only a small number of other variables* $\varphi(y)$. This will in general require a proper choice of block spin and of the splitting procedures, i.e. kernels $\mathcal{A}$, in order to make the couplings between different layers small. Since the density of points increases exponentially with the label j of the layer Λ_j, there could otherwise be strong correlations between a variable $\varphi(x)$, $x \in \Lambda_k$ and many variables $\varphi(y)$ attached to sites y on finer layers Λ_j, $j = k+1, k+2, ..., N$. In this case the system as a whole could fail to be noncritical, cp. the discussion in section I-2 and below. In particular the coupling between block spin Φ^j and fluctuation field ζ^j in the Hamiltonian should be made as small as possible.

As we saw in section I, it is possible to choose the kernel $\mathcal{A}$ so that there is no coupling between ϕ^j and ζ^j in the kinetic part $\mathcal{H}_{free} = -\frac{1}{2}(\phi, \Delta\phi)$ of the Hamiltonian. This is achieved by imposing condition (2.7) which says that for $x \in \Lambda_j$, $\Delta_z \mathcal{A}(z,x)$ should be constant as a function of $z \in$ *base* on cubes $y \in \Lambda_j$. Equivalently, for $x \in \Lambda_j$,

$$\Delta_z \mathcal{A}(z,x) - \Delta_w \mathcal{A}(w,x) = 0 \quad \text{if} \quad z \underline{\in} y, \ w \underline{\in} y \quad \text{for some} \quad y \in \Lambda_j \ . \tag{1.6}$$

This equation can be solved by relaxation. Because of invariance under translations by integer multiples of a_j, it suffices to consider $x = 0$. Let $\chi(z) = 1$ if $z \underline{\in} x$ and 0 otherwise. One starts with $\mathcal{A}(z,x) = a_j^{-4}\chi(z)$; this satisfies the constraint (1.4c). To each z one selects a partner $w \neq z$ in the same cube $y \in \Lambda_j$ and updates $\mathcal{A}(z,x)$ and $\mathcal{A}(w,x)$ by opposite amounts in such a way that eq. (1.6) holds at the particular pair (z,w). This is repeated until convergence is reached. The result is shown for $a_j/a_N = 3$ in table 1a.

One sees that $\mathcal{A}(z,x)$ is very small unless z is in cube x or its nearest neighbours.

The small exponential tails of $\mathcal{A}(z,x)$ are inconvenient for numerical work. Therefore we set $\mathcal{A}(z,x) = 0$ unless z is in cube x or its nearest neighbours, and impose eq. (1.6) only for pairs (z,w) in cube x or its nearest neighbour. [In other words, Δ is replaced by a Laplacian with Dirichlet boundary conditions on the boundary of the desired support of $\mathcal{A}$.] The effective Laplacian obtained with such a kernel $\mathcal{A}$ is shown in table 1b.

So much about the kernel $\mathcal{A}$. It was designed to work for bare coupling strength λ_u which is not too large. In the Ising limit $\lambda_0 = \infty$ one would have to use nonlinear splits. An example of a nonlinear split for gauge theories will be described in subsection II-2.

Table 1a. Kernel $\mathcal{A}(z,0)$ for block size 3, at $z = (\mathbf{z}_1\mathbf{z}_2 00)$ for $\mathbf{z}_1, \mathbf{z}_2 = 0, ..., 7$

0.01	0.00	0.00	0.00	0.00	0.00	0.00	0.00
0.02	0.01	0.00	0.00	0.00	0.00	0.00	0.00
−0.02	−0.02	−0.01	0.00	0.00	0.00	0.00	0.00
−0.15	−0.13	−0.03	−0.01	0.00	0.00	0.00	0.00
−0.08	−0.07	0.00	0.00	−0.01	0.00	0.00	0.00
0.28	0.23	0.09	0.00	−0.03	−0.01	0.00	0.00
1.31	1.10	0.23	−0.07	−0.13	−0.02	0.01	0.00
1.57	1.31	0.28	−0.08	−0.15	−0.02	0.02	0.01

Table 1b. Effective Laplacian for block size 3, at $x = (\mathbf{x}_1\mathbf{x}_2 00)$ for $\mathbf{x}_1, \mathbf{x}_2 = 0, ..., 3$

0.03	0.00	0.00	0.00
−0.27	−0.02	0.00	0.00
2.81	−0.03	−0.02	0.00
−18.85	2.81	−0.27	0.03

Now I turn to a description of the sweeps through the multigrid. Let us again denote by φ^j the restriction of the multigrid field φ to layer Λ_j. One performs sweeps through the layers Λ_j of the multigrid in turn. Typically we performed 5 sweeps through Λ_j before proceeding to Λ_{j-1}, if $j \neq 0$. From Λ_0 we proceeded to Λ_N again. The purpose of the sweeps through Λ_j is to equilibrate φ^j (its equilibrium distribution will of course depend on the actual values of φ^k, $k \neq j$). One could achieve this by updating $\varphi(x)$ for all $x \in \Lambda_j$. Because φ satisfies the constraint that its average over all $x \in \Lambda_j$, which lie inside one cube $y \in \Lambda_{j-1}$, is zero, one would actually have to update pairs of variables $\varphi(x_1), \varphi(x_2)$ attached to sites $x_1, x_2 \in \Lambda_j$ inside the same cube. This is feasible. However it is simpler to equilibrate φ^j by updating the block spin Φ^j *keeping the fluctuation field* ζ^j *fixed.* Φ^j determines φ^j but satisfies no constraint. Φ^j receives contributions from $\varphi^{j-1}, \varphi^{j-2}, ...$ also. Therefore updating Φ^j may also change these. These changes are innocuous because they are made with transition probabilities which respect the stationary probability distribution. The dominant and desired effect is the change of φ^j, which is the highest frequency part of Φ^j. φ^j is expected to equilibrate fast, because it has a correlation length of $\approx L$ lattice spacings only (remember $L = 2$ is best).

The *fast approach to correct conditional probability distributions for fields* φ^j *on individual layers* comes from the fast decay of correlation functions $\langle \varphi^j(x)\varphi^j(y)\rangle$. It *is not sufficient to guarantee that the whole system approaches equilibrium fast*, because $\varphi^j(x)$ may be strongly correlated with many variables $\varphi^k(y)$ with $k > j$ - cp. section I-2. Suppose that the coupling between layers is so strong that φ^j is essentially determined[1] by the φ^k's with $k > j$. Then correct conditional probability distributions for φ^j are trivially attained, but from this one learns nothing about the approach of the whole system to equilibrium. Therefore it is wise to follow the path outlined by successful rigorous analytical work, as we did, and to reduce coupling between layers as much as possible. The simple proposal [19] to change the field ϕ by amounts that are constant on cubes $x \in \Lambda_j$ fails to do so. It leads to couplings between layers j and $k > j$ which are too big by a factor L^{k-j}. Therefore this is not a good choice, notwithstanding arguments which say in effect that it also leads to integrable correlation functions $\langle\varphi^j(x)\varphi^j(y)\rangle$.

I will now explain how the updating of $\Phi^j(x)$ is done. Sweeps through Λ_N are standard Monte Carlo sweeps. For $j \neq N$ one updates Φ^j, holding the fluctuation field ζ^j fixed. Therefore a change $\delta\Phi^j(x)$ modifies the original field ϕ by an amount

$$\delta\phi(z) = \mathcal{A}(z,x)\ \delta\Phi^j(x) \qquad\qquad (x \in \Lambda_j)$$

By our choice of $\mathcal{A}$, this is nonvanishing only for z in cube x in base space and on its nearest neighbors; these cubes have sidelength a_j.

The Hamiltonian $\mathcal{H}(\phi = \mathcal{A}\Phi^j + \zeta^j)$ is split as follows

$$\mathcal{H} = \mathcal{H}_{free} + \widehat{V}_j + [V - \widehat{V}_j] \qquad ; \qquad \widehat{V}_j = a_j^4 \sum_{x\in\Lambda_j} \widehat{\mathcal{V}}_j(\Phi^j(x))$$

where $\mathcal{H}_{free}$ is the exact free Hamiltonian, and $\widehat{V}_j$ is a local approximation to the effective interaction which one would obtain after integrating out ζ^j; it depends only on Φ^j, but not on ζ^j.

$$\widehat{\mathcal{V}}_j(\Phi^j(x)) = \frac{1}{2}m_j^2\Phi^j(x)^2 + \lambda_j\Phi^j(x)^4$$

We found that the choice $\widehat{\mathcal{V}}_j = 0$ is good enough, at least for reasonably small λ_u. Because of our truncation of kernels $\mathcal{A}$, $\mathcal{H}_{free}$ will contain a linear coupling between Φ^j and ζ^j, cp. eq. (I-1.6)

$$\mathcal{H}_{free}(\phi) = -\frac{1}{2}(\Phi^j, \Delta_{eff}^{(j)}\Phi^j) + (\Phi^j, \xi^j) \ + \ \Phi^j\text{-independent terms}$$
$$\xi^j(x) = -a_N^4 \sum_{z\in base} \zeta^j(z)\Delta_z\mathcal{A}(z,x) \qquad\qquad (x \in \Lambda_j)$$

The fundamental field ϕ is held in storage all the time. Before one starts sweeping through Λ_j, the block spin field Φ^j and fluctuation field ζ^j are computed; ξ^j is determined from ζ^j and is held in storage. Then one performs for each $x \in \Lambda_j$ the following steps. $\Phi^j(x)$ is updated by some amount $\delta\Phi^j(x)$ with probability distribution determined by $\mathcal{H}_{free} + \widehat{V}_j$. The corresponding change $\delta\phi(z)$ of the fundamental field on cube x and its nearest neighbours is computed, and the change δV of the exact interaction Hamiltonian $\lambda_u\phi^4 + \frac{1}{2}m_u^2\phi^2$ on these cubes as well. If $\delta V - \delta\widehat{V}_j > 0$

[1] For example, deep in the two phase region the fluctuation field ζ^0 (which is given by φ^k's with $k > 0$) "knows" the state of the system. It determines the (sign of) magnetization $\Phi^0 = \varphi^0$. If there were no coupling between ζ^0 and Φ^0, updating Φ^0 would give a tunneling probability of 50%, because $\mathcal{H}^0(\Phi^0)$ is symmetric. The observed tunneling rate is much lower.

then the proposed change $\delta\Phi^j$ is rejected with probability $1 - \exp(-\delta V + \delta\widehat{V}_j)$. The parameters m_j, λ_j, which are free to begin with, must be chosen so that the rejection rate is low. With the choice $m_j = \lambda_j = 0$ we got acceptance rates of 90% at $\lambda_u \approx 2/3$.

For one sweep through Λ_j, one must compute the interaction for each $z \in$ *base* $2^4 + 1$ times ($= 1 +$ no. of nearest neighbors). This suggests that the amount of computational work for one sweep through a layer Λ_j for $j \neq 0, N$ is about the same as for a standard Monte Carlo sweep through the basic lattice. This turned out to be correct. A standard FORTRAN 77 program was written. Priority was given to transparency, without bothering about vectorization. On a VP 100 machine, it performed as follows for a 12^4 lattice, in comparison with ordinary Monte Carlo sweeps through the finest lattice

1 sweep ordinary Monte Carlo	1.7 sec
same + evaluation of effective action	
(once per 5 sweeps)	2.5 sec
1 sweep through trigrid, brutto	5.7 sec

The time for one sweep through a trigrid includes evaluation of the effective action and splitting of the field into block spin and fluctuation field part (once per 5 sweeps).

To show the efficiency of the multigrid procedure in fighting critical slowing down, we show the total magnetization vs. sweep number for a single component ϕ^4-model in the two phase region slightly above the critical point, compared to a standard Monte Carlo run (figure 2). The effective action $\mathcal{H}^0$ for the same parameter values is shown in figure 3b.

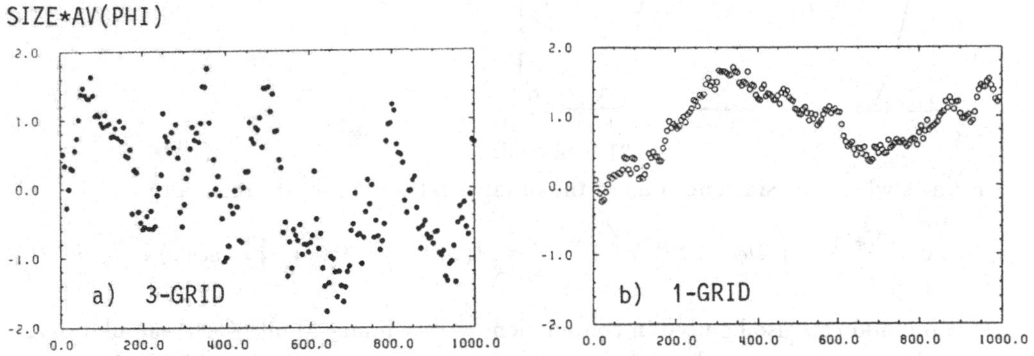

Fig. 2. Magnetization vs. no. of sweeps through a multigrid (a) compared with standard Monte Carlo (single grid) (b) at $g_0 = 4!\lambda_u = 16.376$, $m_u^2 = -1.15$ (cold start).

2. Effective Potential and Conditional Probabilities

I will now explain how to extract the effective action (or potential) $\mathcal{H}_{eff}(\Phi^0) \equiv \mathcal{H}^0(\Phi^0)$ and its derivatives. The index 0 refers to layer Λ_0, which is a single point. By definition the real variable

$$\Phi^0 = \text{magnetization} \tag{2.1}$$

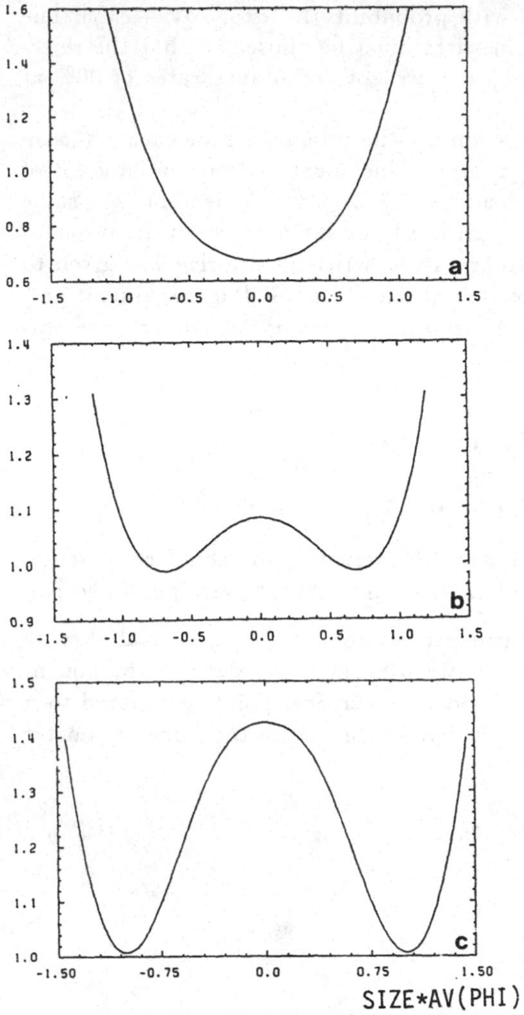

Fig. 3. Effective Potential as a function of the magnetization (in units where sidelength of the lattice = 1) for $4!\lambda_u = 16.376$ and three values of m_u^2 near the critical surface
a) $m_u^2 = -1.14$
b) $m_u^2 = -1.15$
c) $m_u^2 = -1.16$

in units where the sidelength a_0 of the original lattice Λ_N is set to 1, and

$$e^{-\mathcal{H}^0(\Phi^0)} = \int \mathcal{D}\phi \; e^{-\mathcal{H}(\phi)} \; \delta\Big(\Phi^0 - \operatorname*{av}_{z\in\Lambda_N} \phi(z)\Big) \quad ; \quad \mathcal{D}\phi = \prod_{z\in\Lambda_N} d\phi(z) \, . \qquad (2.2)$$

Our method can also be used in conjunction with ordinary Monte Carlo simulations, for polynomial theories like $\lambda\phi^4$. In contrast with other approaches [13,20,21] we do *not* perform separate simulations for individual values of Φ^0.

Instead we regard the original Hamiltonian $\mathcal{H}$ as a function of the magnetization Φ^0 and fluctuation field $\zeta^0(z)$, viz

$$\Phi^0 = \operatorname*{av}_{z\in\Lambda_N} \phi(z) \qquad , \qquad \zeta^0(z) = \phi(z) - \Phi^0 \, . \qquad (2.3)$$

We define fluctuating coupling constants $g_n(\zeta^0)$ by

$$\mathcal{H}(\Phi^0 + \zeta^0) = \sum_{n=0}^{4} g_n(\zeta^0)\,(\Phi^0)^n \qquad (2.4)$$

The Monte Carlo simulation produces a sequence of N_S configurations ϕ_a, this specifies a sequence of fluctuation fields ζ_a. For each of these configurations we evaluate the fluctuating coupling constants

$$g_{n,a} = g_n(\zeta_a) \tag{2.5}$$

We show that, by using these data, one can compute the power series expansion of $\mathcal{H}^0(\Phi^0)$ around any arbitrary point $\widehat{\Phi}$ up to any desired order in

$$\delta\Phi = \Phi^0 - \widehat{\Phi} \,. \tag{2.6}$$

Without loss of generality we assume that an additive constant in $\mathcal{H}$ has been so chosen that $\int \prod d\phi(z) \exp[-\mathcal{H}(\phi)] = 1$. The probability distribution $\exp[-\mathcal{H}(\phi)]\ \mathcal{D}\phi$ of ϕ specifies a probability distribution of Φ^0, viz $\exp[-\mathcal{H}^0(\Phi^0)]\ d\Phi^0$, and a probability distribution $dP(\zeta^0)$ of ζ^0.

We exploit the fact that the probability of Φ^0 equals the probability of ζ^0 times the conditional probability of Φ^0 given ζ^0, integrated over ζ^0. The conditional probability density of Φ^0 for given ζ^0 is

$$\mathcal{Z}(\zeta^0)^{-1} e^{-\mathcal{H}(\Phi^0+\zeta^0)} \quad \text{with} \quad \mathcal{Z}(\zeta^0) = \int d\Phi^0\ e^{-\mathcal{H}(\Phi^0+\zeta^0)} \tag{2.7}$$

The Monte Carlo configurations ζ_a are distributed with probability $dP(\zeta)$, thus $\int dP(\zeta^0)\ f(\zeta^0) = N_S^{-1} \sum_a f(\zeta_a)$. Therefore

$$\begin{aligned} e^{-\mathcal{H}^0(\widehat{\Phi}+\delta\Phi)} &= N_S^{-1} \left\{ \sum_{a=1}^{N_S} \mathcal{Z}(\zeta_a)^{-1} e^{-\mathcal{H}(\widehat{\Phi}+\delta\Phi+\zeta_a)} \right\} \\ &= N_S^{-1} \left[\sum_{a=1}^{N_S} \mathcal{Z}_a^{-1} \exp\left[-\sum_{n=0}^{4} g'_{n,a}(\widehat{\Phi})(\delta\Phi)^n \right] \right] \end{aligned} \tag{2.8}$$

Herein $\mathcal{Z}_a$ is defined by a 1-dimensional integral, and $g'_{n,a}$ are defined by reexpansion, as follows

$$\mathcal{Z}_a = \int d\Phi \exp\left[-\sum_{n=0}^{4} g_{n,a}\Phi^n \right]$$

$$\sum_{n=0}^{4} g_{n,a}\Phi^n = \sum_{n=0}^{4} g'_{n,a}(\widehat{\Phi})(\Phi - \widehat{\Phi})^n$$

Expanding both sides of (2.8) in a power series in $\delta\Phi$ expresses $\mathcal{H}^0(\widehat{\Phi})$ and its derivatives in terms of the fluctuating coupling constants $g_{n,a}$.

This method could be extended to obtain information about effective actions $\mathcal{H}^j$ at other length scales. In our multigrid approach, the fluctuating coupling constants are calculated for the purpose of updating Φ^0 (the block spin at the last layer Λ_0, which is a single point) anyway. The effective action $\mathcal{H}^0$ is therefore obtained essentially for free. Some examples of effective actions in the vicinity of the critical surface are shown in Fig. 3. From the effective potential $\mathcal{H}^0(\Phi^0)$ one can determine renormalized coupling constants and masses if one knows the wave function renormalization constant Z_1. In a particular definition they are 2nd and 4th derivative of $\mathcal{H}^0(Z_1^{1/2}\Phi)$ at the minimum of $\mathcal{H}^0$. In the 2-phase region, it is more convenient to use the position of the minimum (i.e. the vacuum expectation value of the field) to define a coupling constant, cp. Lüscher's lectures at this school. Results will be presented elsewhere [2].

To compute wave function renormalization one may determine an effective action $\mathcal{H}^0(\Phi^0, \Phi^0_{,\mu})$ which depends on the 5 lowest Fourier components of ϕ. Again it is not necessary to do separate sweeps for individual values of $\Phi^0, \Phi^0_{,\mu}$.

3. A Multigrid Procedure for Pure Gauge Theories

In this subsection I will sketch a multigrid procedure for pure gauge theories. At the time of this writing I do not have a finished computer program to implement it yet.

In a pure lattice gauge theory, the fundamental field variables $u(b)$ are attached to links b of a lattice Λ_N; they take their values in the gauge group G. The gauge group shall be a group of matrices, like $SU(2)$ or $SU(3)$. We denote by $\mathcal{G}$ the linear space of all matrices which are linear combinations of elements of G with real coefficients.

To set up the multigrid procedure, one must define a block spin U^j for every layer Λ_j of the multigrid, and a split of the field u into a background field B^j, which is determined by the block spin U^j, and a fluctuation field ζ^j. Both B^j and ζ^j are fields attached to links of the basic lattice. The updating procedure will then consist in sweeps through the layers Λ_j. On each layer, the block spin U^j will be updated keeping the fluctuation field ζ^j fixed. This induces a change in the fundamental field u.

One could use block spins which take their values in the linear space $\mathcal{G}$. This leads to effective Hamiltonians of dielectric type [22]. Here we shall work with block spins $U^j(b')$ which take their values in the gauge group G, using objects $\Phi^j(b') \in \mathcal{G}$ as auxiliary quantities only. Also B^j and ζ^j will take their values in G. Our proposal is motivated in part by rigorous analytical work of Balaban [11] on RG-transformations and UV-stability in pure gauge theories.

The decomposition of u will exhibit it as a product, rather than a sum

$$u(b) = B^j(b)\ \zeta^j_\mu(z) \qquad \text{for} \qquad b = (z + e_\mu, z) \tag{3.1}$$

e_μ = lattice vector in Λ_N in μ-direction, $\mu = 1, ..., 4$. Under gauge transformations, $B^j(b)$ transforms like a lattice gauge field, but $\zeta^j_\mu(z)$ transforms like a matter field in the adjoint representation sitting at site z.

To define a block spin, we imagine that Λ_j has been identified with a sublattice of Λ_{j+1}. In this way, a point $i(x) \in \Lambda_{j+1}$ is assigned to every point $x \in \Lambda_j$. For instance, $i(x)$ may be the block center when x' is regarded as a block = cube in Λ_{j+1}. If $x \in \Lambda_j$ then $\hat{i}(x) \equiv i^{N-j}(x) = i \circ ... \circ i(x)$ will be a site in Λ_N. Under a gauge transformation Ω on Λ_N, the block spin $U^j(b)$ will transform according to

$$U^j(b) \to \Omega(\hat{i}(y))\ U^j(b)\ \Omega(\hat{i}(x))^{-1} \qquad \text{for} \qquad b = (x, y)\ ,\quad x, y \in \Lambda_j \tag{3.2}$$

The block spin is recursively defined. One starts with $U^N = u$ on Λ_N. For $j < N$ one selects classes $C^j(b)$ of paths C on Λ_{j+1} from $i(x)$ to $i(y)$, when $b = (x, y)$ is a link in Λ_j. Set

$$\Phi^j(b) = \sum_{C \in C^j(b)} U^{j+1}(C) \in \mathcal{G} \tag{3.3}$$

$U^{j+1}(C)$ is the parallel transporter along C constructed from the lattice gauge field U^{j+1} on Λ_{j+1}. Finally one performs a generalized polar decomposition [22] of $\Phi^j(b)$ to obtain $U^j(b)$

$$\Phi^j(b) = U^j(b)\ \rho^j_\mu(x) \qquad \text{for} \qquad b = (x + e'_\mu, x)\ . \tag{3.4}$$

If $G = SU(2)$ then $\rho^j_\mu(x)$ is a nonnegative real number. For $G = U(N)$, $N \geq 3$ it is a positive semidefinite $N \times N$ matrix, and for $G = SU(N)$, $N \geq 3$ it is a positive semidefinite $N \times N$ matrix times a complex phase factor $e^{i\theta}$. The decomposition is almost always unique.

A convenient special choice is the $\sqrt{3}$ scheme of Cordery, Gupta, and Novotny [23]. In this scheme, Λ_j is oblique to Λ_{j+1}, and $a_j = \sqrt{3}\ a_{j+1}$. If $b = (x,y)$ is a link in Λ_j, then $i(x)$ and $i(y)$ lie at opposite ends of the space diagonal in a 3-dimensional elementary cube in Λ_{j+1}. $C^j(b)$ consists of all paths of length $3a_{j+1}$ from $i(x)$ to $i(y)$.

Now we turn to the definition of the background fields B^j. The fundamental action $\mathcal{H}(u)$ may for instance be a Wilson action. Given the block spin U^j on Λ_j, one may define a background field B^j on Λ_N to equal that lattice gauge field u which minimalizes some approximation $\widehat{\mathcal{H}}^j(u)$ to $\mathcal{H}(u)$ - for instance $\widehat{\mathcal{H}}^j(u) \equiv \mathcal{H}(u)$ - subject to the constraint that the blockspin on Λ_j which is determined by u equals U^j. This is the analog of the background field $\Psi^j = \mathcal{A}\Phi^j$ of section I. This Ψ^j was a field on Λ_N that extremizes the free massless Hamiltonian $\mathcal{H}_{free}$ subject to the condition that its block spin on Λ_j has the prescribed value Φ^j.

When the background field B^j is defined, as a function of U^j and therefore of u, the fluctuation field ζ^j is also defined as a function of u. Updating the block spin U^j at b while keeping the fluctuation field ζ^j fixed, changes u by

$$\delta u(b') = \delta B^j(b')\ \zeta^j_\mu(x') \qquad\qquad (b' \in \Lambda_N \text{ arbitrary})\ .$$

δB^j is the change of the background field induced by the change of the block spin U^j. With a background field defined as above, a change of U^j at link $b \subset \Lambda_j$ can change $\delta u(b)$ at links $b' \subset \Lambda_N$ very far from b, although usually by very little. This is the analog of the problem of the exponential tails in the kernels $\mathcal{A}$. It is inconvenient for numerical purposes.

I propose to remedy this problem in a similar way as with the kernels $\mathcal{A}$. While updating at a link $b = (x,y) \subset \Lambda_j$ one admits a change of u only at links $b' = (z_1, z_2)$ on Λ_N such that z_1, z_2 are in a finite neighbourhood $\mathcal{N}_j(b) \subset \Lambda_N$. x, y may be regarded as cubes of side a_j in Λ_N. We may for instance take $\mathcal{N}_j(b)$ as union of these cubes and their nearest neighbours. This fixes at the same time a set $\widetilde{\mathcal{N}}_j(b)$ of links in Λ_j.

In numerical work, the purpose of the background field definition is to specify how u changes when the block spin is updated at a link $b \subset \Lambda_j$. While performing such an updating, an instantaneous field configuration u is in storage. We may then admit for the stated purpose a definition of B^j on $\mathcal{N}_j(b)$ which depends not only on the block spin U^j but also on the configuration u outside of $\mathcal{N}_j(b)$ where it will not be affected by the updating. (We do not need B^j outside of $\mathcal{N}_j(b)$ because it will not be changed there by the updating of $U^j(b)$.)

A suitable definition is to take for B^j that lattice field which extremizes $\widehat{\mathcal{H}}^j$ subject to the condition that $B^j = u$ outside of $\mathcal{N}_j(b)$, and its block spins at links $b_1 \in \widetilde{\mathcal{N}}_j(b)$ take their prescribed values $U^j(b_1)$.

The extremum may be determined by relaxation on $\mathcal{N}_j(b)$. While relaxing one changes B^j at pairs of links in $\mathcal{N}_j(b)$ in such a way that the block spins U^j remain unchanged. This is possible if the sets $\mathcal{C}^j$ of paths satisfy suitable conditions [24].

It is not necessary to find the exact extremum. One may define the background field as result of a specified number of relaxation sweeps starting from a specified initial configuration $B^{\circ j}$. The initial configuration $B^{\circ j}$ must be chosen independent of the above mentioned instantaneous field configuration u *inside* $\mathcal{N}_j(b)$, apart from its dependence on the block spins. Otherwise one risks violating detailed balance. For instance one could choose $B^{\circ j}$ as follows on $\mathcal{N}_j(b)$. For links $b' \subset \Lambda_N$ which connect different cubes $x, y \in \Lambda_j$ one sets $B^{\circ j}(b') = U^j(\bar{b})$, $\bar{b} = (x, y)$. For links b' inside a single cube set $B^{\circ j}(b') = 1$. This respects block spins when sets $\mathcal{C}^j$ of paths are suitable chosen (à la Balaban for example).

4. Cautionary Remark

We insisted on producing field configurations ϕ on the basic lattice with exactly the right probability distribution - as given by the action on the basic lattice -, independent of how many sweeps are performed at each visit to a particular layer.

Proposals have been made where this is not true [25]. It was proposed repeatedly to update block spins Φ^j with probabilities given by approximate effective actions $\widehat{\mathcal{H}}^j$, without our subsequent filtering which corrects for the error in the effective action. This produces incorrect probabilities for Φ^j. Relying on the subsequent sweeps through the finest lattice to correct this is very dangerous, because correction of the probability distribution of Φ^j will be hampered by critical slowing down. One will get incorrect probabilities no matter how many sweeps are made altogether, unless one makes infinitely many sweeps *each time* one visits the basic lattice.

III. Towards Computer Simulation of Continuum Theories without UV-Cutoff via Simulation of Polymer Systems on the Multigrid

Suppose we want to take the continuum limit by letting the lattice spacing $a_N = L^{-N} a_0$ of the basic lattice go to zero. If we want to keep the scale factor L in the multigrid fixed, the number N of layers will have to become infinite. Continuum theories can be reformulated as theories on an "infinitegrid"

$$\Lambda_\infty = \Lambda_0 + \Lambda_1 + \Lambda_2 + \ldots \quad , \quad a_j = L^{-j} a_0 \tag{1.1}$$

The number of points per volume a_0^d increases exponentially with j like L^{dj} in d dimensions. Obviously one cannot store in a computer field configurations φ^j on all the ∞ many layers Λ_j. We propose to reformulate the theory as a polymer system on the infinitegrid Λ_∞ and to do simulations for that.

It was proposed by K. Pinn and the author some time ago [3] to do computations for ordinary lattice field theories by simulating polymer systems on the lattice. It was pointed out that one could handle nonlocal Hamiltonians in this way which are difficult to handle otherwise. Meanwhile, such simulations were done with great success for the Ising model by Swendsen and others [26].

Suppose for simplicity[1] that we are only interested in the long distance behaviour

[1] The method can deal with correlation functions that are smeared with arbitrary smooth test functions, but we will not discuss this here.

of the theory in the sense that we consider only observables $\mathcal{O}$ which depend on the fundamental field ϕ through the block spin Φ on Λ_0 only. Then we could compute their expectation values by Monte Carlo simulation on Λ_0 if we knew the exact effective Hamiltonian $\mathcal{H}_{eff}(\Phi)$, also called $\mathcal{H}^0$ in the multigrid context.

$$\langle \mathcal{O} \rangle = Z^{-1} \int [\prod_{x\in\Lambda_0} d\Phi] \ \mathcal{O}(\Phi) \ e^{-\mathcal{H}_{eff}(\Phi)} \tag{1.2}$$

with Z such that $\langle 1 \rangle = 1$. To use this, the effective Boltzmann factor $e^{-\mathcal{H}_{eff}}$ will have to be computed *at the same time* from a suitable polymer representation.

For the purpose of this section, polymers on Λ_∞ are certain finite subsets P of Λ_∞ *which contain at least one point x in* Λ_0. We write Ψ for the background field which is determined by the lattice field on Λ_0

$$\Psi(z) = \int_{x\in\Lambda_0} \mathcal{A}(z,x)\Phi(x) \tag{1.3}$$

Letter z is reserved for arguments on base space, i.e. in the continuum. We think of $\mathcal{H}_{eff}$ as sum of a local part $\mathcal{H}_{loc}$ plus nonlocal terms that are irrelevant in the RG-sense. $\mathcal{H}_{loc}$ will be determined by renormalized coupling constants (including a mass m_0 etc.) = running coupling constants at scale a_0; we regard them as given. For instance

$$\begin{aligned} \mathcal{H}_{loc}(\Psi) &= \mathcal{H}_{kin}(\Psi) + \int dz \ V_{loc}(\Psi(z)) \\ \mathcal{H}_{kin}(\Psi) &= \frac{1}{2}\int dz \ \left[(\nabla_\mu \Psi(z))^2 + m_0^2 \Psi(z)^2\right] \end{aligned} \tag{1.4}$$

The polymer representation of the effective Boltzmannian will be of the form

$$e^{-\mathcal{H}_{eff}(\Psi)} = e^{-\mathcal{H}_{kin}(\Psi)} \sum_{X\subseteq\Lambda_0} \exp\left[-\int_{z\underline{\in} X} V_{loc}(\Psi(z))\right] \cdot \sum_{\mathcal{P}:X+\sum_{P\in\mathcal{P}} P\cap\Lambda_0=\Lambda_0}{}' \ \prod_{P\in\mathcal{P}} R(P|\Psi) \tag{1.5}$$

The inner sum $\sum'$ extends over collections $\mathcal{P}$ of compatible polymers P which intersect both Λ_0 and some other layers in Λ. The subset X together with the intersection of all the polymers P with Λ_0 must exhaust Λ_0. Points $x \in \Lambda_0$ are identified with cubes in base space, we write $z\underline{\in}x$ when z is in such a cube, and similarly for subsets $X \subset \Lambda_0$ (union of cubes). The polymer activity $R(P|\Psi)$ shall be local in the sense that it depends on $\Psi(z)$ only for $z\underline{\in}\ P\cap\Lambda_0$.

There are three different notions of compatibility which are useful in practice, depending on how one sets up the polymer representation

(a) P_1, P_2 compatible if $P_1 \cap P_2 = \emptyset$ (i.e. P_1, P_2 are disjoint).

(b) P_1, P_2 compatible if their $\subseteq$-convex hulls are disjoint. (1.5')

(c) P_1, P_2 compatible if $P_1 \cap \Lambda_0$ and $P_2 \cap \Lambda_0$ are disjoint.

Regarding points $x \in \Lambda_\infty$ as cubes in base space, the $\subseteq$-convex hull of a finite subset $X \subset \Lambda_\infty$ consists of all $y \in \Lambda_\infty$ such that $x_1 \subseteq y \subseteq x_2$ for some cubes $x_1, x_2 \in X$.

The fact that terms $X \neq \Lambda_0$ with R-factors represent irrelevant additions to $\mathcal{H}_{loc}$ is reflected in renormalization conditions on the polymer activities R such as

$$R(P|0) = 0 \quad ;$$
$$\underset{z_1 \in x}{\text{av}} \int dz_2 \ \frac{\delta^2}{\delta\Psi(z_1)\delta\Psi(z_2)} R(P|\Psi)\Big|_{\Psi=0} = 0 \qquad \text{for all } x \in \Lambda_0 \text{ etc.} \tag{1.6}$$

The polymer activities themselves are finitedimensional integrals. Typically there is one integration variable $\varphi(x)$ for each point $x \in P$ which is not in Λ_0.

For the simulation procedure the convergence properties of the sums in the polymer representation of $\exp[-\mathcal{H}_{eff}]$ are crucial. The polymer activities $R(P|\Psi)$ must be sufficiently small when one of the following conditions is fulfilled

1. P contains many points.
2. Points in P are far apart horizontally. (1.7a)
3. P contains a point in Λ_k with k large.

The relevant horizontal distance is the length of the shortest tree on points $x \in P$. The length of the tree's link between $x \in \Lambda_j$, $y \in \Lambda_k$ for $k \leq j$ is the distance between the centers of cubes x, y in units of a_k.

If the sum over polymers in (1.5) is restricted to polymers which do not intersect layers Λ_k with $k > j$, then the result is the effective Boltzmannian of a theory with UV-cutoff a_j^{-1}. Thus condition 3 reflects the desire that this cutoff dependence should be weak.

More precisely, one must insist on UV-convergence in the sense that for any $x \in \Lambda_0$, $j \geq 0$

$$\sum_{P: x \in P, \ P \cap \Lambda_j \neq \emptyset} |R(P|\Psi)| \leq b(j) \tag{1.7b}$$

with $\sum_j b(j) < \infty$. For instance

$$b(j) = b_0 \ L^{-\alpha j} \quad , \quad \alpha \approx 2 \quad , \tag{1.7c}$$

or at least

$$b(j) = b_0 \ j^{-2} \quad , \tag{1.7d}$$

b_0 should better not be too large.

Exponential UV-convergence (1.7c) requires exact fulfilment of renormalization conditions, cp. section IV. It comes about because convergent graphs behave typically like $\int (d^4p/p^4) \ p^{-2}$ and $p \sim a_j^{-1} = L^j a_0$. The graphs in a nonperturbative context are not ordinary Feynman graphs, but have similar properties. This is explained in Pordt's contribution to these proceedings [5].

In the next section I will describe ways to derive polymer representations of the form (1.5) - with compatibility condition of the most convenient form (1.5'c) - whose convergence properties can be proven analytically to hold true at least in some weakly coupled models. (A renormalizable example is massless lattice ϕ^4-theory, whose infrared behaviour can be studied on an infinitegrid $\Lambda_0 + \Lambda_{-1} + \Lambda_{-2} + \ldots$, $a_{-j} = L^j a_0$). Lack of convergence will be very easy to detect in numerical work.

Let us now turn to simulations. Suppose for the sake of the following argument that the activities $R(P|\Psi)$ are all nonnegative. Simulations require a finite volume $|\Lambda_0|$. A state of the polymer system is a pair $(\mathcal{P}, \Phi)$, when $\mathcal{P}$ is a possibly empty collection of compatible polymers on the infinitegrid, and Φ is a field configuration on Λ_0.

Since every polymer $P \in \mathcal{P}$ contains at least one point $y \in \Lambda_0$, and no two compatible polymers may contain the same point $y \in \Lambda_0$, there can only be finitely many of them, and we may conveniently store the coordinates $(j, x^1, ..., x^d)$ of all the points $x \in P$, including the label j of the layer Λ_j in which is x. $\mathcal{P}$ determines X as Λ_0 minus union of the sets $P \cap \Lambda_0$, $P \in \mathcal{P}$. The states are distributed with probability distribution

$$d\text{prob}(\mathcal{P}, \Phi) \propto \Big[\prod_{P \in \mathcal{P}} R(P|\mathcal{A}\Phi)\Big] \exp\Big[-\mathcal{H}_{kin}(\mathcal{A}\Phi) - \int_{z \in X} \mathcal{V}_{loc}(\mathcal{A}\Phi(z))\Big]\, \mathcal{D}\Phi \qquad (1.8)$$

One makes sweeps through Λ_0. While at $x \in \Lambda_0$ one may update $\Phi(x)$, add or remove a point from the polymer P with $x \in P$. If there is no such polymer P in $\mathcal{P}$, it is not possible to remove a point, but one may create a new polymer P which contains one point y in addition to x. Only points y can be added which give rise to a new polymer which is compatible with the others. Appropriate transition probabilities for these steps are given by standard Metropolis rules [3].

In order to add points y to polymers $P \ni x$ one may propose y with an a priori probability, and then filter proposals after computing the change in activity. One may propose a layer j with probability $b(j)/\sum b(k)$, and a point y in Λ_j with probability $\propto \exp(-\text{tree distance of } y \text{ from } P)$.

The convergence properties stated above will ensure that the probability of encountering any polymers with many points, or having points in layers Λ_k with large k, will be small, in spite of the fact that the number of points in Λ_k increases exponentially with k.

Simulation will produce a sequence of N_S states $(\mathcal{P}_a, \Phi_a)$. The expectation value of an observable $\mathcal{O}(\Phi)$ will be given by

$$\langle \mathcal{O} \rangle = N_S^{-1} \sum_a \mathcal{O}(\Phi_a) \qquad (1.9)$$

in the limit $N_S \to \infty$. This follows from eqs. (1.2) and (1.5).

Unfortunately, activities $R(P|\Psi)$ are not usually positive semidefinite. How to deal with this problem effectively is one of the main unsolved problems in this approach. A poor mans way is to use $|R|$ for probabilities and include signs in expectation values. This will only work when there are very few polymers.

When activities are positive, it is legitimate to impose the supplementary condition that there is an even number of polymers P (for large volume) because the presence of a polymer far away will not affect expectation values of local quantities in the absence of long range order. One might be tempted to generalize this to situations with possibly negative activities, restricting attention to states with even numbers of polymers with negative activities. This would make the right hand side of (1.8) positive. But it is incorrect. Is there a correct way to exploit the idea that faraway negative polymers are unimportant? This problem is under study, but I am not prepared to present results yet. The idea is to use polymerization once again to decouple correlations between sign factors.

The expansion (1.5) which I proposed as a basis for polymer simulations can be modified in several ways by using methods of the theory of polymer system to rewrite the expression.

For instance, locality of $\mathcal{H}_{kin}, \mathcal{V}_{loc}$, in Ψ means that they are somewhat nonlocal functions of the blockspin Φ, because of nonlocality of kernel $\mathcal{A}$. The same is therefore true of polymer activities R (We had to deal with such a locality problem in section II already). Polymer activities could be made local in Φ at the expense of introducing polymers which lie entirely in Λ_0. These could have activities which lack the factors of fractional powers of coupling constants that are present in the other polymer activities and make them small in weakly coupled models.

I conclude the section by mentioning an alternative approach. If the compatibility condition is of the convenient form (1.5'c), as will be the case for the expansions described in section IV, then expansion (1.5) can be partially summed. For nonempty finite subsets S of Λ_0 define sums over polymers

$$\widetilde{R}(S|\Psi) = \sum_{P:P\cap\Lambda_0=S} R(P|\Psi) \,. \tag{1.10}$$

Making use of the fact that compatibility condition (1.5'c) involves only $P \cap \Lambda_0$, expansion (1.5) takes the form of a polymer expansion on the simple lattice Λ_0

$$e^{-\mathcal{H}_{eff}(\Phi)} = e^{-\mathcal{H}_{kin}(\Psi)} \sum_{X\subseteq\Lambda_0} \exp\left[-\int_{z\in X} \mathcal{V}_{loc}(\Psi(z))\right] \cdot \sum_{\mathcal{P}:X+\sum_{S\in\mathcal{P}} S=\Lambda_0} \prod_{S\in\mathcal{P}} \widetilde{R}(S|\Psi) \tag{1.11}$$

$\mathcal{P} =$ collection of disjoint finite subsets S of Λ_0. One could think of computing first the activities (1.10) in parametrized form by polymer simulation on an infinitegrid as described before, and then proceed to a polymer simulation on Λ_0 based on (1.11). This would be more in the spirit of a renormalization group approach, except that one expands effective Boltzmannians rather than effective Hamiltonians, for reasons explained in section I. The need for a parametrization of the Ψ-dependence in $\widetilde{R}$ would be a drawback of this alternative approach, though.

IV. Polymer Systems on the Multigrid

1. General Theory of Polymer Systems

In the last section it was proposed to use eq. (III-1.5) as a basis for computer simulation of continuum theories. This equation is an example of a general kind of expansions that are used in statistical mechanics and quantum field theory. Another example are the Feynman graph expansions of perturbation theory. We will only be interested in convergent expansions, though. To develop a general point of view, let us ask the

Question: What do expansion methods of (classical) statistical mechanics really do?

Answer: They express observable quantities (e.g. free energy as function of external fields or sources) of an infinite system in terms of properties of small subsystems.

Expansion = sequence of approximate answers that are determined by partition functions of ("small") subsystems of increasing size.

Example: Virial expansion for a real gas. The leading term (ideal gas) is known if the partition function for a single particle is known. The 1st order correction involves

2-particle clusters and is known if the partition functions of systems of 1 or 2 particles are known, etc.

Different expansions are obtained by different choices of subsystems. For instance, in a lattice gas the subsystem might either consist of $n = 1, 2, 3, \ldots$ particles, or it might consist of $1, 2, 3, \ldots$ sites that could be occupied.

A systematic procedure is provided by the theory of polymer systems [27]. The polymers considered here are a mathematical abstraction. To specify a polymer system one must, first of all, specify a *set* Λ *of sites* which may be occupied by the polymers - for instance the squares of a chessboard. Certain finite nonempty subsets P of Λ are declared to be polymers - for instance unions of squares that can be cut out of card board without falling apart. A polymer $P = \{x\}$ with only a single site x is called a monomer.

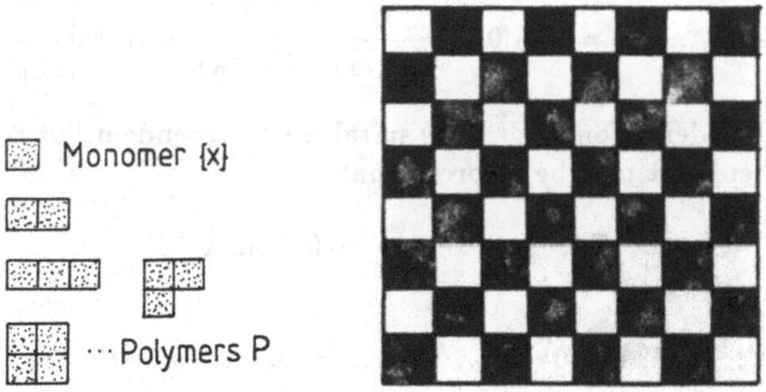

One specifies a notion of compatibility for polymers, so that no polymer is compatible with itself. Until further notice, two polymers P_1, P_2 shall be considered as compatible when they are disjoint. Finally, a (real) activity is assigned to each polymer P.

Once a polymer system is defined in this way, partition functions of finite subsets X of Λ are defined

$$Z(X) = \sum_{X=\sum P} \prod_P A(P) \tag{1.1}$$

Summation is over partitions of X into mutually compatible polymers P. In particular

$$Z(\emptyset) = 1$$

It is important to note that the activity $A(P)$ is uniquely determined by the partition functions $Z(X)$ for $X \subseteq P$. Proof: Suppose true for (no. of sites in P) $\leq n$. Consider a polymer X with $n+1$ sites. Then $A(X) = Z(X) - \sum \prod$ (activities of polymers with $\leq n$ sites) is determined by partition functions $Z(Y)$ for $Y \subseteq X$. $\surd$

There exists a formula for the free energies

$$\ln Z(X) = \sum_{x \in X} \ln A(\{x\}) + \sum_Q a(Q) \prod_{P \in Q} \overline{A}(P) \,. \tag{1.2}$$

The first term involves monomer activities. Sum over Q is over collections of not necessarily distinct polymers P such that the following graph is connected: Draw a vertex for each P and link two such vertices when the polymers are incompatible. $a(Q)$ are combinatorial coefficients [27] and

$$\overline{A}(P) = A(P) / \prod_{x \in P} A(\{x\}) \; .$$

Convergence of the infinite series for $\ln Z(X)$ is assured if the activities $A(P)$ satisfy suitable bounds [27]. The convergence conditions are very restrictive - for instance they may be violated because the monomer activities $A(\{x\})$ are too small - and it is therefore a good idea to avoid taking logarithms by way of expansion (1.2) whenever possible. Instead one may proceed as follows [1].

It is customary and convenient to consider partition functions as functions of external fields or sources Ψ such that desired correlation functions are obtained as derivatives of Z

$$G(z_1, ..., z_n) = Z(\Lambda|0)^{-1} \frac{\delta}{\delta\Psi(z_1)} \cdots \frac{\delta}{\delta\Psi(z_n)} Z(\Lambda|\Psi)\Big|_{\Psi=0} \tag{1.3}$$

By including in the definition of $Z(X|\Psi)$ suitable (X-dependent but Ψ-independent) normalization factors, it may be enforced that

$$Z(X|\Psi = 0) = 1 \qquad \text{for all } X \; . \tag{1.4}$$

This implies

$$A(P|\Psi = 0) = \delta_{1,n} \qquad (\text{n} \; = \; \text{no. sites in } P)$$

It follows now from the polymer representation (1.1) that

$$G(z_1, z_2) = \sum_P \frac{\delta}{\delta\Psi(z_1)} \frac{\delta}{\delta\Psi(z_2)} A(P|\Psi = 0) + \\ + \sum_{P_1, P_2 \; compat.} \frac{\delta}{\delta\Psi(z_1)} A(P_1|\Psi = 0) \; \frac{\delta}{\delta\Psi(z_2)} A(P_2|\Psi = 0) \; . \tag{1.5}$$

etc. Often $Z(X|\Psi)$ depends on $\Psi(z)$ only if $z \in X$. In this case the same is true of $A(X|\Psi)$. It follows that the series for $G(z_1, z_2)$ and for higher correlation functions will converge on an infinite lattice (or multigrid) Λ if for every $z \in \Lambda$ the following series converges absolutely, uniformly for Ψ in a complex neighbourhood of 0

$$\sum_{P: z \in P} |A(P|\Psi)| < \infty \tag{1.6}$$

Derivatives with respect to Ψ may be estimated by use of the Cauchy formula [8].

Summing up, an expansion is specified by selecting a class of finite subsystems X of an infinite (or large) system Λ, and defining partition functions $Z(X)$ for these. The partition functions $Z(X)$ for $X \subseteq P$ specify truncated quantities $A(P)$, called activities. Free energies, correlation functions etc. for the infinite system may be expressed in terms of these polymer activities. When proper normalization conventions are imposed the resulting expansions will converge under condition (1.6).

In applications, partition functions are integrals which would factorize if the integrands would. Therefore it suffices to consider the integrands as partition functions of a polymer system and determine the corresponding activities. This is achieved with the help of Pordt's general tree formula [28] which will now be explained. This formula generalizes tree formulas that had been used for a long time in constructive field theory [14,29].

An n-tree η is a map of integers $[2,...,n] \to [1,...,n-1]$ such that $\eta(i) < i$ for all i. [This specifies a tree graph with root 1 and links $(i, \eta(i))$. Its vertices are labelled in a way which is compatible with the partial order that is specified by the rooted tree graph.] One introduces $n-1$ auxiliary integration variables $(s_1, ..., s_{n-1}) = s$ in the real interval $[0,1]$, and defines

$$f(\eta|s) = \prod_{i=2}^{n} [s_{i-2}s_{i-3}...s_{\eta(i)}] \tag{1.7}$$

Empty products are to be read as 1.

Given a subset X of Λ, write X^* for the set of pairs (x,y) of distinct points in X. Consider partition functions $Z(X)$ of a polymer system on Λ of the form

$$Z(X) = \exp \mathcal{E}(X) \tag{1.8}$$

with $\mathcal{E}(\emptyset) = 0$. One introduces positive real variables $t_{xy} \equiv t_{yx}$ associated with pairs $(x,y) \in \Lambda^*$. Consider interpolating interactions $\mathcal{E}(X|t)$ which depend on the variables $t = (t_{xy})_{(x,y)\in\Lambda^*}$ such that

$$\begin{aligned} &\text{i)} \quad \mathcal{E}(X|t) = \mathcal{E}(X) \text{ when } t_{xy} = 1 \text{ for all } (x,y) \in X^* \\ &\text{ii)} \quad \mathcal{E}(X|t) \text{ depends only on } t_{xy} \text{ with } (x,y) \in X^* \\ &\text{iii)} \quad \mathcal{E}(X_1 + X_2|t) = \mathcal{E}(X_1|t) + \mathcal{E}(X_2|t) \text{ if } X_1 \cap X_2 = \emptyset \\ &\qquad \text{and } t_{xy} = 0 \text{ whenever } x \in X_1, y \in X_2. \end{aligned} \tag{1.9}$$

Retain the notations for trees. Given an ordered sequence of points $x_1, ..., x_n$ define

$$t(s)_{x_j x_i} = s_i s_{i+1} ... s_{j-1} \quad \text{if} \quad i < j\ . \tag{1.10}$$

The activities $A(P)$ associated with partition functions $Z(X)$ equal

$$A(P) = (n-1)!\ \mathcal{S}\left\{ \sum_{\eta} \int ds\ f(\eta|s) \left[\prod_{i=2}^{n} \frac{\partial}{\partial t_{x_i x_{\eta(i)}}} \right] e^{\mathcal{E}(P|t)} \Bigg|_{t=t(s)} \right\} \tag{1.11}$$

if $P = \{x_1, ..., x_n\}$. $\mathcal{S}$ stands for symmetrization with respect to $x_2, ..., x_n$; any element of P can be chosen as x_1. The η-summation runs over all n-trees, and $s_1, ..., s_{n-1}$ are integrated from 0 to 1.

2. Polymer Systems on the Multigrid

Our aim is to derive an expansion of the form (III-1.5) for the Boltzmann factor $\exp[-\mathcal{H}_{eff}(\Phi)]$ of the effective theory on the lattice Λ_0. This is a special kind of a phase cell cluster expansion. Phase cell cluster expansions were introduced by Glimm and Jaffe [15], and further developed by Feldmann, Magnen, Sénéor and Rivasseau [16], Battle and Federbush [17], Pordt and the author [1].

We imagine starting with a fundamental theory defined on base space Λ_N = lattice of lattice spacing $a_N = L^{-N}a_0$, and take the continuum limit $N \to \infty$ later.

In multigrid language, $\mathcal{H}_{eff} = \mathcal{H}^0$ is the last of a sequence of effective Hamiltonians which are given by eq. (I-20a...e). Thus

$$e^{-\mathcal{H}^0(\Phi)} = e^{-\frac{1}{2}\int_{z\in base}[\nabla\Psi(z)]^2} Z_0(\Lambda|\Psi)$$
$$Z_0(\Lambda|\Psi) = \int [\prod_{j=1}^{N} d\mu_{v^j}(\varphi^j)] \exp[-V(\Psi + \sum_{j=1}^{N} \mathcal{A}\varphi^j) + const] \tag{2.1}$$

φ^j is a field on Λ_j, and $\mathcal{A}\varphi^j$ was defined in eq. (I-2.22). The effective interaction Boltzmannian Z_0 depends on the block spin $\Phi \equiv \Phi^0$ on Λ_0 through the background field

$$\Psi(z) = \int_{x\in\Lambda_0} \mathcal{A}(z,x)\Phi(x) \tag{2.2}$$

For suitable choice of the additive constant to V one has

$$Z_0(\Lambda|0) = 1 .$$

We regard Z_0 as partition function of a system on the multigrid $\Lambda = \Lambda_0 + ... + \Lambda_N$. It depends parametrically on Ψ. Using the Fourier representation (I-1.14) of Gaussian measures, the above formula can be written as

$$Z_0(\Lambda|\Psi) = \int [\prod_{j=1}^{N} \prod_{x\in\Lambda_j} \frac{d\varphi^j(x)\, dq(x)}{2\pi}]\ e^{-V(\Psi+\sum_{j=1}^{N}\mathcal{A}\varphi^j)} .$$
$$\cdot \exp\Big\{\sum_{j=1}^{N}[\int_{x\in\Lambda_j} iq(x)\varphi^j(x) \ - \ \frac{1}{2}\int_{x,y\in\Lambda_j} q(x)\mathrm{v}^j(x,y)q(y)]\Big\} \tag{2.3}$$

The propagators v^j on Λ_j were introduced in section I, cp. eq. (I-2.17). They have correlation length $\approx a_{j-1}$ for $j \geq 1$.

The desired expansion (III-1.5) is basically a polymer representation of this partition function, simplified by use of normalization conventions. This will be explained in a moment.

We may define the polymer system by saying which subsets $P \subset \Lambda$ are admitted as polymers, state compatibility conditions for them, and specify partition functions $Z_0(X|\Psi)$ for all polymers X, together with $Z_0(\emptyset|\Psi) = 1$.

To begin with, we will admit as polymers P all finite nonempty subsets of Λ with the following properties. Let $j = j_P$ be the smallest integer so that $P \cap \Lambda_j$ is nonempty - that is, a_j is side length of the largest cubes in P. We require

i) if $y \in P$ then $y \subseteq x$ for some $x \in P \cap \Lambda_{j_P}$

ii) P is $\subseteq$-convex. That is, if $x_1 \in P$ and $x_2 \in P$ then $x_1 \subseteq y \subseteq x_2$ implies $y \in P$ (2.4)

An example is shown in figure 4.

Two such polymers shall be compatible if they do not intersect.

The partition functions $Z_0(X|\Psi)$ will be defined as integrals. Typically there is one integration variable per point $x \in X$, $x \notin \Lambda_0$. The same will be true for corresponding activities $A(P|\Psi)$.

In order to assure that activities for polymers P with a large number of points

Fig. 4. An example of a polymer satisfying conditions (2.4 i,ii)

are very small, a further expansion step shall be performed at the end. One splits $A(P|\Psi) \equiv A_0(P|\Psi)$ into pieces in the form ($j = 0$ for now)

$$A_j(P|\Psi) = \sum_{Q \subseteq P}' R_j(Q|\Psi) \tag{2.5}$$

The split will be such that the smallest sets Q give the dominant contribution. Summation is over those subsets Q of P with $Q \cap \Lambda_{j_P} = P \cap \Lambda_{j_P}$ whose $\subseteq$-convex hull is P. It is then convenient to regard Q as the polymers. They have property (2.4 i) but not (2.4 ii). Q_1 and Q_2 are compatible if their convex hulls do not intersect. Property (2.4 i) implies that this is equivalent to compatibility condition c) of section III if $j = 0$.

In the following text, polymer shall mean a set P which satisfies (2.4 i) *and* (2.4 ii) unless the contrary is explicitly stated.

For definiteness I will consider $\lambda\phi^4$-theory again

$$V(\phi) = \int_{z \in \Lambda_N} \{\frac{1}{2} m_N^2 \phi(z)^2 + \lambda_N \phi(z)^4 + \frac{1}{2}(\mathcal{Z}_N - 1)[\nabla\phi(z)]^2\} \tag{2.6}$$

m_N, λ_N, $\mathcal{Z}_N$ are bare mass, bare coupling constant and wave function renormalization constant. In four dimensions, $\lambda\phi^4$-theory does not have a nontrivial continuum limit, but it is still of interest to consider a large but finite UV-cutoff a_N^{-1}. [Also the infrared limit of the massless $\lambda\phi^4$-theory can be studied with the analytical methods of this section, using a multigrid $\Lambda_0 + \Lambda_{-1} + \Lambda_{-2} + ...$] In less than 4 dimensions, $\lambda\phi^4$-theory is superrenormalizable, at least for weak enough coupling, the continuum limit exists, and one may omit the wave function renormalization counterterm.

Let me now explain how the desired expansion (III-1.5) of the effective Boltzmann factor $\exp[-\mathcal{H}_{eff}(\Phi)]$ is obtained from a polymer representation of Z_0.

The partition functions $Z_0(X|\Psi)$ shall have the following properties, for background fields Ψ of the form (2.2)[1]

$$Z_0(X|\Psi = 0) = 1 \quad \text{for all } X \tag{2.7a}$$

$$Z_0(X|\Psi) \equiv 1 \quad \text{if } X \cap \Lambda_0 = \emptyset \tag{2.7b}$$

$$Z_0(X|\Psi) = \exp\left\{-\int_{z \underline{\in} X} [\frac{1}{2} m_0^2 \Psi(z)^2 + \lambda_0 \Psi(z)^4]\right\} \quad \text{if } X \subseteq \Lambda_0 \tag{2.7c}$$

[1] Remember that points $x \in \Lambda$ can be identified with cubes in base space. $z \underline{\in} X$ means $z \underline{\in} x$ for an $x \in X$, and $z \underline{\in} x$ means that z is contained in cube x. Indices 0 refer to lattice Λ_0.

The following properties of activities $A(P|\Psi)$ follow from this. Let $n =$ no. points in P.

$$A(P|\Psi = 0) = \delta_{1,n} \qquad \text{for all } P \tag{2.8a}$$

$$A(P|\Psi) \equiv \delta_{1,n} \qquad \text{if } P \cap \Lambda_0 = \emptyset \tag{2.8b}$$

$$A(P|\Psi) = \delta_{1,n} \exp\left\{ - \int_{z \in P} \left[\frac{1}{2} m_0^2 \Psi(z)^2 + \lambda_0 \Psi(z)^4 \right] \right\} \quad \text{if } P \subseteq \Lambda_0 \tag{2.8c}$$

Assume finite volume $|\Lambda_0|$ and UV-cutoff N to begin with. The polymer representation of the partition function reads ($\sum$ is union of disjoint sets as always)

$$Z_0(\Lambda|\Psi) = \sum_{\mathcal{P}: \sum_{P \in \mathcal{P}} P = \Lambda} \prod_{P \in \mathcal{P}} A(P|\Psi) \ .$$

We may restrict the sum to $\mathcal{P}$'s which make a nonvanishing contribution. By condition (2.8b) such $\mathcal{P}$ may not contain polymers which do not intersect Λ_0 other than monomers, and the monomers will merely produce factors 1. There will also be no polymers which lie entirely in Λ_0, other than monomers. The monomers in Λ_0 contribute a factor (2.8c). In conclusion

$$Z_0(\Lambda|\Psi) = \sum_{Y \subseteq \Lambda_0} \sum'_{\mathcal{P}: \sum_{P \in \mathcal{P}} P \subseteq \Lambda, \ \sum P \cap \Lambda_0 = Y} \prod_{x \in \Lambda_0 \setminus Y} A(\{x\}|\Psi) \prod_{P \in \mathcal{P}} A(P|\Psi) \tag{2.9}$$

The prime $'$ indicates that all polymers in $\mathcal{P}$ must intersect both Λ_0 and the rest of Λ. We may insert formula (2.8c) for the monomer activities $A(\{x\}|\Psi)$, and perform the final split (2.5) of activities to obtain expansion (III-1.5) as a final result. In the continuum limit $N \to \infty$, arbitrary polymers P on the infinitegrid may appear in $\mathcal{P}$, but their activities R will be very small if they penetrate into layers Λ_j with very large j, as a result of renormalization. The infinite volume limit can also be taken if $\Phi(x)$ is zero except on a finite set Y of points $x \in \Lambda_0$, since polymers P which contain points in Λ_0 far from Y will have very small activities too.

I proceed to the definition of partition functions for subsets X of Λ which satisfy conditions (2.4 i,ii). $Z_0(X|\Psi)$ is generalization of the effective interaction Boltzmannian (2.1) on lattice Λ_0 to theories with a general kind of combined volume and UV-cutoff specified by X. It will be convenient to consider similar generalizations[1] $Z_j(X|\Psi)$ of the interaction Boltzmannian on lattice Λ_j for subsets X of $\Lambda_{\geq j}$

$$e^{-\mathcal{H}_j(\Phi^j)} = e^{-\frac{1}{2}\int_{z \in base} [\nabla \Psi(z)]^2} Z_j(\Lambda_{\geq j}|\Psi) \tag{2.10a}$$

with

$$\Psi(z) = \int_{x \in \Lambda_j} \mathcal{A}(z,x)\Phi^j(x) \tag{2.10b}$$

The partition functions shall be defined for arbitrary fields Ψ on base space, not necessarily of the form (2.10b), but are later only needed for background fields of the form (2.10b). $Z_j(X|\Psi)$ shall be local in Ψ in the sense that

$$Z_j(X|\Psi) \text{ shall depend on } \Psi(z) \text{ only for } z \in X \ . \tag{2.11}$$

[1] I use the following notations for subsets of Λ.
$\Lambda_{\geq j} = \Lambda_j + \Lambda_{j+1} + ... + \Lambda_N$, $\Lambda_{\leq j} = \Lambda_0 + \Lambda_1 + ... + \Lambda_j$;
$X_j = X \cap \Lambda_j$, $X_{\geq j} = X \cap \Lambda_{\geq j} = \sum_{k=j}^{N} X_k$

I will describe a recursive construction of the partition functions because this will make the origin of UV-convergence most intuitive. A supplementary discussion can be found in Pordt's contribution to these proceedings [5].
The basic rule shall be that $Z_j(X|\Psi)$ must be a good approximation of $Z_j(\Lambda_{\geq j}|\Psi)$ for fields $\Psi = \mathcal{A}\Phi^j$ with Φ^j that vanishes outside $X_j \equiv X \cap \Lambda_j$. In other words, there should be little dependence on the combined cutoff X. We imagine splitting the effective interaction Hamiltonian into a local part and an irrelevant part which may contain nonlocal pieces. We require that the local part is independent of the cutoff X. For instance

$$\frac{1}{4!}\int_{x_2,\ldots,x_4} \frac{\delta^4}{\delta\Phi^j(x_1)\ldots\delta\Phi^j(x_4)} \ln Z_j(X|\Psi = \mathcal{A}\Phi^j)\Big|_{\Phi^j=0} = \lambda_j \tag{2.12a}$$

$$\text{for all } X \text{ with } X \cap \Lambda_j \neq \emptyset, \text{ and all } x_1 \in X \cap \Lambda_j\ ,$$

and similarly for mass parameter m_j and wave function renormalization $\mathcal{Z}_j$. Since our blockspin is a block average [viz $\Phi^j(x) = \mathrm{av}_{z\underline{\in}x}\Psi(z)$] the derivatives can also be written as

$$\frac{\delta}{\delta\Phi^j(x)} = a_j^{-d}\int_{z\underline{\in}x}\frac{\delta}{\delta\Psi(z)} \qquad \text{for} \qquad x \in \Lambda_j \tag{2.13}$$

in d dimensions. Consideration of the special case $X = \Lambda_{\geq j}$ of eq. (2.12a) shows that λ_j is a running coupling constant at scale a_j which is related to the effective Hamiltonian by

$$\lambda_j = \frac{1}{4!}\int_{x_2,\ldots,x_4} \frac{\delta^4}{\delta\Phi^j(x_1)\ldots\delta\Phi^j(x_4)} \mathcal{H}^j(\Phi^j) \tag{2.12b}$$

This is independent of x_1 by translation invariance, which we assume.

$Z_j(X|\Psi')$ shall be defined in terms of the polymer representation of $Z_{j+1}(X_{\geq j+1}|\Psi)$. The definition will be such that the recursion relation for effective Boltzmannians (or Hamiltonians) is reproduced in the special case $X = \Lambda_{\geq j}$. This guarantees that $Z_0(\Lambda|\Psi)$ will equal the effective interaction Boltzmannian on lattice Λ_0.

The polymer representation for Z_j is a generalization of expansion[1] (2.9) for Z_0

$$Z_j(X|\Psi) = \sum_{Y\subseteq X_j}\ \sum_{\mathcal{P}:\sum_{P\in\mathcal{P}} P\subseteq X,\ \sum P\cap\Lambda_j = Y}\ \prod_{x\in X_j\setminus Y} A_j(\{x\}|\Psi) \prod_{P\in\mathcal{P}} A_j(P|\Psi)$$

$$\text{for } X \subseteq \Lambda_{\geq j} \tag{2.14a}$$

with monomer activities

$$A_j(\{x\}|\Psi) = \exp\left\{-\int_{z\underline{\in}x}\left[\frac{1}{2}m_j^2\Psi(z)^2 + \lambda_j\Psi(z)^4 + \frac{1}{2}(\mathcal{Z}_j - 1)[\nabla\Psi(z)]^2\right]\right\} \tag{2.14b}$$

For $j = N$, the interaction Boltzmannian is defined by expansion (2.14) with only a single term $Y = \emptyset$, $\mathcal{P} = \emptyset$. This is in agreement with the definition of the model which says that $Z_N(\Lambda_N|\Psi) = e^{-V(\Psi)}$, with $\lambda\phi^4$-interaction (2.6).
We introduce X-dependent running coupling constants $\lambda_j(X|z)$, $m_j(X|z)$, $\mathcal{Z}_j(X|z)$ such that they equal the previously introduced running coupling constants when $X = \Lambda_{\geq j}$.

$$\lambda_j(\Lambda_{\geq j}|z) \equiv \lambda_j \qquad \text{etc.} \tag{2.15}$$

---
[1]Empty products are always equal to 1. $X_j = X \cap \Lambda_j$.

They may be chosen constant on blocks $z \underline{\in} x \in \Lambda_j$. $\mathcal{Z}_j$ is regarded as a matrix. The recursion relation for Z_j shall read for polymer $X \subseteq \Lambda_{\geq j-1}$,

$$Z_{j-1}(X|\Psi) = \mathcal{N}_j(X) \int d\mu_{v^j}(\varphi^j) \sum_{Y \subseteq X_j} Z_{loc}(X_{\geq j}, Y|\Psi + \mathcal{A}_{X_j}\varphi^j) \cdot \sum_{\mathcal{P}: \sum_{P\in\mathcal{P}} P \subseteq X_{\geq j},\ \sum P\cap\Lambda_j = Y} \prod_{P\in\mathcal{P}} A_j(P|\Psi + \mathcal{A}_{X_j}\varphi^j) \quad (2.16)$$

with normalization factor $\mathcal{N}_j$ that ensures normalization (2.7a) and

$$Z_{loc}(X', Y|\Psi) = \exp\Big\{-\int_{z\underline{\in} X'_j\setminus Y} \Big[\frac{1}{2}m_j^2(X'|z)\Psi(z)^2 + \lambda_j(X'|z)\Psi(z)^4 + + \frac{1}{2}\big(\mathcal{Z}_j(X'|z)_{\mu\nu} - \delta_{\mu\nu}\big)\nabla_\mu\Psi\nabla_\nu\Psi(z)\Big]\Big\}$$

for $X' \subseteq \Lambda_{\geq j}$, $Y \subseteq X' \cap \Lambda_j$. Summation runs over collections $\mathcal{P}$ of polymers which intersect both X_j and $X_{\geq j+1}$.

The kernel $\mathcal{A}_{X_j}(z,x)$ is only needed for arguments $x \in \Lambda_j$. It shares the properties of the kernel $\mathcal{A}$ except that

$$\mathcal{A}_{X_j}(z,x) = 0 \qquad \text{unless} \qquad z\underline{\in} X_j\ ,\ x \in \Lambda_j\ . \qquad (2.17a)$$

For the moment we may imagine that this kernel is obtained from $\mathcal{A}$ by multiplication with step functions. A different definition which avoids jumps at the boundary of X_j is mentioned at the end of section IV-3.

The integrand in (2.16) depends on $\varphi^j(x)$ only for $x \in X_j$. Therefore the Gaussian measure $d\mu_{v^j}$ can be replaced by a Gaussian measure $d\mu_{v^j_{X_j}}$ with covariance $v^j_{X_j}$ which involves only integration variables $\varphi^j(x)$ for $x \in X_j$.

$$v^j_{X_j}(x,y) = \begin{cases} v^j(x,y) & \text{if } x,y \in X_j \\ 0 & \text{otherwise} \end{cases} \qquad (2.17b)$$

Such a Gaussian measure has support on fields that vanish outside X_j, since

$$d\mu_{v^j_{X_j}}(\varphi^j) = \int \Big[\prod_{x\in X_j} \frac{d\varphi^j(x)\, dq(x)}{2\pi}\Big] \exp\{-i\int_{x\in X_j} \varphi^j(x)q(x)\} \cdot \cdot \exp\{-\frac{1}{2}\int_{x,y\in X_j} q(x)v^j(x,y)q(y)\} \qquad (2.17c)$$

The integrand of the Gaussian measure in (2.16) equals $Z_j(X_{\geq j}|\Psi)$ in the special case $X = \Lambda_{\geq j-1}$. Therefore the definition reproduces the recursion relation for effective interaction Boltzmannians (2.10a), viz

$$Z_{j-1}(\Lambda_{\geq j-1}|\Psi) = \mathcal{N}_j \int d\mu_{v^j}(\varphi^j)\ Z_j(\Lambda_{\geq j}|\Psi + \mathcal{A}\varphi^j)\ . \qquad (2.18)$$

For general X however, the integrand is not simply Z_j but differs from it by replacement of $\lambda_j \equiv \lambda_j(\Lambda_{\geq j})$ by $\lambda_j(X_{\geq j})$ etc. These running coupling constants are determined by the requirement that the renormalization conditions (2.12a) are fulfilled, with $j-1$ in place of j. Remember that this ensures that $Z_{j-1}(X|\Psi)$ is a good approximation to $Z_{j-1}(\Lambda_{\geq j-1}|\Psi)$. In more detail, the philosophy is as follows.

For the purpose of integrating out the field components φ^j at length scale a_j one may regard λ_j as a bare coupling constant and λ_{j-1} as a renormalized one. As in renormalization theory, one thinks of the renormalized parameters λ_{j-1} etc. as given, and determining the bare parameters. The differences $\delta\lambda_j = \lambda_j - \lambda_{j-1}$ etc. are regarded as a consequence of the interaction. When the interaction is partly switched off by imposing sharper cutoffs (smaller X), then $\delta\lambda_j$ should be adjusted, keeping λ_{j-1} fixed. In this way λ_j becomes X-dependent. The factors $A(P|\Psi)$ with $P \in \mathcal{P}$ will involve coupling constants $\lambda_k(\cdot)$ at smaller length scales a_k, $k \geq j+1$. We leave them unaffected by the size of $X_j = X \cap \Lambda_j$.

This defines the partition functions $Z_j(X|\Psi)$ for all j and all $X \subset \Lambda_{\geq j}$, and therefore also the activities $A_j(P|\Psi)$. The final activities are $A \equiv A_0$.

To compute the activities one needs the running coupling constants λ_j etc., either exactly, or approximately but with increasing accuracy for larger and larger j.

For instance, in the superrenormalizable $\lambda\phi^4$-theory in 3 dimensions, one would retain the recursion relation (2.6), setting in it $\lambda_j(X|z) = \lambda$ independent of j, X, z, and $\mathcal{Z}_j(X|z) \equiv 1$ and $m_j^2(X|z) = m^2 - \delta m_j^2(\mathrm{v}_X)$. Here $\delta m_j^2(\mathrm{v}_X)$ is obtained from the perturbation theoretic (low order) mass counterterm by splitting the free propagator into pieces $v = \sum v^j$ and substituting $v^k_{X_k}$ for v^k if $k > j$, and 0 otherwise. v^k_Y is defined by

$$v^k_Y(z_1,z_2) = \begin{cases} \displaystyle\int_{x_1,x_2\in Y} \mathcal{A}(z_1,x_1)\mathrm{v}^k(x_1,x_2)\mathcal{A}(z_2,x_2) & \text{if } z_1,z_2 \underline{\in} Y \\ 0 & \text{otherwise} \end{cases} \tag{2.19}$$

$$\text{for } \; Y \subseteq \Lambda_k$$

The renormalization conditions (2.12a) are not needed for superrenormalizable models. They will however be fulfilled approximately by the explicit choice of parameters.

In asymptotically free renormalizable theories one may use approximate perturbative expressions for running coupling constants as well. Typically this will lead to slower UV-convergence like an integrable power of j^{-1}, cp. (III-1.7d). When renormalization conditions are exactly fulfilled, convergence is exponential in j, like (III-1.7c), because convergent graphs converge typically like $\int (d^4p/p^4)\, p^{-2}$ and $p \sim a_j^{-1} = L^j a_0$. Something more will be said about UV-convergence in section IV-3 and in Pordt's contribution to these proceedings [5].

From the recursion relation for Z_j one can deduce a recursion relation for activities A_j which can then be iterated. To this end one considers interpolating quantities which depend on decoupling variables t_{xy}, and applies Pordt's general tree formula (1.11). The t-variables are introduced into propagators v^j, kernels $\mathcal{A}(z,x)$, X-dependent couplings, and normalization constants $\mathcal{N}_j(X)$ in such a way that $Z_j(X,t|\Psi)$ depends only on t_{xy} for $x,y \in X$, and

$$\begin{gathered} Z_j(X,t|\Psi) = Z_j(X_1,t|\Psi)\; Z_j(X_2,t|\Psi) \\ \text{if } \; X_1 \cap X_2 = \emptyset,\; X = X_1 \cup X_2,\; t_{xy} = 0 \; \text{ whenever } \; x \in X_1,\; y \in X_2\,. \end{gathered} \tag{2.20}$$

This equation must hold also when X_2 does not intersect Λ_j. For such sets X_2 we use the trivial interpolation

$$Z_j(X_2,t|\Psi) \equiv 1 \qquad \text{for} \qquad X_2 \cap \Lambda_j = \emptyset\,. \tag{2.21}$$

Application of the tree formula (1.11) gives an expression for the activities which involve sums over n-trees η. Such a sum is easy to perform, also numerically, since it amounts to summing the $n-1$ integer variables $\eta(i)$ $(i = 2, ..., n)$ over the ranges $1, ..., i-1$.

Lastly I should describe the final expansion step (2.5). The activity $A_{j-1}(P|\Psi)$ is an integral over variables $\varphi(x)$ attached to sites $x \in P \cap \Lambda_{\geq j}$. The final activities are $A \equiv A_0$.

One introduces auxiliary variables $u(x)$ which are ultimately set to one. Basically the last expansion step consists in introducing u-depending activities by substituting $u(x)\varphi(x)$ for $\varphi(x)$, and doing a Taylor expansion to lowest order with remainder in each variable $u(x)$ according to

$$f(\xi) = f(0) + \int_0^1 du \, \frac{\partial f}{\partial u}(u\xi) \, .$$

In this way, $A_j(P|\Psi)$ becomes a sum of terms associated with subsets $Q \subseteq P$. Q contains those points where one takes the remainder. One shows that the contribution of Q is zero unless P is the $\subseteq$-convex hull of Q [30].

If one were to proceed literally in this way, the interpolating u-dependent activities would not satisfy the renormalization conditions identically in u. We require that they do, in order that the activities R_j satisfy the renormalization conditions. This is achieved as follows. The integrals which define activities A_j depend on propagators $v_{P_{\geq j}}$, and kernels $\mathcal{A}_{X_{\geq j}}(z,x)$ which vanish when $z \notin X_{\geq j}$ or $x \notin X_{\geq j}$. The X-dependent coupling constants are determined as solutions of renormalization conditions, in this way they become functions of such propagators and kernels also. It is convenient to think of their X-dependence as arising in this way. One substitutes $u(x)\varphi(x)$, $u(x)\mathrm{v}(x,y)u(y)$ for field variables $\varphi(x)$ and multigrid propagators $\mathrm{v}(x,y)$ which determine v. Then one performs the Taylor expansion to lowest order with remainder in all variables $u(x)$.

This expansion step assures that the new activity is proportional to a fractional power of the coupling constant for every point $x \in P \cap \Lambda_{\geq j}$. This is used in the convergence proof for weakly coupled models. A similar device was widely used by Magnen and Sénéor [16].

3. Polymer Expansions of Effective Interaction Boltzmannians on a Single Lattice

In this section I will explain how the multigrid setup described so far can also be interpreted as iteration of exact renormalization group transformations $Z_j = e^{-\mathcal{H}^j} \to Z_{j-1} = e^{-\mathcal{H}^{j-1}}$, in which the effective Boltzmannians Z_j are given as partition functions of a polymer system on a *single* lattice Λ_j. This provides a substitute for the exact renormalization group approach of Kupiainen and Gawędzki [8,9] which avoids its most serious complication - the need for a large-field small-field split.

The polymer representation (2.14) of the effective Boltzmannian $\exp[-\mathcal{H}^j(\Phi^j)] = Z_j(\Lambda_{\geq j}|\Psi)\exp[-\frac{1}{2}\int(\nabla\Psi)^2]$ at the length scale a_j involves a sum over polymers P on the multigrid $\Lambda_{\geq j}$. Since the compatibility condition for polymers $P_1, P_2 \subset \Lambda_{\geq j}$ which intersect Λ_j is equivalent to the condition that $P_1 \cap \Lambda_j$ and $P_2 \cap \Lambda_j$ do not intersect, we may again perform a partial summation over polymers to obtain a polymer representation on Λ_j, as in section III.

$$Z_j(\Lambda_{\geq j}|\Psi) = \sum_{X\subseteq\Lambda_j} \prod_{x\in\Lambda_j\setminus X} A_j(\{x\}|\Psi) \sum_{S:\sum_{S\in\mathcal{S}} S=X} \prod_{S\in\mathcal{S}} \widetilde{R}_j(S|\Psi) \tag{3.1a}$$

with

$$A_j(\{x\}|\Psi) = \exp\left\{-\int_{z\in x}\left[\frac{1}{2}m_j^2\Psi(z)^2 + \lambda_j\Psi(z)^4 + \frac{1}{2}(\mathcal{Z}_j - 1)[\nabla\Psi(z)]^2\right]\right\} \tag{3.1b}$$

Summation over $\mathcal{S}$ runs over collections of disjoint nonempty finite subsets of Λ_j whose union is X, and

$$\begin{aligned}\widetilde{R}_j(S|\Psi) &= \sum_{P:P\subseteq\Lambda_{\geq j},\ P\cap\Lambda_j=S}{}' A_j(P|\Psi) \\ &= \sum_{Q:Q\subseteq\Lambda_{\geq j},\ Q\cap\Lambda_j=S} R_j(Q|\Psi)\ .\end{aligned} \tag{3.2}$$

The sum over P runs only over polymers which satisfy conditions (2.4 i) *and* (2.4 ii), while the second formula is based on the further decomposition (2.5) and involves summation over sets Q which need only satisfy (2.4 i).

As a consequence of renormalization conditions (2.7a), (2.12a), the following renormalization conditions will be fulfilled for all activities $A(P|\Psi)$ individually, when P is not a monomer, and therefore also for their sums $\widetilde{R}_j$.

$$\widetilde{R}_j(P|\Psi = 0) = 0$$

$$\int_{x_2,\ldots,x_n\in\Lambda_j} \frac{\delta^4}{\delta\Phi^j(x_1)\ldots\delta\Phi^j(x_n)}\widetilde{R}_j(P|\Psi = \mathcal{A}\Phi^j)\bigg|_{\Phi^j=0} = 0 \quad \text{for } n = 2,4;\ x_1 \in P$$

and a similar further relation from wave function renormalization. The reader is invited to check that this fulfils (2.12a) because $Z_j(X|0)^{-1} = 1$ and only the leading term $X = \emptyset$, $\mathcal{S} = \emptyset$ in (3.1a) will make a contribution to the integrated 2nd and 4th derivative of Z_j.

To establish convergence properties (of Greens functions etc., cp. section IV-1) in weakly coupled models, one derives recursive bounds on the activities $\widetilde{R}_j(S|\Psi)$ which imply that

$$\sum_{S:x\in S} |\widetilde{R}_j(S|\Psi)| < \infty \qquad \text{for all } j\ .$$

Along the way, one derives similar bounds for the contributions from individual polymers $Q \subset \Lambda_{\geq j}$ to the sum (3.2). I will not go into details of their derivation, but will indicate how they look. Bounds on $\widetilde{R}_j(X|\Psi)$ will be stated for fields Ψ which are smooth as appropriate for a field at length scale a_j, viz

$$\Psi(z) = \int_{x\in\Lambda_j} \mathcal{A}(z,x)\Phi^j(x) \tag{3.4}$$

Because of nonlocality of the kernels $\mathcal{A}$, violent behaviour of the blockspin Φ^j outside of X could jeopardize smoothness of the field Ψ for $z\in X$. This presents a problem, which is solved [4] by stating bounds not only on individual activities $\widetilde{R}_j$ but also on products of such, under weaker smoothness conditions. The smoothness conditions are eq. (3.4) together with conditions on the set supp Φ^j where Φ^j may be different from 0. For simplicity I ignore wave function renormalization, assuming $\mathcal{Z}_j = 1$ at first. The bounds read as follows for a single activity

$$|\widetilde{R}_j(S|\Psi)| \leq \widetilde{b}_j(S|\Psi) \qquad \text{if} \qquad \Psi = \mathcal{A}\Phi^j \ \text{ with } \ \text{supp}\ \Phi^j \subseteq S \tag{3.5a}$$

Herein

$$\begin{aligned} \widetilde{b}_j(S|\Psi) &= c_j(S) \exp \int_{z \in S} [-\frac{\mu_j^2}{2} \lambda_j^{1/2} |Re\ \Psi(z)|^2 + 16\lambda_j |Im\ \Psi(z)|^4] \\ c_j(S) &= (C\lambda_j^{1/8})^{n-1} \exp[-\mu_j L_{tree}(S)] \\ &\text{for } n = \text{no. points in } S \geq 1, \text{ while } c_j(\{x\}) = O(\lambda_j^{1/2}) \ . \\ \mu_j &= O(a_j^{-1}) \end{aligned} \tag{3.5b}$$

Bounds are valid for Φ^j in a complex strip such that

$$|Im\ \Psi(z)| \leq \lambda_j^{-1/4} \mu_j \ . \tag{3.5c}$$

C is some constant and $L_{tree}(S)$ is the length of the shortest tree on vertices $x \in S$.

Bounds of this form where used by Kupiainen and Gawedzki for the large field region [9]. We use them in conjunction with the renormalization conditions also in the small field region.

Actually these bounds are established not only for the single kernel $\mathcal{A}$ but for a family of kernels which share its properties (smoothness, decay, and prescribed block averages). Such a family is formed by convex combinations of kernels $\mathcal{A}_P$ which are constructed exactly like $\mathcal{A}$, using however Laplacians with Dirichlet boundary conditions on certain faces of cubes x. A kernel of this type was used in our numerical work - see section II-1.

So far I assumed that wave function renormalization constants $\mathcal{Z}_j = 1$ in eq. (3.1b). This can be achieved by absorbing the wave function renormalization into a factor $(\mathcal{Z}_{j-1}/\mathcal{Z}_j)^{1/2}$ in the block spin definition (I-1.2), thereby eliminating the $(\mathcal{Z}_j - 1)\ (\nabla\Psi)^2$-term in the monomer activity $A_j(\{x\}|\Psi)$. $\mathcal{Z}_j^{-1}$ stays actually finite in massless lattice ϕ^4-theory in the infrared limit $j \to -\infty$, cp. [15,16]. This is easily understood on the basis of perturbation theory with running coupling constants, cp. Pordt's contribution and references given there.

References

[1] G. Mack and A. Pordt, Commun. Math. Phys. **97**, 267 (1985)

[2] G. Mack and S. Meyer, *Multigrid Simulations for ϕ^4-Theory* (in preparation)

[3] G. Mack and K. Pinn, Phys. Lett. **B173**, 434 (1986)

[4] G. Mack and H.-J. Timme, in preparation

[5] A. Pordt, *Renormalization Theory for Use in Convergent Expansions of Euclidean Quantum Field Theory* (contribution to these proceedings) see also

G.Gallavotti , The Structure of Renormalization Theory: Renormalization, Form Factors and Resummation in Scalar Field Theory , in : Critical Phenomena, Random Systems, Gauge Theories , K.Osterwalder and R.Stora (Eds.) , Les Houches 1984.

G.Gallavotti, Renormalization Theory and Ultraviolet Stability for Scalar Fields via Renormalization Group Methods , Rev. Mod. Phys. 57 ,471-561 (1985).

G.Gallavotti and F.Nicolò , Renormalization Theory in Four Dimensional Scalar Fields (I) , Comm. Math. Phys. 100 , 545-590 (1985) .

G.Gallavotti and F.Nicolò , Renormalization Theory in Four Dimensional Scalar Fields (II) , Comm. Math. Phys. 101 , 247-282 (1985) .

[6] M. Göpfert and G. Mack, Commun. Math. Phys. **82**, 545 (1982)

[7] K. Wilson, Phys. Rev. **D3**, 1818 (1971); Rev. Mod. Phys. **55**, 583 (1983)

J. Kogut and K. Wilson, Phys. Reports **12C**, 75 (1974)

[8] K. Gawędzki and A. Kupiainen, Commun. Math. Phys. **77**, 31 (1980)

[9] K. Gawędzki and A. Kupiainen, Commun. Math. Phys. **99**, 197 (1985) and references given there

[10] T. Balaban, Commun. Math. Phys. **89**, 571 (1983)

[11] T. Balaban, Commun. Math. Phys. **109**, 249 (1987) and *Renormalization Group Approach to Lattice Gauge Field Theories: II. Analysis of Effective Densities, and Their Expansions, in the Complete Models*, preprint 87/B205, Harvard University, 1987

[12] T. Balaban, J. Imbrie, and A. Jaffe, *Exact renormalization group for gauge theories* in: Progress in Gauge Field Theories, G. 't Hooft et al. (eds.), Plenum Press, New York 1984

[13] K. Pinn, Ph.D. thesis (in preparation) Hamburg, Spring 1988

[14] J. Glimm and A. Jaffe, Quantum Physics, Springer Verlag, Heidelberg 1981

[15] J. Glimm and A. Jaffe, Fortschr. Physik **21**, 327 (1973)

[16] J. Magnen and R. Sénéor, Ann. Inst. Henri Poincare **24**, 95 (1976)

J. Feldman, J. Magnen, V. Rivasseau, and R. Sénéor, *Infrared* ϕ_4^4, Les Houches lectures 1984; Commun. Math. Phys. **109**, 437 (1987) and references given there

[17] P. Federbush and G. Battle, Ann. Phys. (N.Y.) **142**, 95 (1982), Commun. Math. Phys. **88**, 263 (1983)

P. Federbush, Commun. Math. Phys. **107**, 319 (1986); Commun. Math. Phys. **110**, 293 (1987), and preprints

[18] A. Brandt, Math. Comp. **31**, 333 (1977)

For a review see:

A. Brandt and N. Dinar, in: Numerical Methods for Partial Differential Equations, S.V. Pastes (ed.), Academic Press, 1979

A. Brandt, in: Multigrid Methods, ed. by W. Hackenbusch and U. Trottenberg, Lecture Notes in Mathematics, Vol. 90, Springer Verlag, Berlin 1982

see also

A. Brandt, D. Ron, and D.J. Amit, in: Multigrid Methods II, ed. by W. Hackenbusch and U. Trottenberg, Lecture Notes in Mathematics, Vol. 1228, Springer Verlag, Cologne 1985

G. Parisi, *Prolegomena to any Future Computer Evaluation of the QCD Mass Spectrum* in: Progress in Gauge Field Theory, G. 't Hooft et al. (eds.), Plenum Press, N.Y. 1984

[19] H. Meyer-Ortmanns, Z. Phys. C **27**, 553 (1985)

J. Goodman and A.D. Sokal, Phys. Rev. Lett. **56**, 1015 (1986)

A.D. Sokal, *New Monte Carlo Algorithm for Quantum Field Theory and Critical Phenomena, or How to Beat Critical Slowing Down* in: Proc. 8th Int. Congress of Math. Physics, Marseille, M. Mebkhout and R. Sénéor (eds.), World Scientific, Singapore 1987

[20] J. Kuti and Y. Shen, *Supercomputing the effective action*, preprint UCSD/PTH 87-14

[21] L. O'Raifeartaigh, A. Wipf, and H. Yoneyama, Nucl. Phys. **B271**, 653 (1986)

H. Yoneyama, *The Constraint Effective Potential and Lattice Field Theory* in: Proceedings of the Workshop and Conference Nonperturbative Methods in Quantum Field Theory, Siófok, Hungary 1986, Z. Horvath, L. Palla and A. Patkos (eds.), World Scientific

[22] G. Mack, Nucl. Phys. **B235** [FS11], 197 (1984)

[23] R. Cordery, R. Gupta, and M.A. Novotny, Phys. Lett. **128B**, 425 (1983)

[24] G. Mack and K. Pinn, *Restrictions on Effective Dielectric Lattice Gauge Theory Actions from their Analyticity Properties*, DESY-report DESY 86-127, Oct. 1986

[25] R. Gupta, *Food for Thought: Five Lectures on Lattice Gauge Theory*, Los Alamos preprint 1987, esp. section 5.7

H. Hahn and T. Streit, *Renormalization Group Multi-Grid Monte Carlo*, preprint Braunschweig (submitted to Phys. Rev. B)

K.E. Schmidt, Phys. Rev. Lett. **51**, 2175 (1983)

M. Faas and H.J. Hilhorst, Physica **135A**, 571 (1986)

[26] R.H. Swendsen and J.S. Wang, Phys. Rev. Lett. **58**,86 (1987)

U. Wolff, *Monte Carlo Simulation of a Lattice Field Theory as Correlated Percolation*, DESY-report DESY 87-092, Aug. 1987

I. Montvay, G. Münster, and U. Wolff (in preparation)

[27] C. Gruber and H. Kunz, Commun. Math. Phys. **22**, 133 (1971)

E. Seiler, Gauge Theories as a Problem in Constructive Quantum Field Theory and Statistical Mechanics, Lecture Notes in Physics, Vol. 159, Springer Verlag, Heidelberg 1982

[28] A. Pordt, Z. Phys. C **31**, 635 (1986)

[29] D. Brydges and P. Federbush, J. Math. Phys. **19**, 327 (1973) and references given there

[30] G. Mack, *Exact Renormalization Group as a Scheme for Calculations*, DESY-report DESY 85-111, Oct. 1985

[31] R.C. Brower, K.J.M. Moriarty, E. Myers, and C. Rebbi, *The Multigrid Method for Fermion Calculations in Quantum Chromodynamics*, preprint, Boston University, 1987

[32] K. Binder in: Monte Carlo Methods in Statistical Physics, K. Binder (ed.), Springer Verlag, Heidelberg 1982

INTRODUCTION TO CONFORMAL INVARIANT QUANTUM FIELD THEORY IN TWO AND MORE DIMENSIONS

Gerhard Mack

II. Institut für Theoretische Physik der Universität Hamburg

I. Scale and conformal invariance in d dimensions

II. Conformal field theory lives on a compact space, or why conformal field theory is interesting in d dimensions

III. The stress energy tensor in a two dimensional scale invariant field theory

IV. Analytic continuation to Euclidean space

V. Quantum field theory as representation theory of a group or Lie algebra (in d dimensions)

VI. Semigroups, finite and infinite dimensional

What is conformal invariant quantum field theory? And why does it live on a tube? The first sections of these lectures will try to answer this. In the literature the discussion is usually purely Euclidean. But it is much simpler and more transparent to start from a quantum field theory in Minkowski space. I will review this approach in sections I-III; the discussion is based on joint work with M. Lüscher. It yields the Virasoro algebra for 2-dimensional theories in a straightforward way. At the same time it makes clear that 2 dimensions are not as special as is often believed. Section V will answer the question what is the substitute for the Virasoro algebra in d dimensions.

The relation between quantum field theory in Minkowski space and Euclidean field theory is of course well known, and the results of the Minkowski space analysis are easily transcribed into Euclidean language (section IV). It is less generally known that there exists also a (semi-)group theoretical version of this correspondence which has produced important results and promises more (section VI).

A great deal is known about conformal invariant quantum field theory in $d \geq 2$ dimensions from work which was done in the seventies [3-22]. Space does not permit to review it all, but I would like to mention very briefly some of the results that are not covered here.

Uniqueness of conformal invariant two and three point functions formed the basis of the Migdal Polyakov bootstrap approach which constructed conformal invariant

Greens functions by skeleton graph expanions [10] . These expansions are the iterative solution of the infinite set of coupled Schwinger Dyson (SD) equations for all the Euclidean Greens functions. A group theoretical approach to conformal QFT was started by the author [14] . It used conformal partial wave expansions of Euclidean Greens functions to show that the ∞ many SD-eqs. amount to the requirement of poles in the partial wave amplitudes with factorizable residues. Roughly speaking, operator product expansions are the solution of these constraints. Using these partial wave expansions , the physical principles of QFT (Osterwalder Schrader positivity, invariance and locality) were reduced to a crossing relation for partial wave amplitudes for 4-point functions (and positivity of their real residues).

This crossing relation was in Euclidean space. There exists also a Minkowski space version. They were also studied independently by Ferrara, Gatto, Grillo and Parisi, and by Polyakov. The two crossing relations can be related by use of the semigroup technology (see e.g. [13]).

The mathematical meaning of the Minkowskian crossing relation rests crucially on the convergence of operator product expansions (in Minkowski space) on the vacuum. Ferrara et al. [15] had shown that the contributions of all derivatives $\partial_{\mu_1}\dots\partial_{\mu_n}\phi^k(z)$ of a nonderivative ("quasiprimary") field ϕ^k to Wilson operator product expansions could be summed up with the result

$$\phi^i(x)\phi^j(y)|\Omega>=\sum_k\int dz\phi^k(z)|\Omega>B^{kij}(z;xy).$$

Integration is over Minkowski space and the kernels $B^{kij}(z;xy)$ are determined by conformal invariance up to certain normalization constants, given the dimension of the fields [19]. This holds as asymptotic expansion for arbitrary states $|\Omega>$ of finite energy. The expansions are convergent if $|\Omega>=vacuum$. Convergence was proven in [19] by pointing out that the expansions are nothing but the decomposition of a unitary representation of the quantum mechanical conformal group into its irreducible components. I lectured at length on the use of convergent operator product expansions in conformal QFT at the 1976 Cargese school [17]; here I will refer the reader to ref.[18].

Also omitted is the study of models. A Virasoro algebra for the Thirring model was constructed as early as 1972 by Ferrara, Gatto and Grillo [3].

I. SCALE AND CONFORMAL INVARIANCE IN d DIMENSIONS

In the last 15 years Quantum Field Theory (QFT) and Classical Statistical Mechanics (CSM) have merged into a single discipline [1]. Nonperturbative studies in quantum field theory are nearly always done in the language of classical statistical mechanics. On the other hand, conformal invariance is important for the study of systems of classical statistical mechanics[1] at a critical point, but the crucial object to study is the stress energy tensor of a quantum field theory in Minkowski space.

According to Wilsons renormalization group theory [2] , the long distance behavior of a system at a critical point is goverened by a renormalization group fixed point, and is therefore described by an exactly scale invariant Euclidean Field Theory. If

[1]It may one day become important again for elementary particle physics if $N=4$ supersymmetric Yang-Mills theory describes nature, for instance.

the theory has Osterwalder Schrader positivity (which is true if there is a theory with nearest neighbour interaction and time reversal invariance in the same universality class) then one can analytically continue to obtain a scale invariant quantum field theory in Minkowski space which obeys the usual principles. It can be expected to have a conserved symmetric traceless stress energy tensor $\Theta_{\mu\nu}$ which defines the generators of space time symmetries

$$\Theta_{\mu\nu}(x) = \Theta_{\nu\mu}(x) = \Theta_{\mu\nu}(x)^*, \qquad \partial^\mu \Theta_{\mu\nu}(x) = 0.$$

$$\begin{aligned} P_\mu &= \int d\mathbf{x}\Theta_{\mu 0}(x^0, \mathbf{x}), \\ M_{\mu\nu} &= \int d\mathbf{x}(x_\mu \Theta_{\nu 0} - x_\nu \Theta_{\mu 0}), \\ D &= \int d\mathbf{x}\mathcal{D}_0, \qquad \mathcal{D}_\mu(x) = x^\nu \Theta_{\nu\mu}(x). \end{aligned} \tag{1}$$

If Θ is traceless then D is time independent and scale invariance holds. (The renormalization group picture permits also spontaneous breaking of scale invariance, with an effective potential that is a homogeneous function of the background field. We ignore this possibility.) In two dimensions the converse is also true . In more dimensions, $\Theta_{\mu\nu}$ may have to be redefined to make it traceless [4]. But if Θ is traceless then also the generator K_μ of conformal transformations is conserved.

$$K_\mu = \int \mathcal{K}_{\mu 0}, \qquad \mathcal{K}_{\mu\rho} = 2x_\mu x^\nu \Theta_{\nu\rho} - x^2 \Theta_{\mu\rho} \tag{2a}$$

More generally we have the relation

$$\partial^\rho \mathcal{K}_{\mu\rho} = 2x_\mu \partial^\nu \mathcal{D}_\nu \tag{2b}$$

This relation was first derived in [5,6] for Lagrangean field theories (under specified conditions on possible derivative couplings which were shown to be fulfilled for Yang Mills theories, for instance).

Scale- and conformal transformations in d-dimensional space time act according to

$$\begin{aligned} x^\mu &\mapsto \rho x^\mu, \quad \rho > 0, & ds^2 &\mapsto \rho^2 ds^2 \quad \text{dilatations} \\ x^\mu &\mapsto \sigma(x)^{-1}(x^\mu - c^\mu x^2), & ds^2 &\mapsto \sigma(x)^{-2} ds^2 \quad \text{conf. transf.} \\ \sigma(x) &= 1 - 2c\cdot x + c^2 x^2 \end{aligned} \tag{3}$$

The d-vector c^μ parametrizes conformal transformations . The transformation law of fields with dimension of mass l and arbitrary spin is furnished by the theory of induced representations [6] ($\Sigma_{\mu\nu}$ = spin representation matrices of the Lorentz group)

$$\begin{aligned} [\phi(x), D] &= i(l + x_\nu \partial^\nu)\phi(x) \\ [\phi(x), K_\mu] &= -i(2lx_\mu + 2x_\mu x_\nu \partial^\nu - x^2\partial_\mu - 2ix^\nu \Sigma_{\mu\nu})\phi(x) \end{aligned} \tag{4}$$

for fields which are not derivatives of other fields (nowadays called quasiprimary fields). More generally there can be an extra term $\kappa_\mu \phi(x)$ in the second formula.

Conformal transformations leave the light cone invariant but they can take spacelike x into timelike x . Because of this seeming causality problem, conformal symmetry was regarded with suspicion in spite of its long history [8], or was considered consistent only for free field theories which have commutators concentrated on the light cone. This would rule out anomalous dimensions. The final resolution of this causality puzzle was found by Lüscher and the author [9] and will be discussed in the next section.

II. CONFORMAL FIELD THEORY LIVES ON COMPACT SPACE S^{d-1} or WHY CONFORMAL FIELD THEORY IS INTERESTING IN d DIMENSIONS.

There is no problem with global conformal invariance of *Euclidean* Greens functions, it follows from infinitesimal conformal invariance in Minkowski space. It was shown in [9] that vacuum expectation values $< 0|\phi(x_1)\dots\phi(x_n)|0 >$ of products of field operators (Wightman functions) admit analytic continuation to an infinite sheeted covering space $\tilde{M}$ of Minkowski space M whenever the Euclidean Greens functions of the theory have conformal invariance. The analysis is based on use of a semigroup , see section VI. I. Segal had shown that the quantum mechanical conformal group $\tilde{G}$, which is the infinite sheeted universal covering of $SO(d,2)$ in d dimensions, can act on $\tilde{M}$, and that $\tilde{M}$ admits a $\tilde{G}$-invariant causal ordering [11]. Thus the action of $\tilde{G}$ on $\tilde{M}$ will not interchange timelike and spacelike on $\tilde{M}$. The analytically continued Wightman functions will be globally $\tilde{G}$-invariant and respect locality for relatively spacelike points on $\tilde{M}$. $\tilde{M}$ has the topology of a torus $S^{d-1}\times\mathbf{R}$. Its points can be parametrized by pairs (e,τ) , where $e=(e^1,\dots,e^d)\in S^{d-1}$ is a unit d-vector, and $\tau\in\mathbf{R}$. Minkowski space M is embedded in $\tilde{M}$ as shown in figure 1.

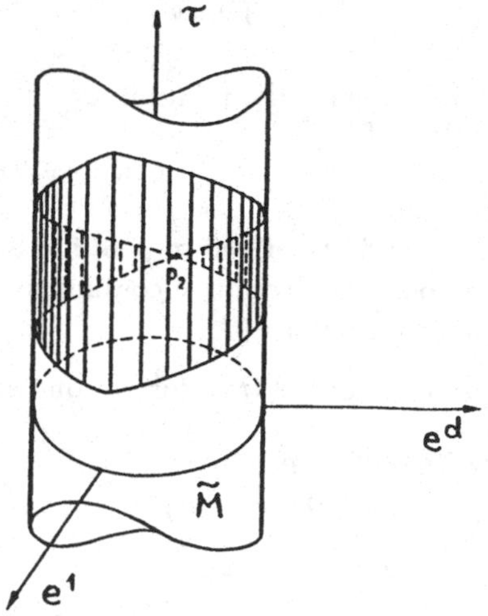

Figure 1. Manifold $\tilde{M}$.
Shaded part is M.
p_2 is the unique point at spatial infinity of $\tilde{M}$.
Drawing for 2 space time dimensions

$$M=\{(e,\tau)\in\tilde{M};-\pi<\tau<\pi,\ e^d>-\cos\tau\} \tag{1}$$

The relation with the usual Minkowski coordinates x^μ is given by

$$x^0=\frac{\sin\tau}{\cos\tau+e^d},\qquad x^i=\frac{e^i}{\cos\tau+e^d},\quad i=1,\dots,d-1 \tag{2}$$

A point (e_1,τ_1) is positive timelike relative to (e_2,τ_2) if

$$\tau_2-\tau_1>Arccos\ e_1\cdot e_2 \tag{3}$$

Arccos is the principal value of arccos.

We see that conformal field theory lives on a compact space. In this assertion, space really means space, not space time. From a spacelike surface Σ in Minkowski space, a compact spacelike surface in $\bar{M}$ is obtained by adding one point at spacelike infinity. The assertion that the theory lives on a compact space means that wave functions etc. have a particular behavior at spacelike infinity, i.e. it imposes boundary conditions.

Time evolution in a general sense fixes the state (Schrödinger wave function) on a spacelike surface Σ, given the state on a neighbouring surface. It is governed by the stress energy tensor. In a conformal theory, it is natural to consider spacelike surfaces τ =const.. The operator

$$H = P^0 + K^0$$

of τ-translations is called the conformal Hamiltonian. H is a positive operator because P^0 is. This follows from

$$H = P^0 + K^0 = P^0 + \mathcal{R}P^0\mathcal{R} \geq 0$$

$\mathcal{R}$ is inverse radius transformation times space reflection in even dimensions d (see below); it was discovered by Kastrup long ago that this is an important special element of the identity component of the conformal group. A similar argument for the positivity of H was first given by I. Segal [11].

On a compact space, the Hamiltonian is expected to have a discrete spectrum, and there should be only a finite number of independent eigenvectors of H with energy below any finite value E. And indeed, the conformal Hamiltonian H has these properties if the theory admits Wilson operator product expansions and if there is only a finite number of (quasiprimary) fields with dimension $l \leq E$. In this case the spectrum of H consists of the values [9]

$$l + n, \quad n = 0, 1, 2, \ldots, \quad l = \text{dimension of a field.}$$

In addition, the vacuum $|0>$ is eigenvector of H to eigenvalue 0.

In conclusion there will be only *a finite number of degrees of freedom effective below a finite energy E* in a conformal theory. Such a theory should be tractable in any number of dimensions d. The use of conformal symmetry is based on the same philosophy as the renormalization group: To treat exactly critical systems is difficult, because for infinite correlation length infinitely many degrees of freedom appear coupled in an essential way. In the renormalization group approach, the treatment of the problem is decomposed into infinitely many similar steps, each of them involves treatment of an auxiliary statistical system which has both an infrared and an ultraviolet cutoff and therefore also finite correlation length , and which one can hope to handle by making approximations with a finite number of degrees of freedom [2,47].

Solving a conformal invariant field theory means constructing field operators ϕ and a Hilbert space of states $\mathcal{H}$ such that all the principles of local relativistic quantum field theory are fulfilled, plus conformal invariance [14].[1] The most difficult

[1] The existence of a "fundamental field" (which satisfies a field equation in the original nonasymptotic theory) can be read off the list of fields by looking for the shadow of a missing member in a tower of fields, see [14] .

is locality [14,15], all other principles can be implemented by use of convergent [19] operator product expansions [16] on the vacuum.[2] This leaves as free parameters the Lorentz spins s_i and dimensions l_i of the fields ϕ^i, and coupling constants g^{α}_{ijk} that multiply the 3-point functions. (In the presence of spin there can be a finite number of linearly independent conformal invariant 3-point functions labelled by α) [10,19]. They would have to be determined from locality.

Let P_E be the projector to eigenvalues of H below E and let $\phi_E \equiv P_E \phi P_E$. Consider only fields of dimension $l \leq E$ (since only for these $\phi_E|0> \neq 0$); there will be finitely many of them. One could try to find approximate solutions by minimizing the violation of locality in the presence of such an energy cutoff, e.g.

$$\sum_{i,j} \int_{x \text{ spacelike}} tr\Big([\phi^i_E(x), \phi^j_E(0)]_\pm ([\phi^j_E(0)^*, \phi^i_E(x)^*]_\pm \Big) = Min \tag{4}$$

for fields on the tube, $\int_x = \int de d\tau$.

Let us return to the quantum mechanical conformal group $\tilde{G}$ = covering group of $SO(d,2)$, and its action on the tube $\tilde{M}$. $\tilde{G}$ has a discrete center [20] Γ,

$$\Gamma = \Gamma_1 \Gamma_2 \approx \mathbf{Z}_2 \times \mathbf{Z} \tag{5}$$

The nontrivial element of Γ_1 is γ_1 = rotation of e by 2π. It acts trivially on $\tilde{M}$. The transformation law under Γ_1 distinguishes bosons and fermions (in $d > 2$ dimensions). Let $\mathcal{R}$ be "conformal inversion", i.e. a rotation of e by π for even dimension d. Then Γ_2 consists of elements γ_2^n, $n \in \mathbf{Z}$ with

$$\gamma_2 = \mathcal{R} e^{i\pi H}$$

The tube $\tilde{M}$ consists of Minkowski space M and its translates under Γ_2 .(In mathematical terminology, M is a "fundamental domain" for Γ_2). We see that

$$e^{2\pi i n H} \in \Gamma \qquad \text{for} \quad n \in \mathbf{Z}.$$

In an irreducible representation of $\tilde{G}$, elements of its center are represented by multiples of the identity (by Schurs lemma). Therefore the spectrum of H in an irreducible representation contains only eigenvalues of the form (l = lowest eigenvalue)

$$l + n, \qquad n = 0, 1, 2, \ldots$$

It turns out that for any (quasiprimary) field ϕ^i , vectors $\phi^i(\cdot)|0>$ span an irreducible representation space for $\tilde{G}$, and the lowest eigenvalue of H is given by the dimension l_i of ϕ^i [19]. ($|0> = vacuum$). Since H generates τ-translations, it follows that

$$\phi(e, \tau + 2\pi)|0> = e^{2\pi i l} \phi(e, \tau)|0>$$

However, one can not expect in general that the fields themselves are periodic functions of τ up to a phase factor [21,22]. If it were the case, then the operator product expansion of fields ϕ_1 and ϕ_2 of dimensions l_1 and l_2 could only contain fields of

[2] I discussed the use of convergent operator product expansions at length in my Cargèse lectures on conformal field theory in 1976, therefore I don't want to repeat it and refer the reader to the article [18].

dimension $l_1 + l_2 + integer$, which is usually not the case in models. For instance [22] , conformal invariant ϕ^3-theory in $d = 6 + \epsilon$ dimensions has a basic field ϕ of dimension $l = \frac{1}{2}(d-2) + \frac{1}{18}\epsilon + ...$, but the operator product expansion of two such fields contains a tower of symmetric tensor fields O_s of rank s with dimensions l_s given by

$$l_s = d - 2 + s + \sigma_s, \qquad \frac{1}{2}\sigma_s = \left[\frac{1}{18} - \frac{2}{3(s+2)(s+1)}\right]\epsilon + \ldots$$

We see that $l_s - s \mapsto 2l$ as $s \mapsto \infty$ but $l_s - s \not\equiv 2l$. Similar results hold for ϕ^4-theory in $4 - \epsilon$ dimensions .

There is an important exception, cf section III. Currents and stress energy tensor in a *2-dimensional* conformal field theory are periodic functions of τ of period 2π. They can be regarded as living on compactified Minkowski space $\bar{M} = \tilde{M}/\Gamma_2$. This is consistent with causality because these fields have commutators which are concentrated on the light cone. Not all local fields in the 2-dimensional conformal models share these properties, though.

Above we considered general dimension d. In two dimensions the group $\tilde{G}$ with center $\mathbf{Z}_2 \times \mathbf{Z}$ is not the universal covering of $SO(2,2)$, and there could exist theories which live not on $\tilde{M} = S^1 \times \mathbf{R}$ but on its simply connected covering. But one chooses not to consider them.

III. THE STRESS ENERGY TENSOR IN 2-DIMENSIONAL SCALE INVARIANT FIELD THEORY

The content of this section is taken from an unpublished old manuscript by Lüscher and the author [23].

In two dimensional Minkowski space it is convenient to use light cone coordinates

$$x_\pm = x^0 \pm x^1 \equiv t \pm r \tag{1}$$

In two dimensions, scale invariance implies tracelessness of the stress energy tensor (see Appendix)

$$\Theta_{\mu\nu} = \Theta_{\nu\mu}, \qquad \Theta^\mu{}_\mu = 0, \qquad \partial^\mu \Theta_{\mu\nu} = 0. \tag{2}$$

A traceless stress energy tensor has only two independent components

$$\Theta_+ = \Theta_{00} + \Theta_{01} \quad \text{and} \quad \Theta_- = \Theta_{00} - \Theta_{01}. \tag{3}$$

The conservation equation $\partial^\mu \Theta_{\mu\nu} = 0$ reads

$$\frac{\partial}{\partial x_-}\Theta_+(x_+, x_-) = 0, \qquad \frac{\partial}{\partial x_+}\Theta_-(x_+, x_-) = 0. \tag{4}$$

Therefore Θ_+ depends only on x_+ , and Θ_- depends only on x_-.

The stress energy tensor is a local field which commutes at spacelike distances

$$[\Theta_\pm(x_+, x_-), \Theta_\pm(y_+, y_-)] = 0 \quad \text{when} \quad (y_+ - x_+)(y_- - x_-) < 0 \tag{5}$$

Given that Θ_+ (Θ_-) depends only on x_+ (x_-), this is only possible if

$$[\Theta_+(x_+), \Theta_-(y_-)] \equiv 0, \tag{6}$$

and the commutator of two + components is localized at $x_+ = y_+$. It follows that

$$[\Theta_+(x_+), \Theta_+(y_+)] = \sum_{l=0}^{3} \delta^{(l)}(x_+ - y_+) O_l(y_+). \tag{7}$$

This is seen as follows. Let us simplify the notation by writing $\Theta(x), (x \in \mathbf{R})$ for $\Theta_+(x_+)$. Consider the bilocal operator

$$F(\xi, y) = [\Theta(x), \Theta(y)], \qquad \xi = x - y \tag{8}$$

$F(\xi, y)$ is an operator valued distribution which vanishes unless $\xi = 0$. Let $f(x)$ be a test function which equals 1 in a neighborhood of zero. Then the quantities

$$O_k(y) = \frac{(-1)^k}{k!} \int d\xi \xi^k f(\xi) F(\xi, y), \qquad k = 0, 1, 2, ... \tag{9}$$

are local operator valued distributions. They do not depend on the particular shape of $f(\xi)$. Let $U(\lambda)$ be the operator which implements scale transformations by λ,

$$U(\lambda)\Theta(x)U(\lambda)^{-1} = \lambda^2 \Theta(\lambda x) \tag{10}$$

since Θ has dimension 2. Therefore

$$U(\lambda) O_k(y) U(\lambda)^{-1} = \lambda^{3-k} O_k(\lambda y). \tag{11}$$

Thus, O_k is a local field of dimension $3 - k$. The dimension cannot be negative. Otherwise, the unique scale- and translation invariant two-point function which satisfies the spectrum condition

$$< 0|O_k^*(x) O_k(y)|0 >= B_k(x - y - i\epsilon)^{2k-6}, \qquad B_k \in \mathbf{C} \tag{12}$$

would not be positive for any B_k.

The matrix elements $< \psi|F(\xi, y)|\chi >$ are tempered distributions localized at $y = 0$, therefore of the form $\sum_{l \geq 0} \delta^{(l)}(\xi) H_l(y)$ [24]. By definition of O_k, $H_l(y) =< \psi|O_l(y)|\chi >$. Since $O_l = 0$ for $l > 3$, because it would have negative dimension, eq.(7) follows as an operator equation.

Next we investigate the consequences of $[\Theta(x), \Theta(y)] = -[\Theta(y), \Theta(x)]$ when inserted in eq.(7). It requires

$$\sum_{l=0}^{3} \delta^{(l)}(x - y) O_l(y) = - \sum_{l=0}^{3} \delta^{(l)}(y - x) O_l(x).$$

Now

$$\delta^{(l)}(y - x) O_l(x) = \sum_{k=0}^{l} (-)^k \binom{l}{k} \delta^{(k)}(x - y) \frac{\partial^{l-k}}{\partial y^{l-k}} O_l(y)$$

Equating coefficients of $\delta^{(k)}(x - y)$ we obtain

$$O_0(y) = - \sum_{l=0}^{3} \frac{\partial^l}{\partial y^l} O_l(y), \tag{13a}$$

$$O_1(y) = \sum_{l=1}^{3} l \frac{\partial^{l-1}}{\partial y^{l-1}} O_l(y), \tag{13b}$$

$$O_2(y) = -\frac{1}{2} \sum_{l=2}^{3} l(l-1) \frac{\partial^{l-2}}{\partial y^{l-2}} O_l(y). \tag{13c}$$

From eq.(12) we observe that O_3 is a constant which can be written as

$$O_3(y) = c\frac{i^3}{6\pi}.$$

Eq.(13c) implies that $O_2(y) = 0$, and (13a) reduces to

$$2O_0(y) = -\frac{\partial}{\partial y} O_1(y).$$

Putting these results together, we have

$$[\Theta(x), \Theta(y)] = c\frac{i^3}{6\pi}\delta'''(x-y) + \delta'(x-y)O_1(y) - \delta(x-y)\frac{1}{2}\frac{\partial}{\partial y}O_1(y).$$

Using translation covariance, eq. (I-1a) requires that

$$\int dx [\Theta(x), \Theta(y)] = -2i\frac{\partial}{\partial y}\Theta(y).$$

Therefore $\frac{1}{2}(\partial/\partial y)O_1(y) = 2i(\partial/\partial y)\Theta(y)$. The requirement of scale invariance, eq.(11), fixes the integration constant, so that $O_1 = 4i\Theta$. Thus we find as a final result for the commutation relations in Minkowski space

$$\begin{aligned}
[\Theta_+(x_+), \Theta_+(y_+)] =& c_+\frac{i^3}{6\pi}\delta'''(x_+ - y_+) \\
&+ 4i\delta'(x_+ - y_+)\Theta_+(y_+) - 2i\delta(x_+ - y_+)\frac{\partial}{\partial y_+}\Theta_+(y_+), \\
[\Theta_-(x_-), \Theta_-(y_-)] =& c_-\frac{i^3}{6\pi}\delta'''(x_- - y_-) \\
&+ 4i\delta'(x_- - y_-)\Theta_-(y_+) - 2i\delta(x_- - y_-)\frac{\partial}{\partial y_+}\Theta_-(y_-), \\
[\Theta_-(x_-), \Theta_+(y_+)] =& 0.
\end{aligned} \tag{14}$$

If parity invariance holds, then

$$c_+ = c_- = c$$

In conclusion, the fact that the components $\Theta_\pm$ of the stress energy tensor depend only on a single variable $x_\pm$ each leads to the result that the commutation relations of the stress tensor with itself are determined (not only at equal times), with only one free constant. In short, $\Theta_{\mu\nu}$ is a Lie field, i.e. it generates a Lie algebra in which $c \cdot$ *identity* appears as only other generator.

We remember now that conformal field theory lives on a tube, whose points may be parametrized by (e, τ), e = unit d-vector. In two dimensions we may set

$$(e^1, e^2) = (\sin\sigma, \cos\sigma) \tag{15a}$$

Fields on the tube must be periodic functions of σ. This parametrization gives for the Minkowski space coordinates

$$x_{\pm} = \tan \frac{\tau \pm \sigma}{2}. \tag{15b}$$

Since the stress tensor on $\tilde{M}$ must be a periodic function of σ, but Θ_+ depends only on $\tau + \sigma$, it must be periodic in τ also. As we noted before, this means that the stress tensor can be thought to live on compactified Minkowski space. Thus, if we define the stress energy tensor on the tube[1] by

$$\mathit{\Theta}_{\pm}(\vartheta) = (\cos \frac{\vartheta}{2})^{-4}\Theta_{\pm}(\tan \frac{\vartheta}{2}) \qquad (\vartheta = \frac{\tau \pm \sigma}{2}) \tag{16}$$

then $\mathit{\Theta}_{\pm}$ are periodic functions of ϑ. We may therefore consider the Fourier components

$$L_k = \frac{1}{8}\int_{-\pi}^{\pi} d\vartheta e^{ik\vartheta}\mathit{\Theta}_+(\vartheta) \tag{17}$$

The commutation relations for the stress energy tensor translate into

$$[L_l, L_k] = (l-k)L_{l+k} + \frac{c}{12}l(l^2-1)\delta_{k+l,0} \tag{18}$$

These are the commutation relations of the Virasoro algebra Vir [25]. Hermiticity of the stress energy tensor translates into

$$L_{-k} = L_k^* \tag{19}$$

The other component Θ_- of the stress tensor gives rise to another Virasoro algebra with generators $\bar{L}_k$ wich commute with the first. The hermitean linear combinations of

$$L_0, L_1, L_{-1} \quad \text{and} \quad \bar{L}_0, \bar{L}_1, \bar{L}_{-1} \tag{20}$$

generate the conformal algebra

$$so(2,2) \approx so(1,2) \times so(1,2).$$

These generators leave the vacuum invariant (because $L_m|0>$ has zero norm for $m = 0, \pm 1$ by by eqs.(24),(26) below). Therefore we have conformal symmetry as expected. The conformal Hamiltonian, which generates τ-translations, is

$$H = L_0 + \bar{L}_0 \tag{21}$$

while $L_0 - \bar{L}_0$ generates rotations of e, i.e. translations of σ.

Let us now imagine that we live in a 2-dimensional conformal invariant world in which the stress energy tensor is the only observable field [7,18]. This means that the algebra $\mathcal{A}$ of observables is generated by the stress energy tensor, and may be taken as universal envelopping algebra of the Virasoro algebras, for instance. The Hilbert space of physical states will then split into superselection sectors[2] [26,27] which carry

[1] The general formula for transforming arbitrary quasiprimary fields to the tube (in any dimension d) can be found in [9]

[2] By definition, observables do not make transitions between different superselection sectors. As a result, relative phases between vectors in different superselection sectors are unobservable.

irreducible *-representations of the algebra of observables, hence unitary irreducible representations of the sum of the two Virasoro algebras. From the fact that $H \geq 0$ in these representations one can deduce that

$$L_0 \geq 0 \qquad \text{and} \qquad \bar{L}_0 \geq 0. \tag{22}$$

This reflects the fact that in 2 dimensions there are two positive generators of translations $P^0 \pm P^1$. (Remember that $H \geq 0$ follows from $P^0 \geq 0$).

Let us find the irreducible representations of the Virasoro algebra with positive energy $L_0 \geq 0$. From the commutation relations

$$[L_0, L_m] = -mL_m \tag{23}$$

we conclude that L_m will take an eigenvector of L_0 to eigenvalue E into an eigenvector to eigenvalue $E-m$. Since L_0 is positive, there must be a lowest eigenvalue. It follows that there is an eigenvector $|h>$ of L_0 such that

$$L_0|h>= h|h>, \qquad L_k|h>= 0 \quad \text{for} \quad k > 0. \tag{24}$$

This is called a lowest weight vector. h is the lowest eigenvalue of L_0 in the irreducible representation of Vir. The vacuum state $|0>$ has weight $h = 0$.

The representation space $\mathcal{H}_h$ will be spanned by vectors of the form

$$L_{k_n} ... L_{k_2} L_{k_1} |h> \equiv |h; k_1, ..., k_n> \tag{25}$$

Their scalar products can be calculated from the Virasoro commutation relations, the action (24) of generators L_k with $k \geq 0$ on the lowest weight vector $|h>$, and the relation $L_{-n} = L_n^*$, as follows. Generators L_n with $n > 0$ are moved to the right until they annihilate $|h>$, and generators $L_n = L_{-n}^*$ with $n < 0$ are moved to the left until they annihilate $<h|$. The remaining generators L_0 give factors h. Thus, for instance

$$\begin{aligned} <h|L_{-m}^* L_{-m}|h> &=<h|[L_m, L_{-m}]|h> \\ &= 2mh + \frac{c}{12} m(m^2-1) \qquad \text{for} \quad m > 0 \end{aligned} \tag{26}$$

if $<h|h>= 1$. In a unitary representation, one must have $<\psi|\psi>\geq 0$; the zero norm vectors are divided out to get $\mathcal{H}_h$. Thus one must have

$$<h|A^* A|h>\geq 0 \tag{27}$$

for any monomial A in the generators L_m. For instance, eq. (26) tells us that $c \geq 0$. By computing the norm $<\psi|\psi>$ of the vector

$$|\psi>= \left[8L_{-9} + 6L_{-7}L_{-2} + 12L_{-6}L_{-3} - 8L_{-2}L_{-5}L_{-2} + 12L_{-3}L_{-4}L_{-2} - 5L_{-3}^3\right]|0>$$

one obtains the restriction

$$c \geq \tfrac{1}{2}$$

or $c = 0$. The theory of one hermitean anticommuting massless free field $\varphi(x)$ yields a model with $c = \frac{1}{2}$. Such a field depends on a single variable $x = x_+$ or x_-.

From (25) and the commutation relations it follows that the spectrum of H in an irreducible positive energy representation of $Vir \oplus Vir$ with lowest weight $(h, \bar{h})$ consists of eigenvalues of the form

$$E = h + \bar{h} + n, \qquad n = 0, 1, 2, \ldots$$

Now we remember that application of a field operator of dimension l to the vacuum yields a representation space of the conformal algebra with eigenvalues of H of the form $l + m$, $m = 0, 1, 2, \ldots$. Therefore, classification of the unitary irreducible representations of Vir yields information on the dimension l of fields in the theory , i.e. on the critical indices of the statistical mechanical system in Euclidean space.

All this was known to Lüscher and the author in 1976 [23]. We tried to evaluate the restriction on c, h from unitarity, with little success. Working out norms of various vectors yielded strange results.[1] The problem was solved later by other authors with the help of the Kac determinant [28] which was not known in 1976. Friedan, Quiu and Shenker [29] showed that either

$$c = 1 - \frac{6}{(m+2)(m+3)}, \qquad 0 \leq m,$$
$$h = \frac{[(m+3)r - (m+2)s]^2 - 1}{4(m+2)(m+3)}, \qquad 1 \leq s \leq r \leq m+1,$$

with m, r, s integer, or

$$c \geq 1, \quad h \geq 0 \quad \text{arbitrary.}$$

Goddard and Olive showed that to each of these pairs there exists a unitary irreducible positive energy representation [30].

It has become costumary to use complex variables $z_\pm$ on the unit circle in place of the angles

$$z_\pm = e^{i(\tau \pm \sigma)} = \frac{1 + ix_\pm}{1 - ix_\pm}.$$

One defines the stress tensor in these coordinates

$$T(z) = 2\pi \left(i \frac{dx}{dz} \right)^2 \Theta(x(z)),$$

where $T = T_\pm$, $z = z_\pm$, $x = x_\pm$ etc. The Fourier expansion (17) reads now

$$T(z) = \sum_{n \in \mathbf{Z}} L_n z^{-n-2}.$$

Consider smeared fields

$$T(f) = \frac{1}{2\pi i} \oint_{S^1} dz T(z) f(z), \qquad z^{-1} f(z) \quad \text{real.}$$

[1] By contrast, this method combined with the theory of induced representations worked beautifully for positive energy representations of the conformal group $\hat{G}$ in $d = 4$ dimensions. It yielded both a complete classification and explicit construction of all of them [20]

The commutation relations read in this language

$$[T(f),T(g)] = T(\mathcal{D}_f g) + \frac{c}{12}\omega(f,g)$$

$$\mathcal{D}_f g = fg' - gf'; \qquad (' = \frac{d}{dz})$$

$$\omega(f,g) = \oint \frac{dz}{2\pi i} f''' g.$$

Thus the adjoint action of the Virasoro algebra defines an action $g \mapsto \mathcal{D}_f g$ of $T(f)$ on functions g on the circle.
$g \mapsto g + \delta t \mathcal{D}_f g$ is of the form of the action of an infinitesimal diffeomorphism [31]

$$z \mapsto z' = z + \delta z, \qquad \delta \ln z = i z^{-1} f(z) \delta t$$

on functions of weight -1 on the circle. By definition, such functions transform under diffeomorphisms $z \mapsto z'(z)$ according to

$$g(z) \mapsto \left(\frac{\partial z'}{\partial z}\right)^{-1} g(z')$$

The extra term $\frac{c}{12}\omega(f,g)$ signals that Vir is a *central extension* of the Lie algebra of the diffeomorphism group $Diff(S^1)$ on the circle. $Diff(S^1)$ has an infinite sheeted covering group [32] whose center is isomorphic to $\mathbf{Z}$ and coincides with the center of one factor $\widetilde{SO(1,2)}$ of the universal covering group of the conformal group in two dimensions. This gives another way of understanding why the stress energy tensor is periodic in τ, i.e. is invariant under the action of the center of the conformal group. It is because the adjoint action of the center of a Lie group on its Lie algebra is trivial.

The results presented above can be generalized to currents j_μ which are conserved together with their axial brothers $j^5_\mu = \epsilon_{\mu\nu} j^\nu$.

$$\partial^\mu j_\mu = 0, \quad \partial^\mu j^5_\mu = 0.$$

In light cone coordinates this gives

$$\frac{\partial}{\partial x_-} j_+(x) = 0, \qquad \frac{\partial}{\partial x_+} j_-(x) = 0.$$

so that $j_\pm$ depend on only one variable $x_\pm$ each. (This observation has been a cornerstone in the construction of 2-dimensional models for decades). The commutation relations can be worked out as for the Virasoro algebra. In the language of complex variables on the circle one has a Fourier expansion for $J \propto j_\pm(x_\pm)$.

$$J(z) = \sum_{n \in \mathbf{Z}} J_n z^{-1-n}.$$

The commutation relations take the form of a Kac Moody algebra, which is the Lie algebra of the central extension of a loop group [32-36]. For instance

$$[J_n, J_m] = n\delta_{n+m,0}, \qquad [J_n, L_m] = nJ_{n+m}$$

for a single current that is associated with a chiral $U(1)$-symmetry. This larger algebra is called the conformal current algebra. It contains Vir. A lowest weight vector for the conformal current algebra is a simultaneous eigenvector of L_0 and J_0 which obeys

$$L_0|q,h> = h|q,h>$$
$$J_0|q,h> = q|q,h>$$
$$L_n|q,h> = J_n|q,h>= 0 \quad \text{for } n > 0.$$

q is the charge. The Sugawara prescription permits to construct a stress energy tensor with the right commutation relations from the current [33]

$$T(z) = \tfrac{1}{2} : J(z)^2 :$$

or, in modes

$$L_n = \tfrac{1}{2} \sum_{l,m:l+m=n} : J_l J_m :$$

Normal ordering :: means that generators J_l with larger l stand to the right. Validity of the Sugawara relation means that the algebra of observables is not simply the universal envelopping algebra of the Lie algebra of current and stress energy tensor, but is obtained from it by factoring out an ideal. (Equating an ideal to zero is the algebraic variant of field equations). Sandwiching the Sugawara relation between lowest weight vectors gives a relation between charge q and h [34],

$$h = \tfrac{1}{2}q^2.$$

IV. ANALYTIC CONTINUATION TO EUCLIDEAN SPACE

In the last section I explained how to compute expectation values $< h|L_{k_1}...L_{k_n}|h >$ of products of Virasoro generators in the vacuum $|0>$ or other lowest weight state $|h>$. Since L_k are the Fourier components of the stress tensor, this yields the vacuum expectation values $< 0|T(z_1)...T(z_n)|0 >$ etc. of products of stress tensors in a two dimensional theory. We moved the generators L_n with $n \geq 1$ to the right, using the Virasoro commutation relations, until they annihilate the lowest weight vector $|h>$. For the vacuum $|0>$ we could do the same for $n \geq -1$, because the vacuum is conformal invariant. The procedure is therefore equivalent to the following. Let $T^{(-)}$ be the annihilation part of T, viz.

$$T^{(-)}(z) = \sum_{n\geq -1} L_n z^{-2-n}, \qquad T^{(-)}(z)|0>= 0. \tag{1}$$

Its commutation relations with the stress tensor are

$$[T^{(-)}(z_0), T(z)] = \frac{c}{2z_{01}^4} + \frac{2}{z_{01}^2}T(z_1) + \frac{1}{z_{01}}T'(z_1), \tag{2}$$

where $z_{01} = z_0 e^{\epsilon} - z_1$, $\epsilon \mapsto +0$. (The familiar $i\epsilon$-prescription for time τ turns into a factor $e^{\epsilon} > 1$ here.) Proceeding as above, i.e. moving the annihilation part of T to the right, yields the recursion relation ($\widehat{\cdot}$ means "omit" $\cdot$)

$$\begin{aligned} < T(z_0)T(z_1)...T(z_n) > =& \frac{c}{2}\sum_{k=1}^{n} \frac{1}{z_{0k}^4} < T(z_1)...\widehat{T(z_k)}...T(z_n)|0 > \\ &+ \sum_{k=1}^{n}\left(\frac{2}{z_{0k}^2} + \frac{\partial}{\partial z_k}\right) < T(z_1)...T(z_n) > \end{aligned} \tag{3}$$

Here $< \quad >$ is short for vacuum expectation values $< 0|...|0 >$ for now.

Vacuum expectation values of local field operators (like the stress tensor) may be analytically continued to Euclidean space. This results in Euclidean Greens functions (=Schwinger functions) which are symmetrical in their arguments. Analytic continuation is also possible for fields which are not local, but pseudolocal as is natural in two dimensions. For a discussion of this case I refer to J. Fröhlichs lectures at this school [37].
Recursion relations like (3) remain true for the Schwinger functions $< T(z_1)...T(z_n) >$ by uniqueness of analytic continuation [31,38]. In the models of interest, Schwinger functions have an interpretation as expectation values of classical statistical mechanical systems, and we may thus interpret the notation $< \quad >$ as such expectation values.

Since the conformal Hamiltonian H (which is generator of translations of the conformal time τ which parametrizes the tube) is positive, we may continue in the variable τ:

$$\tau \mapsto i\tau$$

The complex variables become

$$z_+ = e^{-\tau+i\sigma} \equiv z,$$
$$z_- = e^{-\tau-i\sigma} \equiv \bar{z}.$$

It is customary to write

$$T_+(z_+) = T(z)$$
$$T_-(z_-) = \bar{T}(z)$$

We see that in the Euclidean domain, z_+ and z_- are complex conjugate of each other. The fact that $T_+(z_+)$ is independent of z_-, viz

$$\frac{\partial}{\partial \bar{z}} T(z) = 0 \qquad \text{etc.},$$

translates into the property that Schwinger functions are holomorphic in arguments z of $T(z)$, and antiholomorphic in arguments z of $\bar{T}(z)$.

Belavin, Polyakov and Zamolodchikov introduced the concept of a primary field [31]. This is a field ψ in whose operator product expansion appears no field of lower dimension than the dimension of ψ. Its commutation relations with the stress tensor read, in the z-picture

$$[L_n, \psi(z)] = z^n \left(z\frac{d}{dz} + (n+1)h \right) \psi(z)$$

$\psi(z)$ leads from the vacuum to a superselection sector which carries a representation with lowest weight h of the Virasoro algebra with generators L_n. If ψ is independent of $\bar{z}$ then h is the dimension. (In general the dimension proper is $l = h + \bar{h}$, while $h - \bar{h}$ is the Lorentz spin. $\bar{h}$ is the lowest weight for the other Virasoro algebra). The possibility of finding the commutation relations not only at equal times follows again from the fact that Θ_+ is independent of of x_-. These commutation relations can be used again to relate expectation values $< 0|T(z)A|0 >$ to expectation values

$< 0|A|0 >$, for monomials A in the fields. These relations remain valid under analytic continuation to Euclidean space. For instance [31]

$$< T(z_0)\psi_1(z_1)...\psi_n(z_n) >= \sum_i \left(\frac{h_i}{z_{0i}^2} + \frac{1}{z_{0i}}\frac{\partial}{\partial z_i} \right) < \psi_1(z_1)...\psi_n(z_n) > .$$

In general, a local field ψ in two dimensions does not depend on $z_+ \equiv z$ or $z_- \equiv \bar{z}$ only, but on both. This implies that the Schwinger functions involving $\psi(z)$ will not be holomorphic or antiholomorphic in z.

Schroer and Rehren [39] have presented arguments that it is always possible to build up local fields ϕ from nonlocal components which depend only on one variable z_+ or z_-.

The first step is the decomposition of the local field $\phi(e,\tau)$ on the tube into pieces which transform according to an irreducible representation of the center Γ of the conformal group, following Schroer and Swieca [21]. These pieces are sums of the form

$$\sum_n e^{i\Delta n}\phi((-)^n e, \tau + \pi n) = \phi_\Delta .$$

These Schroer Swieca fields are natural objects to consider *in 2 dimensions*, because they are still relatively local to the observables (stress tensor and currents), although generally not to themselves. Relative locality to the observables follows from the fact that the field ϕ was (relatively) local, and stress tensor and currents are 2π-periodic functions of τ which transform trivially under the center Γ. In more than 2 dimensions this would be no longer true.

The second step is an argument, based on operator product expansions, that these Schroer Swieca fields can be chosen to depend on only one variable. There could be a multiplicity problem; I am not prepared to discuss that. I refer the reader to B. Schroers lectures at this school for further results of this approach.

V. QUANTUM FIELD THEORY AS REPRESENTATION THEORY OF A GROUP OR LIE ALGEBRA

It is sometimes said that conformal invariance cannot be very useful in more than two dimensions because then one does not have the infinite dimensional Lie algebra. I have already tried to argue against this pessimistic point of view in section 2. Moreover, the Virasoro algebra is not Lie algebra of a symmetry group, because it does not leave the vacuum invariant (for $c \neq 0$) - only the generators of the conformal algebra $so(1,2) \times so(1,2)$ do. The Virasoro algebra should properly be regarded as a spectrum generating algebra, or as a Lie algebra associated with an algebra of observables [27]. Such an algebra exists in any local relativistic quantum field theory, and the Lie algebra can be described in a general way.

Borchers proposed to regard local relativistic quantum field theory as *-representation of an algebra of test functions (Borchers algebra) [42]. This idea can be specialized to conformal QFT as follows. The algebra consists of equivalence classes of finite sequences of test functions, with product $\otimes$:

$$f = f^0, f^1_{\mu_1,\nu_1}(x_1), f^2_{\mu_1\nu_1\mu_2\nu_2}(x_1,x_2), ..., f^N_{\mu_1\nu_1...\mu_N\nu_N}(x_1,\ldots,x_N). \tag{1}$$

For simplicity of notation, let us adopt the obvious identification of constants f^0, functions f^1 of a single variable, etc. with sequences f of test functions with only one nonvanishing term. The product $\otimes$ in the algebra is defined by

$$(f^0 \otimes g^1)_{\mu_1\nu_1}(x_1) = f^0 g^1_{\mu_1\nu_1}(x_1)$$
$$(f^1 \otimes g^1)_{\mu_1\nu_1\mu_2\nu_2}(x_1, x_2) = f^1_{\mu_1\nu_1}(x_1) g^1_{\mu_2\nu_2}(x_2), \tag{2}$$
$$\text{etc.}$$

A complex *-algebra is obtained by admitting complex test functions, with complex conjugation as *-operation. The requirement of a *-representation means that the representation operator $\Theta(f)$ of f obeys

$$\Theta(f)^* = \Theta(f^*)$$

Until further notice we shall restrict attention to real f. The hermitean representation operators $\Theta(f)$ are to be interpreted as smeared products of stress tensors.

$$\Theta(f) = f^0 \mathbf{1} + \sum_{k=1}^{N} \int dx_1 ... dx_k f^k_{\mu_1\nu_1...\mu_k\nu_k}(x_1, ..., x_k) \Theta^{\mu_1\nu_1}(x_1) ... \Theta^{\mu_k\nu_k}(x_k) \tag{3}$$

The algebra $\mathcal{A}$ is obtained from the free algebra generated by constants f^0 and functions f^1 of a single variable and $\otimes$-products of such functions, by imposing the following equivalence relations.

TRACELESSNESS, SYMMETRY AND CONSERVATION OF STRESS TENSOR.

$$f^1_{\mu\nu}(x) \equiv f^1_{\mu\nu}(x) + \eta_{\mu\nu} h(x) + [k_{\mu\nu}(x) - k_{\nu\mu}(x)] + \partial_\mu l_\nu(x) \tag{E1}$$

for arbitrary functions $h, k_{\mu\nu}, l_\mu$ of x. $\eta_{\mu\nu}$ is the metric tensor.

LOCALITY.

$$f^1 \otimes g^1 = g^1 \otimes f^1 \text{ if the supports of } f^1 \text{ and } f^2 \text{ are relatively spacelike.} \tag{E2}$$

CONFORMAL SYMMETRY. *Eq.(E3) below.*

The generators J^{AB} of the conformal algebra $so(d,2)$ are linear combinations of the generators $P_\mu, M_{\mu\nu}, D, K_\mu$ introduced in section I ($A, B = 0...d-1, d+1, d+2$). [6]. They are constructed from the stress tensor according to eqs.(I.1)f. Integration is over space, but because the result is time independent, we may also smear over time. Therefore there exist special test functions $f^{AB}(x)$ of a single variable such that the generators take the form

$$J^{AB} = \Theta(f^{AB}).$$

The conformal Hamiltonian is $H = J^{0d+2}$. The transformation law of the stress tensor under conformal transformations takes the form

$$[J^{AB}, \Theta(g^1)] = \Theta(\partial^{AB} g^1)$$

where ∂^{AB} are known differential operators. This translates into an equivalence relation in the Borchers algebra,

$$f^{AB} \otimes g^1 - g^1 \otimes f^{AB} \equiv \partial^{AB} g^1. \tag{E3}$$

This should be read as an equivalence relation between sequences of test functions in which there is only a single nonvanishing term. Note that it is an equivalence between functions with different numbers of arguments (2 on the left, 1 on the right), i.e. *inhomogeneous*. Our additional requirement that the stress tensor generates some symmetries leads thus to a profound change in the structure of the Borchers algebra. It provides for an inhomogeneous relation.

The algebra constructed in this way shall be called the Θ-Borchers algebra. One is interested in positive energy representations of this algebra. The positive energy requirement reads

$$H = \boldsymbol{\Theta}(f^{0d+2}) \geq 0. \tag{4}$$

The result of the analysis of the stress tensor of a scale invariant theory in 2 dimensions (section III) may be rephrased as follows:

The Θ-Borchers algebra in 2 dimensions is isomorphic to the envelopping algebra of the Lie algebra $Vir \oplus Vir$ (two copies of Virasoro algebra).

More generally, one may regard the Θ-Borchers algebra as envelopping algebra of a Lie algebra $\mathcal{V}$ which generalizes $Vir \oplus Vir$. The Lie algebra $\mathcal{V}$ is made of equivalence classes of commutators $f \otimes g - g \otimes f$. Its representation operators are linear combinations of

$$\Theta(f^1), \quad [\Theta(f^1), \Theta(g^1)], \qquad [[\Theta(f^1), \Theta(g^1)], \Theta(h_1)], \quad \ldots$$

where f^1, g^1, h^1 are functions of a single variable; $\Theta(f^1) = \boldsymbol{\Theta}(f^1)$ is the smeared stress tensor.

Since conformal field theory lives on a tube $\tilde{M} = S^{d-1} \times \mathbf{R}$, it is appropriate to regard x as points on $\tilde{M}$, viz $x = (e, \tau)$. The requirement that we want theories with a discrete spectrum of the generator H of τ-translations, as is required by validity of operator product expansions, may be incorporated by admitting as test functions all elements of the dual of the space of quasiperiodic generalized functions of τ. (Quasiperiodic functions are those which admit a Fourier series representation, rather than requiring a Fourier integral.)

The complex Virasoro algebra contains generators L_n, $(n > 0)$ which annihilate the vacuum. (They are not hermitian, therefore only contained in the complex Lie algebra, but they are in the real *Euclidean* algebra - see section VI). The same is true in general for the complex Lie algebra $\mathcal{V}$ which is generated by the stress tensor (see above) smeared with complex test functions f. Let S be the set of real numbers in the spectrum of H, and let S_Δ be the set of differences of numbers in S. Let E_γ be the projector on the eigenvalue γ of H. Decompose

$$\begin{aligned} \mathcal{V} = \sum_{\alpha \in S_\Delta} \mathcal{V}_\alpha, \qquad X = \sum_{\alpha \in S_\Delta} X_\alpha, \quad (X_\alpha \in \mathcal{V}_\alpha) \\ \text{where} \quad X_\alpha = \sum_{\beta, \gamma \in S : \gamma = \beta + \alpha} E_\gamma X E_\beta \end{aligned} \tag{5}$$

Elements of $\mathcal{V}_\alpha$ transfer energy α. $\mathcal{V}_\alpha$ contains $\Theta(f^1)$ for $f^1(\epsilon,\tau) \propto \tilde{f}^1(e)e^{i\alpha\tau}$ etc. More precisely it contains $\lim_{T\mapsto\infty}(2T)^{-1}\Theta(\chi_T f^1)$ where $\chi_T(\tau) = 1$ for $|\tau| < T$, and $= 0$ otherwise. Clearly the spectrum condition implies that

$$X|0>= 0 \quad \text{for } X \in \mathcal{V}_\alpha \quad \text{with } \alpha > 0.$$

The same is true of elements X in $\mathcal{V}_0$ with $< 0|X|0 >= 0$. They form a subalgebra if the vacuum is unique. Clearly

$$[\mathcal{V}_\alpha, \mathcal{V}_\beta] \subseteq \mathcal{V}_{\alpha+\beta}, \quad \text{in particular } [\mathcal{V}_0, \mathcal{V}_\alpha] \subseteq \mathcal{V}_\alpha.$$

It is interesting that in the ϕ^3-model in $6+\epsilon$ dimensions anomalous parts of of the dimensions of all fields are relatively rational, to first order in ϵ. They determine the spectrum of H as we know.

The conformal generators are in $\mathcal{V}_0$ and $\mathcal{V}_{\pm 1}$. A further decomposition of $\mathcal{V}$ can be effected by harmonic analysis [19] on the conformal group $\tilde{G}$.

In conclusion, the situation in d dimensions is much as in two dimensions, except that the representation theory of $\mathcal{V}$ looks less tractable . But remember that the situation in two dimensions was just a little better than this in 1976. We knew the algebra and how to construct positive energy representations by the method of lowest weight, but we were unable to find out which were unitary, except for free field examples, see section III.

There is also a *group theoretical* formulation of the constructions described above. Consider the group **G** which is obtained from the free group on objects $W(f)$ ($f = f_{\mu\nu}(x) real$) by imposing equivalence relations. The representation operator for $W(f)$ is to be interpreted as unitary operator

$$W(f) = e^{i\Theta(f)}, \qquad \Theta(f) = \int d^dx f_{\mu\nu}(x)\Theta^{\mu\nu}(x). \tag{6}$$

The equivalence relations are

$$\begin{aligned} &W(f) = W(-f)^{-1} \\ &W(f)W(g) = W(g)W(f) \quad \text{if } f,g \text{ have relatively spacelike supports} \\ &W(tf^{AB})W(g)W(-tf^{AB}) = W(e^{it\theta^{AB}}g) \quad (t \text{ real}) \end{aligned} \tag{7}$$

and dependence on f should only be through equivalence classes (E1).

The group **G** is the generalization to d dimensions of the central extension of $Diff(S^1) \times Diff(S^1)$ which we encountered in 2 dimensions.

VI. SEMIGROUPS, FINITE AND INFINITE DIMENSIONAL

The results of Lüscher and the author on global conformal invariance, which were reviewed in section II, were derived by use of semigroups [9]. Starting from Greens functions in d-dimensional Euclidean space which are invariant under the Euclidean conformal group G, a contractive representation T of a real subsemigroup S of G (with the same dimension as G) was constructed by appeal to Osterwalder Schrader (OS) positivity [OS-positivity is the physical positivity for quantum field theory [13]]. A nonabelian Hille Yosida theorem was proven [43] which implied that this contractive representation T of S could be analytically continued (through a complex semigroup S_c) to a unitary positive energy representation of the quantum mechanical conformal group in Minkowski space.

The (finite dimensional) conformal group in d-dimensional Minkowski space is the infinite sheeted universal covering G^* of $SO(d,2)$, its Euclidean brother is the twofold covering G of $SO(d+1,1)$. Both groups share a common subgroup

$$G_+ = \widetilde{SO(d,1)} = Spin(d,1).$$

The semigroup S is mapped into itself by the adjoint action of G_+. And

$$\Lambda \in S \quad \text{implies} \quad \theta(\Lambda^{-1}) \in S.$$

θ is Euclidean time reversal, see below. Schematically, the relation between representations is (T contractive means norms $||T(\Lambda)|| \leq 1$)

pseudounitary representation of G with OS-positivity

$\Downarrow$

contractive representation T of S with $T(\theta(\Lambda^{-1})) = T(\Lambda)^*$

$\Downarrow$

unitary positive energy representation of G^*

Pseudo-unitary means that there exists an invariant but not necessarily positive semidefinite scalar product. For irreducible representations it is furnished by the Knapp-Stein intertwining operator [45], it is the same that is used for the complementary series [46] of unitary representations of G. The semigroup can act on an invariant proper subspace ("half") of the representation space of G, with scalar product modified by action of the time reversal operator [9,13].

The real Lie algebras[1] $\mathcal{G}$ and $\mathcal{G}^*$ of the two groups (Euclidean and Minkowski conformal group in our application- but the construction is more general) are related as follows. There is an involution θ of the Lie algebras $\mathcal{G}^*$ and $\mathcal{G}$ - i.e. an automorphism with $\theta^2 = 1$. This furnishes a split

$$\mathcal{G}^* = \mathcal{G}^*_+ + \mathcal{G}^*_-, \quad \text{with } \theta(X) = \pm X \quad \text{for } X \in \mathcal{G}^*_\pm \tag{1a}$$

The real Lie algebra of the Euclidean group is then

$$\mathcal{G} = \mathcal{G}^*_+ + i\mathcal{G}^*_- \equiv \mathcal{G}_+ + \mathcal{G}_- \tag{1b}$$

$\mathcal{G}^*_+$ generates the common (noncompact) subgroup G_+ of G and G^*. If $\mathcal{G}^*$ admits positive energy representations then $\mathcal{G}^*_-$ contains a nontrivial cone V which is invariant under the adjoint action of G_+ [44]. The elements of V are conjugate under G_+ to $-tH$, $t > 0$, where H is hermitean generator of a factor SO(2) of the maximal compact subgroup of (the adjoint representation of) G^* if G^* is simple [44]. In d-dimensional conformal field theory H is the conformal Hamiltonian. A semigroup S^o is generated by $\mathcal{G}^*_+ - iV \subset \mathcal{G}$, it consists of finite products

$$ue^{X_1}...e^{X_n}, \qquad u \in G_+, \quad X_1, \ldots, X_n \in V. \tag{1c}$$

S is obtained from S^o by taking closures.

[1]I use physicists conventions for real Lie algebras: Their generators are hermitean in a unitary representation of the group, i.e. $e^{iX} \in G$ if $X \in \mathcal{G}$.

In two dimensional conformal quantum field theory one has two positive generators $P_{\pm}$ of translations. Therefore one has the possibility of performing analytic continuation in two variables $t \pm r$. Suitable components of the stress tensor depend on only one of them, as we have seen. In this way the problem is reduced to the study of a 1-dimensional theory. The special case $d = 1$ is therefore of particular interest:

$$G = \widetilde{SO(2,1)} \supset S \Rightarrow G^* = \widetilde{SO(1,2)}.$$

In this case, the two groups are locally isomorphic to each other and to $SL(2,\mathbf{R})$. By carrying out the above construction one gets positive energy representations (i.e. representations of the interpolated discrete series) of the infinite sheeted universal covering group of $SL(2,\mathbf{R})$ from representations of $SL(2,\mathbf{R})$ in the complementary series. (A similar analysis was done for other groups by Schrader [48]). This is of interest because Vir contains infinitely many subalgebras $sl(2,\mathbf{R})$.

Here I wish to generalize these constructions to the infinite dimensional case which is of interest in 2-dimensional conformal field theory.

The *real* Lie algebra $Vir \oplus Vir$ is generated by the stress energy tensor in Minkowski space, smeared with real test functions. I will describe the Euclidean semigroup S, and its Lie algebra $\mathcal{G}$, and initiate study of an associated group. They turn out to be interesting objects. Restricting attention to one factor, which comes from one of the two algebras Vir, one finds an algebra $\mathcal{G}$ which behaves very much like an infinite dimensional brother of $sl(2,\mathbf{R})$.

It has both a noncompact and a compact Cartan subalgebra, both 1-dimensional (not counting the center) , as is the case for $sl(2,\mathbf{R})$. It admits an Iwasawa decomposition $\mathcal{G} = \mathcal{K} + \mathcal{A} + \mathcal{N}$ and a Bruhat decomposition $\mathcal{G} = \mathcal{X} + \mathcal{M} + \mathcal{A} + \mathcal{N}$, as are familiar from the theory of finitedimensional noncompact groups.[1] Half of the generators $\tilde{K}_n \in \mathcal{K}$ of the "maximal compact subgroup " K with Lie algebra $\mathcal{K}$ generate compact 1-parameter groups (in the adjoint representation). Together with their commutators they span the Lie algebra $\mathcal{K}$. The Lie algebra $\mathcal{N}$ is nilpotent in the sense that there exists a decreasing sequence of subalgebras $\mathcal{N} = \mathcal{N}_1 \supset \mathcal{N}_2 \supset \dots$ such that

$$[\mathcal{N}_k, \mathcal{N}_m] \subseteq \mathcal{N}_{k+m}, \qquad \text{and} \quad \bigcap_{m=1}^{\infty} \mathcal{N}_m = \emptyset. \tag{2}$$

Let us start with the real Virasoro algebra $\mathcal{G}^* = Vir$. It is generated by

$$T(f) = \frac{1}{2\pi i} \oint dz\, zT(z)f(z), \qquad f(z) \text{ real }, z = ie^{i\tau} \in S^1. \tag{3a}$$

I have changed conventions compared to section III because reality properties will be important here. $T(f)$ is a hermitean linear combination of generators L_n of the Virasoro algebra.

$$T(f) = \sum_n f_n L_n, \quad \text{if } f(z) = \sum f_n z^n. \tag{3b}$$

[1] The Iwasawa decomposition is a cornerstone of modern representation theory of (finite dimensional) noncompact Lie groups, because Harish Chandra's subquotient theorem [49] asserts that every unitary irreducible representation of such a group is subquotient of an induced representation on the homogeneous space $G/KAN = K/M$ (K = maximal compact subgroup, A abelian noncompact, M = centralizer of A in K). The inducing representation of MAN is trivial on the nilpotent subgroup N [46].

For real f we have $\bar{f}_{-n} = f_n$. The Virasoro generators obey the commutation relations and hermiticity condition which were derived in section III

$$[L_n, L_m] = (n-m)L_{n+m} + \frac{c}{12}n(n^2-1)\delta_{n+m,0} \tag{4a}$$

$$L_n = L^*_{-n} \tag{4b}$$

Note that L_n are not in the real Lie algebra $\mathcal{G}^*$, except for $n = 0$, because they are not hermitean. The hermitean generators of the finite dimensional conformal group $\widetilde{SL(2, \mathbf{R})} \subset G^*$ in Minkowski space are

$$\begin{gathered}\frac{1}{2}(L_1 + L_{-1}), \quad \frac{i}{2}(L_1 - L_{-1}) \quad \text{(noncompact)} \\ L_0 \qquad \text{(compact)}\end{gathered} \tag{5}$$

The involution is now defined by

$$\theta(L_n) = -(-)^n L_{-n}, \qquad \theta(c) = -c \tag{6}$$

This extends the above mentioned involution for the finite dimensional conformal algebra $sl(2, \mathbf{R})$. It is easily verified that θ is an automorphism. It gives

$$\theta(T(f)) = T(\theta f) \quad \text{with } \theta f(z) = -f(-z^{-1}) \tag{7}$$

This yields a decomposition (1) of the Virasoro algebra $\mathcal{G}^*$ into even and odd parts $\mathcal{G}^*_+ + \mathcal{G}^*_-$,

$$T(f) \in \mathcal{G}^*_\pm \quad \text{if } \theta f = \pm f, \qquad c \in \mathcal{G}^*_- \tag{8}$$

The Euclidean algebra is defined by $\mathcal{G} = \mathcal{G}^*_+ + i\mathcal{G}^*_-$, as in eq.(1). Thus

$$T(f_+ + if_-) \in \mathcal{G} \quad \text{if } f_\pm(z) \text{ are real, and } f_\pm(-z^{-1}) = \mp f(z) \tag{9}$$

Equivalently

$$T(f) \in \mathcal{G} \quad \text{if } f(-z^{-1}) = -\overline{f(z)}. \tag{9'}$$

In particular, if we set

$$f^n(z) = i\left(\frac{z}{i}\right)^n \tag{10}$$

then

$$T(f^{-n}) = ii^n L_n \equiv \tilde{L}_n \in \mathcal{G}. \tag{11a}$$

In addition, the central charge

$$\tilde{c} \equiv ic \in \mathcal{G}.$$

We see that the generators L_n of the Virasoro algebra, multiplied with i if n is odd, are elements of the real Euclidean Lie algebra $\mathcal{G}$, so they would be hermitean in a unitary representation of $\mathcal{G}$. The commutation relations of the hermitean generators are

$$\begin{gathered}[\tilde{L}_n, \tilde{L}_m] = i\{(n-m)\tilde{L}_{n+m} + \frac{\tilde{c}}{12}n(n^2-1)\delta_{n+m,0}\} \\ \tilde{L}_n = \tilde{L}^*_n, \qquad \tilde{c} = \tilde{c}^*.\end{gathered} \tag{12}$$

$\mathcal{G}$ contains the finite dimensional Euclidean conformal algebra $\mathcal{H}_1 \approx sl(2,\mathbf{R})$. It is the first of an infinite family of algebras $\mathcal{H}_n \approx sl(2,\mathbf{R})$ $(n = 1,2,3...)$. The generators of $\mathcal{H}_n$ are

$$\begin{aligned} &\frac{1}{2n}(\tilde{L}_n + \tilde{L}_{-n}) \quad \text{(compact)} \\ &\frac{1}{n}\tilde{L}_0 + \frac{\tilde{c}}{24}(n - \frac{1}{n}), \quad \frac{1}{2n}(\tilde{L}_n - \tilde{L}_{-n}) \quad \text{(noncompact)} \end{aligned} \tag{13}$$

Note that, compared to the Minkowski space conformal algebra $sl(2,\mathbf{R})$, one compact and one noncompact generator have switched place. The subgroups $\mathcal{H}_n$ are images of $\mathcal{H}_1$ under an endomorphism ρ_n of Lie algebras $\mathcal{G}$ and $\mathcal{G}^*$ which is induced by the map

$$\begin{aligned} \rho_n(L_n) &= \frac{1}{n}L_{nk} + \frac{c}{24}\left(n - \frac{1}{n}\right)\delta_{k,0}, \\ \rho_n(c) &= nc. \end{aligned} \tag{14}$$

The endomorphisms ρ_n intertwine inequivalent positive energy representations of Vir with different central charge.

The Cartan decomposition of the Lie algebra $\mathcal{G}$ of a finite dimensional noncompact group exhibits $\mathcal{G}$ as a sum of a compact subalgebra $\mathcal{K}$ and a noncompact part $\mathcal{P}$ [46].

$$\begin{aligned} &\mathcal{G} = \mathcal{K} + \mathcal{P} \\ &[\mathcal{K},\mathcal{K}] \subseteq \mathcal{K}, \quad [\mathcal{K},\mathcal{P}] \subseteq \mathcal{P}, \quad [\mathcal{P},\mathcal{P}] \subseteq \mathcal{K} \end{aligned} \tag{15}$$

The generators of $\mathcal{K}$ generate compact 1-parameter groups in the adjoint representation. The same is not necessarily true in the group itself. In particular, the compact generator of $sl(2,\mathbf{R})$ generates a noncompact 1-parameter subgroup of the infinite sheeted universal covering of $SL(2,\mathbf{R})$.

A Cartan decomposition exists also for our infinite dimensional Euclidean algebra $\mathcal{G}$. The meaning of "compact" is somewhat different, though. This is not so unexpected. Infinite dimensional Lie groups are not locally compact, therefore it is not a priori clear what the meaning of a "maximal compact subgroup" K should be.

The $sl(2,\mathbf{R})$-subalgebras $\mathcal{H}_n$ span all of $\mathcal{G}$. So one might expect that the generators of $\mathcal{K}$ should be the compact generators of the $sl(2,\mathbf{R})$-subalgebras. But this is inconsistent with $[\mathcal{K},\mathcal{K}] \subseteq \mathcal{K}$. These generators are all odd under the involution θ. So they cannot form a nonabelian proper subalgebra, since their commutators would be even.

Instead I select for $\mathcal{K}$ (resp. $\mathcal{P}$) those generators of $\mathcal{G}$ that are even (resp. odd) under the involution $L_n \mapsto -L_{-n}$, $c \mapsto -c$, that is $\tilde{L}_n \mapsto -(-)^n\tilde{L}_n$, $\tilde{c} \mapsto -\tilde{c}$. The generators of $\mathcal{K}$ are then

$$\tilde{K}_n = \frac{1}{2n}(\tilde{L}_n - (-)^n\tilde{L}_{-n}) \in \mathcal{K}, \quad n = 1,2,... \tag{16}$$

Half of these generators - those with n odd - are compact generators of some $sl(2,\mathbf{R})$-subalgebra. Among them is the compact generator $\tilde{K}_1$ of the finite dimensional Euclidean conformal algebra.

Let us introduce some more noncompact subalgebras of $\mathcal{G}$:

$\mathcal{A}$	generators $\tilde{L}_0$, $\tilde{c}$	commutative
$\mathcal{N}$	generators $\tilde{L}_n$, $n = 1,2,...$	nilpotent
$\mathcal{X}$	generators $\tilde{L}_{-n}, n = -1,-2,...$	nilpotent,

and the centralizer $\mathcal{M}$ of $\mathcal{A}$ in $\mathcal{K}$. One finds

$$\mathcal{M} = 0. \tag{17}$$

Looking at the list of generators, we see that $\mathcal{G}$ admits an Iwasawa decomposition

$$\mathcal{G} = \mathcal{K} + \mathcal{A} + \mathcal{N}. \tag{18}$$

and a Bruhat decomposition

$$\mathcal{G} = \mathcal{X} + \mathcal{M} + \mathcal{A} + \mathcal{N}. \tag{19}$$

The algebras $\mathcal{N}$ and $\mathcal{X}$ are nilpoptent in the sense described earlier in eq. (2), and $\mathcal{X} = \theta(\mathcal{N})$.

Let us finally note that $\mathcal{G}$ admits both compact and noncompact Cartan subalgebras, such as (disregarding the center $\tilde{c}$)

$$\begin{array}{lll} \mathcal{T}_1 : & \text{generator } \tilde{K}_1 & \text{compact} \\ \mathcal{T}_2 : & \text{generator } \tilde{L}_0 & \text{noncompact} \end{array} \tag{20}$$

These are at the same time compact and noncompact Cartan subalgebras of the finite dimensional Euclidean conformal algebra $\mathcal{H}_1$.

Next we turn to the description of the Euclidean semigroup S and the invariant cone $V \subset \mathcal{G}^*_-$ that is associated with it.

The real Virasoro algebra $\mathcal{G}^*$ is the Lie algebra of a central extension of the group $Diff(S^1)$ of diffeomorphisms of the circle (and of its infinite sheeted universal covering group which consists [32] of diffeomorphisms F of $\mathbf{R}$ with $F(x + 2\pi) = F(x)+2\pi$). $T(f) \in \mathcal{G}^*$ for real f; it generates a 1-parameter group of diffeomorphisms of S^1, with infinitesimal transformation

$$\delta \ln z = if(z)\delta t$$

From this we see that the subgroup $\mathcal{G}^*_+$ which is generated by

$$T(f_+) \in \mathcal{G}^*_+, \qquad f_+(z) = -f_+(-z^{-1})$$

is associated with diffeomorphisms $F: \quad S^1 \mapsto S^1$ of the circle which are symmetric with respect to the imaginary axis in the sense that

$$F(-z^{-1}) = -F(z)^{-1}.$$

These diffeomorphisms have (at least) two fixpoints $\pm i$, since necessarily $f_+(\pm i) = 0$. From this it is clear that

$$V = \{T(f_-) \in \mathcal{G}^*_-, \quad f_-(z) < 0\} \tag{21}$$

is a cone in $\mathcal{G}^*_-$ which is invariant under the adjoint action (conjugation) by elements of G_+. If $T(if_-) \in iV$ then

$$\delta \ln z = i(if_-(z))\delta t > 0$$

and this property is invariant under conjugation by diffeomorphisms in $\mathcal{G}_+$.

The elements $T(f_+ - if_-)$, $(f_-(z) < 0)$, of $\mathcal{G}_+^* - i\mathcal{V}$ generate a semigroup S as in the finite dimensional case, eq.(1c). In particular

$$e^{-T(f)} \in S \quad \text{if } f(z) = f(-z^{-1}) > 0.$$

This real semigroup S is subsemigroup of the complex semigroup S_c which is generated by $T(f)$ with complex f obeying $Imf(z) < 0$. It was pointed out by G. Segal [50] that S_c is a Lie semigroup, and unitary positive energy representations of the central extension $\widetilde{Diff(S^1)}$ of the diffeomorphism group define contractive representations of S_c. It follows from the first assertion that S is also a Lie semigroup. Assuming that the nonabelian Hille Yosida theorem proven in [43] for the finite dimensional case remains valid in our infinite dimensional setting, it furnishes a converse of Segals second assertion: Contractive representations of the real Lie-semigroup S can be continued to unitary positive energy representations of the universal covering group of $\widetilde{Diff(S^1)}$. At the Lie algebra level this result is elementary.

The complex semigroup S_c cannot be embedded into a Lie group, because neither $Diff(S^1)$ nor its universal covering possess a complexification. In the finite dimensional case, S_c cannot be embedded into a Lie group either, because also the universal covering of $SO(d,2)$ possesses no complexification, but the real semigroup S can be embedded into a Lie group $Spin(d+1,1)$. This motivates the search for a Lie group G with Lie algebra $\mathcal{G}$ in the infinite dimensional case.

I do not know whether such a Lie group exists. Heuristic arguments suggest that a nonvanishing central charge $\tilde{c} \neq 0$ presents an obstruction. One could begin by constructing a group which is not necessarily a Lie group , by taking finite products of elements of S and the inverse semigroup S^{-1}. With $\Lambda \in S$ also $\theta(\Lambda^{-1}) \in S$. Therefore S^{-1} is image of S under time reflection.

There are reasonable groups N, A, X associated with the subalgebras $\mathcal{N}, \mathcal{A}, \mathcal{X}$ of $\mathcal{G}$. Consider the 1-parameter subgroups of N.

$$n_l(t) = \exp \frac{it}{l}\tilde{L}_l, \qquad l = 1,2,3,...$$

In terms of n-fold commutators

$$n_l(t)\tilde{L}_k n_l(t)^{-1} = \sum_{n\geq 0} \frac{(it)^n}{n!l^n}[\tilde{L}_l,[\tilde{L}_l,...[\tilde{L}_l,\tilde{L}_k]...]]. \tag{22}$$

Remember now that $\tilde{L}_k = T(f^k)$, with f^k defined in eq.(10). Evaluating the multiple commutators one finds that the following equation holds for $f(z) = f^k(z) = i(-iz)^k$, and therefore also for finite sums f of such f^k's.

$$\begin{aligned} n_l(t)T(f)n_l(t)^{-1} &= T(f_t) \qquad \text{where} \\ f_t(i\zeta)/i\zeta = f(i\zeta_t)/i\zeta_t, &\qquad \zeta_t = \zeta(1+t\zeta^l)^{-1/l} \end{aligned} \tag{23}$$

We see that the 1-parameter groups $t \mapsto n_l(t)$ with $l > 0$, which are generated by elements of $\mathcal{N}$, can be regarded as groups of (algebraic) transformations $f \mapsto f_t$ of germs of holomorphic functions at $z = 0$ which are pure imaginary on the imaginary axis:

$$f(z) = \sum_{n\geq 0} f_n z^n, \qquad \sum_{n\geq 0} |f_n|t^n < \infty \quad \text{for some } t > 0. \tag{24}$$

They may be regarded as the elements of the dual of the Lie algebra $\mathcal{N}$. We may thus regard N as a group of analytic transformations of germs of real analytic functions on the imaginary axis at $z = 0$.

Similarly the 1-parameter subgroup $t \mapsto n_l(t)$ with $l < 0$, which are generated by $\mathcal{X}$, act on germs of holomorphic functions at $z = \infty$ which are pure imaginary on the imaginary axis.

Finally, A is a 1-parameter subgroup of the finite dimensional conformal group, it acts by dilation.

$$\begin{aligned} a(t)T(f)a(t)^{-1} &= T(f_t), \qquad \text{where} \\ f_t(i\zeta)/i\zeta &= f(i\zeta_t)/i\zeta_t, \qquad \zeta_t = \epsilon^t\zeta. \end{aligned} \tag{25}$$

APPENDIX (taken from Lüscher and Mack, ref. 23)

Here it is proven that $\Theta_{\mu\nu}$ is traceless in a scale invariant theory in two dimensions: $\Theta^{\mu}{}_{\mu} = 0$. To this end consider the Schwinger two-point function $S_{\mu\nu\rho\sigma}$ of $\Theta_{\mu\nu}$:

$$S_{\mu\nu\rho\sigma}(x) = \text{ analytic continuation of } < 0|\Theta_{\mu\nu}(x)\Theta_{\rho\sigma}(0)|0 > . \tag{1}$$

This function is real analytic for $x \in \mathbf{R}^2, x \neq 0$, and is covariant under rotations and dilations:

$$S_{\mu\nu\rho\sigma}(\lambda x) = \lambda^{-4} S_{\mu\nu\rho\sigma}(x). \tag{2}$$

Moreover, by locality , symmetry, and conservation of $\Theta_{\mu\nu}$ we have:

$$\begin{aligned} S_{\mu\nu\rho\sigma}(x) &= S_{\nu\mu\rho\sigma}(x) = S_{\mu\nu\sigma\rho}(x). \\ S_{\mu\nu\rho\sigma}(x) &= S_{\rho\sigma\mu\nu}(-x) = S_{\rho\sigma\mu\nu}(x). \end{aligned} \tag{3}$$

Thus, $S_{\mu\nu\rho\sigma}$ has six independent components and can be written as follows:

$$S_{\mu\nu\rho\sigma} = (x^2)^{-4} \sum_{i=1}^{6} A_i T^i_{\mu\nu\rho\sigma}, \qquad A_i \in \mathbf{C}, \tag{4}$$

where ($\epsilon_{\mu\nu} = -\epsilon_{\nu\mu}$, $\epsilon_{01} = +1$, $g_{00} = g_{11} = 1$)

$$\begin{aligned} T^1_{\mu\nu\rho\sigma} &= (x^2)^2 g_{\mu\nu} g_{\rho\sigma} \\ T^2_{\mu\nu\rho\sigma} &= (x^2)^2 (g_{\mu\rho} g_{\nu\sigma} + g_{\mu\sigma} g_{\nu\rho}) \\ T^3_{\mu\nu\rho\sigma} &= x^2 (g_{\mu\nu} x_\rho x_\sigma + g_{\rho\sigma} x_\mu x_\nu) \\ T^4_{\mu\nu\rho\sigma} &= x_\mu x_\nu x_\rho x_\sigma \\ T^5_{\mu\nu\rho\sigma} &= x^2 \{ g_{\mu\nu} (x_\rho \epsilon_{\sigma\delta} x^\delta + x_\sigma \epsilon_{\rho\delta} x^\delta) + g_{\rho\sigma} (x_\mu \epsilon_{\nu\delta} x^\delta + x_\nu \epsilon_{\mu\delta} x^\delta) \} \\ T^6_{\mu\nu\rho\sigma} &= (x_\mu \epsilon_{\nu\delta} x^\delta + x_\nu \epsilon_{\mu\delta} x^\delta) x_\rho x_\sigma + (x_\rho \epsilon_{\sigma\delta} x^\delta + x_\sigma \epsilon_{\rho\delta} x^\delta) x_\mu x_\nu . \end{aligned}$$

T^5 and T^6 are odd under parity and are absent in (4) if parity is conserved. The continuity equation $\partial^\mu S_{\mu\nu\rho\sigma} = 0$ now fixes the numbers A_i up to two arbitrary constants A_+, A_-:

$$\begin{gathered} A_1 = 3A_+, \quad A_2 = -A_+, \quad A_3 = -4A_+, \quad A_4 = 8A_+, \\ A_5 = A_-, \quad A_6 = -2A_- \end{gathered}$$

Upon inserting these values into eq.(4) we find

$$S^{\mu}{}_{\mu\rho\sigma}(x) = S^{\mu}{}_{\mu}{}^{\rho}{}_{\rho}(x) = 0.$$

Hence $< 0|\theta^{\mu}{}_{\mu}(x)\theta^{\rho}{}_{\rho}(0)|0 >$ vanishes, which, by the Reeh-Schlieder theorem [51], implies that $\theta^{\mu}{}_{\mu}(x) = 0$.

REFERENCES

[1] A. Jaffe and J. Glimm, *Quantum physics. A functional integral point of view*, Springer Verlag Heidelberg 1981

K. Osterwalder and S. Schrader, *Axioms for Euclidean Greens functions I, II*, Commun. Math. Phys. **31** ,83 (1973); **42**, 281 (1975)

[2] K. Wilson, Phys. Rev. **D3**, 1818 (1971); Rev. Mod. Phys. **55**, 583 (1983)

J. Kogut and K. Wilson, Phys. Reports **12C**, 75 (1974)

[3] S. Ferrara, A.F. Grillo and R. Gatto, *Conformal algebra in two space time dimensions and the Thirring model* Nuovo Cimento **12**, 959 (1972), and in: *Scale and conformal symmetry in hadron physics*, R. Gatto (ed), Wiley Interscience, New York 1973

[4] C.G. Callan, S. Coleman and R. Jackiw, *A new improved energy momentum tensor*, Ann. Phys. (N.Y.) **59**, 42 (1970)

[5] G. Mack, *Partially conserved dilatation current*, Ph.D. thesis, Bern 1967

[6] G. Mack and Abdus Salam, *Finite component field representations of the conformal group*, Ann. Phys.(N.Y.) **53**, 174 (1969)

[7] G. Mack and K. Symanzik, *Currents, stress tensor and generalized unitarity in conformal invariant quantum field theory*, Commun. Math. Phys. **27**, 247 (1972)

[8] E. Cunningham, *The principle of relativity in electrodynamics and an extension thereof*, Proc. London Math. Society **8**, 77 (1909)

H. Batemann, *The transformation of electrodynamical equations*, Proc. London Math. Society **8**, 223 (1909)

P.A.M. Dirac, *Wave equations in conformal space*, Ann. Math. **37**, 429 (1936)

H. A. Kastrup, *Zur physikalischen Deutung und darstellungstheoretischen Analyse der konformen Transformationen von Raum und Zeit*, Ann. Physik **9**, 388 (1962). A historical survey and further references can be found here. For an extensive bibliography up to 1978 see

I. Todorov et al., ref. 12

[9] G. Mack and M. Lüscher, *Global conformal invariance in quantum field theory*, Commun. Math. Phys. **41**, 203 (1975)

[10] Conformal bootstrap based on skeleton perturbation theory

A.M. Polyakov, *Conformal symmetry of critical fluctuations*, Zh. ETF Pis. Red. **12** 538 (1970), Engl. Transl. JETP Letters **12**, 381 (1970)

A.A. Migdal, *Conformal invariance and bootstrap*, Phys. Letters **37**, 356 (1971)

G. Parisi and L. Peliti, *Calculation of critical indices*, Lett. Nuovo Cimento **2**, 627 (1971)

G. Mack and I. Todorov, *Conformal invariant Green functions without ultraviolet divergences*, Phys. Rev. **D8**, 1764 (1973)

G. Mack and K. Symanzik, ref. 7

G. Mack, *Conformal invariance and short distance behavior in quantum field theory*, in : Lecture Notes in Physics **17**, W. Rühl and A. Vancura (eds.), Springer Verlag Heidelberg 1972

E.S. Fradkin and M. Palchik, *Conformal invariant solutions of quantum field equations I, II*, Nucl. Phys. **B99**, 317 (1975),

[11] I. Segal, *Causally oriented manifolds and groups*, Bull. Amer. Math. Soc. **77**, 958 (1971)

T. Go, H.A. Kastrup and D. Mayer, *Properties of dilatations and conformal transformations in Minkowski space*, Rep. Math. Phys. **6**, 395 (1974)

[12] I. Todorov, M.C. Mintchev and V.B. Petkova, *Conformal invariance in quantum field theory*, Publ. Scuola Normale Superiore, Pisa 1978

[13] G. Mack, *Osterwalder Schrader positivity in conformal invariant quantum field theory*, in: Lecture Notes in Physics, vol. **37**, H. Rollnik and K. Dietz (eds.),Springer Heidelberg 1975

[14] G. Mack, *Group theoretical approach to conformal invariant quantum field theory*, J. de Physique (Paris) **34** C1 (supplement au no. 10) 99 (1973), and in *Renormalization and invariance in quantum field theory*, E.R. Caianello (ed.), Plenum press 1974 , (and proceedings of the Karpacz winter school, Feb. 1973.) See also ref. 13, 52.

[15] A.M. Polyakov, *Non-Hamiltonian approach to quantum field theory at short distances*, Zh ETF **66**, 23 (1974), engl. transl. JETP **39**, 10 (1974)

S. Ferrara, A. Grillo and R. Gatto, *Conformal algebra in space time and operator product expansions*, Springer Tracts in Modern Physics **67** (1973)

[16] Operator product expansions

K. Wilson, *Non-Lagrangean Models of current algebra* , Phys. Rev. **179**, 1499 (1969)

S. Ferrara, A. Grillo and R. Gatto, ref.15

G. Mack, ref. 14

A.M. Polyakov, ref. 15

J. Kupsch, W. Rühl and B.C. Yunn, *Conformal invariance of quantum fields in two dimensional space time*, Ann. Phys. (N.Y.) **89**, 115 (1975)

V. Dobrev, V. Petkova, S. Petrova and I. Todorov, *Dynamical derivation of vacuum operator product expansions in Euclidean conformal quantum field theory*, Phys. Rev. **D13**, 887 (1976)

W. Rühl and B.C. Yunn, *Operator product expansions in conformally covariant quantum field theory*, Commun. Math. Phys. **48**, 215 (1976)

M. Lüscher, *Operator product expansions on the vacuum in conformal quantum field theory in two space time dimensions*, Commun. Math. Phys. **50**, 23 (1976)

G. Mack, refs. 18,19

[17] G. Mack, *Conformal invariant quantum field theory* (Cargese lectures 1976) in: *New developments in quantum field theory and statistical mechanics*, M. Levy and P.K. Mitter (eds), Plenum Press, N.Y. 1977

[18] G. Mack, *Duality in quantum field theory*, Nucl. Phys. **B118**, 445 (1977)

[19] G. Mack, *Convergence of operator product expansions on the vacuum in conformal invariant quantum field theory*, Commun. Math. Phys. **53**, 155 (1977)

[20] G. Mack, *All unitary ray representations of the conformal group SU(2,2) with positive energy*, Commun. Math. Phys. **55**, 1 (1977)

[21] B. Schroer and J.A. Swieca, *Conformal transformations of quantized fields*, Phys. Rev. **D10**, 480 (1974)

M. Hortacsu, R. Seiler and B. Schroer, *Conformal symmetry and reverberations*, Phys. Rev. **D5**, 2518 (1972)

B. Schroer, J.A. Swieca and A.M. Völkel, *Global operator product expansions in conformal invariant relativistic quantum field theory*, Phys. Rev. **D11**, 1509 (1975)

[22] G. Mack, *Conformal invariance and short distance behavior in quantum field theory*, in : Lecture Notes in Physics **17**, W. Rühl and A. Vancura (eds.), Springer Verlag Heidelberg 1972

[23] M. Lüscher and G. Mack, *The energy momentum tensor of a critical quantum field theory in 1 + 1 dimensions* (1976) unpublished

[24] I. M. Gelfand and G.E. Shilov, *Generalized functions* vol. 1, Academic Press, N.Y. 1965

[25] M.A. Virasoro, *Spin and unitarity in dual resonance models*, in: *Duality and symmetry in hadron physics*, E. Gotsman (ed.), Weizmann Science Press, Jerusalem 1971

[26] G.C. Wick, E.P. Wigner and A.S. Wightman, *Intrinsic parity of elementary particles*, Phys. Rev. **88**, 101 (1952)

R.F. Streater and A.S. Wightman, *PCT, spin & statistics and all that*, Benjamin, New York 1963, section I-1

[27] R. Haag and D. Kastler, *An algebraic approach to quantum field theory*, J. Math. Phys. **5**, 848(1964)

S. Doplicher, R. Haag and J.E. Roberts, *Fields, observables and gauge transformations I,II*, Commun. Math. Phys. **13**, 1 (1969); **15**, 173 (1969)

[28] V. Kac, *Contravariant form for infinite dimensional Lie algebras and superalgebras*, Lecture Notes in Physics, Springer Berlin 1979, and ref. 36

I. B. Frenkel and V.G. Kac, *Basic representations of affine Lie algebras and dual resonance models*, Invent. Math. **62** , 23 (1980)

B.L. Feigin and D.B. Fuchs, *Verma modules over the Virasoro algebra*, Lecture Notes in Mathematics **1060**, 230, Springer Berlin 1984

[29] D. Friedan, Z. Quiu, S. Shenker, *Conformal invariance, unitarity and critical exponents in two dimensions*, Phys. Rev. Letters **151B**, 37 (1985)

- - -, *Details of the non-unitarity proof for highest weight representations of the Virasoro algebra*, Commun. Math. Phys. **107**, 535 (1986)

[30] P. Goddard and D. Olive, *Kac Moody algebras, Conformal symmetry and critical exponents*, Nucl. Phys. **B257**, 226 (1985)

- -, *Kac Moody and Virasoro algebras in relation to quantum physics*; Int. J. Mod. Phys. **1**, 303 (1986)

P. Goddard, A. Kent and D. Olive, *Unitary representations of the Virasoro and super-Virasoro algebras*, Commun. Math. Phys. **103**, 105 (1986)

[31] A.A. Belavin , A.M. Polyakov and A.B. Zamolodchikov, *Infinite conformal symmetry in two dimensional quantum field theory*, Nucl. Phys. **B241**, 333 (1984)

[32] A. Pressley and G. Segal, *Loop groups*, Oxford Science Publications, Oxford 1986

[33] H. Sugawara, *A field theory of currents*, Phys. Rev. **170**, 1659 (1968)

[34] V. G. Knizhnik and A.B. Zamolodchikov, *Current algebra and Wess Zumino model in two dimensions*, Nucl. Phys. **B241**, 333, (1984)

I. Todorov, *Current algebra approach to conformal invariant two dimensional models*, Phys. Letters **153 B**, 77 (1985)

[35] C. Itzykson and J.B. Zuber, *Two-dimensional conformal invariant theories on a torus*, Nucl. Phys. **B275** [FS 17], 580 (1986)

A: Cappelli, Phys. Letters **B185**, 82 (1987)

A. Cappelli, C. Itzykson and J.B. Zuber,*Modular invariant partition functions in two dimensions*,Nucl. Phys. **280** [FS 18],445 (1987)

R.R. Paunov and I.T. Todorov, *Modular invariant quantum field theory models of U(1) conformal current algebra*, Phys. Letters **B 196**, 519 (1987)

[36] V. Kac, *Infinite dimensional Lie algebras*, Birkhäuser , Boston 1983, 2nd edition: Cambridge University Press, Cambridge (England) 1986

[37] J. Fröhlich, *Statistics of fields, the Yang-Baxter equation, and the theory of knots and links*, (contribution to these proceedings)

[38] For an alternative, purely euclidean ,derivation see e.g.

C. Itzykson, *Invariance conforme et modeles critiques bidimensionels*, cours au DEA de Physique Theorique de Marseille, CPT-86/P.1915 Marseille 1986

[39] B. Schroer, *New methods and results in conformal* QFT_2 *and the string idea*, (contribution to these proceedings)

K.H. Rehren and B. Schroer, *Exchange algebra on the light cone and order-disorder 2n-point functions in the Ising field theory*, Phys. Letters B (in press)

K.H. Rehren, *Locality of conformal fields in two dimensions: Exchange algebra on the light cone*, (submitted to Commun. Math. Phys.)

[40] H.A. Kastrup, *Gauge properties of Minkowski space*, Phys. Rev. **150**, 1189 (1964)

[41] A. Cappelli and A. Coste ,*On the stress tensor of free conformal field theories in higher dimensions*, Saclay preprint 1987

[42] H. J. Borchers, *Algebraic aspects of Wightman field theory*, in: *Statistical mechanics and field theory*, R.N. Sen and C. Weil (eds)

[43] ref 9, Appendix

[44] M. Lüscher, *Analytic representations of simple Lie groups and continuation to contractive representations of holomorphic Lie semigroups*, DESY-report DESY 75/71 (1975) and Ph.D. Thesis Hamburg 1975

[45] A.W. Knapp and E.M. Stein, *Intertwining operators for semi-simple groups*, Ann. of Math. (2) **93**, 489 (1971)

K. Koller, *The significance of conformal inversion in quantum field theory*, Commun. Math. Phys. **40**, 15 (1975)

[46] G. Warner, *Harmonic analysis on semisimple Lie groups, vol I*, Springer Heidelberg 1972

see also

N.R. Wallach, *Harmonic analysis on homogeneous spaces* , Marcel Dekker, New York 1973

I.M. Gelfand, M.I. Graev and N. Ya. Vilenkin, *Generalized functions, vol. 5*, Academic Press, New York 1966

[47] G. Mack, *Multigrid methods in quantum field theory*, (contribution to these proceedings)

[48] R. Schrader, *Reflection positivity for the complementary series of SL(2n,C)*, Publications of RIMS (Kyoto) **22**, 119 (1986)

[49] G. Warner, ref. 46, proposition 5.5.1.5

[50] G. Segal, *The definition of conformal field theory*, in: *Links between geometry and mathematical physics*, workshop on Schloss Ringberg, march 1987, preprint /87-58 Max Planck Institute für Mathematik, Bonn

[51] H.Reeh and S.Schlieder,Nuovo Cimento **22**, 1051 (1961)

[52] V.K. Dobrev, G. Mack, V.B. Petkova, S.G. Petrova and I.T. Todorov, *Harmonic analysis on the n-dimensional Lorentz group and its application to conformal field theory*, Lecture Notes in physics vol.**63**, Springer Heidelberg 1977

CONTINUOUS WILSON RENORMALIZATION GROUP AND THE 2-D O(n) NON-LINEAR σ- MODEL (*)

P.K. Mitter
Laboratoire de Physique
Théorique et Hautes Energies (**)
Université Pierre et Marie
Curie (Paris VI)

T.R. Ramadas (***)
School of Mathematics
Institute for Advanced
Study
Princeton, N.J.

I-INTRODUCTION

The two dimensional O(n) non-linear σ- model with $n \geqslant 3$ is perturbatively renormalisable and asymptotically free [1, 2, 3]. The theory is expected to be characterized by a single phase with a mass gap (a provocative contrary opinion is expressed in [12]. However these non-perturbative issues have not yet been definitively resolved. As a prior step the non-perturbative ultraviolet stability of the model has to be proved. There is no doubt that the most powerful way of studying these problems is via the Wilson Renormalization group method [4]. The purpose of these notes is to present a new way of studying UV cutoff removal in this model via the continuous Wilson renormalization group [4] and to present some results in this

---

(*) Expanded version of lectures of P.K. Mitter at the 1987 Cargèse Summer School on "Nonperturbative Quantum Field Theory".

(**) Laboratoire Associé au CNRS UA 280. Postal address : LPTHE, Université Pierre et Marie Curie, Tour 16, 4 place Jussieu, 75230 Paris Cedex 05, France.

(***) on leave of absence from the Tata Institute of Fundamental Research, Bombay, India.

context. It is based on [5] where further details can be found.

To begin with recall that the 2-d O(n) σ model is based on the Euclidean action

$$S = \frac{1}{2g_0^2} \int_{\mathbb{R}^2} d^2x \, |\nabla \phi|^2 \tag{1.1}$$

with the constraint

$$|\phi(x)|^2 = 1 \tag{1.2}$$

and the corresponding Euclidean functional measure. Here $\phi(x)$ is a $\mathbb{R}^n$ valued scalar field and $|\,.\,|$ is the euclidean norm in $\mathbb{R}^n$. The spins $\phi(x)$ thus live on the sphere S^{n-1} and throughout we take $n \geq 3$. To define the Euclidean measure we have to introduce an ultraviolet cutoff Λ and a volume cutoff $\mathcal{V}$. For example this could be the lattice cutoff with $\mathbb{R}^2$ replaced by $\Lambda^{-1}\mathbb{Z}^2$ with appropriate b.c. on $\partial\mathcal{V}$. If we now choose $g_0(\Lambda)$ by the 2-loop asymptotic freedom formula, and define rescaled field ϕ_R by $\phi = Z^{1/2}(\Lambda)\,\phi_R$ with $Z(\Lambda)$ appropriately chosen then we expect that the $\Lambda \to \infty$ limit to exist (non-perturbatively) for the (invariant) ϕ_R correlation functions. $g_0(\Lambda)$ and $Z(\Lambda)$ are computable from the work in [2].

If we were now to study the UV cutoff removal via Wilson's methods we are immediately faced with the problem that the simplest and most appealing RG transformations destroy the δ- function constraint. Here we adopt the following strategy : drop the constraint (1.2) and replace (1.1) by :

$$S = \frac{1}{2g_0^2}\left[\int d^2x \, |\nabla\phi|^2 + \tilde{\lambda}_0 \int d^2x \left(|\phi(x)|^2 - 1\right)^2\right] \tag{1.3}$$

Naively, as $\tilde{\lambda}_0 \to \infty$ we recover the δ- function constraint. However we shall not work 'naively'. Note that $\tilde{\lambda}_0$ has mass dimension 2 (since ϕ has mass dimension 0). We let λ_0 depend on the cutoff :

$$\tilde{\lambda}_0 = \Lambda^2 \lambda_0 \tag{1.4}$$

where $\lambda_0 > 0$ is a free cutoff independent parameter. In other words : we let the "geometry" fluctuate with the cutoff. As $\Lambda \to \infty$, we are approaching the δ- function. To renormalize the model we choose $g_0(\Lambda)$ as before and scale $\phi \to Z(\Lambda)^{1/2}\phi$. Then we expect the $\Lambda \to \infty$ limit to exist for the O(n) invariant correlation functions.

Note that since we are in d = 2, and the interaction in (1.3) is polynomial we are not obliged to stick to lattice cutoffs. A convenient 'continuum cutoff' version is provided by :

$$S(\Lambda) = \frac{1}{2}\int d^2x \left|\nabla e^{-\frac{\Delta}{2\Lambda^2}}\phi\right|^2 + \left(Z_0(\Lambda)-1\right)\frac{1}{2}\int d^2x\, |\nabla\phi|^2 + \Lambda^2 \frac{\lambda_0}{2g_0^2(\Lambda)}\int d^2x \left(Z_0(\Lambda)\, g_0^2(\Lambda)\,|\phi(x)|^2 - 1\right)^2 \qquad (1.5)$$

where, for convenience, we have rescaled the fields $\phi \to g_0\,\phi$. To define the model we must also put in a volume cutoff $\mathcal{V}$, with b.c. on $\partial\mathcal{V}$.

The success of the above programme was first checked in 1-loop perturbation theory (after adding a magnetic field and working in infinite volume), [6]. The 1-loop renormalization constants to take $\Lambda \to \infty$ limit coincide with those of [2], (except that Z_0 is appropriate to rescaled fields $\phi \to g_0\,\phi$) and $\lambda_0 > 0$ is a free parameter, which disappears in the renormalized theory. This leads us to expect that under the Wilson RG flow λ_0 gets driven to a fixed point λ_* (as in Wilson's theory of critical phenomena [4]). This picture was then verified (without symmetry breaking) in the local approximation to the continuous Wilson RG flow [6]. It was found that λ, as well as other interactions arising under RG iterations, get driven to a fixed point, g is "marginal" and there are no relevant directions [6]. We should mention here that the "local approximation" is similar to the RG iteration of a certain hierarchical 2d - O(n) non-linear σ model whose continuum limit was studied in [7], with a very similar picture.

In these notes we study the full Wilson continuous RG flow with (1.5) as the starting action. Wilson [4] derived a differential equation governing the flow of the effective potential. This equation has been exploited very effectively in [9] in the study of critical phenomena in ε- expansions, in [10] for an elegant proof of perturbative renormalizability of $(\phi^4)_4$ and recently in [11] to derive (or rederive) improved estimates for Mayer expansions. We use this differential equation to prove the existence of the renormalized trajectory (scale dependent renormalized effective potential) in the $\Lambda \to \infty$ limit as a series in the (scale dependent) effective charge and in the fields. In doing this we obtain the following picture :

1) Z, g are the only marginal variables and their flows coincide exactly with those obtained in [2] 2) all other interactions contract to a fixed point 3) there are no relevant directions. This confirms the picture obtained in the local approximation except that we now obtain the correct marginal flows (the wave function renormalization is missed out in the local approximation).

Let us now point out that the above expansion (upto a high enough order in the fields) is just what one needs in the so called "small field region" of rigorous RG theory (see e.g. [8]). For a non-perturbative control of the $\Lambda \to \infty$ limit one needs suitable stability estimates in the "large field region". <u>The large field problem in this model is under investigation</u>. The continuous RG can be viewed as an "adjoint" continuous Markov process on the space of probability measures on $\mathcal{S}'(\mathbb{R}^d)$; the differential equation that we have exploited being the Fokker-Planck equation. At present the differential equation is most effective for organising the small field expansion whereas the adjoint semigroup should be exploited for the a priori stability estimates for large fields.

We have been asked, in this school and elsewhere : in what sense is one picking up a renormalized continuum "non-linear σ- model" ? It is worth emphasizing that as far as the renormalized continuum theory is concerned it is the "universality class" that matters (and not the particular cutoff theory).

The universality class is characterized by symmetries, short and long distance behaviour. We have O(n) symmetry and the short distance behaviour is the same as computed in traditional perturbative approaches in a local chart (the renormalization constants are the same). The conjectured long distance behaviour (mass gap) is as yet unproven. More detailed questions of a geometrical nature can be meaningfully asked. This is because in the family of measures constituting the renormalized trajectory each measure has a built in cutoff (scale) and is thus supported on spaces of highly differentiable functions. Multiplying functions pointwise now gives measurable functions, and non-linear field configurations have non-zero measure. We can ask for the probability of fluctuations around submanifolds of $\mathcal{S}'(\mathbb{R}^2)$, (e.g. the space of smooth maps into S^{n-1}). These questions are not pursued here.

In Sec. II we introduce the Wilson method and explain our basic results. Sec. III is devoted to the local approximation. This supplies both insight as well as a 'boundary condition' exploited later. In Sec. IV we prove the existence of the underlying fixed point in the sense explained earlier and in Sec. V analyze the marginal flow and its stabilization. More details will be found in [5].

II- THE WILSON RG APPROACH TO UV CUTOFF REMOVAL

In this section we define our model, in accord with the introduction, explain the continuous Wilson RG flow and the governing differential equation as well as the strategy for UV cut off removal. We also state our basic result.

Since we shall be working with an O(n) invariant model with $n \geqslant 3$ in $d = 2$ in the "symmetric phase" we must introduce initially a volume cutoff : $\mathcal{V} = (-\frac{\ell}{2}, \frac{\ell}{2})^2$ and an UV cutoff Λ. Let $\Delta_{\mathcal{V}}$ be the Dirichlet Laplacian (dirichlet data on $\partial\mathcal{V}$). Then our cutoff action is : (ϕ: $\mathbb{R}^n$ valued scalar field in $\mathcal{V}$, (.,.) scalar product in $\mathbb{R}^n$, $n \geqslant 3$)

$$S_{\Lambda,\mathcal{V}} = \frac{1}{2}\int_{\mathcal{V}} d^2x \left(\phi(x), (-\Delta_{\mathcal{V}})\, e^{-\frac{\Delta_{\mathcal{V}}}{\Lambda^2}}\phi(x)\right) + \tilde{\mathcal{V}}_{\mathcal{V}}^{(0)} \qquad (2.1)$$

where

$$\tilde{\vartheta}^{(0)}_{\mathcal{V}} = \frac{1}{2}(Z_0-1)\int_{\mathcal{V}} d^2x\,(\phi,(-\Delta_{\mathcal{V}})\phi) + \Lambda^2\frac{\lambda_0}{2g_0^2}\int_{\mathcal{V}} d^2x\,\left(Z_0 g_0^2\,|\phi(x)|^2-1\right)^2 \tag{2.2}$$

Here λ_0 is an arbitrary cutoff independent positive constant. It is dimensionless. When we take the $\Lambda \uparrow \infty$ limit, $Z_0 g_0$ will have to be chosen functions of Λ <u>but λ_0 will be arbitrary</u>: <u>it is driven to a fixed point under RG transformations.</u>

Define :

$$C_{\Lambda,\mathcal{V}} = (-\Delta_{\mathcal{V}})^{-1}\, e^{\Delta_{\mathcal{V}}/\Lambda^2} \tag{2.3}$$

It is the UV cutoff Dirchlet covariance. Its integral kernel (Dirichlet propagator) is given by :

$$C_{\Lambda,\mathcal{V}}(x,y) = \sum_{n_1,n_2\in\mathbb{Z}} \frac{\ell^2 e^{-\frac{(n_1^2+n_2^2)}{\Lambda^2\ell^2}}}{n_1^2+n_2^2}\, u_{\underline{n},\ell}(x)\, u_{\underline{n},\ell}(y) \tag{2.4}$$

where the $u_{n,\ell}$ are normalized eigenfunctions of $\Delta_{\mathcal{V}}$.

Let $\mu_{C_{\Lambda,\mathcal{V}}}$ be the Gaussian measure of mean 0 and covariance $C_{\Lambda,\mathcal{V}}$ on $\mathcal{S}'(\mathbb{R}^2)$. Each $\mathbb{R}^n$ valued field $\phi(x)$ can be written as a vector ($\phi_1(x)$, ..., $\phi_n(x)$) with respect to an orthonormal basis (e_j) of $\mathbb{R}^n$. The ϕ_j are independent Gaussian random variables distributed according to $\mu_{C_{\Lambda,\mathcal{V}}}$. The cutoff measure $\mu_{\Lambda,\mathcal{V}}$ of our model is defined by

$$d\mu_{\Lambda,\mathcal{V}}(\phi) = \prod_{i=1}^{n} d\mu_{C_{\Lambda,\mathcal{V}}}(\phi_i)\, e^{-\tilde{\vartheta}^{(0)}_{\mathcal{V}}(\phi)} \tag{2.5}$$

upto an overall normalization constant.

The generating functional of the Schwinger (correlation) functions is :

$$Z_{\Lambda,\mathcal{V}}(j) = \int d\mu_{\Lambda,\mathcal{V}}(\phi)\, e^{\phi(j)} \tag{2.6}$$

where $j \in \mathcal{S}(\mathbb{R}^2)\otimes\mathbb{R}^n$ is the "source".

We recall that in the Wilson approach the UV cutoff removal involves three basic steps :

A) First, using the scaling property :

$$u_{\underline{n},\ell}(x) = \Lambda\, u_{\underline{n},\Lambda\ell}(\Lambda x) \tag{2.7}$$

in d = 2 of the normalized Dirichlet eigenfunctions, one derives :

$$C_{\Lambda,V}(x,y) = C_{1,\Lambda V}(\Lambda x,\Lambda y) \tag{2.8}$$

where $\Lambda V = (-\Lambda\frac{\ell}{2}, \Lambda\frac{\ell}{2})^2$. We have the corresponding relation for Gaussian random fields :

$$\phi(x) = \Phi(\Lambda x) \tag{2.9}$$

where ϕ is distributed according to $\mu_{C_{\Lambda,V}}$ and Φ according to $\mu_{C_{1,\Lambda V}}$

Using (2.9) and (2.2), (2.5) becomes :

$$\left.\begin{aligned} d\mu_{\Lambda,V} &= d\mu_{C_{1,\Lambda V}}(\Phi)\, e^{-\mathcal{V}^{(0)}_{\Lambda V}(\Phi)} \\ \text{where} \\ \mathcal{V}^{(0)}_{\Lambda V} &= \frac{1}{2}(Z_0-1)\int_{\Lambda V} d^2x\, (\Phi, -\Delta_{\Lambda V}\Phi) \\ &+ \frac{\lambda_0}{2g_0^2}\int_{\Lambda V} d^2x\, (Z_0 g_0^2 |\Phi(x)|^2 - 1)^2 \end{aligned}\right\} \tag{2.10}$$

and for the generating functional (2.6) :

$$\left.\begin{aligned} Z_{\Lambda,V}(j) &= \int d\mu_{C_{1,\Lambda V}}(\Phi)\, e^{-\mathcal{V}^{(0)}_{\Lambda V}(\Phi) + (\Phi, j_\Lambda)_{L^2(\Lambda V)}} \\ \text{where} \\ j_\Lambda(x) &= \Lambda^{-2} j(\Lambda^{-1}x) \end{aligned}\right\} \tag{2.11}$$

The reference measure in (2.10 - 2.11) has a "unit cutoff", but an expanded volume.

B) The <u>next</u> step is to introduce the RG transformation which lowers the cutoff $1 \to e^{-t}1$ and scales back to 1, (t > 0). Define :

$$C^{(t)}_{\Lambda V} \doteq C_{1,\Lambda V} - C_{e^{-t}1,\Lambda V} > 0 \tag{2.12}$$

so that

$$\begin{aligned} C_{1,\Lambda V} &= C_{e^{-t},\Lambda V} + C^{(t)}_{\Lambda V} \\ \Phi_i &= \phi^{(1)}_i + \zeta_i \end{aligned} \tag{2.13}$$

$\phi_i^{(1)}$, ζ_i are independent random fields distributed according to $\mu_{C_{e^{-t},\Lambda V}}$ and $\mu_{C^{(t)}_{\Lambda V}}$ respectively. Then from (2.10), (2.13) and using (2.8 - 2.9) again we get :

$$d\mu_{\Lambda,V} = \prod_{i=1}^{n} \left(d\mu_{C_{1,e^{-t}\Lambda V}}(\Phi) \otimes d\mu_{C^{(t)}_{\Lambda V}}(\zeta_i) \right) \exp\left[- \mathcal{V}^{(0)}_{\Lambda V}\left(\Phi(e^{-t}\cdot) + \zeta \right) \right] \tag{2.14}$$

The continuous RG transformation is defined by :

$$\exp\left[- \mathcal{V}^{(t)}_{e^{-t}\Lambda V}(\Phi(\cdot)) \right] = \int \prod_{i=1}^{n} d\mu_{C^{(t)}_{\Lambda V}}(\zeta_i) \exp\left[- \mathcal{V}^{(0)}_{\Lambda V}\left(\Phi(e^{-t}\cdot) + \zeta \right) \right] \tag{2.15}$$

The generating functional Z(j) of (2.11) can be expressed using the RG flow $\mathcal{V}^{(t)}$. An easy calculation gives for <u>large t</u> (and this will be the case in the future) :

$$\begin{aligned} Z_{\Lambda,V}(j) = \int \prod_{i=1}^{n} d\mu_{C_{1,e^{-t}\Lambda V}}(\Phi_i) \exp\Big[&- \mathcal{V}^{(t)}_{e^{-t}\Lambda V}(\Phi) \\ &+ e^{2t}\left(\Phi, e^{-\Delta} j_\Lambda(e^{-t}\cdot)\right)_{L^2(e^{-t}\Lambda V)} - \frac{1}{2} e^{4t}\left(j_\Lambda(e^{t}\cdot), (e^{-\Delta}-1)(-\Delta)^{-1} j_\Lambda(e^{t}\cdot)\right)_{L^2(e^{-t}\Lambda V)} \Big] \end{aligned} \tag{2.16}$$

Similar formulae were obtained earlier by Wilson [4], except that we are working in finite volume

C) We now consider a sequence of UV cutoffs $\Lambda = \Lambda_N = L^N$, $L \geqslant 2$. (Strictly speaking $\Lambda_N = L^N\mu$, where μ is a fixed scale, but in formulae only Λ_N/μ enters, so we have set $\mu = 1$) $\Lambda_N \to \infty$ as $N \to \infty$. We also choose M fixed with $M \ll N$. Then L^M is our "running scale".

In (2.15) we integrate down to :

$$t = t_{N-M} \equiv (N-M)\log L \tag{2.17}$$

$$e^{-t_{N-M}}\Lambda_N = \Lambda_M = L^M$$

In this case we obtain from (2.16) :

$$Z_{\Lambda_N,V}(j) = \int \prod_{i=1}^{n} d\mu_{C_{1;L^M V}}(\Phi_i)\exp\Big[-\mathcal{V}^{(t_{N-M})}_{L^M V}(\Phi) + L^{-2M}\big(\Phi, e^{-\Delta} j(L^{-M}\cdot)\big)_{L^2(L^M V)} - \frac{1}{2}L^{-4M}\big(j(L^{-M}\cdot),(-\Delta)^{-1}(e^{-\Delta}-1)j(L^{-M}\cdot)\big)_{L^2(L^M V)}\Big] \tag{2.18}$$

Our aim is to show that the UV cutoff can be lifted, i.e.

$$\lim_{N\to\infty} \mathcal{V}^{(t_{N-M})}_{L^M V} = \mathcal{V}^{(*)}_{L^M V} \tag{2.19}$$

exists (M- fixed) (in a sense to be explained below) for a suitable choice of initial conditions $Z_0(N)$, $g_0(N)$ and for any $\lambda_0 > 0$ (see 2.10). In this case the renormalized generating functional is given by :

$$Z^{(*)}_{V}(j) = \int \prod_{i=1}^{n} d\mu_{C_{1;L^M V}}(\Phi_i)\exp\Big[-\mathcal{V}^{(*)}_{L^M V}(\Phi) + L^{-2M}\big(\Phi, e^{-\Delta} j(L^{-M}\cdot)\big)_{L^2(L^M V)} - \frac{1}{2}L^{-4M}\big(j(L^{-M}\cdot),(-\Delta)^{-1}(e^{-\Delta}-1)j(L^{-M}\cdot)\big)_{L^2(L^M V)}\Big] \tag{2.20}$$

and L^M is the running scale.

In this program we must study the large t asymptotics of $\mathcal{V}^{(t)}$ given by the RG transformation (2.15). It is easy to show, starting from (2.15), that $\mathcal{V}^{(t)}$ satisfies the following non-linear differential equation :

$$\frac{\partial}{\partial t} \mathcal{V}^{(t)}_{e^{-t}\Lambda\mathcal{V}} = -\int_{e^{-t}\Lambda\mathcal{V}} d^2x \sum_{i=1}^{n} (x \cdot \nabla \Phi_i(x)) \frac{\delta \mathcal{V}^{(t)}_{e^{-t}\Lambda\mathcal{V}}}{\delta \Phi_i(x)} \tag{2.21}$$

$$- \int_{(e^{-t}\Lambda\mathcal{V})\times(e^{-t}\Lambda\mathcal{V})} d^2x\, d^2y \; K_{e^{-t}\Lambda\mathcal{V}}(x,y) \sum_{i=1}^{n} \left[- \frac{\delta^2 \mathcal{V}^{(t)}_{e^{-t}\Lambda\mathcal{V}}}{\delta \Phi_i(x) \delta \Phi_i(y)} + \frac{\delta \mathcal{V}^{(t)}_{e^{-t}\Lambda\mathcal{V}}}{\delta \Phi_i(x)} \frac{\delta \mathcal{V}^{(t)}_{e^{-t}\Lambda\mathcal{V}}}{\delta \Phi_i(y)} \right]$$

where

$$K_{\mathcal{V}}(x,y) = \sum_{n_1,n_2} e^{-\sum n_i^2/\ell^2} u_{\underline{n},\ell}(x)\, u_{\underline{n},\ell}(y) \tag{2.22}$$

$$\mathcal{V} = \left(\frac{\ell}{2}, \frac{\ell}{2}\right)^2, \quad \Lambda\mathcal{V} = \left(-\frac{\Lambda\ell}{2}, \frac{\Lambda\ell}{2}\right)^2$$

Such an equation (in ∞ volume) was first derived by Wilson [4]. For $\Lambda = L^N$, $t = (N-M)\log L$, $e^{-t}\Lambda\mathcal{V} = L^M \mathcal{V}$.

When $\mathcal{V} \uparrow \mathbb{R}^2$,

$$K_{\mathcal{V}}(x,y) \longrightarrow \int \frac{d^2q}{(2\pi)^2} e^{iq(x-y)} e^{-q^2} \equiv K(x-y) \tag{2.23}$$

which has exponential decay.

Its Fourier transform

$$\tilde{K}(q) = e^{-q^2} \tag{2.24}$$

is C^∞ at $q = 0$. <u>Hence (2.21), the RG flow equation, has no infra-red divergences in the infinite volume limit</u>. This is simply a reflection of the fact that the hard field ζ appearing in (2.15) has as propagator (2.12), and in the infinite vol. limit its Fourier transform is

$$\tilde{C}^{(t)}(q) = \frac{1}{q^2}\left(e^{-q^2} - e^{-e^{2t}q^2}\right)$$

which is regular at $q = 0$. <u>In the following we take advantage of this fact to study the RG flow directly in infinite volume</u>.

When $\mathcal{V} \uparrow \mathbb{R}^2$, the RG flow equation reads :

$$\frac{\partial \mathcal{V}^{(t)}}{\partial t} = -\int_{\mathbb{R}^2} d^2x \sum_{i=1}^{n} \left(x \cdot \nabla \Phi_i(x)\right) \frac{\delta \mathcal{V}^{(t)}}{\delta \Phi_i(x)}$$
$$- \int_{\mathbb{R}^2 \times \mathbb{R}^2} d^2x\, d^2y\, K(x-y) \sum_{i=1}^{n} \left[- \frac{\delta^2 \mathcal{V}^{(t)}}{\delta \Phi_i(x) \delta \Phi_i(y)} + \frac{\delta \mathcal{V}^{(t)}}{\delta \Phi_i(x)} \frac{\delta \mathcal{V}^{(t)}}{\delta \Phi_i(y)} \right] \tag{2.25}$$

with K(x) given by (2.23 - 2.24), with initial condition (see 2.10)

$$\mathcal{V}^{(0)} = \frac{1}{2}\left(Z_0(N) - 1\right) \int_{\mathbb{R}^2} d^2x \left(\Phi, -\Delta \Phi\right)$$
$$+ \frac{\lambda_0}{2 g_0^2(N)} \int_{\mathbb{R}^2} d^2x \left(Z_0(N) g_0^2(N) |\Phi(x)|^2 - 1\right)^2 \tag{2.26}$$

It is convenient to define :

$$\psi = Z_0^{1/2} g_0 \Phi \tag{2.27}$$

and rewrite $\mathcal{V}^{(0)}$ as :

$$\mathcal{V}^{(0)} = \frac{1}{g_0^2}\left[\frac{Z_0 - 1}{Z_0} \cdot \frac{1}{2} \int d^2x\, |\nabla \psi|^2 + U_0(\psi)\right]$$

where (2.28)

$$U_0 = \frac{\lambda_0}{Z} \int d^2x \, \left(|\psi(x)|^2 - 1 \right)^2$$

Note that $U^{(0)}$ has a (degenerate) minimum on constant field configurations

$$\psi(x) = v \, , \quad |v|^2 = 1 \tag{2.29}$$

We now parametrize the evolved potential $\mathcal{V}^{(t)}$ as :

$$\left.\begin{aligned} \mathcal{V}^{(t)} &= \frac{1}{g_t^2}\left[\frac{(Z_t - 1)}{Z_t} \cdot \frac{1}{2} \int d^2x \, |\nabla \psi_t|^2 + U_t \left(\psi^{(t)}; g_t^2, Z_t \right) \right] \\ \psi^{(t)} &= Z_t^{1/2} g_t \, \Phi \end{aligned}\right\} \tag{2.30}$$

and demand :

(i) $\quad \dfrac{\delta U_t}{\delta \psi} = 0 \qquad$ at $\quad \psi = v \, , \quad |v|^2 = 1$

(ii) Define (2.31)

$$u^{[2]}_{ij}(x, y \, ; t) = \left. \frac{\delta^2 U_t}{\delta \psi_i(x) \, \delta \psi_j(y)} \right|_{\psi = v}$$

and

$$u^{[2]}_{ij}(x, y \, ; t) = \int \frac{d^2 p}{(2\pi)^2} \, \tilde{v}^{[2]}_{ij}(p \, ; t) \, e^{ip\cdot(x-y)}$$

impose

$$\left. \frac{d}{dp^2} \right|_{p^2 = 0} \tilde{v}^{[2]}_{ij}(p \, ; t) = c_t \, v_i v_j$$

i.e. no δ_{ij} term.

Conditions (2.31) (i) and (ii) suffice to define the effective coupling g_t and wave function Z_t at scale t.

Define the Taylor coefficients at the minimum $\psi = v$:

$$u^{[m]}_{d_1 \cdots d_m}(x_1 \cdots x_m; t, g_t^2) = \left. \frac{\delta^m U_t}{\delta\psi_{d_1}(x_1) \cdots \delta\psi_{d_m}(x_m)} \right|_{\psi = v} \tag{2.32}$$

m ⩾ 2.

and expand in power series in g_t^2:

$$u^{[m]}_{d_1 \cdots d_m}(x_1 \cdots x_m, t, g_t^2) = \sum_{k \geqslant 0} \frac{(g_t^2)^k}{k!} u^{[m],k}_{d_1 \cdots d_m}(x_1 \cdots x_m; t) \tag{2.33}$$

Then our basic result is the following:

<u>Theorem</u>: Let

$$\beta_2 = -\frac{(n-2)}{2\pi}, \quad \gamma_2 = \frac{1}{2\pi} \tag{2.34}$$

and β_3, γ_3 appropriately chosen (see later). Choose the initial conditions:

$$\begin{aligned} \frac{1}{g_0^2(N)} &= \frac{1}{\bar{g}^2} - \beta_2 \cdot N \log L + \frac{\beta_3}{\beta_2} \log(N \log L) \\ Z_0(N) &= \left(\frac{g_0^2(N)}{\bar{g}^2} \right)^{-\frac{\gamma_2}{\beta_2}} \end{aligned} \tag{2.35}$$

Let $t_{N-M} = (N-M)\log L$, $M \ll N$ and fixed.

Then, the following limits exist (in the sense of formal power series):

$$\begin{aligned} \lim_{N \to \infty} g^2_{t_{N-M}} &= \tilde{g}^2_M = g_0^2(M) + O((g_0^2(M))^2) \\ \lim_{N \to \infty} Z_{t_{N-M}} &= \tilde{Z}_M = 1 + \hat{\gamma}_2(M)\tilde{g}_M^2 + O((\tilde{g}_M^2)^2) \\ \hat{\gamma}_2(M) &= \gamma_2(M \log L) \end{aligned} \tag{2.36}$$

Moreover, for every m ⩾ 2, k ⩾ 0

$$\lim_{t \to \infty} \tilde{v}^{[m],k}_{d_1 \cdots d_m}(p_2 \cdots p_m; t) = \tilde{v}^{[m],k}_{d_1 \cdots d_m, *}(p_2, \cdots, p_m) \tag{2.37}$$

exists uniformly over compacts. Here the v are the Fourier transforms of (2.32), see (2.40).

<u>Hence, in the sense of convergence of Taylor series coefficients</u> :

$$\lim_{N\to\infty} \mathcal{V}_{t_{N-M}} = \mathcal{V}_M^{(*)} \tag{2.38}$$

where,

$$\mathcal{V}_M^*(\Phi) = \frac{1}{2}(\tilde{Z}_M - 1)\int d^2x\, |\nabla\Phi|^2 + \frac{1}{\tilde{g}_M^2}\, U_*\left(\tilde{Z}_M^{1/2}\tilde{g}_M\Phi ; \tilde{g}_M^2\right)$$

with

$$U_*(\psi;\tilde{g}_M^2) = \sum_{m\geq 2} \frac{1}{m!}\int d^2x_1 \ldots dx_m\; u^{[m]}_{d_1\cdots d_m *}(x_1\cdots x_m;\tilde{g}_M^2)\,\cdot$$

$$\cdot\left(\psi_{d_1}(x_1) - v_{d_1}\right)\cdots\cdots\left(\psi_{d_m}(x_m) - v_{d_m}\right)$$

and

$$u^{[m]}_{d_1\cdots d_m *}(x_1\cdots x_m;\tilde{g}_M^2) = \sum_{k\geq 0}\frac{(\tilde{g}_M^2)^k}{k!}\, u^{[m],k}_{d_1\cdots d_m *}(x_1\cdots x_m)$$

It follows that Z_t, $g^2{}_t$ are the <u>only</u> "<u>marginal</u>" directions which get stabilized for large t with the "asymptotic free" initial conditions (2.35) and <u>all other directions are "irrelevant"</u>, corresponding to the contraction (2.37). In particular there are <u>no</u> relevant directions.

Note that the Taylor expansion of U_* can be written in a manifestly invariant form by expanding the coefficient tensors in invariants (see later) and averaging the vector v over O(n).

The proof of the above theorem, together with all details, can be found in [5]. In the remainder of these notes we explain how one arrives at the above result.

We first write the RG flow equation (2.35) directly in terms of U_t (defined by 2.30). We get :

$$\frac{\partial U_t}{\partial t} = -\left(\frac{1}{Z_t}\frac{dZ_t}{dt}\right)\frac{1}{2}\int d^2x\, |\nabla\psi|^2 - g_t^2 \frac{d}{dt}\left(\frac{1}{g_t^2}\right)\cdot U_t \tag{2.39}$$

$$+\frac{1}{2}\left(g_t^2\frac{d}{dt}\left(\frac{1}{g_t^2}\right) - \frac{1}{Z_t}\frac{dZ_t}{dt}\right)\int d^2x \sum_{i=1}^{n}\psi_i(x)\frac{\delta U_t}{\delta\psi_i(x)}$$

$$-\sum_{i=1}^{n}\int d^2x\,(x\cdot\nabla\psi_i(x))\frac{\delta U_t}{\delta\psi_i(x)}$$

$$-Z_t\int d^2x\, d^2y\, K(x-y)\sum_{i=1}^{n}\left\{-g_t^2\left[\frac{(Z_t-1)}{Z_t}(-\Delta_x)\delta(x-y) + \frac{\delta^2 U_t}{\delta\psi_i(x)\delta\psi_i(y)}\right]\right.$$

$$\left.+\left[\frac{(Z_t-1)}{Z_t}(-\Delta_x)\psi_i(x) + \frac{\delta U_t}{\delta\psi_i(x)}\right]\cdot\left[\frac{(Z_t-1)}{Z_t}(-\Delta_y)\psi_i(y) + \frac{\delta U_t}{\delta\psi_i(y)}\right]\right\}$$

By taking successively the first and second (functional) derivatives of (2.39) at $\psi = v$, and using 2.31 (i), (ii) we obtain the flow equations for g_t^2 and Z_t.

Using translation invariance (in infinite vol.) we write the Fourier transforms of (2.32) as:

$$(2\pi)^2\,\delta^{(2)}(p_1+\cdots+p_m)\,\tilde{v}^{[m]}_{j_1\ldots j_m}(p_2\cdots p_m\,;\,t, g_t^2) \tag{2.40}$$

<u>The flow equations for g_t^2 and Z_t then read</u>:

$$\left[\frac{d}{dt}\left(\frac{1}{g_t^2}\right)\right]\left(\sum_{i=1}^{n}\tilde{v}^{[2]}_{ij}(0\,;\,t, g_t^2)\,v_i\right) \tag{2.41}$$

$$= \frac{1}{g_t^2}\cdot\frac{1}{Z_t}\left(\frac{dZ_t}{dt}\right)\left(\sum_{i=1}^{n}\tilde{v}^{[2]}_{ij}(0\,;\,t, g_t^2)\,v_i\right)$$

$$-2Z_t\int\frac{d^2p}{(2\pi)^2}\tilde{K}(p)\sum_{i=1}^{n}\tilde{v}^{[3]}_{jii}(p,-p\,;\,t, g_t^2)$$

and

$$\left(1+\frac{1}{2}g_t^2\sum_i v_i\left(\frac{d}{dp^2}\right)\Big|_{p^2=0}\tilde{v}^{[3]}_{ijk}(p,-p\,;\,t)\right)\Big|_{\delta_{jk}\,\mathrm{cmts.}}\frac{1}{Z_t}\frac{dZ_t}{dt} \tag{2.42}$$

$$= \frac{d}{dt}\left(\frac{1}{g_t^2}\right)(g_t^2)^2\left(\sum_i v_i \frac{d}{dp^2}\Big|_{p^2=0} \tilde{v}^{[3]}_{ijK}(p,-p;t)\Big|_{\delta_{jK}\ \text{contrib.}}\right)$$

$$- Z_t\left[-\frac{g_t^2}{4}\sum_i \int \frac{d^2q}{(2\pi)^2}\,\tilde{K}(q)\,\Delta_p\Big|_{p=0} \tilde{v}^{[4]}_{iijK}(q,p,-p;t)\Big|_{\delta_{jK}\ \text{contr.}}\right.$$

$$\left. + 4\left(\frac{Z_t-1}{Z_t}\right)\tilde{v}^{[2]}_{jK}(0;t)\Big|_{\delta_{jK}\ \text{contr.}} - 2\sum_i \tilde{v}^{[2]}_{ij}(0;t)\,\tilde{v}^{[2]}_{iK}(0;t)\Big|_{\delta_{jK}\ \text{contr.}}\right]$$

In the above, $\Delta_p = \sum_{\mu=1}^{2} \frac{\partial^2}{\partial p_\mu^2}$ is the 2-dim Laplacian, and $\tilde{K}(q)$ is given by (2.24).

III- THE LOCAL APPROXIMATION.

The RG flow equation (2.39) leads to (non-linear) coupled differential equations for the Taylor coefficients (2.32) and their Fourier transforms (2.40). The equations for the latter quantities simplify considerably in the limit of vanishing moments. This is the <u>local approximation</u>. The local approximation can be set up directly from (2.39) by replacing in the <u>non local term</u> involving $\delta U_t/\delta \psi_i(x)$

$$K(x-y) \rightarrow \delta(x-y) \tag{3.1}$$

and $Z_t = 1$, since in this case there is no wave function renormalization. If U_0 is local without derivatives, then so is U_t. The reason for first considering the local approximation is twofold :

(i) it gives a correct qualitative picture, except that we miss the wave function renormalization (which is recovered when we go beyond this approximation as we shall later).

(ii) it gives an important technical input ("boundary condition") which will be exploited later. The local approximation is only the first step of a systematic procedure.

In the local approximation (2.39) becomes :

$$\frac{\partial U_t^{(loc.)}}{\partial t} = - g_t^2 \frac{d}{dt}\left(\frac{1}{g_t^2}\right) \cdot U_t^{(loc)} + \frac{1}{2} g_t^2 \frac{d}{dt}\left(\frac{1}{g_t^2}\right) \int d^2x \sum_i \psi_i(x) \frac{\delta U_t^{(loc)}}{\delta \psi_i(x)}$$

$$- \sum_i \int d^2x \, (x \cdot \nabla \psi_i(x)) \frac{\delta U_t^{(loc)}}{\delta \psi_i(x)} + g_t^2 \sum_i \int d^2x \, d^2y \, K(x-y) \frac{\delta^2 U_t^{(loc)}}{\delta \psi_i(x) \delta \psi_i(y)}$$

$$- \sum_i \int d^2x \, \frac{\delta U_t^{(loc)}}{\delta \psi_i(x)} \frac{\delta U_t^{(loc)}}{\delta \psi_i(y)} \qquad (3.2)$$

Write :

$$U_t^{(loc)} = \int d^2x \, u_t(|\psi(x)|^2) \qquad (3.3)$$

and call $|\psi(x)|^2 = \tau$. Use the fact :

$$\sum_i \int d^2x \, (x \cdot \nabla \psi_i(x)) \frac{\delta U_t^{(loc)}}{\delta \psi_i(x)} = -2 \, U_t^{(loc)} \qquad (3.4)$$

Then from (3.2) we get (primes denote τ partial derivatives)

$$\frac{\partial u_t}{\partial t} = 2u_t - 4\tau (u_t')^2 + \frac{g_t^2}{2\pi} (n u_t' + 2\tau u_t'') \qquad (3.5)$$

$$+ g_t^2 \frac{d}{dt}\left(\frac{1}{g_t^2}\right)(\tau u_t' - u_t)$$

$u_t(\tau)$ has minimum at $\tau = 1$. Using this we derive the local approximation to the g_t^2 flow :

$$\frac{d}{dt}\left(\frac{1}{g_t^2}\right) = - \frac{1}{(2\pi)} \left[(n+2) + 2 \frac{u_t'''(1)}{u_t''(1)} \right] \qquad (3.6)$$

We have :

and

$$\frac{d}{dt}\left(\frac{1}{g_t^2}\right) = b_2 + b_3 g_t^2 + \cdots\cdots \tag{3.7}$$

$$U_t^{(loc)} = \sum_{k\geq 0} (g_t^2)^k \, U_t^{(loc),(k)}$$

To lowest order from (3.5):

$$\frac{\partial u_t^{(0)}}{\partial t} = 2u_t^{(0)} - 4\tau\left(\frac{\partial u_t^{(0)}}{\partial \tau}\right)^2 \tag{3.8}$$

This equation has a unique fixed point :

$$u_*^{(0)}(\tau) = \frac{1}{2}\left(|\tau|^{1/2} - 1\right)^2 \tag{3.9}$$

$u_*^{(0)}(\tau)$ has a convergent Taylor expansion around $\tau = 1$:

$$u_*^{(0)}(\tau) = \sum_{p\geq 2} \frac{\lambda_p^*}{p!}(\tau - 1)^p$$

with

$$\lambda_p^* = (-1)^p 2^{-p} \prod_{\ell=1}^{p-2} (2\ell+1), \quad p \geq 2 \tag{3.10}$$

In particular,

$$\lambda_2^* = \frac{1}{4}, \quad \lambda_3^* = -\frac{3}{8}, \quad \lambda_4^* = \frac{15}{16}$$

It is easy to show that the fixed point $u_*^{(0)}$ is attractive. Write :

$$\sigma = \sqrt{\tau} \tag{3.13}$$

$$u_t(\tau) = \hat{u}_t(\sigma)$$

$\hat{u}_t(\sigma)$ has minimum at $\sigma = 1$. From (3.8) we get :

$$\frac{\partial \hat{u}_t^{(0)}}{\partial t} = 2\hat{u}_t^{(0)} - \left(\frac{\partial \hat{u}_t^{(0)}}{\partial \sigma}\right)^2 \tag{3.14}$$

Taylor expand around $\sigma = 1$:

$$\hat{u}_t^{(0)}(\sigma) = \sum_{p \geqslant 2} \hat{\lambda}_p(t) \frac{(\sigma-1)^p}{p!} \tag{3.15}$$

Then.

$$\frac{d\hat{\lambda}_2}{dt} = 2\left(\hat{\lambda}_2 - (\hat{\lambda}_2)^2\right)$$

and for $p \geqslant 3$ (3.16)

$$\frac{d\hat{\lambda}_p}{dt} = -2\left(p\hat{\lambda}_2 - 1\right)\hat{\lambda}_p + f_p\left(\hat{\lambda}_2, \dots, \hat{\lambda}_{p-1}\right)$$

where the f_p are polynomials.

Clearly

$$\hat{\lambda}_2(t) \underset{t\to\infty}{\longrightarrow} \hat{\lambda}_{2*} = 1$$

For large t, $p \geqslant 3$

$$\frac{d\hat{\lambda}_p}{dt} = -2(p-1)\hat{\lambda}_p + f_p\left(\hat{\lambda}_{2*}, \hat{\lambda}_3, \dots, \hat{\lambda}_{p-1}\right)$$

It follows by induction that for $p \geqslant 3$:

$$\hat{\lambda}_p(t) \underset{t\to\infty}{\longrightarrow} \hat{\lambda}_{p*}$$

This shows that $u_*^{(0)}$ is attractive.

If we consider the order k term in (3.7), then very similarly to the above we can prove :

$$u_t^{(k)} \to u_*^{(k)}$$

in the sense of convergence of Taylor coefficients around $\tau = 1$. (the details are omitted)

Next consider the g_t^2 flow (3.6). To lowest order, for large t

$$\frac{d}{dt}\left(\frac{1}{g_t^2}\right) = -\frac{1}{(2\pi)}\left[(n+2) + 2\,\frac{\lambda_3^*}{\lambda_2^*}\right] + O(g_t^2)$$

where the λ_p^* are given by (3.10). We get :

$$\frac{d}{dt}\left(\frac{1}{g_t^2}\right) = -\frac{(n-1)}{(2\pi)} + O(g_t^2) \tag{3.17}$$

The leading order 'local approximation' coefficient

$$b_2 = -\frac{(n-1)}{(2\pi)}$$

differs from the true value $-\frac{(n-2)}{2\pi}$ (see subsequent sections), for the same reason that $Z_t = 1$ (in this approximation). The local approximation gives <u>the correct picture for the irrelevant terms</u>, <u>but not entirely correct for the marginal flow</u>. Note that there are no relevant directions.

For future purposes define :

$$\tilde{v}^{[m],k}_{j_1\cdots j_m}(t) = \left.\frac{\partial^m u_t^{(k)}(\psi)}{\partial\psi_{j_1}\cdots\partial\psi_{j_m}}\right|_{\substack{\psi = v\\ |v|^2=1}} \tag{3.18}$$

$$\tilde{v}^{[m],k}_{j_1\cdots j_m}(t) \longrightarrow \tilde{v}^{[m],k}_{j_1\cdots j_m\,*}$$

We record here some lowest order coefficients :

$$\tilde{v}^{[2],0}_{ij\,*} = v_i v_j \;,\; \tilde{v}^{[3],0}_{ijk\,*} = (\delta_{ij}\, v_k + \text{perms}) - 3 v_i v_j v_k \tag{3.19}$$

$$\tilde{v}^{[4],0}_{ijk\ell\,*} = (\delta_{ij}\,\delta_{k\ell} + \text{perms}) - 3(\delta_{ij}\, v_k v_\ell + \text{perms}) + 15\, v_i v_j v_k v_\ell$$

etc.

IV- THE FIXED POINT BEYOND THE LOCAL APPROXIMATION

4.1

We return to the exact flow equations (2.39), (2.41) and (2.42). We expand :

$$U_t = \sum_{k\geq 0} (g_t^2)^k U_t^{(k)} \tag{4.1}$$

$$\left.\begin{aligned} \frac{d}{dt}\left(\frac{1}{g_t^2}\right) &= \beta_2 + \sum_{k\geq 1} \beta_{k+2} (g_t^2)^k \\ \frac{1}{Z_t}\frac{dZ_t}{dt} &= \gamma_2 g_t^2 + \sum_{k\geq 2} \gamma_{k+1} (g_t^2)^k \end{aligned}\right\} \tag{4.2}$$

and

$$Z_t - 1 = \hat{\gamma}_2 g_t^2 + \sum_{k\geq 2} \hat{\gamma}_{k+1} (g_t^2)^k \tag{4.3}$$

We assume that the $\hat{\gamma}_j$ exist as $t_{N-M} \to \infty$. This follows from (4.2) provided the initial conditions $g_0^2(N)$, $Z_0(N)$ are correctly chosen (see (2.35)). The flow of marginal parameters will be calculated in Section V.

Using (4.1 - 4.3) we can derive the flow equations for the $U_t^{(k)}$ starting from (2.39). Thus we have :

$$\frac{\partial U_t^{(0)}}{\partial t} = -\sum_{i=1}^{n} \int d^2x \, (x.\nabla\psi_i(x)) \frac{\delta U_t^{(0)}}{\delta \psi_i(x)} - \int d^2x \, d^2y \, K(x-y) \sum_{i=1}^{n} \frac{\delta U_t^{(0)}}{\delta \psi_i(x)} . \frac{\delta U_t^{(0)}}{\delta \psi_i(y)} \tag{4.4}$$

We write out explicitly the equation for $U_t^{(1)}$ to get the feel :

$$\frac{\partial U_t^{(1)}}{\partial t} = -\sum_i \int d^2x \, (x \cdot \nabla\psi_i(x)) \frac{\delta U_t^{(1)}}{\delta\psi_i(x)} - \sum_i 2 \int d^2x \, d^2y \, K(x-y) \frac{\delta U_t^{(0)}}{\delta\psi_i(x)} \cdot \frac{\delta U_t^{(1)}}{\delta\psi_i(y)}$$

$$+ \Bigg[-\gamma_2 \frac{1}{2} \int d^2x \, |\nabla\psi|^2 - \beta_2 U_t^{(0)} + \frac{1}{2}(\beta_2 - \gamma_2) \int d^2x \sum_i \psi_i(x) \frac{\delta U_t^{(0)}}{\delta\psi_i(x)}$$

$$- \int d^2x \, d^2y \, K(x-y) \sum_i \left\{ \frac{\delta^2 U_t^{(0)}}{\delta\psi_i(x)\,\delta\psi_i(y)} + 2\hat{\gamma}_2 \left(-\Delta_x \psi_i(x)\right) \frac{\delta U_t^{(0)}}{\delta\psi_i(y)} \right.$$

$$\left. + \hat{\gamma}_2 \frac{\delta U_t^{(0)}}{\delta\psi_i(x)} \cdot \frac{\delta U_t^{(0)}}{\delta\psi_i(y)} \right\} \Bigg] \tag{4.5}$$

More generally : for $k \geqslant 1$,

$$\frac{\partial U_t^{(k)}}{\partial t} = -\sum_i \int d^2x \, (x \cdot \nabla\psi_i(x)) \frac{\delta U_t^{(k)}}{\delta\psi_i(x)}$$

$$- \sum_i 2 \int d^2x \, d^2y \, K(x-y) \frac{\delta U_t^{(0)}}{\delta\psi_i(x)} \frac{\delta U_t^{(k)}}{\delta\psi_i(y)} + R^{(k-1)}\left(U_t^{(0)}, \ldots, U_t^{(k-1)}\right) \tag{4.6}$$

where R_{k-1} is a polynomial of degree 2 in the $U_t^{(j)}$, $j = 0, \ldots, k-1$, and (functional) derivatives there of.

Our basic result is, [5] :

$$U_t^{(k)} \underset{t \to \infty}{\to} U_*^{(k)} \tag{4.7}$$

in the sense of convergence of Taylor coefficients (2.37).

We will first sketch the proof of (4.7) for $k = 0$. This is the basic case. Once this is understood, the proof of (4.7) for $k \geqslant 1$ by induction starting from (4.6) is very easy.

Start from (4.4). It is then easy to derive the flow equations for the Taylor coefficients (2.32 - 2.33) :

$$u^{[m],0}_{j_1 \cdots j_m}(x_1, \cdots x_m; t) = \left. \frac{\delta^m U_t^{(0)}}{\delta\psi_{j_1}(x_1) \cdots \delta\psi_{j_m}(x_m)} \right|_{\psi = v} \tag{4.7}$$

whose Fourier transform is

$$(2\pi)^2 \delta^{(2)}(p_1 + \cdots + p_m)\, \tilde{v}^{[m],0}_{j_1 \cdots j_m}(p_2 \cdots p_m ; t) \tag{4.8}$$

To write down the flow equation we need some notation :

$$M = (1, 2, ..., m) \ ; \ K, L, ... \subset M \ ; \ |K| = \#(K)$$

$$J_M = (j_1, ..., j_m) \ ; \ p_M = (p_1, ..., p_m) \tag{4.9}$$

$$\underset{\sim}{p}_M = (p_2, ..., p_m).$$

Then we get : (remember $\sum_{i \in M} p_i = 0$)

$$\begin{aligned} & -\frac{\partial}{\partial t} \tilde{v}^{[m],0}_{J_M}(\underset{\sim}{p}_M ; t) \\ & = \left[\left(\sum_{\ell=2}^{m} p_\mu^{(\ell)} \frac{\partial}{\partial p_\mu^{(\ell)}} \right) - 2 \right] \tilde{v}^{[m],0}_{J_M}(\underset{\sim}{p}_M ; t) \\ & + 2 \sum_{i=1}^{n} \sum_{\substack{K \cup L = M \\ |K| \geq |L|}}' \tilde{v}^{[1+|K|],0}_{\{i\} \cup J_K}(p_K ; t)\, \tilde{K}\Big(\sum_{j \in K} p_j\Big)\, \tilde{v}^{[1+|L|],0}_{\{i\} \cup J_L}(p_L ; t) \end{aligned} \tag{4.10}$$

Here the prime on $\sum$ means that for $|K| = |L|$, the same partition is not repeated twice.

Starting from (4.10) we will prove, uniformly in p_M in compacts, that :

<u>Proposition 4.1</u>

$$\lim_{t \to \infty} \tilde{v}^{[m],0}_{J_M}(\underset{\sim}{p}_M ; t) = \tilde{v}^{[m],0}_{J_M}(\underset{\sim}{p}_M)_*$$

exists.

The proof will be by induction. First we will prove the existence of the limit for m = 2, 3 and then consider the general case. The fixed point functions for m = 2, 3 4 are needed later for the computation of the marginal flow, see (2.41, 2.42).

Let us note that if we set $p_M = 0$ in (4.10) then the equations collapse to those of the "local approximation" of Section III. In this case the existence of the limit was proved in Section III. This fact will be exploited as a "boundary condition" when solving (4.10) for large t.

4.2 THE CASE m = 2.

Begin with m = 2 (since $\tilde{v}_j^{[1]} = 0$). For this case, (4.10) reads explicitly :

$$-\frac{\partial}{\partial t}\tilde{v}^{[2],0}_{j_1 j_2}(p;t) = \left(p_\mu \frac{\partial}{\partial p_\mu} - 2\right)\tilde{v}^{[2],0}_{j_1 j_2}(p;t) + 2\sum_{i=1}^{n} \tilde{K}(p)\,\tilde{v}^{[2],0}_{i j_1}(p;t)\,\tilde{v}^{[2],0}_{i j_2}(p;t) \tag{4.11}$$

Decompose into invariants :

$$\tilde{v}^{[2],0}_{j_1 j_2} = \tilde{v}^{[2],0}_{1}\,\delta_{j_1 j_2} + \tilde{v}^{[2],0}_{2}\,v_{j_1} v_{j_2} \tag{4.12}$$

We then get from (4.11)

a) $$-\frac{\partial}{\partial t}\tilde{v}^{[2],0}_{1} = \left(p_\mu \frac{\partial}{\partial p_\mu} - 2\right)\tilde{v}^{[2],0}_{1} + 2\tilde{K}\left(\tilde{v}^{[2],0}_{1}\right)^2$$

(4.13)

b) $$-\frac{\partial}{\partial t}\tilde{v}^{[2],0}_{2} = \left(p_\mu \frac{\partial}{\partial p_\mu} - 2\right)\tilde{v}^{[2],0}_{2} + 4\tilde{K}\,\tilde{v}^{[2],0}_{2}\,\tilde{v}^{[2],0}_{1} + 2\tilde{K}\left(\tilde{v}^{[2],0}_{2}\right)^2$$

From the initial condition

$$U_0^{(0)} = \frac{\lambda_0}{2}\int d^2x\,\left(|\psi(x)|^2 - 1\right)^2$$

it is easy to check :

$$\left.\begin{aligned}\tilde{v}_1^{[2],0}(p; t=0) &= 0\\ \tilde{v}_2^{[2],0}(p;0) &= 4\lambda_0\end{aligned}\right\} \quad (4.14)$$

Since $\tilde{v}_1^{[2],0} = 0$ is a fixed point of 4.13 a), it follows that

$$v_1^{[2],0}(p\,;t) = 0 \quad (4.15)$$

and (4.13 b) becomes

$$-\frac{\partial}{\partial t}\tilde{v}_2^{[2],0} = \left(p_\mu \frac{\partial}{\partial p_\mu} - 2\right)\tilde{v}_2^{[2],0} + 2\tilde{K}\left(\tilde{v}_2^{[2],0}\right)^2 \quad (4.16)$$

with the initial condition (4.14).

For $s \geqslant 0$, define :

$$\left.\begin{aligned}v_2^{[2],0}(s;t) &\equiv \tilde{v}_2^{[2],0}(e^{s/2}p\,; 2t)\\ k(s) &\equiv \tilde{K}(e^{s/2}p) = e^{-p^2e^s}\end{aligned}\right\} \quad (4.17)$$

using (2.24). Then (4.16) reads :

$$-\left(\frac{\partial}{\partial t} + \frac{\partial}{\partial s}\right)v_2^{[2],0} = -v_2^{[2],0} + k(s)\left(v_2^{[2],0}\right)^2 \quad (4.18)$$

$$\left.\begin{aligned}&\text{Writing} \qquad t = u + v\,,\ s = u - v\\ &\text{so that} \qquad u = \frac{t+s}{2}\,,\ v = \frac{t-s}{2}\\ &\text{and} \qquad v_2^{[2],0}(s;t) \equiv \hat{v}(u;v)\end{aligned}\right\} \quad (4.19)$$

(4.18) reads :

$$\frac{\partial \hat{v}}{\partial u} = \hat{v} - k(u-v)\,\hat{v}^2 \quad (4.20)$$

with $u \,\varepsilon\, [\frac{s}{2}, \infty)$. The last equation is easily solved by the substitution $\hat{v} = \frac{1}{\xi}$, since it then becomes linear. The solution written in terms of

$v_2^{[2],0}(s\,;t)$ reads :

$$\frac{1}{v_2^{[2],0}(s\,;t)} = \frac{e^{-t/2}}{v_2^{[2],0}(s-\frac{t}{2}\,;\frac{t}{2})} - \frac{e^{-s}}{p^2}\left(e^{-e^s p^2} - e^{-p^2 e^{s-t/2}}\right)$$

Going back to (4.17) and putting s = 0, we get :

$$\frac{1}{\tilde{v}_2^{[2],0}(p\,;2t)} = \frac{e^{-t/2}}{\tilde{v}_2^{[2],0}(e^{-t/4}p\,;t)} - \frac{1}{p^2}\left(e^{-p^2} - e^{-p^2 e^{-t/2}}\right) \tag{4.21}$$

Now :

$$\lim_{t\to\infty} \tilde{v}_2^{[2],0}(e^{-\frac{t}{4}}p\,;t) = \lim_{t\to\infty} \tilde{v}_2^{[2],0}(0\,;t) = 1 \tag{4.22}$$

Since $v_2^{[2],0}(o\,;t)$ is governed exactly by the equations of the local approximation (Section III) and we have used (3.19). It follows from (4.21 - 4.22) :

$$\lim_{t\to\infty} \tilde{v}_2^{[2],0}(p\,;t) = \tilde{v}_2^{[2],0}(p)_* = \frac{p^2}{1-e^{-p^2}} \equiv f_*(p) \tag{4.23}$$

It is easy to check that the convergence is uniform over compacts.

S 4.3 THE CASE m = 3.

M = (1, 2, 3) and $J_{M}^{i}(j_1 j_2 j_3)$

We decompose into invariants :

$$\tilde{v}_{j_1 j_2 j_3}^{[3],0}(p_2, p_3\,;t) = \left\{ \tilde{v}_1^{[3],0}(p_2, p_3\,;t)\, V_{j_1}\, \delta_{j_2 j_3} + (\text{perms})' \right\}$$

$$+ \; V_{j_1} V_{j_2} V_{j_3}\, \tilde{v}_2^{[3],0}(p_2, p_3\,;t) \tag{4.24}$$

The (perms)' above are just those necessary to ensure the symmetry of

$$\delta(p_1+p_2+p_3)\, \tilde{v}_{j_1 j_2 j_3}^{[3],0}(p_2, p_3)$$

under the transposition $(j_\ell, p_\ell) \rightleftharpoons (j_m, p_m)$, $(\ell, m) \in \{1, 2, 3\}$. Plugging in (4.24) into (4.10) we get, for large t (use 4.23),

$$-\frac{\partial}{\partial t}\tilde{v}_1^{[3],0}(p_2, p_3; t)$$
$$= \Big[\sum_{\ell=2}^{3}\Big(p_\mu^{(\ell)}\frac{\partial}{\partial p_\mu^{(\ell)}}\Big)-2\Big]\tilde{v}_1^{[3],0}(p_2, p_3; t)$$
$$+ 2\tilde{K}(p_2+p_3) f_*(p_2+p_3)\tilde{v}_1^{[3],0}(p_2, p_3; t)$$

$$-\frac{\partial}{\partial t}\tilde{v}_2^{[3],0}(p_2, p_3; t) \tag{4.25}$$
$$= \Big[\sum_{\ell=2}^{3}\Big(p_\mu^{(\ell)}\frac{\partial}{\partial p_\mu^{(\ell)}}\Big)-2\Big]\tilde{v}_2^{[3],0}(p_2, p_3; t) + 2\Big(\sum_{i=1}^{3}\tilde{K}(p_i) f_*(p_i)\Big)\tilde{v}_2^{[3],0}(p_2, p_3; t)$$
$$+ 2\sum_{i=1}^{3}\tilde{K}(p_i) f_*(p_i)\Big(\tilde{v}_1^{[3],0}(p_i, p_j; t) + \tilde{v}_1^{[3],0}(p_i, p_k; t)\Big)_{j\neq k\neq i} \tag{4.26}$$

and $(p_1 + p_2 + p_3) = 0$, f_* is given by (4.23)

Consider first (4.25). Set

$$v_1^{[3],0}(s\,;t) = \tilde{v}_1^{[3],0}(e^{s/2}p\,; 2t)$$
$$k(s) = \tilde{K}(e^{s/2}p_1) \tag{4.27}$$
$$\tilde{f}_*(s) = f_*(e^{s/2}p_1)$$

Then (4.25) reads :

$$\Big(\frac{\partial}{\partial t}+\frac{\partial}{\partial s}\Big)v_1^{[3],0}(s;t) = \big(1-\tilde{k}(s)f_*(s)\big)v_1^{[3],0}(s;t) \tag{4.28}$$

Setting $t = u + v$, $s = u - v$, the linear equation (4.28) is easily integrated to give :

$$v_1^{[3],0}(s;t) = v_1^{[3],0}\Big(s-\frac{t}{2};\frac{t}{2}\Big)e^{\frac{t}{2}}\,\frac{1-e^{-p_1^2 e^{s-\frac{t}{2}}}}{1-e^{-p_1^2 e^{s}}} \tag{*}$$

Setting $s = 0$, and using (4.27) we get from (*) :

(**)

$$\tilde{v}_1^{[3],0}(\underset{\sim}{p};2t) = \tilde{v}_1^{[3],0}(e^{-\frac{t}{4}}\underset{\sim}{p};t)e^{t/2} \cdot \frac{1-e^{-p_1^2 e^{-t/2}}}{1-e^{-p_1^2}}$$

$v_1{}^{[3],0}(o\ ;\ t)$ is governed exactly by the local approximation of Section III.

Using (3.19):

$$\lim_{t\to\infty} \tilde{v}_1^{[3],0}(o;t) = 1$$

Using this we get from (**):

$$\lim_{t\to\infty} \tilde{v}_1^{[3],0}(\underset{\sim}{p};2t) = \tilde{v}_{1*}^{[3],0}(p_2, p_3)$$

$$= \frac{p_1^2}{1-e^{-p_1^2}} = f_*(p_2+p_3) \qquad (4.29)$$

The above convergence is uniform over compacts.

We can now turn to (4.26), and replace the inhomogeneous term by its asymptotic value using (4.29). The resulting linear equation (for large t) is integrated as above and one obtains:

$$\tilde{v}_2^{[3],0}(\underset{\sim}{p};2t) = \tilde{v}_2(e^{-t/4}\underset{\sim}{p};t)e^{t/2}\prod_{i=1}^{3}\left[\frac{1-e^{-p_i^2 e^{-t/2}}}{1-e^{-p_i^2}}\right]$$

$$-\frac{1}{\prod_{i=1}^{3}(1-e^{-p_\ell^2})}\cdot\sum_{i=1}^{3}\Bigg[\Big\{-p_j^2\big(e^{-p_i^2}-e^{-p_i^2 e^{-t/2}}\big)$$

$$+\frac{p_i^2 p_j^2}{p_i^2+p_k^2}\left(e^{-(p_i^2+p_k^2)}-e^{-(p_i^2+p_k^2)e^{-t/2}}\right)\Big\} \qquad (***)$$

$$+\{(j \rightleftarrows k)\}\Bigg]_{j\neq k\neq i}$$

Once again $v_2{}^{[3],0}(o\ ;\ t)$ is governed exactly by the local approximation of

Section III. Using (3.19):

$$\lim_{t\to\infty} \tilde{v}_2^{[3],0}(0;t) = -3$$

Using this we get from (***)

$$\lim_{t\to\infty} \tilde{v}_2^{[3],0}(p_2, p_3; 2t) = \tilde{v}_{2*}^{[3],0}(p_2, p_3)$$

where

$$\tilde{v}_{2*}^{[3],0}(p_2, p_3) = -\frac{1}{\prod_{\ell=1}^{3}\left(1-e^{-p_\ell^2}\right)} \cdot \sum_{i=1}^{3}\left[\left\{-p_j^2 e^{-p_i^2} + \frac{p_i^2 p_j^2}{p_i^2+p_k^2} e^{-(p_i^2+p_k^2)} + \frac{p_j^2 p_k^2}{p_i^2+p_k^2}\right\} + \{j \rightleftarrows k\}\right]_{j\neq k\neq i} \qquad (4.30)$$

It is easy to show from (4.30) by carefully taking limits:

$$\tilde{v}_2^{[3],0}(p,-p) = -\left(2f_*(p)+1\right) \qquad (4.31)$$

We now proceed to the general case i.e. arbitrary 'm'.

4.4 THE GENERAL CASE.

We return to (4.10). Without loss of generality we take m = $2q$ with $q \geqslant 2$. Our inductive assumption is: for m = 2, 3, ..., $2q$ - 1,

$$\lim_{t\to\infty} \tilde{v}_{J_M}^{[m],0}(\underset{\sim}{p}_M; t) = \tilde{v}_{J_M *}^{[m],0}(\underset{\sim}{p}_M) \qquad *$$

exists uniformly in p_M. We have seen that this is satisfied for m = 2, 3.

Separating out explicitly the term on the r.h.s. of (4.10) involving $\tilde{v}^{[2q],0}$ we get: (M = (1, ..., $2q$))

$$-\frac{\partial}{\partial t}\tilde{v}^{[2q],0}_{J_M}(\underset{\sim}{p}_M;t)=\left[\sum_{\ell=2}^{2q}\left(p_\mu^{(\ell)}\frac{\partial}{\partial p_\mu^{(\ell)}}\right)-2\right]\tilde{v}^{[2q],0}_{J_M}(\underset{\sim}{p}_M;t)\ +$$

$$+\,2\sum_i\sum_{\substack{K\cup L=M\\|K|=2q-1}}\tilde{v}^{[2],0}_{\{i\}\cup J_L}(p_L;t)\,\tilde{K}(p_L)\,\tilde{v}^{[2q],0}_{\{i\}\cup J_K}(p_K;t)\ +$$

$$+\,2\sum_i\sum_{\substack{K\cup L=M\\2\le|K|\le 2q-2\\|K|\geq|L|}}{}'\ \tilde{v}^{[1+|K|],0}_{\{i\}\cup J_K}(p_K;t)\,\tilde{K}\Big(\sum_{j\in K}p_j\Big)\,\tilde{v}^{[1+|L|],0}_{\{i\}\cup J_L}(p_L;t)\qquad(4.32)$$

We shall decompose the O(n) tensors into invariants :

$$\tilde{v}^{[2q],0}_{J_M}(\underset{\sim}{p}_{J_M};t)=\Big[\Big(\delta_{j_1j_2}\delta_{j_3j_4}\cdots\delta_{j_{2q-1}j_{2q}}\tilde{v}^{[2q],0}_{1}(\underset{\sim}{p}_{J_M};t)+\text{perms}\Big)$$

$$+\Big(\delta_{j_1j_2}\delta_{j_3j_4}\cdots\delta_{j_{2q-3}j_{2q-2}}v_{j_{2q-1}}v_{j_{2q}}\tilde{v}^{[2q],0}_{2}(\underset{\sim}{p}_{J_M};t)+\text{perms}\Big)$$

$$+\ \cdots\cdots\cdots\cdots$$

$$+\ v_{j_1}\cdots v_{j_{2q}}\tilde{v}^{[2q],0}_{q+1}(\underset{\sim}{p}_{J_M};t)\Big]\qquad(4.33\text{ a})$$

For $|K| = 2q - 2$,

$$\tilde{v}^{[2q-1],0}_{\{i\}\cup J_K}(p_K;t)=\Big[\Big(\delta_{j_1j_2}\cdots\delta_{j_{2q-3}j_{2q-2}}v_i\,\tilde{v}^{[2q-1],0}_{1}(p_K;t)+\text{perms}\Big)$$

$$+\Big(\delta_{j_1j_2}\cdots\delta_{j_{2q-5}j_{2q-4}}v_{j_{2q-3}}v_{j_{2q-2}}v_i\,\tilde{v}^{[2q-1],0}_{2}(p_K;t)+\text{perms}\Big)$$

$$+\ \cdots\cdots\cdots$$

$$+\ v_{j_1}\cdots v_{j_{2q-2}}v_i\,\tilde{v}^{[2q-1],0}_{q}(p_K;t)\qquad(4.33\text{ b})$$

and similarly for lower rank tensors.

Also for large t,

(4.33 c)

$$\tilde{v}^{[2],0}_{i\ell}(p_\ell; t) \longrightarrow v_i v_\ell f_*(p_\ell)$$

with f_* given by (4.23).

Using (4.33 a - c) we get from (4.32):

(4.34 a)

$$-\frac{\partial}{\partial t}\tilde{v}^{[2q],0}_1(p_{J_M}; t)$$
$$= \left[\sum_{\ell=2}^{2q}\left(p^{(\ell)}_\mu \frac{\partial}{\partial p^{(\ell)}_\mu}\right) - 2\right]\tilde{v}^{[2q],0}_1(p_{J_M}; t) + 2\, Q^{[2q-1],0}_1(p_{J_M}; t)$$

and for $j \geq 2$

$$-\frac{\partial}{\partial t}\tilde{v}^{[2q],0}_j(p_{J_M}; t) = \left[\sum_{\ell=2}^{2q}\left(p^{(\ell)}_\mu \frac{\partial}{\partial p^{(\ell)}_\mu}\right) - 2\right]\tilde{v}^{[2q],0}_j(p_{J_M}; t) +$$

$$+2\left\{\sum_{(\text{perms})'} f_*(p_\ell)\,\tilde{K}(p_\ell)\right\}\tilde{v}^{[2q],0}_j(p_{J_M}; t) +$$

$$+2\left\{\sum_{(\text{perms})'} f_*(p_\ell)\,\tilde{K}(p_\ell)\right\}\tilde{v}^{[2q],0}_{j-1}(p_{J_M}; t)$$

$$+ \; Q^{[2q-1],0}_2(p_{J_M}; t)$$

(4.34 b)

In the above $Q^{[2q-1],0}_i$ involve invariants $\tilde{v}^{[m],0}$ for $3 \leq m \leq 2q - 1$, and for these $t \to \infty$ limit exists by the inductive hypothesis. The $\sum_{(\text{perms})'}$ refers to sums over just those permutations necessary to ensure symmetry properties. <u>There are at least two terms in such a sum</u>.

A <u>We begin with (4.34 a)</u>. The equation is linear with smooth coefficients and hence has a unique solution for finite t. To study the asymptotics of $t \to \infty$, we first exploit a consequence of O(n) invariance (and this enables us to go ahead without a detailed knowledge of the structure of $Q_1^{[2q-1],0}$).

By O(n) invariance of $U_t^{(0)}(\psi)$:

$$\frac{d}{d\omega}\bigg|_{\omega=0} U_t^{(0)}\left(\psi_i + \omega\varepsilon_{ij}\psi_j\right) = 0 \tag{*}$$

where ε_{ij} is an antisymmetric tensor.

From (*) it is easy to derive : for any $j \in (2, \ldots, 2q)$

$$\tilde{v}_1^{[2q],0}(p_{J_M}; t)\bigg|_{p_j=0} = \tilde{v}_1^{[2q-1]}(p_{J_K}; t) \tag{4.35}$$

$|K| = 2q - 1$, $K \subset M$, and $q \geqslant 2$.

Define :

$$v_1^{[2q],0}(\lambda; t) \equiv \tilde{v}_1^{[2q],0}(\lambda p_{J_M}; t) \tag{4.36}$$

By (4.35) and the inductive hypothesis :

$$\lim_{t\to\infty} \frac{d^m}{d\lambda^m}\bigg|_{\lambda=0} v_1^{[2q],0}(\lambda; t) \quad \text{exists for } m = 0, 1, 2 \tag{4.37}$$

Note that $v_1^{[2q],0}(\lambda; t)$ satisfies :

$$-\frac{\partial}{\partial t} v_1^{[2q],0}(\lambda; t) = \left(\lambda\frac{\partial}{\partial\lambda} - 2\right) v_1^{[2q],0}(\lambda; t) + Q_1^{[2q-1]}(\lambda; t) \tag{4.38}$$

and $v_1^{[2q],0}(\lambda; t)$ is smooth in λ, since the coefficients of the equation are smooth. We make the Taylor expansion upto second order :

$$v_1^{[2q],0}(\lambda; t) = \sum_{m=0}^{2} \frac{\lambda^m}{m!}\left(\frac{\partial}{\partial\lambda}\right)^m v_1^{[2q],0}(\lambda; t) + \frac{\lambda^3}{3!} R_1^{[2q],0}(\lambda; t) \tag{4.39}$$

and similarly for $Q_1^{[2q-1]}(\lambda; t)$. From (4.38 - 4.39) we get :

$$-\frac{\partial}{\partial t} R_1^{[2q],0}(\lambda;t) = \left(\lambda \frac{\partial}{\partial \lambda} + 1\right) R_1^{[2q],0}(\lambda;t) + \hat{Q}_1^{[2q-1]}(\lambda;t) \tag{4.40}$$

(R_1, $\hat{Q}_1$ are given by the remainder formula for Taylor series).

Because of (4.37) and (4.39) we have to show :

$$\lim_{t\to\infty} R_1^{[2q],0}(\lambda;t) = R_{1*}^{[2q],0}(\lambda) \text{ exists} \quad *$$

for λ pointwise in compact sets.

First consider the special case of (4.40) obtained by setting $\lambda = 0$:

$$-\frac{\partial}{\partial t} R_1^{[2q],0}(0;t) = R_1^{[2q],0}(0;t) + \hat{Q}_1^{[2q-1]}(0;t) \tag{4.41}$$

Since,

$$\lim_{t\to\infty} \hat{Q}_1^{[2q-1]}(0;t) = \hat{Q}_{1*}^{[2q-1]}(0) \text{ exists} \tag{4.42}$$

it follows from (4.41) that

$$\lim_{t\to\infty} R_1^{[2q],0}(0;t) = R_{1*}^{[2q],0}(0) \text{ exists} \tag{4.43}$$

Now we go back to (4.40) and define

$$\left.\begin{aligned} \bar{R}_1^{[2q],0}(s;t) &\equiv R_1^{[2q],0}(e^s\lambda;t) \\ \bar{Q}_1^{[2q-1]}(s;t) &\equiv \hat{Q}_1^{[2q-1]}(e^s\lambda;t) \end{aligned}\right\} \tag{4.44}$$

and then (4.40) reads :

$$\left(\frac{\partial}{\partial s} + \frac{\partial}{\partial t}\right) \bar{R}_1^{[2q],0}(s;t) = -\bar{R}_1^{[2q],0}(s;t) - \bar{Q}_1^{[2q-1]}(s;t) \tag{4.45}$$

For large t, the last term on r.h.s. of (4.45) can be replaced by its asymptotic form (which exists by the inductive hypothesis) $\bar{Q}_{1*}^{[2q-1]}(s)$

We can solve the linear equation (4.45) after the change of variables $t = u + v$, $s = u - v$. We get (for large t):

$$\bar{R}_1^{[2q],0}(s;t) = e^{-\frac{t}{2}}\, \bar{R}_1^{[2q],0}\left(s-\frac{t}{2};\frac{t}{2}\right) - e^{-s}\int_{s-\frac{t}{2}}^{s} ds_1\, \bar{Q}_{1*}^{[2q-1]}(s_1)\, e^{s_1} \qquad (4.46)$$

From (4.42- 4.43):

$$\left.\begin{array}{l} \lim\limits_{t\to\infty} \bar{R}_1^{[2q],0}\left(s-\frac{t}{2};\frac{t}{2}\right) = R_{1*}^{[2q],0}(0) \\ \lim\limits_{s_1\to-\infty} \bar{Q}_1^{[2q-1]}(s_1) = \hat{Q}_{1*}^{[2q-1]}(0) \end{array}\right\} \text{exist} \qquad (4.47)$$

It now follows from (4.46):

$$\begin{aligned} \lim_{t\to\infty} \bar{R}_1^{[2q],0}(s;t) &= -e^{-s}\int_{-\infty}^{s} ds_1\, \bar{Q}_{1*}^{[2q-1]}(s_1)\, e^{s_1} \\ &= \bar{R}_{1*}^{[2q],0}(s) \quad \text{exists} \end{aligned} \qquad (4.48)$$

From (4.39), (4.37) and (4.48) it follows that

$$\lim_{t\to\infty} \tilde{v}_1^{[2q],0}(\underset{\sim}{p}_{J_M};t) = \tilde{v}_{1*}^{[2q],0}(\underset{\sim}{p}_{J_M}) \quad \text{exists} \qquad (4.49)$$

B Next we turn to $\tilde{v}_j^{[2q],0}(\underset{\sim}{p}_{J_M};t)$ governed by (4.34 b), for $j \geqslant 2$. We assume (inductively) that the $t \to \infty$ limit exists for $\tilde{v}_{j-1}^{[2q],0}(\underset{\sim}{p};t)$. The first step is satisfied because of (4.49).

Define:

$$h_{2*}^{(j)}(\underset{\sim}{p}) \equiv \left\{\sum_{(perms)'} f_*(p_\ell)\tilde{K}(p_\ell)\right\} \tilde{v}_{(j-1)*}^{[2q],0}(\underset{\sim}{p}) + \frac{1}{2} Q_{2*}^{[2q-1],0}(\underset{\sim}{p}) \tag{4.50}$$

(which exists by the inductive hypothesis)

$$h_{1*}^{(j)}(\underset{\sim}{p}) = \sum_{(perms)'} \left(f_*(p_\ell)\tilde{K}(p_\ell)\right) \tag{4.51}$$

Define also :

$$\left.\begin{aligned} \bar{v}_j^{[2q],0}(s;t) &= \tilde{v}_j^{[2q],0}(e^{s/2}p\,;2t) \\ \bar{h}_{i*}^{(j)}(s) &= h_{i*}^{(j)}(e^{s/2}p) \end{aligned}\right\} \tag{4.52}$$

Using (4.50 – 4.52) we get from (4.34 b) for large t :

$$\left(\frac{\partial}{\partial t} + \frac{\partial}{\partial s}\right)\bar{v}_j^{[2q],0}(s;t) = \left(1 - \bar{h}_{1*}^{(j)}(s)\right)\bar{v}_j^{[2q],0}(s;t) - \bar{h}_{2*}^{(j)}(s) \tag{4.53}$$

Without l.o.g. set $p_\ell = 1$. As stated after (4.34 b), there are at least two terms in the sum (4.51). Thus :

$$\begin{aligned} \bar{h}_{1*}^{(j)}(s) &= m_j\, f_*(s)\, \tilde{K}(s) \\ &= m_j \frac{e^s}{1-e^{-e^s}}\, e^{-e^s} \end{aligned} \tag{4.54}$$

with $m_j \geq 2$. Changing variables, t = u + v and s = u – v, the linear equation (4.53) can be integrated. We get :

$$\bar{v}_j^{[2q],0}(s;t) = e^{t/2} \frac{\left(1-e^{-e^{(s-t/2)}}\right)^{m_j}}{\left(1-e^{-e^{s}}\right)^{m_j}} \bar{v}_j^{[2q],0}\left(s-\frac{t}{2};\frac{t}{2}\right)$$

$$-\frac{e^{s}}{\left(1-e^{-e^{s}}\right)^{m_j}} \int_{s-\frac{t}{2}}^{s} ds_1\, \bar{h}_{2*}^{(j)}(s_1)\, e^{-s_1} \left(1-e^{-e^{s_1}}\right)^{m_j} \tag{4.55}$$

We have that the limits :

1) $$\lim_{t\to\infty} \bar{v}_j^{[2q],0}\left(s-\frac{t}{2};\frac{t}{2}\right) = \tilde{v}_{j*}^{[2q],0}(0)$$

2) $$\lim_{s_1\to -\infty} \bar{h}_{2*}^{(j)}(s_1) = h_{2*}^{(j)}(0)$$

exist since they coincide with the local approximation of Section III. Moreover

3) $$e^{-s_1}\left(1-e^{-e^{s_1}}\right)^{m_j} \underset{s_1\to-\infty}{\sim} e^{(m_j-1)s_1}$$

Since $m_j \geqslant 2$ we get from (4.55) using (1, 2, 3) :

$\lim_{t\to\infty} \bar{v}_j^{[2q],0}(s;t) = \bar{v}_{j*}^{[2q],0}(s)$ exists. The existence has been proved here for generic momenta. In [5], the limits are taken in the uniform topology.

The proof of the general case is complete, and proposition 4.1 is proved.

4.5 CONVERGENCE OF $U_t^{(k)} \to U_*^{(k)}$, $k \geqslant 1$.

Starting from (4.6) one derives the equations for Taylor coefficients analogous to (4.10). We have :

$$-\frac{\partial}{\partial t}\tilde{v}^{[2],k}_{d_1 d_2}(p;t) = \left(p_\mu \frac{\partial}{\partial p_\mu} - 2\right)\tilde{v}^{[2],k}_{d_1 d_2}(p;t) + \tag{4.56}$$

$$+ 4 \sum_i \tilde{v}^{[2],0}_{i d_1}(p;t)\, \tilde{v}^{[2],k}_{i d_2}(p;t) + Q^{(k-1)}_{d_1 d_2}(p;t)$$

and for m ⩾ 3.

$$-\frac{\partial}{\partial t}\tilde{v}^{[m],k}_{J_M}(\underset{\sim}{p}_M;t) = \left[\sum_{\ell=2}^{m}\left(p^{(\ell)}_\mu \frac{\partial}{\partial p^{(\ell)}_\mu}\right) - 2\right]\tilde{v}^{[m],k}_{J_M}(\underset{\sim}{p}_M;t) + \tag{4.57}$$

$$+ 4 \sum_i \sum_{\substack{K \cup L = M \\ |K| = m-1}} \tilde{v}^{[2],0}_{\{i\}\cup J_L}(p_{J_L};t)\, \tilde{K}(p\;)\, \tilde{v}^{[m],k}_{\{i\}\cup J_K}(p_{J_K};t) + Q^{[k-1],(m-1)}_{J_M}(\underset{\sim}{p}_M;t)$$

Here $Q^{(k-1)}$ involves (as polynomial) in Taylor coefficients of $U_t^{(k-1)}$ and $Q^{(k-1),(m-1)}$ involves Taylor coefficients of $U_t^{(k-1)}$ and those of $U_t^{(k)}$ upto order (m − 1). These equations are analysed exactly as in § 4.4 for $U_t^{(0)}$ coefficients, and the inductive proof of convergence of $U_t^{(k)}$, k ⩾ 1, follows the same pattern. We omit the details. We have :

<u>Proposition 4.2</u>

$$\lim_{t\to\infty} \tilde{v}^{[m],k}_{J_M}(\underset{\sim}{p}_M;t) = \tilde{v}^{[m],k}_{J_M *}(\underset{\sim}{p}_M) \tag{4.58}$$

exists uniformly over compacts.

V THE RG FLOW OF MARGINALS.

We now turn to the g_t^2 and Z_t flows governed by (2.41), (2.42). We first study these flows to lowest order :

$$\left.\begin{aligned} \frac{d}{dt}\left(\frac{1}{g_t^2}\right) &= \beta_2 + O(g_t^2) \\ \frac{1}{Z_t}\frac{dZ_t}{dt} &= \gamma_2 g_t^2 + O((g_t^2)^2) \end{aligned}\right\} \tag{5.1}$$

and compute the coefficients β_2, γ_2.

Replace the coefficients $\tilde{v}^{[m]}_{d_1\cdots d_m}(\cdot;t)$ in (2.41, 2.42) by their lowest order fixed point forms $\tilde{v}^{[m],0}_{d_1\cdots d_m *}(\cdot)$, valid for large t. For m = 2, 3, 4 decompose into invariants. For m = 2, see (4.12), (4.15), (4.23). For m = 3, see (4.24), (4.29), (4.30 - 4.31). For m = 4, take (4.33 a) with $q = 2$.

We then get from (2.41) - (2.42) the formulae for β_2, γ_2 of (5.1)

$$\beta_2 = \gamma_2 - 2\int \frac{d^2p}{(2\pi)^2}\, \tilde{K}(p)\Big[n\,\tilde{v}^{[3],0}_{1*}(p,-p) + 2\tilde{v}^{[3],0}_{1*}(0,p) + \tilde{v}^{[3],0}_{2*}(p,-p)\Big] \tag{5.2}$$

$$\gamma_2 = \int \frac{d^2q}{(2\pi)^2}\, \tilde{K}(q)\,\frac{1}{4}\Delta_p\Big|_{p=0}\Big\{n\,\tilde{v}^{[4],0}_{1*}(q,p,-p) + \tilde{v}^{[4],0}_{1*}(p,q,-p) + \tilde{v}^{[4],0}_{1*}(p,-q,-p) + \tilde{v}^{[4],0}_{2*}(-p,-q,q)\Big\} \tag{5.3}$$

The $\tilde{v}^{[3],0}_{d*}$ are available from (4.29), (4.30), (4.31). We get :

$$\left.\begin{aligned} \tilde{v}^{[3],0}_{1*}(p,-p) &= f_*(0) = 1 \\ \tilde{v}^{[3],0}_{1*}(0,p) &= f_*(p) \\ \tilde{v}^{[3],0}_{2*}(p,-p) &= -\,(2f_*(p)+1) \end{aligned}\right\} \tag{5.4}$$

with $f_*(p)$ given by (4.23).

We also need $\tilde{v}_{j*}^{[4],0}$. These fixed point functions satisfy equations obtained from (4.32) for $q = 2$, after inserting (4.33 a) for $q = 2$. It is easy to show :

$$\left.\begin{aligned} \tilde{v}_{1*}^{[4],0}(q,p,-p) = \tilde{v}_{1*}^{[4],0}(-p,-q,q) = 1 \\ \tilde{v}_{1*}^{[4],0}(p,q,-p) = f_*(p-q) \end{aligned}\right\} \tag{5.5}$$

In particular

$$\begin{aligned} G &\equiv \Delta_p\Big|_{p=0} \tilde{v}_{1*}^{[4],0}(p,q,-p) = \Delta_p\Big|_{p=0} \tilde{v}_{1*}^{[4],0}(p,-q,-p) \\ &\doteq \Delta_q f_*(q) \\ &= f_*'' + \frac{1}{q} f_*' \end{aligned} \tag{5.6}$$

where primes refer to derivatives with respect to $q = (q^2)^{1/2}$, the radial variable.

Finally,

$$\begin{aligned} H &\equiv \Delta_p\Big|_{p=0} \tilde{v}_{2*}^{[4],0}(-p,-q,q) \\ &= \frac{8}{(1-e^{-q^2})^2}\left[(e^{-q^2}-1) + q^2 e^{-q^2} + (q^2)^2 e^{-q^2}\right] \end{aligned} \tag{5.7}$$

Using

$$f_*(q) = q^2\left(1-e^{-q^2}\right)^{-1}, \quad \tilde{K}(q) = e^{-q^2}$$

it is useful to rewrite (5.7) in the form :

$$H = -4\,\frac{(f_*\tilde{K})'}{q\tilde{K}} + 2\,\frac{\tilde{K}' f_*'}{\tilde{K}} \tag{5.8}$$

From (5.3) we have, using (5.5 – 5.8) :

$$\gamma_2 = \frac{1}{4}\int \frac{d^2q}{(2\pi)^2}\, \tilde{K}(q)\,(2G+H)$$

$$= \frac{1}{4}\int_0^\infty \frac{dq}{2\pi}\, q \left(2\left(f_*'' + \frac{1}{q} f_*'\right) + \frac{2\tilde{K}' f_*'}{\tilde{K}} - \frac{4\,(f_* \tilde{K})'}{q \tilde{K}} \right)\tilde{K}$$

$$= \frac{1}{(2\pi)}\cdot\frac{1}{4}\int_0^\infty dq \left(2\frac{d}{dq}\left(\tilde{K} f_*' q\right) - 4\left(f_* \tilde{K}\right)'\right)$$

$$= \frac{1}{2\pi}\, f_*(0)\,\tilde{K}(0) = \frac{1}{2\pi}$$

$$\gamma_2 = \frac{1}{2\pi} \qquad (5.9)$$

From (5.2) we get, using (5.4) and (5.9) :

$$\beta_2 = \frac{1}{2\pi} - (n-1)\cdot 2 \int \frac{d^2p}{(2\pi)^2}\, \tilde{K}(p)$$

$$= \frac{1}{2\pi} - (n-1)\cdot 2\cdot \frac{1}{2}\cdot\frac{1}{(2\pi)}$$

or

$$\beta_2 = -\frac{(n-2)}{2\pi} \qquad (5.10)$$

Starting from (2.41 - 2.42), and using the results of Section IV, the higher order contributions (4.2) with $k \geqslant 1$ to the marginal flows can be calculated. Now choose $g_0^2(N)$ and $Z_0(N)$ as in (2.35). Then it is easy to check (and this is well known) (2.36). <u>The marginal flows get stabilized for $t_{N-M} \to \infty$.</u> The asymptotic freedom formulae (2.35) (bare charge $\to 0$ as $N \to \infty$) agree with [1,2] ; our wave function renormalization corresponds to rescaled fields ($\phi \to g_0^2\, \phi$ with respect to [2]).

ACKNOWLEDGEMENTS

P.K.M. thanks G. Jona Lasinio and the Dipartimento di Fisica of the Università di Roma - La Sapienza, and the INFN, Sezione di Roma, for providing warm hospitality and a stimulating environment in which part of this work was done.

REFERENCES

1. A.M. Polyakov, Phys. Lett. 59 B, 79 (1975)
2. E. Brézin and J. Zinn-Justin, Phys. Rev. B 14, 3110 (1976)
3. E. Brézin, J.C. Le Guillou and J. Zinn-Justin, Phys. Rev. D 14, 2615 (1976)
4. K.G. Wilson and J. Kogut, Phys. Rep. 12 C, 75 (1974)
5. P.K. Mitter and T.R. Ramadas, IAS, Princeton preprint (in preparation)
6. P.K. Mitter and T.R. Ramadas : in, Proceedings of the Ringberg Workshop (1987 February) on "Non-linear field transformations in Quantum Field Theory" to be published in Lecture Notes in Physics(Springer-Verlag) ; ed. P. Breitenlohner, D. Maison and K. Sibold.
7. K. Gawedski and A. Kupiainen : Comm. Math. Phys. 106, 553 (1986)
8. K. Gawedski and A. Kupiainen : in Critical Phenomena, Random Systems, Gange theories, K. Osterwalder and R. Stora, eds. (Elsevier, 1986)
9. F.J. Wegner, in Phase Transitions and Critical Phenomena, vol. 6, Eds. C. Domb and M.S. Green, Acad. Press, London, New-York (1976)
10. J. J. Polchinski, Nucl. Phys. B 231, 269 (1984)
11. D.C. Brydges and J. Kennedy, Journ. of Stat. Phys., Vol. 48, 19 (1987)
12. E. Seiler et al, Max Planck Institut (Munich) preprints MPI - PAE / PTh 76/87 and 75/87.

NUMERICAL SIMULATIONS: OLD AND NEW PROBLEMS

Giorgio Parisi

Dipartimento di Fisica II Universita' di Roma "Tor Vergata"

and INFN, sezione di Roma

I. Introduction

The results obtained for lattice gauge theory simulations in the last years (i. e. after the previous summer school of this series[1,2]) do not look very exiting to the outsider: however these years have been years of hard work in which the results obtained in the past have been consolidated, by putting under controll some of the systematic and statisticals errors. Moreover a serious effort has been done in the construction of home made high speed supercomputers which should run numerical simulations for lattice QCD at full time. Some of these supercomputers have already started to produce interesting results[3,4] and it is likely that in the next years they will improve qualitatively our controll of lattice QCD.

At the present moment we are at the beginning of a new phase of carefull and systematic investigations of lattice QCD: I think that it is better to present in these lectures neither the results obtained from this new supercomputers, or the results obtained by the conventional ones (on this last subject there are two very nice reviews[5,6]). Here I will mainly discuss some theoretical problems, whose solution is not yet compleately satisfactory.

II. General principles

At the present moment string theory has not reached such a mature stage to allow intensive computer simulations, which should have as output the dimensionality of space time and the quark and leptons mass spectrum (or any other quantitity it may be interesting to know). Waiting for such an event, it may be useful to review the theoretical basis of computer simulations.

It is usual nowdays to represent quantum systems as functional integrals (i. e. as integrals over an infinite dimensional functional space); sometimes these integrals may be defined directly, however in many cases it is necessary to approximate these infinite dimensional integrals by finite dimensional integrals: this is usually done by introducing a lattice spacing (to cutoff the high frequency part) and to consider only fields defined in box. If this strategy works, we can say that the functional integral is hyperfinite, i. e. it may be reached as the limit of integrals over finite dimensional manifolds.

The hyperfiniteness of the theory, in its functional integral representation, seems to be a crucial prerequisite for doing computer simulations. The simplest field theories are obviously hyperfinite.

The problems arise when the theory is invariant under a symmetry group (G). It convenient to keep G-invariance also at the level of the finite dimensional approximations, because this procedure garantees that the final theory is authomatically G-invariant: otherwise some unwanted divergences may arise and one has to finely tune some of the parameters to avoid them. In principles instead of integrating over the the φ fields, we could factor out the symmetry group and integrate over only φ/G; however this procedure is dangerous because not all smooth functions of φ/G extend to smooth functions of φ, e. g., if $f(x)$ ($x \in R$) is even and smooth, we must have $f'(0)=0$.

The real problems arise when the group G is an infinite dimensional group which has only (non trivial) infinite dimensional representations (this happens for example in gauge theory and in gravity). It is obvious that in such a case G-invariance cannot be kept, if the functional integration is restricted to a finite dimensional space. The usual solution, at least in gauge theory, is to have the theory invariant under a finite dimensional group ($\tilde{G}$ which is a subgroup of G) such that, when the functional integral over the finite dimensional space approches the infinite dimensional integrals, $\tilde{G}$ must approach (in some sence) G (this is possible only if the symmetry group is hyperfinite).

A second (more subtle possibility) is to have the theory invariant only under a subgroup G^0 of G and proving that in the infinite dimensional limit (under the appropriate conditions) G^0 invariance implies G invariance. This happens for rotational invariance: if higher derivatives terms are absent in the action, invariance under a subgroup of O(3) implies O(3) invariance.

However, after that we have shown that the theory is hyperfinite (and for string theories that has not yet been done), we are not at the end. The number of variables (N), on which we must integrate, is very large ($N=10^6$ is a typical number) and an integral over such an high dimensional space cannot be done easily. More precisely, we need an integration algorithm such that the computer time, needed to get the results with a given accuracy, is proportional to a power of N, not to exp(aN), otherwise the computation would be far outside the capabilities of any present and (quite likely) future computer.

Essentially integrals over such a huge space can be done only when they are of the form

$$\int d\mu(x) f(x), \qquad (1)$$

where $d\mu(x)$ is a probability measure and $f(x)$ is a quantity which is of O(1). In this case it is often possible to find simple algorithms, which generate a sequence x_k such that

$$\lim_{M\to\infty} 1/M \sum_{k=1,M} f(x_k)= \int d\mu(x) f(x), \qquad (2)$$

and the value of M, needed to have reasonable accurate results (i. e. x_M not correlated to x_1), increases as a polynomial in N.

In order to bring a finite dimensional functional integral under the form (1), we must firstly eliminate anticommuting variable and check that the resulting measure ($d\mu$) is positive definite. If $d\mu$ is not positive definite (this happens for example in presence of topological terms for the two dimensional σ-model and four dimensional gauge theories, or more crucially, in presence of a chemical potential for fermions at finite temperature), serious problem arise and it is not clear yet if they have a solution. A possible way out is discussed in section V.

III The Langevin equation

The Langevin equation is very interesting as far as it provides an analytic framework for discussing the properties of numerical simulations[7,8,9]; indeed it is believed that the long time behaviour of a local (non energy-conserving) algorithm is controlled by an appropriate Langevin equation. The convergence of the pseudofermions approach[10] for numerical simulations with fermions (beyond the quenched approximation[11]) can be carefully studied using the Langevin equation[12]. Moreover the Langevin equation itself has been used long time ago[13] in numerical simulations in order to increase the signal to noise ratio.

Generally speaking, the field φ will a function of x and t, where x is the space and t is the simulation time; the time dependence of φ is arbitrary, if the following condition is satisfied:

$$<f>=\lim_{t\to\infty} 1/t \int_0^t d\tau f(\varphi(\tau)), \tag{3}$$

where the brackets <> denotes the equilibrium expectation value.
The simplest form of the Langevin equation is the following:

$$\partial\varphi/\partial t = -K * \delta S/\delta\varphi + K^{1/2} * \eta, \tag{4}$$

where the equilibrium probability distribution is given by

$$d\mu(\varphi) = d\varphi \exp(-S[\varphi]), \tag{5}$$

η is a white noise:

$$\overline{\eta(x,t_1)\eta(y,t_2)} = 2\, \delta(x-y)\, \delta(t_1-t_2), \tag{6}$$

and K is an integral kernel

$$(K * f)(x) = \int dy\, K(x,y)f(y) \tag{7}$$

with positive eigenvalues.

Eq. (4) holds when the field φ is unconstrained, otherwise a more complex Langevin equation should be written.

The key tool for the study of the Langevin equation is the Fokker-Plank equation, which controlls the time evolution of the probability $P[\varphi,t)$, i. e. the probability distribution of the field φ at time t. Using the Fokker-Plank equation it is easy to prove that eq. (3) is satisfied for any choice of the kernel K. (In principle it may be useful also to consider Langevin equations with φ-dependent noise, and this can be easily done by a careful application of the Ito differential calculus).

In any computer simulation the time t and the space x cannot be continous variables and they must be discretized; it is usual to call ε the time step and a the lattice spacing. In the most naive approach the discretized for of the Langevin equation is

$$\varphi(x,(n+1)\varepsilon)= \varphi(x,n\varepsilon)+\varepsilon[-K * \delta S/\delta\varphi + K^{1/2} * b(x,n)] \tag{7}$$

where the b's are random Gaussian variables with variance:

$$\overline{b(x,n\varepsilon)\, b(y,m\varepsilon)} = 2\, \delta_{x,y} /a^D\, \delta_{n,m} /\varepsilon, \tag{8}$$

and ε is small enough: a minimal condition is that the term proportional to ε in the r. h. s. of eq. (7) should be smaller than φ for generic configurations.

The choice of the kernel K is crucial in order to have efficient computer simulations. The simplest, and the mostly widely used, is $K=\delta(x-y)$ (different choices will be discussed in the next section. If we consider for semplicity a scalar theory where

$$\delta S/\delta\varphi = - \quad (-\Delta+\mu^2+g\varphi^2)\varphi, \tag{9}$$

(Δ is the lattice Laplacian), we must have (for this choice of the kernel) that

$$\varepsilon=O(a^2), \tag{10}$$

at least in dimensions less or equal than 4, where the theory is non trivial (neglecting logarithms) in the continuum limit. This means that the computer time to perform a numerical simulation of the Langevin equations (which is naturally proportional to $1/(a^D\varepsilon)$ increases as

$$1/(a^{(D+2)}), \tag{11}$$

in the limit of small lattice spacing.

The presence of the term $1/a^2$ is related to the so called critical slowing down and some effort has been done to suppress it: they will be discussed in the next section.

IV Acceleration

At it was proposed in my Cargese lectures[2] (for an implementation of this proposal see refs. 14,15) the simplest possibility for accelerating the computer simulations and killing the term $1/a^2$ in eq. (11) is to select an appropriate form of the kernel K (in the same notes the multigrid Monte Carlo method was also proposed, we consider here only the Langevin equation approach, which can be investigated analytically).

The choice of the Kernel K is very similar in spirit to preconditioning in the study of the solution of linear differential equations. It is evident that if the kernel $K(x,y)$ can be written as $K(x-y)$, the moltiplication by K is a convolution which can be done quite efficiently in a time proportional to $a^D \ln(1/a)$, or using the convolution theorem for Fourier transform and the Fast Fourier Transform or other fast methods for convolutions[16] (we will neglect the term $\ln(1/a)$ in the rest of these lectures, we note however that the appearence of this logarithmic factor is also common to the multigrid method).

A very natural choice of the kernel K is $1/(p^2+m^2)$ in Fourier space, where m is approximatively the physical mass (which we assume to be finite in the limit $a\to0$). In this way the noise term forces directly correlations on the physical scale.

Unfortunately this method is not very efficient in not too low dimensions; indeed it is evident that the kernel K may tame the kinetical term, it can do nothing against the fact that as soon as $D>2$, $\varphi^3>>\varphi$.

A simple analysis, which neglects the fact that μ is also divergent when a goes to zero, gives you the condition

$$\varepsilon < \min(1,\ a(D-2)/g). \tag{12}$$

This condition may be modified by a careful analysis, which can be done in pertubation theory and it is left to the interested reader, however the final conclusions are that in four dimensions

$$\varepsilon < a^2 / g^k, \tag{13}$$

(k being 1 or 2).

In other words, if the theory is superrenormalizable, the dynamics at short distances is trivial and it can be done nearly analytic. This method does not seem to work for renormalizable theories described by eq. (9).

It should be however noticed that this sad result (eq. 13) may be modified if we change the theory. Let us consider the non-linear sigma model in two dimensions, which is a renormalizable theory. The same analysis as before suggest that, if one takes care of the constraint on the fields, no serious problems should arise here and ε can be taken of O(1) (neglecting logarithms). The main difference between the two theories is the absence (in the sigma model) of an operator as φ^2 which is invariant under all the symmetry groups and has dimensions smaller than the action.

This result should be more carefully studied in pertubation theory, moreover it should be very interesting to understand if the same strategy can be used in 4 dimensional gauge theories, where we must pay a special care to the consequences of gauge invariance.

V Complex probabilities

Obviously probabilities are real. How could we have eqs. (1-2) satisfied when the probability distribution is complex (or equivalently when the action S in eq. (3) is complex)? The solution is in principle very simple: it may be possible that

$$\int d\mu(x) f(x) = \int d\rho(z,\overline{z}) f(z) \qquad (14)$$

where $x \in R$, $z \in C$, μ is non positive distribution on the real line, ρ is a positive measure on the complex plane and f is an analytic functions.

In other terms[17] we are deforming the integration path from the real line to the complex plane and it is possible that the final measure is positive definite.

The real troubles start when we want to find an explicite algorithms for implement eq. (2). An intriguing possibility is to naively use the Lagevin equation with complex action, complex variables and real noise[18,19]. There are proofs of the correctness of the methods, unfortunately such proofs are only formal; moreover it is well known that sometimes the complex Langevin method fails and it is not clear at all how to find an internal criterium to judge the correctness of the results[20].

A solution of this problem would be very interesting, not only for QCD: due to Fermi-statistics, almost any computation with many electrons has to deal with non definite integrals, which are very hard to extimate with numerical techniques. An efficient method for computing non positive definite integrals may be very useful in solid state physics and in chemistry. It should be also clear that the evolution in real time of a few body quantum system is a fascinating problem which at the present moment goes far beyond our possibilities.

References

1. G. Parisi in "Progress in Gauge Field Theory", edited by, G. 't Hooft, A. Jaffe, H. Lehmann, P. K. Mitter, I. M. Singer and R. Stora (Plenum Press), New York 1984.
2. B. Berg in "Progress in Gauge Field Theory", edited by, G. 't Hooft, A. Jaffe, H. Lehmann, P. K. Mitter, I. M. Singer and R. Stora (Plenum Press), New York 1984.
3. N. H. Christ and A. E. Terrano, Phys. Rev. Lett. , 56, 11 (1986)

4. M. Albanese, P. Bacilieri, M. Bernaschi, S. Cabasino, N. Cabibbo, F. Costantini, G. Fiorentini, F. Flore, L. Fonti, M. P. Lombardo, P. Giacomelli, P. Marchesini, E. Marinari, F. Marzano, P. Paolucci, G. Parisi, S. Petrarca, F. Rapuano, E. Remiddi, R. Rusack, G. Salina and R. Tripiccione, Phys. Lett. B, to be published.
5. M. Fukugita in "Lattice Gauge Theory Using Parallel Processors" edited by Li Xiaoyuan, Qiu Zhaoming and Ren Haicang.
6. R. Gupta in "Lattice Gauge Theory Using Parallel Processors" edited by Li Xiaoyuan, Qiu Zhaoming and Ren Haicang.
7. P. C. Hoemberg and D. Halperin, Rev. Mod. Phys. **49**, 435 (1977).
8. G. Parisi and Wu Yongshi, Scientia Sinica, **24**, 483 (1981).
9. G.Parisi "Statistical Field Theory" Benjamin, New York, (1988) and references therein.
10. F. Fucito, E. Marinari, G. Parisi and C. Rebbi, Nucl. Phys. **B180** FS2], 369 (1982).
11. G. Parisi, E. Marinari and C. Rebbi, Nucl. Phys. **B190** [FS3], 734 (1981).
12. F. Fucito and E. Marinari, Nucl. Phys. **B190** [FS3], 237 (1981)
13. G. Parisi Nucl. Phys. **B205** [FS5], 337 (1982), M. Falcioni, E. Marinari, M. Paciello, G. Parisi, F. Rapuano, B. Taglienti and Zhang Yi-cheng, Nucl. Phys. **B215** [FS7], 256 (1983).
14. C. G. Batrouni et. al Phys. Rev. **D32**, 2736 (1985).
15. P. Rossi, C. T. H. Davis and G. P. Lepage, San Diego preprint UCSD-PTH 87/08.
16. H. J. Nussbaumer "Fast Fourier Transform and Convolution Algorithms" Springer (1981).
17. J. Zinn-Justin (private comunication).
18. G. Parisi, Phys. Lett. **131B**, 393 (1983).
19. J. R. Klauder, Phys. Rev. **A29**, 2036 (1984).
20. For recent work on this subject see B. Soderberg, Lund preprint LU TP 86-27, Nucl. Phys. to be published.

THE ROLE OF LOCALITY IN STRING QUANTIZATION

Raymond Stora

LAPP, BP. 909, F-74019 Annecy-le-Vieux Cedex
and
CERN, TH-Division, CH-1211 Geneva 23

INTRODUCTION

Whereas free string quantization is by now part of the common knowledge[1], it still calls for some comments : it is a gauge theory, for the gauge group "Diff × Weyl" ; gauge fixing has then to be performed according to the Faddeev Popov procedure and the ensueing Slavnov symmetry can be found, given local gauge functions. It is customary to work in a Landau type conformal gauge and furthermore to eliminate the Weyl ghost and antighost as well as the multiplier field so that only a pair of diffeomorphism ghost and antighost are left over, besides the string field. These eliminations lead to a loss of nilpotency of the Slavnov symmetry which, in the existing versions only holds modulo the ghost equations of motion. The alternative use of the Hamiltonian formalism[2] not only spoils the world sheet geometry but prevents a systematic use of world sheet locality which is necessary for an unambiguous classification of the anomalies. These questions have been straightened out recently[3),4),5)], including the study of the localized Slavnov symmetry through the corresponding current algebra study[5),6)] and the supersymmetric extension[7]. It is the aim of these notes to review some of these recent constructions.

We shall mostly be concerned with the one orbit theory in the conformal gauge, i.e. assume no global zero mode. In this framework, we describe the Slavnov symmetry and the corresponding Slavnov identity (section II), the covariance under diffeomorphisms, and the corresponding Ward identity (section III), the localized Slavnov symmetry and the corresponding current algebra (section IV). Locality is used throughout and provides unambiguous answers, e.g. on the existence and form of the anomalies.

In general, however, the conformal gauge is not a good gauge. In the present framework, this is signalled by the presence of global zero modes which depend on the boundary conditions and have to be gauge fixed. As an example, the case where the world sheet is a compact Riemann surface of genus $g > 1$ is treated in some details (section V).

Some concluding remarks are gathered in section VI.

II. THE FREE BOSONIC STRING IN THE CONFORMAL GAUGE : THE SLAVNOV SYMMETRY

We start with a string with world sheet Σ mapped into R^D equipped with a metric $(\, , \,)$: the usual (σ,τ) variables will be collectively denoted $x = (x^1, x^2)$ and the string field by $\vec{X}(x) \in R^D$. $g^{\alpha\beta}(x)$ $\alpha,\beta = (1,2)$ is a metric on Σ. Unless otherwise stated we shall work in the euclidean framework, i.e. both $g^{\cdot\cdot}$ and $(\, , \,)$ are euclidean. The string action is

$$S_{inv}(X.g) = \frac{1}{2} \int_\Sigma d^2x \sqrt{g}\, g^{\alpha\beta}(x) \, (\partial_\alpha \vec{X}, \partial_\beta \vec{X})(x) \tag{1}$$

Since $S_{inv}(X,g)$ only depends on the conformal class of the metric g (i.e. is invariant under the Weyl scaling $g \to e^{\varphi} g$) it is convenient to use the following parametrization : z, $\bar{z}$ denote some complex analytic coordinates associated with a background (conformal class of) metric $\overset{\circ}{g}$ and the conformal class of g will be parametrized by the Beltrami differential μ ($|\mu| < 1$) such that

$$ds^2 = \frac{1}{2} g_{\alpha\beta} \, dx^\alpha \, dx^\beta \propto |dz + \mu d\bar{z}|^2 \, , \qquad |\mu| < 1 \tag{2}$$

S_{inv} can then be rewritten as

$$S_{inv}(\vec{X}, \mu, \bar{\mu}) = \int_\Sigma \frac{dz \wedge d\bar{z}}{2i} \, \frac{1}{1-\mu\bar{\mu}} \, (\partial_z - \bar{\mu}\partial_{\bar{z}}\vec{X}, \partial_{\bar{z}} - \mu\partial_z \vec{X}) \tag{3}$$

The geometrical object associated with μ is

$$\hat{\mu} = \mu \, d\bar{z} \otimes \frac{\partial}{\partial z} \tag{4}$$

It is called a Beltrami differential. It is a type $(0,1)$ one form with value a type $(1,0)$ vector field. It is convenient, in order to check the homogeneity of formulae under holomorphic changes of variables (e.g. change of chart on Σ) to think of μ as bearing a lower index $\bar{z}$ and an upper index z ($\mu \sim \mu_{\bar{z}}^{z}$). μ is assumed to be C^∞.

It is easy to work out the transformation of both $\vec{X}$ and μ under an infinitesimal diffeomorphism represented by a vector field

$$\gamma = \gamma^z \partial_z + \gamma^{\bar{z}} \partial_{\bar{z}} = (\gamma \cdot \partial) \tag{5}$$

$$\begin{aligned} \delta_\gamma X &= (\gamma \cdot \partial)\, X \\ \delta_\gamma \mu &= \partial_{\bar{z}} c^z + c^z \partial_z \mu - \mu \partial_z c^z \end{aligned} \tag{6}$$

where

$$c^z = \gamma^z + \mu \gamma^{\bar{z}} \, . \tag{7}$$

All formulae have to be completed through the substitution $\mu \leftrightarrow \bar{\mu}$ $z \leftrightarrow \bar{z}$ $i \leftrightarrow -i$, denoted by "c.c.". It is easy to check that

$$[\delta_\gamma, \delta_{\gamma'}] = \delta_{[\gamma,\gamma']} \tag{8}$$

where $[\gamma,\gamma']$ is the Lie bracket.

The variable c^z (eq. 7) has been discovered by C. Becchi[4] in connection with the question of holomorphic factorization of the Green functional in the conformal gauge. We shall return to this point later.

Turning γ into a Faddeev Popov ghost (6) is transformed into

$$\begin{aligned} s\, \vec{X} &= (\gamma \cdot \partial)\, \vec{X} \\ \hat{s}\, \mu &= \partial_{\bar{z}} c^z + c^z \partial_z \mu - \mu \partial_z c^z \end{aligned} \tag{9}$$

and eq. (8) becomes

$$s\, c^z = c^z\, \partial_z\, c^z \tag{10}$$

Note that in eq. (10), there is <u>no</u> summation over $z, \bar{z}$.

These formulae are easily obtained in terms of complex analytic coordinates Z pertaining to the complex structure parametrized by μ : one has

$$d\, Z = \lambda (dz + \mu d\bar{z}) \tag{11}$$

where the non local integrating factor λ fulfills (as a consequence of $d^2 = 0$):

$$- \partial_{\bar{z}} \lambda + \mu \partial_z \lambda + \lambda \partial_z \mu = 0 \tag{12}$$

The action of diffeomorphisms is easily worked out in terms of the Z variables, after an easy elimination of λ, using eq. (12). Note that in terms of c^z, all formulae are local, as a consequence of the elimination of the only non local item which occurs in the geometry, namely the integrating factor λ.

Of course we have

$$s^2 = 0 \tag{13}$$

since all we have done is to represent the Lie algebra of diffeomorphisms.

Gauge fixing is carried out as usual. The conformal gauge is characterized by the gauge function

$$\mu - \mathring{\mu} \tag{14}$$

where $\mathring{\mu}$ is a prescribed Beltrami differential. If one stays in a Landau type gauge, the multiplier field can be eliminated and μ must be replaced by $\mathring{\mu}$ everywhere. From now on we shall suppress the upper script $\circ$ on $\mathring{\mu}$, remembering that μ is now a classical field.

The gauge fixed action is then

$$\begin{aligned} &S_{gf}(X, \mu, \bar{\mu}, C^z, C^{\bar{z}}, b_{zz}, b_{\bar{z}\bar{z}}) \\ &= S_{inv}(X, \mu, \bar{\mu}) + S_{gh}(\mu, \bar{\mu}, C^z, C^{\bar{z}}, b_{zz}, b_{\bar{z}\bar{z}}) \end{aligned} \tag{15}$$

where b_{zz} is the antighost, bearing the indices of a quadratic differential to insure the homogeneity of the formulae.
S_{inv} is given by eq. (3), and

$$S_{gh} = \int \frac{dz \wedge d\bar{z}}{2i} \; \bar{b}_{zz} \; s\mu + \text{"c.c"} \tag{16}$$

with $s\mu$ as in eq.(9).

Introducing sources coupled to the s-variation:

$$S_{source} = \int \frac{dz \wedge d\bar{z}}{2i} \left[(\mathfrak{X}, sX) + \Gamma \, s \, C^z + \text{c.c.} \right] \tag{17}$$

the Slavnov symmetry eqs (9),(10) extended by

$$\begin{aligned} &s \, b_{zz} = 0 \quad \& \; (\text{"c.c."}) \\ &s \, \mathfrak{X} = s \, \Gamma = 0 \end{aligned} \tag{18}$$

is expressed through the Slavnov identity

$$\int \frac{dz \; d\bar{z}}{2i} \left[\left(\frac{\delta S}{\delta X} , \frac{\delta S}{\delta \mathfrak{X}} \right) + \frac{\delta S}{\delta \Gamma} \frac{\delta S}{\delta C^z} + \frac{\delta S}{\delta b_{zz}} \frac{\delta S}{\delta \mu} + \text{"c.c."} \, .. \right] (z, \bar{z}) \; = \; 0 \tag{19}$$

For

$$S = S_{tot} = S_{inv} + S_{gf} + S_{source} \tag{20}$$

The Slavnov identity is a current algebra Ward identity for the energy momentum tensor components Θ_{zz}, $\Theta_{\bar{z}\bar{z}}$ whose correlation functions are generated by functional differenciation with respect to μ, $\bar{\mu}$ respectively.

Note, in order to compare with the usual formulae, that eq.(19), derived as a consequence of the nilpotency property eq.(11) can also be interpreted as the invariance of S under the transformation $\tilde{s}$:

$$\begin{aligned} &\tilde{s} = s \quad \text{on} \quad X, \mathfrak{X}, \Gamma \\ &\tilde{s}\,\mu = 0 \qquad \tilde{s}\, b_{zz} = \frac{\delta S}{\delta \mu} = \Theta_{zz} \end{aligned} \tag{21}$$

However $\tilde{s}$ is only nilpotent modulo the ghost equation of motion.

Passing to the quantum level amounts to replacing the classical action S by the vertex functional Γ which also takes into account one loop diagrams (and, here, only those) and requiring eq.(19) with S replaced by Γ.

The "holomorphy" anomaly is then found on the right hand side with a coefficient proportional to D-26 and the local form

$$\mathfrak{a} \sim \int \frac{dz \wedge d\bar{z}}{2i}\, \mu\, \partial_z^3 C^z + \text{"c.c."} \tag{22}$$

If several charts are needed to cover Σ eq.(22) is no good. One should then use e.g. eq.(31) of ref. 3) and convert it into the present notations.

III. THE DIFFEOMORPHISM WARD IDENTITY[4),5),12),13)]

Besides the Slavnov symmetry associated with Diffeomorphisms and their gauge fixing, the classical action admits another invariance under the action of diffeomorphisms, which commutes with the Slavnov symmetry. Actually the action of diffeomorphisms is uniquely defined on $\vec{X}$ and μ, but not on C^z, b_{zz} and the other fields. One set of transformations was proposed by C. Becchi[4)], another one by L. Baulieu and M. Bellon[5)]. They are equivalent modulo the equations of motion. We shall limit ourselves to the latter.

The symmetry in question reads:

$$\begin{aligned} \partial_\Lambda X &= (\lambda.\partial)\, X \\ \delta_\Lambda \mu &= \partial_{\bar{z}}\Lambda^z + \Lambda^z \partial_z \mu \quad \mu \partial_z \Lambda^z \\ \delta_\Lambda C^z &= \Lambda^z \partial_z C^z - C^z \partial_z \Lambda^z \equiv [\Lambda, C] \\ \delta_\Lambda b_{zz} &= (\Lambda^z \partial_z + 2\partial_z \Lambda^z)\, b_{zz} \end{aligned} \tag{23}$$

and "c.c.".

The commutation properties are

$$[\delta_\Lambda, \delta_{\Lambda'}] = \delta_{[\Lambda,\Lambda']} \tag{24}$$

with

$$[\Lambda,\Lambda'] = \Lambda^z \partial_z \Lambda'^z - \Lambda'^z \partial_z \Lambda^z \tag{25}$$

We have used the obvious notation

$$\Lambda^z = \lambda^z + \mu\lambda^{\bar z} \Leftrightarrow \lambda^z = \frac{\Lambda^z - \mu\Lambda^{\bar z}}{1 - \mu\bar\mu} \tag{26}$$

The last of eq.(23) is obtained by duality, demanding that S_{gh} be invariant. One may similarly infer the formula for $\delta_\Lambda \Gamma$. $\delta_\Lambda \mathfrak{X}$ is not completely computable by duality. It is only so modulo the ghost equation of motion, which nevertheless allows to write a Ward identity e.g. for the connected Green's functional (the Legendre transform of the vertex functional). Details will appear elsewhere. Either the Ward identity or the Slavnov identity can be used to prove the holomorphic factorization of the connected Green functional[4),5)]:

$$Z^c(J_x, J_c, J_b, \mu, \text{"c.c."}) = Z^c(J_x, J_c, J_b, \mu) + Z^c(J_x\ J_{\bar c}\ J_{\bar b}, \bar\mu) \tag{27}$$

where J_{field} is the source argument corresponding to the field in question. It is in the course of requiring the holomorphic factorization property that C. Becchi discovered that the proper definition of the correlation functions of the Θ_{zz}, $\Theta_{\bar z\bar z}$ components of the energy momentum tensor goes through differenciation with respect to μ, $\bar\mu$ and that the proper ghost fields are C^z, $C^{\bar z}$, rather than γ^z, $\gamma^{\bar z}$.
$Z^c(J_x, J_c, J_b, \mu)$ may be derived from the action eq.(20) by putting $C^{\bar z} = \bar b_{\bar z\bar z} = \bar\Gamma = \bar\mu = 0$ (i.e. forgetting the "(c.c.)" world) which reduces to the FMS action[1)], including terms coupled to the stress energy tensor and to the Slavnov variations, and the S operation reduces to s_{FMS}, except for the interchange $s\mu$, $sb = 0$ into $s_{FMS}\mu = 0$ $s_{FMS}b = \delta S/\delta\mu$. s_{FMS} is nilpotent. One may similarly define a nilpotent $\bar s_{FMS}$. $s_{FMS} + \bar s_{FMS}$ leaves invariant $(S_{inv} + S_{gh})|_{\mu=\bar\mu=0}$ but $(s_{FMS} + \bar s_{FMS})^2$ only vanishes modulo the ghost equations of motion.

IV. LOCALIZED SLAVNOV SYMMETRY AND THE SLAVNOV CURRENT ALGEBRA

The Slavnov symmetry eqs.(9),(10) can be thought of as involving an "infinitesimal space time independent anticommuting parameter". This point of view which is conveniently replaced by the present one where s is a graded derivation, calls for the corresponding generalization: can one construct an extension of s into s_{loc}, involving, besides the former variables, a space time dependent even field $\lambda(x)$ (the localization of the ghost of the "small anticommuting parameter"), together with an odd gauge field

$$\alpha = \alpha_\mu(x)\, dx^\mu \,, \qquad \alpha_\mu(x) \text{ odd} \tag{28}$$

serving as a source of the Noether current corresponding to the Slavnov symmetry, in such a way that $s^2_{loc} = 0$, and that the action S can be extended into an action S_{loc}, depending on α, but not on λ, left invariant by s_{loc}. Both s_{loc} and S_{loc} are assumed to reduce to the "global" Slavnov symmetry and the former S for $\lambda = 1$, $\alpha = 0$.

Besides being a natural question to ask, this sheds a new light on the following little puzzle alluded to in the Introduction : "the anomaly" ($\propto$ D-26) is found in the lack of nilpotency, at the quantum level, of the Q operator obtained by integrating the time component of the Slavnov Noether current[2]. This result relies on a renormalization procedure needed to pass from the classical to the quantum situation, which is performed by using Wick ordering in a way which is not known to be unique and whose ambiguity is not known by virtue of admitted principles.

Formally Q^2 appears as an equal time commutator term for the Slavnov current algebra Ward identity. A study of that Ward identity, using the locality principle which leads to a parametrization of the ambiguities, should therefore show the result that its anomalies are rigidly linked to the "global" holomorphy anomaly, thus explaining unambiguously why the nilpotency anomaly is also proportional to D-26.

One solution[5),6)], which exhibits this phenomenon, has been constructed. This does not completely settle the question, but shows that the Slavnov current algebra cannot have more anomalies. The construction follows the lines of Ref. 8), which deals with the Yang-Mills case. The extension thus obtained extends the holomorphic factorization property to correlation functions involving the Slavnov Noether current.

"Minimal coupling" is used throughout, i.e. the following replacement both in S eqs.(3),(16),(17) and in s eqs.(9),(10):[6)]

$$\mu \to \hat{\mu} = \mu\,(1 + \beta_z \mathfrak{c}^z) - \beta_{\bar{z}} \mathfrak{c}^z \tag{29}$$

$$s \to s_{loc} \qquad S \to S_{loc}$$

together with the adjunction of

$$s_{loc}\, \beta_z = -\, \partial_z \lambda/\lambda \qquad s_{loc}\, \lambda = 0 \tag{30}$$

The correlation functions of the components $J_z\ J_{\bar{z}}$ of the Noether current are obtained by functional differenciation with respect to β_z, $\beta_{\bar{z}}$. (Here, $\mathfrak{c}^z = \lambda C^z \quad \alpha_z = \lambda\beta_z$). One can easily check that $\left.\frac{\delta S}{\delta \beta_z}\right|_{\beta=0}$ is the known Noether current.

The algebra thus constructed is isomorphic with a trivial extension of the Slavnov algebra. The anomaly of the corresponding Noether current Ward identity is thus

$$\mathfrak{a}_{loc} \sim \int \frac{dz\, d\bar{z}}{2i}\, \partial^z\, \mathfrak{c}^z\, \hat{\mu} + \text{"c.c."} \tag{31}$$

and its coefficient is the same as that of $\mathfrak{a}$ (eq.(22)) to which it reduces for $\lambda = 1$, $\beta = 0$.

V. GLOBAL ZERO MODES

So far we have constructed an action which is well defined in the absence of global zero modes, which is the case for field theory $(\Sigma = R^2)$. For other types of surfaces, occuring in string theory, the global zero modes have to be discussed in each particular case, and gauge fixed. We propose to do this, as an exercise, for Σ compact. The constant zero modes for $\vec{X}$ lead to the extension of the s operation according to

$$s\,\vec{X} \;=\; (\gamma.\partial)\,\vec{X} + \vec{c} \qquad s\,\vec{c} = 0 \tag{32}$$

$\vec{c}$ independent of x

A convenient gauge function is $\int \vec{X}\, \rho_{z\bar{z}}\, dz\, d\bar{z}$ where ρ is the density which describes $\overset{\circ}{g}$ in terms of the $z, \bar{z}$ variables. The corresponding Faddeev Popov term is

$$\vec{\bar{c}} \int [(\gamma.\partial)\vec{X} + \vec{c}]\rho_{z\bar{z}}\, dz\, d\bar{z} \tag{33}$$

and the integration over $\vec{\bar{c}}$, $\vec{c}$ yields a determinant factor $(\int \rho\, dz\, d\bar{z})^{D/2}$, whereas the Laplacian has to be restricted to $\vec{X}$ orthogonal to constants with respect to the metric ρ.

More serious are the zero modes occuring in the ghost sector:

$$s\,\mu \equiv \partial_z C^z + C^z \partial_z \mu - \mu \partial_z C^z = 0 \tag{34}$$

describes the conformal Killing vectors for the complex structure defined by μ, as one can easily see by passing to the Z variables (cf. eq.(11)). As one knows the vector space of solutions has complex dimension 3 for genus $g = 0$ $(\Sigma \sim s^2)$, 1 for $g = 1$ $(\Sigma \sim T^2)$, 0 for $g > 1$. We shall henceforth assume $g > 1$.

The antighost zero modes fulfill

$$(\partial_{\bar{z}} - \mu\partial_z - 2\partial_z\mu)\, b_{zz} = 0 \tag{35}$$

as one can easily see by integration by parts in eq.(16),(9). In terms of the Z variables (cf. eq.(11)), one sees that b_{zz} is a holomorphic quadratic differential for the complex structure described by μ. One knows there is none for $g = 0$, 1 for $g = 1$ 3g-3 otherwise. Let ϕ^i, $i = 1,\ldots,3g-3$ be a basis of solutions, functionals of μ.

Applying s (eqs.(9),(10)) to eq.(35), with b_{zz} replaced by ϕ^i, one easily finds that $s\phi^i$ is of the form

$$s\phi^i = (2\partial_z C^z + C^z\partial_z)\phi^i + A^i{}_j\, \phi^j \tag{36}$$

where $A^i{}_j$ is a matrix independent of $z, \bar{z}$, linear in C^z, and otherwise a functional of μ fulfilling

$$s\, A = A^2 \tag{37}$$

because $s^2 = 0$.

One then finds a new nilpotent s operation, including the invariance of the action under translation of b along its zero modes : it is sufficient to keep $s\vec{X}$, $s\mu$, sC^z and take

$$s\, \bar{b}_{zz} = c_i\phi^i + b_i s\phi^i \tag{38}$$

with

$$s\, b_i = -c_i \qquad s\, c_i = 0 \tag{39}$$

$$c_i \text{ even}, \quad b_i \text{ odd}$$

instead of $s\, b_{zz} = 0$.

One still has $s^2 = 0$. That s, extended by eq.(38) leaves S_{gh} invariant, is due to the identity

$$\int s\phi^i\, s\mu = 0 \tag{40}$$

with $s\phi^i$ given by eq.(36).

In order to perform gauge fixing, let us choose a set $\{h_i\}$, smooth in μ, dual to the system $\{\phi^j\}$:

$$\int \phi^j h_i \, dz \, d\bar{z} \equiv (\phi^j, h_i) = \delta^j{}_i \tag{41}$$

and choose as gauge functions

$$(b, h_i) , \quad (b, sh_i) \tag{42}$$

of the fermionic and bosonic type, respectively. In the corresponding Landau gauge, the corresponding Faddeev Popov Lagrangian involving the bosonic antighost $\bar{c}_i$ and the bosonic antighost $\bar{b}_i$, reads:

$$\bar{c}_i (c_i + b_j (s\phi^j, h_i)) + \bar{b}_i (c_j (\phi^j sh_i) + b_j (s\phi^j, sh_i)) \tag{43}$$

with the constraints

$$(b, h_i) = (b, sh_i) = 0 \tag{44}$$

The corresponding Faddeev Popov superdeterminant is

$$^{s}\det \begin{Vmatrix} \delta^j_i & (s\phi^j, h_i) \\ -(s\phi^j, h_i) & (s\phi^j, sh_i) \end{Vmatrix} \tag{45}$$

VI. CONCLUDING REMARKS

We have reviewed recent reinvestigations of string first quantization, taking as an example the free bosonic string. Emphasis was put on locality, together with the construction of nilpotent Slavnov symmetries. These are two principles which lead to a class of "DET'"s resulting from gaussian integrations. The test for the correctness and usefulness of these principles is of course the possibility to construct sufficiently many observables. It is however encouraging to see some puzzles resolved, such as for instance the difficulty to construct an off shell nilpotent Slavnov symmetry[1),2),10)], as well as the tight relationship between the "nilpotency" anomaly[2)] and the trace or holomorphy anomaly.

Of course, the route followed here is a bit strange : the gauge function $\mu-\overset{\circ}{\mu}$ is only good if there is only one orbit under diffeomorphisms[11)]. Otherwise, one should choose an appropriate gauge function, which, for topological reasons, cannot depend on μ alone, but also on $\bar{\mu}$ thus spoiling holomorphic factorization from the start. One thus completely looses contact with holomorphic geometry. One possibility is to choose harmonic type gauges[11]. The procedure proposed here, i.e. to gauge fix the antighost zero modes signalling the inadequacy of the gauge function may not be equivalent. On the other hand, it respects the contact with holomorphic geometry.

ACKNOWLEDGEMENTS

I wish to thank C. Becchi, L. Baulieu, M. Bellon, J.C. Wallet for numerous discussions as well as communicating their results prior to publication. Thanks are also due to L. Alvarez-Gaumé, O. Alvarez, J.M. Bismut, L. Bonora, D. Friedan, C. Gomez, C. Itzykson, C. Reina, P. Windey for keeping me informed on their work.

NOTE ADDED

In section III we have for simplicity described the Diffeomorphism Ward identity proposed by L. Baulieu, M. Bellon. This seems to be a particularity of the "b-c" system, whereas the identity proposed by C. Becchi holds for all conformal systems. It reads:

$$\delta_\lambda \mu = \delta_\Lambda \mu \qquad \text{as in eq.(23)}$$

$$\delta_\lambda \varphi_q = (\lambda.\partial)\varphi_q + q(\partial_z \lambda^z + \mu \partial_z \lambda^{\bar{z}})\varphi_q, \quad \text{and} \quad \text{c.c.}$$

for φ_q of type $(q,0)$ ($q = -1$ for $\mathcal{C}^z$, $q = +2$ for b_{zz}).
It expresses the invariance of the theory under a diffeomorphic change of the background complex structure parametrized by the complex analytic coordinates z.

REFERENCES

1) M.B. Green, J.H. Schwarz, E. Witten, Superstring Theory, Vol. I, Cambridge University Press (1987) and references therein.
(FMS) D. Friedan, E. Martinec, S. Shenker, Nucl. Phys. B271, 93 (1986).
2) M. Kato, K. Ogawa, Nucl. Phys. B212, 443 (1983).
S. Hwang, Phys. Rev. D28, 14 (1983) and Goteborg Dissertation (1986).
3) L. Baulieu, C. Becchi, R. Stora, Phys. Lett. 180B, 55 (1986) and references therein.
4) C. Becchi, "On the Covariant Quantization of the Free String : The Conformal Structure", Genova preprint 1987.

5) L. Baulieu, M. Bellon, "Beltrami Parametrization and String Theory", LPTHE 87-39, Paris, June 1987.

6) J.P. Ader, J.C. Wallet, Phys. Lett. 192B, 103 (1987);
J.P. Ader, J.C. Wallet, in preparation;
R. Marnelius, Phys. Lett. 186B, 351 (1987).

7) L. Baulieu, M. Bellon, R. Grimm, "Beltrami Parametrization for Superstrings", LPTHE 87-43, Paris, July 1987, and L. Baulieu's lecture, these proceedings.

8) L. Baulieu, B. Grossmann, R. Stora, Phys. Lett. 180B, 95 (1986).

9) C. Becchi, in preparation.

10) E. Cohen, C. Gomez, P. Mansfield, Phys. Lett. 180B, 55 (1986);
P. Mansfield, Nucl. Phys. B283, 551 (1987);
P. Mansfield, University of Oxford preprint 46/87.

11) L. Bonora, M. Martellini, C.M. Vialett, CERN Th.4796/87.

12) A.A. Belavin, A.M. Polyakov, A.B. Zamolodchikov, Nucl. Phys. B241, 333 (1984).

13) T. Eguchi, H. Ooguri, Nucl. Phys. B282, 308 (1987).

ON THE BRST STRUCTURE OF THE CLOSED STRING AND SUPERSTRING THEORY

Laurent Baulieu
Laboratoire de Physique Théorique et Hautes Energies
Université Pierre et Marie Curie, Paris.

The Beltrami parametrization for Riemann surfaces has been recognized recently as a usefull tool for string theory quantization [1]. In this lecture I will report further results concerning this parametrization which have been obtained in collaboration with Marc Bellon. I will point out in particular the possibility of gauge fixing in a BRST invariant way the global zero modes occuring in string multiloop amplitudes via the ghost propagator [2] and a supersymmetric generalization of the Beltrami parametrization which is relevant for superstrings [3].

Shortly speaking, the Beltrami parametrization consists into expressing the line element of the string worldsheet under the following factorized form :

$$ds^2 = g_{\alpha\beta}\, dx^\alpha dx^\beta = e^\phi\, [dx^+ + \mu^+ dx^-]\,[dx^- + \mu^- dx^+] \qquad (1\text{-a})$$

This amounts to the following expression of the 2-dimensional metrics $g_{\alpha\beta}$:

$$g_{\alpha\beta} = e^\phi \begin{pmatrix} 1 & \mu^- \\ \mu^+ & 1 \end{pmatrix} \begin{pmatrix} 0 & 1 \\ 1 & 0 \end{pmatrix} \begin{pmatrix} 1 & \mu^+ \\ \mu^- & 1 \end{pmatrix} \qquad (1\text{-b})$$

In complex coordinate notation one would denote $\mu_{\bar z}{}^z - \mu^+$, $\mu_z{}^{\bar z} - \mu^-$, $dz - dx^+$ and $d\bar z - dx^-$. The object $\mu_{\bar z}{}^z d\bar z\, \partial/\partial z$ is called Beltrami differential.

With the Beltrami parametrization of $g_{\alpha\beta}$ the Polyakov string action reads :

$$\mathcal{I}_{cl} = \int_\Sigma d^2z\, (1-\mu^+\mu^-)^{-1} (\partial_+ - \mu^-\partial_-)\vec{X}\cdot(\partial_- - \mu^+\partial_+)\vec{X} \qquad (2)$$

and both non vanishing components of the classical energy momentum tensor are :

$$T^+_{cl} = \delta I/\delta\mu^+ = (\, (\partial_+ - \mu^-\partial_-)\vec{X})/(1-\mu^+\mu^-)\,)^2$$
$$T^-_{cl} = \delta I/\delta\mu^- = (\, (\partial_- - \mu^+\partial_+)\vec{X})/(1-\mu^+\mu^-)\,)^2 \quad . \qquad (3)$$

For a conformally flat metric on the worldsheet (i.e. $\mu = 0$) a factorized form of the Polyakov action and of the energy momentum tensor is well known [5]. Eqs. (2) and (3) show that the use of the variables μ^+, μ^- and ϕ for parametrizing $g_{\alpha\beta}$ proves the persistance of a factorization even for arbitrary metrics $g_{\alpha\beta}$. It is therefore suggested that one should rotate each (2.D) vector field (i.e. each object usually denoted by (1,0)) by the matrix $\begin{pmatrix} 1 & \mu^+ \\ \mu^- & 1 \end{pmatrix}$, with an obvious generalization for each tensor (p,p'). By applying this rotation to the diffeomorphism ghost vector field (ξ^+, ξ^-) one thereby introduces the following redefined ghost field Ξ:

$$\begin{pmatrix} \Xi^+ \\ \Xi^- \end{pmatrix} = \begin{pmatrix} 1 & \mu^+ \\ \mu^- & 1 \end{pmatrix} \begin{pmatrix} \xi^+ \\ \xi^- \end{pmatrix} \quad (4)$$

The use of the ghost variables Ξ^+ and Ξ^- rather than ξ^+ and ξ^- determines a factorized expression for the action of the Weyl x diffeomorphism BRST operator s [1]:

$$\begin{aligned} & sX = \xi^\alpha \partial_\alpha X, \\ & s\mu^+ = \partial_- \Xi^+ + [\Xi^+, \mu^+]_+, \qquad s\mu^- = \partial_+ \Xi^- + [\Xi^-, \mu^-]_-, \\ & s\Xi^+ = \tfrac{1}{2}[\Xi^+, \Xi^+]_+, \qquad s\Xi^- = \tfrac{1}{2}[\Xi^-, \Xi^-]_- . \end{aligned} \quad (5)$$

We have defined the graded brackets $[\ ,\ ]_\pm$ by $[\Xi^\pm, \mu^\pm]_\pm \equiv \Xi^\pm \partial_\pm \mu^\pm - \mu^\pm \partial_\pm \Xi^\pm$ and $\frac{1}{2}[\Xi^\pm, \Xi^\pm]_\pm = \Xi^\pm \partial_\pm \Xi^\pm$. Following the general rule, the grading of each object is defined as the sum of its ghost number and form degree in dx^+ and dx^-. For instance, μ is an even object while Ξ, s, d and $dx^\pm$ are odd. The geometrical origin of the separation between the variables μ^+, Ξ^+ and μ^-, Ξ^- which is seen in eq. (5) is simply the invariance under diffeomorphisms of the null directions on the worldsheet which are independently determined by μ^+ and μ^-. The Beltrami differential components μ^+ and μ^- are Weyl invariant. This is not the case for the scalar part of the metric $\sqrt{g} - e^{\phi}(1-\mu^+\mu^-)$. The BRST transform of ϕ involves the Weyl ghost Ω :

$$\begin{aligned} s\phi &= \Omega + \partial_\alpha \xi^\alpha + \xi^\alpha \partial_\alpha \phi \\ s\Omega &= \xi^\alpha \partial_\alpha \Omega . \end{aligned} \quad (6)$$

One can easily verify that $s^2 = 0$ on all fields, a property which is guaranteed by the equivalence between eqs. (5,6) and the usual expression for the action of the BRST operator on $g_{\alpha\beta}$ and ξ^α, $sg_{\alpha\beta} = L_\xi g_{\alpha\beta} + \Omega\, g_{\alpha\beta}$ and $s\,\xi^\alpha = \xi^\beta \partial_\beta \xi^\alpha$.

A consistent gauge-fixing of the string action corresponds to the addition of an s-exact and thus BRST invariant term to the action. For a worldsheet of genus $g \leq 1$ one may chooses μ and ϕ as gauge functions and gauge fix the 3 components μ^+, μ^-, ϕ of $g^{\alpha\beta}$ equal to given functions, say $\mu^+_0(x^+,x^-)$, $\mu^-_0(x^+,x^-)$ and $\phi_0(x^+,x^-)$. For $g > 1$ such gauge conditions are too stringent. In this case, we shall see shortly that they can improved for taking into account the existence of conformal classes of metrics which are not equivalent modulo diffeomorphisms. This will necessitate the introduction of new constant ghost degrees of freedom [2]. If one considers only the cases $g = 0,1$ one simply needs to introduce antighosts $\bar{\Xi}^\pm$ and $\bar{\Omega}$ and Stückelberg type auxiliary fields $b^\pm$ and b with $s\bar{\Xi}^\pm - b^\pm$, $s\bar{\Omega} - b$, $sb^\pm - 0$, $sb - 0$. This permits the construction of ghost dependent gauge fixed and BRST invariant actions with total ghost number zero. By choosing μ and ϕ as gauge functions, one defines the BRST invariant action $\mathcal{I}$ as [1] :

$$\mathcal{I} = \mathcal{I}_{cl} + \mathcal{I}_{GF},$$

$$\begin{aligned}\mathcal{I}_{GF} &= s\{\int_\Sigma d^2z\ (\bar{\Xi}^+(\mu^+-\mu_0^+)+\bar{\Xi}^-(\mu^--\mu_0^-)+\bar{\Omega}(\phi-\phi_0))\} \\ &= \int_\Sigma d^2z\ (b^+(\mu^+-\mu_0^+)+b^-(\mu^--\mu_0^-)-\bar{\Omega}(\Omega+\partial_\alpha\xi^\alpha+\xi^\alpha\partial_\alpha\phi) \\ &\quad -\bar{\Xi}^+(\partial_-\Xi^+ +[\Xi^+,\mu^+]_+)-\bar{\Xi}^-(\partial_+\Xi^- +[\Xi^-,\mu^-]_-)\}.\end{aligned} \tag{7}$$

The variations with respect to $b^\pm$, b, $\bar{\Omega}$, Ω yield the wanted gauge conditions :

$$\begin{array}{lll} \mu^+=\mu_0^+ & \mu^-=\mu_0 & \phi=\phi_0 \\ \bar{\Omega}=0 & \Omega=\partial_\alpha\xi^\alpha+\xi^\alpha\partial_\alpha c & \end{array} \tag{8}$$

Inserting these equations of motion into $\mathcal{I}$, one gets :

$$\mathcal{I}\sim\mathcal{I}_0[\mu_0,\vec{X},\Xi,\bar{\Xi}]=\int_\Sigma d^2z\ \{(1-\mu_0^+\mu_0^-)^{-1}(\partial_+-\mu_0^-\partial_-)\vec{X}\cdot(\partial_--\mu_0^+\partial_+)\vec{X}$$

$$-\bar{\Xi}^+(\partial_-\Xi^+ +[\Xi^+,\mu_0^+]_+)-\bar{\Xi}^-(\partial_+\Xi^- +[\Xi^-,\mu_0^-]_-) \tag{9}$$

This action is the one studied in ref [1]. $\mathcal{I}_0$ is invariant under the action of the following nilpotent BRST operator s_0 where μ_0 is now a background :

$$s_0\vec{X}=\xi^\alpha\partial_\alpha\vec{X},$$

$$\begin{array}{ll} s_0\mu_0^+=\partial_-\Xi^+ +[\Xi^+,\mu_0^+]_+, & s_0\mu_0^-=\partial_+\Xi^- +[\Xi^-,\mu_0^-]_-, \\ s_0\Xi^+=\frac{1}{2}[\Xi^+,\Xi^+]_+, & s_0\Xi^-=\frac{1}{2}[\Xi^-,\Xi^-]_-, \\ s_0\bar{\Xi}^+=0, & s_0\bar{\Xi}^-=0. \end{array} \tag{10}$$

$\mathcal{I}_0[\mu_0,\vec{X},\Xi,\bar{\Xi}]$ is also invariant under a background diffeomorphism symmetry which is governed by a background ghost vector field Λ :

$$\begin{array}{ll} \sigma\vec{X}=\lambda^\alpha\partial_\alpha\vec{X}, & \Lambda^\pm\equiv\lambda^\pm+\mu_0^\pm\lambda \\ \sigma\mu_0^+=\partial_-\Lambda^+ +[\Lambda^+,\mu_0^+]_+, & \sigma\mu_0^-=\partial_+\Lambda^- +[\Lambda^-,\mu_0^-]_-, \\ \sigma\Lambda^+=1/2[\Lambda^+,\Lambda^+]_+, & \sigma\Lambda^-=1/2[\Lambda^-,\Lambda^-]_- \\ \sigma\Xi^+=\Lambda^+\partial_+\Xi^+ +\Xi^+\partial_+\Lambda^+ & \sigma\Xi^-=\Lambda^-\partial_-\Xi^- +\Xi^-\partial_-\Lambda^- \\ \sigma\bar{\Xi}^+=\Lambda^+\partial_+\bar{\Xi}^+-2\bar{\Xi}^+\partial_+\Lambda^+, & \sigma\bar{\Xi}^-=\Lambda^-\partial_-\bar{\Xi}^{+-}-2\bar{\Xi}^-\partial_-\Lambda^-. \end{array} \tag{11}$$

One has $s_0^2=s_0\sigma+\sigma s_0=\sigma^2=0$ and $s_0\mathcal{I}_0=\sigma\mathcal{I}_0=0$. The transformation law of $\bar{\Xi}$ under σ identifies the diffeomorphism antighost $\bar{\Xi}^\pm$ as a differential of rank 2, which should be in fact denoted as $\bar{\Xi}_{\pm\pm}$. The Ward identities associated to the s_0 or σ invariances can be used as a criterium for defining the theory generated by the action $\mathcal{I}_0$ as a "conformal theory" [1,4,5].

The advantages of considering general backgrounds $\mu_0(x^+, x^-)$ while preserving factorized equations, are twofold. Firstly this permits one to go further and consider worldsheets with $g > 1$ by a rather straighforward generalization. Secondly one can use μ_0^+ and μ_0^- as the sources of the energy momentum $T^\pm$. One has :

$$T^+ = \delta I/\delta\mu^+_0 = ((\partial_+ - \mu^-_0\partial_-)X)/(1-\mu^+_0\mu^-_0))^2 - [\Xi^+,\bar{\Xi}^+]_+$$

$$T^- = \delta I/\delta\mu^-_0 = ((\partial_- - \mu^+_0\partial_+)X)/(1-\mu^+_0\mu^-_0))^2 - [\Xi^-,\bar{\Xi}^-]_- \qquad (12)$$

where $[\Xi^\pm,\bar{\Xi}^\pm] = \Xi^\pm\partial_\pm\bar{\Xi}^\pm - 2\bar{\Xi}^\pm\partial_\pm\Xi^\pm$. By using the known BRST technics for the S_0 and σ invariances one can then deduce the Ward identities for Green functions involving the string field, its ghosts and insertions of the energy momentum tensor. One can then explore the properties of the current algebra of the energy momentum tensor. By taking advantage of the quasi Yang-Mills structure of equations (10), it is also straightforward to derive the corresponding local BRST equations [1,9]. In this way, and by using the technics introduce in ref. [10], one can relate the properties of the energy momentum tensor to those of the BRST Noether current.

Although the fields ϕ, Ω, $\bar{\Omega}$ are not present in the gauge fixed action $\mathcal{S}_0$, their introduction from the requirement of Weyl invariance is usefull. Indeed, when computing the partition function steming from $\mathcal{S}_0$, the Weyl invariance should be broken at least at two stages. One must introduce a Weyl non invariant regulator and consequently one expects a cosmological type counterterm $\sqrt{g} = e^\phi(1-\mu^+\mu^-)$ in the action. Moreover, for $g > 1$ one will end up with an integration over the moduli parameter space which necessitates the computation of objects of the type $\langle f | \ell \rangle = \int d^2x\, [\sqrt{g}]^\lambda f(x^+,x^-)\, \ell(\mu, x^+, x^-)$ where λ stand for a given weight. It is therefore important to control the gauge fixing of $\sqrt{g}$, i.e. of ϕ, as we have done by starting from an action which is BRST invariant not only in the diffeomorphism sector but also in the Weyl sector. In order to convince oneself of the non triviality of the Weyl sector one can for instance consider the following exercice[6]. Instead of the simplest gauge fixing action $\int d^2x(\bar{\Omega}(\phi - \phi_0))$ that we have used above, one might tentatively choose another BRST invariant action $\int d^2x\, s(\bar{\Omega}\Box\phi) = \int d^2x\, (\phi\Box b - \bar{\Omega}\Box\Omega)$. One immediately sees that in this gauge the Weyl ghost propagates and cannot be integrated out from the action. One finds also that the field ϕ (and thus the scalar part of the metrics) propagates together with the Stückelberg field b of the Weyl sector. This gauge choice is of course consistent due to the BRST invariance: When one computes the string critical dimension, b and ϕ contribute each one by one unit to the conformal anomaly coefficient, but both Weyl ghost Ω and antighost $\bar{\Omega}$ contribute negatively each one by one unit and thus all fields of the Weyl sector conspire for giving a vanishing contribution to the conformal anomaly, so that the critical value $D = 26$ is also obtained in this gauge. This exemple, although it is a rather perverse one, indicates the relevance of carefully gauge fixing the Weyl invariance.

The algebraïc structure and thereby the descent anomaly equations of the BRST symmetry of the string are now easily obtained. One observes that eqs. (5) are similar to the BRST equations of a Yang-Mills theory. Therefore if one defines the following quantities [1]

$$\tilde{\mu}^+ \equiv \mu^+dx^- + \Xi^+ \qquad\qquad \tilde{d}_- \equiv dx^-\partial/\partial x^- + s$$

$$\tilde{\mu}^- \equiv \mu^- dx^+ + \Xi^- \qquad\qquad \tilde{d}_+ \equiv dx^+ \partial/\partial x^+ + s \qquad (13)$$

one finds the following compact form for the BRST equations, equivalent to eqs. (5):

$$\begin{aligned} \tilde{d}_- \tilde{\mu}^+ - 1/2\,[\tilde{\mu}^+, \tilde{\mu}^+]_+ &= 0 \\ \tilde{d}_+ \tilde{\mu}^- - 1/2\,[\tilde{\mu}^-, \tilde{\mu}^-]_- &= 0 \end{aligned} \qquad (14)$$

Notice that at ghost number zero eqs. (14) are empty since $dx^+ \wedge dx^+ = dx^- \wedge dx^- = 0$. Moreover, if we introduce the enlarged embedding space $(X, \theta, \bar{\theta})$ which is relevant for string field theory [7] (θ and $\bar{\theta}$ are a pair of additionnal Grassman coordinates), eq. (13) suggests the possibility of defining $s = d\theta\, \partial/\partial\theta|_{\theta=0}$, $\tilde{\mu}^\pm = \mu^\pm dx + \mu_\theta^\pm d\theta$, $\Xi^\pm = \mu_\theta^\pm\, d\theta\,|_{\theta=0}$.

The consistant anomalies, i.e. the solutions modulo s- and d- exact terms of the following equations:

$$\int_\Sigma \Delta^1{}_2 = 0 \qquad\qquad \oint \Delta^2{}_1 = 0 \qquad (15)$$

(Δ'_2 is a 2-form with ghost number 1 and $\Delta^2{}_1$ is a 1-form with ghost number 2), can be computed by using the following identity:

$$\partial\,(\tilde{\mu}\partial\tilde{\mu}\partial^2\tilde{\mu}) = \tilde{\mu}\partial\tilde{\mu}\partial^3\tilde{\mu} \qquad (16)$$

Indeed by combining the identity (16) and eqs. (14) one gets:

$$\begin{aligned} \tilde{d}_-\,(\tilde{\mu}^+ \partial_+{}^3 \tilde{\mu}^+) &= \partial_+(........) \\ \tilde{d}_+\,(\tilde{\mu}^- \partial_-{}^3 \tilde{\mu}^-) &= \partial_-(........) \end{aligned} \qquad (17)$$

Eqs. (17) are analogous to the descent equations which generate the pyramids of cocycles in gauge theories of forms [8]. The part with ghost number 1 in $\tilde{\mu}(\partial)^3\tilde{\mu}$ determines the following expression of the conformal anomaly $\int \Delta^1{}_2$ which only depends on μ and Ξ :

$$\int_\Sigma \Delta^1{}_2 = \chi \int_\Sigma d^2x (\, \Xi^+ \partial_+{}^3 \mu_0{}^+ + \Xi^- \partial_-{}^3 \mu_0{}^-) \qquad (18)$$

The term with ghost number 2 in $\tilde{\mu}(\partial)^3\tilde{\mu}$ determines the second cocycle $\oint \Delta^2{}_1$, i.e. the Schwinger type form of the anomaly:

$$\oint \Delta^2{}_1 = \chi \oint (dx^+\, \Xi^+ \partial_+{}^3 \Xi^+ + dx^-\, \Xi^- \partial_-{}^3 \Xi^-) \qquad (19)$$

The computation of the anomaly coefficient χ can be done by various methods and is proportional to D-26. It is of interest to insert the consistent anomaly (17) at the right hand side of the Ward identity stemming from the BRST invariance of the action . By differentiating this broken Ward identity with respect to $\Xi^\pm$ and using the property that $\mu_0{}^\pm$ is the source of the full energy momentum tensor $T^\pm$ one gets:

$$\begin{aligned} \partial_- T^+ + [\mu_0{}^+, T^+]_+ &= (\,D-26\,)/2\ \partial_+{}^3 \mu_0{}^+ \\ \partial_+ T^- + [\mu_0{}^-, T^-]_- &= (\,D-26\,)/2\ \partial_-{}^3 \mu_0{}^- \end{aligned} \qquad (20)$$

Eqs. (20) indicate that the conformal anomaly occuring for $D \neq 26$ affects symmetrically and independently the left and right part of the theory. Notice that the modifications that we shall introduce in the BRST symmetry for curing the problem of global zero modes for $g > 1$ will not change the form of the consistent anomalous Ward identity (20).

Let us now discuss worldsheets with $g > 1$ [2]. Due to the existence of conformal classes of metrics which cannot be transformed one into each other by diffeomorphisms, the gauge choice $\mu^\pm = \mu_0{}^\pm (x^+, x^-)$ is too strong: If one tries to use for $g > 1$ the action $\mathcal{I}_0$, the signal of the inconsistency is the non existence of a propagator for the diffeomorphism ghosts $\bar{\Xi}^+$ and $\bar{\Xi}^-$

The equations of motion of Ξ^+ and Ξ^- are:

$$\partial_-\Xi^+ - \Xi^+\partial_+\mu_0^+ - 2\mu_0^+\partial_+\Xi^+ = 0$$
$$\partial_+\Xi^- - \Xi^-\partial_-\mu_0^- - 2\mu_0^-\partial_-\Xi^- = 0 \qquad (21)$$

Since the Rieman-Roch theorem indicates the existence of 3g-3 independent zero modes for both eqs.(21), one finds a degeneracy of the action (9) under the following transformations [2]:

$$\delta\Xi^+ = \sum_{1\leq i\leq 3g-3} \epsilon_i^+ h^{i+}(\mu_0, x^+, x^-)$$
$$\delta\Xi^- = \sum_{1\leq i\leq 3g-3} \epsilon_i^- h^{i-}(\mu_0, x^+, x^-) \qquad (22)$$

and thereby the non existence of the ghost propagator. The $\epsilon^\pm_i$'s are constant anticommuting parameters, and the $h^{\pm i}$'s a basis of quadratic differentials solutions of eq. (21). In ref. [2] it is shown how to generally gauge fix in a BRST invariant way such a degeneracy of an action duc to global zero modes. In the present case we introduce two sets of 3g-3 constant commuting ghosts γ^+_i and γ^-_i in correspondence with the ϵ^+_i and ϵ^-_i, together with constant anticommuting Stückelberg fields λ^+_i and λ^-_i. The improved nilpotent BRST symmetry operator s is now defined as:

$$s\vec{X} = \xi^\alpha\partial_\alpha\vec{X},$$
$$s\mu^+ = \partial_-\Xi^+ + [\Xi^+, \mu^+]_+, \qquad s\mu^- = \partial_+\Xi^- + [\Xi^-, \mu^-]_-,$$
$$s\Xi^+ = \tfrac{1}{2}[\Xi^+, \Xi^+]_+, \qquad s\Xi^- = \tfrac{1}{2}[\Xi^-, \Xi^-]_-,$$
$$s\bar{\Xi}^+ = b^+ + \sum_{1\leq i\leq 3g-3}\gamma_i^+ h^i \qquad s\bar{\Xi}^- = b^- + \sum_{1\leq i\leq 3g-3}\gamma_i^- h^{i-},$$
$$s\gamma_i^+ = \lambda_i^+, \qquad s\gamma_i^- = \lambda_i^-,$$
$$s\lambda_i^+ = 0, \qquad s\lambda_i^- = 0$$
$$sb^+ = -\sum_{1\leq i\leq 3g-3}\lambda_i^+ h^i, \qquad sb^- = -\sum_{1\leq i\leq 3g-3}\lambda_i^- h^{i-}. \qquad (23)$$

By introducing functions $F^\pm_i$ which as such that $\det(\langle F^\pm_i | h^{i\pm}\rangle) \neq 0$ we modify the gauge fixing action in the diffeomorphism sector as follows:

$$s(\int_\Sigma d^2z\, [\bar{\Xi}^+(\mu^+ - \mu_0^+) + \bar{\Xi}^-(\mu^- - \mu_0^-)) \rightarrow$$

$$s[\int_\Sigma d^2z\, (\bar{\Xi}^+(\mu^+ - \mu_0^+ + \sum_{1\leq i\leq 3g-3}\gamma_i^+ h^{i+}) + \bar{\Xi}^-(\mu^- - \mu_0^- + \sum_{1\leq i\leq 3g-3}\gamma_i^- h^{i-})\,]$$
$$+ 1/2\alpha_{ij}\lambda_i^+\lambda_j^- \qquad (24)$$

The α_{ij}'s stand for a (3g-3) x (3g-3) symmetrical matrix of real gauge parameters. From the new BRST invariance, eqs.(23), it is possible to control the dependence of the partition function upon different choices of the functions $\mu_0^\pm$ and $F^\pm_i$ as well as of the gauge parameters α_{ij}. Moreover, by expanding eq. (24) and eliminating the $b^\pm$ fields one gets:

$$\mathcal{I} \sim \mathcal{I}_0 [\ \overline{X}, \mu = \mu_0 + \sum_{1 \leq i \leq 3g-3} \gamma_i F_i\ , \Xi\ , \overline{\Xi}\ ]\ +$$

$$1/2 \alpha_{ij} \lambda_i^+ \lambda_j^- \ +$$

$$\lambda^+_i \int_\Sigma d^2x\ \overline{\Xi}^+ F^+_i \ +\ \lambda^-_i \int_\Sigma d^2x\ \overline{\Xi}^- F^-_i \qquad (25)$$

By setting equal to zero the gauge parameters α_{ij}, one finds by variation with respect to λ^+_i and λ^-_i that the antighosts $\overline{\Xi}^+$ and $\overline{\Xi}^-$ are now constrained to remain orthogonal to the global zero modes in a BRST invariant way. This permits a consistent definition of the ghost propagator. Notice that the replacement of μ by $\mu = \mu_0 + \sum \gamma_i F_i$ after elimination of the fields $b^\pm$ identifies the constant ghosts γ_i as moduli parameters. This construction, and the study of Ward identities for $g > 1$ after elimination of fields b and λ_i is displayed in more detail in ref. [2]. In this reference it is shown in particular that the known expression of the measure for the integration over the moduli space can be recovered by choosing a Feynman type gauge for the constant ghosts (i.e. $\alpha_{ij} = 0$) and by integrating out the fields $\lambda_i^\pm$ and the antighosts $\overline{\Xi}^\pm$ after a mode expansion. The gauge fixing of the trivial global zero modes of the type X^μ = cte is also worked out through a BRST invariant procedure. It is in fact an interesting result that that the partition function is fully expressible by following the by now familiar principle of BRST symmetry [8].

In the case of superstrings a "super Beltrami" parametrization can be defined [3]. For describing the spin structure of the superstring, one associates to μ^+ (resp μ^-) a supersymmetric partner $\alpha^\oplus$ (resp α^-). The index $\oplus$ (resp $-$) denotes a positive (resp.negative) helicity for 2-D spinors. The field $\alpha^\oplus$ is obtained by extracting the superWeyl inert spin 3/2 component from the 2-D gravitino $\psi^\oplus = \psi^\oplus{}_\alpha dx^\alpha$ [3]:

$$\psi^\oplus = e^{\phi/2} (\lambda^- ([dx^+ + \mu^+ dx^-] \ + \alpha^\oplus dx^-)$$
$$\psi^- = e^{\phi/2} (\lambda^\oplus ([dx^- + \mu^- dx^+] \ + \alpha^- dx^+) \qquad (26)$$

The spinors $\lambda^\oplus$ and λ^- correspond to the spin 1/2 parts of the gravitino and are the supersymmetric partners of the scalar part of the metric.

The BRST transformation of α and μ is obtained from the supersymmetric generalization of eq. (14):

$$\tilde{d}_- \tilde{\mu}^+ - 1/2\,[\,\tilde{\mu}^+ , \tilde{\mu}^+]_+ - i\tilde{\alpha}^\oplus \tilde{\alpha}^\oplus = 0$$
$$\tilde{d}_- \tilde{\alpha}^\oplus - [\,\tilde{\mu}^+ , \tilde{\alpha}^\oplus]_+ = 0$$

$$\tilde{d}_+ \tilde{\mu}^- - 1/2\,[\,\tilde{\mu}^- , \tilde{\mu}^-]_- - i\tilde{\alpha}^- \tilde{\alpha}^- = 0$$
$$\tilde{d}_+ \tilde{\alpha}^- - [\,\tilde{\mu}^- , \tilde{\alpha}^-]_- = 0 \qquad (27)$$

where $\tilde{\alpha}^\oplus \equiv \alpha^\oplus dx^- + \chi^\oplus$ and $\chi^\oplus$ and χ^- stand for both component of the commuting spinor ghost of local 2-D supersymmetry. The bracket $[\tilde{\mu}, \tilde{\alpha}]$ is defined as $[\tilde{\mu}, \tilde{\alpha}] = \tilde{\mu}\partial\tilde{\alpha} + \frac{1}{2}\tilde{\alpha}\partial\tilde{\mu}$, which identifies $\tilde{\alpha}$ as a differential of rank 1/2. Factorization along + and − directions is still present in eqs. (26). Expansion in ghost number yields:

$$s\mu^+ = \partial_-\Xi^+ + [\Xi^+,\mu^+]_+ + 2i\,\chi^\oplus\tilde{\alpha}^\oplus$$
$$s\alpha^\oplus = \partial_-\chi^\oplus + \Xi^+\partial_+\alpha^\oplus - 1/2\alpha^\oplus\partial_+\Xi^+ + 1/2\chi^\oplus\partial_+\mu^+ - \mu^+\partial_+\chi^\oplus$$

$$s\Xi^+ = \tfrac{1}{2}[\Xi^+,\Xi^+]_+ + i\,\chi^\oplus\chi^\oplus$$
$$s\chi^\oplus = \Xi^+\partial_+\chi^\oplus + 1/2\chi^\oplus\partial_+\Xi^+ \qquad (28)$$

and one has similar equations with + <–> − . The classical superstring action reads as follows [3]:

$$\mathcal{I}_{cl} = \int_\Sigma d^2z\,(1-\mu^+\mu^-)^{-1}\{(\partial_+-\mu^-\partial_-)\vec{X}\cdot(\partial_--\mu^+\partial_+)\vec{X} +$$
$$2i\alpha^-\vec{F}^\oplus(\partial_+-\mu^+\partial_+)\vec{X} - 2i\alpha^\oplus\vec{F}^-(\partial_+-\mu^+\partial_+)\vec{X} + 2\alpha^\oplus\alpha^-\vec{F}^\oplus\vec{F}^-\} -$$
$$\int_\Sigma d^2z\{i\vec{F}^\oplus(\partial_+-\mu^+\partial_+)\vec{F}^- - i\vec{F}^-(\partial_+-\mu^+\partial_+)\vec{F}^\oplus + \vec{W}.\vec{W}\} \qquad (29)$$

A straightforward generalization of the bosonic case can be done for a BRST invariant gauge fixing procedure of this action. The integration over supermoduli space is now found from the necessity of gauge fixing the global zero modes of the diffeomorphism and supersymmetry antighosts.

Acknowledgments

It is a pleasure to thank my collaborators M. Bellon and R. Stora.

REFERENCES

[1] L. Baulieu, M. Bellon, Beltrami Parametrization and String Theory, Phys. Lett. B 196 (1987) 142, C. Becchi, Genova preprint, R. Stora, these proceedings.

[2] L. Baulieu, M. Bellon, "BRST invariant gauge fixing procedure of global zero modes in covariant string theory", LPTHE preprint 1987, submitted to phys. Lett. B.

[3] L. Baulieu, M. Bellon, R. Grimm, LPTHE preprint, June 1987, to appear in Phys. Lett. B. and "Weyl superspace from Lorentz superspace", in preparation.

[4] L. Baulieu, C. Becchi, R. Stora, Phys. Lett. 180B (1986) 55.

[5] D. Friedan, E. Martinec, S. Schenker, Nucl. Phys. B271 (1986) 93.

[6] I thank O. Babelon for drawing my attention on the possible propagation of the Weyl ghost.

[7] A. Neveu, P.C. West, H. Nicolaï, Phys. Lett. 167B (1986) 307 ; W. Siegel, Phys. Lett. 149B (1984) 157 ; W. Siegel, B. Zwieback, Nucl. Phys. B263 (1986) 105 ; L. Baulieu, S. Ouvry, Phys. Lett. 171B (1986) 57.

[8] L. Baulieu, Cargese Lectures, July 1983, "Algebraïc construction of gauge invariant theories" in Perspectives in particles and fields, eds. Basdevant-Lévy, Plenum Press.

[9] J.P. Ader, J.C. Wallet, Phys. Lett. 192B (1987) 103 and in preparation.

THE A-D-E CLASSIFICATION OF CONFORMAL INVARIANT FIELD THEORIES IN TWO DIMENSIONS

Andrea Cappelli
Service de Physique Théorique
CEN-Saclay, 91191 Gif-sur-Yvette Cedex, France
and
INFN, Sezione di Firenze, Largo E. Fermi 2
50125, Florence, Italy

1. Introduction: The classification problem

Two dimensional massless quantum field theory invariant under conformal transformations is a well developed and beautiful subject [1,2,3], which finds applications in string theory [1] and critical phenomena in statistical physics [2]. We shall present results obtained at Saclay, in collaboration with C. Itzykson and J.-B. Zuber [4,5,6] and we shall mainly discuss the applications to critical phenomena.

Let us briefly recall the relevant features of conformal field theories. D. Friedan's lectures in these proceedings provide a larger introduction. These theories have a stress-energy tensor $T_{\mu\nu}$ which is traceless, symmetric and conserved. Its two independent components $T(z) = (T_{11}+iT_{12})/2$, $\partial_{\bar z}T(z)=0$, and $\bar{T}(\bar z) = (T_{11}-iT_{12})/2$, $\partial_z\bar{T}(\bar z)=0$, generate independent analytic transformations of the coordinates $z = x^o+ix^1$ and antianalytic of $\bar z = x^o-ix^1$, respectively. They yield conformal rescalings of the metric.

$$ds^2 = dz d\bar z = \left(\frac{\partial z}{\partial z'}\right)\left(\frac{\partial \bar z}{\partial \bar z'}\right) dz'\, d\bar z' \qquad (1.1)$$

Conformal Ward Identities determine the algebra of transformations [3]: for the Fourier components $T(z) = \sum_n L_n z^{-n-2}$, $\bar T(\bar z) = \sum_n \bar L_n \bar z^{-n-2}$, $n \in \mathbb{Z}$, we have

$$\begin{aligned}[L_n, L_m] &= (n-m)\, L_{n+m} + \frac{c}{12}(n^2-1)\, n\, \delta_{n+m,0} \\ [\bar L_n, \bar L_m] &= (n-m)\, \bar L_{n+m} + \frac{\bar c}{12}(n^2-1)\, n\, \delta_{n+m,0}\end{aligned} \qquad (1.2)$$

These are two Virasoro algebras with central charges $c = \bar{c}$. This charge measures the response of quantum fluctuations to a classical curved background, because it parametrizes the trace anomaly $\langle T^{\mu}_{\mu}\rangle = (c/12)R$ [1,2]. While in string theory c = 0 by unitarity of the theory embedded in a D-dimensional space-time [2], here c can take positive values. It will play the rôle of "principal quantum number" in the following classification.

The states of the conformal theory are described by irreducible representations of the Virasoro algebra [7]. These are built on highest weight states $|h\rangle$, defined by

$$L_o|h\rangle = h|h\rangle \ , \quad L_n|h\rangle = 0 \qquad n > 0 \tag{1.3}$$

A generic state in the representation of weight h is

$$|\psi_h\{n\}\rangle = L_{-n_k}L_{-n_{k-1}}\ldots L_{-n_1}|h\rangle \ , \quad n_k \geqslant n_{k-1} \geqslant \ldots n_1 \geqslant 0 \tag{1.4}$$

There is a one-to-one correspondance between state and fields of the theory: to h.w. states correspond "primary fields" $\Phi_{h,\bar{h}}$ applied to the vacuum $|0\rangle$

$$\Phi_{h,\bar{h}}(0,0)|0\rangle = |h,\bar{h}\rangle = |\bar{h}\rangle|h\rangle \tag{1.5}$$

$\Delta = h+\bar{h}$ is the scaling dimension of $\Phi_{h,\bar{h}}$ and $s = h-\bar{h}$ its spin, as it follows from the two-point function:

$$\langle\Phi_{h,\bar{h}}(z,\bar{z})\Phi_{h',\bar{h}'}(0,0)\rangle = \delta_{hh'}\delta_{\bar{h},\bar{h}'}\, z^{-2h}\, \bar{z}^{\,-2\bar{h}} \tag{1.6}$$

Generic states in eq.(1.4) correspond to descendant fields. They are built by the local product of the primary field and T(z), $\bar{T}(\bar{z})$. The fields of the theory must have a closed algebra under the short distance expansion [2,3]

$$\Phi_{h\bar{h}}(z,\bar{z})\Phi_{h'\bar{h}'}(0,0) = \sum_{h'',\bar{h}''} \frac{C_{hh'h''}C_{\bar{h}\bar{h}'\bar{h}''}}{z^{h+h'-h''}\bar{z}^{\bar{h}+\bar{h}'-\bar{h}''}}\left\{\Phi_{h''\bar{h}''}(0,0)+\text{descendants}\right\} \tag{1.7}$$

$C_{hh'h''}$, $C_{\bar{h}\bar{h}'\bar{h}''}$ are the structure constants of the algebra. Eq.(1.7) makes sense for both sides inserted in a correlation function.

By studing representation theory, the authors of ref.[3] showed that for

$$\begin{aligned} c &= 1-6\,\frac{(p-p')^2}{pp'} \qquad p,p' \in \mathbb{Z} \ \textit{and} \ > 0 \\ h_{r,s} = \bar{h}_{r,s} &= \frac{(pr-p's)^2-(p-p')^2}{4pp'} \ , \quad 0 \leqslant r \leqslant p', \quad 0 \leqslant s \leqslant p \end{aligned} \tag{1.8}$$

the primary fields $\Phi_{h,\bar{h}}$ with $(h,\bar{h}) \in \{(h_{rs}, \bar{h}_{r's'})\}$ have a closed algebra. In these cases we can have a "minimal" theory, which has a finite number of primary fields.

Unitarity is an important requirement in quantum field theory. The theories with $c < 1$ have positive norm states if [8] and only if [9]

$$c = c(m) = 1 - \frac{6}{m(m+1)} \quad , \quad m = 3,4,\dots$$
$$h_{rs} = \frac{((m+1)r-ms)^2-1}{4m(m+1)} \quad , \quad 1 \leqslant r \leqslant m-1, \quad 1 \leqslant s \leqslant m \tag{1.9}$$

Unitary conformal theories for $c < 1$ are only minimal . For $c > 1$, unitarity allows any $h,\bar{h} > 0$. Non-unitary theories are also interesting in critical phenomena, but they can be non minimal. For example, self-avoiding polymers correspond to a non minimal theory at $c = 0$ [10].

Now we have the elements to state the problem of classification of conformal theories. It amounts to finding all possible sets of dimensions $\{(h,h)\}$ and multiplicities of primary fields (i.e. the number of fields for every pair of dimensions), together with structure constants of their algebra, eq.(1.7). This program is almost completed for $c < 1$ unitary theories. The dimensions are fixed in eq.(1.9), multiplicities are determined by constructing the partition function [11], and structure constants are determined by the 4-point functions, which are calculable for minimal theories [3,12]. Here we shall discuss the classification of multiplicities. It applies to all minimal theories, both unitary ($p' = p-1 = m$) and non unitary ones.

2. Modular invariance of the partition function on a torus

Let us consider a finite portion of the complex plane with periodic boundary conditions, i.e. a torus with periods ω_1, ω_2; $\tau = \omega_2/\omega_1$ is the modular ratio (with convention $\mathrm{Im}\tau > 0$). The partition function is the trace of the transfer matrix for the strip of width ω_1

$$\begin{aligned} Z(\tau) &= \mathrm{Tr}\left(e^{-\mathrm{Im}\omega_2 \mathcal{H} + i\mathrm{Re}\omega_2 \mathcal{P}} \right) \\ &= \mathrm{Tr}\left(e^{i2\pi\tau(L_0 - c/24)} \, e^{-i2\pi\bar{\tau}\left(\bar{L}_0 - c/24\right)} \right) \end{aligned} \tag{2.1}$$

$\mathrm{Im}\omega_2$ and $\mathrm{Re}\omega_2$ are the evolution time and space, respectively. The hamiltonian $\mathcal{H}$ an the momentum $\mathcal{P}$ operators in the w-strip are obtained by their forms in the infinite z-plane through the transformation $z = \exp(i2\pi\, w/\omega_1)$ [11,13]:

$$\mathcal{H} = \frac{2\pi}{\omega_1}\left(L_0 + \bar{L}_0 - c/12\right) \quad , \quad \mathcal{P} = \frac{2\pi}{\omega_1}\left(L_0 - \bar{L}_0\right) \tag{2.2}$$

For $c < 1$ minimal theories in eqs. (1.8,9), the trace decomposes in pairs of Virasoro irreducible representations as follows

$$Z(\tau) = \sum_{(h,\bar{h})\in\{(h_{rs},\bar{h}_{r's'})\}} \mathcal{N}_{h\bar{h}}\ \chi_h(\tau)\ \overline{\chi_{\bar{h}}(\tau)} \tag{2.3}$$

$\chi_h(\tau) = \mathrm{Tr}_{(h)}\left(e^{2i\pi\tau(L_0-c/24)}\right)$ is the character of the representation of weight h, which is a known function [8,14]. $\mathcal{N}_{h,\bar{h}}$ is the unknown multiplicity of the field $\Phi_{h,\bar{h}}$ in the theory. It must be a non-negative integer, because it comes from a decomposition of representations; moreover, $\mathcal{N}_{0,0} = 1$ for the unicity of the vacuum.

Cardy showed that $\mathcal{N}_{h\bar{h}}$ can be determined by requiring the invariance of $Z(\tau)$ under reparametrizations of the torus [11], the modular transformations [15]. In fact, the torus is $\mathbb{T} = \mathbb{C}/\Lambda$, where Λ is a lattice of points generated by ω_1,ω_2. Actually, ω_1', ω_2' generates the same lattice

$$\begin{cases}\omega_2' = a\omega_2 + b\omega_1 \\ \omega_1' = c\omega_2 + d\omega_1\end{cases} \tag{2.4}$$

for $a,b,c,d \in \mathbb{Z}$, ad-bc=1. The modular group $\Gamma = PSL(2,\mathbb{Z})$ acts on τ as follows

$$\tau \longrightarrow {}^{A}\tau = \frac{a\tau+b}{c\tau+d}, \qquad A = \pm\begin{pmatrix}a & b\\ c & d\end{pmatrix} \in \Gamma \tag{2.5}$$

The modular invariance conditions on the partition function are

$$Z(\tau) = Z({}^{A}\tau) \ , \quad \forall\, A \in \Gamma \tag{2.6}$$

We shall now discuss the solutions of these conditions. Some technical steps are needed to transform the functional relations (2.6) into linéar equations for the matrix $\mathcal{N}$ [4]. The characters can be arranged in a N-dimensional vector $\vec{\chi} = (\chi_\lambda)$, $\lambda = 1,\ldots,N$, $N = 2pp'$, where $\lambda = rp\text{-}sp'$ in eq.(1.8), and $h_{rs} = (\lambda^2-1)/2N$. Notice however the parity symmetries

$$\chi_\lambda = \chi_{\lambda+N} = \chi_{-\lambda} = -\chi_{\omega_0\lambda} \tag{2.7}$$

where $\omega_0 = \omega_0(N)$ is a known number, $\omega_0^2 = 1 \bmod 2N$.

In the N-dimensional vector space of $\vec{\chi}$, the action of Γ is an unitary linear representation:

$$\begin{aligned}&\chi_\lambda({}^{A}\tau) = U(A)_{\lambda\lambda'}\ \chi_{\lambda'}(\tau) \ , && A \in \Gamma, \\ &U(A)\ U(B) = U(AB) \ , && UU^{\dagger} = \mathbb{1}\end{aligned} \tag{2.8}$$

The solution of the modular invariant conditions in eq.(2.6) will be in two steps:

(A) of finding matrix $\mathcal{N}_{\lambda,\bar{\lambda}}$ commuting with U(A)

$$\mathcal{N} = U(A)\, \mathcal{N}\, U(A)^{\dagger}, \qquad \forall\, A \in \Gamma \tag{2.9}$$

(B) of imposing positivity conditions $\mathcal{N}_{h,\bar{h}} \geq 0$, ($\mathcal{N}_{h,\bar{h}} = 1$ for $h=\bar{h}=0$):

$$\mathcal{N}_{h(\lambda),h(\bar{\lambda})} = \mathcal{N}_{\lambda,\bar{\lambda}} + \mathcal{N}_{-\lambda,\bar{\lambda}} - \mathcal{N}_{\omega_0\lambda,\bar{\lambda}} - \mathcal{N}_{-\omega_0\lambda,\bar{\lambda}} \tag{2.10}$$

Eq.(2.9) follows from (2.3,6,8); eq.(2.10) follows from eq.(2.7), after we have taken an independent set of $h,\bar{h}$ values in the summation eq.(2.3), that is $\lambda,\bar{\lambda} \in \Delta$, where Δ is the fundamental domain of the parity group $\{\pm 1, \pm\omega_0\}$ on integers $\mathbb{Z}/N\mathbb{Z}$.

(A) Modular invariants

The subgroup Γ_{2N}

$$\Gamma \supset \Gamma_{2N} = \left\{ \Gamma \ni \begin{pmatrix} a & b \\ c & d \end{pmatrix} = \pm \begin{pmatrix} 1 & 0 \\ 0 & 1 \end{pmatrix} \bmod 2N \right\} \tag{2.11}$$

acts by a global phase $U(B)_{\lambda,\lambda'} = \delta_{\lambda,\lambda'} \exp i\varphi(B)$, $B \in \Gamma_{2N}$ [4]. Eq.(2.9) is trivial for elements of Γ_{2N}, non trivial conditions follow from invariance under the finite group $M_{2N} = \Gamma/\Gamma_{2N} = PSL(2,\mathbb{Z}/2N\mathbb{Z})$. The general form of the matrix invariant under the adjoint action of a finite group is

$$\mathcal{N} = \frac{1}{|M_{2N}|} \sum_{B \in M_{2N}} U(B)\, \mathcal{D}\, U^{\dagger}(B) \tag{2.12}$$

for any N×N matrix $\mathcal{D}$. By a proper parametrization of $\mathcal{D}$, we find a set of $(N/2)^2$ non zero invariants: $\mathcal{N}_{k,\ell}$, k,ℓ integers modulo N/2. Many of them are equal up to a phase:

$$\mathcal{N}_{k,\ell} = e^{i\varphi(A)}\, \mathcal{N}_{ak+b\ell,\, ck+d\ell} \quad , \qquad A = \begin{pmatrix} a & b \\ c & d \end{pmatrix} \in M_{N/2} \tag{2.13}$$

These two matrices are on the same orbit of the group $M_{N/2}$ in the space of matrices. It follows that independent invariants correspond to different orbits of $M_{N/2}$. A simple analysis shows an orbit for any δ divisor of N/2, $\delta | N/2$ [16,17]. All these solutions were previously found in Ref.[4], by following a different method. Call them $\mathcal{N}_\delta$: for example, when N/2 has no square factors,

$$(\mathcal{N}_\delta)_{\lambda,\lambda'} = \delta_{\lambda,\omega\lambda'} \quad \bmod N \tag{2.14}$$

with $\omega = \omega(\delta)$, $\omega^2 = 1 \bmod 2N$. The dimension of the space of solution $\mathcal{N}_\delta$ is the number of factors of N/2, so it grows like $\log N \sim 2 \log m$ on average, for $c = c(m)$ in eq.(1.9).

(B) Positive invariants, A-D-E Classification

The condition of positive multiplicities is far more restrictive: it only allows two or three invariants for any value of c(m). We insert the linear combinations

$$\mathcal{N}_{\lambda,\lambda'} = \sum_{\delta | N/2} C(\delta)\, \delta_{\lambda',\omega(\delta)\lambda} \qquad (2.15)$$

into eq.(2.10) and look for solutions in $C(\delta) \in \mathbb{Z}$. An explicit analysis up to order $N \sim 10^4$ shows a regular pattern [4]: two solutions for any c(m) and 6 exceptional cases. This is in remarkable correspondance to the Cartan-Dynkin classification of simple Lie algebra of types A, D and E (those having all roots of equal length) [18]. This correspondance is explained by an example: m = 12, c = 35/36. A positive invariant is

$$Z = \frac{1}{2}\sum_{s=1}^{12}\left\{|\chi_{1,s}+\chi_{7,s}|^2+|\chi_{4,s}+\chi_{8,s}|^2 + |\chi_{5,s}+\chi_{11,s}|^2\right\} \qquad (2.16)$$

The diagonal terms $|\chi_{r,s}|^2$ for $h = \bar{h} = h_{r,s}$ are non zero for both indeces r and s in the lists

$$\left.\begin{array}{l} r = 1,4,5,7,8,11 \\ s = 1,2,\ldots\ldots,12 \end{array}\right\}\text{Coxeter exponents of}\left\{\begin{array}{l} E_6 \\ A_{12} \end{array}\right. \qquad (2.17)$$

with the correct multiplicities also. Coxeter exponents caracterize some properties of the root space of the Lie algebras encoded in the Dynkin diagram.

Later, the following theorem was proven:

All the positive modular invariants partition functions of $c < 1$ minimal conformal theories are classified by pairs of Lie algebras of types A, D and E.

The two existing proofs are by inspection and uses arithmetical properties of integers [6,19]. Consider for example a line of the modular sesquilinear form eq.(2.3, 2.15):

$$\bar{\chi}_\lambda \sum_{\lambda'} \mathcal{N}_{\lambda,\lambda'}\chi_{\lambda'} = \bar{\chi}_\lambda\left(C(\delta_1)\chi_{\omega_1\lambda}+\ldots+C(\delta_k)\chi_{\omega_k\lambda}\right) \geq 0 \qquad (2.18)$$

where $\lambda \in \Delta$, $\delta_i | N/2$, $i = 1,\ldots,k$ and we are free of taking $C(\delta_i) \geq 0$. The conditions on a term $C(\delta_i)\bar{\chi}_\lambda\chi_{\omega_i\lambda}$ are:

i) If $\pm\omega_i\lambda \in \Delta$, the restriction to the fundamental domain Δ of λ values eqs.(2.10) yields a positive term for any $C(\delta_i) > 0$;

ii) If $\pm\omega_i\lambda \in \omega_0\Delta$, the restriction yields a negative term. Either $C(\delta_i)=0$ or it exists another positive term in eq.(2.18), say $C(\delta_j)\bar{\chi}_\lambda\chi_{\omega_j\lambda}$, such that it compensate it: $\omega_j\lambda = \pm\,\omega_0\omega_i\lambda \bmod N$ and $C(\delta_j) \geq C(\delta_i)$.

These conditions on the $C(\delta)$'s can be reduced to simpler conditions for the product of two integers modulo N, whose solutions are listed. It

Table I List of partition functions of $c < 1$ minimal conformal theories. $\chi_{r,s}$ are the characters of representations $h_{r,s}$. The unitary theories corresponds to $p' = m-1$, $p = m$ or $p = m-1$, $p' = m$ by exchanging $r \leftrightarrow s$, for $c = 1-6/m(m+1)$, $m = 3,4,\ldots$.

$$\frac{1}{2}\sum_{r=1}^{p'-1}\sum_{s=1}^{p-1} |\chi_{rs}|^2 \qquad (A_{p'-1}, A_{p-1})$$

$$p' = 4\rho+2,\ \rho\geq 1 \qquad \frac{1}{2}\sum_{s=1}^{p-1}\left\{\sum_{\substack{r\ \mathrm{odd}=1\\ r\neq 2\rho+1}}^{4\rho+1} |\chi_{rs}|^2 + 2|\chi_{2\rho+1\ s}|^2 + \sum_{r\ \mathrm{odd}=1}^{2\rho-1}\left(\chi_{rs}\chi^*_{p'-r\ s} + \mathrm{c.c.}\right)\right\} \qquad (D_{2\rho+2}, A_{p-1})$$

$$p' = 4\rho,\ \rho\geq 2 \qquad \frac{1}{2}\sum_{s=1}^{p-1}\left\{\sum_{r\ \mathrm{odd}=1}^{4\rho-1} |\chi_{rs}|^2 + |\chi_{2\rho\ s}|^2 + \sum_{r\ \mathrm{even}=1}^{2\rho-2}\left(\chi_{rs}\chi^*_{p'-r\ s} + \mathrm{c.c.}\right)\right\} \qquad (D_{2\rho+1}, A_{p-1})$$

$$p' = 12 \qquad \frac{1}{2}\sum_{s=1}^{p-1}\left\{|\chi_{1s}+\chi_{7s}|^2 + |\chi_{4s}+\chi_{8s}|^2 + |\chi_{5s}+\chi_{11s}|^2\right\} \qquad (E_6, A_{p-1})$$

$$p' = 18 \qquad \frac{1}{2}\sum_{s=1}^{p-1}\left\{|\chi_{1s}+\chi_{17s}|^2 + |\chi_{5s}+\chi_{13s}|^2 + |\chi_{7s}+\chi_{11s}|^2 + |\chi_{9s}|^2 + \left[(\chi_{3s}+\chi_{15s})\,\chi^*_{9s} + \mathrm{c.c.}\right]\right\} \qquad (E_7, A_{p-1})$$

$$p' = 30 \qquad \frac{1}{2}\sum_{s=1}^{p-1}\left\{|\chi_{1s}+\chi_{11s}+\chi_{19s}+\chi_{29s}|^2 + |\chi_{7s}+\chi_{13s}+\chi_{17s}+\chi_{23s}|^2\right\} \qquad (E_8, A_{p-1})$$

turns out that all positive invariants are labelled by pairs of Lie algebras as above (see Table I). These proofs do not employ properties of Dynkin diagrams, then they do not clarify the rôle of Lie algebras in our problem. We add some comments on this correspondance:

i) Classifications by A, D and E Lie algebras also apply to other subjects, like catastrophe theory and discrete subgroups of SU(2) [20]. Hopefully our problem could be related to the former [21] and the latter [6].

ii) A direct construction of positive invariants is under study, which applies properties of the Dynkin diagrams of the corresponding algebras [22]. The necessity of the construction is also needed for a proof.

Let us notice a posteriori that modular invariance, plus simple physical conditions, has strongly constrained the possible conformal theories. We miss an a priori explanation of the relation between conformal and modular invariance.

In order to complete the classification of $c < 1$ conformal theories, we need the structure constants of the algebra of fields for any A-D-E theory. They were computed in part [12]. Two theories with the same fields but different algebras are possible a priori, but they are not known. In the meantime, known selection rules [3,23] allow to verify the closure of the algebra for all theories in the A-D-E classification [4,16].

3. Applications

3.1 Statistical mechanics

A statistical model on a space lattice, like the Ising model, can have critical points with infinite correlation length. At these points it is appropriately described by a massless field theory in the continuum, which is invariant under infinitesimal (global) translations, rotations and scale transformations.

A conformal transformation is equivalent to a local rotation and scale transformation. The general conformal theory [3] tells that local rotation and scale invariance follows from global invariance if the generators of transformations have a local form, that is the stress-energy tensor exists. Locality of the interaction implies conformal invariance.

Primary fields (or descendants) of the conformal field theory are to be identified to (mutually local) variables of the statistical model, through their critical exponents. The classification of multiplicities yields a classification of universality classes, that is of possible sets of exact critical exponents. We discuss the unitary theories only the simplest example is the Ising model, for $m = 3$ in eq.(1.9), $c = 1/2$. Here we have only one partition function with all scalar fields (Table I). From dimensions $h_{1,1}=0$, $h_{2,1} = 1/2$, $h_{2,2} = 1/16$, we identify the identity operator as $\Phi_{0,0}$, the Ising spin σ as the primary field $\Phi_{1/16,\ 1/16}$, because the critical exponent η is $1/4$, and the energy operator ϵ as

$\Phi_{1/2,1/2}$, because its correlation relates to the specific heat singularity $\alpha = 0$.

For larger values of $c = c(m)$, one statistical model for any universality class is shown in Table II. (This in turns implies that we did not forget any more condition on the theory). The "principal" series (algebras (A,A)) contains generalizations of the Ising model called solid-on-solid models [24]. They have a generical multicritical point: for $c = c(m)$, m-1 phases coexist without any symmetry relating them, the order parameter has m-2 independent components, everyone with its critical exponent. From table I, scalar fields of all possible scaling dimensions h_{rs} appear in this theories.

The "complementary" series (A,D) contains the q-state Potts model, for $3 \leq q \leq 4$, $q = 4\cos^2(\pi/m+1)$, m odd, [25] (for m even they are tricritical version of the same models). The fields have half of the possible scaling dimensions, some have integer spin.

The exceptional cases are fitted by new models found by V. Pasquier [26]. He uses Dynkin diagrams in their construction: in relation to his work we proposed the A-D-E classification. (We also profited by a suggestion of V. Kaç).

Conformal invariant theories with extended symmetries are also A-D-E classified by some technical reasons we shall now discuss.

3.2 Wess-Zumino-Witten model and other models

The Wess-Zumino-Witten model is a two-dimensional σ-model, that is a theory with scalar fields living on the manifold of a compact group G [27]. This model may describe the compactification of the extra dimensions in string theory. Here we consider $G = SU(2)$. Besides conformal invariance, the theory possess a local SU(2) invariance, whose generators are the holomorphic currents $J(z) = J^a(z)t^a, \bar{J}(\bar{z}) = \bar{J}^a(\bar{z})t^a$. The current algebra is the affine Kac-Moody algebra $g = A_1^{(1)}$

$$J^a(z) = \sum J_n^a z^{-n-1}, \qquad \left[J_n^a, J_m^b\right] = i\,\epsilon^{abc}J_{n+m}^c + \frac{k}{2}\,n\,\delta^{ab}\delta_{n+m,0} \tag{3.1}$$

and the same for $\bar{J}^a(z)$ [7]. The semi-direct product of Virasoro and Kac-Moody algebras have highest weight representations $|\ell\rangle$ of "isospin" ℓ: $L_n|\ell\rangle = J_n^a|\ell\rangle = 0$, $n > 0$, $J_0^a|\ell\rangle = t_\ell^a|\ell\rangle$. Unitarity requires the "level" $k \in \mathbb{Z}$ and the compatibility of current and conformal algebra requires a finite number of representations in the theory, whose possessing isospin $0 \leq \ell \leq k/2$. The conformal charge is $c^g = 3k/(k+2)$. The partition function on a torus has the familiar form

$$Z(\tau) = \sum_{\lambda,\lambda'=1}^{k+1} \mathcal{N}^{aff}_{\lambda\lambda'}\, \chi_\lambda^{aff}(\tau)\, \overline{\chi_{\lambda'}^{aff}(\tau)} \tag{3.2}$$

where $\chi_\lambda^{aff}(\tau) = \mathrm{Tr}_{(\ell)}\left(e^{i2\pi\tau(L_0-c/24)}\right)$ is the character of representation with $\ell = (\lambda-1)/2$ [8], and $\mathcal{N}^{aff}_{\lambda,\lambda'}$ are the multiplicities of fields. A modular invariance problem can be consider as in the conformal case [16]. The result is the following theorem:

Table II Examples of statistical models in the universality classes derived by the A-D-E classification.

m	c	Series A-A	Series D-A	Cases E-A
3	1/2	Ising (A_2, A_3)		
4	7/10	Ising tricr. (A_3, A_4)		
5	4/5		Potts 3-states (A_4, D_4)	
6	6/7		Potts tricr. (D_4, A_5)	
⋮				
11	21/22			New Model (A_{10}, E_6)
12	25/26			" " (E_6, A_{11})
⋮				
18		Restricted		" " (A_{17}, E_7)
19		Solid-on-Solid		" " (E_7, A_{18})
⋮			Potts q-states	
29				" " (A_{28}, E_8)
30				" " (E_8, A_{29})
⋮				
∞	1		q=4	

Table III List of theories with an infinite dimensional Lie algebras generalizing the Virasoro algebra. We report conformal charges c and some corresponding statistical models. Theories with a star are A-D-E classified.

	c =	1/2	7/10	4/5	1	3/2	2	3
		Ising	Ising tric.	Potts 3q	Ashkin Teller	....		
*) Minimal Conformal		1/2,	7/10,	4/5,..,1				$c = 1 - \frac{6}{m(m+1)}$
*) Min. Superconformal N=1			7/10,		,1,..3/2			$c = \frac{3}{8}\left(1 - \frac{8}{m(m+2)}\right)$
*) Min. Superconformal N=2					,1...........,3			$c = 3 - \frac{6}{m}$
*) $A_1^{(1)}$ affine algebra					,1...........,3			$c = \frac{3k}{k+2}$
*) $\mathbb{Z}_K$ Parafermionic alg.		1/2,		4/5,	,1.......,2			$c = 2\,\frac{k-1}{k+2}$
$\mathbb{Z}_3$ algebra				4/5,...........,2				$c = 2 - \frac{24}{p(p+1)}$
*) S_3 & spin 4/3 algebra				6/7,........,2				$c = 2 - \frac{24}{p(p+4)}$

Any positive modular invariant of SU(2) current algebra is classified by one Lie algebra A,D or E.

In fact we can prove the relation [4,16]

$$\mathcal{N}^{conf}_{h_{rs},\bar{h}_{rs}} = \mathcal{N}^{aff}_{r,\bar{r}} \otimes \mathcal{N}^{aff}_{s,\bar{s}} \qquad (3.3)$$

between Virasoso invariants $\mathcal{N}^{conf}$ of charge c(m) and pairs of current algebras invariants $\mathcal{N}^{aff}$ of levels k = m-2 and k' = m-1.

These modular properties come from representation theory: the "coset-space construction" [8] builds the Hilbert space of states of Virasoro representations by using $A_1^{(1)}$ Hilbert spaces as follows

$$\text{Virasoro } (c(m)) = \frac{g(\text{level } m\text{-}2) \otimes g'(\text{level } 1)}{h(\text{level } m\text{-}1)} \qquad (3.4)$$

for $g = g' = h = A_1^{(1)}$ (Notice that $c(m) = c^g + c^{g'} - c^h$). It follows that Virasoro and Affine characters are related functions, $\chi^{conf}_{c(m)}$ transforms in τ as the pair $(\chi^{aff}_{m-2}, \chi^{aff}_{m-1})$, thus giving an insight in eq.(3.3).

The SU(2) current algebra Hilbert space is employed in building representations of other infinite Lie algebras. These algebras generalize Virasoro by adding N = 1 [28] or N = 2 [29] two dimensional supersymmetry or $\mathbb{Z}_K$ discrete symmetry [30]. It follows that the A-D-E classification of modular invariant partition functions can been extended similarly [5,17,31]. Table III report these extended algebras, together with their conformal charges. The other one in Table III is probably related to SU(3) current algebra [32]. The search of explicit statistical models or field theories fitting in the classification is actively pursued [33]. Models with c > 1 are interesting non-minimal conformal theories.

Within the non-trivial structures for $c \leqslant 1$ already investigated, we mention the experimental realization of supersymmetry in two dimensional critical phenomena [34]. For c = 7/10 the tricritical Ising model has N = 1 global supersymmetry; it is a model for helium adsorbed on a surface. For c = 1, the Ashkin-Teller model has a critical line, points of it possess N = 1 or N = 2 supersymmetry [35]. Supersymmetry at the critical point implies non-trivial Ward Identities between correlations of fermionic and bosonic fields; however a simple physical picture is still lacking.

I thank my colleagues in this work C. Itzykson and J.B. Zuber and also J. Cardy, P. di Francesco, D. Friedan, I. Kostov, V. Pasquier and H. Saleur for discussions. I acknowledge the Angelo Della Riccia foundation for partial support.

REFERENCES

[1] A.M. Polyakov, Phys. Lett. 103B, 207, 211 (1981);
D. Friedan, in "Recent Advances in Field theory and Statistical Mechanics", Les Houches School 1982, J.B. Zuber and R. Stora editors, North Holland 1984.

[2] J.L. Cardy, in "Phase Transitions and Critical Phenomena", Vol 11, editors C. Domb and J. Lebowitz, Academic Press, London 1986.
[3] A.A. Belavin, A.M. Polyakov and A.B. Zamolodchikov, Nucl. Phys. B241, 333 (1984).
[4] A. Cappelli, C. Itzykson and J.B. Zuber, Nucl. Phys. B280, 445 (1987).
[5] A. Cappelli, Phys. Lett. B185, 82 (1987).
[6] A. Cappelli, C. Itzykson and J.B. Zuber, Saclay preprint 87-59, Comm. Math. Phys. 112 (1987).
[7] For a review, see P. Goddard and D. Olive Int. J. Mod. Phys. A1 (1986) 303;
V.G. Kaç, "Infinite Dimensional Lie Algebras", Cambridge Un. Press, London 1985.
[8] P. Goddard, A. Kent and D. Olive, Comm. Math. Phys. 103, 105 (1986).
[9] D. Friedan, Z. Qiu and S. Shenker, Phys. Rev. Lett. 52, 1575 (1984); in "Vertex Operators in Mathematics and Physics", J. Leposwky, S. Mandelstam and I. Singer eds. Springer, Berlin 1985; Comm. Math. Phys. 107, 535 (1986).
[10] H. Saleur, J. Phys. A20 (1987) 82;
P. Di Francesco, H. Saleur and J.B. Zuber, Nucl. Phys. B285, 454 (1987), J. Stat. Phys. in press.
[11] J.L. Cardy, Nucl. Phys. B270, 186 (1986).
[12] V.S. Dotsenko and V.A. Fateev, Nucl. Phys. B240 (1984) 312, B251 (1985) 691; Phys. Lett. B154 (1985) 291.
[13] C. Itzykson and J.B. Zuber, Nucl. Phys. B275, 580 (1986).
[14] A. Rocha-Caridi, in "Vertex operators in mathematics and physics", op. cit.
[15] R.C. Gunnings, "Lectures on Modular Forms", Princeton Un. press, 1962.
[16] D. Gepner, Nucl. Phys. B287, 111 (1987).
[17] D. Gepner and Z. Qiu, Nucl. Phys. B285, 423 (1987).
[18] J.E. Humpreys, Introduction to Lie Algebras and representation theory, Springer, Berlin 1972.
[19] A. Kato, Mod. Phys. Lett. A2, 585 (1987).
[20] P. Slodowy, in "Algebraic geometry", J. Dolgachev ed. Lecture Notes in Mathematics 108, Springer Berlin 1983.
[21] J. Cardy, private communication.
[22] W. Nahm, private informations. See also the works:
P. Bouwknegt and W. Nahm, Phys. Lett. B184, 359 (1987);
D. Bernard, Ph.D. Thesis, Université de Paris-Sud (Orsay), January 1987;
P. Bouwknegt, Amsterdam preprint ITFA 87/05;
P. Bowcock and P. Goddard, Nucl. Phys. B285, 651 (1987).
[23] P. Christe, Ph. D. Thesis, Bonn. Univ. 1986, IR-86-32, P. Christe R. Flume, Phys. Lett. B188 (1987) 219.
[24] D.A. Huse, Phys. Rev. B30, (1984) 3908;
A.W.W. Ludwig and J.L. Cardy, Nucl. Phys. B285 (1987) 687.
[25] B. Nienhuis, J. Stat. Phys. 34 (1984) 73, J. Phys. A15 (1982) 189.
[26] V. Pasquier, Nucl. Phys. B285, 162 (1987), J. Phys. A20, L221, L217 (1987), Saclay preprint 87/014,062.

[27] E. Witten, Comm. Math. Phys. 92, 455 (1984);
V.G. Knizhnik and A.B. Zamolodchikov, Nucl. Phys. B247 (1984) 83;
D. Gepner and E. Witten, Nucl. Phys. B278 (1986) 493.
[28] D. Friedan, Z. Qiu and S. Shenker, Phys. Lett. 151B, 37 (1985);
M. Bershadsky, V. Zmizhnik and M. Teitelman, Phys. Lett. 151B, 31 (1985
[29] W. Boucher, D. Friedan and A. Kent, Phys. Lett. 172B, 316 (1986);
P. Di Vecchia, J.L. Petersen, M. Yu and H.B. Zheng, Phys. Lett. 174B (1986) 280;
V.G. Kaç and I.T. Todorov, Comm. Math. Phys. 102, 337 (1985);
E.B. Kiritsis, Cal. Tech. preprint 68-1390, 1987.
[30] V.A. Fateev and A.B. Zamolodchikov, Sov. Phys. JETP 62, 215 (1985).
[31] D. Gepner, Princeton preprint March 1987;
S.K. Yang and F. Ravanini, Nordita preprint 87/25-P.
F. Ravanini, Nordita preprint 87/43-P.
[32] V.A. Fateev and A.B. Zamolodchikov, Nucl. Phys. B280, 644 (1987) and Landau Institute preprint 1986;
E. Verlinde et al, Utrecht preprint 1987.
[33] M. Jimbo, T. Miwa and M. Okado, Nucl. Phys. B275 (1986) 517, RIMS preprints 564, 566, 572;
H.J. De Vega and M. Karowski, Nucl. Phys. B285, 619 (1987);
V. Rittenberg, in Paris-Meudon Colloqium; 22-26 September 1986; H. De Vega et al editors.
[34] D. Friedan, Z. Qiu and S. Shenker, Physics Today, Jan. 1987 p.S19.
[35] Von Gahlen and V. Rittenberg, J. Phys. A20, 227 (1987); A19, L1039;
D. Friedan and S. Shenker, Chicago Un. preprint EFI 86-17,
S.K. Yang, Nucl. Phys. B285, 639 (1987), 183;
S.K. Yang and H.B. Zeug, Nucl. Phys. B285, 410 (1987);
H. Saleur, Saclay preprint 87-46, April 1987;

THE BREAKDOWN OF DIMENSIONAL REDUCTION

A. Kupiainen

Helsinki University, TFT
00170 Helsinki, Finland

1. *QFT* WITH RANDOM COUPLINGS

Some of the most interesting encounters of field theory and statistical mechanics have been in recent years in the domain of disordered systems i.e. field theories with random couplings. Apart from occasional speculations about the relevance of such constructs for physics in the Planck scale or beyond, they have been used to model the coupling of impurities to (equilibrium) statistical systems.

Consider for instance the $O(N)$ sigma models with action (on a lattice)

$$S(\phi) = \beta \sum_x [(\nabla \vec{\phi}(x))^2 + \vec{h} \cdot \vec{\phi}(x)] \tag{1}$$

where $\vec{\phi}(x)$ maps Z^d into S^{N-1} and $\vec{h}$ is a constant external field. The usual equilibrium properties of (1) are well known. The principle of universality states that the qualitative properties of the phase diagram (as a function of β and h) as well as the quantitative properties (critical exponents) of the critical point are independent of the particular form of (1), depending only on the symmetry group $O(N)$ and the space dimension d (as long as we stay in the class of short range interactions).

In particular, two invariants depending only on N are the *upper* and *lower critical dimensions* d_u and d_ℓ defined as the smallest dimension above which the critical point is gaussian and the smallest dimension above which there is spontaneous symmetry breaking, respectively. As is well known,

$$d_u(N) = 4 \tag{2}$$

$$d_\ell(N) = \begin{cases} 1, & N = 1 \\ 2, & N > 1 \end{cases} . \tag{3}$$

Moreover, at d_ℓ there is interesting N-dependent behaviour: for $N = 2$ the theory has a massless large β phase whereas for $N > 2$ the mass is positive for all β, $m \sim e^{-\beta \cdot \mathrm{const}}$ as $\beta \to \infty$.

What happens as impurities are coupled to (1)? A large class of impurities is modelled by adding to (1) random couplings

$$\delta S = \sum_{|x-y|=1} \delta J(x,y) \vec{\phi}(x) \vec{\phi}(y) + \sum_x \delta \vec{h}(x) \vec{\phi}(x) \tag{4}$$

with $\delta J(x,y)$, $\delta h(x)$ suitable random variables. (4) includes diluted ferro- and anti-ferromagnets, spin glasses, random field models etc. Far from a general picture exists for (4), neither a general method of attack; only separate studies of separate submodels of (4). Even qualitative structure of the phase diagram is far from established. However, if the randomness δJ and δh are taken small, it turns out that δh dominates: δJ does not change the ferromagnetic transition (i.e. d_ℓ will not change) whereas δh will have drastic effects however small it is.

The most controversial case during the past ten years has been the $N = 1$ model with $\delta J = 0$ i.e. the Random Field Ising Model (RFIM)

$$S(\phi, h) = \beta \sum_x ((\nabla\phi)^2 - h(x)\phi(x)) \tag{5}$$

where $h(x)$ are independent random variables with variance ε^2:

$$\overline{h(x)h(y)} = \varepsilon^2 \delta(x-y) \quad . \tag{6}$$

The mathematical difficulty of RFIM (as well as other random systems) arises from the quenched h-averages, taken only after the statistical ones. E.g. the Green functions are

$$G(x_1, ..., x_n) = \overline{< \phi(x_1), ..., \phi(x_n) >} \tag{7}$$

with

$$< F(\phi) > = \mathcal{N}^{-1} \sum_\phi e^{-S(\phi,h)} F(\phi) \tag{8}$$

and the sum is over $\phi: \ Z^d \to \{-1,+1\}$.

Apart from having several experimental realizations, the RFIM has gained interest because it has been a theoretical puzzle for many years. Not unrelatedly, various theoretical tricks have been applied to it: the replica trick, supersymmetry and dimensional reduction to mention some. We will next review the two main approaches that have been taken to the problem and that lead to conflicting conclusions.

2. TWO WAYS TO USE CLASSICAL SOLUTIONS

The first argument for d_ℓ was given by Imry and Ma [1] in 1975. They considered zero temperature i.e. $\beta = \infty$. Hence for a given h we should find the ground state of (5) ϕ_h and then G_n of (7) are given by

$$\prod_i \phi_h(x_i) \quad . \tag{9}$$

Imry and Ma studied the probability of ϕ_h being disordered as follows. As is well known, an Ising configuration $\phi(x) = \pm 1$, $x \epsilon Z^d$, specifies a set of disjoint closed surfaces $\{S\}$ in (the dual of) Z^d, and the action becomes

$$S = \beta \sum_S A(S) - \beta \sum_{V+} h(x) + \beta \sum_{V-} h(x) \tag{10}$$

where $A(S)$ is the area of S and $V^{\pm}$ are the $\pm$ regions determined by $\{S\}$ (we put say + boundary conditions). For $h = 0$ the surfaces are suppressed, and as temperature is increased, they are excited untill they condense at the critical temperature. The question now is whether a random h has sizable probability of creating surfaces. Thus, consider one surface. Its energy with respect to the $\phi = 1$ state is

$$E(S) = A(S) - 2 \sum_{V(S)} h(x) \equiv A(S) + H(S)$$

with $V(S)$ the volume enclosed by S. $H(S)$ is a sum of independent random variables and thus has variance $4\varepsilon^2 V(S)$. Hence

$$\text{Prob}\ (E(S) < 0) \leq \exp - \frac{A(S)^2}{8\varepsilon^2 V(S)} \leq \exp - \frac{c}{\varepsilon^2} R^{d-2} \tag{11}$$

for a gaussian distribution and a cubic S of side R. Thus, for $d > 2$, the larger the S, the more unprobable it is, whereas for $d < 2$ the situation is reversed. At $d = 2$ there is a finite probability to have surfaces on all scales. Imry and Ma concluded, that $d_\ell = 2$.

Although simple, the Imry-Ma argument has several problems. The first is that of entropy: if we estimate the probability that a given point has a S of area R^{d-1} around it, we find $\exp[-O(1)R^{d-1}]$ surfaces. This number dominates (11), for R large. Of course most such events are not independent, and indeed this counting problem was later solved by Fisher, Fröhlich and Spencer [2] using a coarse graining method, which we, too, will use below. The second problem is that of surfaces within surfaces: if we are at d_ℓ we should expect the ground state typically to have surfaces of all scales within each other. It is not clear at all how to extend the Imry-Ma argument to such configurations.

These problems and the lack of means to perform more detailed analysis led to a wide acceptance of the so called *dimensional reduction* argument [3-6], most elegantly formulated by Parisi and Sourlas [5].

Starting from a general scalar field action in d dimensions

$$S(\phi, h, d) = \int d^d x \left[\frac{1}{2}(\nabla\phi)^2 + V(\phi(x) + h(x)\phi(x)\right] \tag{12}$$

we look again for classical solutions

$$-\Delta\phi + V'(\phi) + h = 0 \quad . \tag{13}$$

This stochastic differential equation may be written, by a Faddeev-Popov type trick, as a supersymmetric field theory, with the action (12) but ϕ replaced by a superfield $\Phi = \phi + \bar{\theta}\psi + \bar{\psi}\theta + a\bar{\theta}\theta$ and h but to zero

$$(13) \cong S(\Phi, 0, d) \tag{14}$$

($\cong$ means Green functions are equal). Finally, (14) can be dimensionally reduced: the generating functional $W_{SS}(J,d)$ of (14) equals that of (12) with $h=0$ and $d \to d-2$:

$$W_{SS}(J,d) = W(j, d-2) \tag{15}$$

provided we take J supported on a $d-2$ dimensional hyperplane in R^d. Thus phase diagram and critical properties of the random field model should coincide with those of the non-random one in two less dimensions. In particular, from (2) and (3) one gets the conjectures $d_u = 6$ and $d_\ell = 3$. This conflict of the value of d_ℓ with that predicted by Imry and Ma was the source of a long controversy. (It should be stressed that Parisi and Sourlas only intended their argument for $d_u = 6$, which is believed to be correct). The decisive question thus is, whether the RFIM in three dimensions is ordered at low temperatures.

An important step in the resolution of the controversy was given by Imbrie [7], who showed, that *at* $\beta = \infty$ and $d = 3$ the model has symmetry breaking. Since this result is still not in conflict with dimensional reduction [8] (the $d = 1$ Ising model has symmetry breaking at $\beta = \infty$ but is disordered for $\beta < \infty$), a separate low temperature analysis in needed.

3. FROM SCALING TO RENORMALIZATION GROUP

I will now scetch a rigorous proof by J. Bricmont and myseld [9] which shows that $d_\ell \leq 2$. The content of the proof is to extend the Imry-Ma scaling idea to an exact renormalization group (RG) analysis, along the lines of [10]. We start from the formula (10) for the action, which allows us to write the partition function

$$\sum_{\phi} e^{-S} = e^{\beta \sum_V h(x)} \sum_{\{S_\alpha\}} \prod_\alpha \zeta(S_\alpha) e^{\beta \sum_{V-} H(x)} \tag{16}$$

where $\zeta(S) = \exp[-A(S)]$, $H = -2h$, and we pulled out the zeroth order gound state energy.

The Imry-Ma argument may now be restated as the potential non-convergence of (16): the effective activity of a single surface once we ignore surfaces within surfaces is $\exp[-\beta A(S) + \beta H(S)]$ and with probability (11) this is not small ($d = 3$).

Thus surfaces on all scales may be bad but the badness is "irrelevant". We are thus naturally led to a RG analysis, which in the Wilsonian form consists of the following steps:

(i) Integrate out from a system with action S degrees of freedom of length scales $\leq L =$ fixed, $O(1)$ number.

(ii) Rescale distances by L, thereby recovering a new action RS describing the rescaled low momentum fluctuations.

(iii) Iterate the RG, hopefully ending up on a manifold of finite dimension in "the space of $S's$". These are the relevant variables of the model.

Let us trace through (i) - (iii) in our case.

(i) "Fluctuations" are surfaces. We start by "summing out" of (16) the small ones of diameter $d(S) < L$. Obviously these will lie in the + or − regions $V^{\pm}$ determined by the large surfaces S_α^L and their sum will contribute free energies $\delta h^{\pm}$ there:

$$Z = e^{\beta \sum (h+\delta h^+)} \sum_{\{S_\alpha^L\}} \prod_\alpha \zeta(S_\alpha^L) e^{\beta \sum_{V^-} (H+\delta H)} \tag{17}$$

(we put $\delta H(x) = \delta h^-(x) - \delta h^+(x)$).

(ii) Before rescaling, we also need to get rid of $\leq L$ fluctuations in the S_α^L: cover $\bigcup_\alpha S_\alpha^L$ by L-cubes and divide the resulting set into components $\bigcup_\alpha LS'_\alpha$. S'_α are the rescaled surfaces, again of all sizes ≥ 1.

Regrouping terms in (17) results in

$$Z = e^{\beta \sum (h+\delta h^+)} Z' \tag{18}$$

$$Z' = \sum_{\{S'_\alpha\}} \prod_\alpha \zeta'(S'_\alpha) e^{\beta' \sum_{V'} H'(x)} \tag{19}$$

with $\zeta'(S')$ a sum of products of $\zeta(S)$ and

$$H'(x) = L^{-2} \sum_y (H(x) + \delta H(x)) \tag{20}$$

with $y \epsilon L$-cube centered at Lx. Finally

$$\beta' = L^2 \beta \quad . \tag{21}$$

This peculiar scaling of β, reflected in the definition of the effective field (20) is chosen because

$$\zeta'(S') \leq e^{-\beta' A(S')} \tag{22}$$

(22) follows, since surfaces obviously scale as L^2. ((22) has to be modified slightly for surfaces which degenerate upon our coarse graining procedure: consider e.g. a long thin tube which scales only with factor L. We ignore such details in the present discussion, see [9].)

(iii) (19) is of the same form as the Z we started with and we may thus iterate the RG. How do the variables flow? The most crucial one is H' of (20). We wish to find its probability distribution, given that of H.

Let us study H' in three steps of approximation

(A) Ignore first δH in (20). Then the variance of H' is

$$(L^{-2} \sum H)^2 = L^{-1} H^2 = L^{-1}(4\varepsilon^2) \equiv 4\varepsilon'^2 \quad , \tag{23}$$

i.e. ε is an irrelevant variable. The flow in this approximation is

$$\varepsilon'^2 = L^{-1}\varepsilon^2, \quad \beta' = L^2 \beta \quad . \tag{24}$$

This is the Imry-Ma scaling, which hence is our *RG* in the zeroth order approximation where surfaces within surfaces, i.e. δH, are ignored. The fixed point is zero temperature, zero field, i.e. we have symmetry breaking.

(B) δH is the free energy of small surfaces, and is a function of H. Assume first that H is nowhere too big; $|H(x)| < \delta$ with δ chosen so that $A(S) >> \delta \cdot \text{Volume}\ (S)$ for S small, $d(S) < L$. Thus δ is $O(1)$, L dependent. Note that this is quite probable: probability for $|H| > \delta$ is $\exp[-O(1/\varepsilon^2)]$. The point is, that for such H, small surfaces are suppressed, δH is computable via a convergent expansion and is almost local in H (with exponential tails) with

$$|\delta H(x)| \leq e^{-O(\beta)} \tag{25}$$

(24) is now replaced by

$$\varepsilon'^2 = L^{-1}\varepsilon^2 + e^{-O(\beta)} \tag{26}$$

and our conclusion about the flow would be unchanged.

(C) Even if original H were bounded as in (B), this would *not* hold upon iteration since H is *relevant* variable: the sum in (2) involves L^3 terms and hence δ expands by a factor of L. Thus we have to embark on the large field problem. The problem of large H is reflected in that, the δH is not a small perturbation; its computation in fact involves *small denominators*. Our solution of this problem is that in the large field region we should not compute δH, i.e. we will not exponentiate the small surface sum. Physically, it is clear, that large fields tend to create domain walls around them, however, this will happen only up to certain scale, namely the one where $A(S) > |H|$ for all S. E.g., given $H(0) = L^n$, after n steps of iteration every surface will dominate this H. Provided all effective fields on all intermediate scales were $\geq \delta$, at this point we can compute $\delta H(0)$ and it will be small. We will loose information on surfaces within the L^n box, but the event $H(0) = L^n$ has probability $\exp[-L^{2n}/\varepsilon^2]$ by far smaller than the free energy in such a box.

A systematic way of organizing this idea for arbitrary field configuration is to iterate probability estimates for the large field variables. They are the $|H(x)|'s$ and they have a recursion analogous to (20) without the δH. The result of our analysis is (20) and (21), (22) together with the estimate

$$\text{Prob}\ (|H^n(x)| \geq \delta) \leq e^{-O(\varepsilon_n^{-2})}$$

(with $\varepsilon_n^2 = L^{-n}\varepsilon^2$).

This flow easily leads to the result on symmetry breaking.

4. CONCLUSION

Dimensional reduction breaks near the lower critical dimension, however we have little clue of the actual mechanism of its breakdown. Approximate *RG* analysis shows [11] that the breaking (for interface divergence) takes formally place arbitrary near *below* the upper critical dimension.

What happens at the lower critical dimension $d_\ell = 2$ (presumably) is a much more subtle question. We have constructed an exactly soluble model (a hierarchical model) which has order for $d > 2$ and disorder for $d = 2$ but simultaneously long range (logarithmic) correlations in $d = 2$. This raises the question whether in the full RFIM at $d = 2$ there is a possibly scale invariant phase for large β. The answer to this question is fully non-perturbative: the corrections to the perturbative small field analysis

$$\begin{aligned} \varepsilon'^2 &= \varepsilon^2 + e^{-O(\beta)} \\ \beta' &= L^2 \beta \end{aligned}$$

which flows to $\beta = \infty$ and $\varepsilon = \varepsilon^* \neq 0$, come solely from the large fields. We believe, although cannot prove, that these corrections drive ε eventually up to the large disorder region, where mass-gap has been shown to exist [12]

References

[1] Y. Imry, S.K. Ma, Phys. Rev. Lett. **35**, 1399 (1975).

[2] D. Fisher, J. Fröhlich, T. Spencer, J. Stat. Phys. **34**, 863 (1984).

[3] A. Aharony, Y. Imry, S.K. Ma, Phys. Rev. Lett. **37** 1367 (1976).

[4] A.P. Young, J. Phys. **C10**, L257 (1977).

[5] G. Parisi, N. Sourlas, Phys. Rev. Lett. **43**, 744 (1979), and Nucl. Phys. **B206**, 321 (1982).

[6] H.S. Hogon, D.J. Wallace, J. Phys. **A14**, L257 (1981).

[7] J. Imbrie, Phys. Rev. Lett. **53**, 1747 (1984), Commun. Math. Phys. **98** 145 (1985).

[8] E. Brezin, H. Orland, Saclay Preprint (1987).

[9] J. Bricmont, A. Kupiainen, Phys. Rev. Lett. **59**, 1829 (1987).

[10] K. Gawedzki, K. Kotetci, A. Kupiainen, J. Stat. Phys. **47**, 701 (1987).

[11] D. Fisher, Phys. Rev. Lett. **56**, 1964 (1986).

[12] J. Imbrie, J. Fröhlich, Commun. Math. Phys. **96**, 145 (1984).

VANISHING VACUUM ENERGIES

FOR NONSUPERSYMMETRIC STRINGS

Gregory Moore

Institute for Advanced Study
Princeton, N.J. 08540, USA

We review a mechanism which has been proposed for the vanishing of the cosmological constant in nonsupersymmetric superstring theories. We discuss in particular how the idea might work at two and three loops, although no concrete model is proposed. One of the main difficulties with the approach is finding explicit models exhibiting the mechanism: only two examples in two uncompactified dimensions are known.

1. INTRODUCTION

The problem of the cosmological constant is an old and vexing one. Recall that in Einstein's equations

$$R_{\mu\nu} - \frac{1}{2}g_{\mu\nu}R = 8\pi G T_{\mu\nu} \tag{1.1}$$

we can't set $T_{\mu\nu} = 0$ in the vacuum. Rather we must calculate

$$\langle 0|T_{\mu\nu}|0\rangle = \hbar\Lambda g_{\mu\nu} \tag{1.2}$$

Dimensional analysis suggests that in quantum gravity we should get

$$\Lambda \sim O(1)(G\hbar)^{-2} \sim 10^{76}(GeV)^4 \tag{1.3}$$

and explicit calculations in a variety of models bear that out. On the other hand, observationally we know that

$$|\Lambda| \sim 10^{\ \ 120}(G\hbar)^{\ \ 2} \tag{1.4}$$

Not long after the invention of supersymmetry it was pointed out [1] that the cancellation of vacuum energies between bosons and fermions in a supersymmetric theory might go a long way towards explaining the observed (near?) vanishing of the cosmological constant. Unfortunately, supersymmetry is not yet observed in nature, and, if it exists at all, it is broken on a scale of roughly at least $\sim 1TeV$. Thus, we would expect $|\Lambda| > 10^{-64}(G\hbar)^{-2}$ after supersymmetry breaking. Even if we use - for example- supersymmetric sum rules to arrange that the vacuum energy is zero after supersymmetry breaking, there will be small corrections

$$\Lambda \sim 10^{-68}(G\hbar)^2 \sim 10^8(GeV)^4 \tag{1.5}$$

from electroweak symmetry breaking and perhaps an additional

$$\Lambda \sim 10^{-78}(G\hbar)^2 \sim 10^{-2}(GeV)^4 \tag{1.6}$$

from QCD effects. In short, there are sources of vacuum energy from every length scale. Something somehow must cancel all these disparate contributions to the vacuum energy.

In these lectures we will review a recently proposed idea for how superstring theories in *nonsupersymmetric* backgrounds might nevertheless have a vanishing cosmological constant [2] . The mechanism is based on the existence of a symmetry of the moduli space of Riemann surfaces, and is therefore truly stringy. At present there are two major difficulties with this approach. First, specific examples illustrating the idea are hard to come by: there are only two known examples, and these are string compactifications to two uncompactified dimensions. [1]Second, the extension to higher loops will be subtle, if not impossible to accomplish. We describe some small steps in this direction in section five.

We begin by describing the general idea. As mentioned, in quantum gravity we are obliged to *calculate* the vacuum energy. In string theory there is a very nice formula for it- at least at one loop - which may be derived as follows.

First, recall that for a free scalar field in D spacetime dimensions the vacuum energy can be calculated from the Euclidean path integral:

$$\begin{aligned} e^{-VE} &= \int D\phi e^{-\int \mathcal{L}} \\ &\equiv Det^{-1/2}(-\partial^2 + m^2) \end{aligned} \tag{1.7}$$

So

$$\begin{aligned} E &= +\frac{1}{2}\hbar \int \frac{d^D k}{(2\pi)^D} log(k^2 + m^2) \\ &= -\frac{1}{2}\hbar \int_0^\infty \frac{ds}{s} \int \frac{d^D k}{(2\pi)^D} e^{-s(k^2+m^2)} \\ &= -\frac{1}{2}\hbar \int_0^\infty \frac{ds}{s} (2\pi s)^{-D/2} e^{-\frac{1}{2}m^2 s} \end{aligned} \tag{1.8}$$

If the field has spin we sum over the polarizations, and if the field is a fermion we change the sign. Thus the contribution of all the particles in string theory is

$$-\frac{1}{2}\hbar(4\pi T)^{D/2} \sum_i (-1)^{F_i} \int_0^\infty \frac{ds}{s} (2\pi s)^{-D/2} e^{-\frac{1}{2}(m_i^2/4\pi T)s} \tag{1.9}$$

[1] There is a difficulty with the four-dimensional model presented in [2] which is described briefly at the end of section three.

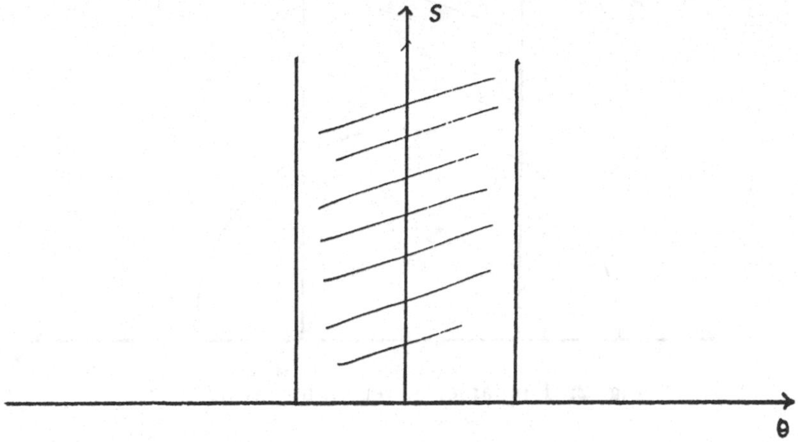

Fig. 1. Fundamental region for field theory

where T is the string tension. We are only summing over the contributions of physical particles, so it is best to use the formulas of light-cone gauge. Then we may express the mass-squared of particles in terms of the Virasoro operators $L_0^{\ell c} = \frac{1}{2}\vec{p}^2 + 4\pi TN + L_0^{int} - \mu$, where $\vec{p}$ is the transverse momentum, N is the number operator for oscillators in uncompactified dimensions, L_0^{int} is the Virasoro operator for whatever conformal field theory cancels the conformal anomaly and μ is a normal ordering constant. The mass-squared is then simply $L_0^{\ell c} + \overline{L}_0^{\ell c}$ at $\vec{p} = 0$ (we work only with closed strings) . The spectrum of a conformal field theory satisfies the constraint that $(L_0^{\ell c} - \overline{L}_0^{\ell c})/4\pi T$ has integral eigenvalues, and physical states are picked out by the zero eigenvalues. Therefore, denoting the spectrum of $N + (L_0^{int} - \mu)/4\pi T$ by $\{h_a\}$ we observe that we can rewrite (1.9) as

$$
\begin{aligned}
&-\frac{1}{2}\hbar(4\pi T)^{D/2}\int_0^\infty \frac{ds}{s}(2\pi s)^{-D/2} \\
&\qquad \int_{-\frac{1}{2}}^{+\frac{1}{2}} d\theta \sum_{a,b}(-1)^F e^{h_a(-\frac{1}{2}s+2\pi i\theta)} e^{\overline{h}_b(-\frac{1}{2}s-2\pi i\theta)}
\end{aligned}
\tag{1.10}
$$

On the other hand, defining $2\pi i\tau = -\frac{1}{2}s + 2\pi i\theta$ and $q = e^{2\pi i\tau}$, we can expand the supertrace

$$
\begin{aligned}
STrq^{L_0^{\ell c}/4\pi T}\overline{q}^{\overline{L}_0^{\ell c}/4\pi T} &= \int \frac{d^{D-2}p}{(2\pi)^{D-2}} \sum_{a,b}(-1)^F q^{p^2/8\pi T + h_a}\overline{q}^{p^2/8\pi T + \overline{h}_b} \\
&= T^{(D-2)/2}(2\pi Im\tau)^{-(D-2)/2}\sum(-1)^F q^h \overline{q}^{\overline{h}}
\end{aligned}
\tag{1.11}
$$

so our final formula is

$$
\Lambda = -\frac{\hbar}{4\pi}T\int \frac{d^2\tau}{(Im\tau)^2} STrq^{L_0^{\ell c}}\overline{q}^{\overline{L}_0^{\ell c}} \tag{1.12}
$$

where $4\pi T$ has been absorbed into $L_0^{\ell c}$.

A crucial question concerns the appropriate region of integration. In field theory we integrate over the region in Fig. 1, and the integral diverges at $s \sim 0$. But in

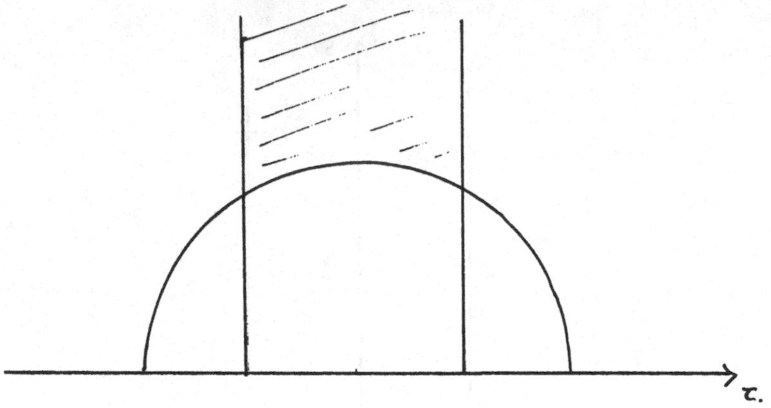

Fig. 2: Fundamental region for string theory

string theory we can interpret the trace as a path integral for a world sheet with the topology of a torus with modular parameter τ. From the invariances of string theory it is "clear" that we should integrate only over a fundamental domain for the modular group in order to avoid double counting, shown in Fig. 2.

This derivation, which is very well known, [3] [4] has several notable features, among these are:

1.) The analysis is completely general, and should apply to any compactification via a conformal field theory, whether or not, for example, the compactified "dimensions" have a spacetime interpretation.

2.) The restriction to the fundamental domain cuts off the ultraviolet divergence at $s \to 0$. As long as no physical partical has $m^2 < 0$ the integral converges. If there is ever a question of the order of integration we integrate $d\tau_1$ first and $d\tau_2$ second.

3.) If there is a spacetime supersymmetry then there are equal numbers of bosons and fermions, so the partition function is zero [2]. In this case the integral obviously vanishes. The nontrivial problem is to explain how the integral can vanish in a nonsupersymmetric theory.

Now that we have an elegant formula for the one-loop cosmological constant let us examine it more closely. Defining $h_{a,p} = p^2/8\pi T + h_a$ the integrand of (1.12) is simply

$$\sum_{h_a, h_b, p} d_{abp} q^{h_{a,p}} \bar{q}^{\bar{h}_{b,p}} \tag{1.13}$$

where d_{ab} gives the degeneracy of primary fields of weight $(h_a, \bar{h}_b)$. D. Friedan and S. Shenker made the wonderful remark that (1.13) resembles a pairing of sections of an infinite dimensional vector bundle over moduli space [5]. Let us see where that philosophy leads us.

[2] Not quite: all we really know is that the integral vanishes *after* integration over τ_1, but *before* integration over τ_2. In explicit examples it always seems to be the case that spacetime supersymmetric partition functions vanish even before the τ_1 integration. This distinction is quite important at higher loops.

We now regard the one-loop cosmological constant as some kind of inner product

$$\Lambda = \langle \psi_L | \psi_R \rangle \tag{1.14}$$

Forgetting, for the moment, that there *are* higher loops we are lead to reformulate the problem of the cosmological constant and ask: Why are ψ_L, ψ_R orthogonal? This question suggests that we might understand the observed vanishing of the cosmological constant as a selection rule.

Selection rules are based on symmetry (e.g. parity, in atomic physics) so we should look for a symmetry of moduli space under which ψ_L and ψ_R transform differently.

In order to implement this idea we will need to undertand (1.13) more concretely. One of the key properties satisfied by

$$(Im\tau)^{-(D-2)/2} \sum_{a,b} d_{ab} q^{h_a} \bar{q}^{\bar{h}_b} \tag{1.15}$$

is that it is modular invariant. It is in this way that modular forms have proved of some use in analyzing one-loop string amplitudes, as we now explain.

Recall that a function $f : \mathcal{H} \to \mathbf{C}$ which is holomorphic and satisfies

$$f(\frac{a\tau + b}{c\tau + d}) = (c\tau + d)^k f(\tau) \tag{1.16}$$

under modular tranformations

$$\begin{pmatrix} a & b \\ c & d \end{pmatrix} \in PSL(2, Z)$$

is a *modular form* of weight k. Actually, there is also a regularity condition at infinity, namely, in an expansion in q, f must be holomorphic at $q = 0$. The quantities which arise in string theory are typically ratios of modular forms, and the denominators are typically combinations of dedekind eta functions. We will also use the term modular form for such objects, i.e., functions f holomorphic on $\mathcal{H}$ and meromorphic at infinity. There exist functions which satisfy the transformation law (1.16)only for transformations lying in some subgroup $\Gamma_1 \subset \Gamma$. These are called modular forms for Γ_1.

The full modular group is generated by

$$S : \tau \to -1/\tau \qquad T : \tau \to \tau + 1 \tag{1.17}$$

subject to the relations $S^2 = 1, (ST)^3 = 1$. Each of the terms in the sum (1.13) is a "modular form" for the subgroup generated by T. But individual terms are not well-behaved under S, nor under combinations of T with S. Thus, in general we cannot expect that (1.15) can be written usefully in terms of modular forms. On the other hand, for some special backgrounds we can regroup the terms into a finite number of summands so that the trace takes the form

$$(Im\tau)^{-(D-2)/2} \sum_i f_i(\tau) \bar{g}_i(\tau) \tag{1.18}$$

such that f_i, g_i are modular forms for some (congruence) subgroup of the modular group. Backgrounds leading to partition functions of the form (1.18) are called "rational backgrounds." We will now explain the motivation for this terminology [2][6] [7].

Consider the partition function of the bosonic string compactified on a torus defined by a d-dimensional lattice Λ :

$$Z = \frac{1}{|\eta^{24}|^2}(Im\tau)^d \sum q^{p_L^2/2}\bar{q}^{p_R^2/2} \tag{1.19}$$

with

$$p_L = p/2 + w$$
$$p_R = p/2 - w$$
$$p \in \Lambda^* \qquad w \in \Lambda$$

The left- and right-moving momenta belong to the set of points $\Gamma = \frac{1}{2}\Lambda^* + \Lambda \subseteq \mathbf{R}^d$. This set of points defines a lattice if and only if the quadratic form of the lattice is rational [3]If Γ is a lattice, then its dual $\Gamma^* = 2\Lambda \cap \Lambda^*$ is a sublattice and we may consider p_L, p_R as vectors in Γ^* shifted by elements of the coset space Γ/Γ^*. Thus we may rewrite

$$Z = \frac{1}{|\eta^{24}|^2}(Im\tau)^d \sum_{\alpha\in\Gamma/\Gamma^*,\beta\in\Gamma/\Gamma^*} \vartheta_{L^*}\begin{bmatrix}\alpha\\0\end{bmatrix} \mathcal{M}_{\alpha\beta}\bar{\vartheta}_{L^*}\begin{bmatrix}\beta\\0\end{bmatrix} \tag{1.20}$$

where

$$\mathcal{M}_{\alpha\beta} = \begin{cases} 1 & \alpha+\beta \in \Lambda^*, \alpha-\beta \in 2\Lambda \\ 0 & \text{otherwise} \end{cases} \tag{1.21}$$

and ϑ_{L^*} is the theta function for the lattice L^*. The moral of the story is that the *rationality* of the background data leads to a partition function of the form (1.18)
The theta functions with characteristics are in fact modular forms because of a theorem of Schoeneberg [8]. These remarks generalize to orbifolds based on rational toroidal compactifications of the bosonic, type II, and heterotic strings. To make this plausible, recall a few basic facts about orbifolds.

Orbifolds are spaces obtained by taking the quotient of a torus by a discrete group action. In string compactification on an orbifold, for every $g \in G$, where G is the orbifold group, there is a twisted sector $\mathcal{H}_g$, and the total Hilbert space of the theory is $\oplus_g \mathcal{H}_g$. The contribution of a twisted sector $\mathcal{H}_g$ to the partition function is

$$tr_{\mathcal{H}_g} P q^{L_0}\bar{q}^{\bar{L}_0} \tag{1.22}$$

where

$$P = \sum_{h\in G} \mathcal{U}(h) \tag{1.23}$$

---

[3] The notion of the quadratic form being rational is sensible because the lattice breaks rotational symmetry to a subgroup of $GL(d, \mathbf{Z})$. Similarity transformations by this group preserve the property of rationality.

with $\mathcal{U}(h)$ the Hilbert space representation of the group element h. Denoting the path integral on a torus with boundary conditions twisted by h, g by $Z(h,g)$, we have the fundamental relation between the path integral and Hamiltonian formalisms:

$$tr_{\mathcal{H}_g} \mathcal{U}(h) q^{L_0} \bar{q}^{\bar{L}_0} \equiv Z(h,g) \tag{1.24}$$

Now consider the modular transformation properties of $Z(h,g)$. Recall that under global diffeomorphisms

$$\begin{aligned} \sigma^1 &\rightarrow d\sigma^1 + b\sigma^2 \\ \sigma^2 &\rightarrow c\sigma^1 + a\sigma^2 \end{aligned} \tag{1.25}$$

the boundary conditions are modified by

$$(h,g) \rightarrow (h^a g^c, h^b g^d) \tag{1.26}$$

Thus, the set of boundary conditions $\{(h,g)\}$ is broken up into orbits. The condition of modular invariance is that $Z(h,g)(a\tau + b/c\tau + d) = Z(h^a g^c, h^b g^d)(\tau)$ [4]. Therefore, if $g^m = 1, h^n = 1$, then the contribution of a twisted sector must be invariant under [9]

$$\Gamma_1(n,m) \equiv \begin{pmatrix} 1 mod n & 0 mod n \\ 0 mod m & 1 mod m \end{pmatrix} \tag{1.27}$$

Indeed, twisted scalars and fermions have partition functions which are expressible in terms of theta functions with characteristics, and the level matching conditions guarantee that these theta functions combine into modular forms for this subgroup[5]. Since the field content of orbifolds for the bosonic, type II, and heterotic strings consists of free bosons and fermions the total partition function

$$\int_{\mathcal{H}/\Gamma} \sum_{g,h} Z(h,g) \tag{1.28}$$

is of the form (1.18) . We conclude that orbifolds based on rational tori are also rational backgrounds.

There is another useful way to think about the integral (1.28) . Using the invariance of $Z(h,g)$ under Γ_1, we see that if we use the coset decomposition

$$\Gamma = \cup_i \Gamma_1(n,m)\gamma_i \tag{1.29}$$

then the terms in a modular orbit are in one-one correspondence with the cosets, and the contribution of a modular orbit to the partition function can be written in two ways:

$$\int_{\mathcal{H}/\Gamma} \sum_i Z(h,g)[\gamma_i] = \int_{\mathcal{H}/\Gamma_1(n,m)} Z(h,g) \tag{1.30}$$

---

[4] and so, I suppose, the Z-functor is a morphism from the "category" of boundary conditions to the "category" of partition functions.

[5] Well, almost. One can have a cancelling modular anomaly between left and right-movers. In this case one obtains modular forms with nontrivial multiplier systems.

Thus, in backgrounds of the kind we have been describing the partition function looks like

$$\Lambda = \sum_{orbits} \int_{\mathcal{H}/\Gamma_1} \bar{g}_i h_{ij} f_j (Im\tau)^k \frac{d^2\tau}{(Im\tau)^2} \qquad (1.31)$$

where h_{ij} is some numerical matrix. The matrix h and the modular forms f, g are functions of the background. Now let us work out the Friedan-Shenker proposal in some more detail in this setting.

Recall that the automorphy factor of a modular form of weight k defines a line bundle

$$\mathcal{L}^k \to \mathcal{H}/\Gamma_1 \qquad (1.32)$$

and we can interpret modular forms of weight k as sections of that line bundle:

$$M_k(\Gamma_1) = \Gamma(\mathcal{L}^k) \qquad (1.33)$$

Moreover, those sections form a vector space on which one may put the Petersson inner product:

$$\langle s_1 | s_2 \rangle = \int_{\mathcal{H}/\Gamma_1} \bar{s}_1 s_2 (Im\tau)^k \frac{d^2\tau}{(Im\tau)^2} \qquad (1.34)$$

In the program of [5] backgrounds correspond to vector bundles, and for rational backgrounds one has the finite dimensional vector bundle [6]:

$$\oplus_{i=1}^N \mathcal{L}^k \to \mathcal{H}/\Gamma_1 \qquad (1.35)$$

Given the Petersson inner product, we get an inner product on sections of this vector bundle, which is the above cosmological constant. According to [5] we should think of ψ_L, ψ_R as sections of this vector bundle [7].

We are finally ready to formulate the idea more precisely. Suppose we have a symmetry, that is, a holomorphic automorphism of the space $\mathcal{H}/\Gamma_1$. We will see in a moment that such spaces do have symmetries which lift naturally to an action on $\Gamma(\oplus\mathcal{L}^k \to \mathcal{H}/\Gamma_1)$, and that the action of the automorphism is unitary in this inner product. Thus, we can decompose $\Gamma(\oplus\mathcal{L}^k)$ into irreducible representations of the symmetry group and, if ψ_L, ψ_R transform in different representations then we have

$$\Lambda = \langle \psi_L | \psi_R \rangle = 0 \qquad (1.36)$$

and we have understood why: it is the fact that left- and right- movers are chiral with respect to this symmetry.

Before ending these general remarks we should note that one might speculate if such moduli space symmetries might provide a general stringy mechanism for producing hierarchies. For example mass corrections may also be regarded as inner products, although of Jacobi forms [10] and one might try to use this observation

[6] It was through the requirement of finite dimensionality of the vector bundle that the special role of rational backgrounds was recognized in [6] .

[7] It is unfortunately not clear that all sections correspond to physical theories. This difficulty will cause problems later.

to construct string models in which massless scalars remain massless at one loop. Attempts of the author, in collaboration with J. Harvey, have so far proved unsuccessful in this direction. The main hindrance has been the lack of four-dimensional models, since mass corrections of massless scalars are tricky and often divergent in two-dimensions.

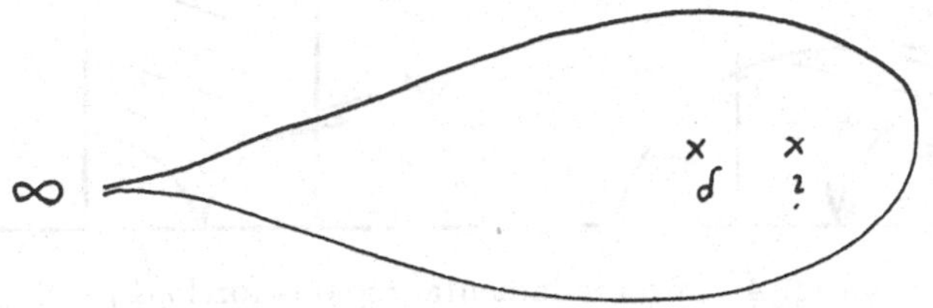

Fig. 3: Moduli space at g=1

2. SYMMETRIES OF MODULAR SURFACES

We must now search for some symmetries of the spaces $\mathcal{H}/\Gamma_1$. These spaces are themselves Riemann surfaces, and are referred to as modular surfaces. We begin with the general remark that an automorphism of a modular surface $\mathcal{H}/\Gamma_1$ for $\Gamma_1 \subset \Gamma$ will lift to an automorphism of $\mathcal{H}$, but the automorphisms of the upper half plane are well known to be just the Mobius transformations. The action of a Mobius transformation ρ on a coset is $[\tau] \rightarrow [\rho \cdot \tau]$ and for this action to be well-defined ρ must lie in the normalizer $N(\Gamma_1)$ for Γ_1 considered as a subgroup of the Mobius group $PSL(2,R)$. One's first thought is to begin with the moduli space $\mathcal{M}_1 = \mathcal{H}^*/\Gamma$, which is illustrated in Fig. 3.

There are three special points on this space which must be left fixed by an automorphism. The only Mobius transformation which fixes three points is the identity, so the moduli space has no symmetries. It would be premature to give up on the idea suggested in the introduction at this point.

For example, we may consider instead a subgroup which fixes an even spin structure, e.g.

$$\Gamma_0(2) = \{ \begin{pmatrix} a & b \\ c & d \end{pmatrix} | c \equiv 0 mod 2 \} \tag{2.1}$$

This subgroup has the nontrivial normalizer $\alpha_2 : \tau \rightarrow -1/2\tau$ which corresponds to the matrix

$$\begin{pmatrix} 0 & -1 \\ 2 & 0 \end{pmatrix} \tag{2.2}$$

Indeed

$$\alpha_2^{-1} \begin{pmatrix} a & b \\ c & d \end{pmatrix} \alpha_2 = \begin{pmatrix} d & -c/2 \\ -2b & a \end{pmatrix} \tag{2.3}$$

It is perhaps worth emphasizing that this transformation exchanges large and small values of $Im\tau$. Two choices of fundamental domain of $\Gamma_0(2)$ are illustrated in Fig. 4.

Of course, we have not used any special properties of two, we could just as well have considered $\Gamma_0(N)$ and $\alpha_N : \tau \rightarrow -1/N\tau$. These transformations are known

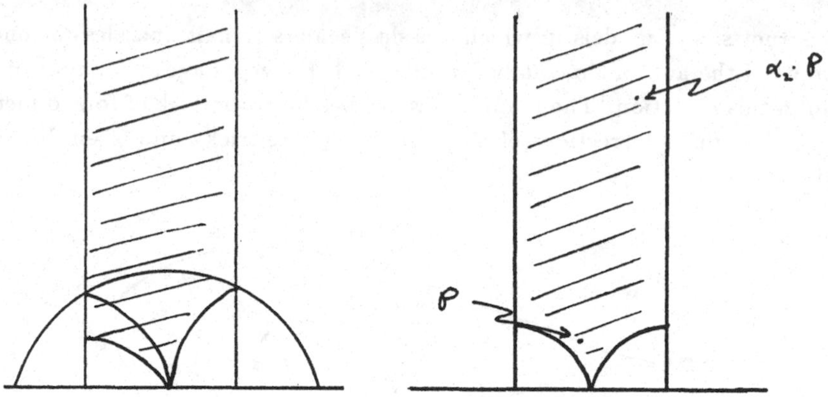

Fig. 4: Two fundamental regions for $\Gamma_0(2)$

as Fricke involutions and are special cases of a more general set of transformations called Atkin-Lehner transformations which are very useful to number theorists in analyzing the structure of the ring of modular forms for $\Gamma_0(N)$ [11]. For general N the full normalizer of $\Gamma_0(N)$ is rather complicated. A nice description of it is given in [12].

From the above remarks we can see that these symmetries will lift to sections of the vector bundles considered in the previous section. If f is a modular form for Γ_1 then so is $f[\rho](\tau) = (det\rho)(c\tau + d)^{-k} f(\rho \cdot \tau)$, when $\rho \in N(\Gamma_1)$, so we define the action to be $f \rightarrow f[\rho]$. Note that the transformation α_N is idempotent, (when k is even) so its eigenvalues are just ± 1.

3. MODEL BUILDING

Now that we have identified an interesting class of symmetries we can set out in search of a model. The following criteria *might* be useful in such a search; they are by no means necessary.

1.) In order for the mechanism to work there are technical conditions on the integrand on moduli space [2] . These conditions basically amount to the condition that there be no tachyon. We thus consider superstrings.

2.) We must be able to examine the transformation of the partition function under transformations of τ which are not modular transformations. Thus the partition function should be exactly solvable, which essentially limits us to orbifolds and compactifications with exactly solvable conformal field theories.

3.) Left and right movers are supposed to transform differently under these symmetries, so we are naturally lead to asymmetric orbifolds.

4.) We must choose N in $\Gamma_0(N)$. Since the relevant identities and ring of modular forms gets more complicated with increasing N the choice $N = 2$ seems a good place to start. In particular, the ring of $\Gamma_0(2)$ forms is generated by combinations of theta functions with half-integral characteristics, so we should look for an orbifold made from a lattice L such that

$$\vartheta_L \in \mathbf{C}[\vartheta_2, \vartheta_3, \vartheta_4] \tag{3.1}$$

and restrict attention to orbifold groups which are powers of $Z/2Z$.

5.) We need to break supersymmetry. The simplest way to do that in light cone gauge is to work with Green-Schwarz fermions and twist by $(-1)^F$ [13].

6.) It is not difficult to see that criteria (4) and (5) require the number of compactified dimensions to be at least 4. Recall that the partition function of a complex fermion is of the form ϑ/η, and by bosonization, the same is true of a left or right-moving compactified boson. Thus, for $Z/2Z$ boundary conditions the contribution to of a given set of boundary conditions is of the form $P/\bar{\eta}^{12}\eta^{24}$ where P is a bihomogeneous polynomial in ϑ and $\bar{\vartheta}$. When we break SUSY with $(-1)^F$ there will be a sector with boundary conditions $(g,1)$ (where g acts by $(-1)^F$ on the GS fermions) which will contribute

$$\frac{\bar{\vartheta}_2^4}{\bar{\eta}^{12}\eta^{24}} P_3 \tag{3.2}$$

where P_3 is again a bihomogeneous polynomial of bidegree $(n, \bar{n})$ and $\bar{n}$ is just the number of compactified dimensions. The whole partition function is supposed to transform into itself (up to a sign) under $\tau \to -1/2\tau$, so we assume that the transformation of the above partition function must be a linear combination of contributions from the other sectors. Using the explicit transformation laws

$$\begin{aligned} 2\vartheta_2(2\tau)\vartheta_3(2\tau) &= \vartheta_2^2(\tau) \\ 2\vartheta_2^2(2\tau) &= \vartheta_3^2(\tau) - \vartheta_4^2(\tau) \\ 2\vartheta_3^2(2\tau) &= \vartheta_3^2(\tau) + \vartheta_4^2(\tau) \\ \vartheta_4^2(2\tau) &= \vartheta_3(\tau)\vartheta_4(\tau) \end{aligned} \tag{3.3}$$

we see that the transform of (3.2) is just

$$\frac{(\bar{\vartheta}_3\bar{\vartheta}_4)^2}{\bar{\eta}^{12}\eta^{24}} \frac{(\bar{\vartheta}_3\bar{\vartheta}_4)^2}{\bar{\vartheta}_2^4} \frac{(\vartheta_3\vartheta_4)^4}{\vartheta_2^8} P_4 \tag{3.4}$$

where P_4 is once again bihomogeneous in the thetas with the *same* bidegree. Comparing with the general form we had before we see that P_4 must be divisible by $\bar{\vartheta}_2^4$ which implies $\bar{n} \geq 4$ as desired.

In fact, if we push this argument a little further we find that the partition function should have the general form

$$\frac{P}{(\bar{\eta}(\tau)\bar{\eta}(2\tau))^4(\eta(\tau)\eta(2\tau))^8} \tag{3.5}$$

therefore we should discuss how to obtain such denominators in conformal field theory.

Recall that an untwisted boson has partition function $Z(1,1) = 1/\eta$, while if we add a $Z/2Z$ twist in the "time" direction we get $q^{-1/24}\prod(1+q^n)^{-1}$. Thus if we have three bosons, one of which is twisted, their partition function is just [8]

$$\frac{1}{\eta(\tau)\eta(2\tau)} \tag{3.6}$$

This seems to be as far as anyone has gone with general guidelines for model building. It is an interesting problem to find sensible methods of constructing models with Atkin-Lehner symmetric partition functions.

We now describe briefly a model in two uncompactified dimensions which has Atkin-Lehner symmetry. Further details are given in appendix A. We compactify the heterotic string to two dimensions, putting the right-movers on the E_8 torus R^8/Γ_8 and the left-movers on $R^{24}/\Gamma_8 \oplus \Gamma_{16}$ and then we construct an orbifold with group $Z/2Z \times Z/2Z$. Let us call the generators α, β, where α contains the twist by $(-1)^F$. By choosing the action appropriately (as in eqs. (A.1)-(A.4)) one can arrange to have only two nonvanishing modular orbits so that the partition function is just

$$Z = Z(\alpha,1) + \epsilon(\alpha,\beta)(Z(\beta,\alpha) + Z(\beta\alpha,\alpha)) + \cdots \tag{3.7}$$

where the ellipsis refers to a sum over $\Gamma_0(2)$ cosets, and $\epsilon = \pm$ is a choice of discrete torsion [9] . Furthermore, the choice of group action (A.1)-(A.4) leads to a partition function of the form

$$\Lambda = \int_{\mathcal{H}/\Gamma_0(2)} (\bar{g}_1 f_1 + \epsilon \bar{g}_2 f_2) \tag{3.8}$$

where $f_1 \leftrightarrow f_2$, $g_1 \leftrightarrow g_2$ under Atkin-Lehner transformations. Thus the choice $\epsilon = -1$ gives a nonsupersymmetric theory with vanishing cosmological constant. In terms of the Friedan-Shenker picture we have

$$\mathcal{L} \oplus \mathcal{L} \rightarrow \mathcal{H}/\Gamma_0(2) \tag{3.9}$$

i.e., a two-dimensional vector bundle. Amongst the various sections of this bundle, we can form the $\pm$ representations

$$\begin{aligned} G_\pm &= g_1 \pm g_2 \\ F_\pm &= f_1 \pm f_2 \end{aligned} \tag{3.10}$$

What we have just asserted is that these may be regarded as sections ψ_L, ψ_R for particular backgrounds. Hence for these backgrounds, the cosmological constant can be written as the inner product

$$\begin{aligned} \Lambda(\epsilon = +1) &= (\,G_+ \quad G_-\,) \cdot \begin{pmatrix} F_+ \\ F_- \end{pmatrix} \neq 0 \\ \Lambda(\epsilon = -1) &= (\,G_- \quad G_+\,) \cdot \begin{pmatrix} F_+ \\ F_- \end{pmatrix} = 0 \end{aligned} \tag{3.11}$$

[8] This is a special case of a trick discovered by E. Witten. See the appendix of [2] for a more general discussion

There are three remarks we should make regarding the theory with $\epsilon = -1$.

1.) After summing over the cosets it turns out the partition function can be written simply in terms of the theta function for the Leech lattice:

$$\begin{aligned} \frac{1}{8}Z = \frac{\vartheta_{Leech}}{\eta^{24}} &= q^{-1} + 24 + 196884q + \cdots \\ &= q^{-1} + 24 + \sum_{n=1}^{\infty} a_n q^n \end{aligned} \tag{3.12}$$

The vanishing of Z when integrated over the fundamental domain has recently been proved from another point of view in [14].

2.) From (3.12) we learn that the number of bosons and fermions differ only at the massless level since the terms proportional to q^n for $n \neq 0$ do not contribute to the physical mass spectrum. In field theory, such a spectrum could not possibly lead to a vanishing cosmological constant since one integrates over the region of Fig. 1. However, in string theory one integrates over the fundamental domain in Fig. 2. The various powers of q all contribute from the integration over the corner regions, the contribution being

$$I_n = \int_{-\frac{1}{2}}^{+\frac{1}{2}} dx \int_{\sqrt{1-x^2}}^{1} \frac{dy}{y^2} e^{2\pi i n x} e^{-2\pi n y} \tag{3.13}$$

Let Σ_n denote the net cosmological constant for levels n and below then we may tabulate the contributions to the cosmological constant as follows:

n	a_n	I_n	$a_n I_n$	Σ_n
-1	1	-11.596	-11.569	-11.569
0	24	$\pi/3$	25.132	13.5367
1	196884	$-.79482 \times 10^{-4}$	-15.6487	-2.112
2	21493760	$.10669 \times 10^{-6}$	2.2932	0.1812
3	864299970	$-.22243 \times 10^{-9}$	-0.1922	-.011

Since $|I_n| < \frac{1}{\pi n} e^{-\sqrt{3}\pi n}$ the series converges rapidly. Graphing the data in the above table is shown in Fig. 5.

Clearly we see here a very stringy effect. It would be interesting to have a spacetime interpretation of this phenomenon. If the partition function is not purely holomorphic but integrates to zero then the cancellations will involve both cancellation from physical particles contributing $q^h \bar{q}^{\bar{h}}$ as well as the above effect.

3.) In [2] a four-dimensional model with AL symmetry was proposed. T. Taylor has pointed out that the choice of signs of modular orbits is not consistent with the existence of an operator formalism. Thus the model probably has an inconsistency. One can change the signs to obtain consistency with an operator formalism, but then the partition function misses being Atkin-Lehner symmetric by a factor of two! Thus, there is no known four-dimensional model exhibiting the phenomenon we are describing.

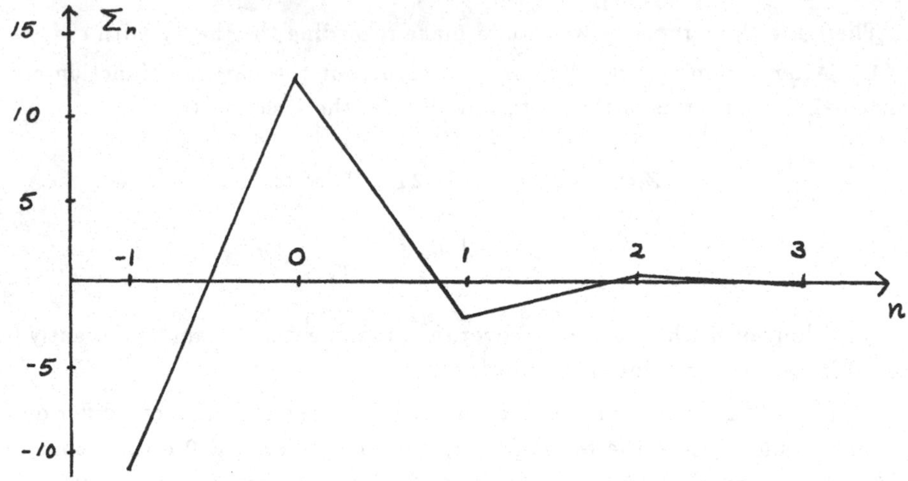

Fig. 5: Graph of the cumulative cosmological constant against level number

This is a serious mistake. The existence of a four-dimensional model is important since gravity is not dynamical in two-dimensions. If there is some kind of obstruction to a four-dimensional model, then, of course, all these considerations are pretty silly. Still, there is a large set of possible symmetries and orbifolds. Moreover, many symmetric string*like* densities were written in [2] some of which were "four-dimensional." Thus, we don't really have a good reason for believing such theories do not exist.

4. THE LEECH THEORY

This section is based on joint work with J. Harvey. One further thing we can learn from the above example is that having a vanishing cosmological constant is compatible with having an extremum-in fact, a minimum- of the vacuum energy in the space of compactification moduli. The point is, that the partition function (3.12) is proportional to the partition function of a toroidal compactification of a nonsupersymmetric string. The fact that the theory we will describe has an extremum was already observed by P. Ginsparg and C. Vafa in [15].

The theory is obtained by compactifying the heterotic string to two spacetime dimensions, with the 24 left-moving bosons moving on the Leech lattice and the 8 right-moving bosons moving on the E_8 lattice. If we use the Green-Schwarz formalism for the right-moving fermions then we must twist by $(-1)^F$ and accompany this with a shift by $\rho = ((\frac{1}{2})^4 0^4)$ in the E_8 lattice. The partition function of the theory is

$$24 \cdot \frac{\vartheta_L}{\eta^{24}}$$

The massless spectrum consists of 8×24 massless scalars

$$|i\rangle_R \otimes \alpha^J_{-1}|0\rangle_L$$

in the untwisted sector, together with 16×24 massless twisted scalars:

$$|\delta\rangle_R \otimes \alpha^J_{-1}|0\rangle_L$$

Here δ is a vector of the form $(p_R + \rho)$ satisfying $\delta^2 = 1$ with $p_R \in \Gamma_8$. There are sixteen possible vectors

$$|((\pm\frac{1}{2})^4, 0^4)\rangle \quad |(0^4, (\pm\frac{1}{2})^4\rangle$$

with an even number of $\pm$ signs.

It is well known that the moduli space of toroidally compactified heterotic string theories is the manifold $SO(8+16,8)/SO(24)\times SO(8)$ modulo some discrete identifications [16] .

Thus we can parametrize theories near the Leech theory by the Lorentz boost

$$exp\begin{pmatrix} 0 & b \\ b^t & 0 \end{pmatrix}$$

where b is a 24×8 matrix, and b^t is the transposed matrix. If we try to evaluate the partition function by inserting vertex operators for the scalar moduli we encounter divergences.

Nevertheless, the partition function is a well defined function of b, at least for small enough b. Therefore, we attempt to give a more direct evaluation of the partition function. We then explain why the vertex operator calculations led to divergences.

For the Leech theory (1.12) becomes

$$\Lambda = -\frac{T}{4\pi}\int \frac{d^2\tau}{(Im\tau)^2}(STr_{\mathcal{H}_1} q^{L_0^{\ell c}}\bar{q}^{\bar{L}_0^{\ell c}} + STr_{\mathcal{H}_\alpha} q^{L_0^{\ell c}}\bar{q}^{\bar{L}_0^{\ell c}}) \tag{4.1}$$

where α is the $Z/2Z$ twist and

$$\begin{aligned} STr_{\mathcal{H}_1} q^{L_0^{\ell c}}\bar{q}^{\bar{L}_0^{\ell c}} &= \frac{1}{2}\frac{\bar{\vartheta}_2^4}{\bar{\eta}^{12}\eta^{24}}\sum q^{\frac{1}{2}\tilde{p}_L^2}\bar{q}^{\frac{1}{2}\tilde{p}_R^2}e^{2\pi i\rho\cdot p_R} \\ STr_{\mathcal{H}_\alpha} q^{L_0^{\ell c}}\bar{q}^{\bar{L}_0^{\ell c}} &= \frac{1}{2}\frac{\bar{\vartheta}_4^4}{\bar{\eta}^{12}\eta^{24}}\sum q^{\frac{1}{2}(\tilde{p}_L+\tilde{\rho}_L)^2}\bar{q}^{\frac{1}{2}(\tilde{p}_R+\tilde{\rho}_R)^2} + (\tau\to\tau+1) \end{aligned} \tag{4.2}$$

The left and right-moving momenta are given by

$$\begin{aligned} \tilde{p}_L &= cosh\sqrt{bb^t}p_L + \frac{1}{b^t}\sqrt{b^t b}sinh\sqrt{b^t b}p_R \\ \tilde{p}_R &= \frac{1}{b}\sqrt{bb^t}sinh\sqrt{bb^t}p_L + cosh\sqrt{b^t b}p_R \end{aligned}$$

where $p_L \in \Gamma_{Leech}$ and $p_R \in \Gamma_8$ and the matrix expressions are defined by their power series. Therefore

$$\begin{aligned} q^{\frac{1}{2}\tilde{p}_L^2}\bar{q}^{\frac{1}{2}\tilde{p}_R^2} &= q^{\frac{1}{2}p_L^2}\bar{q}^{\frac{1}{2}p_R^2} \\ &e^{-2\pi Im\tau(p_L\cdot sinh^2\sqrt{bb^t}\cdot p_L + p_R\cdot sinh^2\sqrt{b^t b}\cdot p_R + p_R\cdot sinh(2\sqrt{b^t b})\sqrt{b^t b}\frac{1}{b}\cdot p_L)} \end{aligned} \tag{4.3}$$

and we therefore find that the full partition function is

$$24\frac{\vartheta_{Leech}}{\eta^{24}} + 24\sum_\delta(e^{-2\pi(Im\tau)\delta\, sinh^2\sqrt{b^t b}\cdot\delta} - 1) + O(q,\bar{q}) \tag{4.4}$$

In the integral (4.1) the first term of (4.4) gives zero, as we have seen; the third term gives corrections which are analytic in the matrix elements of b, for small b. The second term is *not* analytic. Indeed, denoting by E_2 the integral

$$E_2(z) = \int_1^\infty e^{-zx} \frac{dx}{x^2}$$

we see that for small z, E_2 is given by $z log z + 1 + O(z)$. Thus we obtain the vacuum energy

$$V = -12T \sum_\delta (\delta \cdot (sinh^2 \sqrt{b^t b}) \cdot \delta) log(\delta \cdot b^t b \cdot \delta) + analytic \quad in \quad b \tag{4.5}$$

In particular, in all directions in the moduli of toroidal compactifications the vacuum energy behaves as in Fig. 6.

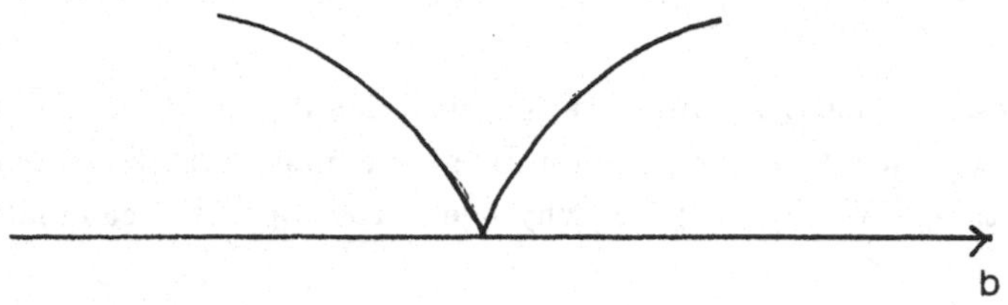

Fig. 6: Vacuum energy near the Leech theory

We now comment on the physical meaning of this answer. Small modifications of the Wilson lines, torsion potential, and metric are equivalent to an $SO(24,8)$ boost by

$$exp \begin{pmatrix} 0 & \frac{1}{2}\delta A & -\frac{1}{2}\delta A \\ -\frac{1}{2}\delta A & -\frac{1}{2}\delta B & \frac{1}{2}\delta B + \delta g \\ -\frac{1}{2}\delta A & -\frac{1}{2}\delta B + \delta g & -\frac{1}{2}\delta B \end{pmatrix}$$

redefining this by an $SO(24) \times SO(8)$ rotation we obtain

$$b = \begin{pmatrix} -\frac{1}{2}\delta A \\ \frac{1}{2}\delta B + \delta g \end{pmatrix}$$

Thus the matrix elements of b are linearly related to small perturbations in A, B, g and hence represent vacuum expectation values for the massless scalar fields ϕ^{iI}. Although the fields ϕ^{iI} have no self-interactions (because they are moduli) they have nonzero interactions with the twisted fields $\phi^{\delta,I}$, and these twisted scalars obtain a mass. From (4.3) it is easy to see that all the twisted scalars obtain a mass shift

$$\delta m^2 = 4\pi T \delta \cdot b^t b \cdot \delta \tag{4.6}$$

Thus, the vacuum energy predicted by the Coleman-Weinberg potential

$$V_{CW} = -\frac{1}{8\pi} \sum m^2 log m^2 \tag{4.7}$$

agrees with the string expression (4.5) to lowest order in the inverse string tension. The two-point functions of vertex operators give infinity because the graph with twisted scalars running around the loop, is divergent.

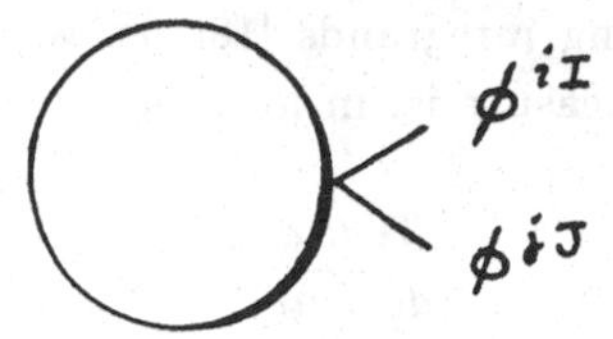

One may also try to derive these mass shifts from a 4pt vertex operator calculation, but here a puzzling thing happens. One would expect that by calculating the correlator

$$\langle V^{iI}(k_1)V^{jJ}(k_2)V^{\delta,K}(k_3)V^{-\delta,L}(k_4)\rangle \tag{4.8}$$

where V^{iI} and $V^{\delta I}$ are the vertex operators of untwisted and twisted scalars, we could determine an effective Lagrangian:

$$\mathfrak{L} = \frac{1}{2}\sum_{i,I}(\partial_\mu \phi^{iI})^2 + \frac{1}{2}\sum_{\delta,I}(\partial_\mu \phi^{\delta,I})^2 - \frac{1}{2}\lambda^{\delta}_{i,j,IJKL}\phi^{iI}\phi^{jJ}\phi^{\delta,K}\phi^{-\delta,L}$$

from which we could obtain δm^2. The correlator (4.8) is easily evaluated to give

$$\delta^i\delta^j(\frac{s}{t}\delta^{IK}\delta^{JL} + \delta^{KL}\delta^{IJ} + \frac{s}{u}\delta^{JK}\delta^{IL}) \tag{4.9}$$

up to an overall constant. One of the peculiarities of two dimensions is that one of s, t, or u is always zero. If we ignore this and work in the $s = 0$ channel we reproduce (4.6) , but this peculiarity should probably not be ignored. One further problem we should mention is that the moduli space of nonsupersymmetric toroidal compactifications contains tachyons [15][16]. Naively we expect that there will be tunneling between the Leech theory and these tachyonic theories [17].

5. HIGHER LOOPS

We have taken pains to phrase things in a language that might generalize to higher loops, but all the concrete statements have been for partition functions for toroidal world sheets. Since we do not even understand (computationally) why strings have vanishing higher loop cosmological constant in R^{10} it might seem a little premature to consider nonsupersymmetric backgrounds. Nevertheless, we can make a few (possibly) relevant comments.

Let us reconsider the mechanism described in section one. There are two key ingredients:

a.) the vacuum energy is an inner product of "wavefunctions" of left and right movers.

b.) the wavefunctions are functions defined on a space with a symmetry and transform in representations of that symmetry.

Statement (a) may be viewed as generalized holomorphic factorization of string integrands. This viewpoint is not altered by the recent remarks on holomorphic factorization of superstring integrands [18], as we now explain. In the bosonic string in flat space the measure is, in local holomorphic coordinates [19] [20] [21] [22],

$$\frac{\prod dt_i d\bar{t}_i}{(det Im\tau)^{13}} |F(t)|^2 \tag{5.1}$$

Already the $(det Im\tau)^{-13}$ could be said to violate holomorphic factorization. Of course, this factor can be removed by introducing g loop momenta $\vec{p}_i$ and writing

$$(det Im\tau)^{-13} \propto \int \prod_{i=1}^{g} \frac{d^{26}\vec{p}_i}{(2\pi)^{26}} e^{i\pi \vec{p}_i \tau_{ij} \vec{p}_j} e^{-i\pi \vec{p}_i \bar{\tau}_{ij} \vec{p}_j} \tag{5.2}$$

The momentum integrals should be considered to be part of the sum over primary fields, and we recover the pairing between holomorphic (antiholomorphic) sections for left and right-movers, which is what is crucial in [5] . Thus, even in rational backgrounds, if there are uncompactified dimensions, the bundle of primary fields is actually infinite dimensional. However, in the bosonic string we can conveniently split compactified and uncompactified contributions and handle the compactified contributions with modular forms.

In the superstring it has been pointed out that, with a particular choice of (δ-function) slice for superteichmuller space, and after integration over the odd moduli, naive holomorphic factorization is spoiled by the correlator of scalar fields in the matter supercurrent [18] . The analog of (5.1) is, in R^{10},

$$\frac{\prod dt_i d\bar{t}_i}{(det Im\tau)^5} \left(F^{(1)}(t)\overline{G}^{(1)}(\bar{t}) - \frac{10}{4\pi} \frac{1}{(Im\tau)^{ij}} F^{(2)}(t)\overline{G}^{(2)}_{ij}(\bar{t}) + \cdots \right) \tag{5.3}$$

As noted in [23] [24] the nonholomorphically factorized terms in (5.3) can easily be written as momentum integrals so that (5.3) becomes

$$\prod dt_i d\bar{t}_i \int \prod_{i=1}^{g} \frac{d^{10}\vec{p}_i}{(2\pi)^{10}} e^{i\pi \vec{p}_i \tau_{ij} \vec{p}_j} e^{-i\pi \vec{p}_i \bar{\tau}_{ij} \vec{p}_j} \left(F^{(1)}(t)\overline{G}^{(1)}(\bar{t}) + p_i \cdot p_j F^{(2)}(t)\overline{G}^{(2)}_{ij}(\bar{t}) + \cdots \right) \tag{5.4}$$

and is once again a pairing of the required kind. The extra terms in (5.3) do indicate that the contributions of compactified and uncompactified dimensions do not split as nicely as before: the $\overline{G}^{(i)}$ will not be modular forms in general. It might be that with a more natural projection from supermoduli to moduli space the $\overline{G}^{(i>1)}$ are zero [25], or that the effects of these terms can be summarized by using the appropriate supersymmetric generalization of τ [23][26]. These possibilities (and others) are currently being investigated by many people. In either case, the usefulness of modular forms will be restored, so we proceed to see what we can do with them.

We now consider the higher genus analog of the Atkin-Lehner transformations. Recall that the automorphisms of the modular surface $\mathcal{H}/\Gamma'$ lift to automorphisms of the upper half plane $\mathcal{H}$. Clearly we may try to generalize this construction by

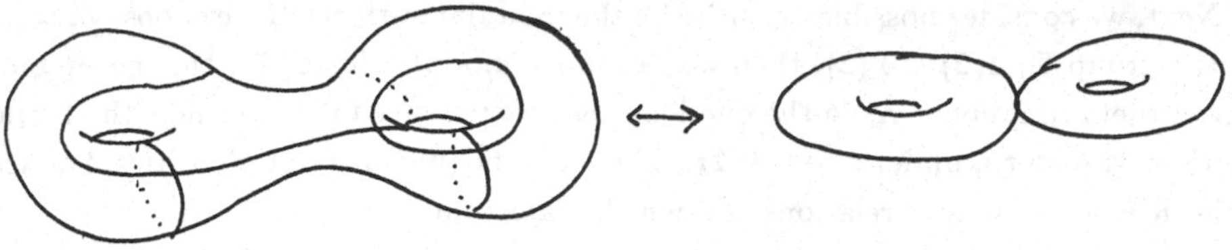

Fig. 7: Funny business at g=2

considering subgroups $\Gamma' \subset \Gamma_g$ of the mapping class group and the corresponding modular "surfaces" T_g/Γ' where T_g is teichmuller space. Again, automorphisms of the modular surfaces will lift to automorphisms of T_g. For $g = 1$ the automorphism group $Aut(T_1) = Aut(\mathcal{H})$ is $PSL(2, R)$, which is much larger than the modular group $PSL(2, Z)$. The Atkin-Lehner transformations are elements of $PSL(2, R) - PSL(2, Z)$. Indeed, if ψ_L, ψ_R transform chirally under the modular group then we have a modular anomaly, and the theory will be inconsistent.

For $g > 1$ a theorem of Royden [27] asserts that the full automorphism group $Aut(T_g)$ is exactly the modular group. Thus, a higher genus extension will be more subtle. One should not give up at this point. There are ways to avoid Royden's theorem. We will describe one of these by considering the situation at $g = 2$. In this case teichmuller space [9]differs from the Siegel upper half space $\mathcal{H}_2$ by the $Sp(4, Z)$ orbit of the diagonal period matrices. The space $\mathcal{H}_2$ has many more autormorphisms than just the symplectic modular group $Sp(4, Z)$. Indeed, among these are the transformations $\tau \to -1/N\tau$ where τ is now the period *matrix*. Therefore, by adding- or if one prefers, subtracting- a set of codimension one to T_2 one obtains a space with automorphisms. The transformation $\tau \to -1/N\tau$ has the weird property that it can relate some (marked) smooth surfaces to surfaces with nodes, as illustrated in Fig. 7.

Note that this funny business only occurs on a codimension one subvariety, which has measure zero. Thus, as long as the string integrand is well-defined on this divisor- which it is if the one-loop cosmological constant is zero- then we can consider the $g = 2$ amplitude to be an integral over a space with automorphisms.

We will now write some string*like* amplitudes to illustrate these remarks. Consider the transformation $\tau \to -1/2\tau$, and consider the series of groups [10]:

$$\Gamma_0(2m, m) \equiv \begin{pmatrix} 1 mod\ m & 0 mod\ m \\ 0\ mod\ 2m & 1\ mod\ m \end{pmatrix} \subset Sp(4, Z) \tag{5.5}$$

for $m = 1, 2, \ldots$. These are all of finite index and normalized by $\tau \to -1/2\tau$. Thus we may consider the modular surfaces $\mathcal{H}/\Gamma_0(2m, m)$ to play the role of the integration region in (1.31) .

[9] Actually, here we use torelli space, obtained by dividing by the diffeomorphisms which fix a marking.

[10] These are rather peculiar groups. For example, the choice $m = 1$ is not the stabilizer of any theta characteristic, and should not be confused with Igusa's group $\Gamma_{1,2}$.

Next, we consider possible densities on the modular surface. If we choose $m = 2$, i.e. the group $\Gamma_0(4,2) \subset \Gamma(2)$, then we can make modular forms for this group out of even theta functions. As in the one-loop case we will need to know how the theta functions transform under $\tau \to -1/2\tau$. There is a beautiful set of identities, known as the Riemann bilinear relations [28] which state that

$$\vartheta\begin{bmatrix}a\\0\end{bmatrix}(z_1|\tau)\vartheta\begin{bmatrix}b\\0\end{bmatrix}(z_2|\tau) = \sum_{\delta\in(Z/2Z)^g} \vartheta\begin{bmatrix}\frac{1}{2}\delta+\frac{1}{2}(a+b)\\0\end{bmatrix}((z_1+z_2)|2\tau)\vartheta\begin{bmatrix}\frac{1}{2}\delta+\frac{1}{2}(a-b)\\0\end{bmatrix}((z_1-z_2)|2\tau) \tag{5.6}$$

These hold for ϑ-functions defined on $\mathcal{H}_g$ for any g. At one loop they reduce (for $z_1 = z_2 = 0$) to (3.3) . At two-loops they give several identities, e.g.

$$\vartheta\begin{bmatrix}\frac{1}{2}&\frac{1}{2}\\0&0\end{bmatrix}^2(\tau/2) = 2\left(\vartheta\begin{bmatrix}00\\00\end{bmatrix}\vartheta\begin{bmatrix}\frac{1}{2}\frac{1}{2}\\00\end{bmatrix} + \vartheta\begin{bmatrix}0&\frac{1}{2}\\0&0\end{bmatrix}\vartheta\begin{bmatrix}\frac{1}{2}&0\\0&0\end{bmatrix}\right)(\tau) \tag{5.7}$$

Finally, recall that the $Sp(4,R)$ invariant measure on $\mathcal{H}_2$ is

$$[d\tau] \equiv \frac{d^2\tau_1 d^2\tau_{12} d^2\tau_2}{(det Im\tau)^3} \tag{5.8}$$

To get a feeling for what kind of densities might arise, recall that at two loops the bosonic string compactified on a rational D-dimensional torus has partition function

$$\int [d\tau](det Im\tau)^{D/2-10} \frac{\sum_{\alpha\in\Gamma/\Gamma^*,\beta\in\Gamma/\Gamma^*} \vartheta_{L^*}[{}^{\alpha}_{0}]\mathcal{M}_{\alpha\beta}\overline{\vartheta}_{L^*}[{}^{\beta}_{0}]}{|\prod_{\epsilon\ even}\vartheta^2[\epsilon]|^2} \tag{5.9}$$

and we expect compactified superstring amplitudes to have a similar flavor. For example, one might expect an integral like

$$\int_{\mathcal{H}_2/\Gamma_0(4,2)} [d\tau](det Im\tau)^2|A(\tau)A(2\tau)|^2$$

$$\frac{(\vartheta[{}^{\frac{1}{2}\frac{1}{2}}_{00}](\tau))^{16} - (\vartheta[{}^{00}_{00}]\vartheta[{}^{00}_{\frac{1}{2}\frac{1}{2}}] + \vartheta[{}^{00}_{0\frac{1}{2}}]\vartheta[{}^{00}_{\frac{1}{2}0}])^8(\tau)}{(\vartheta[{}^{00}_{00}]\vartheta[{}^{\frac{1}{2}\frac{1}{2}}_{00}]\vartheta[{}^{00}_{\frac{1}{2}\frac{1}{2}}](\tau))^{8/3}(\vartheta[{}^{00}_{00}]\vartheta[{}^{\frac{1}{2}\frac{1}{2}}_{00}]\vartheta[{}^{00}_{\frac{1}{2}\frac{1}{2}}](2\tau))^{8/3}} \tag{5.10}$$

where

$$A(\tau) \equiv \frac{\vartheta[{}^{00}_{00}]\vartheta[{}^{\frac{1}{2}\frac{1}{2}}_{00}]\vartheta[{}^{00}_{\frac{1}{2}\frac{1}{2}}](\tau)}{\vartheta[{}^{\frac{1}{2}\frac{1}{2}}_{\frac{1}{2}\frac{1}{2}}](\tau)} \tag{5.11}$$

This integral vanishes nontrivially since it is odd under $\tau \to -1/2\tau$. Furthermore, if it *were* the vacuum amplitude of a compactified string theory, the theory would have two uncompactified dimensions, with tachyons projected out by $L_0 = \overline{L}_0$.

From the factorization expansion we may isolate the terms which would correspond to exchange of massless scalars with vacuum quantum numbers:

$$\int_{\mathcal{H}_1/\Gamma_0(2)} \frac{d^2\tau_1}{(Im\tau_1)^2} \int_{\mathcal{H}_1/\Gamma_0(2)} \frac{d^2\tau_2}{(Im\tau_2)^2} \int^{\infty} ds$$

$$\Big[\frac{\vartheta_2^{16}(\tau_1) + 16(\vartheta_3\vartheta_4)^8(\tau_1)}{(\eta(\tau_1)\eta(2\tau_1))^8} \frac{\vartheta_2^{16}(\tau_2) - 16(\vartheta_3\vartheta_4)^8(\tau_2)}{(\eta(\tau_2)\eta(2\tau_2))^8} \tag{5.12}$$

$$+\frac{\vartheta_2^{16}(\tau_1) - 16(\vartheta_3\vartheta_4)^8(\tau_1)}{(\eta(\tau_1)\eta(2\tau_1))^8} \frac{\vartheta_2^{16}(\tau_2) + 16(\vartheta_3\vartheta_4)^8(\tau_2)}{(\eta(\tau_2)\eta(2\tau_2))^8}\Big]$$

where $\int ds$ is the proper time representation of a zero mass propagator at zero momentum. The expression (5.12) should be compared with (A.7). One can play similar games to write amplitudes which would correspond to $D = 4, 6$ uncompactified dimensions; similar considerations apply at $g = 3$. Indeed, such expressions come by the truckloads: the nontrivial problem is to find the theories (if any) which give such expressions.

The specific symmetries $\tau \to -1/N\tau$ are too simple to extend beyond $g = 3$. Indeed, the following argument, due to J.-B. Bost, shows that if $\tau \to -1/N\tau$ is a symmetry of a dense subset of the Schottky locus [11] $S_g \subset \mathcal{H}_g$, then $g = 1, 2, 3$. Recall that S_g is invariant under modular transformations. These include

$$\begin{aligned} \tau &\to \tau + B \\ \tau &\to -1/\tau \end{aligned} \tag{5.13}$$

for B a symmetric integral matrix. Therefore, applying these transformations we see that if, for almost every $\tau \in S_g$ we have $-1/N\tau \in S_g$, then for all N, B, k and almost all $\tau \in S_g$, $\tau + N^{-k}B \in S_g$. Thus, it follows that the tangent space of S_g at τ is $\frac{1}{2}g(g+1)$ dimensional, and this can only hold if $g = 1, 2, 3$. Whether or not compactifications of modular varieties T_g/Γ' can yield other kinds of new symmetries beyond $g = 3$ remains an open problem. Because of the above argument it seems likely that an appropriate modular surface would involve the torelli group in some nontrivial way.

The most serious drawback of the higher loop extensions we have been describing is that they rely on order by order vanishing of Λ in string perturbation theory. Since the string coupling is given by the vacuum expectation value of a quantum field- the dilaton- this would imply an exact degeneracy in the vacuum energy, which in turn implies we have a massless scalar coupling to matter with the same strength as gravity [29]. This is ruled out by experiment.

[11] The Schottky locus is the subspace of $\mathcal{H}_g$ defined by the period matrices of Riemann surfaces

Still, these ideas might be of some use in a complete solution of the problem. For example the univeral Grassmannian provides an embedding

$$\cup \mathcal{M}_g \hookrightarrow Gr \tag{5.14}$$

and many people have suggested that there might be a density on Gr which satisfies something like

$$\int_{Gr} \psi_R^* \psi_L = \sum_g \int_{\mathcal{M}_g} \mu_g$$

where μ_g is the string measure in some background. Such a thing might be an instanton or semiclassical approximation, or perhaps is the result of a fixed point theorem [12]. The benefits of such a representation would be enormous. Among other things, we know that $Aut(Gr)$ is very large, containing at least $GL(\infty)$. But we need not emphasize Grassmannians overmuch. What we do want to emphasize is that if we formulate string theory in the spirit of the FS proposal then we have a framework in which to address the problem of the cosmological constant. If we have a nonperturbative formula like

$$\Lambda = \int_{\mathcal{M}} \psi_L^a h_{ab} \overline{\psi}_R^b \tag{5.15}$$

where $\mathcal{M}$ is some complex space with automorphisms, then, for some special backgrounds h will be equivariant with respect to these automorphisms. If the world is *chiral* with respect to these symmetries, that is, if $\psi_L \psi_R$ transform in different representations then we can understand two things:

a.) $\Lambda = 0$ because of an underlying stringy symmetry principle.

b.) Only special backgrounds will respect this symmetry, and we might hope that only for special values of the dilaton vev will h be equivariant- which would explain the absence of massless dilatons.

In conclusion, there are several issues that need to be addressed: lack of four dimensional examples, all-loop extensions, and a spacetime interpretation of cancellations from the holomorphic powers q^n. Furthermore, we would not want to have to impose (generalized) Atkin-Lehner symmetry on the ground state, rather, we should deduce it from a more fundamental principle. Finally, if it is a fundamental symmetry of the ground state, then it should do more than give a *post*diction of the vanishing cosmological constant, it should lead to *pre*dictions of phenomena which might violate a naive interpretation of the effective field theory principle. Undoubtedly most readers will regard these as insuperable difficulties with this approach. We hope there are some who regard them as open problems.

ACKNOWLEDGEMENTS

I would like to thank J. Harvey for collaboration that lead to the results in section four, S. Shenker for a useful discussion on modular geometry, and T. Taylor for pointing out the error in the four-dimensional theory in [2] . I would also

[12] This latter possibility is a suggestion of E. Witten.

like to thank J.-B. Bost, C. McMullen, and A. Morozov for very useful discussions on higher loop extensions. I am also grateful to the Institut D'Etudes Scientifique de Cargése for hospitality and to the organizers if the Cargése school on Nonperturbative Quantum Field Theory for the invitation to speak. This work was also supported by DOE contract DE-AC02- 76ER02220.

APPENDIX A. AN EXAMPLE THEORY

We compactify the heterotic string to two dimensions, right-movers compactified by Γ_8 and the left-movers compactified by $\Gamma_8 \oplus \Gamma_{16}$ where Γ_{16} is the root lattice of $Spin(32)/Z_2$. We thus begin with a gauge symmetry $E_8 \times Spin(32)/Z_2$. We consider a $Z_2 \times Z_2$ orbifold where the group generators act by

$$\tilde{\alpha}_{GS} = (-1)^F \qquad \tilde{\beta}_{GS} = e^{2\pi i(\frac{1}{2}J_{56}+\frac{1}{2}J_{78})} = \begin{pmatrix} 1_4 & \\ & -1_4 \end{pmatrix} \tag{A.1}$$

on the Green-Schwarz fermions, while

$$\begin{aligned} \tilde{\alpha} &: p \rightarrow p + ((\tfrac{1}{2})^4 0^4) \\ \tilde{\beta} &: p \rightarrow \begin{pmatrix} 1_4 & \\ & -1_4 \end{pmatrix} p \end{aligned} \tag{A.2}$$

on the right-moving bosons (the form of the transformation is determined by world-sheet supersymmetry up to shift vectors), and

$$\begin{aligned} \alpha &: p \rightarrow -p \\ \beta &: p \rightarrow p \end{aligned} \tag{A.3}$$

for the coordinates on the E_8 maximal torus and finally

$$\begin{aligned} \alpha &: p \rightarrow -p \\ \beta &: p \rightarrow \begin{pmatrix} 0 & 1 \\ 1 & 0 \end{pmatrix} p \end{aligned} \tag{A.4}$$

for momenta conjugate to the coordinates on R^{16}/Γ_{16}. Note that the action of α and β on spinors of $O(8)$ is independent of chirality, a fact which will be important to us momentarily.

The only nonvanishing modular orbits are those associated with the boundary conditions $(\alpha, 1)$ and (β, α) and the partition functions may be evaluated to be

$$Z(\alpha, 1) = \frac{\bar{\vartheta}_2{}^4 (\bar{\vartheta}_3 \bar{\vartheta}_4)^4}{\bar{\eta}^{12}} \frac{(\vartheta_3 \vartheta_4)^{12}}{\eta^{24}} \tag{A.5}$$

and

$$Z(\beta, \alpha) = \frac{(\bar{\vartheta}_3 \bar{\vartheta}_4)^2 (\frac{1}{2}(\bar{\vartheta}_3 \bar{\vartheta}_4)^2 \bar{\vartheta}_2{}^4)}{\bar{\eta}^{12}} \frac{(\vartheta_2 \vartheta_3)^4 \frac{1}{2^3} \vartheta_2^8 (\vartheta_3 \vartheta_4)^4}{\eta^{24}} \tag{A.6}$$

We have used the Jacobi triple product formula in deriving these formulæ .

We may now assemble these partition functions to obtain the $\Gamma_0(2)$ symmetric part of the partition function:

$$Z(\alpha,1)+\epsilon(\beta,\alpha)(Z(\beta,\alpha)+Z(\alpha\beta,\alpha))=2^4\frac{1}{(\eta(\tau)\eta(2\tau))^8}[(\vartheta_3\vartheta_4)^8+\epsilon(\beta,\alpha)\frac{1}{2^4}\vartheta_2^{16}] \tag{A.7}$$

where we have introduced a relative phase ϵ between modular orbits which was called by Vafa "discrete torsion" [9]. In terms of the discussion in section three $g_1=g_2=2^4$, and

$$f_1=\frac{1}{(\eta(\tau)\eta(2\tau))^8}(\vartheta_3\vartheta_4)^8$$

$$f_2=\frac{1}{(\eta(\tau)\eta(2\tau))^8}\frac{1}{2^4}\vartheta_2^{16}$$

As shown in [9] one obtains consistent string theories with either choice of sign for ϵ (the sign of the orbit $(\alpha,1)$ is fixed by spin-statistics). Choosing the plus sign we obtain a partition function with positive AL symmetry, and choosing the minus sign gives a theory with negative AL symmetry.

In order to derive the full partition function we must sum over the coset transforms of $\Gamma_0(2)$. Actually, very little work is involved here since the space of weight twelve forms is two-dimensional and there is only one cusp form. Thus we may write, for example

$$\frac{1}{2}Z_{total}=4\frac{\vartheta_\Lambda}{\Delta}+constant \tag{A.8}$$

where ϑ_Λ is the theta function of the Leech lattice. We must also explain the factor of $\frac{1}{2}$ on the LHS. It has been pointed out in [30] that there is a subtlety when compactifying strings to two dimensions in the light cone gauge. Namely, the infinite Lorentz boost required to go to light-cone gauge projects out one of the $O(8)$ chiralities of the Green-Schwarz fermions. Thus there are two sets of spectra, "up-moving" and "down-moving" obtained by boosting along the $\pm z$ axis. These might differ for the massless states. In our case the action of the group generators on the Green-Schwarz fermions is independent of chirality, so we need only multiply the partition function in (A.7) by two to obtain Z_{total}. Therefore, expanding (A.7) one finds

$$\frac{1}{2}Z_{total}=\frac{4}{q}-2^5\cdot 3\cdot(1+2\epsilon(\alpha,\beta))+\cdots \tag{A.9}$$

Neither choice of sign for the discrete torsion leaves an equal number of bosonic and fermionic massless states. Since the tachyon is projected out by integration across the fundamental domain we can conclude that the choice $\epsilon=-1$ yields a theory with vanishing cosmological constant.

We now briefly describe the massless particles in the theory. In the untwisted sector we must form invariant states out of the states

$$|a \quad or \quad i\rangle\otimes\begin{cases}|p_L\rangle & p^2=2\\ \alpha^i_{-1}|0\rangle & \text{i=1,24}\end{cases} \tag{A.10}$$

where a and i are spinor and vector $SO(8)$ indices. The α and β invariant states are, for example

$$\begin{aligned} &|i=1,4\rangle_R \otimes (|p_L\rangle + |-p_L\rangle) \qquad p\in\Gamma_8 \\ &|a=1,4\rangle_R \otimes (|p_L\rangle - |-p_L\rangle) \qquad p\in\Gamma_8 \end{aligned} \tag{A.11}$$

with a similar set of linear combinations for the Γ_{16} lattice. These states occur in equal numbers for bosons and fermions. In addition there are also fermionic states

$$\begin{aligned} &|a=1,4\rangle_R \otimes \alpha_{-1}^{i=1,8}|0\rangle_L \\ &|a=1,4\rangle_R \otimes (\alpha_{-1}^{i=9,16} + \alpha_{-1}^{i+8})|0\rangle_L \\ &|a=5,8\rangle_R \otimes (\alpha_{-1}^{i=9,16} - \alpha_{-1}^{i+8})|0\rangle_L \end{aligned} \tag{A.12}$$

giving a net number of 96 fermions.

In the β sector we must form states out of $|a,i\rangle_R \otimes |t_R\rangle \otimes |t_L\rangle$. The sign of β on the vacuum $|0\rangle$ is -1 by modular invariance. In terms of the oscillators

$$A_{-1/2}^{i=9,16} = (\alpha_{-1/2}^{i} - \alpha_{-1/2}^{i+8})$$

we have massless states

$$\begin{aligned} &|i=1,2\rangle_R \otimes |t_R=1,2\rangle \otimes A_{-1/2}^{i=1,8}|t_L=1,2\rangle \qquad \epsilon=-1 \\ &|a=1,2\rangle_R \otimes |t_R=1,2\rangle \otimes A_{-1/2}^{i=1,8}|t_L=1,2\rangle \qquad \epsilon=+1 \end{aligned} \tag{A.13}$$

In the $\alpha\beta$ sector we find similarly

$$\begin{aligned} &|i=1,2\rangle_R \otimes |t_R=1,2\rangle \otimes |t_L=1,2^4\rangle \otimes |t_L=1,2\rangle \qquad \epsilon=-1 \\ &|a=1,2\rangle_R \otimes |t_R=1,2\rangle \otimes |t_L=1,2^4\rangle \otimes |t_L=1,2\rangle \qquad \epsilon=+1 \end{aligned} \tag{A.14}$$

Finally, in the α sector there are no massless states. From (A.11)-(A.14) one finds that the first two terms in the net partition function are indeed given by (A.9) .

REFERENCES

[1] B. Zumino, "Supersymmetry and the vacuum." Nucl. Phys. **B89**(1975)535.

[2] G. Moore. "Atkin-Lehner Symmetry," Nucl. Phys. **B293**(1987)139.

[3] R. Rohm, Nucl. Phys. **B237**(1984)553.

[4] J. Polchinski, Comm. Math. Phys. **104**(1986)37.

[5] D. Friedan and S. Shenker, Phys. Lett. **175B**(1986)287; Nucl. Phys. **B281**(1987)509.

[6] D. Friedan and S. Shenker, unpublished.

[7] J. Harvey, G. Moore, C. Vafa, "Quasicrystalline Compactification" HUTP-87/A072; IASSNS/HEP-87/48; PUPT-1068.

[8] B. Schoeneberg, Elliptic modular functions (Springer 1974).

[9] C. Vafa, Nucl. Phys. **B273**(1986)592.

[10] M. Eichler and D. Zagier, The theory of Jacobi forms (Birkhauser, Basel)1985.

[11] A.O.L. Atkin and J. Lehner, "Hecke Operators on $\Gamma_0(m)$," Math. Ann. **185**(1970)134.

[12] J. Conway and Norton, "Monstrous Moonshine," Bull. Lon. Math. Soc. **11**(1979)308.

[13] L. Dixon and J. Harvey, Nucl. Phys. **B274**(1986)93.

[14] W. Lerche, B.E.W. Nilsson, A.N. Schellekens, and N.P. Warner, "Anomaly Cancelling Terms from the Elliptic Genus," Cern preprint, CERN-TH-4765/87.

[15] P. Ginsparg and C. Vafa, Nucl. Phys. **B289**(1987)414.

[16] K.S. Narain, Phys. Lett. **169B**(1986)41;
K.S. Narain, M.H. Sarmadi, and E. Witten, Nucl. Phys. **B279**(1987)369;
P. Ginsparg, Phys. Rev. **D35**(1987)648;
K.S. Narain and M.H. Sarmadi, Phys. Lett. **184B**(1987)165.

[17] C. Vafa, private communication.

[18] A. Morozov and A. Perelomov, preprint ITEP-104.

[19] A.A. Belavin and V.G. Knizhnik, Phys. Lett. **168B**(1986)210; A. Belavin and V. Knizhnik, Sov. Phys. JETP **64**(1986) 214=[ZhETF**91**364; J.-B. Bost and Th. Jolicoer, Phys. Lett. **174B**(1986)273;R. Catenacci,M. Cornalba, M. Martinelli, and C. Reina, Phys. Lett. **172B**(1986)328.

[20] A. Beilinson and Yu. Manin, "The Mumford form and the Polyakov measure in string theory," Comm. Math. Phys. **107** (1986)359.

[21] P. Nelson, "Lectures on Moduli Space and Strings," Harvard preprint HUTP-86/A047.

[22] J.-B. Bost, "Fibrés déterminants, déterminants régularisés et mesures sur les espaces de modules des courbes complexes," Séminaire Bourbaki, 39ème année, 1986-87, no. 676.

[23] M. Baranov and A. Schwarz, "On the multiloop contribution to the string theory" preprint.

[24] E. Verlinde and H. Verlinde, Phys. Lett. Phys. Lett. Phys. Lett. **192B**(1987)95.

[25] This possibility has been emphasized by E. Witten.

[26] E. D'Hoker and D. Phong, Nucl. Phys. **B292** (1987)317.

[27] H.L. Royden, *in* Advances in the theory of Riemann surfaces, ed. L. Ahlfors et a., Ann. of Math. Studies no. 66.

[28] D. Mumford, Tata Lectures on Theta (Birkhauser, Basel, 1983).

[29] M. Green, J.H. Schwarz, and E. Witten, Superstring theory (Cambridge University Press, 1987).

[30] J. Bagger, C. Callan, J. Harvey, Nucl. Phys. **B278** (1986)550.

TORSION CONSTRAINTS AND SUPER RIEMANN SURFACES

Philip Nelson

Physics Department
Boston University
Boston MA 02215, USA

ABSTRACT

Super Riemann surfaces are the arena for two-dimensional superconformal field theory. They can be regarded as smooth 2|2-dimensional supermanifolds equipped with a reduction of their structure group to the group of upper triangular 2×2 complex matrices. The integrability conditions for such a reduction turn out to be (most of) the famous torsion constraints of 2d supergravity. The other torsion constraints are merely conditions to fix some of the gauge freedom in this description, or to specify a particular connection on such a manifold, analogous to the Levi-Civita connection in Riemannian geometry.

Unlike ordinary Riemann surfaces, a super Riemann surface $\widehat{X}$ cannot be regarded as having only one complex dimension; instead it has one commuting and one anticommuting complex coordinate. Nevertheless, in certain important aspects SRS behave as nicely as if they had only one dimension. In particular they possess an analog of the Cauchy-Riemann operator $\bar{\partial}$ on an ordinary Riemann surface X. The latter is a first-order differential operator with values in a canonical line bundle $\omega \equiv \Omega^1 X$ over the surface; this bundle in turn is in a sense a square root of the bundle of volume forms. Similarly super Riemann surfaces have a canonically-defined first-order operator $\hat{\bar{\partial}}$ taking values in the bundle $\hat{\omega} \equiv \mathrm{BER}\, \Omega^1 \widehat{X}$ of half-volume forms. This observation allows us to write a superstring world-sheet action depending only on a "superconformal structure" analogous to a conformal structure on an ordinary 2-surface. Furthermore, the operator $\hat{\bar{\partial}}$ furnishes a short resolution of the structure sheaf of the super Riemann surface, making possible a Quillen theory of determinant line bundles.

The material in this talk, and further discussion, can be found in the following papers:

1. S. Giddings and P. Nelson, "Torsion constraints and super Riemann surfaces," to appear in Phys. Rev. Lett. **59**, 2619 (1987).
2. P. Nelson, "Holomorphic coordinates for supermoduli space," to appear in Comm. Math. Phys. **115**, 167 (1988).
3. S. Giddings and P. Nelson, "The geometry of super Riemann surfaces," Comm. Math. Phys. (1988).

RENORMALIZATION THEORY FOR USE IN CONVERGENT EXPANSIONS OF EUCLIDEAN QUANTUM FIELD THEORY

Andreas Pordt

II. Institut für Theoretische Physik der Universität Hamburg

1. Introduction

Renormalization is necessary for perturbation and cluster expansion methods of Euclidean quantum field theory. Bounds for Feynman graphs show that divergencies for renormalizable theories come from subgraphs with a small number of low frequency external lines [1,2]. Renormalization group suggests a way to avoid such divergencies by introducing running coupling constants. The tree expansion introduced by G. Gallavotti and F. Nicolò [3-6] represents the expansion in powers of running coupling constants instead of renormalized coupling constants. Running coupling constants are related by renormalization group equations and the bare coupling constants have to be chosen such that the renormalized coupling constants are finite if an ultraviolet cutoff is removed.

Nonperturbative effects coming from large field contributions cause the divergence of perturbation expansions. Cluster expansion methods take into account such nonperturbative effects and provide convergent series expansions for Euclidean Greens functions. A polymer system on the multigrid is a natural example of a phase cell cluster expansion (see G. Mack's lecture and [7]).

In section 2 of this paper the use of renormalization with running coupling constants for perturbation theory is discussed. The resummation of Feynman graph expansion in terms of running coupling constants is presented . Section 3 discusses the renormalization for a polymer system on the multigrid.

2. Renormalization with running coupling constants

Consider the following partition function for a d-dimensional scalar field theory

$$Z(\Psi)=\frac{1}{\mathcal{N}}\int\prod_{z\in\mathbf{R}^d}d\Phi(z)\exp\{-\frac{1}{2}(\Phi,v^{-1}\Phi)-V(\Phi+\Psi)\} \tag{1}$$

where the normalization constant $\mathcal{N}$ is determined by $Z(0)=1$. The free propagator v is the Yukawa potential with mass m

$$v=(-\triangle+m^2)^{-1}.$$

Then the free part of the interaction is

$$\frac{1}{2}(\Phi, v^{-1}\Phi) = \frac{1}{2}\int \Phi(z)(-\triangle + m^2)\Phi(z).$$

With the Gaussian measure

$$d\mu_v(\Phi) = (\det 2\pi v)^{-\frac{1}{2}} \exp\{-\frac{1}{2}(\Phi, v^{-1}\Phi)\} \prod_{z\in\mathbf{R}^d} d\Phi(z)$$

we can write

$$Z(\Psi) = \int d\mu_v(\Phi) \exp\{-V(\Phi + \Psi) + e\} \tag{2}$$

where

$$e = -\ln \int d\mu_v(\Phi) \exp\{-V(\Phi)\}$$

$Z(\Psi)$ is the generating functional for free propagator amputated Greens functions. The (connected) n-point free propagator amputated Greens functions $F(z_1, ..., z_n)$ $(F_c(z_1, ..., z_n))$ are defined by

$$\begin{aligned} F(z_1, ..., z_n) &= \frac{\delta^n}{\delta\Psi(z_1)\dots\delta\Psi(z_n)} Z(\Psi)|_{\Psi=0} \\ F_c(z_1, ..., z_n) &= \frac{\delta^n}{\delta\Psi(z_1)\dots\delta\Psi(z_n)} \ln Z(\Psi)|_{\Psi=0}. \end{aligned} \tag{3}$$

$Z(\Psi)$ and $\ln Z(\Psi)$ can be represented as a formal Taylor expansion in Ψ

$$\begin{aligned} Z(\Psi) &= 1 + \sum_{n\geq 1} \frac{1}{n!}\int_{z_1,\dots,z_n\in\mathbf{R}^d} dz_1 \dots dz_n\; F(z_1, ..., z_n)\Psi(z_1)\dots\Psi(z_n) \\ \ln Z(\Psi) &= \sum_{n\geq 1} \frac{1}{n!}\int_{z_1,\dots,z_n\in\mathbf{R}^d} dz_1 \dots dz_n\; F_c(z_1, ..., z_n)\Psi(z_1)\dots\Psi(z_n). \end{aligned} \tag{4}$$

For renormalization group calculations split the free propagator v into propagators v^j which obey the bound

$$|v^j(z_1, z_2)| \leq c_1 L^{j(d-2)} \exp\{-c_2 L^j |z_1 - z_2|\} \tag{5}$$

where $L > 1$ is a fixed scale factor and c_1, c_2 are positive constants. For example

$$v^j(z_1, z_2) = \int_{-\infty}^{\infty} \frac{dp}{(2\pi)^d} \frac{e^{ip(z_1-z_2)}}{p^2}[e^{-L^{-2j}p^2} - e^{-L^{-2(j-1)}p^2}]. \tag{6}$$

Then we obtain for the massless free propagator $v = (-\triangle)^{-1}$

$$v = \sum_{j=-\infty}^{\infty} v^j.$$

For the massive free propagator $v = (-\triangle + m^2)^{-1}$, $m^2 > 0$, we can find similarly

$$v = \sum_{j=0}^{\infty} v^j$$

such that the bound (5) for v^j is fulfilled. Let us consider a partition function with ultraviolet and infrared cutoff

$$Z_k^{(N)}(\Psi) = \int d\mu_{\sum_{j=k}^N v^j}(\Phi) \exp\{-V^{(N)}(\Phi+\Psi) + e_k^{(N)}\}. \tag{7}$$

$V^{(N)}$ is called bare interaction and $e_k^{(N)}$ is fixed by imposing $Z_k^{(N)}(0) = 1$. The effective interaction $V_k^{(N)}$ is defined by

$$Z_k^{(N)}(\Psi) = \exp\{V_k^{(N)}(\Psi)\}. \tag{8}$$

Suppose that the bare interaction $V^{(N)}$ depends on a finite number of parameters $\{\lambda_\alpha^{(N)}\}$. A theory is called renormalizable if one can choose the finite bare parameters $\{\lambda_\alpha^{(N)}\}$ such that the effective theory is well defined when the ultraviolet cutoff is removed ($N \to \infty$), i.e. Greens functions exist for the partition functions

$$Z_k(\Psi) = \lim_{N\to\infty} Z_k^{(N)}(\Psi)$$

for all k.

$Z_k^{(N)}$ and $Z_{k-1}^{(N)}$ are related by the renormalization group equations

$$Z_{k-1}^{(N)}(\Psi) = \int d\mu_{v^k}(\Phi) Z_k^{(N)}(\Phi+\Psi) e^{\delta e_k^{(N)}} \tag{9}$$

where

$$\delta e_k^{(N)} = -\ln \int d\mu_{v^k}(\Phi) Z_k^{(N)}(\Phi).$$

The effective interaction and connected Greens functions are related by

$$V_k^{(N)}(\Psi) = -\sum_{n\geq 1} \frac{1}{n!} \int_{z_1,\dots,z_n\in\mathbf{R}^d} F_{k,c}^{(N)}(z_1,\dots,z_n)\Psi(z_1)\dots\Psi(z_n). \tag{10}$$

If we start with a local bare interaction $V^{(N)}$ we obtain a nonlocal effective interaction $V_k^{(N)}$. But the effective interaction is almost local since we have integrated fields with covariances v^j , $j = k+1, k+2, \dots, N$, and v^j has decay length of order L^{-j}. Therefore the coefficients $F_{k,c}^{(N)}$ of the effective interaction show exponentially decay with decay length of order $L^{-(k+1)}$. Using Taylor expansion for external fields we can represent the effective interaction in a local form

$$V_k^{(N)}(\Psi) = -\sum_{n\geq 1} \frac{1}{n!} \sum_{m_2,\dots,m_n\in\mathbf{N}^d} \int_{z\in\mathbf{R}^d} F_{k,c,m_2,\dots,m_n}^{(N)}(z)\Psi(z) \prod_{a=2}^n \partial_z^{m_a}\Psi(z) \tag{11a}$$

where

$$F_{k,c,m_2,\dots,m_n}^{(N)}(z) = \int_{z_1,\dots,z_n\in\mathbf{R}^d} \frac{(z_2-z)^{m_2}\dots(z_n-z)^{m_n}}{m_2!\dots m_n!} F_{k,c}^{(N)}(z_1,\dots,z_n). \tag{11b}$$

We have used the multiindex notation $m = (m^1, \dots, m^d) \in \mathbf{N}^d$, $m! = \prod_{\mu=1}^d m^\mu!$ and $|m| = \sum_{\mu=1}^d m^\mu$. Define a localization operator $\mathcal{L}$ by

$$\mathcal{L}V_k^{(N)}(\Psi) = -\sum_{n\geq 1} \frac{1}{n!} \sum_{\substack{m_2,\dots,m_n\in\mathbf{N}^d : \\ \sum_{a=2}^n |m_a|\leq K_n}} \int_{z\in\mathbf{R}^d} F_{k,c,m_2,\dots,m_n}^{(N)}(z)\Psi(z) \prod_{a=2}^n \partial_z^{m_a}\Psi(z) \tag{12}$$

with $K_n \in \{-1\} \cup \mathbf{N}$. For example for a theory with reflection symmetry $\Psi \to -\Psi$

$$\mathcal{L}V_k^{(N)}(\Psi) = -\frac{1}{2}\int_z m_k^{(N)}(z)^2\Psi(z)^2 - \frac{1}{4!}\int_z \lambda_k^{(N)}(z)^4\Psi(z)^4 - \frac{1}{2}\sum_{\substack{\mu,\nu=1\\ \mu\leq\nu}}^{d}\int_z \beta_{k,\mu\nu}^{(N)}(z)\Psi(z)\partial_{z^\mu}\partial_{z^\nu}\Psi(z) \quad (13)$$

where

$$m_k^{(N)}(z)^2 = \int_{z_2} F_{k,c}^{(N)}(z,z_2)$$

$$\lambda_k^{(N)}(z) = \int_{z_2,z_3,z_4} F_{k,c}^{(N)}(z,z_2,z_3,z_4)$$

$$\beta_{k,\mu\nu}^{(N)}(z) = \int_{z_2}(z_2^\mu - z^\mu)(z_2^\nu - z^\nu)F_{k,c}^{(N)}(z,z_2).$$

For Euclidean invariant effective actions we obtain $m_k^{(N)}(z)^2 = m_k^{(N)^2}$, $\lambda_k^{(N)}(z) = \lambda_k^{(N)}$, $\beta_{k,\mu\nu}^{(N)}(z) = \beta_k^{(N)}\delta_{\mu\nu}$. Let us consider the nonlocal remainder term

$$(1-\mathcal{L})V_k^{(N)}(\Psi).$$

The 2nd order term in Ψ is

$$\frac{1}{2}\sum_{m\in\mathbf{N}^d:|m|=4}\int_{z_1,z_2}\frac{(z_2-z_1)^m}{m!}F_{k,c}^{(N)}(z_1,z_2)\Psi(z_1)\partial_z^m\Psi(z)|_{z=z_1+\Theta(z_2-z_1)}$$

for some $\Theta \in [0,1]$. Using that $F_{k,c}^{(N)}(z_1,z_2)$ is smaller than a factor $\exp\{-const\cdot L^{k+1}|z_1-z_2|\}$ we gain a factor $L^{-4(k+1)}$ for the remainder term by

$$|z_2-z_1|^m\exp\{-const\cdot L^{k+1}|z_1-z_2|\} \leq O(1)\cdot L^{-4(k+1)}.$$

In next renormalization group steps fields with covariances v^j for $j \leq k$ are integrated. Then external fields Ψ are replaced by $v^j(z,\cdot)$ and the factor $\partial_z^m\Psi(z)$ in (12) is replaced by $\partial_z^m v^j(z,\cdot)$. This gives a factor L^{4j}. Therefore we get a convergence producing factor $L^{4(j-k-1)}$. In summary the effective interactions are split into a local and a nonlocal term. The local term is parametrized by running coupling constants and the nonlocal term gives small contributions because of the convergence producing factors. The renormalization group procedure starts with a local interaction $V^{(N)}$ such that $\mathcal{L}V^{(N)} = V^{(N)}$. For example for the localization operator given by (13) we take

$$V^{(N)}(\Psi) = -\frac{1}{2}m^{(N)^2}\int_z\Psi(z)^2 - \frac{1}{4!}\lambda^{(N)}\int_z\Psi(z)^4 - \frac{1}{2}\sum_{\mu=1}^{d}\beta^{(N)}\int_z\Psi(z)\partial_\mu^2\Psi(z). \quad (14)$$

In the rest of this section I will show the relation of renormalization by running coupling constants and perturbative renormalization. Let

$$v = \sum_{j:j\leq N} v^j$$

be the free propagator with UV-cutoff. For notational simplicity we consider only the case of renormalization for the quartic interaction term. Generalizations are easily obtained. Consider the partition function

$$Z(\Psi) = \int d\mu_v(\Phi) \exp\{-\frac{\lambda_N}{4!} \int_z (\Phi + \Psi)(z)^4 + e_N\}. \tag{15}$$

The partition function for an effective theory is

$$Z_k(\Psi) = \int d\mu_{v^{\leq k+1}}(\Phi) \exp\{-\frac{\lambda_N}{4!} \int_z (\Phi + \Psi)(z)^4 + e_k\}. \tag{16}$$

The running coupling constant is defined by

$$\lambda_k = -\int_{z_2,z_3,z_4} \frac{\delta^4}{\delta\Psi(z)\delta\Psi(z_2)\delta\Psi(z_3)\delta\Psi(z_4)} \ln Z_k(\Psi)|_{\Psi=0}. \tag{17}$$

Define a new partition function by

$$\bar{Z}_k(\Psi) = Z_k(\Psi) \exp\{\frac{\lambda_{k+1}}{4!} \int_z \Psi(z)^4\} \tag{18}$$

for all $k \leq N$. We set $\lambda_{N+1} = 0$. The renormalization group flow for the running coupling constant is given by

$$\delta\lambda_j = \lambda_{j+1} - \lambda_j = \int_{z_2,z_3,z_4} \frac{\delta^4}{\delta\Psi(z)\delta\Psi(z_2)\delta\Psi(z_3)\delta\Psi(z_4)} \ln \bar{Z}_j(\Psi)|_{\Psi=0}. \tag{19}$$

The recursion relation for the partition function $\bar{Z}_k$ is

$$\begin{aligned}\bar{Z}_{k-1}(\Psi) = \int d\mu_{v^k}(\Phi^k) \bar{Z}_k(\Phi^k + \Psi) \exp\{-\frac{\lambda_k}{4!}[\int_z (\Phi^k + \Psi)^4 - \int \Psi^4] - \\ - \frac{\delta\lambda_k}{4!} \int (\Phi^k + \Psi)^4 - \delta e_{k-1}\}\end{aligned} \tag{20}$$

with $\delta e_k = e_{k+1} - e_k$.

The Feynman graph expansion for the partition function is

$$Z(\Psi) = \exp\{\sum_{G \in \mathcal{F}_c} I(G)\}. \tag{21}$$

$\mathcal{F}_c$ is the set of all connected Feynman graphs and $I(G)$ is the Feynman integral for the Feynman graph G. Each line in the Feynman graph corresponds to a free propagator v with UV-cutoff and each vertex corresponds to the negative bare coupling constant $-\lambda_N$. Removing the cutoff ($N \to \infty$) the Feynman integral is no longer convergent for all Feynman graphs G. In standard perturbation theory the bare interaction is splitted in a renormalized interaction and a counterterm. The vertices in the Feynman graphs corresponds to the renormalized coupling constant or to a counterterm insertion. The counterterm is chosen such that the divergencies in each Feynman graph are cancelled. We want to show how to renormalize by using running coupling constants. Assign to each propagator in a Feynman graph an integer $j \leq N$.

Interprete this line in a Feynman integral as propagator v^j. The sum over all assignments reproduce the Feynman integral. Denote by $G^{\geq i}$ the subgraph of G when we omit all propagator lines v^j with $j < i$. $G^{\geq i}$ consists of connected components. Denote by $\mathcal{C}^i_\gamma$ a connected component of $G^{\geq i}$ which contains at least one propagator v^i. The set of all connected Feynman graphs with propagartors $v^N, v^{N-1}, ..., v^{k+1}$ containing at least one propagator is denoted by $\bar{\mathcal{F}}^k_c$. The partition functions are represented by

$$\bar{Z}_k(\Psi) = \exp\{\sum_{G\in\bar{\mathcal{F}}^k_c} \bar{I}(G\} \tag{22}$$

for all $k < N$. The rules for computation of a renormalized Feynman integral $\bar{I}(G)$ are

(i) Vertex rule : From each vertex in G emerges at least one propagator line v^j, $k+1 \leq j \leq N$. If m is the maximal index of all propagator lines emerging from a vertex in G then this vertex corresponds to the negative running coupling constant $-\lambda_m$.

(ii) Subtraction rule : Suppose that the vertices of the Feynman graph G are labelled by 1,...,n. A vertex of a subgraph G' of G is called maximal if all other vertices of G' have larger labels. For a connected component $\mathcal{C}^j_\gamma$ of G with four external lines let $\mathcal{R}(\mathcal{C}^j_\gamma)$ be the graph obtained from $\mathcal{C}^j_\gamma$ by linking external lines which emerge from not minimal vertices to the minimal vertex of $\mathcal{C}^j_\gamma$. Then replace for each connected component $\mathcal{C}^j_\gamma$ by $\mathcal{C}^j_\gamma - \mathcal{R}(\mathcal{C}^j_\gamma)$.

3. Renormalization theory on the multigrid

For an introduction of multigrid methods I refer the reader to G.Mack's lecture. Euclidean quantum field theory can be represented as a polymer system on a multigrid Λ. The field Φ and free propagator v are transformed to the multigrid by the following split

$$\begin{aligned}\Phi(z) &= \int_{x\in\Lambda} \mathcal{A}(z,x)\varphi(x)\\ v(z_1,z_2) &= \sum_j \int_{x_1,x_2\in\Lambda^j} \mathcal{A}(z_1,x_1)\mathrm{v}(x_1,x_2)\mathcal{A}(z_2,x_2) \equiv \sum_j v^j(z_1,z_2)\end{aligned} \tag{23}$$

for $z, z_1, z_2 \in \mathbf{R}^d$. Free propagators v_j obey the bound (5). Suppose that for all finite subsets X of Λ partition functions $Z(X)$ are defined. Then activities $A(P)$ are defined by

$$Z(X) = \sum_{X=\sum P} \prod_P A(P) \tag{24}$$

for all finite subsets X of Λ. The sum runs over all disjoint partitions of X. For convergence properties we need that $A(P)$ is small if P contains blocks $x \in \Lambda^j, y \in \Lambda^k$ with $|j-k|$ large. In the last section we have seen how to gain convergence factors $L^{-\epsilon|j-k|}, \epsilon > 0$, by introducing running coupling constants. In this section I will show how convergence factors for activities $A(P)$ are produced by imposing renormalization conditions.

Feynman graphs are naturally defined on a multigrid. Introduce $\mathcal{A}$-lines which connect points $z \in \mathbf{R}^d$ with blocks $x \in \Lambda$ on the multigrid and v-lines which connect blocks x, y on layers Λ^j of the multigrid. By split (23) each propagator line in a Feynman graph connecting vertices $z_1, z_2 \in \mathbf{R}^d$ can be represented by two $\mathcal{A}$-lines which connect z_1, z_2 by vertices $x_1, x_2 \in \Lambda^j$ respectively and a v-line connecting x_1 and x_2. A Feynman graph on a multigrid is called point connected if the graph is connected when vertices on the multigrid which represent same blocks are identified. A Feynman graph expansion for activities consists of point connected Feynman graphs. For renormalization the following definition is useful. A Feynman graph on a multigrid is called k-vertically irreducible if it is point connected and it is not possible by cutting less than k+1 $\mathcal{A}$-lines to separate vertices $x \in \Lambda^j, y \in \Lambda^{j'}, j \neq j'$, in the Feynman graph. Vertices x, y are separated if there is no path in the Feynman graph which connects x and y. Feynman graphs on a multigrid can be decomposed into k-vertically irreducible components. The point is that a k-vertically irreducible component needs no renormalization if k is large enough (for a renormalizable theory). Correspondingly, activities can be expressed by k-vertically irreducible activities. In the following the definition for k-vertically irreducible activities is given.

Let the partition function for a finite subset X of Λ be of the following form

$$Z(X|\Psi) = \int d\mu_{v_X}(\Phi) \exp\{-V(X|\Phi + \Psi) + \int_{x \in X} j(x)(\varphi + \psi)(x)\} \tag{25}$$

where

$$v_X(z_1, z_2) = \int_{x_1, x_2 \in X} \mathcal{A}(z_1, x_1) \mathrm{v}(x_1, x_2) \mathcal{A}(z_2, x_2).$$

$j(x)$ is an external source and $V(X|\Phi + \Psi)$ an X-dependent interaction. We will use the notations

$$X^j = X \cap \Lambda^j, \quad X^{\geq j} = X^j + X^{j+1} + \dots, \quad X^{<j} = X^{j-1} + X^{j-2} + \dots .$$

Define new partition functions

$$Z_j(X|\Psi) = \int d\mu_{v_{X^{\geq j+1}}}(\Phi) \exp\{-[V(X|\Phi + \Psi) - V(X^{<j+1}|\Phi + \Psi)] + \int_{x \in X^{\geq j+1}} j(x)(\varphi + \psi)(x)\}. \tag{26}$$

We have

$$Z_k(X|\Psi) = \begin{cases} 1, & \text{if } k \geq h(x) = \max\{j | X^j \neq \emptyset\} \\ Z(X|\Psi), & \text{if } k < l(X) = \min\{j | X^j \neq \emptyset\}. \end{cases}$$

Let $\tau : \Lambda \to \mathbf{N}$ ($\mathbf{N}$ = set of all natural numbers) be a function with finite support supp $\tau = \tau^{-1}(\mathbf{N} - \{0\})$. τ can be considered as a finite collection of blocks x with multiplicity $\tau(x)$. We will use the notation

$$\tau! = \prod_{x \in \Lambda} \tau(x)! \,, \quad |\tau| = \sum_{x \in \Lambda} \tau(x) \,, \quad \frac{\delta^\tau}{\delta \psi^\tau} = \prod_{x \in \Lambda} \frac{\delta^{\tau(x)}}{\delta \psi(x)^{\tau(x)}}.$$

For a polymer P and a collection τ with supp $\tau \subset \Lambda^{<l(P)}$ we call $(P|\tau)$ an extended polymer and define a partition function and activity by

$$\begin{aligned} Z(P|\Psi) &= \frac{1}{\tau!} \frac{\delta^\tau}{\delta \psi^\tau} Z_{l(P)-1}(P + \text{supp } \tau | \Psi)|_{\Psi^{<l(P)}=0} \\ A(P|\tau) &= \frac{1}{\tau!} \frac{\delta^\tau}{\delta \psi^\tau} A_{l(P)-1}(P + \text{supp } \tau | \Psi)|_{\Psi^{<l(P)}=0} \end{aligned} \tag{27}$$

where A_j is the activity for the partition function Z_j and $\Psi^{<j}$ is defined by

$$\Psi^{<j}(z) = \int_{x\in\Lambda^{<j}} A(z,x)\psi(x).$$

Partition functions and activities for extended polymers are related by

$$Z(X|\tau) = \sum_{n\geq 1} \sum_{X=\sum_{i=1}^n P_i} \sum_{X=\sum_{i=1}^n \tau_i} \prod_{i=1}^n A(P_i|\tau_i). \tag{28}$$

Let $(P_1|\tau_1), ..., (P_n|\tau_n)$ be extended polymers. We call $[(P_1|\tau_1), ..., (P_n|\tau_n)]$ a k-partition ($k \in \mathbf{N}$) of $(X|\tau)$ into extended polymers if

$$\bigcup_{i=1}^n \text{supp } \tau_i \subset X + \Lambda^{<l(X)} , \; X = \sum_{i=1}^n P_i , \; \tau = \sum_{a=1}^n \tau_a^{<l(X)} , \; |\tau_a|_X| \leq k$$

for all $a \in \{1, ..., n\}$. $B^{[k]}(X|\tau)$ is the set of all k-partitions of $(X|\tau)$ into extended polymers. For extended polymers $(P|\tau)$ k-vertically irreducible activities $A^{[k]}(P|\tau)$ are defined by

$$Z(X|\tau) = \sum_{n\geq 1} \sum_{[(P_1|\tau_1),...,(P_1|\tau_n)]\in B^{[k]}(X|\tau)} N\{\prod_{i=1}^n A^{[k]}(P_i|\tau_i)\frac{\delta^{\tau_i|x}}{\delta j^{\tau_i|x}}\} \tag{29}$$

for all finite subsets X of Λ and collections τ with supp $\tau \subset \Lambda^{<l(X)}$. The symbol $N\{\ldots\}$ means that all derivatives operate inside the bracket. In Feynman graph expansion $A^{[k]}(P|\tau)$ consists of k-vertically irreducible Feynman graphs.

For sets $Q_1, ..., Q_n$ a graph $\gamma(Q_1, ..., Q_n)$ consists of vertices $1, ..., n$ and lines (ij) if $i \neq j$ and $Q_i \cap Q_j \neq \emptyset$. A k-partition $[(P_1|\tau_1), ..., (P_n|\tau_n)]$ is called connected if $\gamma(P_1, ..., P_n, \text{supp } \tau_1|_X, ..., \text{supp } \tau_n|_X)$ is connected. The set of all k-partitions of $(X|\tau)$ into extended polymers is denoted by $B_c^{[k]}(X|\tau)$. Then we obtain the following representation for activities

$$A(X|\tau) = \sum_{n\geq 1} \sum_{[(P_1|\tau_1),...,(P_1|\tau_n)]\in B_c^{[k]}(X|\tau)} N\{\prod_{i=1}^n A^{[k]}(P_i|\tau_i)\frac{\delta^{\tau_i|x}}{\delta j^{\tau_i|x}}\} \tag{30}$$

For extended Polymers $(P_1|\tau_1), ..., (P_m|\tau_m)$ and collections $\sigma_1, ..., \sigma_n$ we call $[(P_1|\tau_1), ..., (P_m|\tau_m)]$ a k-vertically irreducible connected partition of $(R|\tau)$ with respect to $\sigma_1, ..., \sigma_n$ if

(i) $R = \sum_{a=1}^m P_a$, $\tau = \sum_{a=1}^n \sigma_a^{<l(R)} + \sum_{b=1}^m \tau_a^{<l(R)}$ and

$$\bigcup_{a=1}^n \text{supp } \sigma_a \cup \bigcup_{b=1}^m \text{supp } \tau_b \subseteq R + \text{supp } \tau$$

(ii) $\gamma(P_1, ..., P_m, \text{supp } \sigma_1|_R, ..., \text{supp } \sigma_n|_R)$ is connected

(iii) There exists no $I \subseteq \{1, ..., m\}$ such that $(\text{supp } \sigma_b) \cap R' = \emptyset$ for $R' = \sum_{a\in I} P_a$ and all $b \in \{1, ..., m\} - I$ and

$$|\sum_{a\in I} \tau_a^{<l(R')} + \sum_{b\in I:\ (\text{supp } \sigma_b)\cap R'\neq\emptyset} \sigma_b^{<l(R')}| \leq k.$$

The set of all k-vertically irreducible connected partitions of $(R|\tau)$ with respect to $\sigma_1, ..., \sigma_n$ is denoted by $C^{[k]}(R\|\sigma_1, ..., \sigma_n|\tau)$. Define

$$A^{[k]}(R\|\sigma_1,...,\sigma_n|\tau) = \sum_{m\geq 1} \sum_{[(P_1|\tau_1),...,(P_m|\tau_m)]\in C^{[k]}(R\|\sigma_1,...,\sigma_n|\tau)} N\{\prod_{i=}^{m} A^{[k]}(P_i|\tau_i)\frac{\delta^{\tau_i|R}}{\delta j^{\tau_i|R}}\}. \quad (31)$$

Then the activities $A(Q|\tau)$ obey the following recursion relations

$$A(X|\tau) = \sum_{\tau',\tau'':\ \tau=\tau'+\tau''} \sum_{R:\ \emptyset\neq R\subseteq X} \sum_{n\geq 0} \sum_{\substack{\sigma_1,...,\sigma_n: \\ \sum_{i=1}^{n} \sigma_a^{<l(X)}=\tau'}} \sum_{X-R=\sum_{i=1}^{n} Q_a} N\{A^{[k]}(R\|\sigma_1,...,\sigma_n|\tau'')\prod_{i=1}^{n} A(Q_a|\sigma_a)\frac{\delta^{\sigma_a|R}}{\delta j^{\sigma_a|R}}\}. \quad (32)$$

For the first factor in the bracket of Eq. (32) we have enough supression factors if k is chosen sufficiently large. The product in the bracket is small if we impose renormalization conditions. For example impose the renormalization condition

$$\int_{z_2} \frac{\delta^2}{\delta\Psi(z)\delta\Psi(z_2)} A(P|\Psi)|_{\substack{\Psi=0\\ j=0}} = 0 \quad (33)$$

for all polymers P and $z \in \mathbf{R}^d$. Then for a collection σ consisting of blocks x_1 and x_2 we obtain

$$A(P|\sigma)|_{\substack{\Psi=0\\ j=0}} = \frac{1}{\sigma!}\int_{z_1,z_2\in\mathbf{R}^d} \mathcal{A}(z_1,x_1)\mathcal{A}(z_2,x_2)\frac{\delta^2}{\delta\Psi(z_1)\delta\Psi(z_2)}A(P|\Psi)|_{\substack{\Psi=0\\ j=0}} =$$
$$= \frac{1}{2}\frac{1}{\sigma!}\int_{z_1,z_2\in\mathbf{R}^d} [\mathcal{A}(z_1,x_1) - \mathcal{A}(z_2,x_1)][\mathcal{A}(z_2,x_2) - \mathcal{A}(z_1,x_2)]$$
$$\frac{\delta^2}{\delta\Psi(z_1)\delta\Psi(z_2)}A(P|\Psi)|_{\substack{\Psi=0\\ j=0}}. \quad (34)$$

From the differences of the $\mathcal{A}$-kernels we can extract a factor $|z_1 - z_2|^2 L^{j_1+j_2}$ for $x_1 \in \Lambda^{j_1}$, $x_2 \in \Lambda^{j_2}$. Using the exponential decay for funtional derivative of $A(P|\Psi)$ we obtain a factor $L^{-2l(P)}$. Thus a factor $L^{j_1+j_2-2l(P)}$ is gained by using the renormalization condition (33).

Acknowledgement

The author wishes to thank the Deutsche Forschungsgemeinschaft for financial support of the work reported here, and the Hamburgische Wissenschaftliche Stiftung for a grant which made the visit to the Cargèse Advanced Study Institute possible.

References

1. J.Feldman, J.Magnen, V.Rivasseau and R. Sénéor, Bounds on Completely Convergent Euclidean Feynman Graphs, Comm. Math. Phys. 98 , 273-288 (1985).

POWER COUNTING AND RENORMALIZATION IN LATTICE FIELD THEORY

Thomas Reisz

Deutsches Elektronen-Synchrotron DESY, Hamburg

1. Introduction

In an Euclidean quantum field theory a space-time lattice provides a quite natural non-perturbative ultraviolet cutoff. This has lead to a variety of non-perturbative methods. Observable physics has to be searched for in the continuum limit. However, this limit is very difficult to be treated. Very often one is forced to use perturbative methods, and this leads to well known diagrammatic expansions. I will briefly outline the very specific structure of momentum space Feynman integrals with a lattice cutoff and discuss the continuum limit behavior. The well known power counting theorems of Weinberg [1] and Hahn, Zimmermann [2] which state sufficient conditions for the convergence of Feynman integrals do not apply in presence of a lattice cutoff. Nevertheless, a power counting theorem can be given for a wide class of lattice field theories where a new kind of an ultraviolet divergence degree is used [4].

In general, a Feynman integral will be divergent if the cutoff is removed. The existence of a power counting theorem ensures that the combinatorics of subtractions to renormalize a diagram is described by Zimmermann's forest formula [3]. However, in lattice field theory counterterms instead of being polynomials are periodic functions in the external momenta, constructed by so called subtraction operators [5].

At first, we have been concerned solely with the problem of ultraviolet divergencies and have assumed all fields to be massive. Appropriate modifications in the presence of massless fields are given in the last section. Furthermore, we assume four-dimensional space-time.

2. Power Counting Theorem [4]

A typical lattice Feynman integral has the following structure.

$$\widehat{I}(q;\mu,a) = \int_{-\frac{\pi}{a}}^{\frac{\pi}{a}} d^4k_1 \cdots d^4k_m \, \frac{V(k,q;a)}{\prod_{i=1}^{n}[\eta_i(l_i(k,q)a)/a^2 + \mu_i^2]}, \tag{1}$$

where a is the lattice spacing, the masses $\mu_i \neq 0$, $V(k,q;a) = F(ka,qa)/a^{m_0}$, $F \in C^\infty$, $m_0 \in \mathbf{Z}$, so that $P(k,q) = \lim_{a\to 0} V(k,q;a)$ exists, and similarly all $\eta_i \in C^\infty$

are non-negative and $\lim_{a\to 0} \eta_i(l_i a)/a^2 = l_i^2$ (l_i being the line momenta and are linear functions in internal momenta k and external momenta q), i.e., the lattice propagators reproduce the continuum propagators as the lattice spacing a tends to zero. Furthermore, the integrand is assumed to be periodic with the Brillouin zone $[-\pi/a, \pi/a]^4$ in all loop momenta $k_1, \ldots, k_m$.

We want to state very general sufficient conditions for the existence of the continuum limit of (1) as well as its coincidence with the formal limit

$$\int_{-\infty}^{\infty} d^4k_1 \cdots d^4k_m \, \frac{P(k,q)}{\prod_{i=1}^{n}(l_i^2(k,q)+\mu_i^2)}. \tag{2}$$

At first we note that the convergence of (2) does not guarantee the existence of the $a \to 0$-limit of (1) as the following example shows. Let

$$V(t,q;a) = \left(\frac{2}{a}\sin\frac{ta}{2}\right)^4 \left(\cos^2\frac{qa}{2} - \cos^2\frac{ta}{2}\right) + \left(\frac{2}{a}\sin\frac{ta}{2}\right)^2 \left(\frac{2}{a}\sin\frac{qa}{2}\right)^2.$$

Then

$$\begin{aligned} \widehat{J}(q;\mu,a) &= \int_{-\pi/a}^{\pi/a} dt \, \frac{V(t,q;a)}{\left(\frac{4}{a^2}\sin^2\frac{ta}{2}+\mu^2\right)^2} \\ &= \frac{16}{a}\int_{-\pi}^{\pi} dt \, \frac{\sin^6\left(\frac{t}{2}\right)}{(4\sin^2\frac{t}{2}+\mu^2a^2)^2} + O(1) \qquad \text{as } a \to 0 \end{aligned} \tag{3}$$

is divergent in the $a \to 0$-limit, whereas

$$\int_{-\infty}^{\infty} dt \, \frac{t^2q^2}{(t^2+\mu^2)^2} \tag{4}$$

is convergent. Furthermore, we exclude additional poles[1] of propagators in the Brillouin zone besides at zero momentum. If this condition would be violated, the general assumptions about the structure of the integrand of (1), especially its periodicity with the Brillouin zone, would not be sufficient for convergence. In particular, the theorem below does not apply to lattice fermions with propagators having poles on the boundary of the Brillouin zone.

To check the continuum limit behavior, the integration domain of (1) is partitioned in dependence on the configuration of the line momenta l_i. For small l_i and large l_i the corresponding propagator is bounded by its continuum limit and a multiple of a^2, respectively. However, the numerator causes some technical problems. It cannot be estimated by its continuum limit as the above example shows. A more careful analysis is necessary. This can be done by introducing the notion of lattice UV-degrees, defined as follows. Let

$$V(u,w;a) = \frac{F(ua,wa)}{a^{m_0}} \quad , \qquad F \in C^\infty. \tag{5}$$

Then the lattice UV-degree of V with respect to u,

$$\delta_u = \overline{\operatorname{degr}}_{\widehat{u}} V(u,w;a),$$

[1]A "pole" of a propagator $1/\left(\eta(ka)/a^2+\mu^2\right)$ denotes a zero of the η-function.

is defined by the behavior of V as $u \to \infty$ and $a \to 0$ *simultaneously*:

$$V(\lambda u, w; \frac{1}{\lambda}a) = A(u, w; a)\, \lambda^{\delta_u} + O(\lambda^{\delta_u - 1}), \quad \lambda \to \infty, \tag{6}$$

where $A(u, w; a) \not\equiv 0$. This degree satisfies all "usual" degree properties, e.g.

$$\begin{aligned} \overline{\mathrm{degr}}_{\widehat{u}}(V_1 + V_2) &\le \max(\overline{\mathrm{degr}}_{\widehat{u}} V_1, \overline{\mathrm{degr}}_{\widehat{u}} V_2) \\ \overline{\mathrm{degr}}_{\widehat{u}}(V_1 \cdot V_2) &\le \overline{\mathrm{degr}}_{\widehat{u}} V_1 + \overline{\mathrm{degr}}_{\widehat{u}} V_2. \end{aligned} \tag{7}$$

The importance of the lattice degrees rests on the estimate

$$|V(u, w; a) - P(u, w)| \le a^l \sum_{finite} |Q_i(u, w)|,$$

where $P(u, w) = \lim_{a \to 0} V(u, w; a)$, Q_i are polynomials finite in number, $l \in \mathbf{N}$ and

$$\begin{aligned} \overline{\mathrm{degr}}_u P(u, w) &\le \overline{\mathrm{degr}}_{\widehat{u}} V(u, w; a) \\ \overline{\mathrm{degr}}_u Q_i(u, w) &\le \overline{\mathrm{degr}}_{\widehat{u}} V(u, w; a) + l. \end{aligned} \tag{9}$$

$\overline{\mathrm{degr}}_u P(u, w)$ is the degree of the polynomial P with respect to u. Note that the relations (9) are inequalities. For instance, if V is the numerator of (3), $\delta_t = \overline{\mathrm{degr}}_{\widehat{t}} V = 4$, whereas $\lim_{a \to 0} V(\lambda t, q; a) = \lambda^2 t^2 q^2$. In general, if $P(u, w) \not\equiv 0$,

$$\overline{\mathrm{degr}}_{uw} P(u, w) = \overline{\mathrm{degr}}_{\widehat{uw}} V(u, w; a) = m_0,$$

i.e., w.r.t. *all* momenta the lattice degree of V coincides with the polynomial degree of the $a \to 0$-limit of V.

Actually, the estimate (8) must be generalized for the various choices of partitions (u, w). Nevertheless, the importance of the lattice degrees rests on the validity of (9), and this allows to formulate a power counting theorem in terms of lattice UV-divergence degrees.

Consider the integral (1). For a given arbitrary linear parametrization $k = k(u, v, q)$, $det(\partial(u, v)/\partial(k)) \neq 0$, the variable u and the constant v define an affine subspace H of the integration momenta. We define a UV-divergence degree of (1) with respect to H by

$$\widehat{\omega}(H) = d + \overline{\mathrm{degr}}_{\widehat{u}} V(k(u, v, q), q; a) - \overline{\mathrm{degr}}_{\widehat{u}} \prod_{i=1}^{n} \left(\frac{\eta_i(l_i(u, v, q)a)}{a^2} + \mu_i^2 \right), \tag{10}$$

where d denotes the dimension of H. Note that (10) is independent of the values of v and q. Using this notion, we formulate a power counting theorem for massive lattice field theories (for a precise formulation see [4])[2].

[2] Actually, in the following theorem it will be sufficient to assume the power counting conditions (11a) only for a finite number of classes of special affine subspaces H, so called Zimmermann subspaces [4].

THEOREM. *Assume a Feynman integral $\widehat{I}$, (1), is given, such that the integrand is periodic with the Brillouin zone in $k_1, \ldots, k_m$, every propagator has only one pole in the Brillouin zone (at vanishing momentum), and the set $\{l_i\}$ of line momenta is natural. If for all affine subspaces H*

$$\widehat{\omega}(H) < 0\,, \tag{11a}$$

the continuum limit of $\widehat{I}$ exists and is given by

$$\lim_{a \to 0} \widehat{I}(q; \mu, a) = \int_{-\infty}^{\infty} d^4k_1 \cdots d^4k_m \, \frac{P(k,q,\mu)}{\prod_{i=1}^{n}[l_i(k,q)^2 + \mu_i^2]}. \tag{11b}$$

We have to explain what naturalness means. The line momenta l_i are of the form

$$l_i = \sum_{j=1}^{m} C_{ij} k_j + q_i \quad , \quad i = 1, \ldots, n.$$

Naturalness means that $C_{ij} \in \mathbf{Z}$, $rank(C_{ij}) = m$, and whenever one chooses another set of independent line momenta, $k'_1, \ldots, k'_m$ say, as integration variables, then this condition will also hold for the k in terms of the k': $k_i = \sum_{j=1}^{m} A_{ij} k'_j$, $A_{ij} \in \mathbf{Z}$. For a Feynman integral this condition is satisfied if all loop momenta $k_1, \ldots, k_m$ coincide with momenta of lines. Naturalness is important in connection with the periodicity. Both conditions ensure that only a small neighborhood of zero momentum contributes to the continuum limit, i.e., that the limit is given by (11b).

In the definition of the function classes to which the integrand belong we have always assumed infinitely often differentiability of the numerator and denominator with respect to momenta. Actually, the denominator needs to be differentiable only in a small neighborhood of zero momenta, and globally continuous. In the case of renormalization, the whole integrand has to be differentiable to a degree depending on the divergence degrees.

3. Renormalization [5]

The existence of a power counting theorem ensures that the combinatorics of renormalizations is given by the forest formula of Zimmermann [3]. In a lattice field theory counterterms have a very special property: they are periodic functions. They result from so-called subtraction operators. A subtraction operator $\widehat{t}_q^{\delta}$ of the order δ is defined by

$$(\widehat{t}_q^{\delta} F)(k,q;a) = \sum_{n=0}^{\delta} \frac{1}{n!} \sum_{i_1,\ldots,i_n=0}^{s} P_{n,i_1\cdots i_n}(q_1, \ldots, q_s; a) \cdot \cdot \left(\frac{\partial}{\partial q_{i_1}} \cdots \frac{\partial}{\partial q_{i_n}} F(k,q;a) \right)_{q=0},$$

where $P_{n,i_1\cdots i_n} \in \mathcal{C}_n^c$ are totally symmetric in $i_1, \ldots, i_n$ and $(2\pi/a)$-periodic in every component $q_1, \ldots, q_s$, $\lim_{a\to 0} P_{n,i_1\cdots i_n}(q_1, \ldots, q_s; a) = q_{i_1} \cdots q_{i_n}$, and that divergencies should in fact be subtracted:

$$[(1 - \widehat{t}_q^{\delta})F](k, \lambda q; \mu, a) = O(\lambda^{\delta+1}), \; \lambda \to 0.$$

F represents a Feynman integrand and is a periodic function of the loop momenta k and external momenta q. The simplest example of a subtraction operator is a Taylor polynomial in lattice momenta, e.g. $\sin(qa)/a$. The order δ is determined by the UV-divergence degree of the diagram to which $\widehat{t}_q^{\delta}$ is applied.

Using this notion we can state the lattice version of the BPHZ subtraction procedure. The combinatorics of subtractions are described by Zimmermann's forest formula, and the Taylor polynomials of the continuum version are to be replaced by subtraction operators. A lattice diagram renormalized by this method is convergent in the continuum limit, and the limit is independent of the lattice action chosen [5]. This means that perturbation theory is universal.

As an example consider the massive $\Phi^4 + a^2\Phi^6$-theory

$$S(\Phi)_{int} = a^4 \sum_{n\in\mathbf{Z}^4} [g\Phi^4(na) + \lambda a^2\Phi^6(na)\,]. \tag{12}$$

The divergence degree of an arbitrary diagram γ having $E(\gamma)$ external lines is given by

$$\widehat{\omega}(\gamma) = 4 + \sum_V (\widehat{\omega}(V) - 4) - E(\gamma),$$

where the sum is over all vertices V of γ, and $\widehat{\omega}(V)$ is the UV-divergence degree of the vertex V. There are two kinds of vertices. The Φ-field has a UV-dimension equal to one, hence $\widehat{\omega}(\Phi^4) = 4$ and $\widehat{\omega}(a^2\Phi^6) = 6 - 2 = 4$. Note that every power of the lattice spacing a decreases the UV-divergence degree by one. We get

$$\widehat{\omega}(\gamma) = 4 - E(\gamma),$$

i.e., the two-point function has a quadratic and the four-point function a logarithmic divergence, as is the case for a continuum Φ^4-theory. All other functions are finite in the continuum limit, without overall subtractions. In particular, λ gets no contributions from renormalization. The continuum limit of the renormalized theory of (12) is the (BPHZ) renormalized Φ^4-theory

$$S_c(\Phi)_{int} = \int d^4x\, g\Phi^4(x).$$

In summary, a power counting theorem can be given for a wide class of lattice field theories satisfying the pole condition, where the notion of a UV-degree must be modified. Renormalization means application of subtraction operators by the combinatorics of the forest formula, and the counterterms are always periodic. They can always be chosen in such a way that they result from local counterterm contributions to the lattice action [5]. Furthermore, the continuum limit of a renormalized theory is universal, i.e., it does not depend on a specific choice of the lattice action.

4. Generalization to massless fields [6,7]

The program can be generalized to apply to massless field theories also. Whereas the lattice provides an ultraviolet cutoff, infrared singularities should be quite the same as in the continuum. The UV-power counting conditions are to be supplemented by IR-power counting conditions which are on the lattice the same as in a continuum field theory [11,12]. Renormalization must respect these additional constraints. Subtractions at vanishing external momenta of a diagram are no longer convergent. A program that works is the lattice version of the auxiliary mass method of Lowenstein and Zimmermann [8,9]. The mass dependence of propagators is modified to be

$$\mu^2 \to \mu^2 + (s-1)^2 M^2 \quad , \quad \mu^2 + M^2 > 0,$$

where μ is the bare mass (may be zero) and M an auxiliary mass. Counterterms are now constructed for $s = 0$, hence being IR-finite. However, to subtract UV-divergencies of a diagram, one must differentiate with respect to external momenta and the "mass parameter" s. This means that counterterms result from generalized subtraction operators, which are combinations of subtraction operators and Taylor polynomials in s. Furthermore, additional finite renormalizations are necessary for two- and three-point functions to avoid infrared singularities by insertion of these graphs as subdiagrams. For instance, inserting a self-energy graph into a massless line yields a singularity which is integrable only if the two-point diagram vanishes at zero momentum.

These conditions are satisfied if one applies the renormalization prescription of the proceeding section, where every subtraction operator $\widehat{t}_q^\delta$ is replaced according to

$$1 - \widehat{t}_q^\delta \;\to\; (1 - \widehat{t}^{\rho-1}_{q(s-1)})\,(1 - \widehat{t}^\delta_{qs}), \tag{13}$$

where the $\widehat{t}$ on the right hand side are generalized subtraction operators [7], and ρ is given by the IR-divergence degree of the corresponding diagram. In most applications, $\rho \geq 1$ only for diagrams of two- and three-point functions, so that these are the only basic field Green functions which are affected by the additional finite renormalizations. This is the lattice version [7] of the BPHZL renormalization prescription [8,9]. Having applied all subtractions, we set $s = 1$ and get a theory which is ultraviolet as well as infrared finite if the IR-divergence degrees of the vertices satisfy

$$\widehat{r}(V) \geq 4 \qquad \text{for all vertices } V, \tag{14}$$

and momentum space diagrams are convergent for non-exceptional external momenta [10]. Massless bare fields remain massless after renormalization. Universality of the continuum limit holds as in the massive case, and counterterms are also periodic. Other renormalization schemes can be implemented by additional finite renormalizations satisfying the IR-constraints (14). For instance, by appropriate finite counterterms the auxiliary mass dependence vanishes and we get Green functions normalized at non-exceptional momenta apart from finite renormalizations for two- and three-point functions so that these amplitudes vanish at zero external momenta.

As an example, consider now the massless $\Phi^4 + a^2\Phi^6$-theory (12). The IR-dimension of the Φ-field is equal to one. Hence, $\widehat{r}(\Phi^4) = 4$ and $\widehat{r}(a^2\Phi^6) = 6$, not being influenced by powers of the lattice spacing. Hence, the massless model (12) is IR-finite renormalizable. Note that whereas the renormalized theory is IR-finite, this does not necessarily hold for the bare theory. The conditions (14) restrict the class of IR-finite renormalizable theories. For instance, a massless Φ^3-theory is IR-divergent in four dimensions, $\widehat{r}(\Phi^3) = 3$. A three-point vertex needs a lattice momentum factor, as is the case in a gauge theory.

Working also for massless theories, the lattice renormalization procedure will apply to lattice gauge field theories. Here in addition a special symmetry argument is needed, i.e., it must be shown that having chosen an appropriate gauge fixing, there exists a BRS-like symmetric lattice theory which is finite in the continuum limit.

REFERENCES

1. S.Weinberg, *High Energy Behaviour in Quantum Field Theory*, Phys. Rev. **118** (1960), 838 – 849.
2. Y.Hahn, W.Zimmermann, *An Elementary Proof of Dyson's Power Counting Theorem*, Commun. math. Phys. **10** (1968), 330–342.
3. W.Zimmermann, *Convergence of Bogoliubov's Method of Renormalization in Momentum Space*, Commun. math. Phys. **15** (1969), 208-234.
4. T.Reisz, *A Power Counting Theorem for Feynman Integrals on the Lattice*, Commun. math. Phys., to be published.
5. T.Reisz, *Renormalization of Feynman Integrals on the Lattice*, Commun. math. Phys., to be published.
6. T.Reisz, *A Convergence Theorem for Lattice Feynman Integrals with Massless Propagators*, Commun. math. Phys., to be published.
7. T.Reisz, *Renormalization of Lattice Feynman Integrals with Massless Propagators*, Commun. math. Phys., to be published.
8. J.H.Lowenstein , W.Zimmermann, *On the Formulation of Theories with Zero-Mass Propagators*, Nucl. Phys. B86 (1975) 77.
9. J.H.Lowenstein, *Convergence Theorems for Renormalized Feynman Integrals with Zero-Mass Propagators*, Commun. math. Phys. 47 (1976) 53-68.
10. K.Symanzik, "Small-Distance Behaviour in Field Theory", Springer Tracts in modern Physics 57 (1971), 222.
11. J.H.Lowenstein, W.Zimmermann, *The Power Counting Theorem for Feynman Integrals with Massless Propagators*, Commun. math. Phys. 44 (1975) 73 – 86.
12. G.Bandelloni, C.Becchi, A.Blasi, R.Collina, *Renormalization of Models with Radiative Mass Generation*, Commun. math. Phys. 67 (1978) 147 - 178.

OPEN STRINGS AND THEIR SYMMETRY GROUPS

Augusto Sagnotti

Dipartimento di Fisica, Università di Roma II, "Tor Vergata"

Via Orazio Raimondo, 00173 Roma ITALY, and INFN, Sezione di Roma

The last three years have seen a large amount of progress in String Theory [1], and the subject itself has undergone a change of scope. This is well reflected in the content of the talks that have preceded this one.

Much of the recent progress in String Theory can be traced to a precise strategy: a careful study of the few models known since the beginnings of the subject, and the abstraction from them of basic properties that one would like to demand from other models. This could be termed a set of "model-building rules". The approach corresponds to the fact, often a source of embarassment to specialists, that String Theory, born as a set of rules rather than as a set of principles, has long resisted attempts to reduce it to a logically satisfying structure.

The first property is **two-dimensional conformal invariance.** This is the motive behind the machinery that has become known as **Two-Dimensional Conformal Field Theory** [2,3]. Conformal invariance entered the subject long ago, as the cure to the unitarity problems of the bosonic model, by necessity defined in terms of oscillators of Minkowski, rather than Euclidean, signature. Conformal invariance is responsible for a transverse spectrum of states, all of positive norm, and embodies the projective invariance (duality) of the originally known amplitudes. Conformal invariance is reflected, in the context of the **closed** bosonic string, in the invariance under two mutually commuting Virasoro algebras. Conformal Field Theory is an algebraic, and thus non-perturbative, framework for the description of all models that share this property of invariance under two distinct Virasoro algebras. The algebraic description proceeds via "primary fields" (tensors of the conformal group), and entails the encoding of the two-dimensional dynamics in the coefficient functions of the operator product algebra. The operator product coefficients are somewhat reminiscent, in their role, of structure constants in ordinary Lie algebras. Like structure constants, they are subject to quadratic constraints. These embody the duality properties of amplitudes. In principle, these data are suffcient to reconstruct the correlators of the fields, and thus the scattering amplitudes of String Theory. The **closed bosonic string** in 26 dimensions is built out of a particular two-dimensional conformal field theory, one consisting of 26 copies of a free massless bosonic field. This model has a distinctive feature: the value of the central charge is 26 for each of the two independent Virasoro algebras. In modern terms, this compensates the contribution (equal to -26) of the conformal field theory of the reparametrization ghosts, to give a resulting model which is **exactly conformally invariant**, i.e. characterized by a total central charge equal to zero [4]. Alternatively, this property is linked to the absence of local anomalies in the two-dimensional field theory. Generalizations of the bosonic model are to retain this property, if they are to describe a space time of Minkowski signature, and a suitable generalization is needed to deal with (possible) super-partners of the string coordinates, in the form of **super-conformal invariance**.

The second property is **modular invariance**. It is essentially the statement that the theory be a theory of closed strings. This property was identified long ago [5], again in the context of the bosonic model. In modern language, following Polyakov [4], one would correct the tree amplitudes of String Theory by the addition of terms where the Conformal Field Theory lives on surfaces of increasing genera. For simplicity, let us restrict attention to the vacuum amplitude, since this is supposed to capture the essence of the matter anyway [3], and let us consider the first correction to the "tree" amplitude. In this case the parameter space has the topology of the torus. The complex structure of the torus, the one datum that a truly conformally invariant theory feels, can be described via a two-dimensional lattice, of sides 1 and Ω, where one can take $\mathrm{Im}(\Omega)$ to be positive (fig. 1). Ω is the "period matrix" of the torus. The crucial point is that doubly periodic (elliptic) functions, such as the bosonic string coordinates, have no way to distinguish between the two sides of the cell, and in general between choices of fundamental cell related by the familiar redefinition

$$\Omega \longrightarrow \Omega + 1 \qquad \text{and} \qquad \Omega \longrightarrow -1/\Omega \quad . \qquad (1)$$

In this respect, the bosonic model is a bit too simple. In proceeding to the **closed superstring**, one is forced to consider conformal fields that are no more doubly periodic along the two homology cycles of the torus (in modern terms, they are sections of line bundles, rather than functions, on the torus). So, if one starts with proper (i.e. doubly antiperiodic, or Neveu-Schwarz) fermions, and one insists on the **symmetries of a theory of closed strings** (i.e. modular invariance) , one is forced to add more contributions. The way to do so is not unique, but a very suggestive possibility is to complete separately the contributions of the left and right movers (i.e. the portions depending analytically and anti-analytically on Ω). The result is the GSO projection [6], which leads to the ten-dimensional closed superstrings. So, at times one has to work harder to attain modular invariance, and this occurs precisely when one is dealing with sections of nontrivial bundles.

It was the great contribution of ref. [7] (see also ref. [8]) to extend the consideration of sections of nontrivial bundles also to the case of bosons. The resulting constructions, recognized as orbifolds of toroidal models, have turned into a fundamental way of exploring the structure of two-dimensional conformal field theory. It is remarkable that all known constructions based on free fields can be understood in these terms. Indeed, this possibility for the closed superstrings was pointed out in ref. [7], and served as a motivation for the orbifold construction.

Armed with the two principles of **conformal invariance** and **modular invariance**, one can proceed to explore the set of conformal field theories. Even restricting oneself to just twisting boundary conditions of free fields, one finds a huge number of possibilities, and the long-standing limitations on the dimensionality of space time in String Theory fall apart [9].

The extent of the confidence in the properties of **closed strings**, and the corresponding amount of progress, have to be contrasted with the situation for **open strings**. Again, the content of the preceding talks makes this point hardly in need to be stressed. Apart for some occasional mention, open strings have been completely left out. This is rather peculiar, because on the one hand String Theory was born in the form of the Veneziano amplitude for open strings, and on the other hand the resurgence of a wide interest in the subject was triggered by the discovery of the Green-Schwarz mechanism, originally motivated by an analysis of open-string amplitudes. The excuses that have been given over the times for the neglect of open strings can be traced to two main points. First of all, the way symmetry groups originally entered open string models is via the so-called Chan-Paton ansatz [10]. This consists in multiplying amplitudes by traces of suitable matrices (including those of the fundamental representations of the O(N), U(N) and USp(2N) Lie algebras [11]). The whole thing looks rather ad hoc, to be contrasted with the neat role played by Kac-Moody algebras [12] in the construction of the heterotic string [13]. Moreover, the open (and closed) bosonic model is rather complicated, and thus somewhat clumsy to deal with. It involves many more diagrams than the closed (extended Shapiro-Virasoro) model, and often delicate divergence cancellations between them. Furthermore, it is usually felt, not without regret, that modular invariance is lost in this case, and that to check for the consistency of open-string models all one can do is appeal to anomaly cancellations, whenever possible. This last point is made particularly dubious by the recent recognition that, in analogy with the special role played by the group SO(32) in the type-I superstring in 10 dimensions, the group SO(8192) selects a special bosonic model in 26 dimensions [14,15].

The rest of this talk is devoted to remeding to these inconveniences. My aim will be convincing the reader, as well as I hope to have convinced the listener, that **the known open-string models in 26 and 10 dimensions have to be understood as parameter-space orbifolds of corresponding left-right symmetric closed models**. The Z_2 twist involved mixes left and right movers. In more geometrical terms, it symmetrizes between the two sides of the parameter surfaces. So, the open Veneziano model in 26 dimensions is seen to descend from the closed (extended) Shapiro-Virasoro model. The same pattern is followed in ten dimensions, where the type-I superstring can be seen to descend from the chiral type-IIb superstring. It should be noticed that the orbifold construction requires symmetry between left and right waves. This means a chiral spectrum for the superstring, since the two Ramond vacua must have the same chirality. The second point I will try to make is that **the size of the Chan-Paton group is determined by modular invariance**, by which I mean that the orbifold construction, applied to the surfaces with automorphisms that admit it, fixes the weights of the diagrams, and explains the very emergence of open strings. **Open strings are the "twisted sector"** of the construction. Therefore, their vertex operators sit at the fixed points. This is familiar stuff. After all, we all knew for ages that open strings are emitted from boundaries! The powers of two that build up the "magic numbers", 32 and 8192, can be related to sizes of the fixed-point sets. Actually, for these models, Neil Marcus and I showed that the group theory can be generated by means of free fermions valued on the boundaries of the parameter surfaces [15]. This corresponds to the long-held picture of "quarks at the ends of strings". From the point of view advocated here, one should keep in mind the analogy with the zero modes of the Ramond sector in the orbifold construction of the superstring.

Let me start by reviewing how Chan-Paton factors [10] originally entered the game. The crucial observation was that the cyclic symmetry of "tree" open-string amplitudes (what we would now call the "disk contribution") is compatible with multiplication by a trace of matrices. The cyclic symmetry (planar duality) is the remnant, in the open-string case, of the total symmetry (non-planar duality) of closed-string amplitudes. Demanding that these (Chan-Paton) factors respect the factorization properties of amplitudes imposes severe constraints. These are somewhat relaxed if, following Schwarz [16], one takes into account the twist symmetry, i.e. the simple behavior of open-string amplitudes under world-sheet parity. The result is an infinite number of constraints, solved long ago by Neil Marcus and myself [11], by appealing to a classic result in the theory of Algebras. This classifies the simple associative algebras over the real numbers [17]. These arguments lead to exclude exceptional groups from the original open-string models in 26 and 10 dimensions [11].

The Chan-Paton ansatz can actually be replaced by a dynamical construction in terms of currents [15], somewhat reminiscent of the corresponding construction for the heterotic string. However, open string currents are valued on the boundaries of the parameter surfaces. They are described, in the simplest of ways, in terms of one-dimensional free fermions, quantized antiperiodically along boundaries. This implements the old intuitive idea of "quarks at the ends of strings", but for a few subtleties. First of all, in order to attain a degeneracy of order 2**[D/2], one needs but D "quarks". Second, these "quarks" have no space-time attributes. Thus, modifying the usual actions by adding

$$S_{group} = \frac{i}{4}\int_{\partial\Sigma} ds\ \beta^I\ \frac{d\beta^I}{ds} \tag{2}$$

produces all the right multiplicities for empty boundaries. A corresponding modification of the vertex operators includes the ß fields, and produces the trace factors. This is all good and well, but it would be nice to predict the number of ß fields, especially since it turns out one needs as many of them as the space-time coordinates. Indeed, ten fermions give SO(32), whereas 26 give SO(8192).

It has been known for a while that boundaries affect the conformal properties of two-dimensional models. For instance, in ref. [18] it is shown that further divergences are introduced, proportional to the lengths of the boundaries, and that the theory "feels" the geodesic curvature of the boundaries. In ref. [15] we noticed that the "smooth doubling" of surfaces forces the boundaries to be geodesics, which can be taken as a boundary condition on the intrinsic metric. Thus, for each boundary, one is left with a (non-logarithmic) divergence proportional to the length, with a coefficient proportional to the number of space-time coordinates, and of the right sign to cancel against the divergence introduced by the boundary fermions. The divergence being not logarithmic, its cancellation is fraught with ambiguities. Moreover, showing that the cancellation takes place requires adding contributions coming partly from the parameter surface and partly from its boundary. This involves standard ways of dealing with integrals of Dirac′ s delta functions over half of the real axis. This was all known to Neil Marcus and myself at the time of writing ref. [15], but it is not stressed there, since it can capture in different amounts one′s interest, due to the ambiguities mentioned above. Still, if one takes the cancellation seriously, and if the boundaries are taken to be geodesics, **the open-string models with proper groups (SO(32) and SO(8192)) exhibit the same divergence structure as closed string models.** Then, are they really to be regarded as closed-string models all the way?

A related observation is that the order of both SO(32) and SO(8192) is a power of two. More simply, empty boundaries contribute a factor of two for each space-time coordinate. Such factors are familiar. There are at least two instances where they arise. One is the Ramond sector of the superstring. In this case one has gamma matrices and, after all, the quantization of the one-dimensional fermions of ref. [15] also gives gamma matrices. The other case is apparently quite different. It is the Z_2 orbifold of a "square" torus, described by Jeff Harvey at this

School. There the powers of two can be traced to the size of the fixed-point set of the involution that defines the orbifold. This encourages one to look for the same structure in the known open-string models in 26 and 10 dimensions.

As usual, it is simple and instructive to consider the genus-one contribution. For simplicity, I will do so for the bosonic string in 26 dimensions. There are then four diagrams, with parameter surfaces having, respectively, the topology of the torus, the Klein bottle, the annulus and the Möbius strip. They can be conveniently described in terms of lattices in the complex plane (figs. 1 and 2). In addition, the latter three surfaces are conveniently described in terms of their "doubles", which are all tori [19]. Ω is the "period matrix" of the doubles. It is purely imaginary for both the Klein bottle and the annulus, but not for the Möbius strip. It should be noticed that the torus contribution is the same as for the closed bosonic string, apart from a factor of two. Thus, it is modular invariant by itself. Moreover, the Klein bottle contribution is seen by inspection of figure 2 to be invariant under $\Omega \rightarrow \Omega + 2$. The resemblance between what one has so achieved and the untwisted (GSO projected) sector of the superstring, or the untwisted sector of the Z_2 orbifold described by Jeff Harvey at this School, is striking. The tricky point is that now the twist affects the parameter surface. Thus, in looking for the analogue of the $\Omega \rightarrow -1/\Omega$ transformation, one better think of what this transformation is meant to achieve. Then it is seen from figure 2 that the annulus contribution is precisely what is needed, since the twist is rotated by 90 degrees with respect to the Klein bottle. Finally, the Möbius strip symmetrizes the twisted sector. Actually, I have gone a bit too fast here. First of all, the two involutions that lead to the Klein bottle and to the annulus are different. One results into two cross-caps, and the other into two boundaries. Moreover, the "shift" of Ω that leads from the annulus to the Möbius strip is just $\frac{1}{2}$, not 1. The clue is in noticing that the two involutions act in the same way on the homology basis, and this is all the string integrand "feels". On the other hand, the zero modes result in the same contribution **only if one works in terms of the modulus of the double**. However, the "proper time" for the Klein bottle is half the modulus of its covering torus, which is again half the "proper time" for the annulus. Expressing amplitudes in terms of "proper time" exhibits the spectrum, and gives a relative factor of 2**D between Klein bottle and annulus. This is the square of the multiplicity factor when the group has order 2**[D/2]! In the same way, symmetrization in the twisted (open) sector requires adding $\frac{1}{2}$ to Ω, which refers to the double, and generates precisely the modulus for the Möbius strip. The oscillator description accomodates the orbifold idea very nicely. Open strings take values over "one half" of the parameter surfaces, and **are closed modulo the doubling of the parameter surfaces**.

Actually, the preceding discussion has left out an important point. This is the choice of projection in the ground state of the twisted sector. In ref. [15] it was pointed out that a twist-even ground state, and thus the group SO(8192) rather than USp(8192), leads to the elimination of some divergences via a principal part prescription. The divergences manifest themselves in the small-Ω limit of the amplitudes corresponding to figure 2. We were inspired by a similar phenomenon discovered by Green and Schwarz in the four-point amplitudes of the SO(32) superstring, and responsible for both finiteness and anomaly cancellation. This last result is discussed at length in ref. [1]. Actually, even for the superstring the same can be seen to occur directly at the level of the partition function, if one refrains from using the "aequatio" of Jacobi, which sets to zero the contributions of the individual diagrams, due to supersymmetry. The cancellations found by Weinberg [20] in the scattering amplitudes of the SO(8192) theory can be traced to the same phenomenon.

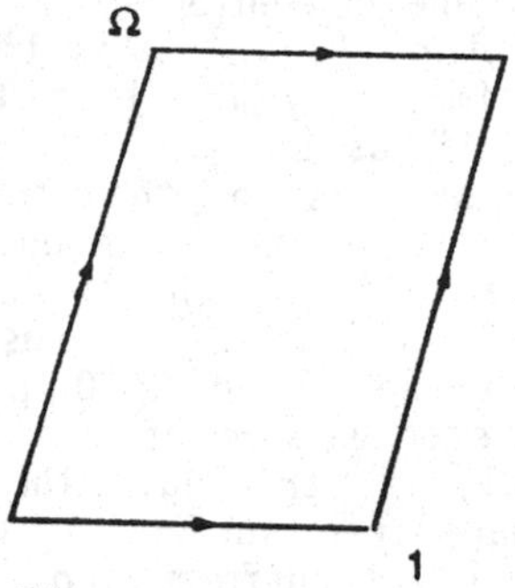

Figure 1. The Torus

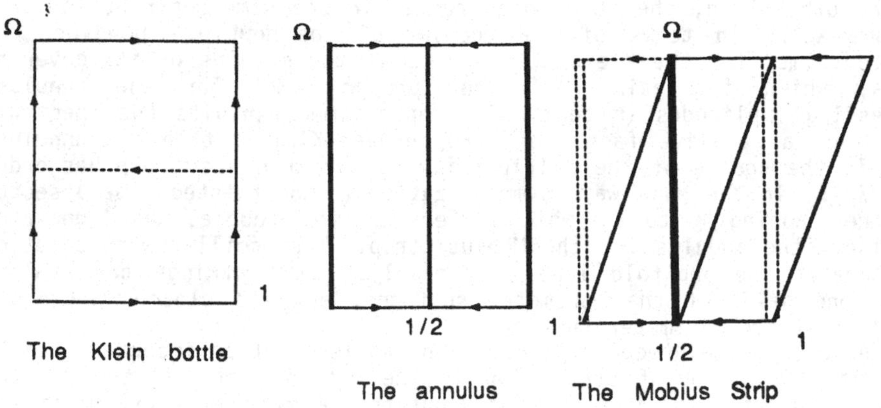

Figure 2.

The picture that emerges from the foregoing discussion has several facets. On the one hand, it is particularly satisfying to see the structure of Conformal Field Theory at work again. On the other hand, one looses the need to consider open-string models as separate entities (or oddities). Everything fits into theories of closed strings, once one allows for the possibility of **twists mixing left and right-movers**, which have been systematically avoided in discussions of orbifolds so far. The "magic rule" of modular invariance is recovered, and this should allow model building with open strings as well. These points clearly deserve a fuller discussion, which will be presented elsewhere.

I am grateful to Giorgio Parisi and to the Organizers for making it possible for me to come to Cargese. I am also grateful to Dan Friedan for a stimulating discussion, and for his interest in the ideas presented in this talk. Finally, the participants to the joint Mathematics-Physics Seminar of the two Universities of Rome, and in particular Massimo Bianchi, Emili Bifet and Gianfranco Pradisi, contributed to making my stay in Rome both pleasant and fruitful.

References

[1] A comprehensive review is: M.B. Green, J.H. Schwarz and E. Witten, "Superstring Theory" (Cambridge Univ. Press, 1987).

[2] A.A. Belavin, A.M. Polyakov and A.B. Zamolodchikov, Nucl. Phys. **B241** (1984) 333;
D. Friedan, in "Recent Advances in Field Theory and Statistical Mechanics", eds. J.B. Zuber and R. Stora (Elsevier, 1984);
D. Friedan, E. Martinec and S. Shenker, Nucl. Phys. **B271** (1986) 93.

[3] D. Friedan and S. Shenker, Phys. Lett. **175B** (1986) 287; Nucl. Phys. **B281** (1987) 509.

[4] A.M. Polyakov, Phys. Lett. **103B** (1981) 207, 211.

[5] J. Shapiro, Phys. Rev. **D5** (1972) 1945. The idea of using modular invariance as a building principle first appeared in the work of W. Nahm, Nucl. Phys. **B114** (1976) 174.

[6] F. Gliozzi, J. Scherk and D. Olive, Nucl. Phys. **B122** (1977) 253.

[7] L. Dixon, J.A. Harvey, C. Vafa and E. Witten, Nucl. Phys. **B261** (1985) 678; **B274** (1986) 285.

[8] S. Hamidi and C. Vafa, Nucl. Phys. **B279** (1987) 465;
L. Dixon, D. Friedan, E. Martinec and S. Shenker, Nucl. Phys. **B282** (1987) 13;
K.S. Narain, M.H. Sarmadi and C. Vafa, preprint **HUTP**- 86/A089 (1986).

[9] I. Antoniadis, C. Bachas and C. Kounnas, preprint **LBL**-22709 (1986);
H. Kawai, D.C. Lewellen and S.-H. Tye, preprints **CLNS** 86/751 (1986), 87/760 (1987);
M. Mueller and E. Witten, Phys. Lett. **182B** (1986) 28;
K.S. Narain, M.H. Sarmadi and C. Vafa, in ref. [8].

[10] J.E. Paton and H.M. Chan, Nucl. Phys. **B10** (1969) 516.

[11] N. Marcus and A. Sagnotti, Phys. Lett. **119B** (1982) 97.

[12] M.B. Halpern, Phys. Rev. **D12** (1975) 1684;
I. Frenkel and V.G. Kac, Invent. Math. **62** (1980) 23;
P. Goddard and D. Olive, in "Vertex Operators in Mathematics and Physics", eds. J. Lepowski, S. Mandelstam and I.M. Singer (Springer-Verlag, 1984).

[13] D.J Gross, J.A. Harvey, E. Martinec and R. Rohm, Nucl. Phys. **B256** (1985) 253.

[14] M.R. Douglas and B. Grinstein, Phys. Lett. **183B** (1987) 52.

[15] N. Marcus and A. Sagnotti, Phys. Lett. **188B** (1987) 58.

[16] J.H. Schwarz, in "Current Problems in Particle Theory", Proc. Johns Hopkins Conference **6**, Florence, 1982.

[17] Actually, the classification of simple associative algebras over the complex numbers is due to Th. Molien (Math. Annalen **41** (1893) 83). Somewhat later, E. Cartan arrived independently at Molien's results, and classified the simple associative algebras over the real numbers as well (C. R. Acad. Sci. **124** (1897) 1217, 1296; Ann. Fac. Sci. Toulouse **12B** (1898) 1). It is amusing to note that Cartan came to consider associative algebras after completing the classification of Lie algebras over the complex numbers, but before completing the classification of the real forms of Lie algebras. J.H.M. Wedderburn is is responsible for a later generalization to associative algebras over arbitrary fields (Proc. London Math. Soc. **(2)6** (1908) 77). I am grateful to E. Bifet for pointing this out to me.

[18] O. Alvarez, Nucl. Phys. **B216** (1983) 125.

[19] See, e.g., M. Schiffer and D.C. Spencer, "Functionals of Finite Riemann Surfaces" (Princeton Univ. Press, 1954). The use of double covers was introduced in String Theory in V. Alessandrini, Nuovo Cimento **2A** (1971) 321, V. Alessandrini and D. Amati, Nuovo Cimento **4A** (1971) 793.

[20] S. Weinberg, Phys. Lett. **187B** (1987) 278.

NEW METHODS AND RESULTS IN CONFORMAL QFT_2 AND THE "STRING IDEA"

Bert Schroer and FU Berlin

Institut für Theorie der Elementarteilchen

Arnimallee 14, 1000 Berlin 33

Abstract Causal fields in conformal QFT_2 yield a new algebra structure: the exchange algebra in which "braid" matrices (special Yang-Baxter structures) appear. They are directly related to the possible dimensional spectra of local fields. The Virasoro structure, i.e. the central extension charge (=Casimir energy) is part of the representation theory. An extension of Unruh's idea allows us to calculate correlation functions of arbitrary conformal $QFT's$ on higher genus Riemann surfaces directly in terms of the flat space correlations. As a generalization of Hawking's temperature, the new positive definite states are characterized by $3g-3$ "temperatures".

I. Introduction and History

Conformal QFT was introduced[1] in the 60^s. However, the field theory community in those days largely ignored or rejected conformal invariance since apart from free fields it seemed to be at odds with the principle of Einstein causality. In those days the consequences of relativistic causality combined with assumptions on the energy-momentum spectrum were the main research topic. Apart from certain structural theorems[2,3] such as TCP and the connection of spin with statistics, the main experimentally testable results were dispersion relations and certain spectral sum rules.

At the beginning of the 70^s, when the intimate relation between relativistic quantum field theory and statistical mechanics through the method of analytic continuation ("Euclidean quantum field theory") became widely appreciated, the issue of conformal invariance reappeared on the statistical mechanics side, where the "causality paradox" created less of a nuisance. In this spirit of statistical mechanics Migdal[4] and Polyakov[5] proposed the so-called "conformal bootstrap", i.e. the study of the Euclidean Schwinger-Dyson equations with conformal invariant boundary conditions. This programme was mathematically refined with the help of group theoretical ideas and techniques.[6]. However, even with all these refinements it remained too structureless as far as non-perturbative solutions are concerned. On the conceptual side the already mentioned paradox remained a sore, particularly in view of the failure of all attempts (even up to our present days) to soften the postulate of Einstein causality i.e. to replace it by a less stringent (macro) causality postulate. The causality paradox of global conformal transformation was clearly formulated in 1972[7]. After its partial understanding in a very limited setting[8], it was finally solved in a series of papers[9,10,11].

The discovery of a classification scheme for conformal QFT_2 by Belavin, Polyakov and Zamolodchikov[12] (BPZ) with additional important later contributions[13,14,15] did not rely on any of the global operator methods of the 70^s and therefore one may ask why should one be

concerned with the "old stuff" . The answer is very simple. It has recently turned out that the old methods which are "global" in nature are mathematically as well as conceptually more powerful than the methods[16] of Kac-Feigin-Fuchs-BPZ which are based on an analysis of the Virasoro algebra and differential equations for the Euclidean correlation functions[15]. Not only can one rederive known results concerning, i.e. the quantization of c values for $c < 1$, the affiliated dimensional spectrum and the 3-point normalizations with ease[17] (no Kac-determinant needs to be calculated!), but one finds a new algebra, the "exchange algebra" of the "intertwiners"[18,19] of the irreducible representation sectors of the energy-momentum tensor (a similar statement holds for loop groups if one uses the current operators). The unitary (positivity) of the field theory turns out to be equivalent to the reality of the (computable) 3-point normalization[17]. This was of course already conjectured in the days of the "conformal bootstrap". At this Summer School I learned from Jörg Fröhlich and Alain Connes that the "exchange matrices" appearing in this new algebra have the structure of "braids" related to Yang-Baxter matrices and that the eigenvalues of these matrices are connected with cohomology-invariants of certain infinite dimensional "exchange" C-algebra[51].
I find it particularly pleasing that all this mathematical and physical wealth appears straight out of the resolution of the "conformal Einstein causality paradox" thus underlining again the great value of deep physical paradoxes.

Most of the analytical insight into quantum field theory, viz. d=2 integrable systems, did not come from the Euclidean functional integral approach. The beauty of the Feynman integral and its suggestive power for approximations remains in stark contrast with its analytic unmanageability. This is a regrettable and perhaps only transitory fact of life in theoretical physics. We therefore propose the thesis that also the "string idea" should be investigated by the global operator method rather than by functional integrals. As a prelude to an operator string approach I will discuss the recently proposed technique of constructing expectation values on Riemann surfaces via the "Hawking-Unruh" averaging[19,20] using Fuchsian groups. This method is not only simpler than the Euclidean approach based on infinite Grassmannians[21] and Valatin-Bogoliubov-transformations, but it also seems to give more physical insight.

I. The "Causality paradox" of conformal QFT and its resolution

The causality paradox in conformally invariant QFT is the statement that the simple-minded equivariant conformal transformation law which is obtained by formally integrating up the infinitesimal (Lie-algebra) transformation is in contradiction with Einstein causality of relativistic QFT.

Theorem[7]: In any local field theory with "reverberation" i.e. non-vanishing timelike commutators (for fermions: anticommutators) the global transformation law:

$$U(b)\,A(x)\,U^{+}(b) = (\sigma(b,x))^{-d_A}\,A\left(\frac{x-bx^2}{\sigma(b,x)}\right) \tag{I.1}$$

$$\sigma(b,x) = 1 - 2bx + b^2x^2$$

(equivariance under special conformal transformations) is inconsistent with causality:

$$[A(x), A(y)] = 0\,, \qquad (x-y)^2 < 0 \tag{I.2}$$

For the proof one takes a special conformal transformation which transforms the spacelike (I.2) into a timelike separation (a continuous path of such transformations goes through a zero of $\sigma(b,x)$. Since the transformed field apart from its changed argument only contains a c-number in front (or matrix of c-numbers in case of a general field), there is a contradiction with the "reverberation" assumption

To make a long story short[9,10,11], the resolution comes from the recognition that the conformal group in Minkowski space is not simply connected. There is a smallest not contractible path c and the corresponding centre element is represented by a non trivial unitary operator, which apart from the special case of the energy momentum tensor, does not commute with the local field:

$$
\begin{aligned}
Z^n A(x) Z^{+n} &\doteq A(x,n) = A_n(x) \cdot \\
Z &= \mathcal{U} \text{ (minimal non-contr. path)}
\end{aligned}
\tag{I.3}
$$

G. Mack introduced the picturesque terminology "heavens and hells" for the index n which ranges over positive and negative integers. A good mathematical formalism is obtained by diagonalizing the unitary operator[9] (the "line bundle approach")

$$
\operatorname{spec}(Z) = \{ e^{-2\pi\theta i} \}
\tag{I.4}
$$

With

$$
A^d_\xi(x) = \sum_\xi Z^n A^d(x) Z^{+n} e^{i\pi n(d-2\xi)}
$$

we obtain the global decomposition of the causal field:[9]

$$
A^d(x) = \sum_\xi A^d_\xi(x)
\tag{I.5}
$$

The notation for the phase normalization convention is that of reference 9. The irreducible components obey the equivariant law:

$$
U(b) A^d_\xi(x) U^+(b) = \left(\frac{1}{\sigma_+(b,x)}\right)^{d-\xi} \left(\frac{1}{\sigma_-(b,x)}\right)^{\xi} A^d_\xi\left(\frac{x-bx^2}{\sigma(b,x)}\right)
\tag{I.6}
$$

Here the suffix $\pm$ indicates the boundary value obtained via the Wightman prescription of adding a small positive imaginary time-like ε-component to the vectors x and b and then taking the boundary value $\lim_{\varepsilon\to 0}\ldots$. From the spectrum condition it immediately follows that:

$$
A^d(x)|0\rangle = A^d_{\xi=0}(x)|0\rangle \,, \quad A^{d+}_0 = A^d_{\xi=d}
\tag{I.7}
$$

For the free fields[8] these two ξ-components are just the (non-local) creation and annihilation parts. Therefore it is reasonable to view the ξ-decomposition (I.5) as a subtle generalization of creation and annihilation operators provided by the global aspects of the conformal group. In order to learn more about the physical interpretation of ξ we consider 3-point functions of A with two other (composite) fields.

$$<0|C\,A\,B|0> \;=\; <0|C_{d_c}\,A_\xi\,B|0> \tag{I.8}$$

and obtain[9]:

$$\xi = \tfrac{1}{2}\,(d_A + d_B - d_c) \tag{I.9}$$

.e. the ξ-spectrum contains all the linear combinations of dimensions of those conformally equivariant composite fields which communicate with A via non-vanishing 3-point functions. We will call these fields A, B, C... "quasiprimary"[12]. Assuming that these composite fields form a complete set in the Hilbertspace yields the statement that formula (I.9) exhausts the possible ξ-spectrum. In higher point functions the global decomposition (I.5) leads to a sum over ξ-values, i.e.:

$$<ABCD> \;=\; \sum_{\xi} <A_{d_A}\;B_{\hat{\xi}(\xi)}\;C_\xi\;D_0> \tag{I.10}$$

In order to obtain a more detailed insight into the conformal decomposition theory (1.5) as a result of the correct global equivariance of local fields, we now specialize to two dimensions in which case the group theory of the finite-dimensional conformal group simplifies. Adopting the standard notation:

$$u = x + t \qquad\qquad v = x - t \tag{I.11}$$

for the two lightcone co-ordinates, the 6-dimensional conformal group:

$$PSL(2,R)_L \otimes PSL(2,R)_R \tag{I.12}$$

acts as a SL(2,R) - linear fractional transformation (Möbius-transformation) on the u-v light-cone co-ordinates. Here we ignore all reflection including the physical reflection (T,P,TCP) and the conformal reflection :$x \rightarrow \frac{x}{x^2}$ which intertwine between the two light-cones. On each side one conveniently uses the Iwasawa decomposition[10]:

$$\begin{gathered} SL(2,R) = M = KAN \\ K = \left\{ \begin{pmatrix} \cos\frac{\alpha}{2} & \sin\frac{\alpha}{2} \\ -\sin\frac{\alpha}{2} & \cos\frac{\alpha}{2} \end{pmatrix} \right\} \\ A = \left\{ \begin{pmatrix} \lambda & 0 \\ 0 & \lambda^{-1} \end{pmatrix} \right\} \quad , \quad N = \left\{ \begin{pmatrix} 1 & \omega \\ 0 & 1 \end{pmatrix} \right\} \end{gathered} \tag{I.13}$$

The existence of the maximal compact subgroup K with its non-trivial universal covering: (not representable by a matrix group)

$$\tilde{K} = \{ \alpha , \alpha \in R \} \tag{I.14}$$

was the origin of the clash between causality and the "naive" equivariance law (I.1).

The correct equivariant transformation law for local fields is most elegantly written in terms of the parametrization in which the light-cone co-ordinates are obtained by applying the K subgroup to the origin on each light-cone i.e.:

$$u = \frac{\cos\beta/2 \cdot 0 + \sin\beta/2}{-\sin\beta/2 \cdot 0 + \cos\beta/2} = \tan\beta/2 \quad , \quad v = \tan\bar{\beta}/2 \tag{I.15}$$

With this notation the correct equivariance law under $\tilde{K}_L$ reads as follows:

$$U(\alpha)\, A^{d,\bar{d}}(\beta,\bar{\beta})\, U^{+}(\alpha) = \left| \frac{\cos^2\frac{\beta}{2}}{\cos^2\frac{\alpha+\beta}{2}} \right|^{d} A^{d\bar{d}}(\beta+\alpha,\bar{\beta}) \tag{I.16}$$

Here the field at the new variable $\beta, \bar{\beta}$ is just an abbreviation for (1.3), i.e.

$$A^{d,\bar{d}}(\beta,\bar{\beta}) \doteq Z_L^{[\beta]}\, Z_R^{[\bar{\beta}]}\, A^{d,\bar{d}}(u,v)\, Z_L^{+[\beta]}\, Z_R^{+[\bar{\beta}]} \tag{I.17}$$

$[\beta]$ = biggest integern in representation:

$$\beta = n\pi + \beta_{red.} \qquad , |\beta_{red}| < \pi$$

The equivariance law under $U(\bar{\alpha})$ is analogous. Note that the non-trivial "reverberation" aspect is now encoded in the non-triviality ($Z \neq \lambda 1$) of the central generator (I.3). The non-trivial part of the conformal equivariance law (1.14a) together with the trivial laws (which we only write for the left light-cone)

$$\begin{aligned} U(\lambda)\, A^{d,\bar{d}}(u,v)\, U^{+}(\lambda) &= \lambda^{d}\, A^{d\bar{d}}(\lambda u, v) \\ U(\omega)\, A^{d,\bar{d}}(u,v)\, U^{+}(\omega) &= A^{d\bar{d}}(u+\omega, v) \end{aligned} \tag{I.18}$$

yield upon diagonalization of Z (I.5) the irreducible conformal equivariance properties:

$$Z\, A^{d\bar{d}}_{\xi\bar{\xi}}(u,v)\, Z^{+} = e^{-2i\pi(d-\xi)}\, A^{d\bar{d}}_{\xi\bar{\xi}}(u,v). \tag{I.19a}$$

$$U(\lambda)\, A^{d\bar{d}}_{\xi\bar{\xi}}(u,v)\, U^{+}(\lambda) = \lambda^{d}\, A^{d\bar{d}}_{\xi\bar{\xi}}(\lambda u, v) \tag{I.19b}$$

$$U(\omega)\, A^{d\bar d}_{\xi\bar\xi}(u,v)\, U^+(\omega) = A^{d\bar d}_{\xi\bar\xi}(u+\omega, v) \tag{I.19c}$$

$$U(b(\gamma))\, A^{d\bar d}_{\xi\bar\xi}(u,v)\, U^+(b(\gamma)) = (1+\gamma u)^{-(2d,\xi)} A^{d\bar d}_{\xi\bar\xi}\left(\frac{u}{1+\gamma u}, v\right) \tag{I.19d}$$

where

$$(1+\gamma u)^{(2d,\xi)} = (1+\gamma u + i\varepsilon)^{-2d+\xi} (1+\gamma u - i\varepsilon)^{-\xi}$$

Note that the dimension appearing in the formula (I.6) for the scalar field ($d = \bar d$) is the total dimension $d + \bar d = 2d$.

It is clear that the Einstein causality requirement (1.2) in the spacelike region u.v. > o will severely restrict the algebraic light-cone structure of the $A^{d,\bar d}_{\xi\bar\xi}$(u, v) appearing in the global decomposition.

$$A^{d\bar d}(u,v) = \sum_{\xi\bar\xi} A^{d\bar d}_{\xi\bar\xi}(u,v) \tag{I.20}$$

In a moment we will see that these irreducible fields $A^{d,\bar d}_{\xi,\bar\xi}$ factorize, i.e.

$$A^{d\bar d}_{\xi\bar\xi}(u,v) = A^{d}_{\xi}(u) \,.\, A^{\bar d}_{\bar\xi}(v) \tag{I.21}$$

The factors will turn out to satisfy a new algebra: the "exchange algebra" in which representation matrices of "braids" will appear. This new algebraic structure is not only sufficient to ensure Einstein causality, but it also seems to be necessary.

We already know the physical interpretation of the new field labels ξ and $\bar\xi$. Specializing the two- and three-point analysis of (1.7) and (1.8) to the two-dimensional light-cone split situation one easily obtains:

$$\xi = \tfrac{1}{2}(d_A + d_B - d_c) \quad ; \quad \bar\xi = \tfrac{1}{2}(\bar d_A + \bar d_B - \bar d_c) \tag{I.22}$$

where the $d_{B,C}(\bar d_{B,C})$ are the (left-right) dimensions of arbitrary (composite) fields which communicate with the field A via non-vanishing 3-point functions. Again the value $\xi, \bar\xi = o$ occurs as a special value of the ξ-spectrum. This value characterizes the "creation" part of the new set of operators. The $A^{d,\bar d}_{o,o}$ with its simplified transformation property ($d = h \equiv d - \xi$ in (1.16)) creates the conformal blocks of BPZ and constitutes the only operator which has been used in the Kac-Feigin-Fuchs infinitesimal Virasoro- (or loop)-algebra representation approach. In order to see that all ξ values occur inside composite 3-point functions, we had to assume that the "quasi-primary" (i.e. Möbius) fields (1.16) generate a complete set of states. Under the same assumption one obtains the validity of the global vacuum expansion for quasiprimary fields:

$$A^{d_1\bar d_1}_{\xi,\bar\xi}(u_1,u_2)\, A^{d_2\bar d_2}_{0,0}(v_1,v_2)|0\rangle = \tag{I.23}$$

$$\sum_{d_3\bar d_3} 2 \exp i\pi(d_3 - \bar d_3)\; C_{312} \int du\, dv\, K(d_i, \bar d_i; u, v)\, A^{d_3\bar d_3}_{00}(u_3,v_3)|0\rangle$$

where the sum extends over all quasiprimary fields with dimensions

$$d_3 = d_1 + d_2 - \xi \quad (\mathrm{mod}\ \mathbb{Z}) \quad ; \quad \bar{d}_3 = \bar{d}_1 + \bar{d}_2 - \bar{\xi} \quad (\mathrm{mod}\ \mathbb{Z}) \tag{I.24}$$

and the kernels have the factorized form

$$K(d_i; \bar{d}_i; u_i, v_i) = K(d_i; u_i) \cdot \bar{K}(\bar{d}_i, v_i) \tag{I.25}$$

$$K(d_i; u_i) = \frac{\Gamma(2d_3)\,\Gamma(\lambda_3)\,\Gamma(1-\lambda_2)}{2\pi\,(u_1-u_2-i\varepsilon)^{\lambda_1}(u_1-u_3-i\varepsilon)^{\lambda_2-1+2d_3}(u_2-u_1+i\varepsilon)^{1-2d_3}(u_2-u_3+i\varepsilon)^{\lambda_3}} \tag{I.26}$$

$$\lambda_1 = d_1 + d_2 + d_3 - 1 \quad , \lambda_2 = d_1 - d_2 - d_3 + 1 \, , \lambda_3 = -d_1 + d_2 - d_3 + 1$$

With this choice of normalization the global vacuum expansion reproduces the 3-point function:

$$\langle A^{d_3\bar{d}_3}(u_3, v_3)\, A^{d_1\bar{d}_1}(u_1, v_1)\, A^{d_2\bar{d}_2}(u_2, v_2) \rangle = \tag{I.27}$$

$$\frac{C_{312}}{(u_3-u_2-i\varepsilon)^{d_3+d_1-d_2}(u_3-u_2-i\varepsilon)^{d_3+d_2-d_1}(u_1-u_2-i\varepsilon)^{d_1+d_2-d_3}} (v, \bar{d})$$

where $(v.\bar{d})$ indicates similar factors referring to the v-light-cone. In the calculation we use the 2-point normalization:

$$\langle A_1(u_1 v_1)\, A_2(u_2, v_2) \rangle = \delta(A_1 = A_2 = A^{d\bar{d}}) \frac{e^{-i\pi(d-\bar{d})}}{(u_1-u_2-i\varepsilon)^{2d}(v_1-v_2+i\varepsilon)^{2\bar{d}}} \tag{I.28}$$

Within the assumed completeness assumption of quasi-primary fields, this calculation of the reproducing property of the 3-point function is at the same time the proof for the validity of the global expansion. In contradistinction to the Wilson expansion (which also holds "off vacuum"), it has the group theoretical status of a Plancheral formula[10]. Note that as a result of (I.21), which in turn results from the u-v factorization of the 3-point function, the n-point function of the irreducible fields (not the local field!) $A^{d,\bar{d}}_{\xi,\bar{\xi}}$ factorize. This follows from an iterative application of the global decomposition (I.17). This in turn means that the operator factorization (I.18) holds. Note that here and again the power of the Wightman framework for structure arguments as opposed to any ordered expectation values (which are useful in perturbation theory) becomes evident. To the educated reader it should also be clear that the sectors generated by diagonalizing the global operators Z_L and Z_R are identical to the conformal blocks of BPZ. However in obtaining them by the global analysis of the "Einstein causality paradox", we receive a vast amount of new insight which hitherto was still missing. Let us first try to reproduce and extend known results. The quasiprimary fields come in families[17]; because one quasiprimary field globally expands with the energy momentum tensor T(u), T(v) à la (I.20) will automatically generate higher quasiprimary fields $A^{(n)}$

$$T(u_1)\, A(u_2,v_2)\, |0\rangle = \sum_n \int du'\, K^{(n)}(u_1-u', u_2-u')\, A^{(n)}(u',v)|0\rangle \quad \text{(I.29)}$$

The beauty and simplicity of the global expansions is that they only deal with quasiprimary fields (and not with their derivatives). It is reasonable to assume that there exists a lowest quasiprimary field which we will later show to be the primary field of BPZ. Hence we write:

$$c_{312} = c_{312}^{prim}\, N_{312}^{n_3 n_1 n_2}\, \bar{N}_{312}^{\bar{n}_3 \bar{n}_1 \bar{n}_2} \quad \text{(I.30)}$$

where the N's are the relative normalizations within one family. The hypergeometric techniques which lead from the global expansion to the 4-point formula:

$$\langle A_1(x_1)A_2(x_2)A_3(x_3)A_4(x_4)\rangle = \frac{\left(\frac{u_2-u_4}{u_1-u_4}\right)^{d_1-d_2}\left(\frac{u_1-u_3}{u_2-u_3}\right)^{d_4-d_3}}{(u_1-u_2)^{d_1+d_2}(u_3-u_4)^{d_3+d_4}} \times \frac{\left(\frac{v_2-v_4}{v_1-v_4}\right)^{\bar{d}_1-\bar{d}_2}\left(\frac{v_1-v_3}{v_2-v_3}\right)^{\bar{d}_4-\bar{d}_3}}{(v_1-v_2)^{\bar{d}_1+\bar{d}_2}(v_3-v_4)^{\bar{d}_4+\bar{d}_3}} \times \quad \text{(I.31)}$$

with $$\sum_{h_0,\bar{h}_0} \exp i\pi(h_0-\bar{h}_0)\, c_{120}^{prim}\, c_{034}^{prim}\, F_{h_0}(x)\, \bar{F}_{\bar{h}_0}(\bar{x})$$

$$F_{h_0}(x) = x^{h_0} \sum N_{120}^{n_1 n_2 0}\, N_{034}^{n_0 n_3 n_4}\, (-x)^{n_0}\, {}_1F_2(d_2-d_1+d_0,\, d_3-d_4+d_0; 2d_0)$$

$$x = \frac{(u_1-u_2)(u_3-u_4)}{(u_1-u_3)(u_2-u_4)} \qquad d_i = (h_i+n_i) \qquad i=1,2,3,4 \quad \text{(I.32)}$$

Such formulae are quite old. For special cases (vertex-operator) they appear already in an Appendix of reference 9. They were derived in BPZ and subsequent papers by resummation techniques[13] starting from Wilson expansion.

Specializing these formulae to the 4-point function: and using the relation

$$\langle A^{h+m,\bar{d}}(x_1)\, T(u_2)T(u_3)\, A^{h+n,\bar{d}}(x_4)\rangle \quad \text{(I.33)}$$

$$T(u_2)T(u_3) = \frac{c/2}{(u_1-u_3)^4} + \frac{T(u_1)+T(u_3)}{(u_2-u_3)^2} \quad \text{(I.34)}$$

$+$ reg. terms at $u_2 = u_3$

one obtains[17] the statement

$$\text{rat. funct}_N(h) \prod_{pq=N} (h - h_{p,q}(c)) \geq 0 \qquad \text{all } N>0 \tag{I.35}$$

Applying the Friedan, Qiu, Shenker[13] argument to this quantity[17] (which is simpler than the Kac determinant), one obtains the desired quantization of c as well as the Kac formula for the dimensions. Furthermore these global techniques allow us to show the absence of ghosts at these quantized values. They also lead to the determination of the multiplicities of quasiprimaries as well as their relative normalization N. For the details of the calculation we refer to a recent publication[17]. If all 4 fields are non-canonical, the generalized global expansion away from the vacuum state replaces the formula (I.34). The discussion becomes more subtle since the form of this new global expansion depends on the structure of the exchange algebra and will be presented in a forthcoming paper.

The derivation of the c-quantization resulting from unitarity based on the global expansion technique is in the spirit of an unpublished (1976) paper by Lüscher and Mack as quoted by Todorov[22], in which only the region $c < \frac{1}{2}$ was excluded and as yet no complete quantization picture was derived. In passing we would like to mention that in the work of Parisi et al.[23] one finds precursors of the global technique which we used in this derivation.

II. The role of the Diffeomorphism group of the circle and the relation of the operator approach to the BPZ formulation

One of the marvellous discoveries of BPZ[12] is the suggestion that there exist "primary" fields, i.e. local fields which have rather simple (infinitesimal) transformation properties under $Diff_L(R) \times Diff_R(R)$. Consider a conformally invariant classical action, i.e.

$$\frac{1}{2}\int \partial_\mu \varphi \partial^\mu \varphi \; d^2x = \int \partial_u \varphi \, \partial_v \; du \, dv \tag{II.1}$$

This is apparently invariant under diffeomorphism:

$$u \longrightarrow f(u) \quad , \quad v \longrightarrow g(v)$$

Barring anomalies and a spontaneous breakdown à la Nambu and Goldstone, in which case the infinitesimal generators would not exist as a result of bad large-distance behaviour (which is certainly absent in the present case), one should expect the symmetry to be realized by unitary operators in Hilbert space à la Wigner and Weyl. There is no guarantee that the vacuum state will be invariant. In fact, we know that in conformal theories the maximum invariance group of the vacuum is the Möbius subgroup of the ∞ dimensional diffeomorphism group. Therefore this large symmetry is somewhat hidden, i.e. not directly visible in the ground state expectation values. Representation theory of groups in the quantum Hilbert space always has two aspects. One must either study the representation theory of the universal covering group (example: rotation group) or of a suitably central extension (example: Galilei group). The most prominent infinite dimensional groups in two-dimensional physics are the loop group and the diffeomorphism group of the light-cone. The former is simply connected and has a non-trivial central extension, whereas the latter has non-trivial covering as well as a central extension. In both cases it is convenient to use the circular instead of the light-cone description which is obtained by the isomorphism:

$$u \longrightarrow z = \frac{1 - iu}{1 + iu} \qquad v \longrightarrow \bar{z} = \frac{1 - iv}{1 + iv} \tag{II.2}$$

It is physically quite obvious, and can be rigorously proved[23], that the only culprit for the non-simple connectedness is still the rigid Abelian subgroup K contained in the diffeomorphism group. Therefore the tilde in $\widetilde{diff}$ is physically related to causality and the subtle generalization of annihilation and creation operators in (I.5) i.e. to the phase spectrum (dimensional spectrum mode Z). But what is the physical interpretation of the central extension of the diffeomorphism group? Consider the outer automorphism of $\widetilde{SL(2,R)}$[19]

$$\tilde{G}_n : h_n^{-1} \widetilde{SL(2,R)} h_n \tag{II.3}$$

with $h_n : z \rightarrow z^n$ (in circular language) being the outer automorphism of level n. For the generators of $\widetilde{SL(2,R)}$ we now use the standard notation

$$L_o = \tfrac{1}{2}(P_L + S_L) \; ; \; L_\pm = \frac{1}{2}(S_L - P_L) \pm i D_L \tag{II.4}$$

where S_L is the light-cone component of the generator of the special conformal transformation (I.6), and D_2 generates the left-handed Lorentz transformation (=scale transformation). L_o is the generator of the compact subgroup $\tilde{K}$. Defining

$$L_0^{(n)} = U^+(h_n)\, L_0\, U(h_n) \tag{II.5a}$$

$$\hat{L}_n = U^+(h_n)\, L_1\, U(h_n) \tag{II.5b}$$

one finds from the required maximal Abelianess of $L_o, 1$:

$$L_0^{(n)} = \frac{1}{n} L_0 + c_n \mathbb{1} \tag{II.6}$$

i.e. the outer automorphism h_n to a "Casimir energy" of the ground state. With

$$\hat{L}_n = \frac{1}{n} L_n \tag{II.7}$$

one easily finds (from Jacobi identities) that the Casimir effects on different levels can be related. The result is the existence of the Virasoro algebra.

$$[L_n, L_m] = (n-m) L_{n+m} + \frac{c}{12\pi} n(n^2-1)\mathbb{1} \tag{II.8}$$

The L_n's thus constructed generate the energy momentum tensor:

$$T(u) = \frac{1}{8\pi} \left(\frac{1}{u+i}\right)^4 \sum_n L_{-n} \left(\frac{u-i}{u+i}\right)^{n+2} \tag{II.9}$$

The c-number functions are a complete set of eigenfunctions of $L_o = \frac{1}{2}(P+S)$ on the u-line; here L_o is considered as a ("first quantized") differential operator. Only for $c < 1$ are the tilde in $\widetilde{diff}_{cent}$, the dimensional spectrum and the Casimir effect corresponding to the suffix "cent" rigidly related. Later we will see that the dimensional spectrum in a local conformal

QFT_2 is directly related to the algebraic structure of an "exchange algebra". The value of c on the other hand is related to the representation theory. Our "global technology" of the previous section did not use directly the existence of $\widetilde{diff}_{cent}$ and the Virasoro algebra. However, to obtain a better understanding of the concepts behind this formalism, it is indispensable to have $diff_{cent}$ at the back of one's mind. Only by being aware of the existence of objects which under *all* G_n transform equivariantly (these are the primary fields[19]

$$U(g_n)\, A^{d\bar{d}}_{\xi\bar{\xi}}(u,v)\, U^{+}(g_n) = |g_n'|^{d}\, e^{-2i\pi(d-\xi)\theta}\, A^{d\bar{d}}_{\xi\bar{\xi}}(g_n(u),v) \tag{II.10}$$

where $\theta(u, g_n)$ is a step function which is one in the region where g is negative) does one obtain a profound conceptional insight. Quasiprimary fields on the other hand have additional additive contributions involving lower quasiprimary fields as well as polynomial expressions in the Schwarz derivative of g multiplied by ordinary derivatives of g. Typical examples have been worked out in reference 18. They define the tensor calculus of the central extension, whereas the primary fields do not "feel" this extension.

In order to make contact with the analytic formalism of BPZ[12], it is helpful to recollect the analytic properties of Poincaré invariant quantum field theories and their conformal extension. For d=2 this is particularly simple[3]. As a result of the ordinary spectrum condition, the vacuum expectation values of Lorentz-covariant fields $A_i(u, v)$

$$W_{i_1 \dots i_{n+1}}(u_1 v_1; u_2 v_2 \dots; u_{n+1} v_{n+1}) = \langle A_i(u_1 v_1) \dots A_{i_{n+1}}(u_{n+1} v_{n+1}) \rangle \tag{II.11}$$

are boundary values of analytic functions:

$$W_{i_1 \dots i_{n+1}}(u_1 v_1; \dots ; u_{n+1} v_{n+1}) = \lim_{\varepsilon, \bar{\varepsilon} \to 0} W_{i_1 \dots i_{n+1}}(\xi_1 \bar{\xi}_1; \dots ; \xi_n \bar{\xi}_n)$$

with

$$\begin{aligned} \xi_i &= u_i - u_{i+1} & \quad Im\, \xi_i &= \varepsilon_i > 0 \\ \bar{\xi}_i &= v_i - v_{i+1} & \quad Im\, \bar{\xi}_i &= -\bar{\varepsilon}_i < 0 \end{aligned} \tag{II.12}$$

The analytic domain (II.3) is called the "primitive tube" . It contains some of the Euclidean points $x_i = (ix^o, \vec{x}_i)$. It can be extended by complex Lorentz transformations to the Bargmann-Hall-Wightman domain[3]

$$\mathcal{T}_n' = \bigcup_{A \in L_n(C)} A\, \mathcal{T}_n \tag{II.13}$$

which also has the name "extended tube". The analytic functions are still univalued in this region. In the case at hand the complex $L(C)$ transformations just rotate the upper u-plane by an angle, whereas the lower v-plane is acted upon by the opposite rotation[3]. Einstein Causality (local (anti)commutativity) enlarges the analytic domain to "permuted extended tube". This domain in which the functions are still univalued is generally not a natural domain: it has a non-trivial holomorphy envelope[2]. The boundary values obtained by approaching the real boundary via different "primitive tubes" belong to different operator orders inside the vacuum expectation values. Note also that the extension via the use of local commutativity leads to regions which do not factorize any more in u- and v-variables. The set of Euclidean points at which quantum field theory makes contact with statistical mechanics constitutes a lower dimensional subset contained in the permuted extended tube. The analytic setting of quantum field theory allows for a slight generalization in d=2. Instead of local fields one may work with the more general "dual" fields or "Yang-Baxter" fields. Instead of vanishing (anti)commutors they are characterized by the space-like algebra[3]:

$$\psi_\gamma(y)\,\psi_\delta(x) = R^{\alpha\beta}_{\gamma\delta}\,\psi_\alpha(x)\,\psi_\beta(y) \tag{II.14}$$

where $x < y$ (left spacelike). For $x > y$ the R matrix is the inverse of (II.14). The Yang-Baxter trilinear relations for R turn out to be a consistency relation of this algebraic structure. Jörg Frölich has pointed out[25] that this algebra arises naturally if one tries to construct the intertwining fields in the sense of Haag el al.[26] for "observable algebras" in two space-time dimensions. This algebra structure leads to the same "permuted extended tube" domain of local quantum field theory, but this time with a violation of univaluedness (i.e. monodromy around analytic ramification points). The boundary values, at the physical points, which are related to the Hilbert-space interpretation, remain however univalued and physically consistent.

Now we are well prepared to study the "conformal extension". Starting again from the primi tive tube, the use of complex scale- and Lorentz-transformation gives the titled upper(lower) halfplanes in (II.13) but this time the tilting may be done with two independent angles. The existence of the proper conformal transformation, or equivalently (using another basis), of the compact conformal subgroup $K_{L,R}$ belonging to the "conformal Hamiltonian" L_o yields another extension. According to our previous discussion, the net result of these extensions may be conveniently described in terms of the two semi-groups[19] $\widetilde{SL}_c^{(L,R)}(2,R)$, the complex extension of the Iwasawa decomposition to $Im\ \alpha > o$ and $Im\ w > o$. This causes a branching in the physical region which forced us to change our picture of space-time to the conformal heavens and hells or introduce irreducible global conformal fields in the L-R worlds which carry a "phase" subscript. As will be seen in the next section, the local fields (this also applies to "dual" or "Yang-Baxter" fields (II.14)) are sums of products of the global irreducible fields. To put it in a more picturesque language: the irreducible global light-cone fields, which also may be considered as a subtle generalization of creation and annihilation operators, form the "raw material" for the construction of local as well the soliton fields.

Besides the standard description of conformal quantum field theory (the light-cone picture) there are two further equivalent descriptions.

1) The compact picture

One transforms the u- and v-light-cones onto circles via the (complex) Möbius isomorphism (II.3). The operators live in the same Hilbert space as in the light-cone picture. The global irreducible ("phase") components fulfil quasi-periodic boundary conditions.

2) The interval (cylinder) picture

Via the logarithmic map one transforms the $z(\bar{z})$ variable of the unit circle to a periodic R mod $2\pi Z$ variable. This is an outer automorphism leading to another Hilbert space (again with positive definite Wightman functions) and a non-vanishing Casimir ground state energy.[27]. The analytic continuation makes contact with statistical transfer matrices on a strip. The new vacuum has lost the $L_{\pm 1}$ invariance.

The compact picture is most typical for d=2 conformal quantum field theories. In earlier works on d=2 quantum field theories, such as for example, Klaiber's partial bosonization[28] of the massless Thirring model and the subsequent full bosonization of Mandelstam[29] and Coleman[30], it was not used. The prize for not using it is the somewhat awkward infra-red aspect of a free massless boson field. This compact picture appears for example in the work of Frenkel and Kac[31]. The boson field lives in a Hilbert space which is the direct product of the Fock space with the quantum-mechanical space generated by the zero modes. Although the boson fields live in a (positive definite) Hilbert space, it has no status as a conformal tensor[8]. Dual models and genus g=o string theory are usually formulated in the interval (cylinder) picture.

The light-cone and compact pictures, although different in interpretation of their variables, have identical analytic functions[23]. Their operator formulae for the Virasoro L_n charges are, however, different. The compact picture yields the simple Fourier formula

$$L_n = \frac{1}{2i\pi} \oint z^{-n-2}\, T(z) \frac{dz}{z} \tag{II.15}$$

whereas the light-cone formula for L_n is slightly more complicated as the result of the orthonormal system of u-eigenfunctions appearing in (II.9). It is well known from general quantum field theory that commutators of conserved charges with local fields can be represented inside Wightman functions by differences of spatial integrals with the relative time component of the current and the field being once positive and once negative imaginary. This follows from our previous remarks concerning the connection between the signs of Im ς (Im$\bar{\varsigma}$) and the operator orderings. For d=2 the two contributions can be combined to yield contour-integrals in ξ and ξ . These contours may be shifted and in this way the BPZ formulae on (analytically continued) Green's functions emerge. In order to maintain an operator interpretation of these analytically continued expectation values in the process of circulating with one variable around, some authors (see Friedan's lectures) insist on introducing an ordering (time or radial ordering) inside the analytic domain. This is like defining a "tube ordering" for the analytic Wightman functions in general quantum field theory. According to my best knowledge this has never been done before. We will also not use it here because it would create havock in connection with the "exchange algebras" of the next section. We will also not employ the somewhat misleading terminology of "radial quantization" for the compact picture. General experience tells us that one should be very economical with the word "quantization". If one wants to use it in this context it should rather be called "angular quantization" since the conformal Hamiltonian L_o is like an angular momentum but without the univaluedness requirement on wave functions.

III. Exchange algebras

The loop groups as well as the circular diffeomorphism group have a new property which distinguishes them from previously studied groups in physics. Their irreducible representations form families which want to be together. In the infinitesimal Kac-Peterson[32] - Feigin-Fuchs[16] - Rocha-Caridi[33] approach to the modular transformation properties of the characters for a fixed c-value are the clearest signal of their relation. In field theoretical language, this property is related to the existence of intertwiners, i.e. field a(u), b(u)... which intertwine between these irreducible sectors. A simple, but unfortunately rather trivial illustration of this phenomenon is supplied by the vertex operator. In the compact picture we define

$$\begin{aligned} V_\alpha(z) &= :\exp i\alpha\varphi(z): \\ \varphi(z) &= q + ip\,\mathrm{Ln}\, z + i\sum_n a_n \frac{z^{-n}}{n} \end{aligned} \tag{III.1}$$

Choosing a torus quantum mechanics and suitably adjusting the real number α the vertex operator will be defined on a finite covering. Although the vertex operator decomposes according to[19]

$$V_\alpha(z) = \sum_n V_\alpha^{(n)}(z) \tag{III.2}$$

where $V_\alpha^{(n)}$ is the contribution acting on the eigenspace of p labelled by n, this additive "charge" structure still yields a trivial situation for the 4-point- (more generally n-point-)function. The ξ-sum in correlation functions always consists of just one term as for free fields. The algebra formed by the vertex operators reads in light-cone co-ordinates $(u_1 > u_2)$

$$\begin{aligned} V_\alpha(u_2)\,V_\alpha(u_1) &= e^{i\pi\alpha^2}\,V_\alpha(u_1)\,V_\alpha(u_2) \\ V_\alpha^+(u_2)\,V_\alpha(u_1) &= e^{-i\pi\alpha^2}\,V_\alpha(u_1)\,V_\alpha^+(u_2) \end{aligned} \tag{III.3}$$

A less trivial example is provided by the Ising model[18,19]. The interwiners we are looking for are precisely the objects which appeared in the global decomposition theory as a subtle generalization of creation and annihilation operators. In the Ising case the centre element Z leads to 3 sectors. In addition to the vacuum sector one obtains the $h = \frac{1}{16}$ and $h = \frac{1}{2}$ sector[12]. The local field which generates the $h = \bar{h} = \frac{1}{16}$ sector is the disorder operator. Its global decomposition reads:

$$\mu(u,v) = a(u)\,a(v) + h.c. + \ldots \tag{III.4}$$

where the a and a^+ are the only surviving contributions on the ket- resp. bra-vacuum.

$$\mu_{0,0}(u,v) = a(u)\,a(v)\ , \quad \mu_{1/16\,1/16} = \mu_{00}^+ \tag{III.5}$$

As a consequence of: $< m \; \mu \, \mu > \neq 0$
where $m = \bar{\psi}\psi$ is the mass operator, there must exist another irreducible component b(u) which has the same dimension as a but a different phase:

$$\begin{aligned} b(u) &= b_{-3/8}^{1/16}(u) \\ \text{i.e.} \qquad Z\,b(u)\,Z^+ &= \exp - \frac{2i\pi Z}{16}\; b(u) \end{aligned} \tag{III.6}$$

Such a field cannot appear in the Kac-Feigen-Fuchs-BPZ approach since

$$b(u)\,|0> = 0 \tag{III.7}$$

However, b as well as a^+ can be applied after a has acted on the vacuum:

$$a^+(u_1)\,a(u_2)\,|0> \;\in\; \text{Vacuum sector of } Z \tag{III.8a}$$

$$b(u_1)\,a(u_2)\,|0> \;\in\; e^{-\frac{2i\pi}{2}} \text{ sector of } Z\,| \tag{III.8b}$$

Two creators a applied to the vacuum vanish. The exotic rules distinguishes the non-trivial conformal creation operators from ordinary free field operators which belong to the representation $c = \frac{1}{2}$. Since no other sector besides $\frac{1}{16}$, $\frac{1}{2}$, 0 exists, there is no need to introduce more light-cone fields. So one expects on each light-cone two 4-point functions:

$$F = \langle a^+(u_1)\, a(u_2)\, a^+(u_3)\, a(u_4) \rangle \tag{III.9a}$$

$$G = \langle a^+(u_1)\, b^+(u_2)\, b(u_3)\, a(u_4) \rangle \tag{III.9b}$$

and a more general 2^{n-1} 2n-point function.
The knowledge of the mixed (order-disorder) correlation function on one line[18] is sufficient to determine all mixed n-point functions of the light-cone conformal creation and annihilators a, a^+, b, b^- and to derive their "exchange algebra" ($u > u'$)

$$\begin{pmatrix} a^+(u') & a(u) \\ b(u') & a(u) \end{pmatrix} = \text{Vertex phase} \begin{pmatrix} 1 & 0 \\ 0 & e^{-i\frac{\pi}{2}} \end{pmatrix} \begin{pmatrix} a^+(u) & a(u') \\ b(u) & a(u') \end{pmatrix}$$

$$\begin{pmatrix} a(u') & a^+(u) \\ b^+(u') & b(u) \end{pmatrix} = \text{Vertex phase} \begin{pmatrix} e^{-i\frac{\pi}{4}} & e^{i\frac{\pi}{4}} \\ e^{i\frac{\pi}{4}} & e^{-i\frac{\pi}{4}} \end{pmatrix} \begin{pmatrix} a(u) & a^+(u') \\ b^+(u) & b(u') \end{pmatrix}$$

$$b(u')\,b^+(u) = \text{Vertex phase}\;\; b(u)\,b^+(u') \tag{III.10}$$

Here the vertex is the phase appearing in (III.4) for dim V= $\frac{1}{16}$. The exchange algebra serves to relate any 2n point function in a non-canonical order into the "ordered" function. The last relation in (III.10) only becomes effective in higher than 4-point functions. It is fairly easy to see that the calculated 2n-point function for the a-b operators of the Ising model are positive definite[19]. It is very plausible that exchange algebras (different from Virasoro or loop algebras) have only one irreducible vacuum representation. A rigorous argument will be presented in a future publication.

Let us now try to generalize the algebraic aspect of the Ising model to the Wess-Zumino-Witten model for SU(2). The suggestion from the analysis of the current algebra is that there are k+1 sectors being distinguished by the lowest angular momentum j with $2j < k$[34]. Furthermore, these sectors correspond precisely to the different centre sectors Z which feature in our global Einstein causality discussion. The local WZW field is therefore of the form $(\alpha, \beta = \pm\frac{1}{2})$:

$$g^{\alpha,\beta}(u,v) = \sum_{j,\sigma,\mu} a^{\alpha}_{(j,\sigma)}(u)\; \tilde{a}^{\beta}_{(j,\mu)}(v) + \ldots \tag{III.11}$$

Here the sum extends over all half-integers $j \leq 2k$ and σ and μ stands for $\pm$. The interpretation is that the operator $a^{\alpha}_{(j\pm)}$ (a) acts only on the Hilbert space H_j belonging to the jth sector and it maps into $H_{j\pm\frac{1}{2}}$ according to the sign$\pm$. Time-reversal invariance requiers us to identify the $\tilde{a}$'s with the time-reversed a's. It is evident that the exchange relations

$(u > u, o < j < 2k)$

$$a^{\beta}_{(1/2,-)}(u')\, a^{\alpha}_{(0,+)}(u) = \eta\, a^{\alpha}_{(1/2,-)}(u)\, a^{\beta}_{(0,+)}(u') \tag{III.12a}$$

$$a^{\beta}_{(j+1,+)}(u')\, a^{\alpha}_{(j,+)}(u) = \eta\, \omega_j\, a^{\alpha}_{(j+1,+)}(u)\, a^{\beta}_{(j,+)}(u') \tag{III.12b}$$

$$\begin{pmatrix} a^{\beta}_{(j-1/2,+)}(u') & a^{\alpha}_{(j+1/2,-)}(u) \\ a^{\beta}_{(j+1/2,-)}(u') & a^{\alpha}_{(j+1/2,+)}(u) \end{pmatrix} = \eta \begin{pmatrix} \alpha_j & \beta_j \\ \beta_j & \delta_j \end{pmatrix} \begin{pmatrix} a^{\alpha}_{(j,+)}(u) & a^{\beta}_{(j+1/2,-)}(u') \\ a^{\alpha}_{(j+1,-)}(u) & a^{\beta}_{(j+1/2,+)}(u') \end{pmatrix} \tag{III.13}$$

$$a^{\beta}_{(\frac{k}{2}-\frac{1}{2},+)}(u')\, a^{\alpha}_{(\frac{k}{2},-)}(u) = \eta\, \tilde{\omega}\, a^{\alpha}_{(k/2-1/2,+)}(u)\, a^{\beta}_{(k/2,-)}(u') \tag{III.14}$$

with $\tilde{a}$ satisfying the analogous relation with the complex conjugate factors and matrices (from time reversal) are equivalent with causality:

$$g^{\alpha\dot{\beta}}(u,v)\; g^{\alpha'\dot{\beta}'}(u',v') = g^{\alpha'\dot{\beta}'}(u',v')\; g^{\alpha\dot{\beta}}(u,v) \tag{III.15}$$

for $(u - u')(v - v') > 0$.

if and only if

$$|\eta| = 1 = |\omega_j| = |\hat{\omega}|$$

$$\begin{pmatrix} \alpha_j & \beta_j \\ \gamma_j & \delta_j \end{pmatrix}^* \begin{pmatrix} \alpha_j & \beta_j \\ \gamma_j & \delta_j \end{pmatrix} = \mathbb{1} \tag{III.16}$$

These conditions just follow by projecting the causality relation into the j respectively the Z sectors. The triple consistency relations ("braid" or Yang-Baxter relations) between three a's are:

$$A_{12}A_{23}A_{12} = A_{23}A_{12}A_{23} \tag{III.17}$$

where the various possible sectors yield $o < j < k$:

$$1)\quad A_{12} = \omega_j \quad , \quad A_{23} = \omega_{j-1/2} \tag{III.18}$$

$$2)\quad A_{12} = \begin{pmatrix} \alpha_{1/2} & \beta_{1/2} \\ \gamma_{1/2} & \delta_{1/2} \end{pmatrix} \quad A_{23} = \begin{pmatrix} 1 & \\ & \omega_{1/2} \end{pmatrix} \tag{III.19}$$

3) $$A_{12} = \begin{pmatrix} \omega_{j-1/2} & & \\ & \alpha_j & \beta_j \\ & \gamma_j & \delta_j \end{pmatrix} \qquad A_{23} = \begin{pmatrix} \alpha_{j-1/2} & \beta_{j-1/2} & \\ \gamma_{j-1/2} & \delta_{j-1/2} & \\ & & \omega_j \end{pmatrix} \tag{III.20}$$

4) $$A_{12} = \begin{pmatrix} \omega_{k/2-1/2} & \\ & \tilde{\omega} \end{pmatrix} \qquad A_{23} = \begin{pmatrix} \alpha_{k/2-1/2} & \beta_{k/2-1/2} \\ \gamma_{k/2-1/2} & \delta_{k/2-1/2} \end{pmatrix} \tag{III.21}$$

The non-trivial general solution of there relations is:

$$\omega_j = \omega \quad , \quad \begin{pmatrix} \alpha_j & \beta_j \\ \gamma_j & \delta_j \end{pmatrix} = D_j R_j D_j^{-1} \tag{III.22}$$

where D_j is an arbitrary real diagonal matrix and R_j is the special unitary matrix:

$$R_j = \frac{1}{S_j(\omega)} \begin{pmatrix} -(-\omega)^{2j+1} & , \sqrt{-\omega\, S_{j-1/2} S_{j+1/2}} \\ \sqrt{-\omega\, S_{j-1/2} S_{j+1/2}} & 1 \end{pmatrix} \tag{III.23}$$

with

$$S_j(\omega) = \sum_{i=0}^{2j} (-\omega)^i \quad \text{and} \quad S_{k/2+1/2}(\omega) = 0 \, , \, \omega = -e^{-\frac{2i\pi p}{k+2}} \tag{III.24}$$

where p and k+2 have no common divisors.

In the two-point function the isoscalar contribution yields:

$$< a^{1/2}_{(1/2,-)}(u)\, a^{-1/2}_{(0,+)}(u') > = - < a^{-1/2}_{(1/2,+)}(u)\, a^{1/2}_{(0,+)}(u') > \tag{III.25}$$

whereas the dimension $d_{\frac{1}{2}}$ of the a shows up in the phase:

$$< a^{1/2}_{(1/2,-)}(u')\, a^{-1/2}_{(0,+)}(u) > = e^{2i\pi d_{1/2}} < a^{1/2}_{(1/2,-)}(u)\, a^{-1/2}_{(0,+)}(u') > \tag{III.26}$$

Hence the η in the exchange algebra fulfill

$$\eta = -\exp i 2\pi d_{\frac{1}{2}} \tag{III.27}$$

Let $d_{1/2}$ be the dimension of the composite field which links the vacuum sector with the $j = \frac{1}{2}$ sector. According to the Z selection rule for the 3-point function of two elementary with one composite field we have:

$$\eta \omega = \exp i\pi (2 d_{1/2} - d_1) \qquad \text{(III.28)}$$

More generally one obtains for the jth sector composite field the phase rule:

$$(\eta \omega)^{2j-1} = \exp i\pi (d_{j-1/2} + d_{1/2} - d_{j+\frac{1}{2}}) \qquad \text{(III.29)}$$

The solution of this recursion gives:

$$d_j = j(j+1) \frac{p}{k+2} \qquad \text{(III.30)}$$

We obtain dimensional trajectories labelled by p. At this point we do not know to which value of the Virasoro charge our trajectory belongs. Comparison with the standard representation theory yields of course p=1. It would be nice if one could prove on purely algebraic grounds that the case $p \epsilon \mathcal{N}, p > 1$ belongs to a direct product of SU(2) with a flavour space. An analogous derivation of the Kac formula for the minimal families from the exchange algebra has been given elsewhere by Rehren.[35] A complete understanding of the minimal models is more subtle than in the SU(2) case. Whereas for SU(2) the value p appears to have the interpretation of a flavour-quantum number, the p for the minimal models appear to label soliton sectors. The understanding of this corresponding c-value goes one step in the direction of the representation theory of the exchange algebra. One needs to construct the energy-momentum tensor as a composite field from its characteristic commutation relations with the exchange fields. Its "strength", which is determined by these commutation relations, will determine c. It is perfectly conceivable that the same braid matrices (e.g. SU(2) and the minimal model) will lead to different c-values. There exists an interesting connection with the analysis of modular invariance of Cardy[27] and Capelli, Itzykson and Zuber[48]. The relevant question in connection with our exchange algebra construction is whether there exist other possibilities of selecting a's from the (k+2) sectors (say a subset of a's), such that causality still holds with a different formula from (III.13) and a different exchange algebra (III.14-16). In this way the very mysterious ADE classification[48] may correspond to a different exchange algebra with different "braid" matrices. We intend to address this problem in a forthcoming publication. Here we content ourselves with the general observation that dimensions are directly related to algebra whereas the Virasoro charge only shows up in the representation. The structure of the exchange algebra has repercussions for the form of the global operator expansion "off vacuum" which one needs in order to complete the bootstrap programme. This question will also be dealt with in a separate publication. Note that the eigenvalues of the diagonal and non-diagonal exchange matrices are equal.

The non-trivial spectral value is expected to have a relation to the cohomology of the exchange algebra. The only well-known studied algebras are those generated by causal fields restricted to, say, a compact space-time region. These local von Neumann algebras are known[36] to be factors of type III in the terminology of Murray - von Neumann as refined by

Powers and Connes[37]. Since the local fields are sums of products of the exchange algebra fields, viz. in the Ising model:

$$\mu(x) = a(u)\,a(v) + h.c. + b(u)\,b(v) + h.c. \tag{III.31}$$

if follows that the exchange algebra fields restricted to a u-interval are also of type III, since algebras of the form $\sum I \times I$ or $\sum II \times II$ cannot yield type III local algebras.

Confessing my almost complete ignorance of these matters, I find it nevertheless tempting to speculate that the restricted exchange algebras may be of type III_λ with a value λ outside the positive Powers set[35]:

$$\lambda = \exp 2i\pi \frac{1}{m+1} \tag{III.32}$$

IV. The possible physical destiny of two-dimensional conformal invariant QFT and the "string idea"

Two-dimensional conformal QFT and its systematic classification with the help of non-Lagrangian operator methods certainly has a mathematical depth which is not easily matched by any other known scheme in theoretical physics. Diffeomorphism groups (and loop groups), their representations and interwiners certainly lead to surprising new concepts. However, the only present-day physical application as a framework of classifying two-dimensional universality classes of critical systems of condensed matter is somewhat limited. One interesting slight generalization is to obtain a classification of integrable d=2 QFT by leading the critical point. Apart from some subtleties, one would expect such an approach to be within reach. The continuous part of the diffeomorphism group breaks down, but discrete subgroups[19] of the centre $Z_L \times Z_R$ as well as the spacelike Yang-Baxter structures of (non-observable) intertwining fields (as discussed in Jörg Fröhlich's lecture) should remain. The S-matrix of integrable models certainly fulfill all the properties of an S-matrix bootstrap, i.e. fits into the conceptional framework which leads to the Veneziano model and its string theoretical reinterpretation. The main difference with the to Veneziano-type S-matrix lies in the fact that zero mass particles are completely absent in the spectrum. Can a generalized string idea with a, d=2 target space perhaps produce such S-matrices?

A more exciting possibility was hinted at in Polyakov's work[36]. D=2 conformal quantum field theories combined with "some sort of string idea" may lead to an analytic understanding of certain higher dimensional non-trivial quantum field theories such as for example the d=3 Ising model. Again there is no gravitation in such models. This possibility is in my opinion much more realistic and interesting than hunting down a theory of everything (Toe) at the scale of the Planck mass (which, as we heard at this summer school, has meanwhile changed into a "quasiclassical approximation" of something). It should be explored by all means. Judging from past experience with attempts to obtain analytic classification schemes (see remarks made in the introduction), a flexible operator (i.e. algebraic) approach seems to be most promising for preliminary studies of this idea.

There appears to be agreement among physicists that a good understanding of conformal theories on Riemann surfaces and their behaviour under Teichmüller deformations is helpful for an understanding of the "string idea". One approach to conformal QFT and Riemann surfaces is to start from the Polyakov functional integral. The mathematical formalism is that of, say, a free field interacting with an external (Euclidean) mertic field. The general formalism of dealing with such external field problems was developed about 20 years ago[39]. Through a correspondence with Fulling this formalism, which leads to Valatin-Bogolubov transformed "vacua", became popular in semiclassical quantizations of general relativity in particular in connection with the Hawking effect. Mathematicians recently refined certain aspects of the formalism, in particular its application to Riemann surfaces. The details can be found in L. Alvarez-Gaumé's lectures[38]. We believe that this Euclidean formalism is mathematically complicated and physically not very revealing. In order to obtain what we believe is a more adequate description, it is interesting to draw some parallels of the Hawking-Unruh acceleration transformation with the Riemann-surface situation. Unruh found[39] that in order to understand flat space quantum field theories in a uniformly accelerated system it is not necessary to solve the QFT in the corresponding changed external metric, but rather to reinterpret the flat-space correlation function via the acceleration transformation in a Rindler half-space, with the, say 0-3 light-cone "wedge" region being the horizon. One finds that the transformed correlation functions restricted to the Rindler world are temperature correlation functions with the Hawking temperature[40]. Of course such a procedure would be meaningless for a true black hole; in that case one has to quantize the matter theory in the Schwarzschild background. It turns out that the Riemann surface case is more like the Unruh situation[20] than like that of a black hole. In the Polyakov functional integral formulation this analogy is only indirectly indicated through the fact that the world-sheet metric was at the beginning just a Lagrangian parameter. Let us look at the operator formalism in the simple case of a free Abelian current I(u) (a (1,0) primary conformal field). Define in the Fock-space the following Unruh-like Fuchsian average[20]

$$J^G(u) = \lim_{N\to\infty} \frac{1}{\sqrt{N}} \sum_{\substack{g_i \in G \\ i=1}}^{N} U(g_i)\, J(u)\, U^{+}(g_i) \qquad \text{(IV.1)}$$

Here G is a discontinuous Fuchsian subgroup of the Möbius group belonging to genus $g > 1$. There are simple arithmetic examples of such groups which are suitable for explicit calculations. For operators with scale dimension $d < 2$ as in the case at hand, the resulting series for the two-point functions:

$$\begin{aligned} \langle J^G(u)\, J^G(u') \rangle &= \lim_{N\to\infty} \sum_i \langle J(u)\, U(g_i) J(u') \rangle \\ &\doteq \langle J(u) J(u') \rangle_G \end{aligned} \qquad \text{(IV.2)}$$

are only conditionally convergent. The case g=1 requires the use of a translation semi-group; by combining terms pair-wise, and using in addition to the $(1,\tau)$ lattice also the "mirror" lattice, one can easily see positivity and express the result in terms of the Weierstraß p - function[20]. The Fuchsian case $g > 1$ is conceptually simpler since one works with the unitary group representers. Positivity of the expression (IV.1) before taking the lines is manifest. The only problem is a clever grouping of terms in order to obtain convergence for the Selberg-type representation of the resulting destribution (hyperfunction). The sum for Fuchsian groups belonging to non-compact genus g Riemann surfaces is conceptually and computationally simpler than that of compact surfaces. In the former case, the Dirichlet region representing the Riemann surface has a free line on the u-light cone (the "infinity" of the non-compact surface) on which positivity holds. In the lattice case the points of the light-cone represented by (hyperbolically equidistant) sequences in the upper half plane transplanted by the elements of the Fuchsian group to the Dirichlet region give limiting point sets. (Julia sets) of possible Hansdorff dimension larger than one. Therefore in this case positivity does not hold on a submanifold but rather on such a general set. This set cannot be shifted by a symmetry operation as in the g=1 case, because on general Riemann surfaces there are no continuous symmetries. The higher point functions retain their free field structure: they are sums over products of the two-point functions. Using an argument of Verlinde and Verlinde[43], one finds that the analytically continued two-point functions are the same as thos obtained as ratios of θ-functions with the help of the Jacobian map[40]. Meanwhile other uses of the Unruh idea have appeared in the literature[44]. Corresponding transformations in those cases cannot be interpreted as global deformations within the symmetry group of the vacuum, which in the conformal case consists of the two Möbius groups. In such cases one does not know to use the flat space quantum field theory and it is highly questionable whether they give rise to global Hawking temperatures. In doing these Fuchsian averages with spinors, one has to recall two facts.

1.) Flat space free spinors permit two natural expectation values

$$\text{a)} \quad \langle 0 | \psi(u)\, \psi^{+}(u') | 0 \rangle \tag{IV.3a}$$

$$\text{b)} \quad \langle \sigma | \psi(u)\, \psi^{+}(u') | \sigma \rangle \tag{IV.3b}$$

where

$$| \sigma \rangle = \sigma(t=i; x=0) | 0 \rangle$$

is a state obtained by the application of the (dis)order operator $(\mu)\sigma$. On the cylinder these expectation values are called Ramon-Schwarz expectation values.

2.) Spinors see part of the covering group $\tilde{M}$ (and therefore $\tilde{G}$) of the Möbius-group.
Taking both (1) and (2) into account, one ends up with 2^{2g} different expectation values (spin structures) from the Fuchsian averages. It is reasonable to express this result in the following way: there exists only one spinor operator, the "Wigner spinor" of the field theoretic textbook; however, on Riemann surfaces one has to consider 2^{2g} different expectation values on the Wigner spinor.

In order to continue, we need some more mathematical concepts. For finite temperature correlation functions (i.e. g=1), the Gelfand-Segal-Naimark[44] reconstruction of the Hilbert space and the operators (in our case by I(u)) has a non-trivial, rather big commutant (for g=o the commutant is just $\lambda 1$), which turns out to be anti-isomorphic to von Neumann algebra[43]. As a result of the compactness of the conformal Hamiltonian L_o, the v. Neumann algebra turns out to be a factor of type I. If one wants the Virasoro algebra, in particular L_o, to have anything to do with the remaining symmetry for g=1, the correct formula is not

$$L_0 = \frac{1}{2\pi i} \oint z^{-2}\, T(z)\, \frac{dz}{z} \tag{IV.4}$$

but rather

$$L_0 = \frac{1}{2\pi i} \oint z^{-2}\, T(z)\, \frac{dz}{z} \quad \text{— (similar contribution from commutant)} \tag{IV.5}$$

This makes the representation of L_o be unbounded below, i.e. not of Feigin-Fuchs type. I do not expect that for $g > 1$ this situation is going to improve.

Now consider some very important subgroups and subsets of the flat space $\widetilde{diff}_{cent}$. The already-mentioned "level n-groups" G_n have the property ($u_2 > 0$, $f(+\infty) = +\infty$).

$$\frac{1}{c_n} < \frac{f(u_1+u_2) - f(u_1)}{f(u_1) - f(u_1-u_2)} < c_n \tag{IV.6}$$

where the smallest c_n measures the distance to a Möbius transformation and $c_n = 1$ for Möbius transformations. The c_n increases with n. Transformations with this bound are called quasi-symmetric. The quasi-symmetric subgroup of diff can be formally generate by the G_n

$$f(u) = \prod_n g_n(u) \quad , \quad g_n \in G_n \tag{IV.7}$$

in a suitable topology. This subgroup forms (if subjected to normalization see below) what is known in mathematics as the universal Teichmüller space of Bers[44]. Q. Teichmüller-spaces belonging to Riemann surfaces are 3g-3 parametric subsets of *normalized* quasi symmetric transformations (in addition to $f(+\infty) = +\infty$ also $f(o) = o$, $f(1) = 1$

$$T_g = \{ f \text{ normalized} \mid f G f^{-1} : \text{Möbius} \} \tag{IV.8}$$

They form a contractible (as a result of the normalization) 3g-3 parametric complex subspace T_g which, loosely speakind, depends only on th egenus of G (the $T's$ of different G's with the same genus are biholomorphic equivalent) and admits the inner antomorphism of G as a "modular" symmetry group.

All these rich mathematical structures are "virtually" there in the flat space conformal field theory, but they only become "activated" through the Fuchsian average and the subsequent Teichmüller deformations. The latter can in principle be put in Virasoro-algebra-like calculations; the equation (IV.8) with the subgroup Ansatz (IV.7) is infinitesimally a Virasoro-algebra-like equation involvingelements of arbitrary high level. To do Teichmüller variations on Fuchsian groups via Virasoro techniques should be interesting for mathematicians.

It is clear that our operator formalism after Teichmüller variations will give expectation values which depend on the Teichmüller parameters (they present themselves as a generalization of the Hawking temperature) and which will have equivariant transformation properties under modular transformations.

These modular properties just follow from the definition of $M_g < T_g$. As an illustrative example let us return to (IV.2) which after application of the Teichmüller variations reads:

$$W(u, u'; \tau_1, \ldots \tau_{3g-3}) = \langle J(u)\, J(u') \rangle_{G(\tau_1 \ldots \tau_{3g-3})} \tag{IV.9}$$

Here the subscript denotes the Unruh average for the Teichmüller deformed Fuchsian group.

The elements of M_g describe an inner automorphism of G and hence lead to modular invariance:

$$\langle J(u) J(u') \rangle_{G(u_1,\dots,u_{3g-3})} =$$

$$\sum_i \langle J(u) U(f(u_1 \dots u_{3g})) U(g_i) U^+(f(u_1 \dots u_{3g}))(u') \rangle \tag{IV.10}$$

$$= \langle J(u) J(u') \rangle_{G(u_1' \dots u_{3g-3}')}$$

if the two Teichmüller-deformed Fuchsian groups are identical sets, i.e. $Mod_g \simeq Aut(G_g)$.

In order to reconstruct the partition function of a conformal field theory on a Riemann surface one follows the procedure of Verlinde and Verlinde[43]. One first represents the conformal energy-momentum tensor as a composite field in terms of the local fields of the model. This requires the construction of a "normal product" via a space-time limiting procedure[43] for the "would be" partition function. Its modular invariance is guaranteed as a result of the modular transformation properties of the 1-point functions. The special case g=1, which requires an averaging using a semi-group, is of considerable interest. In this case one hopes to make contact with the amazing results of Capelli, Itzykson and Zuber[48]. These authors found that the use of modular invariance of partition functions represented as sums over products of characters allows for a classification in terms of A D and E Dynkin diagrams. Partition functions are more universal than local fields. They correspond in some sense to a maximal set of (bosonic) local fields, i.e. a set which is stable under "soliton completion". Let us illustrate this concept in an example. For the Ising model, the operators m(x) and T(u), T(v) form a set which is closed under Wilson expansions as well as under the global expansions considered in this paper. They are, however, not "maximal" since the Hilbert space can be enlarged by adding another (soliton) sector. The enlarged space is irreducible with respect to the operator-algebra generated by the (dis)order operator $(\mu)\sigma$. This new operator together with m (and the identity operator) yields a maximal set of primary fields. Besides the trivial possibility of forming tensor products with other independent local fields, there is no further natural enlargment. Maximal finite families of primary local fields only exist for $c < 1$.[27]. Trying to understand the CIZ classification within the framework of flat space local field theory amounts to comprehending the existence of a finite number of possibilities of maximal local fields which can be formed from the same "raw material" given in terms of the global irreducible light-cone fields. Their "exchange algebra" structure regulates the possibilities of forming maximal local sets. In terms of local von Neumann algebras, one has to construct type III_1 local algebras from type III "exotic" (possibly III with a phase) von Neumann algebras. From this "raw material" one may also construct dual fields or Fröhlich's intertwining fields which would again belong to "exotic" local von Neumann algebras. From the viewpoint of modular properties, the partition functions of "string theories" are distinguished within the set of all conformal partition functions. The left- and right-hand pieces of the total string partition functions are characterized as being modular forms of specific weights with no anomalies[49] (these phases are not "anomalous" within conformal field theories). This absence of anomalies is related to the statement that in Polyakov's[50], an understanding of the "Liouville issue" and its related quantization seems to be crucial in order to use the functional integral for strings as a means to obtain analytic insight into higher-dimensional quantum field theory. Polyakov proposed that the 3d Ising model should be analysed as a string theory. The metric free operator approach proposed here seems to be ideally suited for exploring the "string idea" in quantum field theory. The Riemann surface formalism only makes contact with the string idea in the g-perturbation expansion. In this approximation one expects the Liouville issue to appear if one looks at the modifications which are necessary in order to liquidate the "projective" left-right anomalies which occur in the Unruh averages of the (Teichmüller-deformed) Fuchsian groups. In order to obtain a non-perturbative starting point, a "universal" averaging which does not distinguish a special genus seems to be necessary. This may either consist of using the groups of all quasi-symmetric diffeomorphisms (the Ber's universal Teichmüller space) or performing an ascending sequence $T_g > T_{g+1} > T_{g+2} > ...$ (say starting from g=2) of averages. An understanding of the relations between Teichmüller deformed Fuchsian groups for ascending geni g and the possibility of splitting such groups (corresponding to the pinching of surfaces) is clearly asked for. The objects which should make contact with the higher dimensional quantum field theory are expected to be spatial integrals over two dimensional fields of dimension one (vertex operators in the standard string approach). The presently-known string picture associates scattering amplitudes in "target space" with such quantities. In terms of the original two-dimensional "source space" these objects are non-local diffeomorphism invariants upon integration over the (ascending) Teichmüller parameters. The hope is that these universal quantities may turn out to be asymptotes (in the sense of scattering theory) of a new local field theory in target space. An operator approach, as suggested here, is not only capable of permitting a formulation of the desired locality properties in target space, but, judging from experience with systematic classifications of integrable and conformal quantum field theories, may lend itself to a classification of local "target theories". Algebraic approaches, as we already mentioned in the introduction, seem to be presently more powerful than those based on Euclidean functional integrals and Lagrangians.

The somewhat vague circumscription in the use of the word string as "the string idea" in this work has been chosen on purpose. I want to stay in complete harmony with Witten's statement that the "core ideas about strings remain elusive". On the physical side one hope may be that the string idea may help to obtain non-perturbative analytical insight into (some) non-trivial higher dimensional quantum field theory. Stringlike ideas may turn out to be the correct vehicle by which the amazing analytic understanding and classification of d=2 quantum field theories can be transported into a higher dimensional "target" spaces. We certainly do not intend to suggest to use it for a "Toe" (theory of everything) at 10^{19} GeV. In this connection I would like to recollect an episode of more than 14 years ago. At that time Gursey and Radicati had the modest idea that the group SU(6) may be useful in the spectral classification and transition amplitudes of hadronic physics. Within a short time this idea turned into a hot fashion. At the height of this fashion, when I was visiting Feza Gursey in Ankara, a press conference was given at an APS meeting in New York. Various physicists talked about the SU(6) idea as being an important discovery in combining Relativity with Quantum Physics. There appeared a half-page report on this press conference in Time Magazine, of which I happened to take a copy with me. Two days later Tullio Regge appeared in Ankara. In greeting him, Feza immediately asked: "Please tell me what is the important discovery which was the main topic of the press conference?" Tullio: "some gentlemen made a very important discovery indeed". Feza (very excited): "What is it?" Tullio: "They discovered that Pauli is dead".

I am convinced that the string idea is a more important discovery than that in this episode. This I maintain even in plain view of the physical disaster of the "mechanical string". By mechanical string I mean that string which came from the attempt to find a nice geometrical scheme for the S-matrix guessed by Veneziano and its subsequent unitarization. These mechanical string ideas always produce zero mass objects which were later identified with gravitons and gauge particles. Many people believed that in this way one obtained a unique quantum theory including quantum gravity. As we heard recently, this theory is neither unique (there are millions of "theories of everything") and there are serious doubts about whether it solves "quantum gravity" in target space (it seems to do this in the d=2 source space) although it may yield classical restriction for the target metric in the form of an Einstein equation. I think that the real physical destiny of the (suitably generalized, i.e. non-mechanical) string idea lies in the classification of local quantum field theory. As we saw in section III, Einstein causality in conformal QFT_2 allows for a beautiful, non-perturbative and very non-trivial "exchange algebra" classification in terms of "braid" matrices. Anybody who worked on the principles of general quantum field theory cannot help but be impressed by this later vindication of the ideas of the fathers of non-Lagrangian nonperturbative QFT. At least in d=2, quantum field theories do not like to be classified and understood by Lagrangians and Euclidean functional integrals. The insight rather comes through representation theory of new operator algebras. Wouldn't it be nice if a string idea could be used to carry this deep non-perturbative insight to obtain a classification of S-matrices of Einstein causal quantum field theories? The crucial test for such an idea would be the connection of conformal QFT_2 with known S-matrices fulfilling the boot-strap principles in d=2. Many such S-matrices are known and they have a more complicated form than the Veneziano amplitude. In particular they do not have tachyons nor zero mass particles. I intend to return to these fascinating questions in future publications.

I am indebted to M. Karowski, K.H. Rehren, T.T. Truong, and R. Stora for interesting and helpful discussion. I thank M. Jacob for the hospitality extended to me during a short visit to the CERN Theoretical Physics. Division.

References

1 J.E. Wess, Nuovo Cimento **18**, 1086 (1960),
H.A. Kastrup, Ann. Phys. **7**, 388 (1962),
L. Castell, Nucl. Phys. **B4**, 343 (1967).

2 R.F. Streater and A.S. Wightman, "PCT, Spin and Statistics, And All That", Mathematical Physics Monograph Series, Benjamin Inc. (1964).

3 R. Jost, "General Theory of Quantized Fields", Americ. Math. Soc. Publication (1963).

4 A.A. Migdal, Phys. Lett. **37B**, 98 (1971),**37B**, 383 (1971).

5 A.M. Polyakov, Zh. Eksp. Teor. Fiz. Pis'ma Red. **12**, 583 (1970); Sov. Phys. - JETP Lett. **12**, 381 (1970).

6 G. Mack and K. Szymanzik, Commun. Math. Phys. **27**, 247 (1972).

7 M. Hortacsu, R. Seiler and B. Schroer, Phys. Rev. **D2**, 2519 (1972).

8 J.A. Swieca and A.H. Völkel, Commun. Math. Phys. **29**, 319 (1973).

9 B. Schroer and J.A. Swieca, Phys. Rev. **D10**, 480 (1974),
B. Schroer, J.A. Swieca and A.H. Völkel, Phys. Rev. **D11**, 1509 (1975).

10 M. Lüscher and G. Mack, Commun. Math. Phys. **41**, 203 (1975),
M. Lüscher Commun. Math. Phys. **50**, 23 (1976)
G. Mack, Commun. Math. Phys. **53**, 155 (1977).

11 J. Kupsch, W. Rühl and B.C. Yunn, Ann. Phys. (N.Y.) **89**, 115 (1975).

12 A.A. Belavin, A.M. Polyakov and A.B. Zamolodchikov, Nucl. Phys. **B241**, 333 (1984).

13 D. Friedan, Z. Qiu and S. Shenker, Phys. Rev. Lett. **52**, 1575 (1984).

14 P. Goddard, A. Kent and P. Olive, Commun. Math. Phys. **103**, 105 (1986).

15 V.J.S. Dotsenko and V.A. Fateev, Nucl. Phys. **B240**, 312 (1984).

16 V.G. Kac, Lecture Notes in Physics **94**, 441 (1979).
B.L. Feigin and D.B. Fuchs, Funct. Anal. Appl. **16**, 144 (1982), ibid. **17**, 241 (1983).

17 K.H. Rehren and B. Schroer, "Quasiprimary Fields: An Approach to Positivity of 2D Conformal Quantum Field Theory", FU Berlin preprint (July 1987).

18 K.H. Rehren and B. Schroer, "Exchange Algebra on the Light-Cone and Order/Disorder 2n-Point Function in the Ising Field Theory", FU Berlin preprint (June 1987), to be published in Phys. Lett. B.

19 B. Schroer ,"Operator Approach to Conformal Invariant Quantum Field Theories and Related Problems", CERN-TH 4566/87, (1987) Nucl. Phys. **B295** [FS21],4 (1988).

27 J. Cardy, Nucl. Phys. **B270** (FS16), 186 (1986).

28 B. Klaiber, Lecture Notes in Theoretical Physics, Boulder Lectures (1967), Gordon and Breach N.Y. .

29 S. Mandelstam, Phys. Rev. **D11**, 3026 (1975).

30 S. Coleman, Phys. Rev. **D11**, 2088 (1975).

31 J.B. Frenkel and V. G. Kac, Invent. Math. **62**, 23 (1980).

32 V.G. Kac and D.H. Peterson, Adv. in Mathematics **53**, 125 (1984)

33 A. Rocha-Caridi, in: "Vertex Operators in Mathematics and Physics", eds. J. Lepowski, S. Mandelstam and I. Singer, M.S.R.I. **3**, 451, Springer, Berlin – New-York (1984).
34 E. Witten, Comm. Math. Phys. **92** (1984) 455,
D. Gepner and E. Witten, Nucl. Phys. **B278**, 493 (1986),
V.G. Kinshnik and A.B. Zamolodchikov, Nucl. Phys. **B247**, 83 (1984)
35 K.H. Rehren, "Locality of Conformal Fields In Two Dimensions: Exchange Algebra On The Light Cone", FUB preprint (Sept. 87).
36 R. Longo, "Algebraic and modular structure of von Neumann algebras of physics", in: Proceedings of Symposia in Pure Mathematics **38**, 2, Providence; Am. Math. Society (1982).
37 A. Connes, "Classification des factures", in: Proceedings of Symposia in Pure Mathematics **38**, 2, Providence; Am. Math. Society (1982).
38 A.M. Polyakov, Phys. Letters **103B**, 207 (1981).
39 B. Schroer, R. Seiler and J.A. Swieca, Phys. Rev. **D2**, 12 (1970) 2927.
40 L. Alvarez-Gaumé, Lectures given at this school.
41 W.G. Unruh, Phys. Rev. **D14**, 870 (1976).
42 S.W. Hawking, Phys. Rev. **43**, 199 (1975).
43 E. Verlinde and H. Verlinde, "Chiral Bosonization, Determinants and the String Partition Function", University of Utrecht preprint (1986).
44 M. Martinelli and N. Sanchez, Phys. Lett. **B192**, 361 (1987).
45 M.A. Neimark: Normierte Algebra, Berlin V.E.B. Deutscher Verlag der Wissenschaften 1959.
46 R. Haag, N.M. Hugenholts and M. Winnink, Commun. Math. Phys. **5**, 214 (1979).
47 L. Bers, "Riemann Surfaces", Lecture Notes, New York University, Inst. of Math. Sciences, Lecture Notes 1957.
48 A. Capelli, C. Itzykson and H.B. Zuber, Nucl. Phys. B, to appear.
49 E. Witten, in: "Geometry, Anomalies, Topology", W.A. Bardeen and A.R. White (eds.), World Scientific (1985).
50 A.M. Polyakov, Phys. Lett. **103B**, 211 (1981).
51 B. Schroer, "on Non-Commutative Geometry And Its Possible Role In Physics", PUC Rio de Janeiro preprint, March 88.

ON V. PASQUIER A-D-E MODELS

Roland Seneor

Centre de Physique Théorique
Ecole Polytechnique
91128 Palaiseau Cedex - France

0 - INTRODUCTION

From the work of Beliavin, Poliakov and Zamolodchikov [1] and of Friedan, Qui and Shenker [2] it follows that for d=2 unitary conformal theories the central charge c, with c < 1, has its values parametrized by an integer $m \geq 2$, given by

$$c = 1 - \frac{6}{m(m+1)}$$

Friedan, Qui and Shenker remarked in particular that m = 3 corresponds to the Ising model and m = 5 to the Q = 3 states Potts model. They suggested that there should exist models interpolating the known q-state Potts models for m = 3, 5 and ∞ (i.e. q = 2, 3 and 4) with q given by

$$q = (2\cos\frac{\pi}{m+1})^2$$

On the other hand it was shown by CARDY [3] that by looking at partition functions on a torus it is possible to recover the conformal content of a given conformal invariant theory. He showed that under the constraint of modular invariance the partition function Z takes the form

$$Z = \Sigma\, N_{h,h'}\, \chi_h(\tau)\chi_{h'}(\tau)$$

where $\chi_h(\tau)$ are the characters of the Virasoro algebra of central charge c and highest weight h, τ being the ratio ω_2/ω_1 of the periods of the torus and $N_{h,h'}$ are positive integers.

In studying the classification of these modular invariant partition functions, Cappelli Itzykson and ZUBER [4] showed that they are related to the families of simple laced Lie algebras $\{A_n\}$, $\{D_n\}$, E_6, E_7 and E_8. V. Pasquier [5] solved the reverse question, i.e. to each of these Lie algebras he associated a statistical model and those models exhaust the whole sequence of allowed values of c.

I - CONSTRUCTION OF PARTITION FUNCTIONS

The partition functions are given on a square lattice $\mathcal{L}$ which is for convenience tilted by 45^0 as shown on Figure 1. There are 2N lines and 2M columns. We define the 2-column j, j=0,...M as the set of points labelled by $\{2i+1,2j\}\cup\{2i,2j+1\}\cup\{2i+1,2j\}$, i=0,...,N-1. The partition function is written in term of transfer matrix T acting on 2-columns. One defines similarly 2-lines. Periodic boundary conditions are imposed in both the vertical and the horizontal directions, i.e. the 2-columns 0 and M and the 2-lines 0 and N are identified. Other boundary conditions can be imposed on 2-lines. At each site (i,j) there is a spin variable $\sigma(i,j)$ (or $\sigma(a)$ with a=(i,j)) which takes its values according to the Dynkin diagram of one of the A, D or E Lie algebras (see Fig.2).

Let us choose from now on one of the A, D or E diagrams and note it $\mathcal{A}$. Let $\Omega_{\mathcal{A}}$ be the set of sites of $\mathcal{A}$. Let $G_{\mathcal{A}}$ (or G for short) be its incidence matrix, i.e. the matrix element G_{ij} is 1 if i and j are neighbors on the Dynkin diagram $\mathcal{A}$ and 0 otherwise. G is an irreducible matrix with non negative elements, thus the Frobenius theorem applies and there is a unique maximal eigenvalue ß with eigenvector S, normalized to 1, whose components are positive. The A, D and E diagrams are the only Dynkin diagrams for which ß is less than 2.

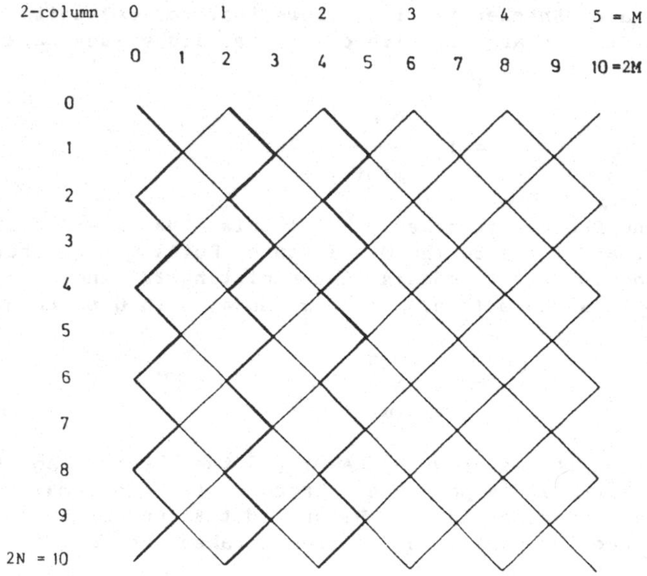

Figure 1

The spin at a point (i,j) of the lattice takes its values in $\Omega_{\mathcal{A}}$. An allowed configuration of spins is a mapping $\sigma : \mathcal{L} \to \Omega_{\mathcal{A}}$ compatible with the boundary conditions and such that if two sites (i,j) are neighbors on the lattice then their values correspond to neighbor points on $\mathcal{A}$. If (i,j) and (i',j') are neighbors on $\mathcal{L}$ and have allowed values we use the notation $\sigma(i,j) \approx \sigma(i',j')$. We introduce the Hilbert space $\mathcal{H}$ of 2-columns $\{|\sigma(0)\ldots\sigma(2N-1)\rangle\ ;\sigma(i)\approx\sigma(i+1)$ for $i=0,\ldots,2N-1\}$ endowed with the usual scalar product.

The transfer matrix is written in term of the identity operator and of operator labelled by lines e_i, $i=0....,2N-1$ and the partition function is given by the usual formula

$$Z_{\mathcal{A}} = \mathrm{tr}_{\mathcal{H}}\, T^M$$

with $T = UV$ where $U = (1+x_1 e_0)\dots(1+x_1 e_{2N-2})$ and $V = (x_2+e_1)\dots(x_2+e_{2N-1})$.

Remark that we can, up to a factor x_2^{2NM}, write the elementary terms as $1 + xe_i$ where $x=x_1$ if i is even and $x=[x_2]^{-1}$ if i is odd.

The e_i are defined on $\mathcal{H}$ by

$$e_i|\sigma(0)\dots\sigma(i-1)\sigma(i)\sigma(i+1)\dots\sigma(2N-1)> =$$

$$= \sum_{\sigma'(i)\approx\,\sigma(i-1)} \delta(\sigma(i-1),\sigma(i+1))\left[\frac{S_{\sigma(i)}S_{\sigma'(i)}}{S(\sigma(i-1))^2}\right]^{1/2} |\sigma(0)..\sigma(i-1)\sigma'(i)\sigma(i+1)..>$$

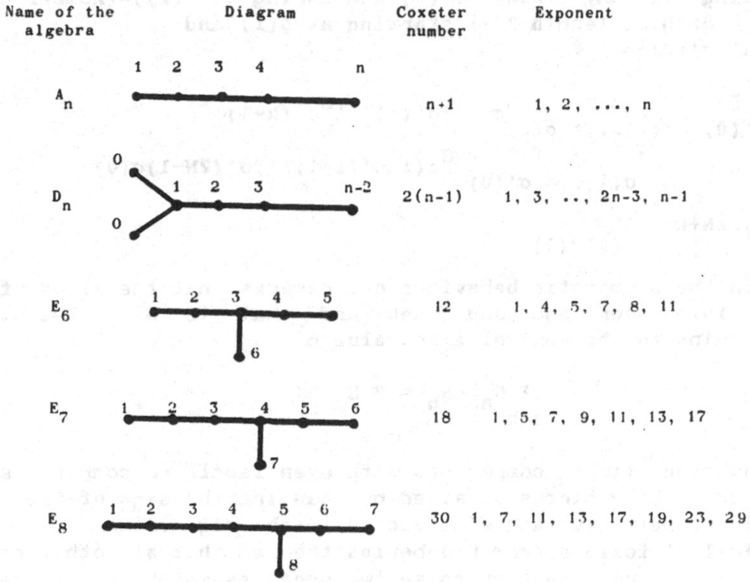

Name of the algebra
Diagram
Coxeter number
Exponent
A n
1 2 3 4 n
n+1
1, 2, ..., n
D n
0 0 1 2 3 n-2
2(n-1)
1, 3, .., 2n-3, n-1
E 6
1 2 3 4 5 6
12
1, 4, 5, 7, 8, 11
E 7
1 2 3 4 5 6 7
18
1, 5, 7, 9, 11, 13, 17
E 8
1 2 3 4 5 6 7 8
30
1, 7, 11, 13, 17, 19, 23, 29

Figure 2

We also introduce orthogonal projectors $P_{|\sigma(k)\sigma(k+1)..\sigma(l)>}$, where $2N-1 \geq l \geq k \geq 0$ and $\sigma(k)\approx \sigma(k+1)\dots\approx \sigma(l)$, by

$$P_{|\sigma(k)\sigma(k+1)..\sigma(l)>}|\sigma'(0)..\sigma'(k)..\sigma'(l)..\sigma'(2N-1)> =$$

$$= \delta(\sigma(k)\sigma'(k))..\delta(\sigma(l)\sigma'(l))\ |\sigma'(0)..\sigma'(k)..\sigma'(l)..\sigma'(2N-1)>$$

In order to deal with objects well defined if $N\to\infty$, we introduce a new trace Tr. It is defined through its action on the projectors P of $\mathcal{H}$

$$\mathrm{Tr}\,P = \frac{\dim \mathrm{Im}P}{\dim\mathcal{H}}$$

Lemma

$$\mathrm{Tr}P_{\sigma(k)\ldots\sigma(l)} = \frac{(G^{2N+k-l})_{\sigma(k)\sigma(l)}}{\mathrm{tr}\,G^{2N}}$$

and

$$\lim_{N\to\infty} \mathrm{Tr}P_{\sigma(k)\ldots\sigma(l)} = \beta^{-(l-k)}$$

Proof:1) $\dim\mathcal{H}$ = number of ways of drawing on $\mathcal{A}$ closed paths of length 2N starting anywhere =

$$= \sum_{\sigma(0)\approx\sigma(1)\ldots\approx\sigma(2N-1)\approx\sigma(2N)\equiv\sigma(0)} G_{\sigma(0)\sigma(1)}\cdots G_{\sigma(2N-1)\sigma(2N)} = \mathrm{tr}\,G^{2N}$$

2) $\dim\mathrm{Im}P = \Sigma_{\sigma'(0)}$ {number of ways of drawing on $\mathcal{A}$ paths of length k starting at any value $\sigma'(0)$ and ending at $\sigma(k)$}×{number of ways of drawing a path of length 2N-1 starting at $\sigma(l)$ and ending at $\sigma'(0)$}=

$$= \sum_{\sigma'(0)} \sum_{\sigma'(0)\ldots\approx\sigma(k)} G_{\sigma'(0)\sigma'(1)}\cdots G_{\sigma'(k-1)} \times$$
$$\times \sum_{\sigma(l)\ldots\approx\sigma'(0)} G_{\sigma(l)\sigma'(l+1)}\cdots G_{\sigma'(2N-1)\sigma(0)}$$
$$= (G^{2N+k-l})_{\sigma(k)\sigma(l)}$$

To obtain the asymptotic behaviour one remarks that the sites of $\mathcal{A}$ can be divided into even and odd ones and that if S is the eigenvector corresponding to the maximal eigenvalue β

$$\sum_{a\approx b} G_{ab}\, S_b = \beta\, S_b$$

This equation links components with even labels to components with odd ones thus, the vector s obtained by changing the sign of the components with odd labels is an eigenvector with eigenvalue $-\beta$. Since G is symmetrical it follows from Froebenius'theorem that all other eigenvalues are smaller in modulus than these two ones. Asymptotically, one has that G behaves as $\beta(|S\rangle\langle S| - |s\rangle\langle s|)$, thus for fixed k and l

$$\mathrm{tr}\,G^{2N} \underset{N\to\infty}{\to} \mathrm{tr}\beta^{2N}(|S\rangle\langle S|+|s\rangle\langle s|) = 2\beta^{2N}$$

and

$$(G^{2N+k-l})_{\sigma(k)\sigma(l)} \underset{N\to\infty}{\to} \beta^{2N+k-l}(S_{\sigma(k)}S_{\sigma(l)} + (-1)^{2N+k-l}s_{\sigma(k)}s_{\sigma(l)})$$
$$= 2\beta^{2N+k-l}$$

For any finite N identifying e_0 and e_{2N} one has

1) $e_i^2 = e_i$

2) $e_i e_{i+1} e_i = e_i e_{i-1} e_i = \beta^{-2} e_i$

3) $[e_i, e_j] = 0$ if $|i-j| > 1$

Moreover for any finite p and $i_1 > .. > i_p \geq 0$

4) $\lim_{N\to\infty} \mathrm{Tr}\, e_{i_1} \ldots e_{i_p} = \beta^{-2p}$

which is the Jones' trace condition.

Remark that 4) cannot be used to compute the partition function since this will involve, for example, to compute terms like $e_1 \ldots e_N$ involving an infinite number of operators for N going to infinity. In this case the behavior of G is not given by the largest eigenvalues and therefore will differ from what obtained by using 1), 2), 3) and 4).

However in the infinite volume limit, i.e. for N and M going to infinity, the e_i's satisfy the 4 conditions, then according to a theorem by Jones [6] this infinite algebra can only be realized, for $\beta<2$, by quantified vales of β given by

$$\beta = 2\cos(\pi/(m+1))$$

Those values are exactly the ones corresponding to the largest eigenvalues of the incidence matrix of the A-D-E Dynkin diagrams.

We now give another expression of the partition function.

II - EXPANSION OF THE PARTITION FUNCTIONS

For each term $1 + xe_i$ acting on the 2jth 2-column, we can choose either the first or the second term. Draw in the diamond whose most left vertex is (i,2j) an horizontal segment for the choice of 1 and a vertical one for the choice of xe_i as shown in Fig.3a). Then Z can be expressed as the sum over all possible diagrams one can draw on the periodic lattice the value of spins at nearest neighbor points should corresponding to nearest neighboring points on the Dynkin diagram. There are 2^{2MN} possible diagrams. Each diagram is made of connected subdiagrams. Each vertex on a given connected part has the same value. Nearest neighbor diagrams correspond to nearest neighbor values on the Dynkin diagram and they have opposite parity. We now give the weights associated to a vertex with value a. There is

1 if no vertical segment passes through it or is closed to it

$(x/\beta)^{1/2} S_a^{-1/2}$ per vertical segment ending at the vertex

$(x/\beta)^{1/2} S_a^{1/2}$ per vertical segment neighboring the vertex

Here x stands for either x_1 or x_2 according to the parity of the segment.

<u>Structure of the diagrams</u>

A connected subdiagram is called a cluster. We set the following definitions [14].

Def.1 A cluster is contractible if it can be reduced to a point. This gives for contractible cluster a notion of interior points.

Def.2 Two cluster are neighbors if they have two sites which are nearest neighbors.

Proposition 1. Any contractible cluster has at least one external point

Proof: By abstract non sense using the topological properties of the torus.

Proposition 2. A contractible cluster Γ has at most one external neighboring cluster Γ'.

Proof: Define the external boundary of Γ as the nearest neighboring vertices of Γ which are not internal. It is easy to see that these vertices have a parity opposite to Γ are connected.

Def.3 Γ' is the father of Γ and Γ the son of Γ'.

Proposition 3. There are cluster which are not fathers of no contractible cluster. They are topologically equivalent to 1-cycle or to 2-cycle on the torus.

Proposition 4. A given configuration exhibits 2 possible topological structure for non contractible cluster:

1) a 2-cycle cluster and no other non contractible cluster

2) 1-cycle cluster in even number and they are of the same type horizontal or vertical

Let us now draw the diagram of connectivity of a configuration. It is obtained by representing each cluster (indexed by its value) by points and two nearest neighbors cluster are connected by a link. The 2 possible structures for the connectivity diagrams are shown on Fig.3b).

Using the weights per cluster defined above it is now possible to sum over all values attached to the trees of the connectivity diagram. In fact suppose there is a link connecting a leave (the extremity of a tree or equivalently a contractible cluster having no son), representing a cluster Γ_a of value a, to another cluster Γ_b of value b. Since the leave has no son its weight is S_a and we have to sum over all a such that $a \approx b$. This gives a factor βS_b. If Γ_b is a non contractible cycle then we add this factor to it. If Γ_b is a contractible cluster and is linked to another cluster Γ_c, its new factor is $(S_b)^{1-1} \times \beta S_b = \beta S_b$ and we can now sum over $b \approx c$.

Finally we get that the sum over the compatible values of a given configuration of horizontal or vertical segments gives

$$(x/\beta)^n \beta^s \sum_{a_1 \approx \ldots \approx a_{2p}} 1 = (x/\beta)^n \beta^s \sum_{a_1 \approx \ldots \approx a_{2p}} G_{a_1 a_2} \ldots G_{a_{2p-1} a_{2p}}$$

$$= (x/\beta)^n \beta^s \operatorname{tr} G^{2p}$$

where n is the total number of vertical links, s the number of sons of all cluster in the configuration and 2p the number of non contractible 1-cycles.

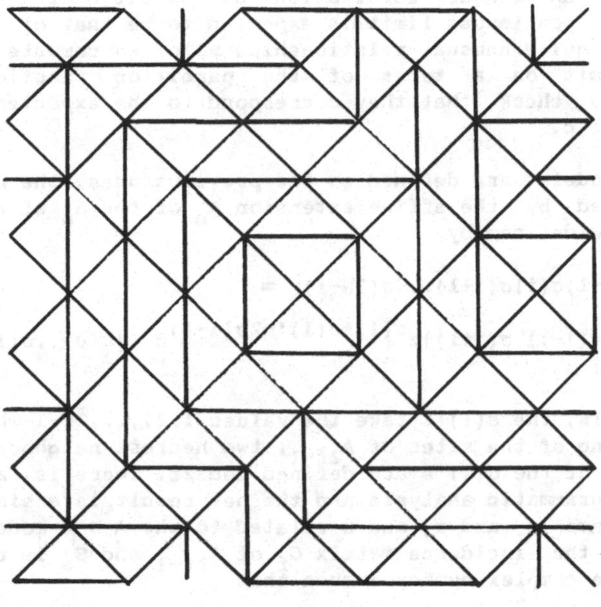

Figure 3a

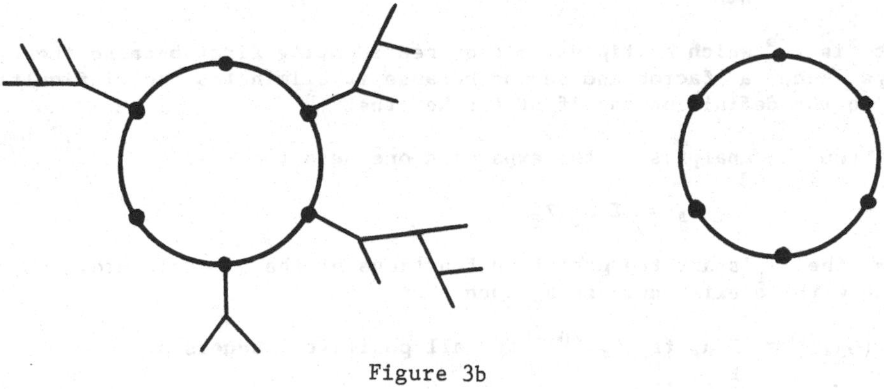

Figure 3b

III - PARTITION FUNCTIONS OF f-MODELS

One of the main problem in defining models is to study their mathematical and physical content. Using the diagrammatic expansion defined above, Pasquier showed that the new model partition functions can be expressed as linear combination of partition functions of simple models whose continuous limit is expected to be that of the Coulomb gas models. These quite unusual relationships allow to compute explicitly the continuum limit on a torus of the partition functions $Z_{\mathcal{A}}$ and in particular to check that they correspond to the expected values of the central charge c.

The f-models are defined as the previous ones, the Dynkin diagrams being replaced by the affine extension $\hat{A}_n$ of the A_n algebra's. However the e_i are now defined by

$$e_i|\sigma(0)\ldots\sigma(i-1)\sigma(i)\sigma(i+1)\ldots\sigma(2N-1)> =$$

$$= \sum_{\sigma'(i)\approx\sigma(i-1)} \delta(\sigma(i-1),\sigma(i+1))z^{\sigma(i)+\sigma(i)'-2\sigma(i-1)}\beta^{-1}|\sigma(0)..\sigma(i-1)\sigma'(i)\sigma(i-1)..>$$

In the f-models, the $\sigma(i)$'s take the values 1,2,...,2f-1 which correspond to a labelling of the sites of A_{2f-1}, two nearest neighbors differing by 1. The values of the $\sigma(i)$'s are defined mod 2f. There is, as in the A-D-E models a diagrammatic analysis and the net result is a similar expansion (with the same x_1 and x_2 and β related to the A-D-E model) where $G_{\mathcal{A}}$ is replaced by the incidence matrix G_f of A_{2f-1} and S_a is replaced by z^a. One choose the complex number z such that

$$z + z^{-1} = \beta$$

where β is the largest eigenvalue of $G_{\mathcal{A}}$. This choice is done in order that the result obtained by summing over the leaves is the same whatever is f; i.e.

$$z^{-b} [\sum_{a\approx b} z^a - \beta z^b] = 0$$

there is z^{-b} which multiply the required identity first because there is always such a factor and second because it eliminates the difficulties dues to the definition mod 2f of the heigths.

From the analysis of the expansion one sees that

$$Z_{\mathcal{A}} = \sum_f a_f Z_f$$

where the Z_f's are the partition functions of the f-models provided one can show there exist numbers a_f such that

$$\mathrm{tr}\,(G_{\mathcal{A}})^{2p} = \sum_f a_f\, \mathrm{tr}\,(G_f)^{2p} \quad \text{for all positive integers p}$$

First this relation should be true for p=0. This leads to

$$n = \sum_f a_f \sum_{j=0}^{2f-1} 1 = \sum_f a_f\, 2f$$

with $n = m$ for $\mathcal{A} = A_m$ or D_m and n = 6, 7 or 8 for E_6, E_7 or E_8. For the general case we multiply both side of the above equation by $-(-1)^p\, t^p$ and sum over p. This gives

$$\mathrm{tr}\ \ln (1 + G_{\mathcal{A}}^2) = \sum_f a_f \ln (1 + G_f^2)$$

Let λ_m and λ_p^f be respectively the eigenvalues of $G_{\mathcal{A}}$ and G_f, then taking account of the symmetry of the set of eigenvalues one gets

$$\underset{m\le h/2}{}(1 + t\,\lambda_m^2)^{n(m)} = \overset{[(f-1)/2]}{\underset{p=0}{}}(1 + t(\lambda_p^f)^2)^{4a_f}$$

where $n(m) = 0$ or 2 for the A-D-E Lie algebras. This relation has to be true for all t and leads to $f\le h$ and to

$$n(m) = \sum_{f|h|fm} 4\, a_f$$

where $f|h|fm$ means that f divides h and h divides fm.

One gets finally that $Z_{\mathcal{A}}$ equal

$1/2\ (Z_{n+1} - Z_1)$	for $\mathcal{A} = A_n$
$1/2\ (Z_{2(n-1)} - Z_{n-1} + Z_2 - Z_1)$	for $\mathcal{A} = D_n$
$1/2\ (Z_{12} - Z_6 - Z_4 + Z_3 + Z_2 - Z_1)$	for $\mathcal{A} = E_6$
$1/2\ (Z_{18} - Z_9 - Z_6 + Z_3 + Z_2 - Z_1)$	for $\mathcal{A} = E_7$
$1/2\ (Z_{30} - Z_{15} - Z_{10} - Z_6 + Z_5 + Z_3 + Z_2 - Z_1)$	for $\mathcal{A} = E_8$

These equalities are true whatever are x_1 and x_2. However we restrict ourself to the critical theories. This correspond to [10] $x_1 x_2 = 1$.

IV - THE CONTINUUM LIMIT OF THE PARTITION FUNCTIONS

According to Cardy [3] informations concerning critical systems can be obtained by looking at their continuum limit on a torus with periodic boundary conditions. It is expected that the continuum limit partition functions behave as

$$\exp(-F\Lambda)\times\{\text{conformal invariant part}\}$$

where F is the free energy per unit surface, Λ is the area of the torus and because of the periodic conditions there is no boundary terms (such a behavior has been shown rigorously by Fisher and Ferdinand [7]). Therefore if one can show that in the equalities of §3 each partition functions has the same free energy they will lead to identities between the conformally invariant parts and we shall see below that the conformally invariant partition functions of f-models are known.

To show the equality of the free energies we follow [8] the analysis of Baxter [9,10]. First it is easy to show that the f-model partition functions Z_f are proportional to the partition function of a 6-vertex model with weights (from now one we stay at the critical point and $x=x_1$) given by

$$a = 1 \qquad b = x/\beta \qquad c = 1 + zx/\beta \qquad d = 1 + z^{-1}x/\beta$$

The free energy only depends on a, b and the product $cd = 1 + x + (x/\beta)^2$. Remark that because of the arithmetical identities of §3 Z_1 have to be interpreted as a f-model with incidence matrix

$$\begin{bmatrix} 0 & 2 \\ -2 & 0 \end{bmatrix}$$

One thus sees that, because in a given equality of §3 the z's are independent of f and are fixed by the value of the Coxeter number of the Dynkin diagram of the left hand side, all free energies of the f-models of the right hand side are the same.

The free energy of the Pasquier models can be analyzed in the same way. One can show [11] that these models belong to a class of generalized restricted SOS models (see [9]). Defining "Boltzmann weight" for a diamond, $W(l,m'|l',m)$, one can show that they are given

$$W(l,l+1|l-1,l) = \alpha_l = 1$$

$$W(l+1,l|l,l-1) = \beta_l = x/\beta \, (S_{l+1}S_{l-1}/S_l^2)^{1/2}$$

$$W(l+1,l|l,l+1) = \gamma_l = 1 + x/\beta \, S_{l+1}/_l$$

$$W(l-1,l|l,l-1) = \delta_l = 1 + x/\beta \, S_{l-1}/_l$$

They satisfy a Yang-Baxter equation provided the parameters x, x' and x" are related by $x + x'' - x' + xx''\beta - xx'x'' = 0$.

One can then parameterize them and show that their largest eigenvalue is the same as that of an eight vertex model with weight given by

$$a = 1 \qquad b = x/\beta \qquad c = \sin\mu/\sin((\mu-v)/2) \qquad d = 0$$

$$x/\beta = \sin((\mu+v)/2)/\sin((\mu-v)/2) \qquad \mu = \pi/\{\text{Coxeter number}\}$$

for which one get that the free energy is the same as the one for the f-models.

This shows that one can, in the equalities of §3, factorize out the free energy contribution and therefore get an equality between the conformally invariant part.

One then introduces a lattice spacing δ and take the limit in which $\delta \to 0$ and at the same time $\delta 2N \to \omega_2$ and $\delta 2M \to \omega_1$. In this limit the f-models are expected [12] to correspond to Gaussian models with partition functions

$$Z_f = \int [D\varphi_f] \exp\left(- \frac{g}{4\pi} \int_T (\nabla\varphi_f)^2 d^2x \right)$$

where the x-space integration is over the torus T with period ω_1 and ω_2 and the fields φ_f are defined mod f.

Introducing $q = \exp -2\pi(\omega_2/\omega_2)$ and the Dedekind function $\eta(q)$, the modular invariant partition function are given by

$$Z_f(g) = \frac{1}{\eta(q)\,\eta(\bar{q})} \sum_{\substack{e\in\mathbb{Z}/f \\ m\in\mathbb{Z}f}} q^{-1/4(eg^{-1/2}+mg^{1/2})^2} \, \bar{q}^{-1/4(eg^{-1/2}-mg^{1/2})^2}$$

Pasquier proposes to choose for the A-D-E diagrams $g = (h-1)/h$. If one computes the small q behavior of the right hand side expression for the partition functions, one gets that the dominant behavior is given by

$$(q\bar{q})^{-c/24}$$

where c is given by the first formula of §0 despite the fact that all f-models have c=1.

In these models fields can also be naturally introduced allowing to define correlation functions. We only give the definition and refer to Pasquier [13] for more details. One can introduce the following objects:

$$\varphi(m)(i,j) = \sum_a (S_a^m/S_a)\, P_{\sigma(i,j)=a}$$

where the S^m's are the set of orthonormalized eigenvectors of G with eigenvalues $2\cos(\pi m/h)$, m running through the exponents and h being the Coxeter number of the Lie algebra $\mathfrak{h}$; the subscript a labels the components of the vectors and S stands for the Frobenius eigenvector with eigenvalue ß.

An unnormalized expectation value is given as usually by

$$<\varphi(m_1)(r_1)\ldots\varphi(m_n)(r_n)> = \mathrm{tr}\, T^{i1}\varphi(m_1)(r_1)T^{i2}\varphi(m_2)(r_2)\ldots$$

if r_1 is a lattice point on the 2-column i_1 etc.. . In term of the diagrammatic expansion this means that if points r_i, $i \in \Omega$, are, in a given configuration, on a cluster of value a then the factors attributed above to the cluster are multiplied by (S_a^{mi}/S_a).

V - CONCLUSION

Pasquier results are impressive, they enlarge considerably the class of known statistical models. Some of his ideas have been presented in this talk with some companion way of understanding them in a more rigorous way. Many interesting questions can be asked as:

1) formulation of the models on a Riemann surface of arbitrary genus

2) extraction of the Virasoro algebras and the links if any with the Jones algebra or more generally with the underlying Hecke algebra and the Yang-Baxter equations

3) the operator structure of the models and its connection with the Beliavin - Poliakov and Zamolodchikov approach

4) extension to c>1 models

I would also thank G. Perrin, D. Hansel and P. Roche for useful comments.

References

[1] A.A. Beliavin, M.M. Poliakov and A.B. Zamolodchikov, J. Stat. Phys. 34 (1984); Nucl. Phys. B241 (1984) 333

[2] D. Friedan, Z. Qui and S. Shenker, Phys. Rev. Lett. 52 (1984) 1575; in Vertex Operators in Mathematics and Physics, Proc. Conf., November 17, 1983, ed. J. Lepowski, S. Mandelstam and I. M. Singer (Springer, New-York 1984) 419

[3] J.L. Cardy, Nucl. Phys. B270 (1986) 186

[4] A. Cappelli, C. Itzykson and J.-B. Zuber, Nucl. Phys. B280 (1987) 445

[5] V. Pasquier, Lattice Derivation of Modular Invariant partition Functions on the Torus, Saclay Preprint SPht/87-62

[6] V. Jones, Invent. Mth. 72 (1983) 25

[7] A.E. Ferdinand and M.E. Fisher, Phys. Rev. 185, 2 (1969) 832

[8] D. Hansel, P. Roche and R. Seneor to be published

[9] G.E. Andrews, R.J. Baxter and P.J. Forrester, J. Stat. Phys. 35:193 (1984)

[10] R.J. Baxter, Exactly Solved Models in Statistical Mechanics (Academic, London, 1982)

[11] V. Pasquier J. Phys. A: Math. Gene. 20 (1987) 1217

[12] P. Di Francesco, H. Saleur and J.-B. Zuber, Saclay Preprint SPhT-86/184 and references therein.

[13] V. Pasquier, Operator content of the A-D-E lattice models, Saclay Preprint SPhT/87014

[14] G. Perrin, Modèles A-D-E sur réseau, Option E. Polytechnique (1987).

GENERALIZATION OF THE SUGAWARA CONSTRUCTION

Jean Thierry-Mieg

CNRS and Royal Society, European exchange program
Department of Applied Mathematics and Theoretical Physics
University of Cambridge, Silver Street
Cambridge, CB3 9EW, U.K.

Introduction

An outstanding problem in the theory of Kac Moody algebras is to try and generalize as many properties of the finite dimensional Lie algebras as possible to the affine case. The beauty of the theory is that this is often feasible and yields a great wealth of results in arithmetics and quantum field theory.

The question that we wish to address here is the generalization of the Racah operators, the generators of the center of the universal enveloping algebra U(G). In other words, we wish to extend the Sugawara construction of the Virasoro algebra [1], which is based upon the quadratic Casimir invariant, to the higher invariants introduced by Chevalley and Racah [2]. The resulting algebra is a composite current algebra of the type defined by Zamolodchikov a few years ago [3].

Indeed, the remarkable connection found by Fateev and Zamolodchikov [4] between the affine algebra $A_2^{(1)}$ and their spin 2 and 3 composite algebra motivated our work and is partially explained by our formalism.

The very same questions were discussed independently by Peter Bouwknegt at this conference [5] using a complementary technique, currents rather than modes, and a less symmetric normal ordering prescription. We have verified that all our results are compatible.

1. The Racah Operators.

Consider a simple compact finite dimensional Lie algebra $\mathcal{G}$. Choose a basis J_a such that the structure constants ${f_{ab}}^c$ are real :

$$[J_a,\, J_b] \;=\; i\, {f_{ab}}^c\, J_c \;\;, \qquad J_a \;=\; J_a^{\dagger} \;. \tag{1.1}$$

Let l denote the rank of this algebra, h and g its Coxeter and dual Coxeter numbers.

The Killing metric g_{ab} is defined as :

$$g_{ab} = -\frac{1}{2g} f_{ac}{}^{d} f_{bd}{}^{c} \ . \tag{1.2}$$

When $\mathcal{G}$ is compact, g_{ab} is regular. We denote by g^{ab} its inverse and use it to raise and lower the indices. Now consider the universal enveloping algebra U(G) It is well known that its center is generated by l operators of degree $e_i + 1$. The e_i are called the exponents of the algebra. The lowest invariant is the quadratic Casimir operator :

$$C = R^{(2)} = \frac{1}{2} g^{ab} J_a J_b \ . \tag{1.3}$$

The higher ones may be written in the form :

$$R^{(i)} = \frac{1}{i!} d^{a_1 \ldots a_i} J_{a_1} \ldots J_{a_i} \ , \tag{1.4}$$

where the d tensors are fully symmetric and pairwise orthogonal under full contraction. In particular, they are traceless.

The exponents of the exceptional algebras are notoriously hard to compute. They were found only around 1950 by Yen Chih Ta, Chevalley and Racah [2] and later Borel and Chevalley [6]. They satisfy interesting identities.
a) The product of the $e_i + 1$ is the order of the Weyl group of $\mathcal{G}$ [7].
b) The lowest and highest exponents are :

$$e_1 = 1 \ , \quad e_l = h - 1 \ . \tag{1.5}$$

c)The sum of the exponents satisfies :

$$N_1 = \sum_i e_i = \frac{1}{2} l h \ . \tag{1.6}$$

This is equivalent to the fact that the $2e_i + 1$ are the Betti numbers of $\mathcal{G}$ and their sum is the exterior degree of the Haar volume form.
e) In addition to these well known identities, we noticed during this conference in collaboration with Peter Bouwknegt the additional relations, true when $\mathcal{G}$ is of type A-D-E :

$$N_2 = \frac{1}{2} \sum_i e_i(e_i - 1) = \frac{1}{3!} l h (h-2), \tag{1.7}$$

$$N_3 = \frac{1}{3!} \sum_i e_i(e_i - 1)(e_i - 2) = \frac{1}{4!} l h (h-2)(h-3), \tag{1.8}$$

It turns out that N_1, N_2, N_3 are respectively the number of subgroups of type A_1, A_2, A_3 in the Weyl group of $\mathcal{G}$. If $\mathcal{G}$ is of type A_n, these identities can be generalized to all orders since $e_i = 1, 2, 3, \ldots n$. They re-sum to the Taylor expansion around $q = 1$ of :

$$\sum_i q^{e_i} = \frac{l}{h-1} \frac{q}{q-1} (q^{h-1} - 1). \tag{1.9}$$

These identities fail at level $p+1$ when $\mathcal{G}$ contains non conjugated subroups of type A_p. Similar equations appear at the end of [8].

Reciprocally, given l and h , one may compute the exponents of every simply laced algebra of level up to 8 by solving the diophantine system (1.5-1.8). This is an indirect but easy way to compute the order of the Weyl group of E_8.

The finite dimensional irreducible representations of $\mathcal{G}$ can be labelled in two different ways. Either, one chooses a Cartan subalgebra $\mathcal{H}$ of $\mathcal{G}$ and gives the Dynkin weights, these numbers are integral. Or one specifies the eigenvalues $\Delta^{(i)}$ of the Racah operators. The later method may seem quite cumbersome since these eigenvalues are complicated polynomials of degree e_i+1 in the Dynkin weights. However it is in some sense more intrinsic. The Racah weights are explicitly frame (i.e. $\mathcal{H}$) independent. They are the true physical observables of the problem. For example, in the case of $A_1 = SU(2)$, the Casimir operator $C = \frac{1}{2}J^2$ is the rotational energy with the well known spectrum $\frac{1}{2}m(m+1)$, $m \in \mathbf{N}$.

In the case of $A_2 = SU(3)$, in the normalisation $d_{abc} = -i\,Tr(J_a\{J_b, J_c\})$, we have :

$$R^{(2)} = \frac{1}{2} g^{ab}\, J_a J_b = \frac{1}{3}(m^2 + n^2 + mn + 3(m+n)\,)$$

$$R^{(3)} = \frac{1}{6} d^{abc}\, J_a J_b J_c = \frac{1}{27}(m-n)(2m+n+3)(m+2n+3)$$

$\Lambda =$ 

(1.10)

2. The generalized Sugawara construction.

Consider the affine Kac-Moody algebra $\mathcal{G}^{(1)}$ in its homogeneous gradation :

$$\begin{aligned} [J_a(m),\, J_b(n)] &= i\, f_{ab}{}^{c}\, J_c(m+n) \,+\, m\,\kappa\, g_{ab}\, \delta_{m+n,0}\;, \\ [J_a(m),\, \kappa] &= 0\;, \qquad m, n \in \mathbf{Z}. \end{aligned} \tag{2.1}$$

and a highest weight module V_Λ with highest weight vector Λ :

$$J_a(m)\,\Lambda = 0, \qquad m > 0\,. \tag{2.2}$$

We define the normal order of 2 currents as :

$$:J_a(m)J_b(n): \;:= \begin{cases} J_a(m)J_b(n) & m < n\;, \\ \frac{1}{2}(J_a(m)J_b(n) \,+\, J_b(n)J_a(m)) & m = n\;, \\ J_b(n)J_a(m) & m > n\;. \end{cases} \tag{2.3}$$

and the normal order of a product of p currents by writing them in a non decreasing order and fully symmetrizing currents with equal mode number. With this convention, any composite local operator, i.e. any convolution of currents, is automatically symmetric in its group indices.

$$\sum_p \;:J_a(m+p)J_b(-p)\ldots: \;=\; \sum_p \;:J_b(m+p)J_a(-p)\ldots:\,. \tag{2.4}$$

Furthermore, such operators evaluated on any given element of V_Λ only involve a finite number of non vanishing terms and are therefore well defined. It is well known [1] that the operators

$$L(m) = \frac{g^{ab}}{2(g+\kappa)} \sum_p : J_a(m+p) J_b(-p) : \tag{2.5}$$

satisfy the Virasoro algebra

$$\begin{aligned} [L(m),\, J_a(n)] &= -n\, J_a(m+n)\,, \\ [L(m),\, L(n)] &= (m-n)\, L(m+n) \;+\; \frac{m(m^2-1)}{12}\, c\, \delta_{m+n,0} \end{aligned} \tag{2.6}$$

where g is the dual Coxeter number of $\mathcal{G}$, κ is the level of the representation and $c = \frac{\kappa d}{g+\kappa}$.

We wish to extend this construction to the higher Racah operators. Consider the local operator with i currents :

$$R^{(i)}(m) \;=\; \frac{1}{i!} d^{a_1 a_2 \ldots a_i} \sum_{p_2 \ldots p_i} J_{a_1}(m+p_1+\ldots+p_i) J_{a_2}(-p_1) \ldots J_{a_i}(-p_i)\,. \tag{2.7}$$

Observe, first of all, that this operator needs no normal ordering since the d tensors are traceless and symmetric. It follows that the Racah currents are primary fields of conformal weight i :

$$[L(m),\, R^{(i)}(n)] = ((i-1)m - n)\, R^{(i)}(m+n)\,. \tag{2.8}$$

It is also clear that the Kac-Moody vacuum is a vacuum for the Racah operators and that on this state the eigenvalue of the zero mode of the Racah current coincides with the finite dimensional Racah weight. It is therefore possible to trade the Dynkin labels, i.e. the vacuum eigen values of the Cartan subalgebra $\mathcal{H}$ and κ for the central charge c of the Virasoro subalgebra and the vacuum eigen values of the l Racah currents.

$$R^{(e_i+1)}(0) = \Delta^{(e_i+1)}\, \Lambda\,, \qquad i = 1, \ldots l\,. \tag{2.9}$$

The hope would be to reexpress all the observables of the theory, i.e. the group invariant quantities, in term of these operators without ever explicitly using the J_a currents. Unfortunately, this is probably not possible as we shall see. Using the techniques of Cvitanović [9] to handle the group indices, it is straightforward to verify that :

$$[J_a(m),\, R^{(i)}(n)] = (\kappa+g)\, m\, R_a^{(i-1,1)}(m+n) \tag{2.10}$$

where $R^{(i,j)}$ denotes an unsaturated current with j symmetrized group indices :

$$R^{(i,j)}_{a_1 a_2 \ldots a_j}(m) \;=\; \frac{1}{i!}\, d_{a_1 a_2 \ldots a_{i+j}}\, (J^{a_{j+1}} J^{a_{j+2}} \ldots J^{a_{i+j}})(m) \;. \tag{2.11}$$

More generaly :

$$[J_a(m),\, R^{(i,j)}_{bc\ldots}(n)] = f_{a[b}{}^{e}\, R^{(i,j)}_{ec\ldots]}(m+n) \;+\; m\,(\kappa + \frac{i-1}{i+j-1} g)\, R^{(i-1,j+1)}_{abc\ldots}(m+n) \;. \tag{2.12}$$

This equation reduces to (2.1) when $i = j = 1$ and to (2.6) when $i = 2, j = 0$. In the critical case $\kappa + g = 0$ the whole current algebra, including its Virasoro subalgebra, is Abelian and commutes with the Kac-Moody currents.

3. The generalized Virasoro algebra

We call the algebra generated by the Racah currents (2.7) the generalized Virasoro algebra. This algebra is very involved and we shall treat explicitly only the $SU(3)$ case. Consider the 2 currents :

$$L(m) = \frac{g^{ab}}{2(\kappa+g)} \sum_n \;:\; J_a(m+n) J_b(-n) \;: \;, \tag{3.1}$$

$$R(m) = \frac{1}{6}\, d^{abc} \sum_{n,p} J_a(m+n+p) J_b(-n) J_c(-p) \;. \tag{3.2}$$

The $L(m)$ satisfy the Virasoro algebra (2.6) and the R(m) are primary fields of spin 3 (2.8). We just need to compute the commutator of 2 R currents. This is already a complex calculation, but fortunately it can be split into several simpler ones. Let us first deal with the group indices. To compute the commutator of $R(m)$ and $R(n)$, we first expand one of them as a product of 3 J and then we use (2.10). This implies and overall factor $(\kappa+g)$. Next, we have to reorder the factors recursively. Each partial reordering is of the form (2.12). It follows that the successive singular terms in the OPE of 2 R, or equivalently the successive terms of the commutator will contain a factor $(\kappa+g)$, $(\kappa+g)(\kappa+\frac{g}{2})$, $(\kappa+g)(\kappa+\frac{g}{2})\kappa$.

To complete the calculation, it is sufficient to keep in the commutator (2.1) the terms proportional to κ and work out the usual Wick contractions of the J_a as if they were free bosons. The result is :

$$\begin{aligned}[R(m),\, R(n)] = {} & \frac{1}{3}(m-n)\,(\kappa+g)\; A(m+n) \\ & + \frac{\delta}{4!}\,(m-n)\,\phi(m,n,p)\,(\kappa+g)(\kappa+\frac{g}{2}) \;:\; J^a(m+n+p) J_a(-p) \;: \\ & + \frac{\delta}{6!}\, m(m^2-1)(m^2-4)\; \kappa D (\kappa+g)(\kappa+\frac{g}{2})\; \delta_{m+n,0} \end{aligned} \tag{3.3}$$

where :

$$A(m) = \frac{3}{8}\, d^{abe} d^{cd}_{\;\;e} \;:\; J_a J_b J_c J_d \;:(m) \;=\; (\kappa+g)^2 \;:\; L(m+p)L(-p) \;: \;\;, \tag{3.4a}$$

$$\phi(m,n,p) \;=\; (m+p)^2 + (m+p)(n+p) + (n+p)^2 - 1\,, \tag{3.4b}$$

$$d_{acd} d_b^{cd} = \delta\, g_{ab} \quad , \quad \delta \;=\; 10/3\,. \tag{3.4c}$$

Let us compare this commutator with the commutator of 2 spin 3 operators of the Zamalodchikov algebra :

$$\begin{aligned}[V(m),\, V(n)] = & \;\frac{16}{22+5c}\,(m-n)\,\Lambda(m+n) \\ & + \;(m-n)(\frac{1}{15}(m+n)^2 - \frac{mn}{6} - \frac{4}{15})\,L(m+n) \\ & + \;\frac{c}{360}\,m(m^2-1)(m^2-4)\,\delta_{m+n,0}\;. \end{aligned} \tag{3.5}$$

where the operator Λ is the regularized square of L :

$$\begin{aligned}\Lambda(m) &= \sum_p \;\ddagger\, L(m+p)L(-p)\,\ddagger\; - \frac{1}{20}(m^2 - \alpha(m))\,L(m) \\ \alpha(2\mathbf{Z}) &= \;4, \quad \alpha(2\mathbf{Z}+1) \;=\; 9\;\;, \end{aligned} \tag{3.6a}$$

and satisfies :

$$[L(m),\, \Lambda(n)] = (3m-n)\,\Lambda(m+n) \;+\; \frac{22+5c}{30}\,m(m^2-1)\,L(m+n)\;\;. \tag{3.6b}$$

The operator Λ differs from A in the definition of the normal ordering. In Λ, the $L(m)$ are reordered as a whole, whereas in A, one reorders the J's implicit in the A's. If we commute L and A, it follows from (2.11) that :

$$[L(m),\, A(n)] \;=\; (3m-n)\,A(m+n) \;+\; \frac{\delta}{2}\,(\kappa+g)(\kappa+\frac{g}{2})\,m(m^2-1)\,L(m+n)\;. \tag{3.7}$$

The fact that the r.h.s. of (3.6) and (3.7) do not match, although $\frac{A}{(\kappa+g)^2}$ and Λ only differ in their normal ordering, shows the existence of a third "quasi-primary" field M transforming like A and Λ :

$$\begin{aligned} M(m) \;&=\; \frac{1}{2(\kappa+g)}\sum_p (\frac{3}{2}m^2 - 5p^2 - 1)\;:\; J^a(m+p)J_a(-p) \;: \;, \\ [L(m),\, M(n)] \;&=\; (3m-n)\,M(m+n) \;-\; 2\,m(m^2-1)\;L(m+n)\;. \end{aligned} \tag{3.8}$$

Thus, we can construct 2 equal primary fields [5]:

$$\begin{aligned}\frac{\tilde{A}(m)}{(\kappa+g)^2} &= \frac{A(m)}{(\kappa+g)^2} \;+\; \frac{\delta}{4}\frac{(\kappa+\frac{g}{2})}{(\kappa+g)}\,M(m) \\ &= \;\tilde{\Lambda}(m) = \Lambda(m) \;+\; \frac{22+5C}{60}\;M(m)\;\;. \end{aligned} \tag{3.9}$$

If we rescale R to

$$\tilde{R}(m) = \sqrt{\frac{2}{\delta(\kappa+g)^2(\kappa+\frac{g}{2})}}\, R(m) \tag{3.10}$$

and use (3.5) and (3.9), we can rewrite (3.3) as :

$$[\tilde{R}(m), \tilde{R}(n)] = [V(m), V(n)] \quad + (m-n)\,\lambda\,\tilde{\Lambda}(m+n)\,. \tag{3.11}$$

$$\lambda = \frac{2(\kappa+g)}{3\delta(\kappa+\frac{g}{2})} - \frac{16}{22+5c} \quad = -\frac{18(\kappa+3)^2}{5(2\kappa+3)(31\kappa+33)} \tag{3.12}$$

The Casimir algebra of $SU(3)$ differs from the Zamolodchikov spin 2 and 3 algebra by the occurence of the extra primary field $\tilde{\Lambda}$. This fields contains a term of the form $\sum_p p^2 \; : J^a(m+p)J_a(-p) :$ and, as an operator, cannot be reexpressed in terms of the Racah currents and their derivatives. In certain cases, however, $\tilde{\Lambda}$ is represented as zero, i.e. it vanishes weakly.

4. Coset constructions

Let $\mathcal{H}$ denote the Cartan subalgebra of $\mathcal{G}$. Let R denote a Racah current of $\mathcal{G}$ and $\hat{R}$ the restriction of this operator to $\mathcal{H}$, obtained by restricting the summation over the a_i in (2.7) to $\mathcal{H}$. Using (2.12), we wish to show by induction that when $\mathcal{H}$ is conformal ($\mathcal{G}$ is of type A-D-E level one) R an d $\hat{R}$ are proportional.

Assume that $\hat{R}^{(i-1,j+1)}$ and $R^{(i-1,j+1)}$ are proportional when their $(j+1)$ explicit indices are in $\mathcal{H}$. This is obviously true when $i=1$ since $R^{(0,j+1)}$ is just a d tensor with all its indices in $\mathcal{H}$.

Consider now a current $R^{(i,j)}$ with all its explicit indices in $\mathcal{H}$. Commute it with an $\mathcal{H}$ current. By (2.12), since $\mathcal{H}$ is abelian, we just get a contraction term of the form $R^{(i-1,j+1)}$ with $(j+1)$ $\mathcal{H}$ indices. The same is true of $\hat{R}$. Thus, by the induction hypothesis, there exists a linear combination $R^{(i,j)} \; + \alpha\,\hat{R}^{(i,j)}$ which is invariant under $\mathcal{H}$. Therefore this combination commutes with $\hat{L}$, the Virasoro algebra of $\mathcal{H}$. Furthermore, if $\mathcal{H}$ is conformal in $\mathcal{G}$, $L = \hat{L}$.

On the other hand, we know that $R^{(i,j)}$ is a primary field of L of spin i, and $\hat{R}$ of $\hat{L}$. Thus :

$$\begin{aligned} 0 \;=\; [\hat{L}(m), (R+\alpha\hat{R})^{(i,j)}(n)] \;&=\; ((i-1)m-n)(R+\alpha\hat{R})^{(i,j)}(m+n) \\ \Rightarrow \quad (R\;+\;\alpha\hat{R})^{(i,j)}(m+n) \;&=\; 0\,. \end{aligned} \tag{4.1}$$

A similar result is proved in [5], but only in the case of $A_2^{(1)}$ level one, by an explicit construction of R using the Frenkel Kac Segal vertex operators.

Let us now reconsider the results of section 3. We can repeat all the calculation using $\hat{R}$. The only modification in (3.3 to 3.10) is that we now have

$$c=2,\ \kappa=1,\ g=0,\ D=2,\ and\ \delta=\frac{4}{3} \tag{4.2}$$

The normalisation (3.10) implies that $\hat{\tilde{R}} = \tilde{R}$, since we know that they are proportional and that they have the same central charge $\frac{1}{180}$. However, if we compute using (3.12) the new value of λ we immediatly find $\hat{\lambda} = 0$. The comparison of our 2 calculations, using R and $\hat{R}$, thus shows that when $\kappa = 1$, $\Lambda(m) = 0$ in (3.11).

This result was found in [4], since their free field construction of V coincides with our $\hat{R}$, and in [5] using an explicit vertex operator construction.

By (2.10), the horizontal algebra $J_a(0)$ commutes with the Zamolodchikov algebra. This has 2 implications concerning the representations of the Z algebra when c=2. Let Λ denote again the Kac-Moody highest weight and μ an element of its Weyl orbit [1]. μ is automatically a Zamolodchikov highest weight. Furthermore, there exists an element ν in the $\mathcal{G}$ orbit of μ sitting in the $\mathcal{H}$ Verma module of Λ. This vector is also a Z highest weight. We obtain in this way an infinite number of degenerate representation of the Zamolodchikov algebra with c=2.

The same method can be applied to the coset space

$$A_2^{(1)} \; level \; \kappa, \;\; A_2^{(1)} \; level \; \kappa' \quad modulo A_2^{(1)} \; level \; \kappa + \kappa',$$

when $\kappa' = 1$, we can replace the second A_2 by its Cartan subalgebra. The comparison of the 2 generalized G.K.O. constructions [10] shows that one obtains a pure Zamolodchikov algebra with $c = 2(1 - \frac{12}{(\kappa+3)(\kappa+4)})$ [5] explaining several key results of [4].

5. Discussion

The Zamolodchikov algebra [3] generalizes the Virasoro algebra in a very natural way. Its unitary representations are not completely known but seem [4] to follow a similar pattern : a discrete sequence with $c = 2(1 - \frac{12}{m(m+1)})$ followed by ionisation at $c = 2$ and a continuum. It admits a BRS operator with critical central charge $c = 100$ and intercept $\alpha = -4$ [11]. At last, as we have shown here, the Zamolodchikov composite algebra offers the correct framework to generalize the Sugawara construction [1] to the higher Racah operators [2]. It is therefore natural to study the Racah current algebra associated to an arbitrary affine Kac Moody algebra. The subject is developing rapidly and several conjectures have been formulated [5], but a proper generalization of the Goddard Kent Olive construction [10] is still lacking.

Acknowledgements

It is a pleasure to thank the organisers of this very nice summer school, Peter Bouwknegt for discussions during the school about the exponents and the primary field $\tilde{\Lambda}$, and Peter Bowcock, Peter Goddard and Hugh Osborn for very many suggestions and comments.

REFERENCES

1. V.G.Kac, Infinite dimensional Lie algebras, chapter 2,
   Cambridge Univerity Press, 1985.
   P.Goddard & D.Olive, Int.J.Mod.Phys. **A 1** (1986) 303.

2. Yen Chih Ta, Comptes Rendus Acad.Sc.Paris **228** (1949) 628-630.
   C.Chevalley, Proc.Int.Congress Math., Cambridge Mass. 1950 II (1952) 21-24.
   G.Racah, Lincei Rend.Sc.Fis.Mat.Nat. **8** (1950) 108-112.

3. A.B.Zamolodchikov, Theor.Math.Phys. **65** (1985) 1205.

4. V.A.Fateev & A.B.Zamolodchikov, Nucl.Phys.B **280** (1987) 644.

5. F.A.Bais, P.Bouwknegt, K.Schoutens & M.Surridge, Amsterdam preprints ITFA 87-12 & 87-21, 1987.

6. A.Borel & C.Chevalley, Mem.Am.Math.Soc. **14** (1955) 1.

7. E.Witt, Abh.Math.Sem.Hansischen Univ. **14** (1941) 289.

8. C.Itzykson, Int.J.Mod.Phys. **A 1** (1986) 65.

9. P.Cvitanović, Group Theory, NORDITA classics illustrated, 1984.

10. P.Goddard, A.Kent & D.Olive, Phys.Lett. **B 152** (1985) 88.

11. J.Thierry-Mieg, Phys.Lett. **B 197** (1987) 368.

CONFORMAL FIELD THEORY AT C=1

Robbert Dijkgraaf, Erik Verlinde and Herman Verlinde

Institute for Theoretical Physics
Princetonplein 5, P.O. Box 80.006
3508 TA Utrecht, The Netherlands

1. C=1 MODELS

1.1. Introduction

Conformal field theory [1] is a subject of great interest to various disciplines in physics. Although we are still far from a complete understanding, partial results seem to indicate that a beautiful, deep mathematical structure lies at its roots. The situation for central charge $c<1$ is by now very well understood [2]. The case $c=1$ however forms in many aspects a natural boundary. Here we meet the new features of an infinity of primary fields and the existence of marginal operators and deformations. Furthermore, the $c=1$ models allow a natural interpretation as a string compactification. As such they can serve as an instructive example of what is to be expected at higher c values.

The basic model with $c=1$ is the gaussian model which is described by a free massless scalar field ϕ compactified on a circle of radius R and with action

$$S = \frac{1}{2\pi} \int d^2z \; \partial\phi\bar{\partial}\phi \tag{1.1}$$

All other known $c=1$ models are orbifold generalizations, i.e. they are constructed by dividing out some discrete symmetry of the quantum theory [3]. In particular we will pay attention to the $\mathbb{Z}_2$ orbifold models obtained by identifying ϕ with $-\phi$.

The operator content of the gaussian model can be decomposed in representations of the U(1) current algebra as generated by the chiral currents $\partial\varphi$ and $\overline{\partial\varphi}$. The U(1) primary fields are given by the set of vertex operators

$$V_{p\bar{p}}(z,\bar{z}) = \exp\left[ip\varphi(z)+i\bar{p}\bar{\varphi}(\bar{z})\right] , \tag{1.2}$$

where we used the equation of motion to write the operator identity $\phi(z,\bar{z}) = \varphi(z)+\bar{\varphi}(\bar{z})$. The operators $V_{p\bar{p}}$ have dimension $(h,\bar{h}) = (\frac{1}{2}p^2,\frac{1}{2}\bar{p}^2)$. Locality implies that the spin $h-\bar{h}$ is integer. In combination with invariance under $\phi \to \phi + 2\pi R$ this restricts the momenta $p,\bar{p}$ to be elements of the lattice

$$\Gamma_R = \left\{(p,\bar{p}) = \left(\frac{n}{R} + \frac{1}{2}mR,\frac{n}{R} - \frac{1}{2}mR\right) ; n,m \in \mathbb{Z}\right\} . \tag{1.3}$$

The $\mathbb{Z}_2$ orbifold models are obtained by first extending the model to a nonlocal model by including twist fields σ_1,σ_2 [4]. These operators have conformal weights (1/16,1/16) and create branch cuts in the $\partial\varphi$ correlators. Projection on states even under $\phi \to -\phi$ yields the local model. In particular the chiral current $\partial\varphi$ is projected out, which breaks the global $O(2)\times O(2)$ symmetry down to a discrete D_4 symmetry.

1.2. Quantum equivalences

In two dimensions many models with a different classical description turn out to be equivalent at the quantum level. A well-known example is the free Dirac fermion which by bosonization is equal to the gaussian model at R=1. With some modifications one can also show that the corresponding orbifold model describes two Majorana fermions, i.e. two decoupled Ising systems [5].

A further example is the electric-magnetic duality that relates the gaussian and orbifold models at R and 2/R via the transformation [6,7]

$$\begin{aligned} V_{p,\bar{p}} &\Leftrightarrow V_{p,-\bar{p}} \\ \sigma_1,\sigma_2 &\Leftrightarrow \frac{1}{\sqrt{2}}\,(\sigma_1 \pm \sigma_2) \end{aligned} \tag{1.4}$$

At the self-dual point $R=\sqrt{2}$ both models have an enhanced symmetry. The gausian self-dual point is known to be equivalent to the k=1 SU(2) WZW model. The k=1 $A_1^{(1)}$ Kac-Moody algebra is generated by the chiral currents $j_1=\cos\sqrt{2}\varphi$, $j_2=\sin\sqrt{2}\varphi$ and $j_3=i\sqrt{2}\partial\varphi$. The spectrum of the self-dual orbifold model also contains the chiral current j_1 and consequently has an $O(2)\times O(2)$ symmetry. Indeed this model is equivalent to a toroidal model, namely at $R = \frac{1}{2}\sqrt{2}$. This can be seen as follows [3]. Consider the twist operators

$$\theta_k = \exp\left(\frac{1}{2} \oint dz\ j_k(z) \right) \qquad k=1,2,3 \tag{1.5}$$

which satisfy $\theta_k^2 = 1$, $\theta_2\theta_1 = \theta_2\theta_1 = \theta_3$ etc. These operators transform the field φ as

$$\begin{aligned} \theta_1\varphi\theta_1 &= -\varphi \\ \theta_2\varphi\theta_2 &= -\varphi + \tfrac{1}{2}\sqrt{2}\pi \\ \theta_3\varphi\theta_3 &= \varphi + \tfrac{1}{2}\sqrt{2}\pi \,. \end{aligned} \tag{1.6}$$

Twisting with $\theta_1\bar{\theta}_1$ or $\theta_3\bar{\theta}_3$ gives the orbifold respectively torus model. The equivalence of both models is now a direct consequence of the SU(2) symmetry which relates θ_1 and θ_3.

1.3. Classification?

Besides $\mathbb{Z}_2$ one can also use other discrete subgroups of SU(2) to construct orbifold models [8]. Only for the three polyhedral subgroups of SU(2) this leads to new models. Thus we arrive at the following picture of c=1 moduli space [7-9].

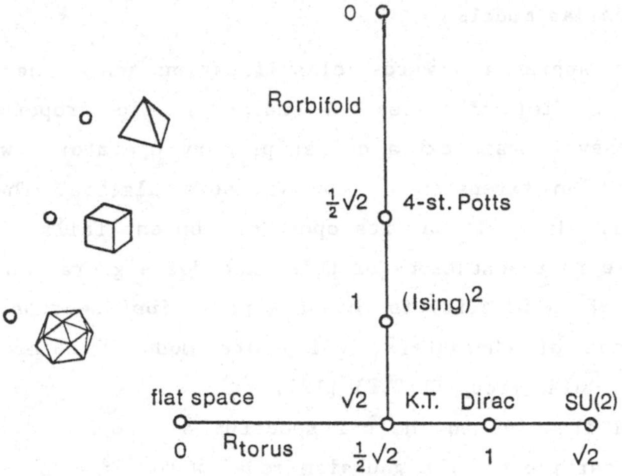

Figure 1

One way to investigate the structure of this moduli space is by using marginal, i.e. dimension (1,1), operators to deform a given conformal model [10]. The intersection point of the orbifold and gaussian line allows three independent marginal deformations. Such points where the number of inequivalent integrable marginal operators jumps to a larger value are called multicritical. Careful analysis shows that there are no other multicritical points. Consequently no new models can be obtained by a continuous deformation of the theories depicted in fig. 1 [7].

One feature all these c=1 models have in common is that their one-loop partition functions can be written as a finite sum of gaussian ones:

$$Z = \sum c_i Z(R_i) \ , \qquad \text{with} \qquad \sum c_i = 1 \ , \tag{1.7}$$

$$Z(R) = |\eta(q)|^{-2} \sum_{(p,\bar{p}) \in \Gamma_R} q^{\frac{1}{2}p^2} \bar{q}^{\frac{1}{2}\bar{p}^2} \tag{1.8}$$

We expect this property to hold for any modular invariant c=1 model. Granted this assumption one can show the completeness of the above picture. First, the presence of a weight (1,0) operator leads uniquely to a gaussian model, since it gives conserved U(1) charges. So the only other linear combinations we have to consider are such that, when the partition function is decomposed into irreducible c=1 Virasoro characters, one finds no chiral currents. If one further imposes non-negative, integer multiplicities, then all possible linear combinations turn out to give precisely the partition functions of the other models in fig. 1 [11].

1.4. Rational gaussian models

A different approach towards classification uses the concept of rational conformal field theories. A characteristic property of these models is that they contain extra chiral primary operators, which can be used to construct an extension of the Virasoro algebra. The theory is called rational if, in addition, its operator content falls into a finite set of irreducible representations of this extended algebra. An alternative characterization of rationality is that the partition function is built up from a finite set of characters, which are modular forms under some subgroup of the modular group $SL(2,\mathbb{Z})$ [12].

The rational theories in the c=1 spectrum are found at the rational values of R^2. In particular, the gaussian model with $\frac{1}{2}R^2 = m/n$ contains the chiral vertex operators $V_\pm(z) = \exp\left(\pm i\sqrt{N}\varphi(z)\right)$ of spin $\frac{1}{2}N = mn$. The representations of the corresponding extension of the current algebra are built on the chiral primary fields $\psi_k = \exp\left(\frac{2\pi ik}{\sqrt{N}}\varphi\right)$, where $k \in \mathbb{Z}_N$. For these values of the momenta the operator products with $V_\pm$ are local. The fusion rules are simply

$$[\psi_k]\cdot[\psi_{k'}] = [\psi_{k+k'}] \ . \tag{1.9}$$

The corresponding 1-loop characters are given by

$$\chi_k(\tau) = \frac{\vartheta\begin{bmatrix} k/N \\ o \end{bmatrix}(\tau)}{\eta(\tau)} \tag{1.10}$$

The field content consists of the operators $\psi_k(z)\psi_{\bar{k}}(\bar{z})$, where $\bar{k}$ is the unique element of $\mathbb{Z}_N$ determined by $k+\bar{k} = 0(\text{mod}2n)$, $k-\bar{k} = 0(\text{mod}2m)$.

A similar structure is found for the orbifold models at rational R^2. The theta functions in (1.10) are modular forms of weight $\frac{1}{2}$ under the congruence subgroup $\Gamma(N)$ of $PSL(2,\mathbb{Z})$. The mathematical result that the space of modular forms of weight $\frac{1}{2}$ is spanned by theta functions [13] seems to indicate that fig. 1. contains at least all rational c=1 conformal field theories.

2. HIGHER LOOP PARTITION FUNCTIONS

2.1. Gaussian partition functions

For the gaussian models (and toroidal compactifications in general) the situation on higher genus Riemann surfaces is well understood. This is due to the fact that these models can be completely analysed in terms of the U(1) current algebra instead of the Virasoro algebra. For an arbitrary conformal field theory we expect the generalized characters, i.e. the holomorphic building blocks of the partition function on a genus g surface, to be labelled by 3g-3 parameters corresponding to representations of the Virasoro algebra. In the gaussian model this reduces, as a consequence of the additivity of U(1) quantum numbers, to g independent labels, namely the loop momenta $(p_i,\bar{p}_i) \in \Gamma_R$ $(i=1,\ldots,g)$ running through the g handles of the surface. The partition function Z_R can accordingly be written as a sum over the momentum lattice Γ_R^g. The dependence on the moduli of the Riemann surface is through the period matrix τ of the abelian differentials. After a calculation that is by now quite standard one finds [14]

$$Z_R(\tau,\bar{\tau}) = \sum_{(p,\bar{p}) \in \Gamma_R^g} \mathcal{Z}_p(\tau)\, \mathcal{Z}_{\bar{p}}(\bar{\tau}) \,, \tag{2.1}$$

$$\mathcal{Z}_p(\tau) = (\det\bar{\partial}_o)^{-1/2} \exp\left[i\pi(p\cdot\tau\cdot p)\right] \,. \tag{2.2}$$

In the rational cases this can be written as a finite sum of characters. For $\frac{1}{2}R^2 = m/n$ and $N = 2mn$ we find

$$Z_R(\tau,\bar{\tau}) = |\det\bar{\partial}_o|^{-1} \sum_{k \in \mathbb{Z}_N^g} \vartheta\begin{bmatrix}k/N\\ o\end{bmatrix}(0|N\tau)\, \overline{\vartheta\begin{bmatrix}\bar{k}/N\\ o\end{bmatrix}(0|N\tau)} \,. \tag{2.3}$$

The integers k_i label the contributions of the primary fields ψ_{k_i} in the i-th loop.

The knowledge of the partition function on an arbitrary Riemann surface is a very strong result, since - as emphasized by Friedan and Shenker [15] - all correlation functions can be obtained by analyzing the partition function at the compactification divisor of moduli space, i.e. on degenerated surfaces. This can be explicitly done for the gaussian models [7] because the factorization properties of the abelian differentials ω_i and their period matrix τ are known [16].

2.2. Orbifolds and double covers

As for the orbifold models we will have 2^{2g} distinct topological sectors corresponding to the different possible boundary conditions of the field ϕ. The partition function is consequently written as the sum over all these sectors. If we choose a homology basis a_i, b_i $(i=1,\ldots,g)$ on the surface, the different sectors can be labeled by a twist structure $\begin{bmatrix}\epsilon\\ \delta\end{bmatrix} = \begin{bmatrix}\epsilon_1 . . \epsilon_g\\ \delta_1 . . \delta_g\end{bmatrix}$ $(\epsilon_i, \delta_i = 0, \frac{1}{2})$, where the field configurations have a branch cut along the cycle $\Sigma_i 2(\delta_i a_i + \epsilon_i b_i)$. All twist structures with $\begin{bmatrix}\epsilon\\ \delta\end{bmatrix} \neq \begin{bmatrix}0\\ 0\end{bmatrix}$ are permuted by the modular group, so we have two modular orbits: twisted or untwisted.

For a particular twist it is natural to consider the double cover $\hat{\Sigma}$ of the Riemann surface on which the field ϕ is again single-valued [17]. The surface $\hat{\Sigma}$ is obtained by taking two copies of the surface Σ, cutting each open along the branched cycle and then pasting the two copies together. The genus is easily seen to be 2g-1. The field ϕ is imposed to be odd under the involution $\iota: \hat{\Sigma} \to \hat{\Sigma}$ which interchanges the two sheets. With respect to the transformation ι the 2g-1 holomorphic one-forms split in g even ones, which project onto the abelian differentials ω_i of Σ, and g-1 odd ones, which are called Prym differentials [16]. The period matrix of the latter will be denoted by Π. Classical solutions for ϕ can now be expressed in terms of the Prym differentials and consequently the classical contribution to the partition function is given in terms of Π. One finds [18,7]

$$Z_R(\Pi,\bar{\Pi}) = \sum_{(p,\bar{p}) \in \Gamma_R^{g-1}} \mathcal{Z}_p(\Pi)\, \mathcal{Z}_{\bar{p}}(\bar{\Pi}) \tag{2.4}$$

$$\mathcal{Z}_p(\Pi) = (\det{}_T \bar{\partial}_o)^{-1/2} \exp\left(i\pi(p\cdot\Pi\cdot p)\right) \ . \tag{2.5}$$

The quantum contribution, given by the chiral determinant of the $\bar{\partial}$-operator on twisted functions, can be calculated using the methods of [14]. Alternatively, one can use the fact that it is independent of the compactification radius and determine it at the multicritical point. Here the orbifold partition function can be set equal to the gaussian one. For the twist $\begin{bmatrix}\epsilon\\ \delta\end{bmatrix} = \begin{bmatrix}0\ldots0\\ 0\ldots\frac{1}{2}\end{bmatrix}$ this leads to the following result

$$(\det{}_T\bar{\partial}_0)^{-\frac{1}{2}} = (\det\bar{\partial}_0)^{-\frac{1}{2}} \frac{\vartheta\begin{bmatrix}\gamma\\ 0\end{bmatrix}(0|2\Pi)}{\vartheta\begin{bmatrix}\gamma & 0\\ 0 & \frac{1}{2}\end{bmatrix}(0|2\tau)} . \tag{2.6}$$

The fact that the right-hand side is independent of the characteristic $\begin{bmatrix}\gamma\\ 0\end{bmatrix} = \begin{bmatrix}\gamma_1\ldots\gamma_{g-1}\\ 0\ldots0\end{bmatrix}$ can be understood from the quantum equivalence of these particular orbifold and gaussian models and is known in the mathematical literature as one of the Schottky relations.

Pinching untwisted cycles produces vertex operator correlators. For each vertex operator on Σ one obtains a conjugate pair of operators on the double cover with opposite momenta, one on each sheet. In this picture it is clear that in the orbifold model momentum conservation is lost. In particular vertex operators can have non zero vacuum expectation values. There is however some restriction due to the discrete D_4 symmetry: only for $(p,\bar{p}) \in 2\Gamma_R$ is the expectation value $\langle V_{p\bar{p}}\rangle$ nonvanishing.

Correlation functions of twist fields are produced by factorization of the partition function at a twisted cycle. The twist correlators are equal to the partition function on the ramified double cover $\hat{\Sigma}$ of Σ branched at the positions of the twist fields.

2.3. Torelli group

All c=1 models that we consider are consistent nonchiral conformal field theories and as such their partition functions are by construction modular invariant. It will however be interesting to discuss the modular properties of the generalized chiral characters.

The mapping class group, i.e. the group of diffeomorphisms not continuously connected to the identity, is generated by Dehn twists around homotopically nontrivial cycles. A Dehn twist D_c can be represented by cutting the Riemann surface along the cycle c and gluing it back together after rotating one of the boundaries over 2π. The mapping class group acts on the gaussian characters through its action on the elements d of the homology group

$$D_c:\ d \to d + \#(d,c)c\ . \tag{2.7}$$

Since these transformations leave the intersection product invariant, they will be represented as elements of $Sp(2g,\mathbb{Z})$. In terms of the period matrix this corresponds to a linear fractional transformation $\tau \to (A\tau+B)(C\tau+D)^{-1}$. Consequently the gaussian characters will be invariant under the Torelli group, which is generated by the Dehn twists around the zero-homology cycles.

This is not the case for the orbifold models. The reason is that a trivial homology cycle, with an appropriate choice of twist structure, will be lifted to a homologically nontrivial cycle on the double cover. This is illustrated for the two loop case in fig. 2.

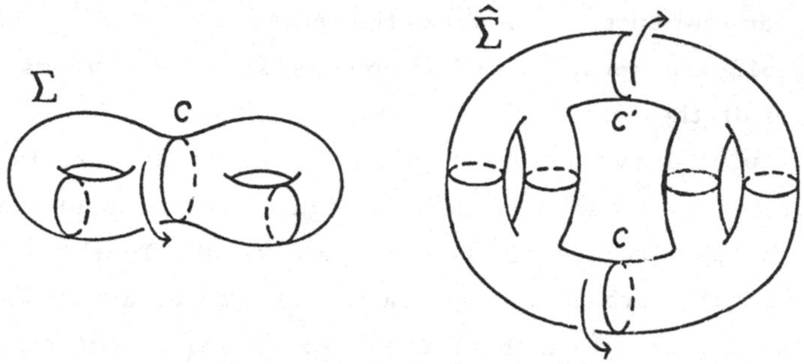

Figure 2

The Dehn twist D_c can be seen to induce the transformation $\Pi \to \Pi + 4$. This result can also be obtained by a factorization argument. If we pinch the cycle c the surface degenerates into two tori, the chiral characters $Z_p(\Pi)$ behave as

$$Z_p(\Pi) = (\det_T \bar{\partial}_o)^{-1/2}\, e^{i\pi p^2 \Pi} \ \underset{=}{\overset{q\to o}{}}\ \langle e^{i2p\varphi}\rangle_1\, q^{2p^2} \langle e^{-i2p\varphi}\rangle_2 + (\text{desc.}) \qquad (2.8)$$

where $\langle\ \rangle_i$ denotes the unnormalized expectation values on the two tori. The Dehn twist D_c sends q to $e^{2\pi i}q$, which indeed reproduces the transformation $\Pi \to \Pi + 4$. Furthermore, since the 1-point functions are nonvanishing and in general $2p^2$ is noninteger, the chiral characters $Z_p(\Pi)$ will transform nontrivially under the Torelli group.

3. OPERATOR FORMALISM ON RIEMANN SURFACES

3.1. Loop momentum operators and modular transformations

The gaussian model on a Riemann surface can be described in terms of the Hilbert space of the theory. More precisely, it is possible to construct a state $|\Sigma\rangle$ which encodes all information of the partition and correlation functions [19]. To this end one chooses a point Q on the surface and a coordinate z in the neighbourhood. The state $|\Sigma\rangle$ is then defined by the requirement that (unnormalized) correlation functions can be computed by taking matrix elements between the vacuum state and $|\Sigma>$. Thus

$$< \prod_k A_k(z_k,\bar{z}_k)> = <0|\prod_k A_k(z_k,\bar{z}_k)\ |\Sigma> \tag{3.1}$$

for any set of local operators A_k. To specify $|\Sigma>$ further we introduce the following operators

$$a[f] = \oint \frac{dz}{2\pi i} f(z)\partial\varphi(z) \tag{3.2}$$

where the contour encircles the point Q. These operators satisfy the commutation relation

$$\left[a[f],a[g]\right] = \oint \frac{dz}{2\pi i} f(z)\partial g(z)\ . \tag{3.3}$$

An immediate consequence of the definition (3.1) is that $|\Sigma>$ is annihilated by those modes a[f] for which the function f(z) is analytically extendible to the whole surface Σ except for possible poles at Q. Since for any basis element $<\psi|$ of the Fock space $J(z) = <\psi|\partial\varphi(z)\ |\Sigma>$ is analytic on Σ-Q, the contour in (3.2) can be pulled off the surface without leaving any residues.

This property, however, does not uniquely fix the state $|\Sigma>$. The reason is that the annihilation modes a[f] are not a maximal set of commuting operators. To see this, consider the class of functions g(z) for which $\partial g(z)$ is extendible to a holomorphic 1-form on Σ-Q. The corresponding operators a[g] all commute with the annihilation operators a[f]. On the other hand $a[g]|\Sigma>$ will in general not be zero, since the functions g need not be single valued on Σ, but may have constant shifts around the non-trivial cycles. The space of such functions g modulo the annihilation modes f is dual to the space of cycles on Σ and thus has dimension 2g. We can associate to any cycle c a function $g_c(z)$, having a branch cut at c with transition function $g_c \rightarrow g_c + 1$. Generalizing the contour deformation argument to the operators $a[g_c]$, we deduce

$$a[g_c] = \oint_c \frac{dz}{2\pi i}\ \partial\varphi(z) \equiv p(c)\ . \tag{3.4}$$

So they measure, when acting on $|\Sigma>$, the momentum running through the cycle c. With the help of these momentum operators we may now decompose the state $|\Sigma>$ into chiral components with definite loop momenta, similarly as we did for the partition functions. Intuitively, it is clear that we can only simultaneously assign loop momenta to non-intersecting cycles. Indeed, the operators p(c) are found to satisfy the commutation algebra [7]

$$[p(c),p(d)] = \frac{i}{2\pi}\#(c,d) \ . \tag{3.5}$$

Hence, a maximally commuting subset of momentum operators is given by $\left\{p(a_i);i=1,\ldots,g\right\}$. Specifying their eigenvalues selects the chiral states.

$$p(a_i)|\Sigma,p\rangle = p_i|\Sigma,p\rangle \ . \tag{3.6}$$

The operators $p(b_i)$ are their canonical conjungates:

$$p(b_i)|\Sigma,p\rangle = \frac{1}{2\pi i}\frac{\partial}{\partial p_i}|\Sigma,p\rangle \ . \tag{3.7}$$

We now like to discuss modular transformations in this formalism. The loop momenta form a representation of the homology group. Modular transformations act as $Sp(2g,\mathbb{Z})$ on the homology cycles, and accordingly as canonical transformations on the loop momentum operators. For the Dehn twist D_c we have

$$D_c\colon\ p(d) \to p(d) - 2\pi i\,[p(d),p(c)]p(c) \ . \tag{3.8}$$

This can be written as a unitary transformation $p(d) \to U_c p(d) U_c^\dagger$, where U_c is given (upto an undetermined phase) by

$$U_c = \exp\left(i\pi p^2(c)\right) \ . \tag{3.9}$$

By this observation we are able to derive the modular properties of the chiral states $|\Sigma,p\rangle$ and the generalized characters $\mathcal{Z}_p = \langle 0|\Sigma,p\rangle$ without using their explicit expressions. For example, under the Dehn twist around $c = \mathbf{r}\cdot\mathbf{a} + \mathbf{s}\cdot\mathbf{b}$ the characters are readily seen to transform as (again up to a phase)

$$D_c\colon|\Sigma,p\rangle \to \exp\left\{i\pi\left[\mathbf{r}\cdot\mathbf{p} + \frac{1}{2\pi i}\,\mathbf{s}\cdot\frac{\partial}{\partial \mathbf{p}}\right]^2\right\}|\Sigma,p\rangle \ . \tag{3.10}$$

The state $|\Sigma\rangle$ of the full theory is a sum over the loop momenta $\mathbf{p},\bar{\mathbf{p}}$:

$$|\Sigma\rangle = \sum_{\mathbf{p},\bar{\mathbf{p}}\,\in\,\Gamma_R^g} |\Sigma,p\rangle \times \overline{|\Sigma,p\rangle} \ . \tag{3.11}$$

It can easily be checked with (3.10) that this state is modular invariant.

It is possible to derive an explicit expression for $|\Sigma >$. It is obtained as a generalized Bogoliubov transformation of the standard vacuum. This is discussed in refs. [19,7].

3.3. Rational models and generalizations

An important question is whether it is possible to generalize the operator formalism on Riemann surfaces to an arbitrary conformal field theory. In general there will be no chiral U(1)-current, so at first one may try to use the stress-energy tensor to determine the state $|\Sigma >$. However, such a generalization would be very difficult both due to the conformal anomaly and the nonabelian nature of the Virasoro algebra. Therefore we propose another -and to our opinion more fruitful - approach to this problem. The basic idea is to work with the algebra of the primary fields instead of the relevant symmetry algebra. This idea is motivated by the following reformulation of the operator formalism for the rational gaussian models.

For these models the state $|\Sigma >$ has a similar structure as the partition function (2.3), i.e. it is given by a finite sum

$$|\Sigma > = \sum_{k \in \mathbb{Z}_N^g} |\Sigma,k\rangle \times \overline{|\Sigma,\bar{k}\rangle} . \tag{3.12}$$

We can select the chiral constituents in the following way. Instead of working with loop momenta we assign "primary fields" to each cycle:

$$\psi_k(c) = \exp\left[2\pi i \frac{k}{\sqrt{N}} p(c)\right] \qquad k \in \mathbb{Z}_N \tag{3.13}$$

satisfying the algebra:

$$\psi_k(c)\psi_{k'}(c) = \psi_{k+k'}(c)$$

$$\psi_k(c)\psi_{k'}(d) = \psi_{k'}(d)\psi_k(c) \exp\left[-2\pi i\frac{kk'}{N} \#(c,d)\right] . \tag{3.14}$$

The operators $\psi_k(c)$ form a representation of $H_1(\Sigma,\mathbb{Z}_N)$ and measure, if c is a generator of the homology, which representation of the extended algebra contributes at c. The chiral state $|\Sigma,k\rangle$ is defined as the common eigenstate of the operators assigned to the a-cycles. The action of the b-cycle operators permutes these states and reflects the fusion rule (1.9)

$$\psi_{k'}(a_i)\ |\Sigma,k\rangle = \exp\left[2\pi i\ \frac{k'k_i}{N}\right]\ |\Sigma,k\rangle \tag{3.15}$$

$$\psi_{k'}(b_i)\ |\Sigma,k\rangle = |\Sigma,k - k'e_i\rangle\ . \tag{3.16}$$

The action of the modular group can also be represented in terms of the "primary fields" $\psi_k(c)$. The unitary transformation induced by the Dehn twist D_c is given by

$$U_c|\Sigma,k\rangle = (-iN)^{-1/2} \sum_{k'\in \mathbb{Z}_N} \exp\left[-i\pi\ \frac{k'^2}{N}\right]\psi_{k'}(c)\ |\Sigma,k\rangle\ . \tag{3.17}$$

Combining this with (3.15-16) we obtain the modular behaviour of the states $|\Sigma,k\rangle$.

More or less as a corollary to this result we find the modular transformation rules of theta functions. An illustrative example is the R=1 model, which is equivalent to a Dirac fermion. For this case we can define, although not completely in line with the presentation upto now, chiral states $]\Sigma,\alpha,\beta\rangle$ satisfying

$$\langle 0|\Sigma,\alpha,\beta\rangle = (\det\bar{\partial}_o)^{-1/2}\ \vartheta\begin{bmatrix}\alpha\\ \beta\end{bmatrix}(\tau) \qquad \alpha,\beta \in (\tfrac{1}{2}\mathbb{Z}/\mathbb{Z})^g\ . \tag{3.18}$$

They are determined by the conditions

$$e^{2\pi i p(a_i)}\ |\Sigma,\alpha,\beta\rangle = e^{2\pi i \alpha_i}\ |\Sigma,\alpha,\beta\rangle \tag{3.19}$$

$$e^{2\pi i p(b_i)}\ |\Sigma,\alpha,\beta\rangle = e^{2\pi i \beta_i}\ |\Sigma,\alpha,\beta\rangle\ . \tag{3.20}$$

One now easily deduces using the CBH-formula

$$e^{2\pi i p(r\cdot a+s\cdot b)} = e^{2\pi i r\cdot p(a)}\ e^{2\pi i s.p(b)}\ e^{i\pi r\cdot s} \tag{3.21}$$

that under the Sp(2g,$\mathbb{Z}$)-transformation

$$\begin{bmatrix}a\\ b\end{bmatrix} \to \begin{bmatrix}D & C\\ B & A\end{bmatrix}\begin{bmatrix}a\\ b\end{bmatrix} \tag{3.22}$$

the spinstructure changes as

$$\begin{bmatrix}\alpha\\ \beta\end{bmatrix} \to \begin{bmatrix}D & -C\\ -B & A\end{bmatrix}\begin{bmatrix}\alpha\\ \beta\end{bmatrix} + 1/2\ \begin{bmatrix}\mathrm{diag}(CD^t)\\ \mathrm{diag}(AB^t)\end{bmatrix} \tag{3.23}$$

a result that, if one uses the explicit form of the theta functions, can only be obtained after a tedious calculation.

REFERENCES

[1] A.A. Belavin, A.M. Polyakov and A.B. Zamolodchikov, Nucl. Phys. B241 (1984) 333.

[2] D. Friedan, Z. Qiu and S. Shenker, Phys. Rev. Lett. 52 (1984) 1575.
J.L. Cardy, Nucl. Phys. B270 [FS16] (1986) 186.
D. Gepner, Nucl. Phys. B287 (1987) 111.
A. Capelli, C. Itzykson and J.B. Zuber, Nucl. Phys. B275 [FS17] (1987) 445.

[3] L. Dixon, J.A. Harvey, C. Vafa and E. Witten, Nucl. Phys. B261 (1985) 620; Nucl. Phys. B274 (1986) 285.

[4] L. Dixon, D. Friedan, E. Martinec and S. Shenker, Nucl. Phys. B282 (1987) 13.

[5] S. Elitzur, E. Gross, E. Rabinovici and N. Seiberg, Nucl. Phys. B283 (1987) 431.

[6] P. Di Francesco, H. Saleur and J.B. Zuber, Nucl. Phys. B285 [FS19] (1987) 454.

[7] R. Dijkgraaf, E. Verlinde and H. Verlinde, C=1 Conformal Field Theories on Riemann Surfaces, Utrecht preprint, THU-87/17.

[8] P. Ginsparg, Curiosities at c=1, Harvard preprint, HUTP-87/A068.

[9] K. Bardacki, E. Rabinovici and B. Säring, String models with c < 1 components, preprint CERN-TH 4760/87.

[10] L.P. Kadanoff, Ann. Phys. 120 (1979) 39.
L.P. Kadanoff and A.C. Brown, Ann. Phys. 121 (1979) 318.

[11] E. Verlinde, unpublished.

[12] J. Harvey, G. Moore and C. Vafa, Quasicrystalline Compactification, Harvard preprint, HUTP-87/A072.

[13] J.-P. Serre, Lecture Notes in Mathematics 672 (Springer, 1977), 29-68.

[14] L. Alvarez-Gaumé, G. Moore and C. Vafa, Comm. Math. Phys. 106 (1986) 1.
L. Alvarez-Gaumé, J.B. Bost, G. Moore, P. Nelson and C. Vafa, Phys. Lett. B178 (1986) 41; Comm. Math. Phys. 112 (1987) 503.
E. Verlinde and H. Verlinde, Nucl. Phys. B288 (1987) 357.

[15] D. Friedan and S. Shenker, Nucl. Phys. B281 (1987) 509.

[16] J. Fay, Theta functions on Riemann surfaces, Springer Notes in Mathematics 352 (Springer, 1973).

[17] S. Hamidi and C. Vafa, Nucl. Phys. B279 (1987) 465.

[18] K. Miki, Phys. Lett. 191B (1987) 127.
D. Bernard, Z_2-twisted fields and bosonization on Riemann surfaces, Meudon preprint 1987.

[19] N. Ishibashi, Y. Matsuo and H. Ooguri, Tokyo preprint, UT-499, 1986.
L. Alvarez-Gaumé, C. Gomez and C. Reina, Phys. Lett. 190B (1987) 55; New methods in string theory, preprint CERN TH-4775/87.
C. Vafa, Phys. Lett. 190B (1987) 47.

GEOMETRIC REALIZATION OF CONFORMAL FIELD THEORY ON RIEMANN SURFACES

Yasuhiko Yamada

Physics Department
Nagoya University
Nagoya, Japan

INTRODUCTION

Recently it has been recognized that conformal field theory (CFT) on Riemann surfaces of arbitrary genus plays an essential role in understanding the profound mechanism of the string theory.

In this talk* I shall discuss one way to construct a CFT (corresponding to charged, free, chiral fermion system) on a family of Riemann surfaces. Motivated by string theory we are interested in the following points:

1) the moduli dependence of the correlation functions,

2) the role of Virasoro and Current algebras on arbitrary Riemann surfaces,

3) unification of the geometric (path integral) approach and the algebraic (representation theory) approach to the string theory.

Our framework can be summarized as the following diagram,

$$\begin{array}{ccccccc} & & & & & & H^{\times} \\ & & & & & & \downarrow \\ \hat{X} \text{ or } \hat{C} & \longrightarrow & UGM & \longrightarrow & \mathbf{P}(F) & \longrightarrow & \mathbf{P}(H). \end{array}$$

(with an arrow from $\hat{X}$ or $\hat{C}$ directly to $H^{\times}$)

Here $\hat{X}$ or $\hat{C}$ is a set of all geometrical data, and the space is embedded by the Krichever map Γ, to the grassmannian, which we call universal grassmann manifold (UGM) following M. Sato.

The UGM is, then, embedded to the projective fermion Fock space $\mathbf{P}(F)$ by the Plücker embedding P. And the fermion Fock space F is isomorphic to the boson Fock space H by the bosonization map B.

On projective Fock space $\mathbf{P}(F)$ or $\mathbf{P}(H)$, there is a natural C^*-bundle $F^{\times}$ or $H^{\times}$,

---

*** The talk is based on the work done in collaboration with N. Kawamoto, Y. Namikawa and A. Tsuchiya[1].**

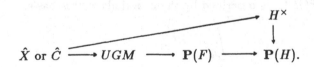

and we can construct a holomorphic section from $\hat{C}$ to $H^{\times}$, which we call τ-function.

In our point of view, $\hat{X}$ (or $\hat{C}$) is an infinite dimensional generalization of the moduli space of surfaces. Although it has a very complicated structure, we can treat it easily when it is embedded in $H^{\times}$ by the τ-function.

The characteristic of our theory is the use of filtration method to controle the infinite dimensional objects like $\hat{X}, \hat{C}$. This method has closer relation to Algebraic geometry (and number theory) than the Hilbert space method of Segal-Wilson[2]. But this difference is a very technical problem, so I will not pursue it further here.

KRICHEVER MAP AS THE PERIOD MAP

Let us start with the ordinary period map $\imath$ (the Torelli map),

$$\imath : m_g \longrightarrow S_g = H_g/\mathrm{Sp}(2g, \mathbf{Z}),$$

which is a map from moduli space of surfaces m_g to the Siegel modular space S_g(=Siegel upper half plane H_g divided by symplectic modular group Sp($2g$,$\mathbf{Z}$)).

The map $\imath$ can be described as follows: Let R be a topological surface (compact connected and oriented) of genus g, and consider its 1st cohomology $H^1(R, \mathrm{C})$ with complex coefficient, which is complex vector space of complex dimension $2g$ and generated by canonical bases of real cohomology, a^i and b^i, the Poincaré dual of the canonical cycles, α_i and β_i, $(1 \leq i \leq g)$.

Now suppose that a complex structure is given on R, then we can divide this space $H^1(R, \mathrm{C})$ into the direct sum of holomorphic part and antiholomorphic part ,

$$H^1(R, \mathrm{C}) = H^0(R, \Omega) \oplus H^0(R, \bar{\Omega}).$$

We call this decomposition a polarization of $H^1(R, \mathrm{C})$. The subspace $H^0(R, \Omega) \subset H^1(R, \mathrm{C})$ moves as parametrized by moduli, which is nothing but the image of the Torelli map. More precisely, we can choose a basis of $H^0(R, \Omega) = \{\{\omega^1, ..., \omega^g\}\}$ as ,

$$\int_{\alpha_i} \omega^j = \delta_{ij},$$

then the subspace $H^0(R, \Omega)$ is described by the β-periods of this basis:

$$\int_{\beta_i} \omega^j = \Omega_{ij},$$

which is the period matrix of the Riemann surface.

Now consider the infinite dimensional generalization of the map $\imath$, which is the Krichever map $\Gamma : \hat{X} \rightarrow UGM$. $\hat{X}$ is a set of all geometrical data consisting of 5-plets $X = (R, Q, L, u, t)$, where R is a Riemann surface and $Q \in R$ is a reference point and L is a line bundle over R, and u and t are local uniformization and trivialization specifying the choise of local coordinate and gauge near Q,respectively.

UGM is a set of finite charge closed subspace of $\hat{K}$, where $\hat{K} = \mathrm{C}((\varsigma))$ is an infinite dimensional vector space of formal Laurent series.

To define the map Γ, we must define a subspace $U(X) \in \hat{K}$ for each data X. We define the subspace $U(X) = H^0(R, L(*Q))_{u,t}$, where $H^0(R, L(*Q))$ is a space of meromorphic section of L holomorphic except for Q, and subscript $u.t.$ means that these sections are expressed as formal Laurent series by the specified coordinate and gauge.

Like the finite dimensional case, this subspace moves as parametried by the geometrical data $X \in \hat{X}$. The important thing here is that for these maps i and Γ, the Torelli type theorems hold, i.e., these maps are injective; we can recover all the geometrical data from its image.

MAYA DIAGRAMS

To express the gross structure of the subspace U of $\hat{K}$, it is convenient to introduce the Maya diagram $M(U)$, which is a semi infinite subset of $\mathbf{Z}$.

When we can choose a basis of $U = \{\{\xi_{m_1}, \xi_{m_2}, \xi_{m_3}, ...\}\}$, such that

$$\xi_{m_i} = \varsigma^{m_i} + \text{(higher order terms)},$$
$$m_1 > m_2 > m_3 > ...,$$

we define $M(U) = \{m_1, m_2, m_3, ...\}$, and display it by the diagram in Fig.1-(a). The charge of the Maya diagram M is defined as follows:

$$charge(M) = N_p(M) - N_h(M),$$

where

$$N_p = \#\{m \in M : m \geq 0\} : \text{particle number},$$
$$N_h = \#\{m \notin M : m < 0\} : \text{hole number}.$$

The UGM is defined precisely as

$$UGM = \{U \subset \hat{K} : \text{closed subspace, s.t. } N_p, N_h \text{ :finite}\}.$$

When $U(X) \subset \hat{K}$ is associated with the geometrical data $X = (R, Q, L, u, t)$, by Riemann-Roch's theorem, we get

$$\begin{aligned} charge(M(U(X))) &= N_p - N_h \\ &= dim H^0(R, L) - dim H^1(R, L) = \chi(L), \end{aligned}$$

which actually implies, $U(X) \in UGM$.

Let us see a few examples: When $L = \Omega$(1-form case), we can take a basis of $H^0(R, \Omega(*Q))$ as

$$H^0(R, \Omega(*Q)) = \sum_{i=1}^{g} \mathbf{C}\omega^i \oplus \sum_{n=1}^{\infty} \mathbf{C}\omega_Q^{(n)},$$

where $\omega_Q^{(n)}$ is a meromorphic 1-form, which has a pole only at Q (the order is n+1). So the corresponding Maya diagram looks like as Fig.1-(b). For $L = O$ (function case), in the positive powers, there are only 1 particle=global holomorphic function (constant), in the negative powers, there are g-holes=Weierstrass gaps (Fig.1- (c)). Note that dual structure of these diagrams (b) and (c), which corresponds to the charge conjugation in field theory and the Serre duality $L \leftrightarrow L^* = \Omega \otimes L^{-1}$ in Algebraic geometry.

We also display the typical Maya diagram for $L = \Omega^2$ (2-form case), where we find 3g-3 holomorphic quadratic differential (Fig.1-(d)). Correspondingly, for $L = \Theta =$

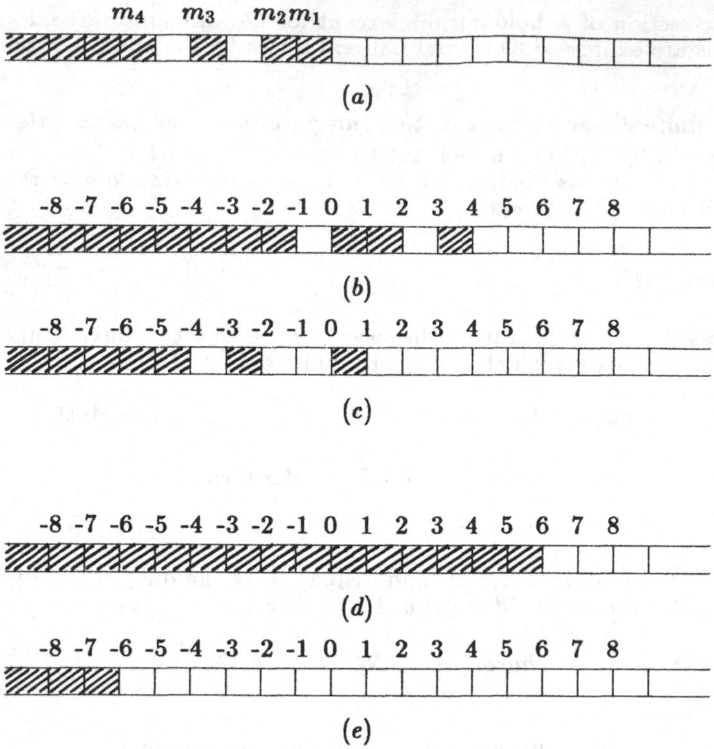

Fig.1 The maya diagrams for (a) : $M = \{m_1, m_2, m_3, ...\}$, (b) : $L = \Omega$, (c) : $L = O = \Omega^0$, (d) : $L = \Omega^2$ and (e) : $L = \Theta = \Omega^{-1}$. $(b) - (e)$ are typical examples for $g = 3$.

Ω^{-1} (vector field case), there are 3g-3 gaps (Fig.1-(e)). We will see that these gaps correspond to the moduli deformation.

DETERMINATION OF THE τ-FUNCTION

Now I explain how to determine the τ-function. In the following discussions we concentrate on the case of $L = \Omega^{1/2}$ (half differential case), which is self dual ($L = L^*$) and has neutral charge.

In our method we use the Sato-Krichever theory for the KP-equations. There are two original works on the KP equations. One is Japanese school's work, which is developed by Sato[3], Hirota, and Date et al[4]. The basic concepts of them are the τ-function and an orthogonal pair of wave functions $\Psi, \bar{\Psi}$. And the Sato theory gives general solution of the KP equations.

The other is Russian school's work developed by Baker, Akhiezer, Krichever and others[5], and it gives the compact solutions of KP equations which are associated with Riemann surfaces.

I cannot explain details of the Sato theory here, but the basic idea and fundamental results can be shortly summarized as follows. The basic result is one to one correspondence between the general solutions of KP equations and points in *UGM*.

As mentioned before, the *UGM* is embedded into projective Fock space $\mathbf{P}(F)$ or $\mathbf{P}(H)$;

$$P : UGM \longrightarrow \mathbf{P}(F) \simeq \mathbf{P}(H) = \mathbf{C}[[t_1, t_2, t_3, \ldots]]$$

The problem here is to characterize the image of this embedding. As is the same as in the ordinary finite dimmensional case, this characterization of the image is given by the Plücker relation, which can be expressed neatly by fermion field operators $\psi(z), \bar{\psi}(z), (z = \varsigma^{-1})$ as follows:

$$Res_Q \psi(z)|U > \otimes \bar{\psi}(z)|U >= 0,$$

or in the Bosonized form,

$$Res_Q \Psi(z, \mathrm{t}, U)\bar{\Psi}(z, \mathrm{t}', U) = 0.$$

Here the wave functions Ψ and $\bar{\Psi}$ are bosonization of 1- particle states, and they are given by letting the vertex operator operate on the τ-function as follows:

$$\Psi(z, \mathrm{t}, U) =< -1|\exp(\sum_{n=1}^{\infty} J_n t_n)\psi(z)|U >=: e^{-Q(z)} : \tau(\mathrm{t}, U),$$

$$\bar{\Psi}(z, \mathrm{t}, U) =< 1|\exp(\sum_{n=1}^{\infty} J_n t_n)\bar{\psi}(z)|U >=: e^{Q(z)} : \tau(\mathrm{t}, U),$$

where

$$\tau(\mathrm{t}, U) =< 0|\exp(\sum_{n=1}^{\infty} J_n t_n)|U > (= \text{bosonization of}|U >),$$

$$Q(z) = \sum_{n=0}^{\infty} n t_n z^n + \log z \frac{\partial}{\partial t_0} - \sum_{n=1}^{\infty} \frac{z^{-n}}{n} \frac{\partial}{\partial t_n}.$$

If we rewrite this bilinear relation in terms of τ- function, then we get the Hirota's bilinear differential equations which are equivalent to the KP equations.

On the other hand, Russian school's work (especially Krichever theory) tells us that there is a one to one correspondence between the geometrical data and normalized Baker-Akhiezer functions, and this normalized BA-function is nothing but the normalized wave function for the compact solution. Moreover, using the Riemann Θ-function, we can give an explicit formula of this normalized BA function corresponding to the data $X = (R, Q, \Omega^{1/2}, u, \sqrt{du})$ as follows:

$$\Psi(z, \mathrm{t}, X) = f(z)\exp(-\sum_{n=1}^{\infty} t_n \omega_Q^{(n)}) \frac{\Theta(I(\mathrm{t}) + I[z]|\Omega)}{\Theta(I(\mathrm{t})|\Omega)},$$

where $f(z)$ is a meromorphic half form which has a single pole at Q of residue 1, and otherwise holomorphic, $I[z] = \int_Q^z \omega$ is Abel- Jacobi map, and

$$I(\mathrm{t}) = \sum_{n=1}^{\infty} I_n t_n : \omega = -d(\sum_{n=1}^{\infty} I_n \frac{z^{-n}}{n}).$$

From this explicit expression we can derive the expression for the τ-function as follows:

$$\tau(\mathrm{t}, X) = e^{\frac{1}{2}q(\mathrm{t})}\Theta(I(\mathrm{t})|\Omega),$$

where

$$q(\mathbf{t}) = \sum_{m,n=1}^{\infty} q_{mn} t_m t_n : \ \omega_Q^{(n)} = d(z^n - \sum_{m=1}^{\infty} q_{mn} \frac{z^{-m}}{m}).$$

This result has been already given by several authors[6,7,8]. Here it should be noted that in this expression there is an ambiguity of normalization constant which can depend on the data X. In the next section we will give an argument to fix this ambiguity.

THE DIFFERENTIAL EQUATION

Here I will discuss about the differential equation satisfied by the τ-function, which is a generalization of the Ward identity in field theoretical view, and is the Gauss-Manin connection in algebro-geometrical view.

First we note that the space $\hat{C} = \{(R, Q, u)\}$ is an infinitesimally homogeneous space on which Lie algebra $g = \sum_{n \in \mathbf{Z}} \mathbf{C}\varsigma^{n+1} \frac{d}{d\varsigma}$ acts as holomorphic vector field. The analogous statement holds for the space $\hat{X}$ on which the algebra $g \oplus (\sum_{n \in \mathbf{Z}} \mathbf{C}\varsigma^n)$ acts. But I only mention $\hat{C}$ here.

More precisely the following theorem holds:

Theorem (Beilinson-Manin-Shehtman)[9]

$$\exists\theta : l \in g \to \theta(l) \in H^0(\hat{C}, \Theta) : \text{Ring homomorphism}$$

$$s.t., \ \ker \theta = B(X) = H^0(R, \Theta(*Q)),$$

$$i.e., \text{ the tangent space } T_X\hat{C} \text{ is isomorphic to } \frac{g}{B(X)}.$$

The meaning is explained in the Fig.2. The space of meromorphic vector fields on R is decomposed into 3 subspaces $(a) \oplus (b) \oplus (c)$, as is indicated in the Maya diagram. A vector field belonging to the part (a) is holomorphic near Q, so it defines coordinate transformation which acts as a vector field on $\hat{C}$ along the fiber. A vector field belonging to the part $(b) \oplus (c)$ has a pole at Q, therefore the equation

$$e^{tl} : \varsigma \to \eta = \varsigma + tl(\varsigma)$$

cannot be considered as coordinate transformation. But this equation still has a meaning as transition function of coordinates between U_∞(n.b.h. of Q) and $U_0 = R - Q$. Then as a result we get a one parameter family of Riemann surfaces R_t, and corresponding vecter field $\theta(l) = \frac{d}{dt}$ deforms the complex structure. Furthermore, if $l \in B(X)$, it can be viewed as coordinate transformation on U_0, but the data $X = (R, Q, u)$ do not know what coordinate is used outside of U_∞, so $\theta(l), l \in B(X)$ deforms nothing, i.e., $B(X)$ is the kernel.

On the other hand, the space $H = \mathbf{C}[[t_1, t_2, ...]]$ (formal power series ring) on which the vacuua are expressed, is a representation space of the $U(1)$-current algebra: $J_n = \frac{\partial}{\partial t_n} (n \geq 0), J_{-n} = nt_n (n > 0)$, so that Virasoro algebra (which is the central extension of g) acts on it by

$$T(z) = \frac{1}{2} : J(z)J(z) : \ .$$

The important thing here is that the action of g on $\hat{C}$ and action of Virasoro algebra on H is compatible (up to anomaly). As a result of this compatibility we get the following differential equation:

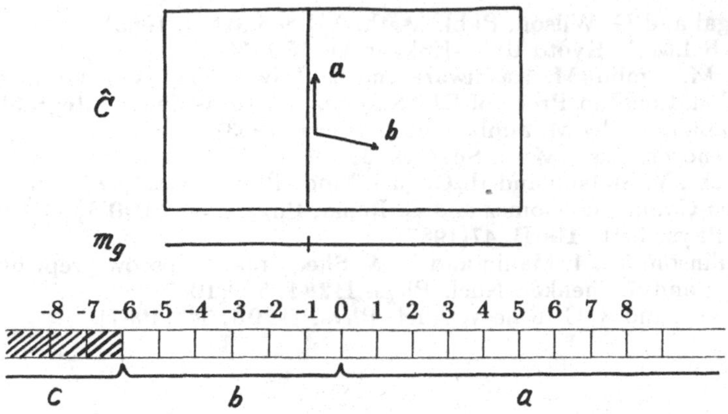

Fig.2 The decomposition of the algebra $g = (a) \oplus (b) \oplus (c)$, and the corresponding vector fields on $\hat{C}$. The gaps (b) deform the moduli.

Theorem [1]

$$[\theta(l) + a(l, X)]\tau(\mathbf{t}, X) = Res_Q(l(z)T(z))\tau(\mathbf{t}, X), \ {}^{\forall}l \in g,$$
$$a(l, X) = -\frac{1}{12} Res_Q(l(z)S(z, X)).$$

Here $\theta(l)$ acts on variable X and $T(z)$ acts on variables $\mathbf{t}$, and $a(l, X)$ is the anomaly term given by projective connection $S(z, X)$ on Riemann surface R as indicated. This equation plays an essential role in characterizing the τ-function uniquely (up to constant)[1].

CONCLUDING REMARK

It seems that string theory includes some interesting mathematics of harmonic (or Hodge) theory on infinite dimensional spaces (e.g., loop space, the universal moduli space, the grassmannians, etc.)[10,11]. For study of them, differential equations would provide an important method as in the finite dimensional case.

We hope that by studying differential equations, we can get important information on the theory, that is, from singularity structure we will get boundary behavior and from monodoromy structure we will get modular property and so on. I think our theory gives the simplest example for such differential equations.

ACKNOWLEDGMENTS

It is a pleasure to thank A. Tsuchiya, Y. Namikawa and N. Kawamoto for an enjoyable collaboration on the topics discussed here. I also thank participants in the school for discussions and the organizers for giving me an opportunity to have a seminar in such congenial circumstances.

REFERENCES

1. N. Kawamoto, Y. Namikawa, A. Tsuchiya and Y. Yamada, preprint, Nagoya Univ. Dept. Math. and Kyoto Univ. KUNS #880 HE(TH) 87/14 (1987)

2. G. B. Segal and G. Wilson, Publ. Math. I.H.E.S.,**61**, 5(1985)
3. M. Sato, R.I.M.S., Kyoto Univ.-Kokyuroku **439**, 30(1981)
4. E. Date, M. Jimbo, M. Kashiwara and T. Miwa, "Transformation Groups for Soliton Equations",in Proc. of RIMS Symp. on Non-Linear Integrable Systems, Kyoto, Japan, ed. by M. Jimbo and T. Miwa (1983)
5. I. M. Krichever, Russ. Mass. Surveys, **32**, 185(1977)
6. N. Ishibashi, Y. Matsuo and H. Ooguri, Mod. Phys. Lett. **A2**, 119(1987)
7. L. Alvarez-Gaumé, C. Gomez and C. Reina, Phys. Lett. **190B**, 55(1987)
8. C. Vafa, Phys. Lett. **190B**, 47(1987)
9. A. A. Beilinson, Yu. I. Manin and Y. A. Shechtman, Moscow preprint (1986)
10. D. Friedan and S. Shenker, Nucl. Phys. **B281**, 509(1987)
11. M. J. Bowick and S. G. Rajeev, Nucl. Phys. **B293**, 348(1987)

INDEX

Printed in the United States
By Bookmasters

Printed in the United States
By Bookmasters